D1153911

Biofilms

WILEY SERIES IN ECOLOGICAL AND APPLIED MICROBIOLOGY

Edited by ***Ralph Mitchell***

MICROBIAL LECTINS AND AGGLUTININS: Properties and Biological Activity
David Mirelman Editor

THERMOPHILES: General, Molecular, and Applied Microbiology
Thomas D. Brock Editor

INNOVATIVE APPROACHES TO PLANT DISEASE CONTROL
Ilan Chet Editor

PHAGE ECOLOGY
Sagar M. Goyal, Charles P. Gerba, Gabriel Bitton Editors

BIOLOGY OF ANAEROBIC MICROORGANISMS
Alexander J. B. Zehnder Editor

BIOFILMS
William G. Characklis and Kevin C Marshall Editors

Biofilms

Edited by

WILLIAM G. CHARACKLIS
Institute for Process Analysis
Montana State University
Bozeman, Montana

KEVIN C. MARSHALL
School of Microbiology
University of New South Wales
Kensington, New South Wales, Australia

A WILEY-INTERSCIENCE PUBLICATION
JOHN WILEY & SONS, INC.
New York • Chichester • Brisbane • Toronto • Singapore

Library of Congress Cataloging-in-Publication Data:

Biofilms/edited by William G. Characklis, Kevin C. Marshall.
p. cm. — (Wiley series in ecological and applied
microbiology)
"A Wiley interscience publication."
Bibliography: p.
Includes index.
ISBN 0-471-82663-4
1. Biofilms. I. Characklis, William G. II. Marshall, Kevin C.
III. Series.
QR100.8.B55B56 1989 89-5681
576'.15—dc19 CIP

Printed in the United States of America

10 9 8 7 6 5 4 3 2 1

This book is dedicated to my lifetime teachers:
Nancy, Greg, and Erin;
for Anna and, especially, Mom

—W. G. C.

CONTRIBUTORS

RUNE BAKKE Hogskolesenteret i Rogaland, Stavanger, Norway

E. J. BOUWER Department of Geology and Environmental Engineering, The Johns Hopkins University, Baltimore, Maryland

JAMES D. BRYERS Duke University, School of Engineering, Durham, North Carolina

WILLIAM G. CHARACKLIS Institute for Process Analysis, Montana State University, Bozeman, Montana

BJØRN CHRISTENSEN Norwegian Institute of Technology, Trondheim, Norway

A. B. CUNNINGHAM Department of Civil and Agricultural Engineering, Montana State University, Bozeman, Montana

ANDREAS ESCHER Basel, Switzerland

WILLI GUJER Swiss Federal Institute for Water Resources and Water Pollution Control, Dubendorf, Switzerland

WHONCHEE LEE Institute for Process Analysis, Montana State University, Bozeman, Montana

BRENDA LITTLE Naval Ocean Research and Development Activity, NSTL, Mississippi

KEVIN C. MARSHALL School of Microbiology, University of New South Wales, Kensington, N.S.W., Australia

GORDON A. MCFETERS Department of Microbiology, Montana State University, Bozeman, Montana

MUKESH H. TURAKHIA Houston, Texas

P. A. WAGNER Naval Ocean Research and Development Activity, NSTL, Mississippi

OSKAR WANNER Swiss Federal Institute for Water Resources and Water Pollution Control, Dubendorf, Switzerland

AN-I YEH Graduate Institute of Food Science and Technology, National Taiwan University, Taipei, Taiwan, R.O.C.

NICHOLAS ZELVER Sandwich, Massachusetts

SERIES PREFACE

The Ecological and Applied Microbiology series of monographs and edited volumes is being produced to facilitate the exchange of information relating to the microbiology of specific habitats, biochemical processes of importance in microbial ecology, and evolutionary microbiology. The series will also publish texts in applied microbiology, including biotechnology, medicine, and engineering, and will include such diverse subjects as the biology of anaerobes and thermophiles, paleomicrobiology, and the importance of biofilms in process engineering.

During the past decade we have seen dramatic advances in the study of microbial ecology. It is gratifying that today's microbial ecologists not only cooperate with colleagues in other disciplines but also study the comparative biology of different habitats. Modern microbial ecologists, investigating ecosystems, gain insights into previously unknown biochemical processes, comparative ecology, and evolutionary theory. They also isolate new microorganisms with application to medicine, industry, and agriculture.

Applied microbiology has also undergone a revolution in the past decade. The field of industrial microbiology has been transformed by new techniques in molecular genetics. Because of these advances, we now have the potential to utilize microorganisms for industrial processes in ways microbiologists could not have imagined 20 years ago. At the same time, we face the challenge of determining the consequences of releasing genetically engineered microorganisms into the natural environment.

New concepts and methods to study this extraordinary range of exciting problems in microbiology are now available. Young microbiologists are increasingly being trained in ecological theory, mathematics, biochemistry, and genetics. Barriers between the disciplines essential to the study of modern microbiology are disappearing. It is my hope that this series in Ecological and Applied Microbiology will facilitate the reintegration of microbiology and stimulate research in the tradition of Louis Pasteur.

There is an increasing realization among scientists and engineers that microbial biofilms play an essential role in engineering processes as diverse as biological waste treatment and energy production. Neither the dynamics of biofilm development nor their biochemical activities on surfaces are well understood. This volume provides a comprehensive review of the physics, chemistry, and biology of biofilm developmet. It presents modelling approaches to biofilm behavior and explains engineering applications of biofilm processes. It should be invaluable to a wide audience of students and researchers in engineering and microbiology.

RALPH MITCHELL

Cambridge, Massachusetts
April 1987

PREFACE

Biofilms is intended to provide a basis for interdisciplinary research, process development, and industrial production in a critical area of environmental biotechnology: interfacial microbial processes. Interfacial properties and processes affect virtually every aspect of engineering and biotechnology. In environmental biotechnology, biological processes at interfaces are most pronounced and persistent. *Biofilms* addresses conceptual and technological needs prominent in environmental biotechnology, a field that focuses on processing of high-volume/low-value products related to more efficient production of energy, materials and chemicals, a cleaner environment, and improved agriculture.

Biofilm science/technology is a relatively new technical discipline, which has emerged in response to the tremendous opportunities and significant costs resulting from pervasive microbial activity at interfaces. *Biofilms* represents an interdisciplinary research area that focuses on understanding and modulating the combination of biological and chemical reactions as well as transport and interfacial transfer processes affecting microbial accumulation and activity at surfaces. Biofilm processes are far more complex and fascinating from a systems viewpoint than processes occurring in other types of microbial systems. For example, biofilms or microbial aggregates at surfaces appear to have much in common with multicellular organisms and are, if anything, more adaptable. Both structure and dynamics of these aggregates need to be systematically and quantitatively examined, a complex challenge addressed in *Biofilms.*

Biofilm processes profoundly impact industrial productivity and competitiveness. Because of their pervasive effects on water quality, power generation, energy efficiency, human and animal health, and product quality, the negative effects of biofilms are observed in most industries. While these processes can be extremely costly to industry and harmful to health, they can be highly beneficial in other applications. As a result, opportunities presented in process innovation, instrumentation development, biotechnological advances, and environmental benefits are of tremendous interest.

Biofilms focuses on the following collective topics: (1) analysis of biological phenomena at surfaces and interfaces and development of biological paradigms for life on surfaces, (2) the relationship between interfacial structure and biological function at interfaces, and (3) advancement of process engineering models for fundamental biological/chemical interactions at surfaces.

A barrier to further development of biofilm science/technology has been an inability to integrate the diverse disciplines required to understand microbial processes at surfaces, especially in large industrial and environmental systems. The interdisciplinary nature of interfacial microbial processes precludes mounting a coherent attack unless microbiologists, chemists, and process engineers integrate their knowledge to address these complex systems.

Biofilms is the first systematic presentation of interfacial microbial processes covering material from fundamentals through applications in technology. In addition to consolidating data from over 680 published references, a substantial amount of previously unpublished data is presented. Analytical methods, microbial reaction stoichiometry, and properties of biotic components in biofilm have been obtained from microbiology. Reaction stoichiometry and kinetics, electrochemistry, interfacial chemistry, and biochemistry topics have their origin in physics and chemistry. The conservation equations (momentum, energy, and mass), the basis of mathematical models derived in engineering, unify the interdisciplinary approach.

Process analysis, an exercise frequently lacking in the biological sciences, unites the diverse activity necessary to advance this discipline. Process analysis encompasses the following: (1) examination of the complex, interconnected processes occurring within interfacial microbial systems through coherent, insightful models; (2) combination of essential thermodynamic and transport properties with process characteristics (e.g., kinetics and stoichiometry) to describe biological metabolism and other behavior at interfaces; (3) a template for structuring the seemingly diverse topics of *Biofilms;* and (4) a basis for engineering design and industrial application.

Engineers and scientists frequently analyze problems on a continuum of scales ranging from microscale to macroscale. Engineering analyses of biofilm processes have generally been accomplished at the *mesoscale* (10^{-3}–10 m). Studies at cellular or molecular dimensions (i.e., the *microscale,* $\leq 10^{-3}$ m) have generally been the domain of biologists and chemists. Biofilms in environmental systems (e.g., an alpine stream) and industrial systems (e.g., a recirculating cooling tower system), on the other hand, are *macroscale phenomena* ($>$ 10 m).

Biofilms integrates the microscale and mesoscale phenomena and introduces methods for "scaling up" these processes to the macroscale. For example, the book addresses the way local structure (at the microscale) of the biofilm system is influenced by initial conditions (i.e., starting materials and process conditions) under which the system was created. The book relates the properties and functional characteristics of the biofilm at the mesoscale (e.g., heat exchange tube) to the cell and interface properties at the microscale (e.g., the biofilm), the contacts and connectivity of the cells with the interface, extracellular components, and the surrounding environment. Critical factors in the design and operation of biofilm reactors at the macroscale are discussed. Methods are presented to design and control the interfacial cellular transport and transformation processes at the mesoscale to achieve a desired structuring for commercial or environmental gain (macroscale). Finally, models are presented to integrate the properties and phenomena observed at the micro- and mesoscale with those of critical importance at the macroscale.

Biofilms can be used for advanced undergraduate and graduate courses in engineering and microbiology. The subject material provides an exceptional topic for combining momentum, energy, and mass transport in a biologically reactive system. For example, professionals in bioprocessing, water and wastewater treatment, petroleum production and refining, and power generation; aquatic ecologists; and environmental engineers will find it a valuable reference because of the focus on relevant fundamental principles in biology and engineering, with repeated emphasis on the importance of their synthesis. The book gives readers a strong foundation for future practice. *Biofilms* also offers a consistent terminology and notation, to make communication more efficient.

Biofilms is divided into five parts. Part I introduces biofilms and process analysis techniques. In addition, experimental biofilm systems are also characterized. Part II reports properties of biofilms and their two major constituents: cells and extracellular polymeric substances. The stoichiometry of microbial biofilm reactions is presented. Part III describes physical, chemical, and microbiological processes (and their rates) that may contribute to biofilm accumulation and activity; it also integrates them into a typical progression of biofilm buildup. Part IV presents modeling approaches for simulating biofilm system behavior. Finally, Part V presents several applications of biofilm processes in the context of the fundamentals previously presented.

Biofilms combines the work of several authors and, at the same time, attempts to use a consistent terminology and notation. However, errors may have occurred. We apologize for any errors and encourage the reader to notify us regarding any inconsistencies in terminology, notation, or presentation.

The senior author (W. G. C.) acknowledges the following persons, in addition to coauthors, for their help and/or support in the preparation of this book: A. W. Busch, my mentor; C. M. Balch, G. F. Bennett, K. J. DeWitt, J. C. Geyer, D. F. Gibson, L. D. Jensen, J. W. Jutila, T. E. Lang, C. E. Renn, J. T. Sears, C. H. Ward, and T. T. Williams for extraordinary administrative and technical support; W. A. Corpe, J. W. Costerton, W. A. Hamilton, R. L. Irvine, D. Jenkins, Z. Lewandowski, J. V. Matson, L. V. McIntire, B. F. Picologlou, H. A. Videla, and P. A. Wilderer for collegial support; B. Batchelor, R. S. Cahoon, P. A. D'Alessandro, M. S. McCaughey, M. J. Nimmons, G. Norrman, S. L. Raef, F. L. Roe, B. M. Thomson, and M. G. Trulear, who contributed significantly to my learning while they were students or staff in my laboratory; the Swiss Federal Institute for Water Resources and Water Pollution Control, where the idea for the book first took shape; the contributions of many colleagues and program managers who served with me on projects supported by the Institute for Biological and Chemical Process Analysis Industrial Associates (Montana State University), National Science Foundation, Office of Naval Research, Environmental

Protection Agency, Electric Power Research Institute, and U.S. Geological Survey; members of the International Association for Water Pollution Research and Control Task Group on Modelling of Biofilm Processes for many stimulating discussions on biofilm topics; copy editor Joe Fineman for improving the manuscript; production editor Isabel Stein for her aid in manuscript revision; Marcia Bottomly, Anne Camper, Gregory Characklis, Diane Doede, Jamie Herigstad, and Wendy Wesen for invaluable help in preparing text, figures, and supporting materials for the manuscript. Thanks to Joyce, George M., George K., Harriet, Nick, and Jerry.

WILLIAM G. CHARACKLIS
KEVIN C. MARSHALL

Bozeman, Montana
Kensington, New South Wales, Australia
May 1989

CONTENTS

Biofilms

PART I

INTRODUCTION

BIOFILMS: A BASIS FOR AN INTERDISCIPLINARY APPROACH

WILLIAM G. CHARACKLIS
Montana State University, Bozeman, Montana

KEVIN C. MARSHALL
University of New South Wales, Kensington, New South Wales, Australia

1.1 BIOFILMS

The first serious study of biofilms was perhaps that of Zobell (1943), who suggested a two step process for microbial colonization, an initial reversible step followed by an irreversible binding of the cell to the surface. Zobell also recognized that macromolecular conditioning films may modify surfaces prior to microbial colonization. Progress in biofilm research was relatively slow until the early 1970s when researchers in various fields became aware of almost universal association of microorganisms with surfaces or with each other (Marshall, 1976). Research on, and hence understanding of, biofilm processes has progressed rapidly in the last decade and is summarized succinctly in the proceedings of a recent workshop (Marshall, 1984).

One of the ultimate aims in studying biofilms is to evolve the means for manipulating these processes for technological and ecological advantage. Thus, we wish to control fouling of a ship hull or heat exchange surface, control dental plaque or periodontal disease, control cholera and other intestinal diseases, enhance biofilm formation in fixed film fermenters, or modify microbial associations in natural environments to establish more effective consortia.

1.1.1 Definition

Microbial cells attach firmly to almost any surface submerged in an aquatic environment. The immobilized cells grow, reproduce, and produce extracellular polymers which frequently extend from the cell forming a tangled matrix of fibers which provide structure to the assemblage termed a *biofilm*.

A biofilm consists of cells immobilized at a substratum and frequently embedded in an organic polymer matrix of microbial origin.

Biofilms sometimes provide a uniform coverage of the wetted surface and in other locations are sometimes quite "patchy." Biofilms can consist of less than a monolayer of cells or can be as thick as 300–400 mm, as in algal mats. The same biofilm can provide a variety of microenvironments for microbial growth. In fact, the clear distinction between biofilms and other microbial systems is the heterogeneity in its microenvironment, which makes transport processes and gradients so important. For example, a thick biofilm can contain both aerobic and anaerobic environments due to oxygen diffusion limitations within the biofilm. In the case of algal mats, the aerobic and anaerobic regions change diurnally because of the production of oxygen in photosynthetic processes during daylight hours.

A biofilm is a surface accumulation, which is not necessarily uniform in time or space.

In the simplest case, biofilms are composed of microbial cells and their products (e.g., extracellular polymers). Such a biofilm generally is a very adsorptive and porous ($\geq$ 95% water) structure. As a result, biofilms observed in many waters consist of a large fraction of adsorbed and entrapped materials such as solutes and inorganic particles (e.g., clay, silt). Many such deposits are found to contain less than 20% volatile mass, suggesting that the organic content is of minor importance, a presumption which may be far from the truth. The organic matrix may be necessary to bind the inorganic components into a coherent deposit.

A biofilm may be composed of a significant fraction of inorganic or abiotic substances held together by the biotic matrix.

1.1.2 Biofilm Systems

A biofilm system consists of the biofilm, the overlying gas and/or liquid layer, and the substratum on which the biofilm is immobilized. A biofilm is composed of microorganisms immobilized at a substratum (i.e., the support surface) generally in association with an organic polymer matrix (Figure 1.1). However, the biofilm may also contain degradable and/or inert particles and sometimes may

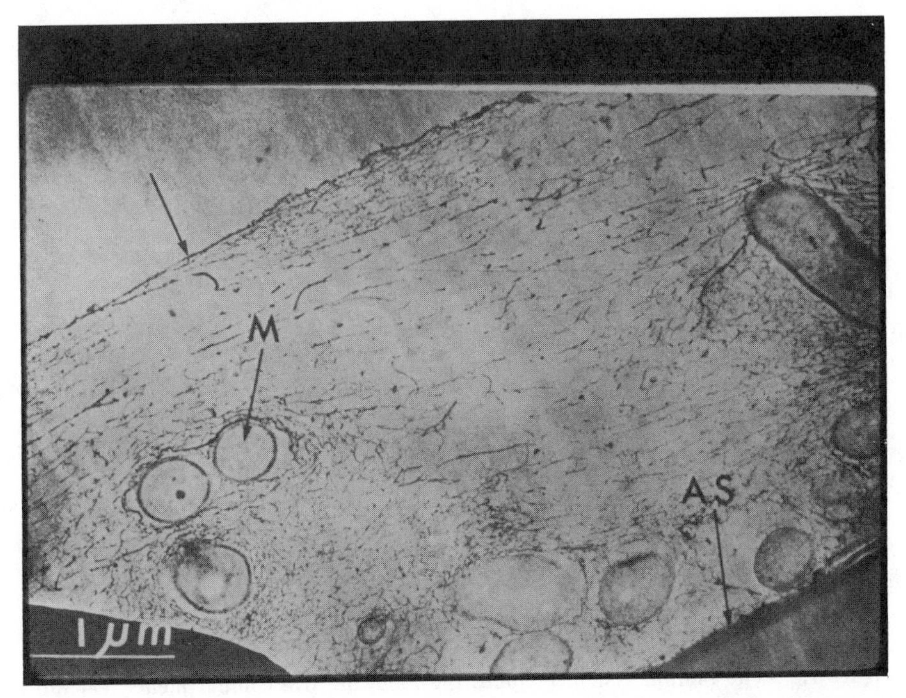

Figure 1.1 A transmission electron micrograph of a biofilm consisting of an undefined microbial population. The biofilm has been treated with ruthenium red, which preferentially stains polysaccharides and demonstrates the significant volume fraction of the extracellular polymer substances. (Reprinted with permission from Jones et al., 1969.)

include macroorganisms such as worms. The biofilm system can be classified in terms of *phases* and *compartments*.

1.1.2.1 System Compartments As many as five compartments can be defined in a biofilm system (Figure 1.2): (1) the substratum, (2) the base film, (3) the surface film, (4) the bulk liquid, and (5) the gas. Each compartment is characterized by at least one phase (gas, liquid, or solid), using the term *phase* in the strict thermodynamic sense. Thus, each compartment can be described in terms of its thermodynamic and transport properties as well as by the transport and transformation processes that dominate within the compartment (Figure 1.3). The system includes all of the "compartments," the phases, the process components (properties and processes), and the geometry of the system (e.g., a rotating biological contactor or a mountain stream).

The substratum plays a major role in biofilm processes during the early stages of biofilm accumulation and may influence the rate of cell accumulation as well as the initial cell population distribution. The substratum generally is an impermeable, nonporous material such as a metal. However, biofilm processes on

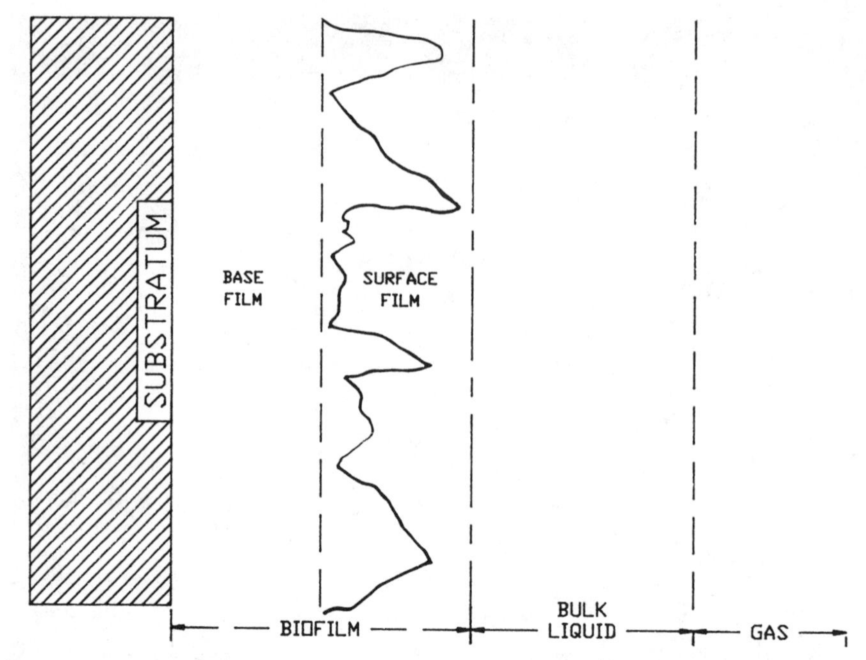

Figure 1.2 The biofilm system includes the following five compartments: (1) substratum, (2) base film, (3) surface film, (4) bulk liquid, and (5) gas. The base film and surface film constitute the biofilm.

semipermeable membranes are of medical and industrial significance. In addition, the substratum can sometimes also serve as the *substrate* (the rate-limiting nutrient for growth), as illustrated by microbial biofilm attack on wood structures or degradation of hydrocarbon droplets.

The biofilm contains two compartments: (1) the base film and (2) the surface film. The base film consists of a rather structured accumulation, having relatively well-defined boundaries. Molecular (diffusive) transport dominates in the base film. The surface film provides a transition between the bulk liquid compartment and the base film. Gradients in biofilm properties in the direction away from the substratum (e.g., a decrease in biofilm density with distance from the substratum) are most important in the surface film. In some cases, the surface film may extend from the bulk liquid compartment all the way to the substratum, especially if a preponderance of filamentous or sheathed (e.g., Gallionella) microorganisms are present. In other cases, the surface film may not exist at all, as in certain monoculture biofilm systems (Figure 1.4). On the other hand, the surface film may be the dominant compartment of the biofilm (Figure 1.5). Advective transport dominates the surface film. The biofilm compartment contains at least two phases (Figure 1.6):

1. A continuous liquid (water) phase which fills a connected fraction of the biofilm volume and contains different dissolved and suspended particulate

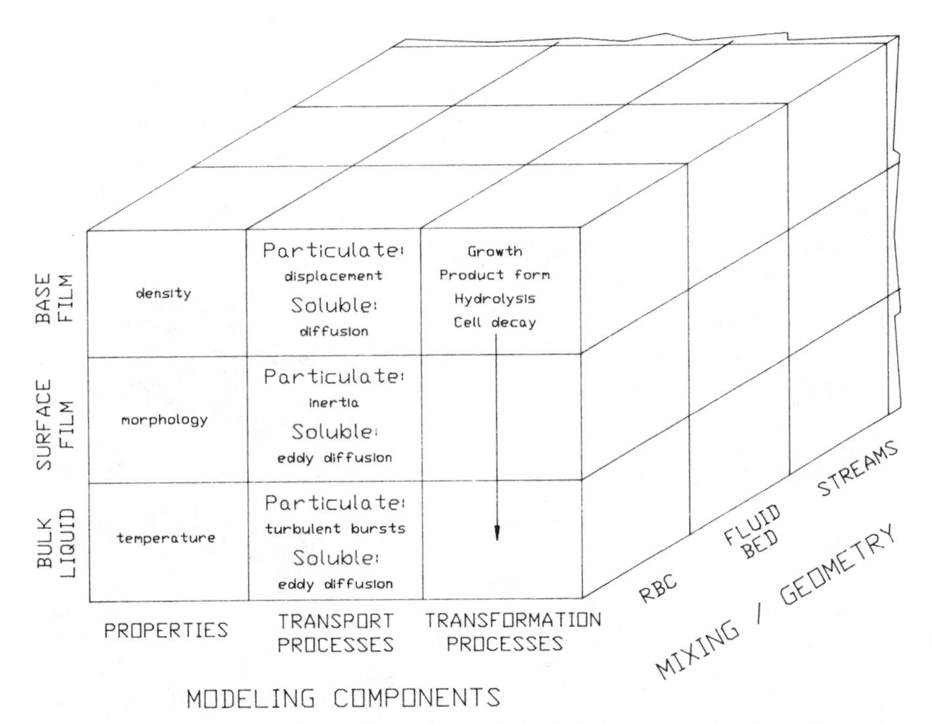

Figure 1.3 The biofilm system modeling matrix includes compartments (e.g., base film, surface film, bulk liquid), modeling components (e.g., properties, processes), and mixing/geometry in the system (e.g., fluidized bed, rotating biological contactor, streams).

materials. The suspended material consists of particles which can move in space independently of one another.

2. A series of solid compartments each composed of specific particulate material, such as a species of microorganisms, extracellular material, or inorganic particles. The solids cannot move freely in space, because they are attached to each other. Movement of attached particles within one solid compartment causes displacement of neighboring particles. Thus, each type of attached solid constitutes a different solid phase, which in addition may contain other components (e.g., sorbed components, components stored within microbial cells).

The transfer of a suspended particle from the liquid phase to a solid phase within the biofilm (e.g., attachment process) has characteristics of a reaction (transformation) process, since educt and product do not belong to the same phase. Thus, the interfacial transfer processes must be distinguished from the transport processes within the biofilm compartment (see Chapter 2 for definitions of transfer, transport, and transformation processes).

Bulk liquid compartment processes affect biofilm system behavior primarily as a result of the mixing or flow patterns resulting from the system's geometry.

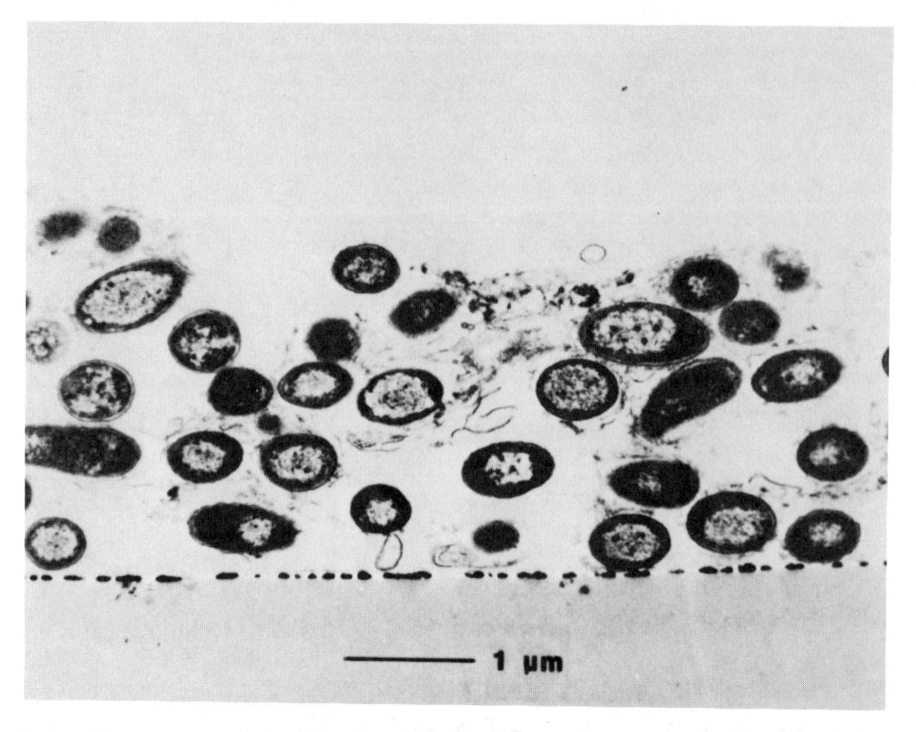

Figure 1.4 A transmission electron micrograph of a *Pseudomonas aeruginosa* biofilm consisting almost entirely of a base film (Bakke et al., 1984).

Figure 1.5 A photomicrograph of a mixed population biofilm consisting largely of a surface film. This very rough biofilm–bulk water interface can significantly influence mass transfer processes between the bulk water and the biofilm. (Courtesy Maria Fruhen, Technische Universität Hamburg-Harburg.)

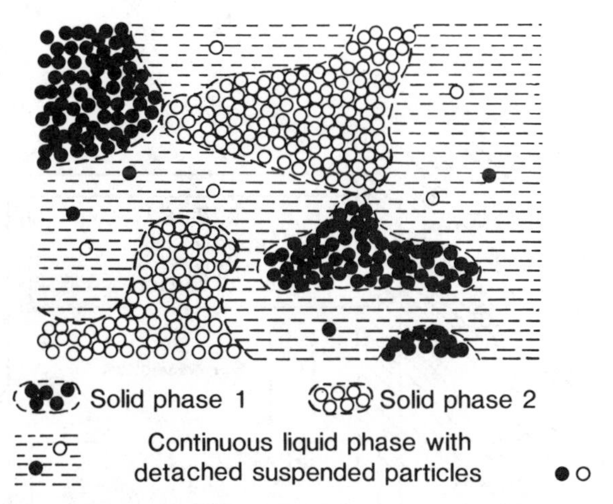

Figure 1.6 Schematic representation of the different phases within the biofilm compartment. The continuous liquid (water) phase may contain dissolved as well as suspended particles. Various solid phases (e.g., different microbial species, inorganic particles, polymeric aggregates) are bound to each other, and individual attached particles cannot move independently of each other (Gujer and Wanner, this volume).

Mass and heat transfer from the bulk liquid compartment to the biofilm compartment is dependent on the fluid dynamic regime: mass transfer in laminar flow will be much slower than in turbulent flow systems. The reactor or system geometry also influences mixing and, consequently, transport processes. Mass transfer will be considerably different in a rotating biological contactor for wastewater treatment and in a packed bed reactor. If microbial activity in these systems is rate-limited by mass transfer, then performance in the two reactors will be significantly different. Thus, the reactor geometry and flow regime frequently determine the progression of biofilm accumulation (Chapter 3).

The gas compartment, absent in some biofilm environments, provides for aeration and/or removal of gaseous microbial reaction products (e.g., CO_2, CH_4).

1.1.2.2 Interaction between System Compartments The interaction between the various compartments in a biofilm system occurs via transport and interfacial transfer processes (Figure 1.7).

Biofilms are distinguished from suspended growth microbial systems primarily by the critical role of transport and transfer processes, which generally are rate-controlling in biofilm systems.

The transport and transfer processes of concern may occur in the bulk liquid compartment as well as in the biofilm (Chapter 9). Bulk liquid compartment transport processes deliver substrate (particulate or dissolved) and microorgan-

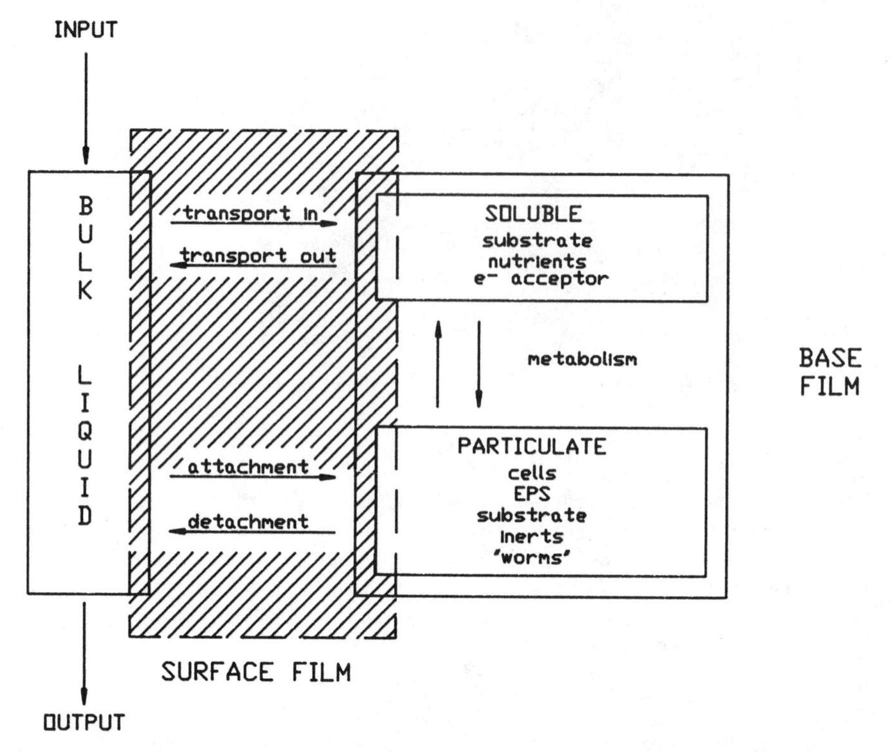

Figure 1.7 A biofilm system, illustrating the interaction between the various compartments through various transport processes (e.g., transport of soluble components in from the bulk liquid compartment to the biofilm) and interfacial transfer processes (e.g., attachment and detachment of particulate components to the biofilm).

isms to the surface film while carrying away detached cells and other products of metabolism. Transport between surface film and base film occurs primarily by molecular diffusion for dissolved components and by volumetric displacement for particulate components. For example, a microbial cell within the film reproduces and the daughter cell "pushes" overlying components up through the film and eventually out into the bulk liquid compartment. There may be other means of transport for particulate components within the biofilm, such as cell motility and macroorganism grazing. Chapter 7 discusses numerous transport and interfacial transfer processes operative in biofilm systems.

1.2 RELEVANCE OF BIOFILMS

Biofilm processes are manifest in many forms and are studied by researchers from a wide variety of disciplines. Some of the more prominent applications of biofilm science and technology are listed in Table 1.1.

TABLE 1.1 Effects and relevance of biofilm processes.

Process	Effects
Biofilm accumulation on heat exchanger tubes and cooling tower fill	Increased heat transfer resistance, energy losses; reduced performance
Biofilm accumulation in water and wastewater conduits, in porous media (e.g., groundwater), and on ship hulls	Increased fluid frictional resistance, energy losses; reduced performance
Accelerated corrosion due to microbial processes at the biofilm–substratum interface	Material deterioration, reduced equipment lifetime
Biofilm formation on sensors, submarine periscopes, and sight glasses	Reduced performance
Biofilm accumulation in drinking water distribution systems	Decrease in water quality and increased health risks
Detachment of microorganisms from biofilms in cooling towers (e.g., *Legionella*)	Health risk
Biofilm accumulation on teeth and gums, urinary tract, and intestines	Health risks
Extraction and oxidation of organic and inorganic contaminants from water and wastewater (e.g., rotating biological contactors, biologically aided carbon adsorption)	Reduced pollutant load to the receiving waters
Benthal biofilm stream activity	Maintenance of instream water quality
Extraction of minerals from ores mediated by biofilms (biohydrometallurgy)	Increased product yield
Immobilized microorganisms in biotechnological industries	Improved productivity and stability of processes

Biofilms serve beneficial purposes in the natural environment and in some modulated or engineered biological systems. For example, biofilms are responsible for removal of dissolved and particulate contaminants from natural streams and in wastewater treatment plants (fixed film biological systems such as trickling filters, rotating biological contactors, and fluidized beds). Biofilms in natural water, called *mats,* frequently determine water quality by influencing dissolved oxygen content and by serving as a sink for many toxic and/or hazardous materials. These same mats may play a significant role in the cycling of chemical elements. Biofilms provide opportunity for syntrophic and other community interactions between microorganisms and a means for survival of microorganisms in natural habitats. Biofilm reactors are used in some common fermentation processes (e.g., the "quick" vinegar process) and may find considerably wider application in the near future.

More recently attention has focused on the nuisance role of biofilms. *Fouling* refers to the undesirable formation of inorganic and/or organic deposits on surfaces, which results in unsatisfactory equipment performance or reduces equipment lifetime. Biofilms cause fouling of industrial equipment such as heat exchangers, pipelines, and ship hulls, resulting in reduced heat transfer, increased fluid frictional resistance, and/or increased corrosion. Fouling is of commercial concern in the manufacture of microelectronics, rolled steel, and paper. In the medicine and dentistry, biofilms are of interest in that they cause health problems.

1.3 GOALS OF THE BOOK

Our goal is to advance the science and engineering of biofilms by providing useful and practical models (physical, conceptual, and mathematical) for them. Our intention is to provide a framework for understanding the accumulation of biofilm and its interaction with the immediate environment. Biofilm systems are described in terms of (1) biofilm properties (Chapters 4 and 5), (2) selected fundamental rate processes (Chapters 6–9), and (3) environmental parameters. These components are integrated via *process analysis* into mathematical models (Chapters 11–13), which serve at least two purposes: (1) facilitating engineering design and (2) helping to understand complex systems and guiding experimental research on them. Much of the presentation in this book is based on the laws of conservation and reactor engineering principles (Chapter 2) and is intended for both engineers and scientists.

The focus of the book is on biofilms accumulated on *abiotic* surfaces. However, the approach and analyses presented are deemed as suitable starting points for biofilms on biotic surfaces as well.

Another goal of the book is to provide the reader with a state of the art review of biofilm science and technology. In doing so, many questions arise concerning our understanding of the mechanisms by which biofilm processes proceed and the usefulness of methods being used to elucidate the mechanisms. Thus, the book may also provide the direction for future research and development related to biofilm processes.

One of the barriers to advances in biofilm science and technology has been the inability to integrate the diverse disciplines that focus on the topic. Different approaches to the study of biofilm phenomena have developed, depending on whether the work has been conducted by microbiologists, physical chemists, or engineers. Each profession has a narrow perspective of the problem and could benefit from exposure to other views.

The apparent complexity of the system also discourages some. For example, the types of substrata available for biofilm formation are only outnumbered by the array of microorganisms participating in biofilm processes.

In our book, the chapter authors are from various disciplines, including process engineering, microbiology, and chemistry, but the material has been integrated to provide a comprehensive treatment of biofilm processes. The contributions of many disciplines are especially evident in Part V, which concentrates on various specific applications of biofilm science and technology.

A barrier to rapid progress in biofilm research is failure of communication, partly due to the multidisciplinary approach to the study of biofilms. Thus, no consistent terminology or notation has been developed for presenting theories and hypotheses. Clearly, more consistency is desirable for improving communication between the disciplines involved in the science and engineering of biofilms.

1.4 TERMINOLOGY

Physical chemists, microbiologists, environmental engineers, and chemical engineers are all involved in research and development related to biofilms. All have their own familiar terminology, which is not always consistent with that used in other disciplines. We have "pooled" some of the terms to arrive at a consistent set of definitions for this book. The combining of terms also reflects the cellular, as opposed to molecular, focus of the book. The resulting definitions will eventually evolve into a more convenient and useful set of terms suitable and satisfactory to all.

The terms *attachment, adhesion,* and *adsorption* are sometimes used casually in the literature, and in some cases synonymously, in reference to the intimate interaction of a cell with the substratum. These terms also serve as a useful illustration of the difficulties encountered in preparing a consistent terminology for an interdisciplinary topic.

We begin with the term *sorption,* which we define as a process in which a *molecule or cell* moves from one phase to be accumulated in another, particularly when the second phase is a solid. Sorption is a general term that includes both adsorption and absorption.

Absorption is the penetration of molecules or cells nearly uniformly among those of another phase to form a "solution" with the second phase. Absorption is a three dimensional process and is best suited to describing the interaction of water phase components (e.g., particulate substrate) with biofilm, which usually has a more accessible three dimensional structure than the substratum. Therefore, absorption is related to attachment (defined below).

Adsorption is the interphase accumulation or concentration of molecules or cells on a substratum or interface. Adsorption is a two dimensional process. Two types of adsorption can be differentiated on a semiquantitative basis: (1) physical adsorption and (2) chemisorption.

Physical adsorption is a reversible or equilibrium adsorption involving primarily physical forces (e.g., van der Waals, hydrogen bonds, protonation, coordination bonds, and water bridging), is characterized by a low heat of adsorption per chemical bond (20–50 kJ mol^{-1}), and exhibits low specificity between the adsorbent and the adsorbate. The net rate of physical adsorption is frequently described by saturation kinetics.

Chemisorption is generally irreversible adsorption and is characterized by a high heat of adsorption per chemical bond (40–400 kJ mol^{-1}) and a more definitive chemical interaction (e.g., ionic or covalent bond). Chemisorption generally exhibits a high specificity of the adsorbate for the adsorbent and usually results in only single layer adsorption. In cell–surface (substratum) interactions, chemisorption is also referred to as *adhesion*, because it refers to more or less irreversible adsorption frequently mediated by an adhesin (Marshall, 1984) such as an extracellular polymer. Definition of chemisorption frequently requires reference to an assay technique to reduce the concept to an operational level (especially in batch or closed systems).

Desorption is the reverse of adsorption and refers to the movement of molecules or cells from the substratum back into the bulk liquid compartment.

Attachment is defined as the capture and/or entrapment of cells in a biofilm. It refers to the interaction of bulk liquid compartment components with the biofilm components, in contrast to adsorption, which occurs at the liquid–substratum interface.

Detachment is the reverse of attachment and is the movement of cells from a biofilm into the bulk liquid compartment. Detachment is loss of components from the biofilm, in contrast to desorption, which is loss of components from the substratum.

The relationship between these processes is illustrated in Figure 1.8. All of these processes are discussed in more detail qualitatively in Chapter 7 and quantitatively in Chapters 8 and 9. Models incorporating these processes are presented in Chapters 11–13.

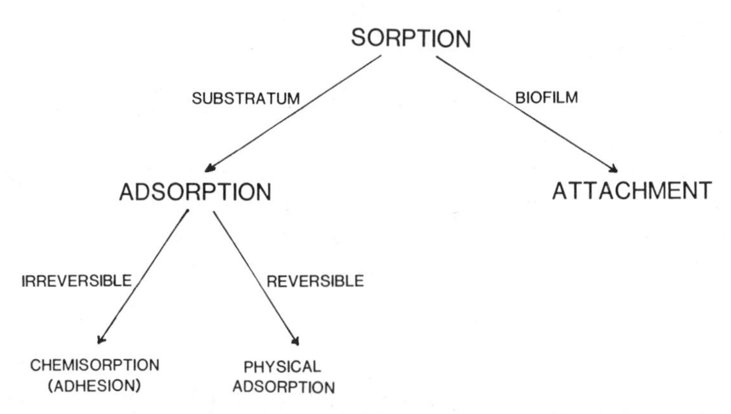

Figure 1.8 The relationship between definitions used to describe the interaction of bulk liquid compartment cells with the substratum or the biofilm.

1.5 NOTATION

Providing consistent notation and units of measure has been challenging. For example, chemical engineers generally use μ as a symbol for viscosity. However, μ also serves as the symbol for microbial specific growth rate. Thus, choices had to be made, usually based on subjective criteria. Similar problems were encountered with units of measure. For example, cell number concentration are generally reported as number per liter, per milliliter, or per 100 milliliters. Yet, neither the liter nor the milliliter is an SI unit. The units and notation recommended by the International Association on Water Pollution Research and Control (IAWPRC) and the Commission on Water Quality of the International Union of Pure and Applied Chemistry (IUPAC) have been adopted whenever possible for biofilms in order to improve communication among biofilm scientists and engineers. A list of all symbols and their units of measure are provided in the Symbols and Notation appendix at the end of the book.

REFERENCES

Bakke, R., M.G. Trulear, J.A. Robinson, and W.G. Characklis, *Biotech. Bioeng.*, **26**, 1418–1424 (1984).

Jones, H.C., I.L. Roth and W.M. Sanders, *J. Bact.*, **99**, 316 (1969).

Marshall, K.C., *Interfaces in Microbial Ecology,* Harvard University Press, Cambridge MA, 1976.

Marshall, K.C. (ed.), *Microbial Adhesion and Aggregation,* Springer-Verlag, Berlin, 1984.

Zobell, C.E., *J. Bacteriol.*, **46**, 39–56 (1943).

PROCESS ANALYSIS

WILLIAM G. CHARACKLIS
Montana State University, Bozeman, Montana

SUMMARY

Process analysis offers a systematic method for the recognition and definition of biofilm-related problems and the development of procedures for their solution. This generally requires (1) development of a mathematical model, (2) experimental testing of the model, and (3) synthesis and systematic presentation of results to ensure full understanding. This chapter presents the basis for process analysis: the equations for conservation of mass, energy, and momentum. The most important results, from the viewpoint of a process analysis of a reacting system, are expressions which quantitatively describe the rate and extent of the fundamental processes contributing to the performance of the overall system. The expressions relevant to biofilms will be discussed in more detail in Chapters 8, 9, and 11. Process analysis permits the formulation of biofilm models that describe their behavior in a wide range of environmental conditions. These models are invaluable for engineering design of biofilm reactors, operation of biofilm reactors, and analysis of biofilm ecosystems.

Mathematical modeling is also an integral part of experimental research because models require that ideas be systematically and rigorously formulated and definitions to be precise prior to mathematical formulation. In addition, mathematical description of a process often offers considerable insight into an experimental design. A more integrated modeling approach, which incorporates microbial physiology and reactor engineering, is needed to understand, control, and exploit biofilm processes. Thus, assumptions and other specific considerations for microbial reactors (e.g., physiology, genetics, ecology) are also presented. Process analysis is a major tool needed for this integrative, interdisciplinary approach. The goal is a predictive model for biofilm accumulation and activity useful for a wide variety of reactor geometries and environments.

2.1 INTRODUCTION

Process analysis refers to the application of systematic methods to the recognition and definition of problems and the development of procedures for their solution. This generally requires (1) development of a mathematical model, (2) experimental testing of the model, and (3) synthesis and systematic presentation of results to ensure full understanding. The *process* denotes an actual series of operations or treatment of materials, as contrasted with the *model,* which is a mathematical description of the process (Himmelblau and Bischoff, 1968). The basis for process analysis rests in the conservation equations: mass, energy, and momentum. The most important results, from the viewpoint of a process analysis of a reacting system, are expressions that quantitatively describe the rate and extent of the fundamental processes contributing to the performance of the overall system.

Much of the past emphasis on biofilms in microbiology laboratories has been on mechanism (e.g., mechanism of adsorption). In engineering research, overall observed processes such as substrate removal have been the major concern, frequently without resort to fundamental rate and stoichiometric concepts.

A more integrated approach, which incorporates microbial physiology and reactor engineering, is needed to understand, control, and exploit biofilm processes. Process analysis is a major tool needed for this integrative, interdisciplinary approach.

There are three main applications of biofilm process analysis with which we are ultimately concerned: (1) operation of existing biofilm reactors or other technical equipment, (2) design of new or modified biofilm reactors and technical equipment, and (3) analysis of natural biofilm processes (e.g., aquatic ecosystems). In the area of operations, both control and optimization of performance stand out as two of the main functions which require sophisticated analysis of biofilm processes. For example, computers must be programmed so that relations describing individual steps of biofilm processes can be combined into an overall biofilm control system, basic parameters in those relations (e.g., kinetic and stoichiometric coefficients) must be evaluated, and qualitative observations (e.g., biofilm physical properties) must be quantified. For these and many related reasons, effective control and optimization of biofilm systems rest on sound process analysis.

The second task, biofilm reactor design, is more difficult. Actual biofilm process plant data are frequently *not* available beforehand, and the engineer must rely on intuition. On the other hand, when modifying existing equipment or designing new equipment similar to that already built, the engineer can draw more heavily on experience. As a consequence, construction of theoretical or semitheoretical mathematical models of the biofilm process frequently is a necessary prelude to design.

The final task, analyzing natural biofilm processes, has become extremely important. We frequently wish to exploit natural processes and, at the same time, minimize human impact on them. Natural biofilm processes frequently pose a greater challenge than industrial or technological biofilm processes because of their oscillatory nature and the large number of variables that significantly affect them.

2.2 MODELING

Modeling is an iterative process. First, a model is formulated in mathematical terms. This model is usually too simple to be realistic; however, it serves as a hypothesis, and a set of experiments is designed to test it. If the results of experiments differ significantly from its predictions, the first model is modified and a new hypothesis (a second model) is formulated (Figure 2.1).

A model never perfectly describes reality, nor is that a necessity. However, the model must describe enough of reality to answer the questions (hypotheses) that have been posed.

2.2.1 Why Model Microbial Systems?

Mathematical modeling attempts to describe aspects of real system behavior with a set of mathematical equations. Mathematical modeling goes hand in

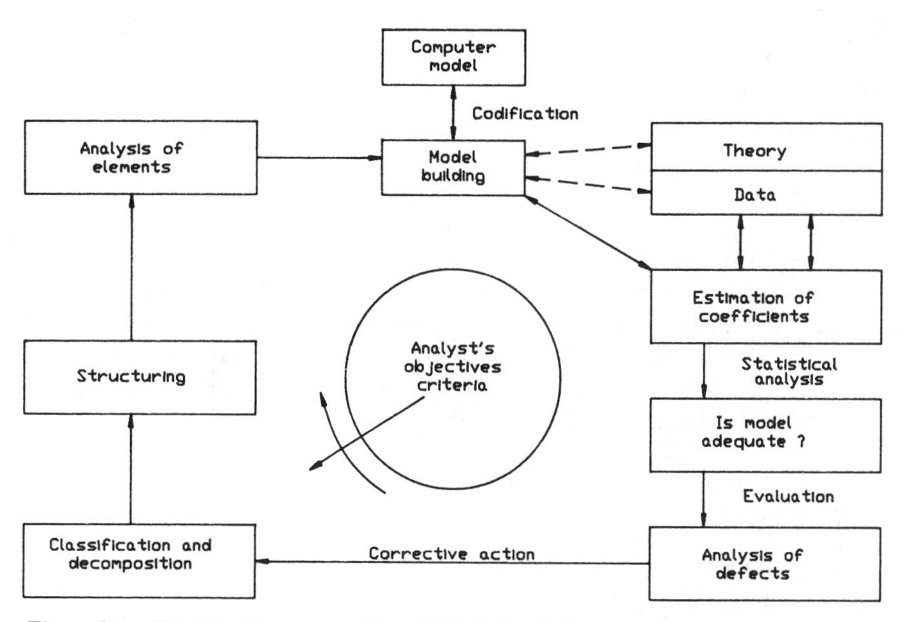

Figure 2.1 The iterative nature of model building (Himmelblan and Bischoff, 1968).

hand with laboratory or field experimental research, because models require ideas that are systematically and rigorously formulated and definitions that are precise prior to mathematical formulation. However, mathematical description of a process often offers considerable insight into an experimental design. Salmon and Bazin (1988) have listed these and other benefits of mathematical modeling microbial ecosystems:

1. Statement of hypotheses in the form of a mathematical equation is usually more precise, more concise, and less ambiguous than the equivalent verbal statement. Model construction tends to impose on the investigator a rigor sometimes lacking in ecology and other life science research.
2. Modeling permits the application of the vast field of mathematical theory to develop *experimentally verifiable* predictions from hypotheses.
3. Important variables and parameters can be identified, thus leading to a more efficient and practical experimental design.
4. Various phenomena can be usefully classified with generality sufficient to permit analogies between widely differing phenomena.

With these potential advantages in mind, our goal is a balance of intuitive understanding of microbial biofilm processes and their quantitative description through simple mathematics.

2.2.2 Types of Models

Three types of models and their combinations can be prepared:

1. Transport phenomena models
2. Population balance models
3. Empirical models

The phenomenological equations of change (conservation of mass, momentum, and energy) are examples of transport phenomena models. Residence time distributions and other age distributions are examples of population balance models. Finally, examples of typical empirical models are those polynomials used to fit empirical data by the method of least squares. Most of our modeling will focus on transport phenomena models and, to a limited extent, on population balance and empirical models.

2.2.3 Simplifying Assumptions

Williams (1972) emphasized the importance of certain procedures in developing models for biological systems:

1. Making complete and rigorous verbal statement of assumptions, with care that relationships to empirical reality are made explicit. This procedure delineates the portion of reality to be considered.
2. Beginning with the simplest possible biological assumptions.
3. Introducing further biological assumptions only when necessary for clearly defined gains in empirical reality. This procedure complicates the system but, at the same time, moves the model closer to reality. Model building is always a compromise between two conflicting demands: (1) the need to pursue reality (wholeness) and (2) the need to limit the complexity of the problem.
4. Beginning with the simplest possible mathematics consistent with the biological assumptions.
5. Introducing further mathematical complexity only when necessary for implementing new biological assumptions.
6. Maintaining a clear distinction between two types of assumption:
   a. The necessary simplifying assumptions for mathematical and experimental tractability, and
   b. The *anacalyptic* (Williams, 1971), or biologically explanatory assumptions, i.e., the important ones that will be tested experimentally. Salmon and Bazin (1988) suggest that anacalyptic assumptions are synonymous with hypotheses.

The growth of even a single population of microorganisms is a very complex process. The task of mathematically modeling such processes is limited by time, insight, and instrument capability, which restrict the number of interactions that can reasonably be considered between a cell and its environment or between a cell population and its environment. Hence, simplifying assumptions are made, which introduce inaccuracies, since the model will not describe all of the details; only continued experimental testing can lead to more realistic mathematical models of these processes.

Many biofilm or microbial growth models make certain simplifying assumptions without stating them explicitly. Such omissions partly result from the undesirable division of tasks between experimenter and modeler. The more common assumptions will be stated explicitly (after Frederickson et al., 1970) and are categorized as *spatial, temporal,* or *functional* simplifications (Savageau, 1976).

2.2.3.1 Distribution of States—Functional Homogeneity within the Population Biological populations are composed of individual organisms, and the individuals are physiologically, morphologically, and genetically different. Individuals differ in size, shape, staining properties, and perhaps in other characteristics, such as motility. Examination for a period of time teaches that individuals differ in age; for a cell we define the *age* as the chronological time since it was formed by fission of its parent. More refined analyses would perhaps reveal even more fundamental, if less obvious, distinctions between individuals.

In relation to biofilms, the age of adsorbed cells in relation to time of fission may vary because of the *daughter–daughter* relationship or the *mother–daughter* relationship:

Daughter–Daughter. The parent cell grows at the substratum and divides by binary fission to give two apparently identical daughter cells remaining at the substratum.

Mother–Daughter. The mother grows at the substratum (adsorbed perpendicularly) and divides by apparent binary fission to give a daughter cell that is motile and departs from the substratum while the mother continues to grow and produce more daughter cells (Kjelleberg et al., 1982).

In summary, individuals of a population do not all exist in the same *state*, but rather represent a distribution of states.

The first simplifying assumption in modeling microbial processes is that the distribution of states within a population of a given species can be ignored and the properties of the culture can be adequately described in terms of a typical individual, whose behavior represents an average over the distribution of states. Certainly this has been the case in most models describing organisms within biofilms, even though concentration, pH, and temperature gradients in the film are known to exist and create environments varying with biofilm depth. Such assumptions at once lead to uncertainty about the validity of the model, since they imply that a whole host of parameters of the population are not important in determining the properties or activities of the population.

There are three reasons for making these assumptions; one involves a conceptual difficulty, and the other two involve practical difficulties. The conceptual difficulty is: What do we mean by the "state" of an individual organism? Can we use some obvious parameter associated with a cell as a measure or index of its state? Would the age of a cell (as defined above) serve this purpose? Or can we use the size of a bacterial cell as an index of its state? Or is some more general notion of state, such as the biochemical constitution of the cell, needed? In some experimental studies of microbial adsorption, growth rates in a chemostat have been used as an indicator of the physiological state of the population (Bryers and Characklis, 1982; Nelson et al., 1985). In each case, the growth *rate* history has significantly influenced the adsorption rate of the microorganisms.

Finally, the equations resulting from models recognizing a distribution of states quickly become mathematically intractable. For these reasons, models where the distribution of states among individual organisms is neglected will be most common in this text.

2.2.3.2 Segregation—Spatial Homogeneity within the System In unicellular microorganisms, life is *segregated* into structurally and functionally discrete units, i.e., cells. Hence, the *number* of individual organisms present in a population must be an important parameter for the description of the population. Such quantities as the biomass of the population must also be included, because application of conservation equations (in this case, mass conservation) is important. Nevertheless, number must be a quantity of prime importance, since the biological characteristics of a population composed of $2n$ organisms having total biomass m are not the same as those of a population composed of n organisms having total biomass m.

In spite of the foregoing arguments, many models make the assumption that segregation of life into discrete units can be ignored (while others ignore mass of cells or size of cells). With such an assumption, number of organisms is not admitted as a parameter to be described by the model and, in effect, the model views the population as biomass *distributed continuously* throughout the culture. Models based on such neglect of segregation are called *nonsegregated* or *distributed*. Biofilm models have frequently taken this form, especially in the engineering literature.

Can a distributed model have any success in the description of unicellular growth or microbial cell adsorption to a substratum? There may be no *practical* need for knowing the numbers of organisms present (biomass may be the quantity of practical importance), but, in general, increase in number (proliferation) and increase in biomass (growth) are coupled processes, so that one cannot really omit the one from a model purporting to describe the course of the other. A possible explanation for the success of some distributed models is that they have been applied to *balanced growth,* or nearly balanced growth, situations. Under balanced growth conditions, growth and proliferation are *proportional,* so that biomass is directly related to number of organisms.

There will be instances where segregated models are essential for a useful biofilm process model. A good example is the early events in the formation of a biofilm (Chapter 12). A small number of individual cells are transported to the substratum, where adsorption results in a "patchy" or nonuniform distribution. In this case, a distributed model does not describe the process sufficiently, because of the low cell concentration. Gujer (1987) has speculated about the influence of "patchiness" on biofilm activity in later stages of accumulation and surmises that intercellular distances of 10 μm can result in transport control of cross feeding between species. Siebel (1987) has observed clear segregation between two species as they accumulate in a biofilm. In a biofilm, the activity of the cells clearly varies with depth in the biofilm in response to gradients in chemical concentrations (e.g., dissolved oxygen, substrate).

2.2.3.3 *Stochastic Phenomena—Temporal Homogeneity* It is not possible to predict the behavior of individual microbial cells with certainty. Thus, the generation times of individual bacterial cells are not all the same, but rather show random deviations about mean values. Models for microbial propagation generally make the simplifying assumptions that these stochastic phenomena can be neglected and that growth can be treated as a deterministic process. If this assumption cannot be made, then nonsegregated models cannot be used either; if cells divide at random times, the number of cells present must be a random variable in the model.

Stochastic population models (so-called "birth-and-death processes") very quickly lead to formidable mathematical difficulties, even when one attempts to model only very simple biological phenomena. Hence, such models have been avoided whenever possible. Avoidance of these models has usually been possible because the enormous number of cells present results in the random deviations being canceled out. There are cases, however, where random deviations are important. These cases always deal with populations where the *total number* of cells is small. Early events in biofilm formation find a very small number of cells adsorbed on the substratum. The prediction of sloughing from the biofilm and other detachment phenomena (Chapter 7) may well require models with a stochastic component. Random variations in environmental parameters in natural systems will frequently dictate biofilm behavior, and hence make stochastic models useful.

2.2.3.4 Biological Structure—Functional Homogeneity within the Cell Two microorganisms having the same biomass and inhabiting the same environment may nevertheless have widely different properties and activities. This is the problem of *state* again: the two organisms have different states. If the model recognizes the existence of a distribution of states, it should also recognize that the distribution may change in response to changes in the environment. If the model does not recognize the existence of a distribution of states, it should at least recognize the possibility that the state of the average or typical organisms (which is all most models consider) can change in response to changes in the environment. This means that parameters *in addition* to population number and population biomass must, in general, be important for the description of population behavior.

Many microbial process models currently used do not recognize this; in most of these models, population biomass is the sole variable employed for describing the population. Since this procedure regards organisms and population biomass as featureless, structureless entities, we shall call such models *unstructured*.

Williams (1967) described a simple structured model to explain some of the observed discrepancies in microbial growth models (e.g., Monod's equation). Williams suggested that the cell could be subdivided into two compartments: (1) the synthetic portion, S, and (2) the structural/genetic portion, G. He suggested that substrate entered the S-compartment first and was then passed on to the G-compartment. Williams obtained remarkable qualitative agreement between his model and experimental observations. Roels and Kossen (1978) and Harder and Roels (1982) have improved Williams's model, but experimental data useful for testing have not been obtained. Part of the reason is that the S and G compartments were not identified as real, measurable components.

So structured models require a vector quantity for describing the biotic component of the system being modeled. Robinson et al. (1984) and Bakke et al. (1984) have described a structured model for a pseudomonad in a chemostat and in a biofilm that separates the biomass into cellular material and extracellular material, i.e., the capsule or slime material. The structured model can predict certain cellular behavior in a biofilm from process coefficients obtained in a chemostat (Chapter 13). Without the added structure (i.e., separating cellular and extracellular components), the predictions are poor. The model is useful because the two components are separately measurable.

2.2.3.5 *Other Temporal Simplifications* There is a vast difference in the time constants that characterize the dynamic responses of microbial systems. The range of differences in cellular control systems is discussed in Section 5.8.3: mass action (milliseconds), enzyme activity (seconds to minutes), enzyme synthesis (generations). This temporal separation of phenomena can often simplify the analysis of complex systems. The variables that respond much faster than the phenomena of interest can be assumed to be at their steady state values; those variables responding more slowly than the phenomena of interest can be assumed to be constants or parameters with given trajectories. The operation of a continuous flow reactor offers another illustration of a temporal simplification. The steady state assumption does not rigorously require that the derivative with respect to time (time rate of change) equal zero. Steady state (or pseudosteady state) can also be assumed if the time derivative is much smaller than some other term in the material balance equation, e.g., the dilution rate.

Model equations for biofilm behavior are frequently derived in a form which permits dynamic solution. Dynamic solutions, however, are difficult to generalize because they are time-dependent and incorporate the entire history of the biofilm. Therefore, clearly defined steady state or pseudosteady state solutions may often be of more practical value. Pseudosteady state modeling is necessary because characteristic times for substrate conversion, microbial growth, and shifts in microbial population composition are orders of magnitude different and result in "stiff" differential equations leading to complex numerical techniques for their solution. Kissel et al. (1984) give a thorough comparison of characteristic times for specific microbial processes. More generally, the following ranges for characteristic times can be given:

Process	Time Range
Biofilm sloughing	Instantaneous
Solute diffusion into biofilm	Sec–min
Biomass growth and erosion	Days
Changes in species distribution in biofilm	Weeks–months

This wide span of characteristic times implies that short term variations of bulk substrate concentrations may be modeled with fixed microbial distributions over the depth of the biofilm. Frequently, shifts in microbial composition can be modeled with (pseudo)steady state substrate profiles. A biofilm is only in a true steady state if all quantities, including nutrient concentrations in the bulk liquid and biofilm, are constant with time and the production of biofilm microorganisms equals losses by erosion to the bulk liquid.

2.3 MODELING COMPONENTS

The process of mathematical modeling is an iterative one in which careful study of a simple model leads to a more realistic, but more complex, model, which is itself the precursor of a better model. Williams's (1967) structured model is an excellent example. The test of each model, however, requires experiments in which the model variables are measured quantities, a weakness in Williams's model. Hence, it is important to determine which variables should be considered and which variables can be measured in the modeling of microbial processes. The variables of concern can be divided into two categories: *biotic* and *abiotic* variables.

The biotic variables of importance depend on the complexity of the microbial process. If the process involves a monoculture, then the number of individuals (or population density), biomass (or biomass concentration), and distribution of physiological states within the population become quantities of significant interest. If the process involves mixed cultures, then an additional class of variables that describe community structure must be considered. Such quantities include the numbers and biomasses of the various populations of the community, in addition to the distribution of physiological states within each population. The physiological state, quite important in modeling any microbial process, is still somewhat difficult to quantify.

Abiotic variables describe the physical and chemical environment in which the microbial process is occurring. Abiotic variables describe the *state* of the abiotic phase or phases of the reaction system. The *chemical state* describes the concentrations of various chemical substances in the liquid and biofilm, such as substrate, nutrients, dissolved oxygen, and pH. Another class of variables describes the *physical state* of the system, which frequently exerts control of the rate of the biological and chemical processes occurring in the reactor. For example, dilution rate is an important variable controlling microbial processes in a continuous fermenter (chemostat). Another example is mechanical shear, which may alter the physiological state of biofilm organisms by influencing lysis rate, altering membrane processes, or influencing mass transfer rates of nutrients to cells.

Process analysis combines these variables in mass, energy, and elemental balance equations with the various stoichiometric, reaction rate, and transport rate *coefficients* to form predictive models (Chapter 11). These coefficients vary with environmental factors, termed *parameters*, such as temperature, pH, inlet substrate concentration, and substratum composition which are generally controlled by the experimenter.

2.4 FUNDAMENTALS REGARDING RATE

The fundamental relationships that underlie transport phenomenon models are the equations for conservation of mass, momentum, and energy, which result from the laws of thermodynamics and Newton's laws of motion. The conservation equations are generally expressed in terms of intensive variables (i.e., variables that are independent of system mass) such as concentration, velocity, and temperature. The conservation equations also include physical properties such as thermodynamic properties, stoichiometric coefficients, transport rates, and chemical reaction rates (Table 2.1).

The thermodynamic properties (e.g., density, heat capacity, and chemical equilibrium constants) can be estimated with reasonable confidence from mechanistic models and can be measured with reasonable accuracy. The transport and transformation coefficients (e.g., viscosity, thermal conductivity, diffusivity, and chemical reaction rate coefficients) can rarely be predicted and are difficult to measure. Even the definitions of these latter quantities are somewhat arbitrary.

The equations for the conservation of mass, energy, and momentum equate the rate of accumulation to the net rate of input by transport and the net rate of input by various rate processes such as chemical reaction, diffusion, radiation, convection, and viscous dissipation. The conservation equations or balance equations can be expressed in words as follows:

Net rate of accumulation in control volume

= rate of transport in through control surface

− rate of transport out through control surface + rate of generation in control volume

− rate of consumption in control volume [2.1]

TABLE 2.1 Intensive Factors, Physical Properties, and Their Relation to the Conservation Equation

Conservation equation	Momentum	Energy	Mass
Intensive variable	v	T	C
Thermodynamic properties	ρ	c_p	K_{eq}
Constitutive properties:			
Transport	ν	α	D
Chemical reaction	—	—	k, a_{ij}
Microbial reaction	—	—	μ_{max}, K_s, Y

Balances are accomplished on extensive parameters. The change of some extensive parameter E (an extensive quantity is proportional to the mass of the system) with time can be caused by two mechanisms:

1. E can be transformed (i.e., generated or consumed) within the control volume by some process (e.g., chemical reaction) occurring at a rate r_E per unit volume (r_E is negative if E is consumed).
2. E can be transported into or out of the control volume through the control surface exposed to the external environment.

2.4.1 Process Time and the Importance of Rate

Physical, chemical, and biological transformations are completed in a certain period of time. For example, the removal of soluble organics in a biological wastewater treatment process occurs in a specified period of time, the hydraulic residence time within the reactor. Another example is a fouling biofilm that accumulates on a heat exchanger surface over a period of time reducing heat transfer until a critical point is reached when the process unit must be shut down for cleaning operations. The rate of biofilm accumulation on a tooth surface is reduced by brushing the tooth twice a day. The time required for these specified changes to occur is inversely proportional to the rate at which the process occurs.

Rate is the most important quantity in process analysis.

The *extent* of the process is quite distinct from its rate. For a chemical reaction, the extent of the reaction is generally referred to as its stoichiometry. The stoichiometry of microbial reactions is frequently expressed in terms of yields (Chapter 6).

2.4.2 The Anatomy of a Balance Equation

Eq (2.1) can be reworded more simply as follows:

$$\text{net rate of accumulation} = \text{net rate of input by transport} + \text{net rate of production by transformation} \quad [2.2]$$

The net rate of accumulation and the net rate of input by transport are *rates of change*. These rates of change are easily measured and observed quantities, but they may be the result of several process rates. Rates of change are extensive quantities and cannot be easily correlated or generalized with factors that describe the environment. The net rate of accumulation, although easily measurable, rarely provides useful, fundamental information.

The *process rates,* on the other hand, are fundamental intensive quantities in that they can be generalized and correlated with factors such as temperature, pressure, composition, velocity, and geometry which describe the environment.

Process rates describe the velocities of various phenomena such as chemical reaction, microbial cell adsorption, diffusion through the biofilm, or viscous dissipation in fluids.

Rates of change *should not be confused with* process rates. *Rates of change generally are measured or observed quantities in a system from which process rates are inferred. A process rate is the most fundamental of the rate quantities.*

2.4.3 Mass Balance Equations

Mass balance equations are necessary tools for analysis of microbial processes within biofilm systems. Consequently, transport and transformation rates (Eq. 2.2) will be encountered frequently. These rates can be classified as follows:

- Transport processes
  - Bulk transport
  - Intraphase transport
  - Interphase transport or
    - Interfacial transfer processes (Chapters 9 and 11)
- Transformation processes
  - Microbial transformations
  - Chemical transformations

Bulk transport, sometimes termed *advective transport,* refers to the movement of material from one location to another as a result of flows. For example, bulk transport carries planktonic microbial cells in a river water through a heat exchanger as a result of fluid flow.

Interphase transport, or interfacial transfer, is transport through a system boundary. The term mass transfer is usually reserved for interphase (e.g., solid–liquid or gas–liquid) transport processes. The transfer of oxygen from the liquid phase to the liquid–biofilm interface is a mass transfer process. Adsorption is considered an interfacial transfer process because the cell is transported from one phase to another but no molecular rearrangement occurs.

Intraphase transport (e.g., diffusion) is transport within one phase. For example, the diffusion of oxygen from the bulk liquid–biofilm interface through the biofilm compartment is an example of an intraphase, or molecular transport, process occurring in the liquid phase of the biofilm compartment. The system and its boundaries must be specifically defined to analyze these processes. The liquid–biofilm interface is frequently difficult to characterize, so defining the boundaries is not always simple.

One of the important criteria in defining transport processes is that molecular structure of the material must remain unaltered. Transport processes are described by rather simple phenomenological equations containing either a concentration or a concentration difference (for transport through the *system* boundaries—e.g., the liquid–biofilm interface) or concentration gradients (for transport within the *system*—e.g., the biofilm). In a biofilm system that consists of two or more compartments, mass balances for each compartment (e.g., the bulk liquid and the biofilm) may be necessary for a thorough analysis.

Chemical and *microbial transformation processes* refer to changes in composition as a result of chemical reactions or microbial activity. For example, the conversion of glucose to CO_2 and cells in a biofilm is a microbial transformation process, because it results in changes in molecular structure. Transformation is metabolism and is dependent on physiology as influenced by the organism genotype and phenotype. Transformation processes are described by *rate equations* or, more specifically, constitutive or kinetic equations of the following general form:

$$r = r(C_1, C_2, \ldots, C_n) \qquad [2.3]$$

where C_1 = concentration of the ith reacting component

Transport and transformation processes can be combined in a general model based on the conservation equations as follows:

$$\underset{\text{transport process}}{\text{in-out}} + \underset{\text{transformation process}}{\text{generation}} = \text{accumulation} \qquad [2.4]$$

If an experimental system is consistent with a mass balance (approximately 100% recovery), the model describing the system is correct. The usefulness of the model, however, depends on the accuracy of the rate expressions chosen to describe the rate processes.

The anatomy of a mass balance equation, as described by Eqs. 2.1, 2.2, and 2.4, is summarized in Table 2.2.

2.5 REACTORS

Many biochemical systems involve *batch* growth of microorganisms. After *seeding*, or inoculating, a liquid *medium* of appropriate composition with living cells, nothing is added to the *culture* (except possibly oxygen) or removed from it (except possibly CO_2) as growth proceeds. The batch reactor is generally well mixed, i.e., no spatial gradients exist. In a batch process, nutrients are depleted and products accumulate.

However, most technologically and ecologically relevant biofilm processes occur in *continuous flow* reactors, open systems in which reactants continuously flow into the reactor while products are continuously removed.

A stable biofilm cannot be maintained in a batch or closed system.

TABLE 2.2 Summary of the Mass Balance Equations

$$\begin{pmatrix}\text{net rate of}\\ \text{accumulation}\\ \text{in system}\\ \text{volume}\end{pmatrix} = \begin{pmatrix}\text{rate of}\\ \text{transport}\\ \text{in through}\\ \text{system surface}\end{pmatrix} - \begin{pmatrix}\text{rate of}\\ \text{transport}\\ \text{out through}\\ \text{system surface}\end{pmatrix} + \begin{pmatrix}\text{rate of}\\ \text{generation}\\ \text{in system}\\ \text{volume}\end{pmatrix} - \begin{pmatrix}\text{rate of}\\ \text{consumption}\\ \text{in system}\\ \text{volume}\end{pmatrix}$$

$$\begin{pmatrix}\text{net rate of}\\ \text{accumulation}\end{pmatrix} = \begin{pmatrix}\text{net rate of input}\\ \text{by transport}\end{pmatrix} + \text{process rate}$$

$$\text{accumulation} = \text{in–out} + \text{conversion}$$

		transport processes		transformation processes

Two types of continuous flow reactors will be important in further discussions: continuous flow stirred tank reactors (CFSTR) and plug flow reactors (PFR).

2.5.1 Residence Time Distribution of Fluid in Vessels

How can we describe the flow characteristics of fluid in a vessel precisely enough to yield information useful in the design of reactors? The traditional approach is to find out how long individual molecules stay in the reactors by using population balance models. The distribution of ages of molecules or particles in the exit stream or the distribution of residence times of molecules within the vessel can be found easily and directly by a widely used experimental technique, *the stimulus-response* technique. The most common stimulus is a tracer (e.g., salt or dye) that is injected (e.g., pulse injection) into the reactor. The response is determined by measuring the effluent concentration of salt or dye. This information can then be used to account for the influence of nonideal flow behavior or mixing of fluid in a chemical flow reactor on reactor performance (Levenspiel, 1972). The stimulus-response technique is illustrated in Chapter 3 for a RotoTorque biofilm reactor as well as a cascade of four RotoTorques in series.

In developing the terminology for reactor mixing, consider the steady state flow, without reaction and without density change (true for our liquid compartment systems), of a single fluid through a reactor. Under these conditions,

$$\Theta = \frac{V}{Q} = \text{mean residence time (holding or space time)} \qquad [2.5]$$

$$D = \frac{Q}{V} = \text{dilution rate (or space velocity)} \qquad [2.6]$$

where V = reactor volume (L^3)
Q = volumetric flow rate ($L^3\ t^{-1}$)

For liquid phase continuous flow systems at steady state, the mean residence time of any system component, i, can be defined as follows:

$$\Theta_i = \text{mean residence time} = \frac{\text{amount of component } i \text{ in the system}}{\text{flow rate of component } i \text{ out of the system}} \qquad [2.7]$$

2.5.2 Continuous Flow Stirred Tank Reactor (CFSTR)

A *continuous flow stirred tank reactor* can be visualized as a well-stirred tank into which there is a continuous flow of reacting material, and from which the (partially) reacted material and product pass continuously. The important characteristic distinguishing the CFSTR from the PFR is the mixing.

2.5.2.1 *Mixing in the CFSTR* In the ideal CFSTR, agitation is assumed to be so vigorous that mixing is complete. If a drop of dye is injected into an ideal CFSTR, it disperses instantaneously. No real reactor can be perfectly mixed, but in practice the behavior of a real reactor can be made to approach that of an ideal CFSTR very closely. The criteria of complete or perfect mixing may be stated in different, but equivalent, ways.

The usual criterion for a CFSTR is simply that the composition of the liquid stream leaving the reactor is the same as that of any sample drawn from the reactor.

From a statistical point of view, mixing is perfect if (1) the probability that a *particle* (organism or molecule) will be in a given subvolume in the culture is the same as the probability that it will be in any other subvolume of equal size, regardless of location in the reactor liquid, and (2) the particles move independently through the vessel. If particles move together in aggregates through the reactor, the mixing is said to be segregated (Levenspiel, 1972).

Let C be the concentration (amount per unit volume of culture) of some substance, biotic or abiotic, in the reactor liquid. Then, application of the principle of conservation of mass to the system in the CFSTR yields the following differential equation (Figure 2.2):

$$\underset{\text{accumulation}}{\underset{\text{rate of}}{\frac{V\,dC}{dt}}} = \underset{\text{transport out}}{\underset{\text{net rate of}}{Q\,(C_i - C)}} + \underset{\text{production}}{\underset{\text{rate of}}{V\,r}} \qquad [2.8]$$

where C = reactor concentration (ML^{-3})
C_i = feed concentration (ML^{-3})
Q = volumetric flow rate (L^3t^{-1})
r = rate of production per unit volume ($ML^{-3}t^{-1}$)
V = liquid volume in the vessel (L^3)

The units of each term in Eq. 2.8 are mass per unit time (Mt^{-1}).

Suppose a steady state ($dC/dt = 0$) has been attained and a substance is consumed or destroyed by reactions within the vessel. Then

$$\underset{\text{transport out}}{\underset{\text{net rate of}}{D(C_i - C)}} = \underset{\text{production}}{\underset{\text{rate of}}{-r}}$$

where $D = Q/V$ = dilution rate (t^{-1})

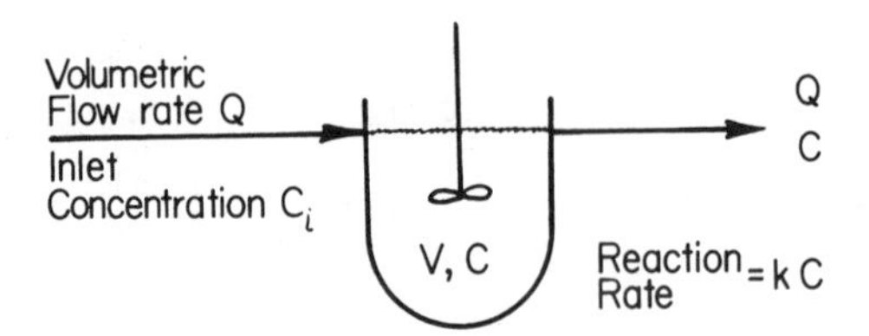

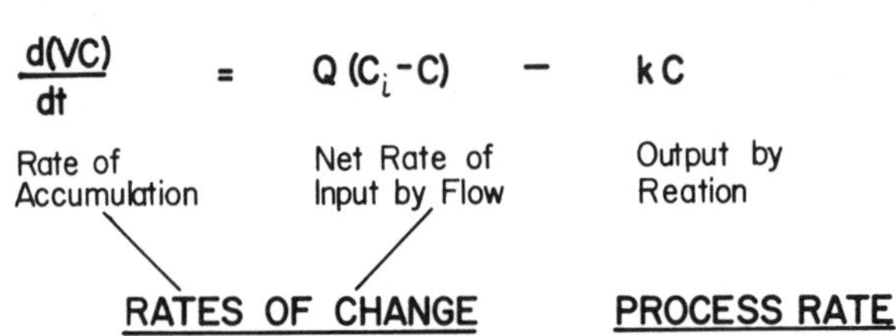

Figure 2.2 A continuous flow stirred tank reactor (CFSTR). The material balance for this reactor is described by Eq. 2.8.

The composition of the feed stream undergoes a discontinuous change when the stream enters the CFSTR (i.e., composition changes from C_i to C). Thus, if the feed to the CFSTR contains organisms, they will experience a discontinuous change, or an environmental shock, upon entering the reactor. More often than not, chemostat experiments are conducted with sterile (or negligible biomass concentration) feeds and the problem of shock is avoided.

A second point of importance regarding CFSTRs is that the residence time of a particle is not fixed but is subject to statistical fluctuation (Frederickson et al., 1970). The density of the distribution of residence time is

$$De^{-Dt} \qquad [2.10]$$

The most probable residence time is zero. The mean residence time is D^{-1}, which is also the standard deviation of the residence time. If a cascade (series) arrangement of equal volume CFSTRs is used (Figure 2.3), say m of them, then the residence time follows a gamma distribution, with density function

$$mD\,\frac{(mDT)^{m-1} - mDt}{(m - 1)!} \qquad [2.11]$$

The mean is again D^{-1} (V is the *total* volume of all m tanks), but the standard deviation is smaller than in the one vessel case, and is $D^{-1}m^{-1/2}$. As the number of tanks in the cascade becomes larger and larger, while D is held constant (thus implying that the individual tanks become smaller and smaller), the standard deviation becomes smaller and smaller. In the limit of very large m, the resi-

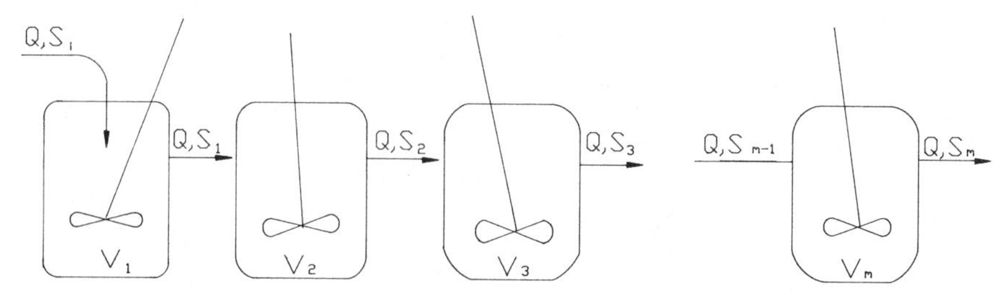

Figure 2.3 A cascade of continuous flow stirred tank reactors (CFSTR) that closely approximates the behavior of a plug flow reactor (PFR). Each reactor or mixing component is thoroughly mixed. As the number of reactors increases for a constant total volume V, the mixing characteristics of the reactor system approach those of the ideal plug flow reactor (Eq. 2.11).

dence time is no longer subject to statistical fluctuations, and is always D^{-1}. This arrangement of reactors is quite effective for simulating a plug flow reactor of volume V.

2.5.2.2 *The Chemostat (Suspended or Planktonic Cells)*

The *chemostat* is a CFSTR in which microbial growth occurs in suspensions; it is an effective technique for attaining and sustaining balanced microbial growth.

Balanced growth *occurs when there is an orderly increase in all cell components of the organism.*

When these conditions exist, the rate of growth is proportional to the logarithm of the cell mass and is termed *logarithmic* or *exponential* growth. All intracellular components are increasing at the same rate. Then, cellular reproduction rate equals RNA production rate equals protein production rate, etc. One of the more significant consequences of continuous culture is that the the planktonic, or suspended, cell population is sustained in balanced growth. This approximates conditions in many natural and engineered environments much more closely than batch cultures do.

A chemostat is illustrated in Figure 2.4; it consists of a vessel with an overflow that maintains a constant volume. Sterile medium is continuously pumped into this vessel and immediately stirred into the culture, displacing some of the stirred culture through the overflow. The culture medium is constituted so that one essential nutrient is in a limiting concentration (i.e., the organisms are nutrient-limited). It is usual to limit the carbon supply, but it is also possible to limit N, Mg, P, S, or any other essential nutrient. By this method, the bacterial concentration (*extent* of bacterial growth) may be controlled, but even more important, by controlling the rate of liquid addition, the *rate* of bacterial growth may be regulated. The volumetric rate of nutrient addition (Q) is usually expressed as the dilution rate ($D = Q/V$).

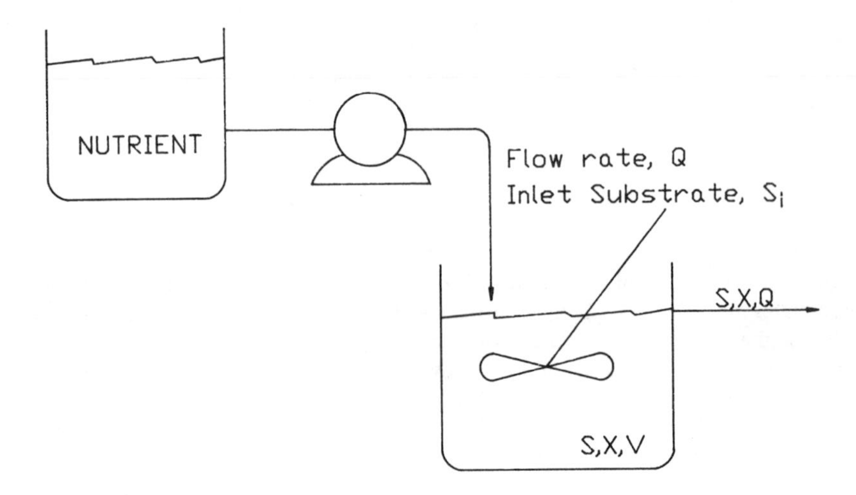

Figure 2.4 A chemostat system consists of a reservoir of nutrients, including the substrate, which are pumped continuously (flow rate Q) through a CFSTR containing microorganisms. Microorganisms are produced in the CFSTR at the expense of substrate and other nutrients. The depleted nutrient mixture and new cells leave the reactor continuously in the effluent.

The dilution rate determines the time a bacterial cell remains in the chemostat. A balance on cell mass in the chemostat yields the following (see Eq. 2.8):

$$\underset{\text{rate of accumulation}}{\frac{dX_c}{dt}} = \underset{\text{rate of growth}}{\mu X_c} - \underset{\text{rate of dilution}}{D X_c} \qquad [2.12]$$

where X_c = biomass concentration (ML_X^{-3})
μ = specific growth rate (t^{-1})

In a chemostat (or any CFSTR), the probability that a cell (particle) in the vessel at some time will be washed out in some subsequent time interval is independent of such factors as the size of the particle or its residence time in the CFSTR. Since washout is equivalent to death so far as the cell population in the CFSTR is concerned, the flow through the vessel imposes a *nonselective* death rate on the population. This point becomes most important when two or more populations are growing and interacting in a CFSTR.

A chemostat, run at one dilution rate, soon establishes steady state conditions. There is no change in bacterial numbers, since bacterial growth exactly balances the bacterial loss by dilution. Thus

$$\mu X_c = D X_c, \qquad \text{or} \quad D = \mu \qquad [2.13]$$

In reactor engineering terms, the chemostat is a CFSTR operated at steady state.

The chemostat is a valuable tool for the determination of bacterial growth rate coefficients, since the dilution rate equals the specific growth rate under steady state conditions. The *mean generation time* g, the time interval required for a cell to divide or the population to double, may be calculated from μ (Section 8.2.1):

$$\mu = \frac{\ln 2}{g} \qquad [2.14]$$

The highest bacterial concentration occurs as $D \rightarrow 0$ (presuming no cell decay processes) and corresponds to that concentration of cells that is produced in a batch culture (a closed system) when substrate (limiting nutrient) is exhausted. There is a very gradual decline in steady state bacterial concentration and a very gradual increase in steady state nutrient concentration as the dilution rate is increased from one experiment to another, until the dilution rate exceeds the highest possible growth rate (Figure 2.5). At this point, D_c, the rate at which bacteria are diluted out, exceeds the rate at which they divide, resulting in a rapid decline in bacterial numbers (washout). In simple terms, the following occurs: at the lower dilution rates, bacteria are starved for a nutrient. As this nutrient is pumped into the vessel, the bacteria utilize it and divide. As the rate of pumping is increased, there is a slight increase in the concentration of this nutrient, as the bacteria are not able to assimilate all of it immediately. There is also a slight decrease in bacterial concentration, as some bacteria are carried out of the vessel before they have an opportunity to divide. This effect is greatly magnified as the dilution rate approaches the maximum growth rate and then exceeds it. More bacteria are washed out than are produced by cell division, and the bacterial concentration falls.

Biofilm accumulation can have a significant effect on the performance of monopopulation chemostats, because detachment of cells from the biofilm serves as a continuous source of bacteria (i.e., a continuous inoculation) for the liquid phase. The result of this *wall growth* is a significant deviation from the mathematical models describing chemostat behavior, especially at high dilution rates (Topiwala and Hamer, 1971; Wilkinson and Hamer, 1974; Molin, 1981). Biofilms in mixed culture chemostats also affect population dynamics in a chemostat (Bryers, 1984, 1986).

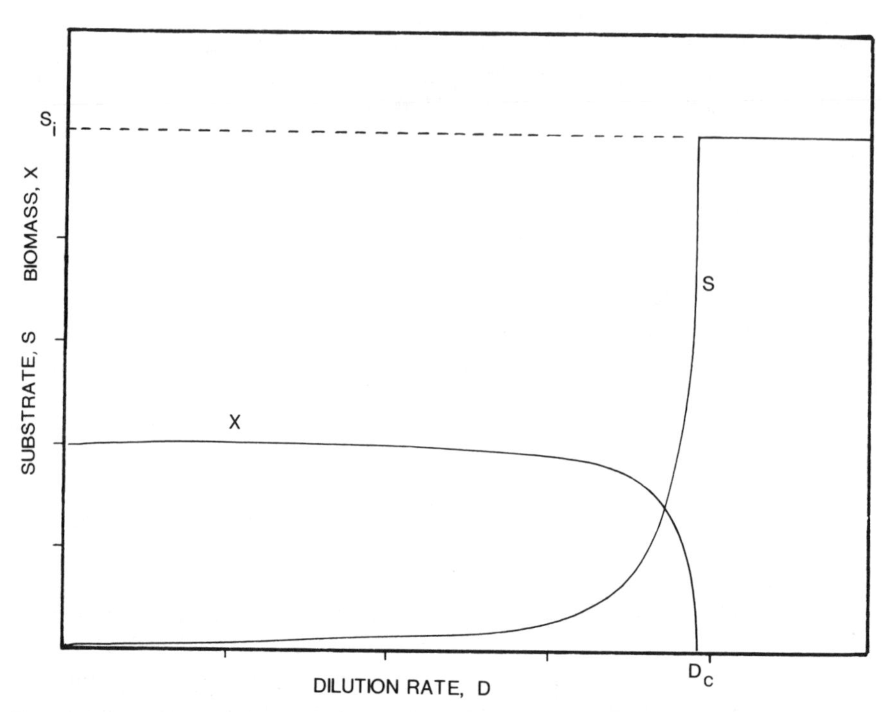

Figure 2.5 Effluent substrate and biomass concentration vary as a function of dilution rate D (reciprocal of residence time Θ) in a chemostat. Increasing D results in increased concentration of substrate in the effluent, since less time is available for its consumption by the microorganisms. At high D, microorganisms do not have sufficient time to reproduce, and their numbers diminish while the effluent substrate concentration increases to the inlet concentration. The coefficients used for this graph are for *P. aeruginosa* (Robinson et al., 1984).

2.5.3 Plug Flow Reactor (PFR)

In a PFR, culture medium and/or organisms are continuously fed to the reactor, where growth and other processes occur. The liquid within the reactor is not stirred, and elements of liquid move progressively through the fermenter without mixing in the direction of flow (Figure 2.6).

In the ideal plug flow reactor, adjacent elements of liquid are assumed to move progressively through the tube without exchange of material between such elements. In addition, the composition of the liquid is assumed to be uniform over any cross section, though the composition obviously changes with distance through the reactor.

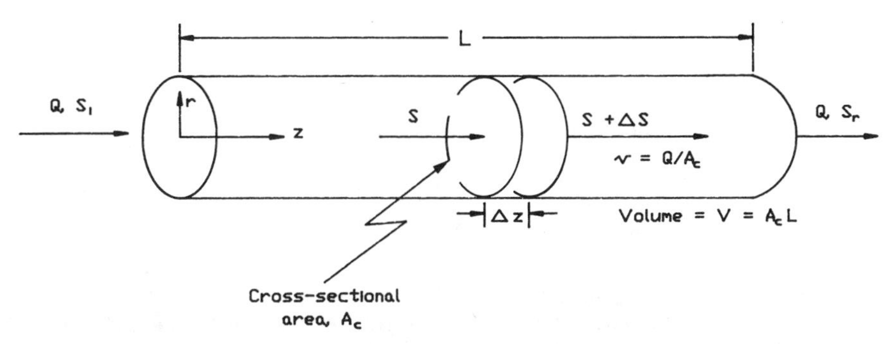

Figure 2.6 A plug flow reactor (PFR) is characterized by deficiency of mixing in the direction of flow, i.e., the fluid entering the reactor moves through the reactor as a "plug." The linear velocity of flow, v, equals the volumetric flow rate Q divided by the cross sectional area of the reactor, A_c.

No reactor can exhibit ideal PFR behavior, but in some cases the behavior can be made to approach the ideal. For example, a series of CFSTRs is quite effective in simulating plug flow (Figure 2.3). A packed bed reactor is another way of simulating plug flow effectively.

Let C be the concentration of some substance in the liquid. Application of the principle of conservation of mass to a system of infinitesimal length (Δz) moving with the flow velocity v through the tube yields the partial differential equation:

$$\underset{\substack{\text{rate of}\\\text{accumulation}}}{\frac{\partial C}{\partial t}} + \underset{\substack{\text{net rate of}\\\text{transport by}\\\text{advection}}}{v\frac{\partial C}{\partial z}} = \underset{\substack{\text{rate of}\\\text{production}}}{r} \qquad [2.15]$$

where z = axial distance from the inlet (L)
v = fluid velocity (Lt^{-1})

In this case, the residence time of a particle is not subject to statistical fluctuations; a particle at position z has been in the vessel for a time z/v, and if the reactor has length L, the transit time will be L/v.

In many cases, mixing in the direction of flow may be important. This is included most easily (though only approximately) by introducing an effective axial diffusion or dispersion coefficient, D_S. The material balance, in this case (a differential material balance described in more detail in Chapter 9), leads to a more complicated equation:

$$\underset{\text{rate of accumulation}}{\frac{\partial C}{\partial t}} + \underset{\text{net rate of transport by advection}}{v\frac{\partial C}{\partial z}} = \underset{\text{rate of transport by diffusion}}{D_S\frac{\partial^2 C}{\partial z^2}} + \underset{\text{rate of production}}{r} \quad [2.16]$$

where D_S = dispersion coefficient (L^2t^{-1})

If steady state can be assumed, the mathematical description can be simplified considerably. D_s is not necessarily the diffusion coefficient used in studies of transport phenomena. D_S is dependent on the flow regime in the reactor and is relatively independent of the usual (molecular) diffusion coefficient, especially if the flow is turbulent.

The most important feature of the PFR from the microbial physiology standpoint is the progressive *change* in environmental conditions seen by an organism traversing the reactor. This is in marked contrast to the constant conditions seen by an organism traversing an ideal CFSTR. In fact, the situation in the ideal PFR is the same as that in the batch reactor, with residence time in the PFR, z/v, replacing the batch reactor holding time. The biofilm organisms in a PFR also see different environmental conditions depending on their location in the PFR.

2.5.4 Batch Reactor

A batch reactor consists of a well-stirred tank which contains no input or output flows. Agitation is vigorous and mixing is complete.

Let C be the concentration of some substance in the reactor liquid (Figure 2.7). Then application of the principle of conservation of mass to the system in the batch reactor yields the following:

$$V\frac{dC}{dt} = Vr \quad [2.17]$$

or

$$\frac{dC}{dt} = r \quad [2.18]$$

Comparing Eq. 2.18 with Eq. 2.8, we see that the batch reactor is a CFSTR without any input or output flows. Comparing Eq. 2.17 with Eq. 2.15, we see that the batch reactor is a PFR without any spatial gradients. The batch reactor, like the PFR, is characterized by *changing* environmental conditions. In the PFR, the change is generally a function of spatial coordinates, while in the batch reactor, changes occur with time.

The batch reactor, in the form of agitated flasks or test tubes, has been one of

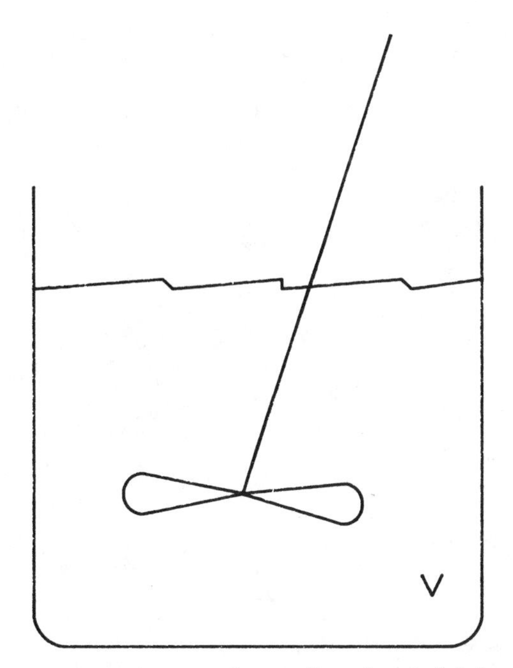

Figure 2.7 A batch reactor has no flow of materials in or out.

the major tools for the microbiologist and fermentation engineer in the past and will continue to be for some time in the future. However the batch reactor, unlike continuous reactors, has little application for biofilm systems, since a stable biofilm cannot be maintained in a closed system after the substrate (energy source) is exhausted.

2.5.4.1 Microbial Growth in Batch Reactors Assume that a single bacterium has been inoculated into a sterile medium and multiplies regularly at a rate consistent with its generation time. If we calculated the theoretical number of bacteria which should be present at various intervals of time and then plotted the data in two ways (logarithm of number of bacteria and actual number of bacteria versus time), we would obtain the curves shown in Figure 2.8. However, this does not represent the entire pattern of growth, but rather one selected portion of the normal growth curve, namely the logarithmic phase of growth (commonly referred to as the log phase). Here, the population increases regularly, doubling at regular time intervals (the generation time) during incubation. In reality, when we inoculate a fresh medium, we obtain a curve of the type illustrated in Figure 2.9. There is an initial period of slow or no growth followed by rapid growth, then a leveling off, and finally a decline in the viable population. Between each of these phases there is a transitional period (curved portion). This represents the time required for all the cells to enter the new phase.

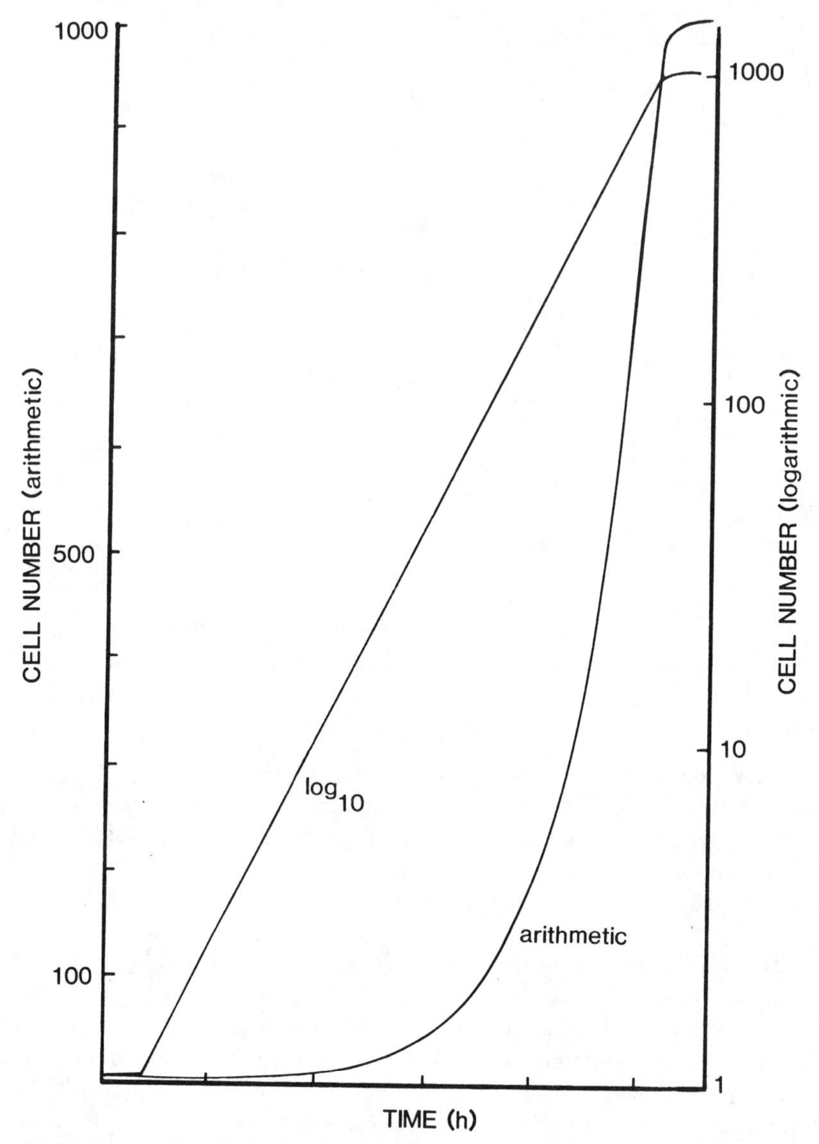

Figure 2.8 The progression of cell numbers in a batch experiment. The linear portion of the semilogarithmic plot is termed logarithmic or log growth.

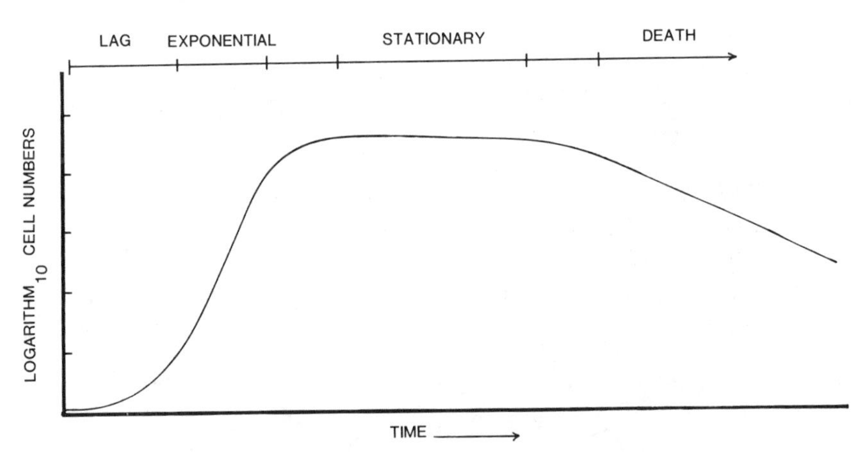

Figure 2.9 The progression of cell numbers in a batch experiment sometimes termed the "growth cycle" or "bacterial growth curve." The beginning of the stationary phase usually coincides with the depletion of the growth-limiting nutrient.

This is typical for most organisms taken from an old culture and inoculated into a fresh medium. Let us examine what happens to the bacterial cells during each of these growth phases.

2.5.4.2 The Lag Phase The addition of inoculum to a new medium is not followed by a doubling of the population according to the generation time. Instead, the population remains temporarily unchanged numerically, as illustrated in the normal growth curve. But this does not mean that the cells are quiescent or dormant; on the contrary, during this stage the individual cells increase in size beyond their normal dimensions. Physiologically they are very active and are synthesizing new cellular constituents. The bacteria introduced into this new environment may be deficient in enzymes or coenzymes, which must first be synthesized in amounts required for optimal operation of the chemical machinery of the cell. Time for adjustments in the physical environment around each cell may also be required. The organisms are actively metabolizing, but there is a lag in the process of cell division.

At the end of the lag phase, cell division commences. However, since not all organisms complete the lag period simultaneously, there is a gradual increase in the population until the end of this period, when all cells are capable of dividing at regular intervals.

2.5.4.3 The Logarithmic Phase During this period the cells divide steadily at a rate determined by their specific physiological capabilities in that growth environment. The logarithm of the number of cells plotted against time results in a straight line, since the generation time remains nearly constant (Figures 2.8 and 2.9). Under optimal conditions, the growth rate is maximal during this phase. During this phase the entire population is most nearly uniform in terms of chemical composition of cells, metabolic activity, and other physiological characteristics. The logarithmic growth phase is another method of achieving *balanced growth*.

2.5.4.4 *The Stationary Phase* The logarithmic phase of growth begins to taper off after several hours, again in a gradual fashion represented by the transition from a straight line through a curve to another straight line of zero slope (Figure 2.9). This trend toward cessation of growth can be attributed to a variety of circumstances, among which are the exhaustion of some nutrients and, less often, the production of toxic products during growth. The population remains constant for a time, perhaps as a result of complete cessation of division or the balancing of reproduction rate by an equivalent death rate.

2.5.4.5 *The Phase of Decline or Death* Following the stationary period the bacteria may die at a rate that exceeds the rate of production of new cells (if, indeed, any cells are still reproducing). The actual causes of death during this phase of the culture are undoubtedly various, depending on the conditions and the bacterial type. Generally, a depletion of essential nutrients and/or an accumulation of products to an inhibitory level (e.g., acids) are sufficient to account for this development.

2.5.4.6 *Summary* The population proceeds gradually from one phase of growth to the next (Figure 2.9), so that not all the cells are in identical physiological condition toward the end of a given phase of growth. The accumulation of cells from the log phase through the stationary phase can be expressed in the form of the logistic equation (Figure 2.9):

$$\frac{dX_c}{dt} = kX_c\left[1 - \frac{X_c}{X_{max}}\right] \qquad [2.19]$$

where X_{max} = maximum biomass concentration (ML_x^{-3})
k = rate coefficient (t^{-1})

When X_c is small enough, $dX_c/dt = kX_c$ and exponential growth is observed. At the stationary phase, $dX_c/dt = 0$ and $X_c = X_{max}$. The logistic curve is a useful empirical model for growth in batch reactors and is also useful in describing the progression of biofilm accumulation (Chapter 8).

An appreciation of the pattern of growth for microorganisms is of great importance for many academic and practical problems. For example, one can use it to predict the approximate time required for a particular bacterial culture to double under specified conditions of cultivation. Thus, the study and use of microorganisms necessitates familiarity with the population changes that occur during the growth of a culture. The restrictiveness of the growth curve, however, was summarized succinctly by Herbert (1961):

> The above sequence of events is usually called the growth cycle, a term which, I suggest, should speedily be abandoned, since it conveys a quite misleading impression that this sequence is a necessary and inevitable feature of bacterial growth, whereas it is in reality a sequence forced upon the organisms by sequential environmental changes which are inevitable when growth occurs in a closed system. Unfortunately, bacteriologists have been inoculating flasks of culture media for so long now that they have come to regard this as part of the natural order of events, instead of as a convenient but highly artificial experimental procedure.

It is important to realize that growth in a closed system, as described above, occurs in a constantly changing environment. Herbert's statement refers to the fact that useful deductions regarding the effect of environment on microbial growth can seldom be drawn from experiments unless they are conducted in such a manner as to obtain balanced growth.

2.6 REACTION STOICHIOMETRY

Stoichiometry is the application of the law of conservation of mass and the chemical laws of combining weights to chemical processes. In its broadest sense, stoichiometry is a *system of accounting* applied to the mass and energy participating in a process involving chemical or physical change. It permits a surprisingly large amount of information to be obtained from a seemingly small number of facts.

Stoichiometry provides information concerning the types of changes and maximum extent of changes that can occur in a reaction process. In an abiotic system, thermodynamic calculations allow the determination of the equilibrium constant for a reaction and hence the actual yield or reaction products for given conditions. Such calculations require knowledge of the exact nature of the reactants and the products, which are not always determinable in a biotic environment. In biotic systems, equilibrium constants are irrelevant, since equilibrium is only attained after all cells are dead. Nevertheless, stoichiometric principles are still very useful for making approximations in microbial systems.

In order to define the stoichiometry of a reaction, the stoichiometric coefficients in the reaction equation must be determined. For example, consider the chemical oxidation of glucose:

$$C_6H_{12}O_6 + 6O_2 \rightarrow 6CO_2 + 6H_2O \qquad [2.20]$$

The stoichiometric coefficients for glucose, oxygen, carbon dioxide, and water are -1, -6, 6, and 6, respectively. The units of the stoichiometric coefficients are *moles*. More generally, the relationship is expressed with a mass balance equation as follows:

$$a_1M_1 + a_2M_2 + \cdots + a_kM_k = 0 \qquad [2.21]$$

where M_i = molecular weight of the ith component (M mol^{-1})
a_i = stoichiometric coefficient of the ith component (mol)

It follows that the stoichiometric relationships also provide a convenient method for comparing rates of reaction for the various reaction components. If the production rate of component i is $r^{(i)}$, then for the reaction described in Eq. 2.33,

$$r^{CO_2} = r^{H_2O} = -0.17r^{\text{glucose}} = -r^{O_2} \quad [2.22]$$

For elementary reactions, the overall process rate r is the same regardless of the reaction component measured. Elementary reactions have rate equations that correspond to their stoichiometric equation (e.g., $A + B \rightarrow C$ and $-r_A = kC_AC_B$). Therefore, for a reaction involving k components

$$r = \frac{r^{(1)}}{a_1} = \frac{r^{(2)}}{a_2} = \frac{r^{(3)}}{a_3} = \cdots = \frac{r^{(k)}}{a_k} \quad [2.23]$$

This relationship indicates that overall conversion rate can be determined by following the rate of appearance or disappearance of the component that is most easily detectable, even if it is not the one of major interest. An example of microbiological relevance is the microbial oxidation of glucose. Busch (1971) has examined this process in a batch reactor, and Turakhia (1986) in a biofilm reactor. Both have observed the following stoichiometry:

$$C_6H_{12}O_6 + 2.5O_2 + 0.7NH_3 \rightarrow 3.5CH_{1.4}O_{0.4}N_{0.2} + 2.5CO_2 + 4.6H_2O \quad [2.24]$$

or, expressed in terms similar to those of Eq. 2.21,

$$3.5M_1 + 2.5M_2 + 4.6M_3 - M_4 - 2.5M_5 - 0.7M_6 = 0 \quad [2.25]$$

where M_1 = biomass (M mol^{-1})
M_2 = carbon dioxide (M mol^{-1})
M_3 = water (M mol^{-1})
M_4 = glucose (M mol^{-1})
M_5 = oxygen (M mol^{-1})
M_6 = ammonia (M mol^{-1})

and $CH_{1.4}O_{0.4}N_{0.2}$ is an empirical formula for the biomass.

The progress of this reaction can be followed by monitoring the concentration of any of the reactants or products. However, oxygen can be measured conveniently, easily, and accurately and is frequently used to determine rates of aerobic microbial reactions. From Eq. 2.24 we see that the ratio of glucose removal rate (molar ratio) to oxygen removal rate is approximately 0.4.

When materials react to form products, it is necessary to determine, after examining the stoichiometry, whether a single reaction or a number of reactions are occurring. When a single stoichiometric equation and a single rate equation represent the progress of the reaction, it is termed a *single reaction*. When more than one stoichiometric expression describes the observed changes, then more than one kinetic expression is needed to follow the changing composition of the reaction components, which results in a *multiple reaction*. Multiple reactions may be classified simply as follows:

$$A \rightarrow B \rightarrow C \qquad \text{series reaction}$$

$$A \begin{cases} \rightarrow B \\ \rightarrow C \end{cases} \qquad \text{parallel reaction}$$

More complicated schemes are possible.

For a single, elementary chemical reaction, it is sufficient to determine the relative amounts of reactant and product at any one time during the reaction in order to obtain the stoichiometric coefficients. However for multiple reactions, a more detailed and cautious procedure must be used. Consider the following illustrative example. *Nitrification* refers to the oxidation of ammonium nitrogen to nitrate nitrogen.

$$NH_4^+ + 2O_2 \rightarrow 2H^+ + H_2O + NO_3^- \qquad [2.26]$$

However, the conversion is better described as a multiple reaction mediated by two different microbial species:

$$NH_4^- + 1.5O_2 \rightarrow 2H^+ + H_2O + NO_2^- \qquad [2.27]$$

$$NO_2^- + 0.5O_2 \rightarrow NO_3^- \qquad [2.28]$$

Using the notation from Eq. 2.21, the reaction stoichiometry is as follows:

$$\begin{array}{rrrrrrl} & +\,2M_2 & +\,M_3 & -\,M_4 & -\,1.5M_5 & +\,M_6 & = 0 \qquad [2.29] \\ M_1\,+ & & & & -\,0.5M_5 & -\,M_6 & = 0 \qquad [2.30] \\ \hline M_1 & +\,2M_2 & +\,M_3 & -\,M_4 & -\,2M_5 & & = 0 \qquad [2.31] \end{array}$$

where M_1 = nitrate ion (M mol^{-1})
M_2 = hydrogen ion (M mol^{-1})
M_3 = water (M mol^{-1})
M_4 = ammonium ion (M mol^{-1})
M_5 = oxygen (M mol^{-1})
M_6 = nitrite ion (M mol^{-1})

The course of nitrification in a batch or plug flow reactor is described in Figure 2.10. At any time during the process, the amount of NH_4^+ consumed results in production of NO_2^- and NO_3^- in a proportion that depends not only on the overall stoichiometry, but also on the rate of the individual reactions. The amount of nitrite N is exaggerated. Very little nitrite appears under natural conditions, except at high pH where NH_3 predominates over NH_4^+. The NH_3 is inhibitory to *Nitrobacter,* so that nitrite oxidation is diminished until NH_3 concentration is reduced to low levels.

2.7 REACTION KINETICS

The rate of reaction is characterized by a *rate equation,* which is generally the result of an empirical curve-fitting procedure. The values of the rate coefficients must be found by experiment, even if the form of the rate equation is suggested by a theoretical analysis or mechanistic model. The determination of a useful rate equation usually requires a study to determine the influence of concentration, followed by those of pH, temperature, etc., on the reaction rate coefficients.

2.7.1 The Rate Equation

The *nth order equation* has been used to great advantage in chemical reactor theory. The equation is a descriptive, two parameter model and is useful over restricted ranges of experimental data:

$$r = kC^n \qquad [2.32]$$

where k = rate coefficient ($t^{-1}\ M^{1-n}\ L^{3(n-1)}$)
n = reaction order (dimensionless)

The equation has proven useful because of its small number of parameters and its simplicity when used in mass balance equations.

Another form of rate equation, used extensively in microbial reactors, is the *saturation rate equation:*

$$r_c = \frac{k_1 C}{k_2 + C} \qquad [2.33]$$

where k_1 = rate coefficient ($ML^{-3}\ t^{-1}$)
k_2 = saturation coefficient (ML^{-3})

From a practical standpoint, the numerical values of the rate coefficients are most important and will depend on the units of measurement. For example, the

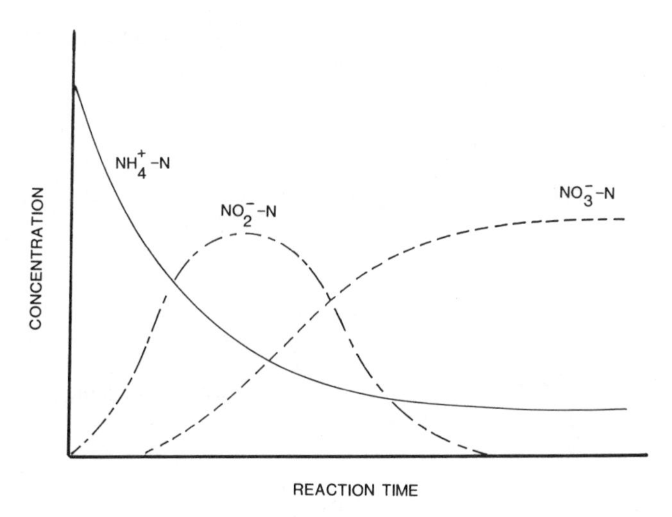

Figure 2.10 A hypothetical progression of nitrogen species during nitrification in a batch reactor. NH_4^+ is oxidized by *Nitrosomonas* to NO_3^-, which is oxidized to NO_2^- by *Nitrobacter*.

glucose removal rate in a microbial reactor can be described by Eq. 2.22, where C can be expressed as either the glucose or the organic carbon concentration. When carbon concentration is the measured variable, k_1 and k_2 will be 40% of the values obtained using glucose measurements since glucose is 40% carbon by weight. Similarly, confusion can result when molar concentration units are compared with mass concentration units. Molar units are often awkward in biofilm systems, since microbial cell mass is one of the products but cannot be readily characterized by a "molecular weight."

2.7.2 Rate Equation versus Balance Equation

It is critically important to distinguish between a *rate equation,* which describes a transformation, and a *balance equation,* which considers both transport and transformation processes. In a batch reactor, where transport processes are not important, the balance equation for a first order reaction is expressed as follows:

$$\underset{\substack{\text{rate of}\\ \text{accumulation}}}{V\frac{dC}{dt}} = \underset{\substack{\text{process}\\ \text{rate}}}{kCV} \qquad [2.34]$$

where k = rate coefficient (t^{-1})

This expression can be confusing, especially if it is referred to as a rate equation. The rate equation for this process is correctly expressed as

$$r = kC \qquad [2.35]$$

The balance equation for the batch reactor can unambiguously be expressed by Eq. 2.23 or as

$$\frac{dC}{dt} = kC \qquad [2.36]$$

A material balance for a continuous reactor (Figure 2.2), using a similar rate expression, can be expressed as follows:

$$V\frac{dC}{dt} = Q(C_i - C) + rV \qquad [2.37]$$

$$V\frac{dC}{dt} = Q(C_i - C) + kCV \qquad [2.38]$$

rate of accumulation	net rate of input by flow	process rate

Equation 2.38 reduces to the balance equation for the batch reaction process (Eq. 2.36) if the flow rate into the reactor is zero ($Q = 0$).

If Eq. 2.34 is mistakenly substituted into Eq. 2.38 and the process rate and the accumulation rate are both expressed as derivatives, then the following results:

$$V\frac{dC}{dt} = Q(C_i - C) + V\frac{dC}{dt} \qquad [2.39]$$

At steady state, all time derivatives go to zero and

$$Q(C_i - C) = 0 \qquad [2.40]$$

This is clearly erroneous and not a general solution. It illustrates the confusion that can result from not distinguishing between rate equations and balance equations.

2.7.3 Steady State versus Equilibrium

In batch reactors, changes in composition occur with time regardless of whether spatial uniformity exists. The changes occur from moment to moment until thermodynamic equilibrium is reached (or until one reactant is depleted). The continuous reactor is different in that any part of the system tends toward a time-invariant state, i.e., steady state. The movement toward a time-invariant state requires constant feed conditions, constant rate of heat removal, etc.

It is important to note that steady state *is not* an equilibrium state. The term equilibrium should be reserved for the time-invariant state of closed systems. The steady state of an open system, such as a continuous reactor, depends on the flow rate, reaction rate, and the size of the system. In the theory of nonequilibrium thermodynamics, the time-invariant or steady state in an open system possesses the same meaning as the equilibrium state in equilibrium thermodynamics which describes the closed systems. A continuous flow of energy is needed to maintain the continuous reactor system at a state away from equilibrium. Thus, a biofilm reactor requires a continuous feed of substrate to maintain the biofilm in a nonequilibrium state, i.e., a state where biofilm organisms are active.

The steady state is the most orderly, efficient, and economical state of an open system.

2.7.4 Classification of Reactions

A useful method of classifying reactions is according to the number and types of phases (or, in some cases, compartments) involved. A phase is a part of a system that is physically and chemically uniform throughout. A reaction is *homogeneous* if it takes place in one phase alone. A reaction is *heterogeneous* if it requires the presence of at least two phases to proceed at its characteristic rate. Consequently, the choice of classification is sometimes difficult and depends on which description or model is more useful. Rigorously, all microbial reactions are heterogeneous, since the biomass constitutes a solid phase and the substrate being consumed is generally soluble, i.e., in the liquid phase. In the same way, a biofilm constitutes a solid compartment. Although segregated models may appear useful, at high substrate concentrations and/or low concentrations of *well-dispersed biomass,* distributed models are frequently satisfactory and microbial reactions can be considered homogeneous.

In heterogeneous reaction systems such as biofilms, reactants must be transported from one phase to another, and the rate of transport (i.e., the mass transfer rate) may control the overall process rate observed. The transport rate is dependent on mixing intensity and system geometry. In any case, if the process consists of a number of rate processes in series (e.g., transport and then reaction), the slowest step of the series exerts the greatest influence and *controls* the overall rate. Consequently, the rate of heterogeneous processes may be controlled either by transport or by reaction.

The major feature distinguishing biofilms from planktonic microbial processes is the rate-controlling influence of transport processes.

2.7.5 Expression of Rates

Process rates are conveniently expressed as intensive measures (i.e., independent of the system mass), so that results can be used in analyzing systems of varying size and composition. Thus the process rate R can be expressed per unit reactor volume:

$$r = \frac{R}{V} \tag{2.41}$$

where V = reactor volume (L^3)
R = conversion rate (Mt^{-1})
r = reaction rate per unit reactor volume ($ML^{-3}t^{-1}$)

The conversion rate can also be based on unit biomass and termed *specific* reaction rate:

$$r_X = \frac{R}{X} \tag{2.42}$$

where X = biomass (M_X)
r_X = reaction rate per unit biomass ($MM_X^{-1}t^{-1}$ or possibly t^{-1})

For heterogeneous systems such as biofilms, the area of reaction surface is the basis for expressing rates:

$$r'' = \frac{R}{A} \tag{2.43}$$

where A = reactive surface area (e.g., biofilm surface area)(L^2)
r'' = reaction rate per unit reaction surface ($ML^{-2}t^{-1}$)

2.8 MICROBIAL REACTIONS VERSUS CHEMICAL REACTIONS

The methods for determining the stoichiometric and kinetic parameters for a given microbial conversion evolved from application to abiotic reactions. Biochemical conversions mediated by viable organisms differ from abiotic chemical conversions in several ways.

1. Microbial reactions are generally *irreversible*. The stoichiometric endpoint generally refers to the exhaustion of one of the reactants. Thus, the equilibrium state is a trivial one from the microbial standpoint.
2. All microbial reactions are *heterogeneous*. They require at least two phases ("solid" microbial cells and water).

3. Low concentrations of reactants and products, in addition to the heterogeneous characteristic, increase the potential for *mass transfer limitations*.
4. The reaction, though frequently represented by one stoichiometric equation, consists of a network of enzymatic reactions occurring in the metabolic region of the cell.
5. The reactions are generally *autocatalytic*, i.e., biomass and related enzymes increase as the reaction proceeds.
6. One of the reactants, biomass, has a *structure* that can influence the stoichiometry and kinetics of the reaction.

REFERENCES

Bakke, R., M.G. Trulear, J.A. Robinson and W.G. Characklis, *Biotech. Bioeng.*, **26**, 1418–1424 (1984).

Bryers, J.D., *Biotech. Bioeng.*, **26**, 948–958 (1984).

Bryers, J.D., *Bioprocess Eng.*, **1**, 3 (1986).

Bryers, J.D. and W.G. Characklis, *Biotech. Bioeng.*, **24**, 2451–2476 (1982).

Busch, A.W. *Aerobic Biological Treatment of Waste Waters,* Gulf Publishing Co., Houston, TX, 1971.

Frederickson, A.G., R.D. Megee, and H.M. Tsuchiya, *Adv. Appl. Microbiol.*, **13**, 419 (1970).

Gujer, W., *Wat. Sci. Tech.*, **19**, 495 (1987).

Harder, A. and J.A. Roels, *Adv. Biochem. Eng.*, **21**, 55 (1982).

Herbert, D., *Symp. Soc. Gen. Microbiol.*, **11**, 391 (1961).

Himmelblau, D.M. and K.B. Bischoff, *Process Analysis and Simulation,* John Wiley and Sons, New York, 1968.

Kissel, J.C., P.L. McCarty, and R.L. Street *J. Env. Eng.*, **110**, 393 (1984).

Kjelleberg, S., B.A. Humphrey, and K.C. Marshall, *Appl. Env. Microbiol.*, **43**, 1166–1172 (1982).

Levenspiel, O., *Chemical Reaction Engineering,* 2nd ed., John Wiley and Sons, New York, 1972.

Molin, G., *Appl. Microbiol. Biotech.*, **13**, 102 (1981).

Nelson, C.H., J.A. Robinson, and W.G. Characklis, *Biotech. Bioeng.*, **27**, 1662–1667 (1985).

Robinson, J.A., M.G. Trulear, and W.G. Characklis, *Biotech. Bioeng.*, **26**, 1409–1417 (1984).

Roels, J.A. and N.W.F. Kossen, *Prog. Ind. Microbiol.*, **14**, 95–203 (1978).

Salmon, I. and M.J. Bazin, "The role of mathematical models and experimental ecosystems in the study of microbial ecology," in J.W.T. Wimpenny (ed.), *Handbook of Laboratory Model Systems for Microbial Ecosystems,* Vol. II, CRC Press, Boca Raton, FL, pp. 235–252, 1988.

Savageau, M.A., *Biochemical Systems Analysis,* Addison-Wesley Publishing Co., Reading, MA, 1976.

Siebel, M.A., "Binary population biofilms," unpublished doctoral dissertation, Montana State University, Bozeman, MT, 1987.

Topiwala, H.H. and G. Hamer, *Biotech. Bioeng.*, **13**, 919 (1971).

Turakhia, M.H. "The influence of calcium on biofilm processes," Ph.D. thesis, Montana State University, Bozeman, MT, 1986.

Wilkinson, T. and G. Hamer, *Biotech. Bioeng.*, **16**, 251 (1974).

Williams, F.M., *J. Theor. Biol.*, **15**, 190 (1967).

Williams, F.M., "Mathematics of microbial populations, with emphasis on open systems," in E.S. Dewey (ed.), *Growth by Intrassusception: Ecological Essays in Honor of G. Evelyn Hutchinson,* Connecticut Academy of Sciences, New Haven, 1972.

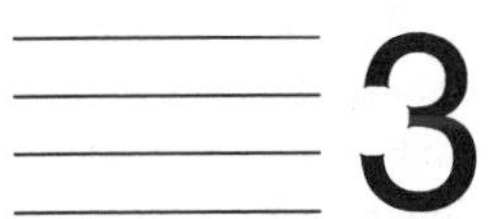

LABORATORY BIOFILM REACTORS

WILLIAM G. CHARACKLIS

Montana State University, Bozeman, Montana

SUMMARY

Some of the more useful laboratory model systems suitable for observing biofilm processes are described. The choice of system depends on the specific biofilm process under consideration and also the convenience of sampling and biofilm manipulation. Some of the laboratory systems have also been adapted for field investigation.

Methods available for monitoring biofilm accumulation and activity have been classified as follows:

- Direct measurement of biofilm quantity (mass and thickness)
- Indirect measurement of biofilm quantity using specific biofilm constituents
- Indirect measurement of biofilm quantity by determining microbial activity within the biofilm
- Indirect measurement of biofilm quantity: effects on biofilm transport properties (see Chapter 14 for more detail)

A combination of methods will be required in most cases; attention should be given to methods that permit the completion of a material balance.

3.1 INTRODUCTION

Experimentation with biofilm processes demands a fundamental approach to establish a systematic framework for analysis of biofilm structure and activity, irrespective of the specific biofilm environments or their particular engineering application. This chapter reviews experimental laboratory systems and methods for measuring biofilm properties and intensive quantities such as reaction rates. The laboratory methods have frequently been extended to field applications.

3.2 EXPERIMENTAL SYSTEMS

3.2.1 System Components

A biofilm experimental system must include certain components, whatever the application, if quantitative data regarding biofilm processes are desired for extrapolation to other systems (Figure 3.1):

1. The *reactor,* where the substratum for biofilm accumulation is located. The reactors can take several geometric forms, which are described below.
2. The *reactor feed,* which requires a continuous water supply. The water must contain all cellular growth requirements (energy, carbon, trace elements) at appropriate concentrations. pH and temperature should be maintained at desired levels. Typically, when large volumetric water flow

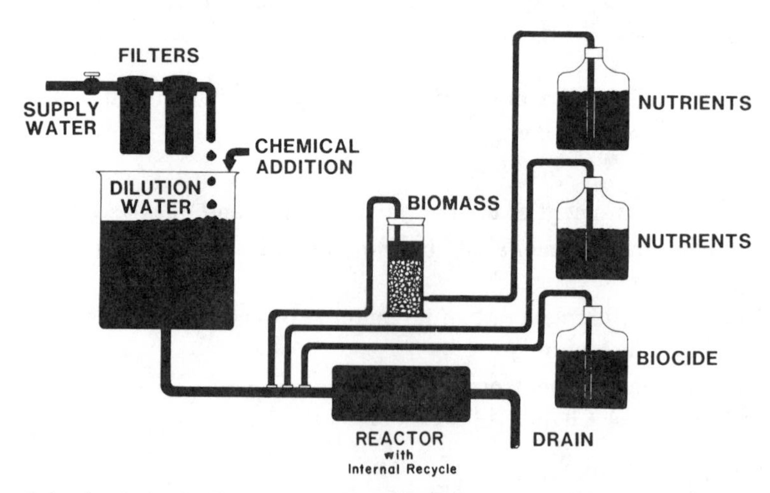

Figure 3.1 A schematic diagram of a typical biofilm reactor system, including the reactor (sometimes with internal recycle), the biomass reactor for continuous inoculation of microorganisms, and the chemical (nutrients, biocide) feed system.

rates are needed, tap water or water from some other source is treated continuously (e.g., to remove residual organic carbon, chlorine, or suspended solids) and the necessary nutrients are added as shown in Figure 3.1. A continuous flow of microorganisms may also be required for certain experiments. A simple packed bed reactor is suitable for providing a continuous, stable supply of microbial biomass.

3. Various appurtenances, including a *pumping system,* a method for controlling flow rates, and instruments for measuring the input and recycle flow rate(s).

3.2.2 Experimental Variables

Laboratory systems permit the observation of biofilm behavior with strict environmental control of a number of variables, including the following:

1. Chemical
   Substrate type
   Substrate concentration
   pH
   Inorganic composition
   Dissolved oxygen
   Microbial inhibitors
2. Physical
   Temperature
   Fluid shear stress
   Heat flux
   Surface composition
   Surface texture (roughness)
   Hydraulic residence time
3. Biological
   Organism type (pure or mixed culture)
   Organism concentration

In laboratory systems, the effect of any one of the variables can be observed by maintaining the other variables constant. This is not usually possible in the field. Nevertheless, field experiments (such as those described in Chapters 14 and 15) are still useful and necessary (Zelver et al., 1984; Matson and Characklis, 1983).

3.2.3 Other Important System Characteristics

3.2.3.1 Open versus Closed Systems An open system is one with inputs and outputs of mass and/or energy (Chapter 2). Since few closed biofilm systems exist in nature or in technological systems, open systems are emphasized over closed (batch) systems in this book. The open reactor system should, however, permit the necessary sampling to accomplish a material balance, including the sampling of the biofilm. The material balance is the starting point for any model or hypothesis verification.

3.2.3.2 *Gradients* Biofilm systems differ from planktonic systems in the importance of spatial gradients. There are good reasons to investigate gradients in velocity, temperature, and concentration in biofilm systems. Microelectrodes are available for determining concentration profiles (e.g., dissolved oxygen and pH) in biofilms, but access for microelectrodes in many reactor systems is difficult, if not impossible. Measuring velocity gradients can also be tedious, but by choosing a conventional geometry, the velocity or shear stress at the biofilm fluid interface can be calculated from the flow rate and/or measurements of the pressure drop (in a tube) or torque (in a rotating geometry).

3.3 REACTORS

Three ideal reactor types are available for the experimental system (Chapter 2, Section 2.5): *continuous flow stirred tank reactor* (CFSTR) in which no concentration gradients exist within the liquid volume; *plug flow reactors* in which reactants and products move as a "plug" from inlet to outlet; *batch reactors* with no inputs or outputs.

The CFSTR provides significant advantages for observing, separating, and evaluating the kinetics and stoichiometry of each biofilm process:

1. The bulk liquid phase is uniform, which makes sampling, chemical analysis, and mathematical modeling simple. Intensive quantities, such as reaction rates, are determined easily because the environmental conditions are fairly uniform in the bulk liquid volume as well as across the reactive surface area.
2. The steady state condition is convenient and reproducible.
3. Biofilms developed in CFSTRs with constant shear stress at the walls are fairly uniform.

Many reactor geometries can be operated as a CFSTR. For example, a tubular reactor is generally perceived as a plug flow reactor. However, if part of the effluent flow is recycled, the system becomes a CFSTR. A number of different reactor geometries can be inserted in the recycle flow system described in Figure 3.2:

1. RotoTorques (rotating annular reactors) are compact and easy to operate. In most cases, RotoTorques are be operated with internal recycle.
2. Circular tubes are the prevalent geometry in heat exchangers, and a wide variety of alloys are available in this form. The fluid dynamics in this ge-

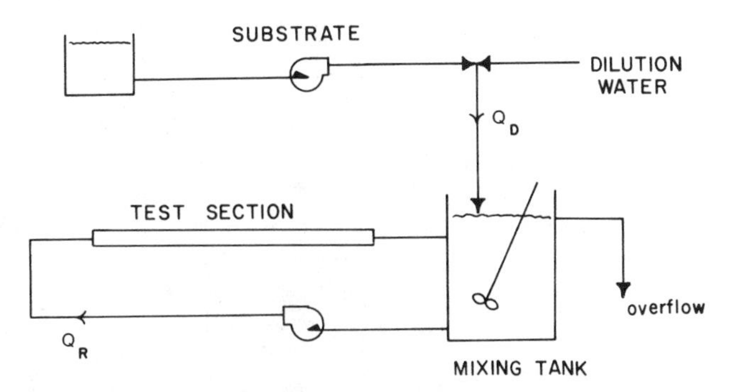

Figure 3.2 A schematic diagram of an experimental reactor system with volume V, influent volumetric flow rate Q_D, and recycle flow rate Q_R.

ometry are well defined. Rectangular tubes (Bakke, 1986) and capillaries (Rutter and Leech, 1980; Powell and Slater, 1983; Escher, 1986) have similar advantages and have been used with success.

3. Open channel systems are a variation of tubular systems except that the flow is exposed to a gas phase (e.g., the atmosphere). These systems have been used to simulate stream bed conditions (Escher, 1983) where gas transfer and radiation (for photosynthesis) must be controlled or measured.
4. Radial flow and rotating disk reactors provide a defined range of fluid shear stress conditions simultaneously at the substratum.
5. Packed or fluidized bed reactors increase the ratio of surface area to volume for biofilm formation through the addition of specific packing material to the reactor.

Many biofilm systems of industrial and ecological concern occur in plug flow reactors. For example, rigorous modeling of biofouling in heat exchanger tubes and pipelines must incorporate longitudinal gradients. In the same way, simulating biofilm processes in natural streams also requires that plug flow be considered.

3.3.1 RotoTorque (Rotating Annular Reactor)

The RotoTorque (IPA, Montana State University), or rotating annular reactor, is an excellent laboratory system for monitoring biofilm development because of its sensitivity, particularly to changes in fluid frictional resistance. The RotoTorque consists of two concentric cylinders: a stationary outer cylinder and a rotating inner cylinder (Figure 3.3). Removable slides (4–12 of them) form an integral part of the inside wall of the outer cylinder and permit sampling of the biofilm so that thickness, mass, and/or biofilm chemical and microbial composition can be determined.

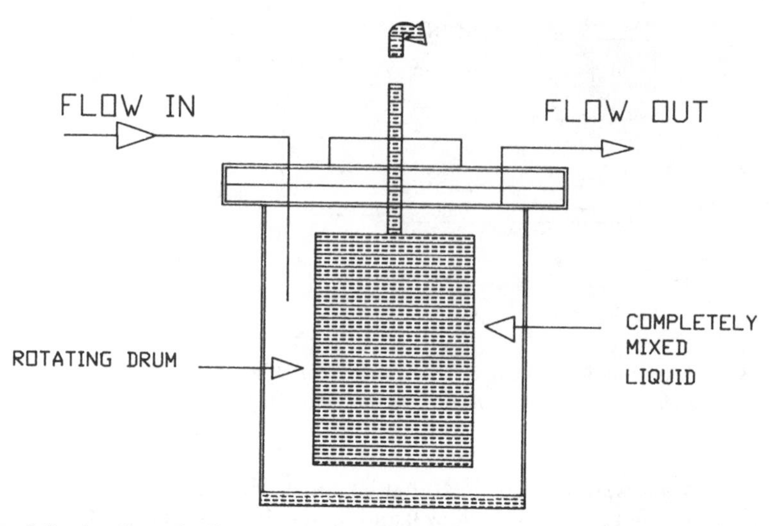

Figure 3.3 A schematic diagram of the RotoTorque, consisting of a rotating inner drum within a stationary outer drum. The RotoTorque is operated as a continuous flow reactor, the liquid in the annular space being completely mixed.

3.3.1.1 Mixing The reactor bulk liquid phase is completely mixed (i.e., it is a CFSTR) by virtue of draft tubes bored eccentrically through the solid inner cylinder (Figure 3.4). The draft tubes are positioned at angles so that the rotation of the inner cylinder pumps the fluid through the tubes. By virtue of the complete mixing, effluent liquid samples represent the reactor liquid composition. Results of dye studies in which the dye was pulsed to the RotoTorque indicate that the reactor represents a CFSTR (Figure 3.5).

In some cases, an external recycle flow has been used to enhance mixing (Trulear, 1983; Turakhia, 1986). The external recycle flow has also been used with the RotoTorque for other purposes:

1. A gas diffuser in the recycle flow increases oxygen input or continuously deaerates the reactor for anaerobic experiments (Trulear, 1983).
2. A sidestream filter continuously removes suspended (planktonic) microorganisms from the reactor liquid (Schaftel, 1982).
3. A sidestream packed column continuously leaches biocide into the reactor liquid.

The recycle flow rate is independent of the dilution flow rate.

The fluid shear stress at the wall is a function of the rotational speed. The mean liquid residence time depends on the dilution flow rate through the reactor. Thus, fluid shear stress and residence time can be varied independently. The reactor residence time can be short (e.g., 10 min) so that planktonic growth

Figure 3.4 The RotoTorque, with eccentrically drilled draft tubes and removable slides. Tubing fittings in the top and one in the bottom permit continuous feed and overflow for nutrients.

is negligible, and all reactor activity can then be attributed to the biofilm. The reactor has a high surface area to volume ratio (300–350 m^{-1}), and most of its surface area is exposed to a uniform shear stress.

Several RotoTorques can be placed in series to simulate a plug flow biofilm reactor (Chapter 2, Section 2.5.2). Theoretically, a step increase in dye concentration will result in an exponential increase in dye concentration in the first reactor. The succeeding reactors will display a lag, and the exponential increase will be slower (Figure 3.6a). Results of a dye test indicated very good agreement between theoretical calculations and measurements for the dye concentration in the effluent from the last of four RotoTorques in series (Figure 3.6b).

3.3.1.2 Torque A torque transducer can be mounted on the shaft between the cylinder and the motor drive to monitor the drag force on the surface of the inner cylinder (Figure 3.7). Fluid frictional resistance can then be calculated from rotational speed and torque measurements. Thus, torque also permits a determination of fluid shear stress when the reactor surface is clean, i.e., under initial conditions. Torque provides a continuous measurement of biofilm accumulation, thus permitting rather effortless, though indirect, observation of biofilm behavior (Characklis and Roe, 1984). For example, torque has been

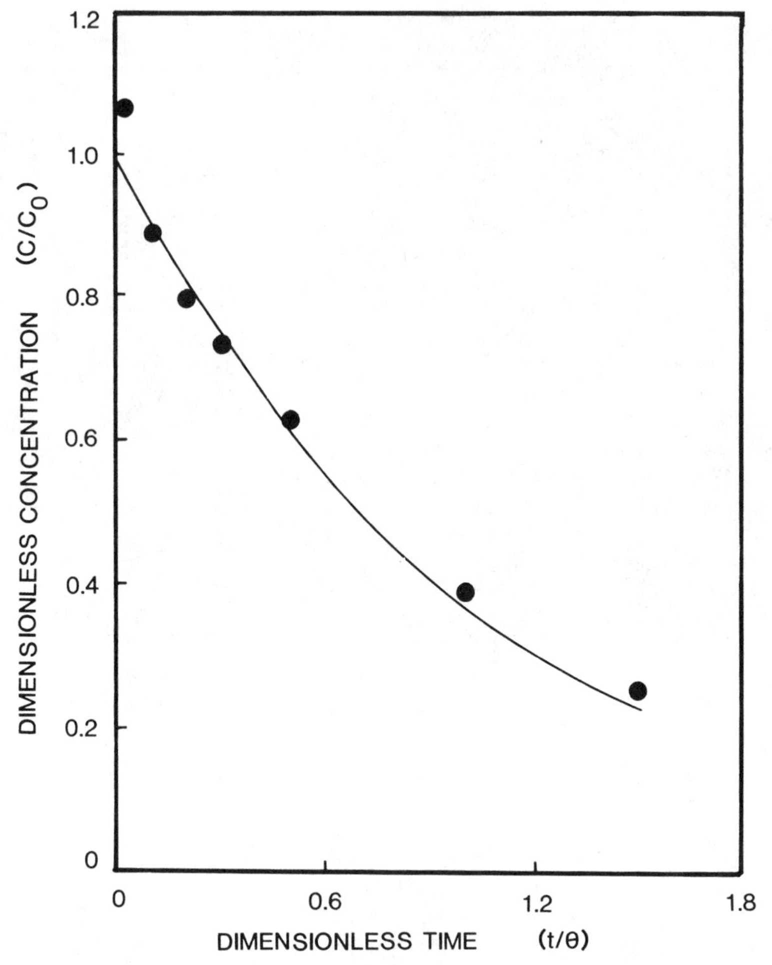

Figure 3.5 Dimensionless dye concentration in the effluent of a RotoTorque as a function of dimensionless time. C_0 is the initial concentration of dye, and Θ is the mean hydraulic residence time. The dye was pulsed to the reactor at time = 0. The line represents the theoretical calculation. (Reprinted with permission from Trulear, 1983.)

used to estimate the effectiveness of chemical additives (e.g., chlorine) for biofilm removal (Figure 3.8).

3.3.1.3 Material Balances Since the RotoTorque is a CFSTR, material balances are conveniently accomplished, permitting useful kinetic analyses of biofilm rate processes including substrate removal, growth, production formation, and detachment (Trulear and Characklis, 1982; Bakke et al., 1984). Through material balance calculations, the reactor permits nonintrusive, nondestructive measurement of biofilm activity normalized over the entire reactor surface area. Specific activities (e.g., substrate removed per cell per unit time) can be determined by removing a slide and determining the cell mass or number per unit area. Oxygen balances are also convenient since oxygen need only enter the reactor with the feed water.

3.3.2 Tubular Reactor

Tubular reactors can be operated as CFSTRs if internal recycling is used (Figure 3.2). This system has advantages when simulating heat transfer tubing or water supply conduits. Often, the experimental system may use tubing of the same composition and diameter as that being simulated. Advantages of the tubular configuration include the following:

1. At high recycle rates ($Q_R \gg Q_D$), the reactor contents are completely mixed and no longitudinal concentration gradients exist in the liquid phase. This simplifies mathematical descriptions and sampling. It also provides a relatively uniform biofilm in the recycle section while allowing

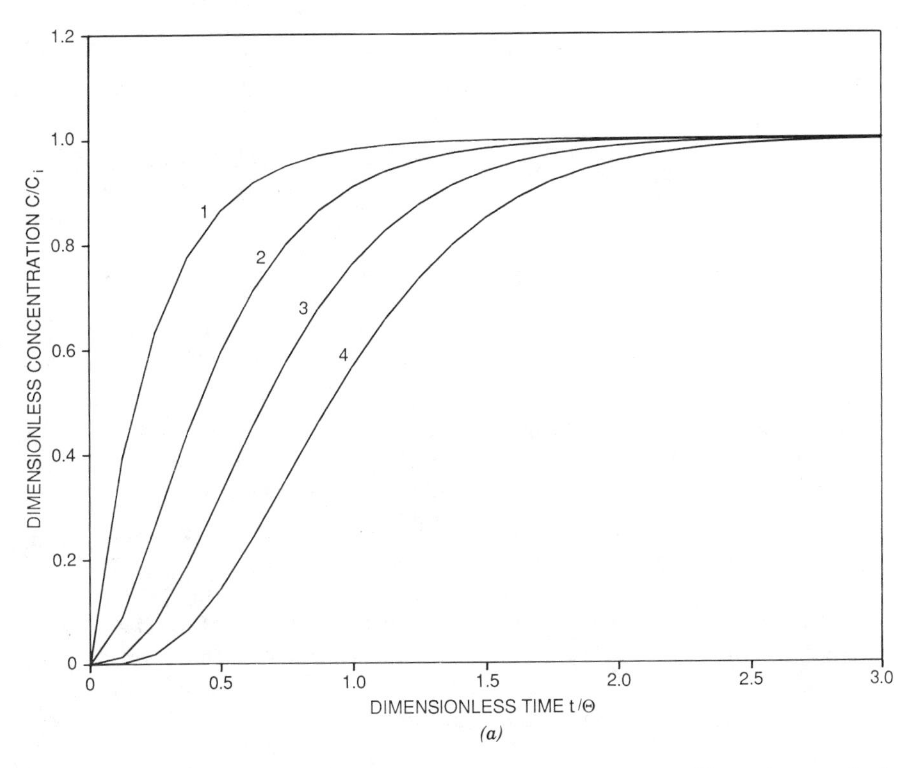

Figure 3.6 (a) Calculated dimensionless dye concentration in the effluent from each of four RotoTorques operated in series. The dye was input to RotoTorque 1 at time 0.

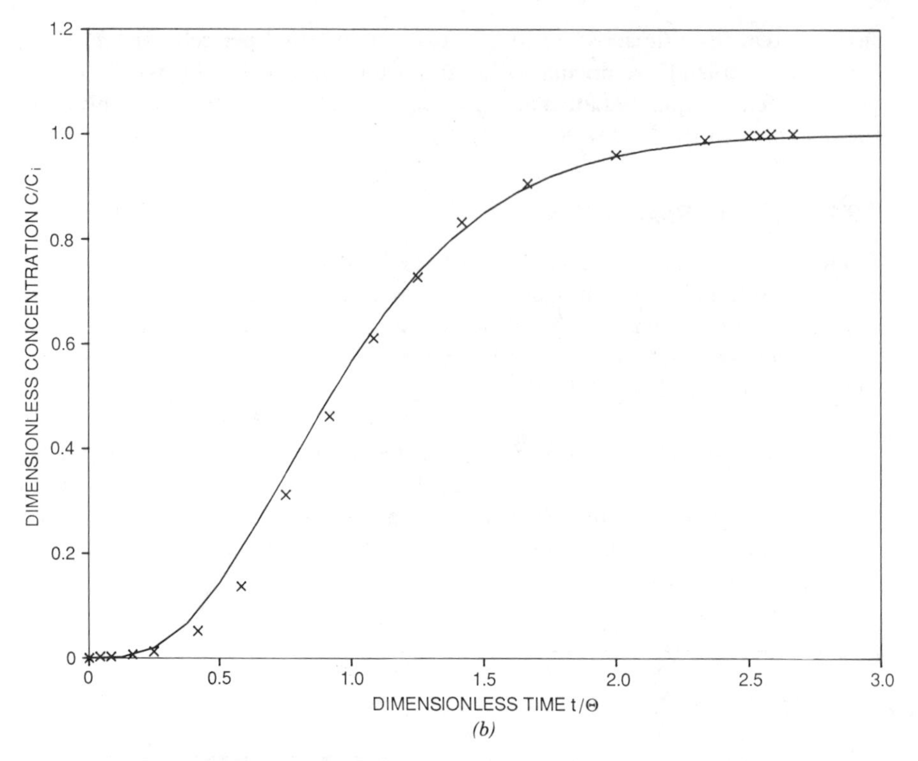

Figure 3.6 (b) Calculated (solid line) and measured (data points) dye concentration in the effluent of the fourth RotoTorque in the cascade (van der Wende, unpublished results). The dye input was begun at time 0 in RotoTorque 1 with an inlet concentration of C_i.

simple control of pH and temperature. From a practical standpoint, this system minimizes the consumption of water, microbial nutrients, and other chemical additives.

2. If desirable, a short hydraulic residence time can be maintained in the system, which minimizes biomass activity in the bulk fluid and restricts microbial activity in the system to the reactor surfaces.
3. The fluid shear stress at the wall in the recycle loop is independent of the mean residence time in the reactor system.

The tubular system can be operated as a plug flow reactor if the flow is once through and the tube is long enough or the flow rate is low enough to establish concentration gradients. Each tubular reactor system may incorporate one or more of the following types of tubular sections (Figure 3.9):

1. A tubular test section in which the pressure drop is monitored by a manometer or pressure transducer during biofilm development. The fluid

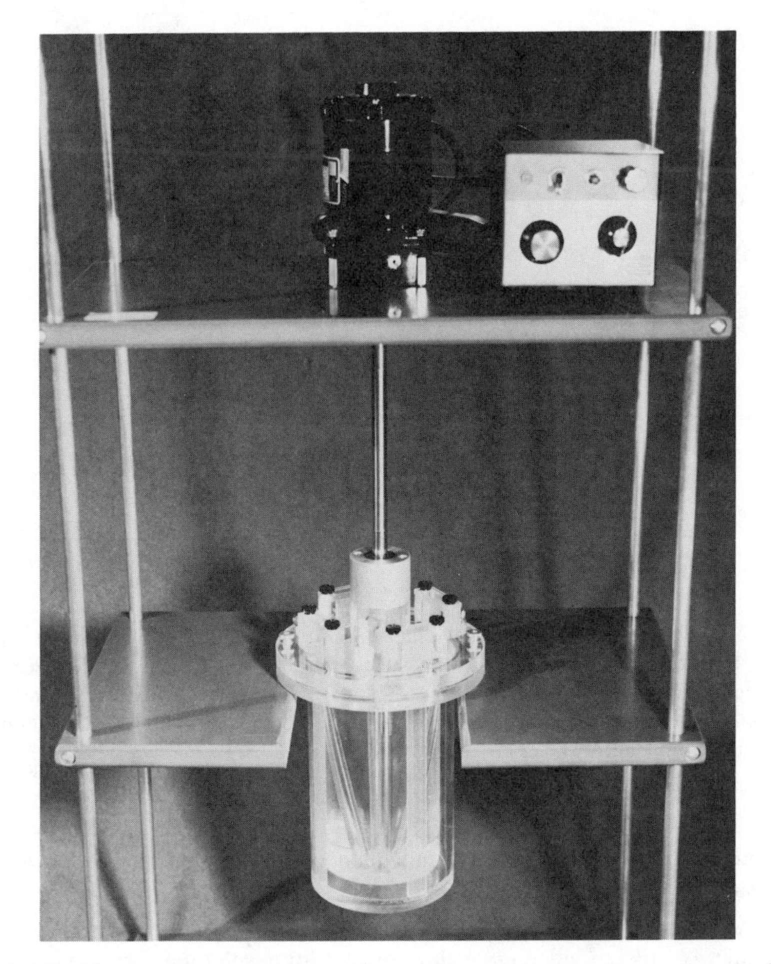

Figure 3.7 The RotoTorque system with variable speed motor for turning the inner cylinder.

frictional resistance can be calculated from flow rate and pressure drop measurement.

2. A test heat exchanger section in which changes in heat transfer resistance are monitored as a function of biofilm development.
3. A tubular section or sleeve, which contains removable sections of tube that can be cut to give a large sampling area. These sample tubes are removed periodically for determining the biofilm thickness or mass, or for biofilm chemical analysis. The Robbins device serves a similar purpose (McCoy et al., 1981), but generally has a reduced sampling surface area limited by the pipe diameter. The biofilm deposit is frequently patchy (Chapter 11),

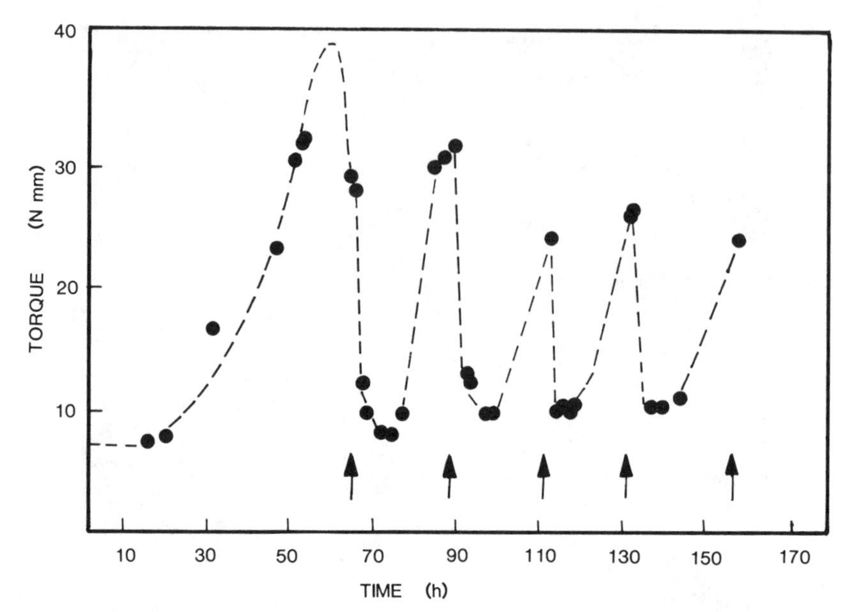

Figure 3.8 The influence of periodic chlorination on biofilm accumulation in a Roto-Torque can be observed indirectly by measuring the torque. Chlorine addition results in an immediate drop in torque (frictional resistance), which corresponded with biofilm detachment. The arrows indicate chlorine additions (12.5 g m^{-3} for 0.5 h). (Reprinted with permission from Characklis et al., 1980.)

and so a large sampling area may be necessary to achieve a representative sample.

4. A glass tubular section for visual or microscopic observations. Periodic micrographs or video recordings of biofilm development may be obtained using such a transparent section.
5. A rectangular duct section may be used when samples of biofilm developing on a flat plate are needed. This is particularly desirable when periodic access to a deposit is required, for instance, using microelectrodes to obtain chemical profiles within a deposit. Pedersen (1982) describes a rectangular duct reactor for biofilm studies which incorporates a large number of removable slides for observing or analyzing biofilm characteristics.

Another form of tubular reactor incorporates image analysis methods and has been used to separate and independently examine individual processes contributing to the colonization of cells at the substratum (Escher, 1986). For laminar flow experiments, microbial cells are grown in a chemostat and continuously flow through a rectangular glass capillary (Figure 3.10). The chemostat operation minimizes variations in cell physiological state and cell concentration throughout an experiment. The inner wall of the rectangular capillary is the sub-

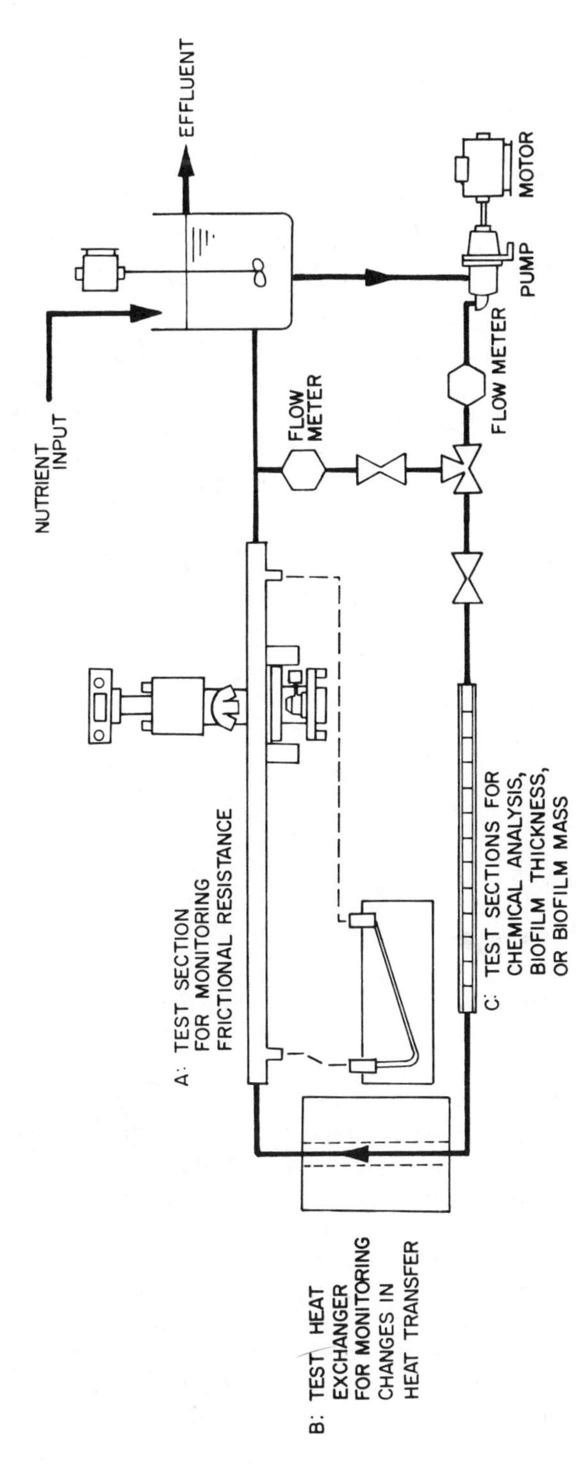

Figure 3.9 Composite diagram of a tubular reactor incorporating the important features of three tubular reactors. The recycle flow rate is much larger than the influent flow rate, and so the reactor is a CFSTR.

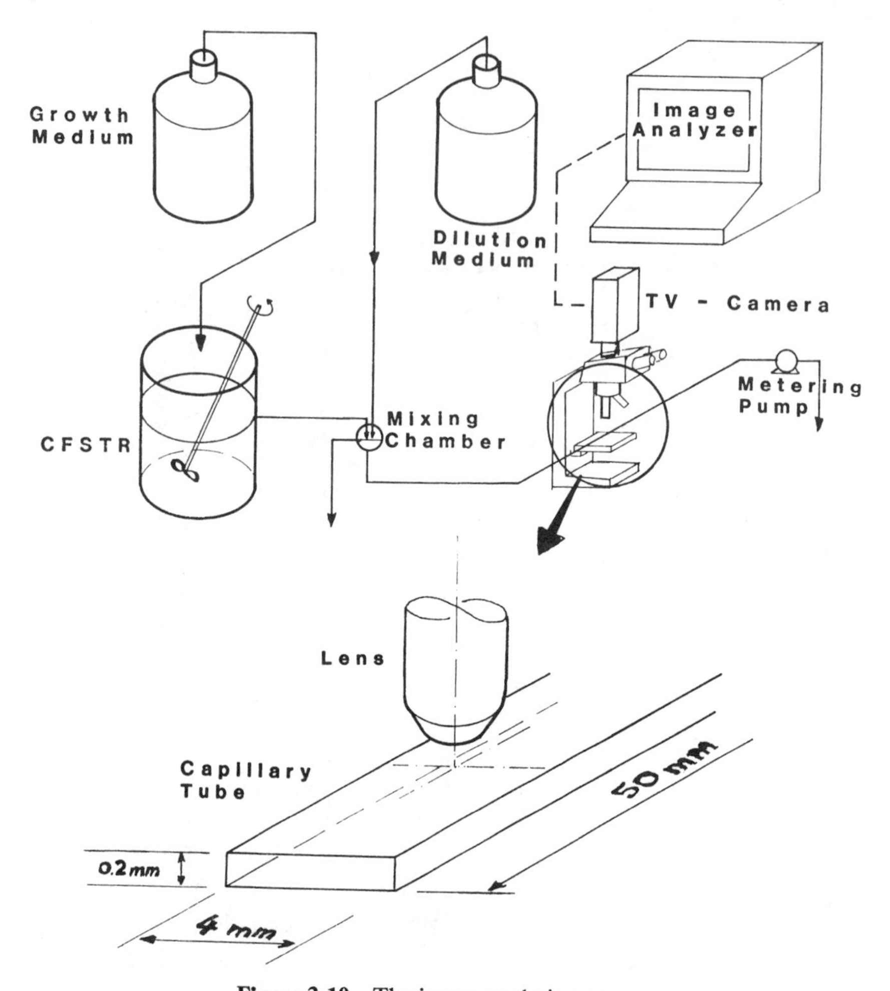

Figure 3.10 The image analysis system.

stratum, and the processes occurring at the substratum are monitored continuously by a high resolution video camera mounted on a microscope, which is focused on the inner wall of the capillary (i.e., the substratum). The video signal is transmitted to an image analyzer, which converts the grey image into a binary image. A single binary image of the specimen is stored on disk at constant time intervals (e.g., every 5 min) for later analysis. The capacity for storing images is dependent on the computer hardware. Escher (1986) could store up to 60 images. The stored images are analyzed, and the events of adsorption, desorption, and multiplication can be observed and enumerated as a function of exposure time. The system also determines the spatial coordinates of each adsorbed cell, substratum area covered by cells, and the frequency distribution of cell characteristics (length, width, roundness). The resulting data permit calculation of the residence time distribution for cells at the substratum. An analysis of the initial events in the colonization of a glass substratum by *P. aeruginosa* using these image analysis techniques is described in Chapter 12.

3.3.3 Open Channel

The open channel geometry is ideal for simulating streambeds, aqueducts, or culverts. An important difference between this geometry and a closed tube geometry is that gas transfer between the atmosphere and water occurs throughout the length of the open channel reactor, whereas the tubular reactor cannot be operated with atmospheric contact. The open channel is very convenient for conducting microelectrode measurements, sampling the biofilm, and generally observing the behavior in the system.

3.3.4 Rotating Disk and Radial Flow Reactor

These two reactor designs (Figures 3.11 and 3.12) are useful because a range of fluid shear stresses can be tested simultaneously for its effect on microbial adsorption (Duddridge et al., 1982).

3.3.4.1 Radial Flow Reactor The radial flow reactor, as described by Fowler and McKay (1980), consists of two parallel disks separated by a narrow gap, typically around 500 μm (Figure 3.11). Culture fluid is pumped in at the center of one of the disks at a constant volumetric flow rate and flows out radially between the disks to a collection manifold. As the cross sectional area available for flow increases with increasing radius, the linear velocity and fluid shear stress decrease. Thus, high shear forces are present near the inlet, and lower ones near the outlet. This geometry is only useful for observing initial events of biofilm accumulation, because of the narrow spacing between the disks. The plug flow characteristics in the gap may have an influence on spatial colonization pattern on the substratum, since detached cells have an opportunity to colonize the substratum at a downstream location. Once cells adsorb at one location, subsequent flow patterns around the adsorbed cells, and hence downstream surface colonization, may be affected. Some deficiencies in the calculation of shear stress have been discovered, which may influence the interpretation of earlier data (Fryer et al., 1984).

3.3.4.2 Rotating Disk Reactor The rotating disk reactor consists of a rotating disk placed in a solution (Figure 3.12). This reactor has been used to study the effect of fluid shear stress on biofilm development (Loeb et al., 1984; Abbott et al., 1983) because the fluid shear stress varies along the radius of the disk, with the highest values at the outer edge and lower ones toward the center. Gulevich et al. (1968) used this system to study the rate of substrate removal by biofilms because the mass transfer boundary layer is constant over the entire disk.

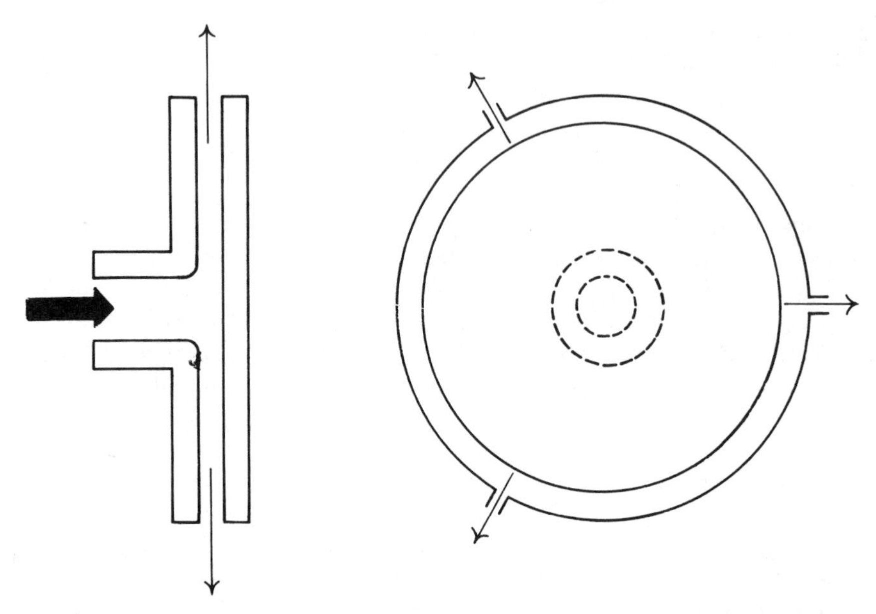

Figure 3.11 The radial flow reactor consists of two parallel disks with a small distance ($\simeq 500\ \mu m$) of separation. (Reprinted with permission from Fowler and McKay, 1980.)

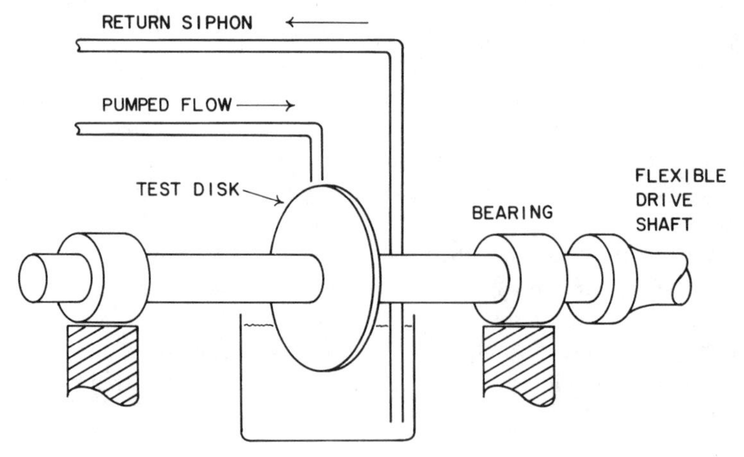

Figure 3.12 A rotating disk reactor of the type employed by Loeb. (Reprinted with permission from Loeb et al., 1984.)

3.3.5 Packed and Fluidized Beds

By adding aggregates to a CFSTR, the surface area for biofilm formation can be greatly increased. In a packed bed, aggregates are packed or are sufficiently heavy that they are not fluidized under flow condition. The packed bed restricts mixing in the axial direction, which may result in the bed operating more as a plug flow reactor than as a CFSTR. A high recycle rate will minimize this effect. A fluidized bed consists of aggregates that are suspended and therefore are well mixed so long as sufficient turbulence is maintained. In both packed and fluidized beds, the aggregates are maintained within the reactor and cannot be washed out (Figure 3.13). The increased surface area may be desirable when high biofilm mass concentration is required.

3.4 OTHER EXPERIMENTAL SYSTEM COMPONENTS

3.4.1 Reactor Feed

The reactor feed consists of three components: dilution water, nutrient solution, and microbial inoculum.

3.4.1.1 Dilution Water Tap water, distilled water, or various treated waters can be used for dilution water, depending on the specific needs of the experiment. The water can be treated before entering the experimental system in various ways including:

1. Flow through an inline 5 μm filter to remove particles
2. Flow through a carbon adsorption column to remove residual chlorine
3. Storage, temperature adjustment, pH adjustment, and aeration in a tank
4. Flow through an inline 0.2 μm filter to remove bacteria

3.4.1.2 Nutrient Solution Nutrient solutions are generally prepared in concentrated form to keep storage and sterilizing volumes small. Nutrient solutions can generally be autoclaved, whereas dilution water is sterilized by filtration. The nutrients are blended with the dilution water before entering the reactor.

3.4.2 Microbial Inoculum

When undefined mixed cultures are selected, a standard inoculum can be prepared to minimize the effects of population distribution differences in laboratory experiments (Characklis et al., 1982). Mixed liquor from a domestic wastewater treatment plant is settled, and the concentrated sludge mixed with glycerol to approximately 25% v/v glycerol. 10 ml aliquots of the resulting suspension are

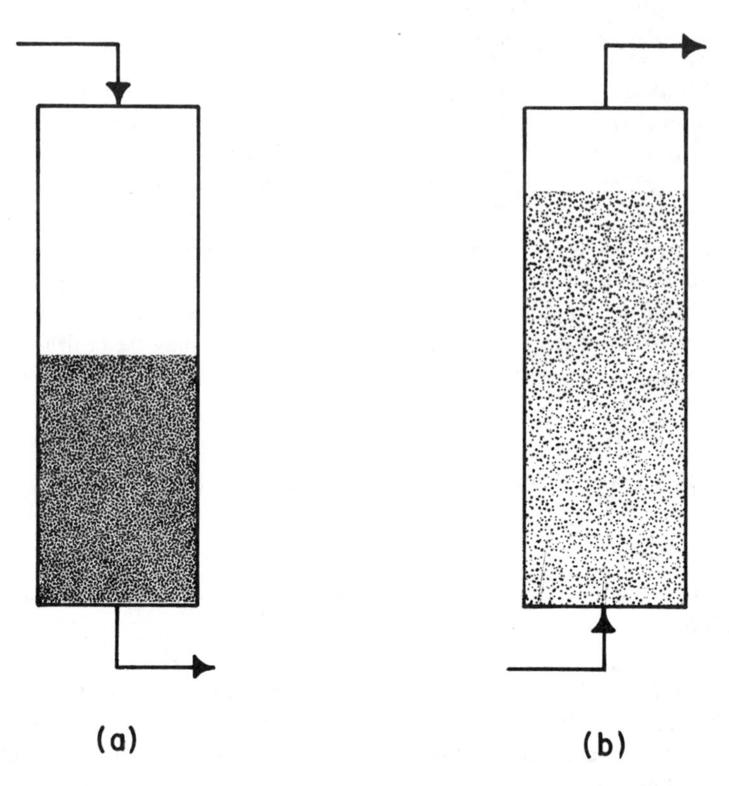

Figure 3.13 The packed bed reactor (a) is generally operated with a downward flow of liquid, while the fluidized bed reactor (b) is operated with an upward flow at a rate which causes bed expansion or fluidization.

transferred to glass ampules, which are quick frozen in liquid nitrogen and stored at −70°C. Growth rate tests on the standard inocula, conducted periodically, indicate no significant changes in overall growth rate over one year.

A method for providing a continuous inoculum employs a packed bed reactor with or without recycle (Characklis et al., 1983). Sterile nutrients are pumped continuously at a constant rate through a packed bed that has been inoculated with a mixed culture. Organisms adsorb to the support medium and begin to grow. After about two or three weeks (for an aerobic reactor), the effluent cell concentration is stable with respect to cell numbers. Limited results indicate that the population distribution is also relatively stable.The ratio of inlet substrate concentration to dissolved oxygen concentration (Chapter 6) determines whether the reactor is aerobic or anaerobic.

When experimenting with single species populations, aseptic technique is used. Regular (e.g., daily) contamination checks of the reactor effluent should be conducted.

3.5 ANALYTICAL METHODS

3.5.1 Direct Measurement of Biofilm Quantity

The only two direct measures of film quantity are thickness (L_f) and mass (Table 3.1). The two quantities are related by the film density (ρ_f), i.e., $\rho_f = X_f''/L_f$, where X_f'' is the biofilm mass per unit area of substratum.

3.5.1.1 Biofilm Mass Wet biofilm is scraped from sample tubes (tubular reactors), slides (RotoTorque), or a sample of the support medium (packed or fluidized beds) and dried to constant weight at an elevated temperature (100–105°C). For substrata unaffected by heat, the biofilm is left on the substratum and the entire section is placed in the oven. After drying, the combined weight of biofilm and substratum is obtained. The substratum is then cleaned, dried, and weighed again. The difference in the two measurements is the dry biofilm mass. The surface area available for growth is generally known, so an areal mass density can be determined. If biofilm thickness or biofilm volume has been measured, the volumetric biofilm density (ρ_f) can be determined. The volumetric biofilm density has units of dry mass per unit wet volume. The areal film density, X_f'', has units of dry mass per unit of substratum area.

3.5.1.2 Biofilm Thickness

RotoTorque. Biofilm thickness has been determined using various methods that differentially locate the biofilm–fluid interface and the biofilm–substratum interface. Trulear and Characklis (1982) used an optical microscope method to determine apparent biofilm thickness in the RotoTorque. The technique was adapted from Sanders (1966) and requires biofilm growth on a thin acrylic plastic slide, which forms an integral part of the RotoTorque wall. The slide is withdrawn from the reactor and placed on a microscope stage. The 10× objective (100× magnification) is lowered until the biofilm surface is in focus and the fine adjustment dial setting is recorded. The objective is then lowered further until the inert plastic growth surface is in focus (Figure 3.14). The difference in fine adjustment settings is compared with a calibration curve, which is generally linear, and the thickness obtained. The sample standard deviation of the measurement is approximately 10–12 μm on biofilms ranging 11–130 μm in mean thickness. The variation includes any irregularities in the biofilm surface. The results of the method must be corrected for the refractive index of the wet biofilm, which is on the order of 1.3 (Bakke and Olsson, 1986). Hence, the term *apparent* biofilm thickness.

Tubular reactor. Bakke (1986) and Bakke and Olsson (1986) have reported an accurate method for *in situ* simultaneous determination of biofilm thickness and refractive index in a transparent rectangular tube. The apparent thickness

TABLE 3.1 Various methods for measurement of biofilm accumulation.

Classification	Analytical Method	References
A. Direct measurement of biofilm quantity	Biofilm thickness	Trulear and Characklis, 1982; Bryers and Characklis, 1980; Hoehn and Ray, 1973; Trulear, 1983.
	Biofilm mass	Trulear and Characklis, 1982; Picologlou et al., 1980; Zelver, 1979.
B. Indirect measurement of biofilm quantity: Specific biofilm constituent	Polysaccharide	Bryers and Characklis, 1980.
	Total organic carbon	Bakke et al., 1984; Little and Lavoie, 1979.
	Chemical oxygen demand	Bryers and Characklis, 1980.
	Protein	Bryers and Characklis, 1980; McCoy, 1979.
C. Indirect measurement of biofilm quantity: Microbial activity within the biofilm	Viable cell count	Costerton and Colwell, 1979; Corpe, 1973; Gerchakov et al., 1977.
	Epifluorescence microscopy	Bakke et al., 1984; Geesey et al., 1978.
	ATP	LaMotta, 1974; Little and Lavoie, 1979; Bobbie et al., 1979.
	Substrate removal rate	Costerton and Colwell, 1979; Trulear and Characklis, 1982; Bakke et al., 1984.
D. Indirect measurement of biofilm quantity: Effects of biofilm on transport properties	Friction resistance	Trulear and Characklis, 1982; Picologlou et al., 1980; Norrman et al., 1977.
	Heat transfer resistance	Characklis et al., 1981; Fetkovich et al., 1977; Knudsen, 1980.

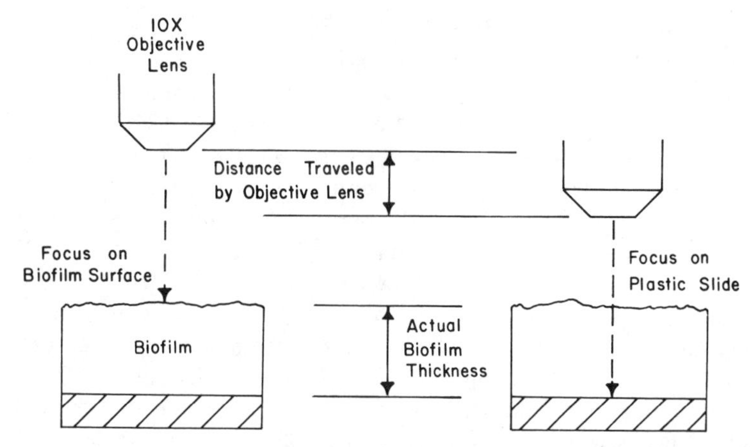

Figure 3.14 The determination of wet biofilm thickness with an optical microscope requires two measurements: (1) the objective lens is lowered until the top of the biofilm is in focus, and the height on the fine calibration adjustment is noted; (2) the objective is lowered further until it focuses on a mark placed on the substratum under the biofilm. The second height is subtracted from the first to obtain the apparent biofilm thickness. The true biofilm thickness is obtained by correcting for the refractive index of the biofilm.

is determined microscopically by observing the biofilm in plan view (i.e., from above) as described for the RotoTorque. Then, another microscopic thickness measurement is made in cross section through the rectangular tube. A 5 μm biofilm thickness can be detected by this method.

Biofilm thickness has also been determined by volumetric displacement in tubular reactor systems (Picologlou et al., 1980). Small sample tubes (12.7 mm in inside diameter and 50 mm in length) were inserted as an integral part of the tubular reactors. The sample tubes were inserted end to end in an acrylic plastic test section (19 mm in inside diameter and 760 mm in length) as shown in Figure 3.15. A modified design for the test section was used more recently to study the effects of roughness on biofouling (Turakhia and Characklis, 1983). The test section is connected to the recycle loop with pipe unions to provide easy access to the sample tubes. At designated intervals, a sample tube is removed from the

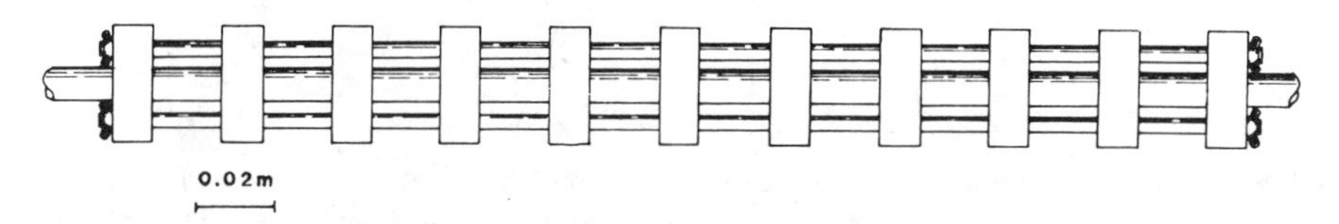

Figure 3.15 The tubular test section consists of a number of short sections of tube in a sleeve for convenient removal. The tube sections can be cut to various lengths to accommodate analytical precision and accuracy needs. The sample tubes are removed periodically for determining biofilm thickness or mass, or for biofilm chemical analysis.

reactor and a clean sample tube inserted in its place. The sample tube with biofilm is then drained to reduce excess water in the biofilm. The drainage time is generally 10 min, but in some cases is reduced to 2.5 min when the biofilm appears to be drying around the end of the sample tube. The volumetric displacement of the biofilm is determined using a displacement cell (Figure 3.16). The latter is filled with a water-surfactant solution (0.3% v/v Turgitol). The initial liquid level before immersing the sample tube is measured by lowering the conductive probe with a micromanipulator until contact is made with the water surface. Contact is indicated by visual observation or by deflection of an ammeter connected in series with the cell, and a 1.5 V power source. A sample tube with biofilm is then immersed in the cell, and the new liquid level (and hence displacement) due to the sample tube and biofilm is determined. The sample tube is cleaned, and its volumetric displacement is determined in the same way. The difference between the displacements of the fouled and clean sample tubes is the biofilm volume. The biofilm thickness is determined by dividing the biofilm volume by the inner surface area of the sample tube. The overall sample standard deviation for this method of determining biofilm thickness is approxi-

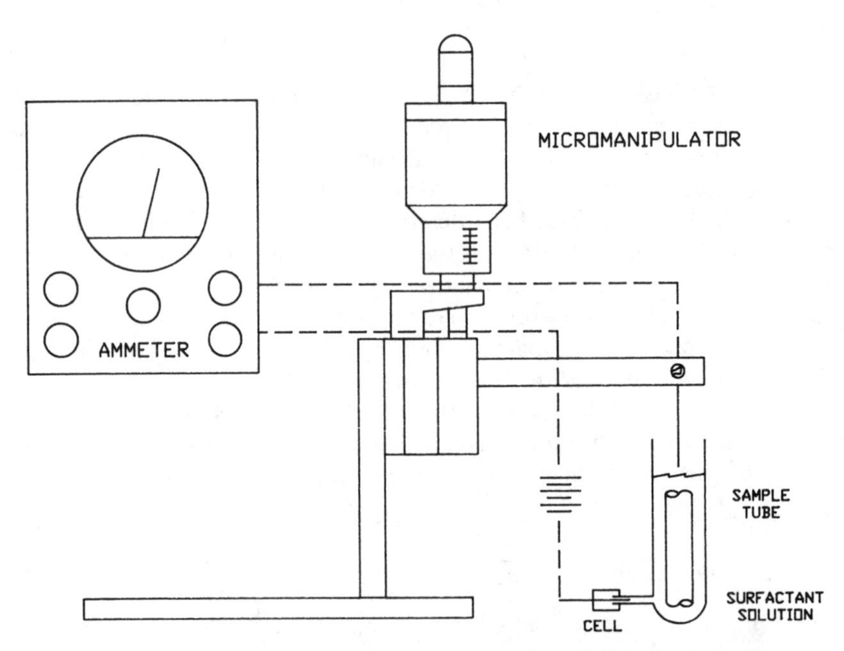

Figure 3.16 The apparatus for measuring wet biofilm volume consists of a test tube containing a surfactant solution in which a tubular test section is immersed. The micrometer lowers a conductive needle until it contacts the fluid, resulting in a spike of current. The tubular test section is removed, cleaned, and reinserted in the test tube for a second height determination. The difference in the height determinations yields a biofilm volume. (Reprinted with permission from Zelver, 1979.)

mately ± 10 μm based on five replicate measurements on 73 samples with mean thicknesses of 50–300 μm.

Norrman et al. (1977) located the biofilm–fluid and biofilm–substratum interfaces in a tubular reactor by means of electrical conductance using a technique adapted from Hoehn and Ray (1973), which uses an apparatus consisting of a steel needle mounted on a micromanipulator. A cylindrical test section provides the substratum for biofilm accumulation and the geometry for the thickness measurement (Figure 3.17). The nonmetallic test section consists of six measurement points, each one being a stainless steel rod (3 mm OD) mounted flush with the inside tube wall. Opposite each rod is a threaded hole, which can be sealed with a screw and O-ring. The thickness of the biofilm on the stainless steel rod surface is measured using the electrical conducting probe depicted in Figure 3.18. The probe and one of the stainless steel rods are connected to an electrical circuit completed by an electrometer. To obtain a measurement, the test section is removed from the reactor, the screws are withdrawn, and the test section is drained for two minutes. The probe is then lowered into the test section through the threaded hole until contact is made with the biofilm surface; a current of about 10^{-8} A is detected and the depth noted on the micromanipulator (Figure 3.18a). Next, the probe is lowered until contact is made with the stainless steel rod surface, when a greater current flow of around 3×10^{-5} A is registered, and the depth is again noted (Figure 3.18b). The difference in depth is the biofilm thickness. The procedure is repeated at the other five locations. The precision is about $\pm 6\%$, and the accuracy, compared to vernier micrometer measurements, is within 5%. If the test section is metal, the stainless steel studs are superfluous.

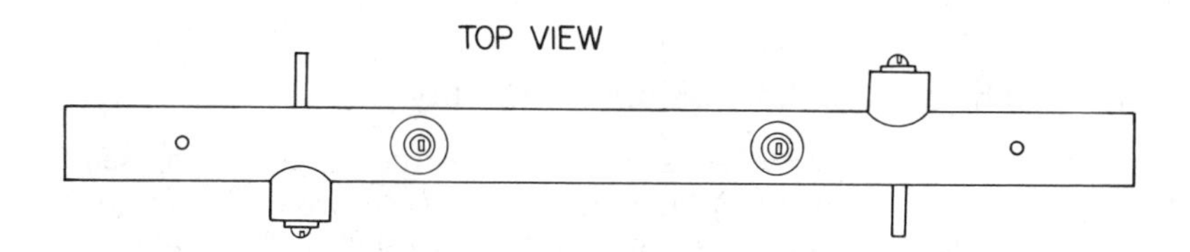

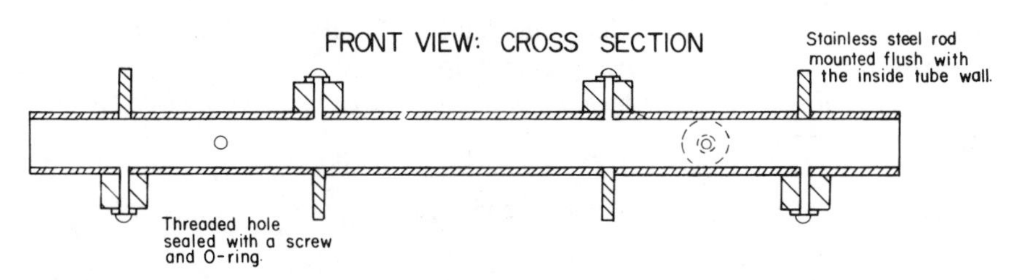

Figure 3.17 The tubular test section used in measuring wet biofilm thickness by electrical conductance. The test section was made of acrylic plastic; the studs were stainless steel. (After Norrman et al., 1977.)

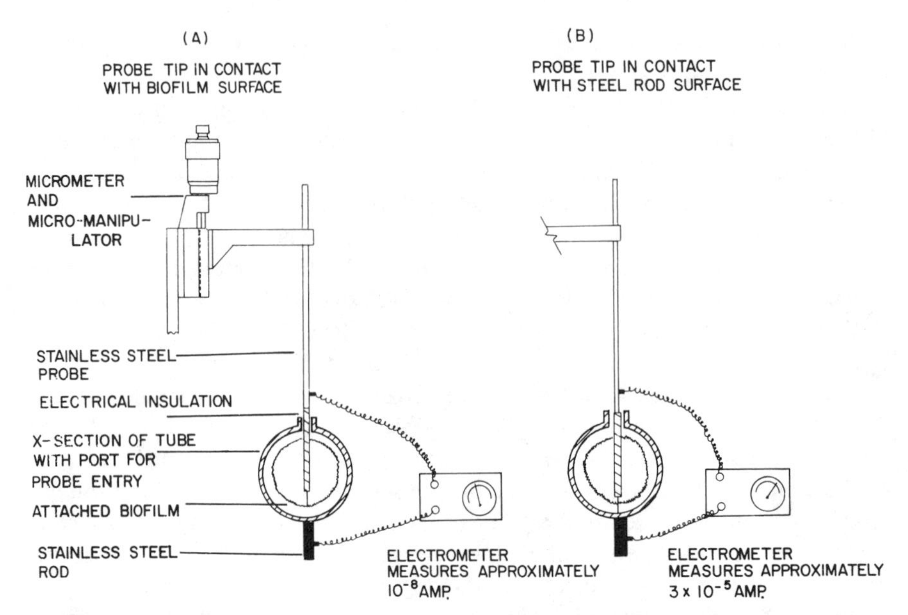

Figure 3.18 The apparatus for measuring wet biofilm thickness by electrical conductance consists of a micrometer that lowers a conducting needle through the holes in the test section (Figure 3.17). The tube is drained for a predetermined interval prior to the measurements. The needle is lowered until it contacts the biofilm, resulting in a spike of current (a). The micrometer reading is noted, and the needle is lowered further until it contacts the stainless steel rod (b). At that point, another spike of current is observed and the micrometer reading is noted. The difference in micrometer readings is the biofilm thickness. (After Norrman et al., 1977.)

3.5.2 Indirect Measurement of Biofilm Quantity

Various specific elemental or molecular biofilm constituents have been used to monitor film development (Table 3.1), because mass measurements are not as sensitive as some other techniques. A calibration curve relating the specific constituent and biofilm mass is needed if material balances are desired. This is less important if a conservative quantity like carbon is used for measuring biomass, substrate, and product formation. The detection limits and standard errors for the various biofilm analyses are presented in Table 3.2.

Two chemical procedures have been used by us as indirect measurements of biofilm formation:

1. Total biofilm organic carbon or chemical oxygen demand
2. Total biofilm polysaccharide

3.5.2.1 Organic Carbon or Chemical Oxygen Demand Indirect quantitation of biofilm is possible by measuring oxidizable organic biofilm material,

TABLE 3.2 Sensitivity and precision of various biofilm accumulation measurement techniques.

Method	Sensitivity		Precision (±)		References
	Measured Quantity	Equivalent COD[a] ($g\ m^{-2}$)	Measured Quantity	Equivalent COD[a] ($g\ m^{-2}$)	
Biofilm thickness (μm)	9	256	9	256	Zelver, 1979
	10	285	10	285	LaMotta, 1974
	10	285	9	256	Trulear, 1980
Biofilm mass ($g\ m^{-2}$)	1.1	1.3	0.1	0.1	Zelver, 1979
Biofilm COD ($g\ m^{-2}$)	0.06	—	0.001	—	Bryers, 1980; Bryers and Characklis, 1980
Biofilm TOC ($g\ m^{-2}$)	0.02	0.057	0.0045	0.0128	Trulear, 1983

[a]Calculated from biofilm thickness values assuming biofilm density 25 kg m^{-3}, COD 1.14 g COD per gram, and biofilm carbon content = 0.4 g carbon per gram of biofilm. (Bryers, 1980; Trulear, 1980, 1983).

which contains significant amounts of organic carbon. A modified chemical oxygen demand (COD) analysis (tubular reactor) provides good sensitivity [60 (mg COD) m^{-2}] and precision [$\pm$1.0 (mg COD) m^{-2}] (Characklis et al., 1982). Modifications to the (*American Public Health Association,* 1976) COD procedure consisted of diluting the dichromate oxidant and ferrous ammonium sulfate titrant.

Bakke et al. (1984) have monitored biofilm accumulation in the RotoTorque using total organic carbon (TOC). Sensitivity and precision of the method are less than 20 (mg TOC) m^{-2} and $\pm$4.5 (mg TOC) m^{-2}, respectively. Carbon, being an atomic element, is a conservative quantity useful for material balances and related kinetic analyses in microbial reactors.

The organic carbon in the biofilm can be segregated into cellular and extracellular (EPS), especially in monopopulation biofilms. Trulear (1983) developed the method reported by Robinson et al. (1984) for a *P. aeruginosa* biofilm. The method requires the cell size of the organisms, a quantity that can now be easily determined by image analysis techniques. The size of the cells leads to a value for cell carbon which is subtracted from total biofilm organic carbon to give extracellular carbon (Figure 3.19). The use of this technique for analysis of a *P. aeruginosa* biofilm system is presented in Chapter 13. First, the biofilm sample is dispersed by homogenizing, and the total organic carbon (TOC) is determined.

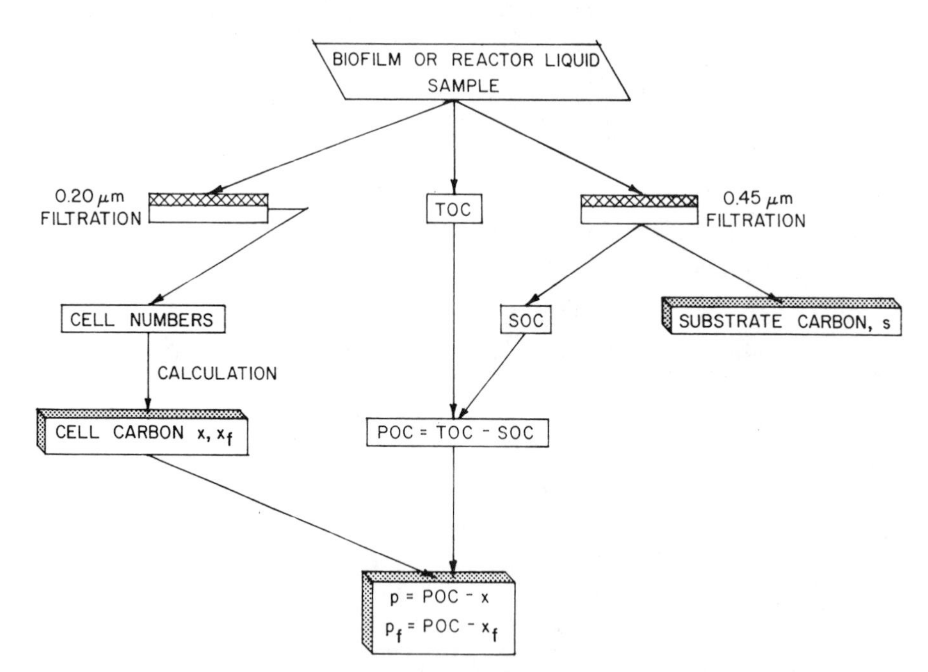

Figure 3.19 Carbon analysis of a biofilm sample can distinguish between cellular carbon in the biofilm (X_f) and extracellular polymeric substance carbon in the biofilm (P_f) (Trulear, 1983). The analysis is also useful for distinguishing these quantities in the bulk water phase (X and P).

An aliquot of the homogenized sample is filtered (0.2 μm), and the cell number and cell volume in the filtered material are determined. The cell mass is determined from data on the cell density, and from it the biofilm cell carbon X_f is calculated (see Chapter 5). The difference between TOC and cell carbon is the EPS carbon in the biofilm, P_f.

3.5.2.2 *Total Polysaccharide* Biofilms contain large quantities of polysaccharide (Costerton and Colwell, 1979; Jones et al., 1969). One method for measuring polysaccharide concentration is based on the reaction of carbohydrate reducing ends (ketones or aldehydes) with strong nonoxidizing acid to yield hydroxymethyl furfural plus other by-products. Condensation between these activated aldehydes and phenolic compounds like resorcinol, naphthol, anthrone, or phenol leads to the formation of colored compounds as a direct function of polysaccharide concentration.

The phenol-H_2SO_4 method of Dubois et al. (1956) has been used as an indirect assay for biofilm during the early stages of its development. The increase in attached polysaccharide with time in a tubular reactor is demonstrated in Figure 3.20 (Bryers, 1980). The problem with this method is that EPS can contain many

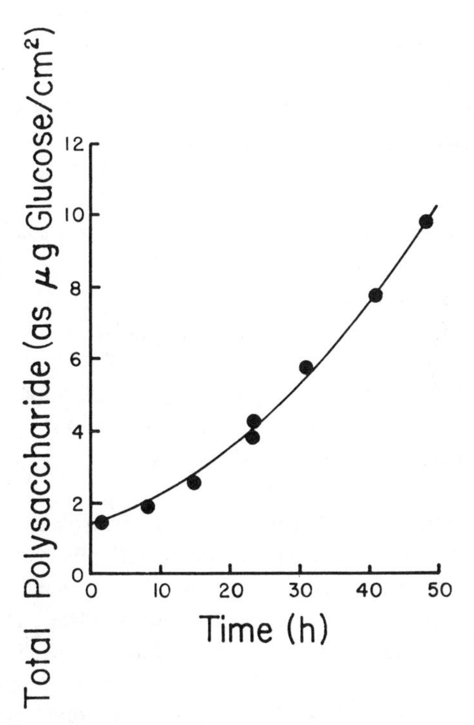

Figure 3.20 Increase in biofilm polysaccharide content during a typical experiment in a tubular reactor. The polysaccharide content is based on a glucose standard curve. (Reprinted by permission from Bryers, 1980.)

different sugar residues, though calibration usually uses one sugar (e.g., glucose). This can lead to errors in determining actual amounts present.

3.5.2.3 *Other Indirect Methods* Siebel, Bakke, and Characklis (unpublished results) have developed an *in situ* light absorbance technique of very high sensitivity, which has been tested in the RotoTorque. Infrared light, absorbed by biochemical compounds in the biofilm, is transmitted by a light emitter from the outside of the clear polycarbonate RotoTorque reactor. A light sensor is mounted on the other side of the RotoTorque, facing the emitter. The differential current between the emitter and sensor is an indirect indicator of biofilm accumulation and can be monitored continuously and nondestructively. Other light-sensitive techniques will undoubtedly be developed soon for biofilm applications, especially techniques using fiber optics technology. The accuracy of light sensitive methods will depend on measures of refractive index or absorbance characteristics of the biofilm.

Sound waves may also be used to monitor biofilm accumulation. However, the accuracy depends on methods to measure the speed of sound in the biofilm.

3.5.3 Microbial Activity

Several analytical methods (currently used or proposed) for the indirect monitoring of biofilm, using metabolic activity as an indicator, are given in Table 3.3. White (1984) has reviewed these and many other sophisticated biochemical techniques for these measurements.

3.5.3.1 *Direct Enumeration* Many investigators have used microscopic techniques to conduct *total cell counts* as a measure of sessile microbial activity in the field as well as in the laboratory (e.g., Corpe, 1973; Bakke et al., 1984). In addition, conventional plating techniques have been used to enumerate *viable cells* per unit area of biofilm. For example, Gerchakov et al. (1977) used viable cell counts to monitor biofilm development on a variety of metal surfaces exposed to tropical seawater. Results indicated an exponential increase in viable cell counts with exposure time for all materials. No direct measure of biofilm quantity was made.

Epifluorescence techniques for *total cell counts* involve disrupting large aggregates of attached biofilm, recovering bacterial cells by filtration, staining them with acridine orange (a fluorescent stain), and counting them directly by epifluorescence microscopy (Zimmerman and Meyer-Riel, 1974). Bakke et al. (1984), using direct counts by epifluorescence microscopy, have counted up to 5.6×10^{11} cells per square meter of biofilm in the RotoTorque reactor. Geesey et al. (1978), using direct counts by epifluorescence, demonstrated the importance of attached bacteria in alpine streams. Nelson et al. (1985) have used epifluorescence techniques to observe initial adsorption of *Ps. aeruginosa* to glass surfaces in the laboratory.

TABLE 3.3 Various measurement techniques for microbial activity within a biofilm.

Technique	Application	Reference
Direct enumeration	Total and viable cell count	Costerton and Colwell, 1979; Corpe, 1973; Gerchakov et al., 1977).
	Total cell count by epifluorescence microscopy	Bakke et al., 1984; Geesey et al., 1978.
Cellular component Chemical analysis	Adenosine triphosphate (ATP)	LaMotta, 1974; Bobbie et al., 1979; Little and Lavoie, 1979.
	Total proteins	White et al., 1979; Bryers and Characklis, 1980; McCoy, 1979.
	Alkaline phosphates	White et al., 1979.
	Muramic acids	Bobbie et al., 1979; White et al., 1979.
	Poly-β-hydroxybutyrate	Bobbie et al., 1979; White et al., 1979.
Substrate removal	Direct measurement	Trulear and Characklis, 1982; Bakke et al., 1984; Costerton and Colwell, 1979.
	Heterotrophic potential	White et al., 1979; Geesey et al., 1978.

Measurement of the adenosine triphosphate (ATP) content of biofilms has been widely used as a chemical assay for biofilm activity. ATP is found only in living organisms: no residual ATP is detectable after cell death (Hamilton and Holm-Hanson, 1967). ATP activity appears limited to the upper layers of the biofilm. For example, LaMotta (1974) observed an increase in ATP content with biofilm thickness up to approximately 320 μm in a RotoTorque. However, the ATP content remained constant as the biofilm thickness increased beyond 320 μm, suggesting the presence of an "active" thickness or layer of biofilm (see Section 3.5.3.2). This may also suggest analytical limitations on the ATP extraction technique, though this has not been confirmed. Attached ATP per unit area has been measured by Little and Lavoie (1979) and Bobbie *et al.* (1979) in two separate field heat exchange units.

White and coworkers (1979, 1984) have experimented with many measurements of biofilm activity including (1) alkaline phosphatase, a bacterial cell wall component, (2) muramic acid, a unique procaryotic cell wall material, (3) poly-β-hydroxybutyrate, and (4) phospholipids.

3.5.3.2 Substrate Removal Rate Microbial activity within the biofilm can be determined by measuring the removal rate of a particular substrate, nutrient, or electron acceptor. Using the RotoTorque, Trulear and Characklis (1982) observed an increase in glucose removal rate with increasing biofilm accumulation up to a critical biofilm thickness termed the *active* thickness. Similar observations have been reported by many others (LaMotta, 1974; Atkinson and Daoud, 1970; Harremoes, 1977; Kornegay and Andrews, 1967). The active thickness is the presumed depth of penetration of substrate, nutrient, and/or electron acceptor before it is exhausted.

Heterotrophic potential is a similar test developed to measure microbial activity in natural waters (Costerton and Colwell, 1979). The test essentially measures the removal of glutamic acid (or other substrate) due to microbial activity in the sample. The heterotrophic potential test may provide spurious results, since the sample is subjected to some rather drastic environmental changes during sampling and processing.

Substrate removal in a CFSTR may be reported in many ways, including the following:

Substrate mass removed per unit wetted substratum area:

$$\frac{Q(S_i - S)}{A} \qquad [3.1]$$

Substrate mass removed per unit biofilm volume:

$$\frac{Q(S_i - S)}{AL_f} \qquad [3.2]$$

Substrate mass removed per unit biofilm mass:

$$\frac{Q(S_i - S)}{AX_f''} \qquad [3.3]$$

Substrate removed per unit biofilm cell:

$$\frac{Q(S_i - S)}{An_f''} \qquad [3.4]$$

where $Q(S_i - S)$ = substrate removal rate ($M_s t$)
A = substratum area (L^2)
L_f = biofilm thickness (L)
X_f'' = biofilm mass (ML_x^{-2})
n_f'' = biofilm cell numbers (# L^{-2})

The quantities are related through the biofilm thickness, biofilm cell, density, and overall biofilm density, since

$$X_f'' A = \frac{\rho_f}{L_f} \qquad [3.5]$$

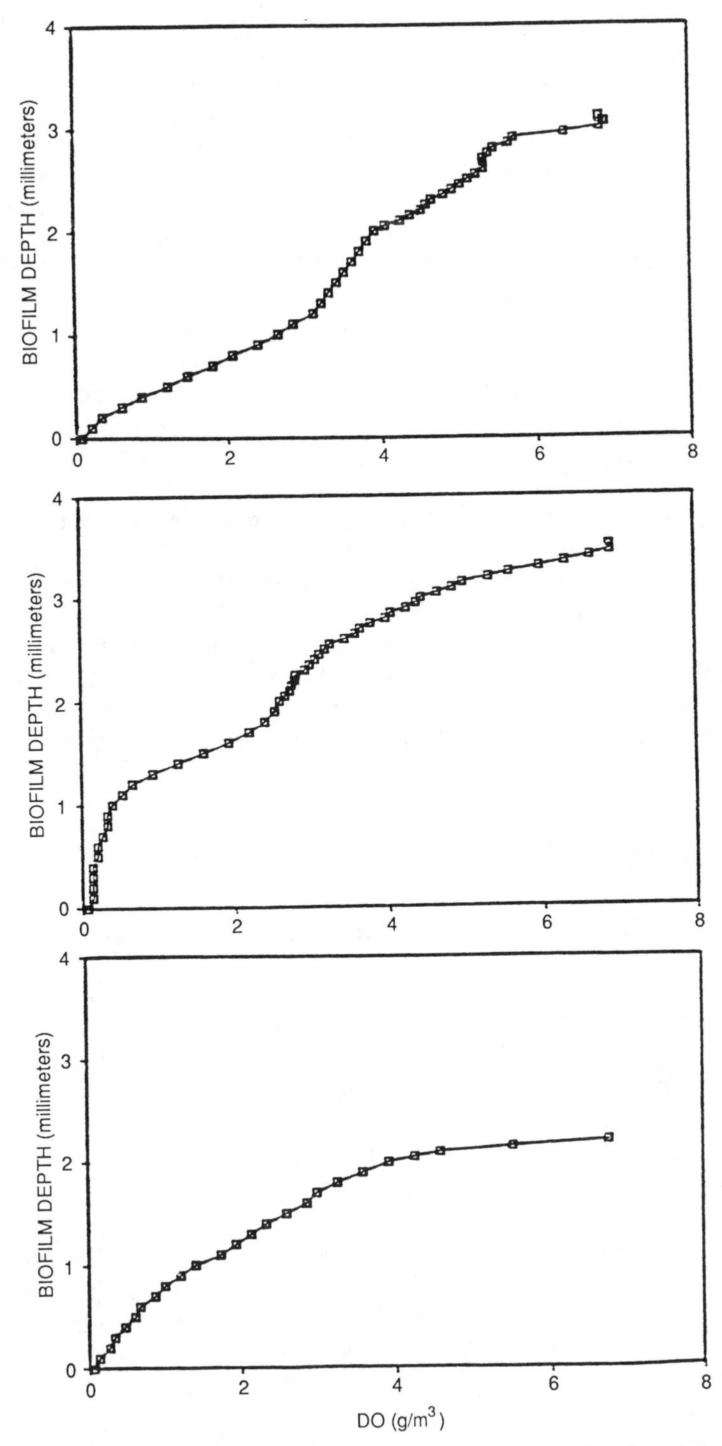

Figure 3.21 Dissolved oxygen profiles in a laboratory biofilm at three locations less than 2 mm apart on the biofilm-water interfacial surface. Patchiness is clearly evident in the various profiles from the same microenvironment. (Lewandowski, unpublished results.)

3.5.4 Effects of Biofilm on Transport Properties

The development of a biofilm in a flowing fluid system generally increases fluid frictional resistance (momentum transport) in the tubular reactor (Picologlou et al., 1980; Zelver, 1979) and in the RotoTorque (Trulear and Characklis, 1982). Thus by monitoring the pressure drop in a tube or torque in a rotating device, the progression of biofilm accumulation can be monitored (Chapter 9). Biofilm accumulation also influences heat transfer resistance (Characklis et al., 1981) in the tubular reactor. Frictional resistance and heat transfer resistance instruments are available commercially (Chapter 9). The influence of biofilms on transport processes is addressed in more detail in Chapters 9 and 14.

3.5.5 Microelectrodes: Biofilm Gradients

Concentration gradients in biofilms can be measured by microelectrodes, a technology introduced into biofilm processes by Bungay et al. (1969). The measured gradients represent measures of metabolic activity within the biofilm. For example, the dissolved oxygen gradient can be used to calculate the biofilm respiration rate. Microelectrode measurements can also demonstrate the patchiness of microbial activity in the biofilm (Figure 3.21). The recent work of Revsbech and Jorgensen (e.g., Nelson et al., 1986; Jorgensen et al., 1986; Cohen et al., 1986) has refined these methods and indicated their value for more detailed biofilm process analysis in the laboratory and in the field.

REFERENCES

Abbott, A., P.R. Rutter, and R.C.W. Berkeley, *J. Gen. Microbiol.*, **129**, 439–445 (1983).

American Public Health Association, *Standard Methods for the Examination of Water and Wastewater,* 14th ed., APHA, Washington, 1976.

Atkinson, B. and Daoud, I.S., *Trans. Inst. Chem. Engrs.*, **48**, 245 (1970).

Bakke, R., "Biofilm detachment," unpublished doctoral dissertation, Montana State University, Bozeman, MT, 1986.

Bakke, R. and P.Q. Olsson, *J. Microbiol. Meth.*, **5**, 93–98 (1986).

Bakke, R., M.G. Trulear, J.A. Robinson, and W.G. Characklis, *Biotech. Bioeng.*, **26**, 1418–1424 (1984).

Bobbie, R.J., Nickels, J.S., and Davis, W.M., "Measurement of microfouling mass and community structure during succession in OTEC simulators," in *Proc. OTEC Biofouling, Corrosion and Materials Workshop,* Rosslyn, VA, 1979, p. 101.

Bryers, J.D., "Dynamics of early biofilm formation in a turbulent flow system," Ph.D. Dissertation, Rice University, Houston, TX, 1980.

Bryers, J.D. and Characklis, W.G., *Water Research,* **15**, 483 (1980).

Bungay, H.R., W.J. Whalen, and W.M. Saunders, *Biotech. Bioeng.*, **11**, 765 (1969).

Characklis, W.G. and F.L. Roe, "Monitoring buildup of fouling deposits on surfaces of fluid handling systems," U.S. Patent 4,485,450, 27 November 1984.

Characklis, W.G., M.J. Nimmons, and B.F. Picologlou, *J. Heat Transfer Eng.*, **3**, 23 (1981).

Characklis, W.G., M.G. Trulear, N.A. Stathopoulos, and L.C. Chang, "Oxidation and destruction of biofilms," in R.L. Jolley, (ed.), (*Water Chlorination*), Ann Arbor Science Publishers, Ann Arbor, MI, 1980, pp. 349-368.

Characklis, W.G., M.G. Trulear, J.D. Bryers, and N. Zelver, *Wat. Res.*, **16**, 1207-1216 (1982).

Characklis, W.G., N. Zelver, C.H. Nelson, R.O. Lewis, D.E. Dobb, and G.K. Pagenkopf, "Influence of biofouling and biofouling control techniques on corrosion of copper-nickel tubes," presented at CORROSION/83, Symposium: Corrosion Induced by Bacteria, Anaheim, CA, 1983.

Cohen, Y., B.B. Jorgensen, N.P. Revsbech, and R. Poplawski, *Appl. Env. Microbiol.*, **51**, 398-407 (1986).

Corpe, W.A., "Microfouling: the role of primary film-forming marine bacteria," in *Proc. 3rd International Congress Marine Corrosion and Fouling*, Northwestern University Press, Evanston, IL, 1973, pp. 598-609.

Costerton, J.W. and R.R. Colwell (eds.), Native Aquatic Bacteria: *Enumeration, Activity and Ecology*, ASTM Press, Philadelphia, 1979.

Dubois, M., D.A. Giles, J.K. Hamilton, P.A. Rebers, and I. Smith, *Anal. Chem.*, **28**, 350 (1956).

Duddridge, J.E., C.A. Kent, and J.F. Laws, *Biotech. Bioeng.*, **24**, 153-164 (1982).

Escher, A., "Algal-bacterial interactions within biofilms," unpublished Master of Science thesis, Montana State University, Bozeman, MT, 1983.

Escher, A. "Bacterial colonization of a smooth surface: An analysis with image analyzer", unpublished doctoral thesis, Montana State University, Bozeman, MT, 1986.

Fetkovich, J.G., G.N. Granneman, L.M. Mahalingam, and D.L. Meier, "Studies of Biofouling in OTEC Plants," in *Proc. 4th Conference of the OTEC*, New Orleans, 1977, p. VIII5.

Fowler, H.W. and A.J. McKay, "The Measurement of Microbial Adhesion", in R.W. Berkeley, J.M. Lynch, J. Melling, P.R. Rutter, and B. Vincent (eds.), *Microbial Adhesion to Surfaces*, Ellis Horwood, Chichester, U.K., 1980, p. 143.

Fryer, P.J., N.K.H. Slater, and J.E. Duddridge, "On the analysis of biofouling data with radial flow shear stress cells," AERE-R11295, United Kingdom Atomic Energy Authority, Harwell, 1984.

Geesey, G.G., R. Mutch, J.W. Costerton, and R.B. Green, *Limnol. Oceanogr.*, **23**, 1214 (1978).

Gerchakov, S.M., D.S. Marszalek, F.J. Roth, B. Sallman, and L.R. Udey, "Observations on microfouling applicable to OTEC systems," in *Proc. OTEC Biofouling and Corrosion Symposium*, Seattle, 1977, p. 63.

Gulevich, W., C.E. Renn, and J.C. Liebman, *Env. Sci. Tech.*, **2**, 113 (1968).

Hamilton, R.D. and O. Holm-Hanson, *Limnol. Oceanogr.*, **12**, 310 (1967).

Harremoes, P., *Vatten*, **2**, 123 (1977).

Hoehn, R.C. and A.D. Ray, *J. Wat. Poll. Cont. Fed.*, **46**, 2302 (1973).

Janda, J. and E. Work, *FEBS Lett.*, **16**, 343 (1971).

Jones, H.C., I.L. Roth, and W.M. Sanders, *J. Bact.*, **99**, 316 (1969).

Jorgensen, B.B., Y. Cohen, and N.P. Revsbech, *Appl. Env. Microbiol.*, **51**, 408–417 (1986).

Knudsen, J.G., "Apparatus and techniques for measurement of fouling of heat transfer surfaces," in J.F. Garey, R.M. Jorden, A.H. Aitken, D.T. Burton, and R.H. Gray (eds.), *Condenser Biofouling Control,* Ann Arbor Science, Ann Arbor, MI, 1980, p. 143.

Kornegay, B.H. and J.F. Andrews, "Characteristics and kinetics of fixed-film biological reactors," Final Report, Grant WP-01181, *Federal Water Pollution Control Administration,* U.S. GPO, Washington, 1967.

LaMotta, E.J., "Evaluation of diffusional resistances in substrate utilization by biological films," doctoral dissertation, University of North Carolina at Chapel Hill, 1974.

Little, B. and D. Lavoie, "Gulf of Mexico OTEC biofouling and corrosion experiment," in *Proc. OTEC Biofouling, Corrosion and Materials Workshop,* Rosslyn, VA, 1979, p. 60.

Loeb, G.I., D. Laster, and T. Gracik, "The influence of microbial fouling films on hydrodynamic drag of rotating discs," in J.D. Costlow and R.C. Tipper, (eds.), *Marine Biodeterioration: An Interdisciplinary Study,* Naval Institute Press, Annapolis, MD, 1984, p. 88.

Matson, J.V. and W.G. Characklis, *J. Cooling Tower Inst.*, **4**, 27 (1983).

McCoy, W.F., "Immunofluorescence as a technique to study marine biofouling bacteria," Directed Research Project M699, University of Hawaii, 1979.

McCoy, W.F., J.D. Bryers, J. Robbins, and J.W. Costerton, *Can. J. Microbiol.*, **27**, 910–917 (1981).

Nelson, C.H., J.A. Robinson, W.G. Characklis, and D. Goodman, *Biotech. Bioeng.*, **27**, 1662 (1985).

Nelson, D.C., N.P. Revsbech, and B.B. Jorgensen, *Appl. Env. Microbiol.*, **52**, 161–168 (1986).

Norrman, G., W.G. Characklis, and J.D. Bryers, *Dev. Ind. Microbiol.*, **18**, 581 (1977).

Pedersen, K., *Appl. Env. Microbiol.*, **43**, 6 (1982).

Picologlou, B.F., N. Zelver, and W.G. Characklis, *J. Hyd. Div. ASCE,* **106**, 733–746 (1980).

Powell, M.S. and N.K.H. Slater, *Biotech. Bioeng.*, **25**, 891–900 (1983).

Robinson, J.A., M.G. Trulear, and W.G. Characklis, *Biotech. Bioeng.*, **26**, 1409–1417 (1984).

Rutter, P.R. and Leech, R., *J. Gen. Microbiol.* **120**, 301–307 (1980).

Sanders, W.M., 3rd, *Int. J. Air Wat. Poll.*, **10**, 253 (1966).

Schaftel, S.O., "Processes of aerobic/anaerobic biofilm development," unpublished Master of Science thesis, Montana State University, Bozeman, MT, 1982.

Trulear, M.G., "Dynamics of biofilm processes in an annular reactor," unpublished Master of Science thesis, Rice University, Houston, TX, 1980.

Trulear, M.G., "Cellular reproduction and extracellular polymer formation in the development of biofilms," unpublished doctoral thesis, Montana State University, Bozeman, MT, 1983.

Trulear, M.G. and W.G. Characklis, *J. Wat. Poll. Contr. Fed.*, **54**, 1288 (1982).

Turakhia, M.H. "The influence of calcium on biofilm processes," unpublished doctoral thesis, Montana State University, Bozeman, 1986.

Turakhia, M.H. and W.G. Characklis, *Can. J. Chem. Eng.*, **61**, 873–875 (1983).

White, D.C., "Chemical characterization of films," in K.C. Marshall (ed.), *Microbial Adhesion and Aggregation* (Dahlem Konferenzen), Springer-Verlag, Berlin, 1984, pp. 159–176.

White, D.C., R.J. Bobbie, J.S. Herron, J.D. King, and S.J. Morrison, "Biochemical measurements of microbial mass and activity from environmental samples," in J.W. Costerton and R.R. Colwell, (eds.), *Native Aquatic Bacteria: Enumeration, Activity and Ecology,* ASTM STP 695, ASTM, 1979, p. 69.

Zelver, N., "Biofilm development and associated energy losses in water conduits," Master of Science thesis, Rice University, Houston, TX, 1979.

Zelver, N., W.G. Characklis, J.A. Robinson, F.L. Roe, Z. Dicic, K. Chapple, and A. Ribaudo, "Tube material, fluid velocity, surface temperature and fouling: A field study," CTI Paper TP-84-16, Cooling Tower Institute, Houston, TX, 1984.

Zimmerman, R. and L.A. Meyer-Riel, *Kiel Meeresforsch.*, **30**, 24 (1974).

PART II

CONSTITUTIVE PROPERTIES OF BIOFILM SYSTEMS

PHYSICAL AND CHEMICAL PROPERTIES OF BIOFILMS

BJØRN E. CHRISTENSEN
Norwegian Institute of Technology, Trondheim, Norway

WILLIAM G. CHARACKLIS
Montana State University, Bozeman, Montana

SUMMARY

Biofilms are composed of two major components: microbial cells and EPS. The microbial species and their morphology as well as the EPS composition largely determine the physical properties of the biofilm. Thus, the biofilm can be considered as an organic polymer gel with living microorganisms trapped in it. The gel has properties that influence the transport of momentum, heat, and mass at the substratum. Changes in transport rates can affect the performance of industrial equipment (fouling), reduce the effectiveness of biocide treatment (e.g., chlorination in water distribution systems), and create unique niches within the biofilm for the proliferation of a variety of microbial species.

The properties of extracellular polymers in biofilms are not known at present, nor is the role of the polymer matrix in biofilm ecology. This chapter is an attempt to encourage further research in the areas of polymer structure and function in biofilms.

Many bacteria produce extracellular polymers, whether grown in suspended cultures or in biofilms. The polymers appear as a highly hydrated capsule attached to the cell or as a viscous, soluble slime. However, the extent and composition of these polymers may vary with physiological state of the organism, and (presumably) so do the physical properties of the polymers. The physical properties of these polymers, primarily polysaccharides, in gels may be a critical to the understanding the physical and physiological behavior of biofilms. Thus, more attention must be focused on the interaction between cell physiology, EPS properties, and the resultant effect on bulk biofilm properties. The characterization of these polymer matrices is necessary because, in many cases, EPS is used to explain otherwise unexplained phenomena.

The properties of biofilm vary with environmental factors such as population distributions in the film, nutrient loading rate, and hydrodynamic shear stress. Acceptable empirical expressions for predicting properties such as density or diffusivity in a biofilm would be very useful in numerous technological and ecological applications. Perhaps through such empirical relationships, a method can be found for integrating fundamental information on extracellular polymers (presumably responsible for the integrity of the biofilms) with bulk biofilm properties. Unfortunately, few successful attempts have been made to isolate and characterize the polymers thought to be important in these biofilm processes.

Microbial extracellular and cell surface polymers are also important in biofilm phenomena during the initial interaction between a bacterial cell and the substratum (Chapter 7). Presumably, irreversible adsorption or adhesion is largely determined by the physical properties of the macromolecules at the cell surface.

The prediction of biofilm properties based on phenomenological models is a major challenge. Biofilm properties may have far-reaching implications in biotechnology, environmental engineering, and microbial physiology.

4.1 INTRODUCTION

The macroscopic properties of a biofilm frequently control the rate and extent of the processes occurring within it. For example, biofilm development may result in unusually high fluid frictional resistance losses in water conduits (Chapter 14) because the rheology and morphology (e.g. roughness) of the biofilm influence the fluid frictional resistance in a conduit. Biofilms accumulating on surfaces of heat transfer tubes significantly reduce the heat transfer rate (Characklis et al., 1981) because the thermal conductivity of the biofilm is significantly less than that of metal heat transfer surface materials. Finally, metabolic activity within the biofilm may be rate-controlled by diffusion of substrate, electron acceptor, nutrients, or inhibitors to individual cells, so biofilm diffusivity is a relevant property. Hence, the rate of transport of momentum (fluid flow), heat, and mass within or through a biofilm is dependent on the properties of the biofilm.

The physical, chemical, and biological properties of the biofilm are also dependent on the particular environment in which the biofilm has accumulated. The physical and chemical components of the aquatic and substratum environment influence the selection of the prevalent biofilm organisms. The predominant organisms, in turn, modify the microenvironment of the substratum in a manner specific to their metabolic activity.

The chemical and physical properties of biofilms are related to the chemical and physical properties of their main components, i.e., the microbial cells and the extracellular macromolecules, which are primarily polysaccharide. This chapter focuses on properties of extracellular polymeric substances (EPS) and macroscopic biofilm properties including biofilm chemical composition and biofilm physical properties. EPS is a useful term if the molecular composition of the polymer matrix is unknown. The extracellular polymer matrix can account for as much as 50 to 90% of the biofilm organic carbon (Bakke *et al.*, 1984). With the currently available analytical techniques for polymer separation and analysis, very little information is available on the precise chemical structure of the wide range of EPS found within a single biofilm. Consequently, the discussion of EPS composition below is limited mainly to the structure of various EPS obtained from single species planktonic populations or from single species biofilms.

4.2 BACTERIAL EXTRACELLULAR POLYMERS (EPS)

Most bacteria produce extracellular polymers, whether they are grown in suspended cultures or in biofilms. In the former case, the polymers appear as a highly hydrated capsule attached to the cell or as viscous, soluble "slime." Such bacterial polymers, almost exclusively polysaccharides, have been studied for many years due to the industrial application of microbially produced gums (Sandford and Baird, 1983) and the involvement of microbial polysaccharides in a variety of specific biological interactions. Despite this knowledge, the physical properties of polysaccharides have received little attention among other scientists. As a result, polysaccharides appear in biofilm and microbial sorption literature as "magic" substances that mediate otherwise unexplained phenomena.

Microbial extracellular and cell surface polymers are important in biofilm phenomena for at least two reasons. First, the interaction between a bacterial cell and the substratum leading to irreversible adhesion is largely determined by the physical properties of the macromolecules at the cell surface. Secondly, extracellular polymers are frequently observed (by electron microscopy) as the extracellular matrix responsible for the integrity of the biofilms. In such cases, the polymers appear as condensed fibers extending from the cell (Figure 4.1). Sample preparation for TEM involves dehydration, which can lead to aggregation and possibly increased adsorption of hydrophilic biopolymers to solid surfaces. The polymeric nature of these molecules is still evident. In fact, electron microscopy has been successfully applied in studies of the physical properties of some biopolymers (Mikkelsen et al., 1985; Lamblin et al., 1979). By using preparations stabilized by antibodies, the problem of collapse or aggregation of the extracellular polymers can be circumvented (Costerton et al., 1981).

Few attempts have been made to isolate and describe the polymers thought to be important in biofilm processes. Extracellular polysaccharides from some bacteria isolated from biofilms are of similar chemical composition to those typical of bacterial polysaccharides isolated from suspended or planktonic bacteria

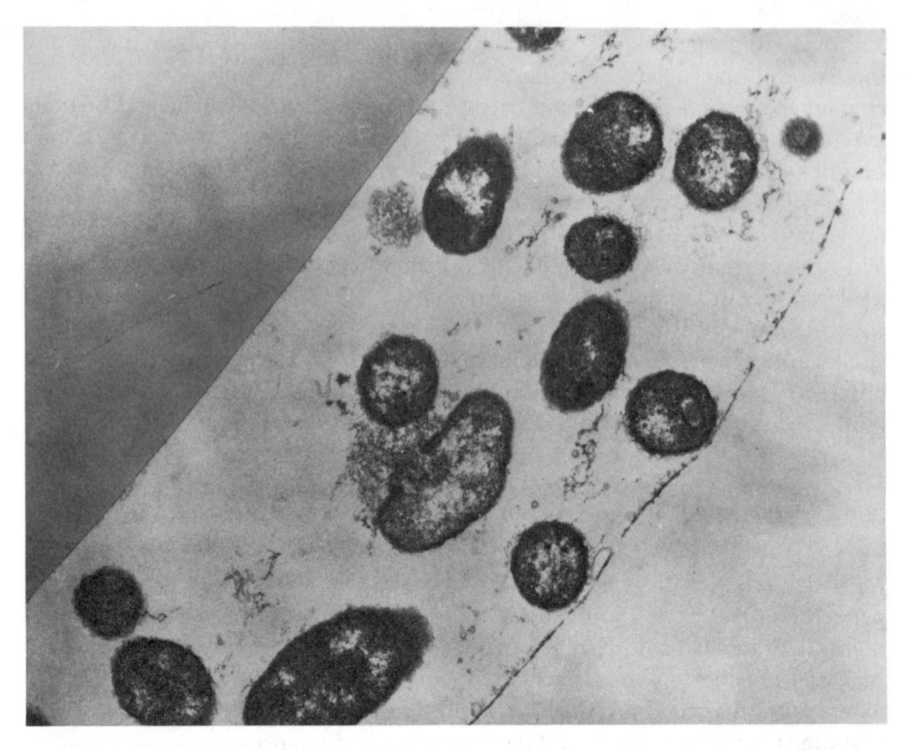

Figure 4.1 Transmission electron microphotograph of a *Ps. aeruginosa* biofilm from a rectangular duct reactor grown on a glucose and mineral salts medium (courtesy R. Bakke). Glucose was the limiting nutrient.

(Corpe, 1970; Uhlinger and White, 1983; Christensen et al., 1985). Thus, chemical and physical properties of the better-characterized polysaccharides isolated from planktonic bacteria probably contribute significantly to the structure and hence the properties of biofilm. In this chapter, we shall discuss the chemical composition of bacterial polysaccharides and emphasize the relationship between chemical composition and physical properties of polymer solutions—in particular, gels, which may be the dominant influence on biofilm properties. Several reviews are available for more details (Aspinall, 1982, 1983; Sutherland, 1977; Ward and Berkeley, 1980; Brant and Burton, 1981).

4.3 MOLECULAR STRUCTURE OF BACTERIAL EXTRACELLULAR POLYMERS (EPS)

The vast majority of bacterial EPS are polysaccharides. Based on their chemical composition, antigenic specificity, and mode of biosynthesis, the bacterial polysaccharides are naturally divided into two groups, namely specific and nonspecific polysaccharides (Kenne and Lindberg, 1983).

4.3.1 Specific Polysaccharides

Special polysaccharides, sometimes called polysaccharide antigens (Jann and Jann, 1977), are specific to individual bacterial strains. Common sugars such as glucose, galactose, mannose, rhamnose, N-acetylglucosamine, glucuronic acid, and galacturonic acid are typical constituents of the specific polysaccharides. The polysaccharides isolated from two marine biofilm bacteria *Pseudomonas atlantica* (Corpe, 1970; Uhlinger and White, 1983) and *Pseudomonas sp*. Strain NCMB 2021 (Christensen et al., 1985) probably belong to this group (Table 4.1). In addition to the common sugars in Table 4.1, several unusual sugars have been found among the specific polysaccharides (Sutherland, 1977; Kenne and Lind-

TABLE 4.1 Relative Molar Ratios of the Sugars and Substituents of the Extracellular Polysaccharides of Two Marine, Biofilm Pseudomonads

	Relative Molar Ratio[a]			
	Pseudomonas sp. NCMB 2021		*Pseudomonas atlantica*	
Substance	PS-A (Christensen et al., 1985)	PS-B (Christensen et al., 1985)	(Uhlinger and White, 1983)[b]	(Corpe, 1970)
Rhamnose	—	—	0.09	—
Fucose	—	—	tr	—
Arabinose	—	—	tr	—
Xylose	—	—	0.03	—
Mannose	—	—	0.36	+
Glucose	1.00	—	1.00	+
Galactose	0.81	—	0.67	+
Glucuronic acid	0.42	—	0.31	+
Galacturonic acid	0.32	—	0.35	—
Mannuronic acid	—	—	0.10	—
N-acetylglucosamine	—	1.00	—	—
KDO (2-keto-3-deoxy-octulosonic acid)	—	1.00	—	—
6-deoxyhexose (not identified)	—	1.00	—	—
O-acetyl groups	+	0.95	—	—
Pyruvate	—	—	—	+

[a] Abbreviations: tr, traces (less than 0.05); +, present, but not quantified; —, not detected.

[b] The data for *Ps. atlantica* have been calculated by the authors from the original data presented by Uhlinger and White (1983).

berg, 1983), which may serve as useful probes for polymer analysis. However, no single type of sugar, including the commonly found uronic acids, can be used as a universal probe for bacterial polysaccharides, due to the large variations in chemical structure encountered within this group.

The typical specific polysaccharide is composed of oligosaccharide repeating units, linear or branched, assembled into a polymer from sugar nucleotides via lipid-linked intermediates (Sutherland, 1977). The oligosaccharide repeating unit may contain two to nine monosaccharides, and additional organic substituents such as pyruvate and acetate are commonly found. Inorganic substituents such as sulfate seem to be less common in bacterial polysaccharides than in algal polysaccharides.

Despite the apparently regular structural pattern of polysaccharides consisting of oligosaccharide repeating units, structural variations and consequently variations in the solution properties (Section 4.4.1) are commonly encountered. Both chain length and substitution pattern can vary, depending on the bacterial strain, growth conditions, and physical-chemical processing of the sample (Sutherland, 1977).

A well-known example of a specific polysaccharide is xanthan gum, the extracellular polysaccharide of *Xanthomonas campestris* (Figure 4.2). It contains a five sugar repeating unit and contains both pyruvate and acetate as substituents (Jansson et al., 1975). This particular polysaccharide has very useful solution

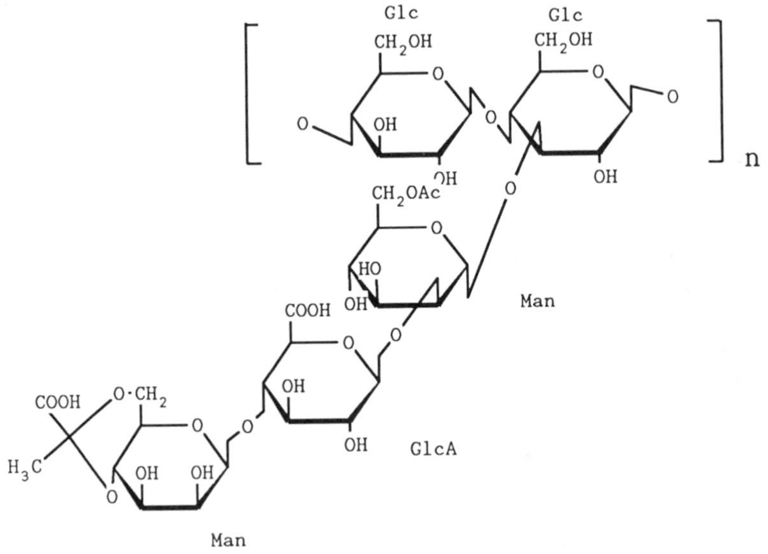

Figure 4.2 Structure of the oligosaccharide repeating unit of xanthan gum, the extracellular polysaccharide of *Xanthomonas campestris*. Note the acetyl substituent in the 6-position of the mannose unit closest to the main chain, and the pyruvate ketal substituent linked in the 4- and 6-positions of the terminal mannose unit. Abbreviations: Glc: D-glucose, Man: D-mannose, GlcUA: D-glucuronic acid, Ac: O-acetyl group, Pyr: pyruvate. All sugars are in the pyranose form.

properties, including high viscosity over a wide range of temperatures and salt concentrations at low shear rates, but is considerably less viscous at high shear stresses (shear thinning). These properties are the basis for the successful commercialization of xanthan gum, a polymer used in many industrial processes, including enhanced oil recovery (Sandford and Baird, 1983).

Specific polysaccharides are sometimes called *polysaccharide antigens* (Jann and Jann, 1977), because they exhibit specific immunological properties. Lipopolysaccharides and teichoic acids, which are components of the cell walls but are not classified as extracellular polysaccharides, are also antigenic. The antigenic specificity of these polymers resides not only in the individual monosaccharides, but also the position and configuration of the glycosidic linkages. Therefore, even a limited number of different monosaccharides can give rise to a very large number of antigens, i.e., a large number of polymers with chemically and physically different properties. Serological techniques to detect antigenic differences may be used for very precise structural determinations of polysaccharides (Jann and Jann, 1977), which otherwise are difficult even with sophisticated spectroscopic and chemical methods.

4.3.2 Nonspecific Polysaccharides

The nonspecific polysaccharides are found in a variety of bacterial strains and are structurally different and generally simpler than the specific polysaccharides. Many of them are homopolysaccharides, containing only one monomer. For example, extracellular cellulose is produced by several *Acetobacter* species, and is chemically similar to cellulose in higher plants, i.e. β-1,4-linked unbranched glucans. β-1,3-linked glucans are also found in several bacterial strains; the best-known example is probably curdlan, which is produced by a variant of *Alcaligenes faecalis* (Harada et al., 1968). Dextrans are polysaccharides with an α-1,6-linked glucan backbone, usually containing glucan branches. Dextrans and related α-1,3-linked glucans are produced by several bacterial species, but those of oral streptococci are of particular interest, since these organisms may form biofilms in the oral cavity, including the teeth. The role of extracellular α-glucans in these processes has been discussed elsewhere (Abbot et al., 1980). Fructans are produced by several bacteria. Levan is essentially β-2,6-linked fructan, but the chains are more or less branched (Lindberg et al., 1973). Inulin is another fructan, mainly composed of β-2,1-fructofuranosyl residues.

Bacterial alginate is produced by several strains of *Pseudomonas aeruginosa* (Linker and Jones, 1966), *P. medocina* (Hacking et al., 1983) and *Azotobacter vinelandii* (Gorin and Spencer,1966). These polymers, very similar to algal alginate, are composed of two monosaccharides: D-mannuronic acid (M) and its 5-epimer L-guluronic acid (G) (Figure 4.3). In addition, *O*-acetyl groups are usually present. The ratio between D-mannuronic acid and L-guluronic acid may vary considerably, and the two sugars may be found as pure poly-M or poly-G sequences, or in alternating (. . . MGMGMG . . .) sequences. All three se-

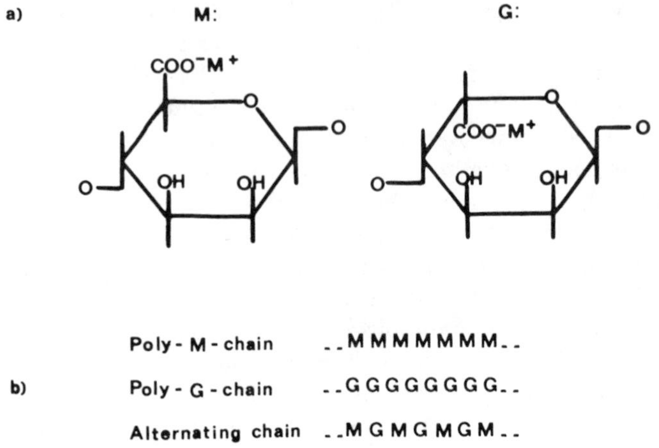

Figure 4.3 (a) Chemical structure of D-mannuronic acid (M) and its 5-epimer L-guluronic acid (G). (b) Different blockwise arrangements of M and G in the alginate chain.

quences may be found in the same polymer. The occurrence and distribution of the various sequences have a remarkable effect on the physical properties of the alginate (see Section 4.4.1). The biosynthesis of alginate is unique in that bacteria first synthesize chains of only D-mannuronic acid. Some of the units are then epimerized at the polymer level into L-guluronic acid (Haug and Larsen, 1971).

4.3.3 Production of More Than One Polysaccharide or Other Extracellular Polymers

An organism may produce two different polysaccharides. In batch culture, the marine *Pseudomonas* sp. NCMB 2021 produces a typical heteropolysaccharide (PS-A) during exponential growth, whereas another polysaccharide (PS-B) is released during decelerating growth and in the stationary phase (Christensen et al., 1985) (Figure 4.4). PS-A contains glucose, galactose, glucuronic acid, and galacturonic acid (molar ratio 1.0:0.81:0.42:0.32), but these sugars are only 71% of the dry weight, and nuclear magnetic resonance studies strongly indicate the presence of other organic substituents. PS-A forms thermoreversible gels at concentrations above 1%, and precipitates with several multivalent cations. PS-B contains KDO (2-keto-3-deoxy-*manno*-octulosonic acid), *N*-acetylglucosamine, and a 6-deoxyhexose in exact equimolar ratios. KDO, often associated with lipopolysaccharides, also occurs in extracellular polysaccharides (Bhattacharjee et al., 1974; Jann et al., 1983), and the high proportion of KDO in PS-B suggests that it is not lipopolysaccharide. Fatty acids were not detected (B. Christensen, unpublished results). PS-B forms nonviscous aqueous solutions and is soluble in 80% ethanol or methanol despite its high molecular weight.

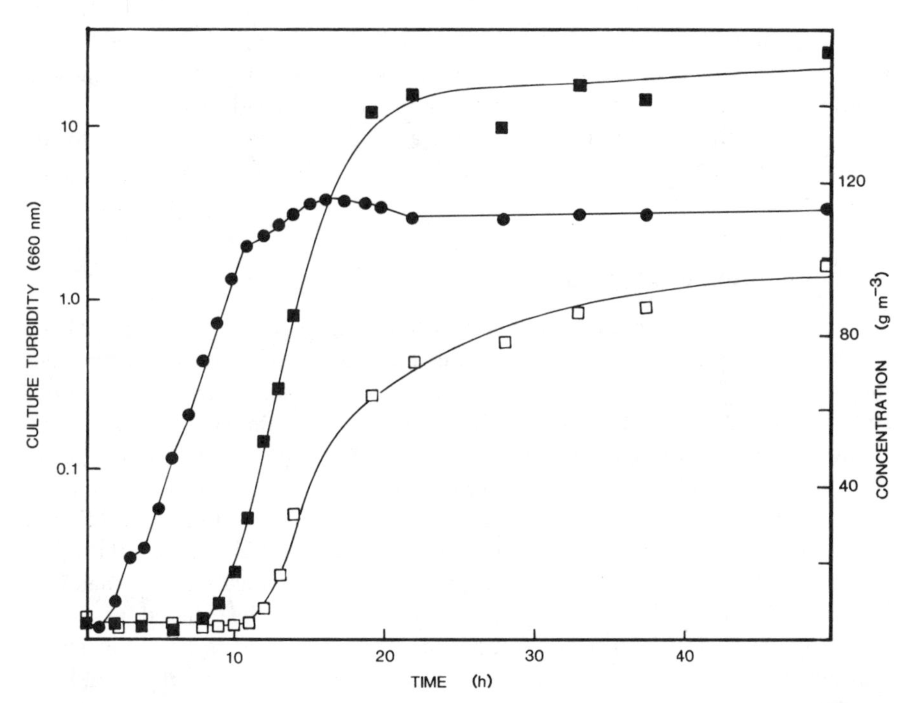

Figure 4.4 Growth curve (culture turbidity at 660 nm) (●) and the production profiles of polysaccharide A (■) and polysaccharide B (□) in a batch culture of *Pseudomonas* sp. strain NCMB 2921 grown at 17°C. Samples (30 to 50 ml) were taken at regular intervals, centrifuged, filtered, and dialyzed. The content of polysaccharide A was measured by the carbazole-sulfuric acid method, and the content of polysaccharide B was analyzed by the cysteine-sulfuric acid method.

Thus, results with PS-B demonstrate that precipitation of polysaccharides from aqueous solutions by addition of alcohols may leave some polymers undetected.

Polymers other than polysaccharides are often found in culture supernatants and in biofilms. Proteins seem to be present, either as a secretion without any known function or as extracellular enzymes, but they are usually not examined in detail unless specific enzymes are being sought. Leakage of intracellular polymers (such as nucleic acids) and cell wall polymers (such as lipopolysaccharides or teichoic acids) are commonly observed as well, but usually in much smaller quantities than the true extracellular polysaccharides.

4.4 PHYSICAL PROPERTIES OF BACTERIAL EXTRACELLULAR POLYMERS (EPS)

A biofilm is an aggregation of cells embedded in a gel matrix. The physical properties of the gel matrix are responsible for many of the observed biofilm phenomena. In this section, some basic properties of the EPS relevant to biofilms will be discussed, namely solution properties, gel formation, and adsorption to surfaces. The solution properties of EPS may seem irrelevant in connection with biofilms, where the polymers occur in a more gel-like or adsorbed state. However, the principles of gel formation and adsorption are closely linked to the solution properties, which demonstrate the major differences between biopolymers and small molecules.

4.4.1 Solution Properties

The term *higher order* (secondary, tertiary, and quaternary) *structures* has been traditionally reserved for proteins and nucleic acids. However, polysaccharides may also adopt a variety of different shapes, including single stranded and multiple stranded helices (Rees et al., 1982). The shape of the polymer has a dramatic influence on its physical properties and is dependent on the primary structure and solvent properties. Changes in solvent composition (e.g., addition of an organic solvent to water), changes in pH, changes in temperature, or addition of inorganic salts may induce conformational transitions that alter physical properties. Examples of transitions include the order–disorder transition upon heating, but the addition of salts increases the thermal stability of the ordered form. The nature of the ordered form of xanthan has been suggested to be either single stranded (Norton et al., 1984) or double stranded (Saito et al., 1984; Stokke et al., 1986).

There are three basic polymer shapes—a sphere, a stiff rod, and a random coil—and their intermediate forms (Figure 4.5). The polysaccharides are usually stiff rods or random coils. The random coils are typically single stranded, linear chains. Stiff rods are formed either by restricted motion (rotation) of the glycosidic bonds, large electrostatic repulsion between equally (usually negative) charged groups along the chain, or formation of helices. Intermediate forms between random coils and stiff rods, for instance *stiff coils* or *wormlike chains*, are also common forms among the polysaccharides.

Aqueous solutions of bacterial polysaccharides are typically viscous, and non-Newtonian behavior is common, i.e., the viscosity is dependent on the shear rate. The solution properties are best described by the intrinsic viscosity $[\eta]$ which is the specific viscosity divided by the concentration at infinite dilution (obtained by extrapolation). The relation between the intrinsic viscosity, polymer shape, and molecular weight is given by the Mark-Houwink equation:

$$[\eta] = KM^{\alpha} \qquad [4.1]$$

where M = molecular weight
α = an exponent dependent upon the stiffness of the chain.

For rigid rods, α equals 1.8, while random coils have α-values between 0.5 and 0.8. Stiff coils have α = 0.8 to 1.8. Ideal spherical molecules have an α-value of 0, i.e., the intrinsic viscosity is independent of the molecular weight and

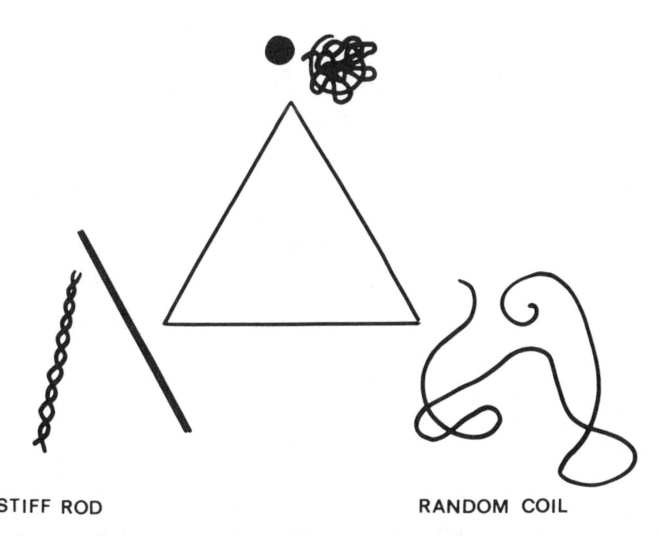

Figure 4.5 Schematic representation of the three basic shapes of polymers. Intermediate forms will be located on the sides of the triangle.

equals 2.5 ml g^{-1} (Tanford, 1961). The presence of charged groups in a bacterial polymer, typically uronic acids or pyruvate, strongly influences its physical properties. The solubility is generally greater for charged polymers in polar solvents such as water. In addition, the nature and concentration of inorganic salts may affect solution properties of the polymer. At low pH, neutralization of acidic polymers occurs, which may lead to precipitation, as in the case of alginate. Several polymers interact strongly with certain cations, leading to precipitation or gel formation. However, salts that do not interact specifically with the polymer may also affect solution properties. Unless severe rotational restrictions exist along the chain, the addition of salts like NaCl will reduce the viscosity of charged polysaccharides because the fixed charges on the chain are shielded against mutual electrostatic repulsion. The polymer chain contracts from an extended, stiff form into a smaller and more flexible form. The variation in the intrinsic viscosity obtained in solvents with different ionic strengths can be used to determine the molecular flexibility of charged polymers (Smidsrød and Haug, 1971). Deviation from Newtonian behavior is typical for stiff rods. At very low shear rates, the polymer chains will be randomly oriented and exhibit Newtonian behavior. However, at higher shear rates, stiff rods will align in the direction of the flow which results in lower viscosity. This property, called *shear thinning*, is useful when viscous solutions have to be pumped. Stiff molecules tend to reduce fluid frictional resistance in turbulent (but not laminar) flow, because the chains resist rapid alignment in localized eddies, thus damping energy transfer between eddies.

Bacterial polysaccharides often have extremely high molecular weights, typically in the range of millions. In this range, even a stiff chain behaves more like a random coil (Holzwarth, 1981). Since the motion between the chain segments is never totally restricted, two structural units of a chain will be randomly oriented relative to each other when the distance between them along the chain is large enough.

The concentration of extracellular polymers in a biofilm will typically be in the range of 1-2% w/v, and at this concentration a very high viscosity is expected. The relationship between the specific viscosity (η_{sp}) and the polymer concentration is given by Tanford (1961):

$$\eta_{sp} = [\eta]C + k[\eta]^2C^2 \qquad [4.2]$$

where C = polymer concentration
k = coefficient

The coefficient k depends on the shape of the polymer. At high concentrations, the interactions between individual chains will sometimes be strong enough to form a gel.

4.4.2 Gel Formation

Biopolymer gels are formed when the molecular chains are partly hydrated and partly cross-linked or aggregated, resulting in a three dimensional network. In addition to the high content of solvent (usually water), gels are characterized by a yield point. The elastic modulus is also much higher than the viscous (loss) modulus. However, the viscous modulus was significantly greater than the elastic modulus for *in situ* measurements on an experimental biofilm (Section 4.6.2). Thermodynamically stable gels of synthetic polymers (e.g., polyacrylamide) are known and may exhibit interesting phase transitions (Tanaka, 1981). However with the possible exception of *i*-carrageenan, biopolymer gels are kinetically, not thermodynamically, stabilized (O. Smidsrød, personal communication). Hence, the induction of gel "collapse" or other phase transition is unlikely.

The potential for gel formation is closely linked to polymer solution properties, but the mechanisms by which intermolecular junctions (cross-links) are formed may vary. In most cases, junctions form within regular sequences along the chains, i.e., sequences where the repeating units are identical. "Structural interruptions" in the regular sequences, such as the presence of a side chain or the presence of another (or modified) monomer, result in increased solubility.

The mechanism of gel formation has been most extensively studied with non-bacterial polymers such as agarose, alginate (from seaweeds), and carrageenans. However, gelling agents of bacterial origin, i.e., extracellular polysaccharides, are well known and have received increased attention lately. These include curdlan (from *Alcaligenes faecalis*), gellan gum (from *Pseudomonas elodea*), and bacterial alginate (from *Azotobacter vinelandii*) (Sandford and Baird, 1983). The gelation of alginate, including bacterial alginate, occurs by the addition of calcium ions, but not monovalent cations like sodium or potassium. The intermolecular junctions are formed between pure poly-L-guluronic acid sequences, and the role of the calcium ions is to bridge the two chains by electrostatic forces (Smidsrød, 1974). The geometry (conformation) of these sequences in the gel state gives rise to an "egg box" arrangement of the chains and the calcium ions (Rees et al., 1982). Interruption of polyguluronic acid sequences by polymannuronic acid or mixed sequences terminates the junctions and gives rise to the "soluble" part of the network.

Recent findings with bacterial alginates indicate that only alginates from *Azotobacter vinelandii* contain polyguluronic acid sequences (Skjåk-Bræk et al., 1986). Several strains of *Pseudomonas* did not contain polyguluronic acid sequences implying that they cannot form classical alginate gels. Ion-dependent gelation also occurs with some carrageenans, but the mechanism is somewhat different and is also less well understood. *i*-carrageenan forms ordered conformations in solution and forms gels when potassium, rubidium, cesium, or ammonium salts are added. Rees et al. (1982) have proposed a model for *i*-carrageenan which the ordered conformation in solution is a formation of interchain double helices where interchain junctions are formed between different double helices by ion bridging. This mechanism has been questioned by Smidsrød (1980).

Gelation occurs in agarose, the primary constituent of commercial agar, by a less controversial mechanism. Agarose forms firm gels when heated solutions cool and the presence of salts is not required. Interchain junctions consist of double helices, and the junction zones are terminated by altered structure of the polysaccharide backbone: the presence of galactose instead of 3,6-anhydrogalactose (Rees et al., 1982).

A true bacterial polysaccharide, gellan gum, forms gels upon the addition of several cations, the most effective being calcium and magnesium. The mechanism of gelation is not known.

Curdlan, which is insoluble in water at room temperature, may form gels on different treatments, including heating, dialysis of alkaline solutions, cooling of heated solutions in 0.2–0.63 M dimethyl sulfoxide, and the addition of calcium ions to weak alkaline solutions (Sandford and Baird, 1983). Curdlan is an example of a neutral polymer that forms gels, and the mechanism of interchain junction formation must necessarily be different than the one for alginate. Saito (1981) suggests, on the basis ^{13}C NMR experiments, that curdlan interchain junctions are the same in gels as in the solid state, namely multiple stranded helices.

The intercellular matrix or gel of biofilms will usually contain several polymers, even in a pure culture system. Although one polymer may dominate, nucleic acids and proteins will probably be present. The presence of other polymers will undoubtedly make the physical properties of the biofilm differ from those of a one component gel system. Such mixed polymer interactions have indeed been studied, and some of the findings may also have relevance to biofilms. A synergistic effect is observed on mixing xanthan gum with a group of plant polysaccharides called galactomannans. Typically, strong gels are formed at concentrations well below those necessary for the individual polysaccharides. It has been suggested that unsubstituted regions in the mannan backbone associate with xanthan chains (in their ordered conformation) into large aggregates (Morris et al., 1977). This interaction has also been suggested in the specific recognition between the plant pathogen *Xanthomonas campestris* and the vascular system of its hosts (Morris et al., 1977). If similar interactions are present between the extracellular polymers of different organisms in a biofilm, a physical stabilization of the biofilm matrix is to be expected.

Phase separation can also occur in polymer mixtures, as illustrated by a mixture of dextrans and polyethylene glycol, a synthetic polymer (Albertsson, 1970). Due to the high molecular weight of polymers, the gain in entropy on mixing two or more polymer solutions is usually much smaller than for small molecules, because for a given weight there are many more small molecules. Instead, the enthalpy of interaction between the chains will dominate the free energy of interaction and will determine whether the polymers will mix or separate into different phases. If the interaction between the same type of chains is energetically more favorable than between unlike chains, then the polymer mixture will separate into two phases, each enriched in one of the polymers. This effect may be observed in biofilms containing two or more organisms producing different polymers and lead to a selection pressure otherwise determined mainly by relative growth rates.

Water content also influences chemical configurations (Dudman, 1977), since dissolved molecules cannot overlap with the EPS gel network. Thus, geometrical restrictions are imposed on the conformation, position, and orientation of dissolved molecules, reducing the number of permissible states. As a consequence, the free energy or chemical potential of the penetrating solute molecules in the EPS is increased with respect to bulk solution of these molecules. Thus, precipitation processes may occur in the biofilm gel even when the solubility product has not been reached in the overlying water. In gel domains adjacent to surfaces, such as observed with bacterial capsules or biofilms, phase transitions of the type that biological macromolecules appear to undergo (e.g., random coil–helix transitions) are predicted to be sharpened (DiMarzio and Bishop, 1974). Thus, changes in microbial activity in biofilms due to rapid changes in the environment (e.g., pH, temperature, soil content) may be in part due to changes in polymer gel properties.

4.4.3 Adsorption of Polymers to Surfaces

Both theoretical considerations (Marshall, 1980; Rutter and Vincent, 1980; Robb, 1984) and electron microscopy studies (Fletcher and Floodgate, 1973; Marshall and Cruickshank, 1973) suggest that cell surface polymers may assist in binding microbial cells to surfaces. The physicochemical principles underlying the nonspecific adsorption of polymers to surfaces are closely related to solution and gelling properties.

Almost any water-soluble polymer tends to adsorb to a surface, sometimes only as a monolayer. The factors which determine the adsorption characteristics of a polymer include the following:

1. Enthalpy factors
   a. The interaction between polymer segments and the substratum
   b. The interaction between polymer segments and the solvent
   c. The interaction between the solid surface and the solvent
   d. The interaction between polymer segments of adjacent chains in solution, at the substratum, or in the solid state
2. Entropy factors
   a. The more flexible the polymer is in the soluble and in the adsorbed state, the higher its entropy
   b. The ordering of solvent around the polymer segments in solution and at the substratum will decrease the entropy of the solvent.

Except for the surface-segment interaction (1a) and the surface-solvent interaction (1c), these factors also determine the solubility of a polymer. If the polymer is soluble in water, 1a and 1c will largely determine its adsorption characteristics. The adsorption of polymers to a substratum is illustrated schematically in Figure 4.6. Because of the large number of segments in a polymer (bacterial polysaccharides may contain several hundred to several thousand monosaccharides), even very weak interactions between polymer segments and the substratum may lead to significant adsorption. Robb (1984) demonstrated that a random coil polymer with each segment-surface interaction as low as 0.01kT still favors the adsorbed state, since there are so many polymer segments.

The main contribution of the polymer chain to the entropy is its flexibility. An increase in flexibility increases entropy, favoring solubility over adsorption. Typ-

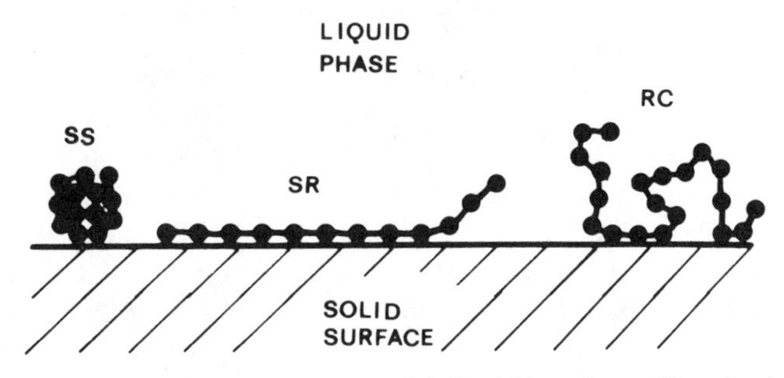

Figure 4.6 Adsorption of polymers to a solid-liquid interface. SS: spherical polymer; RC: randomly coiled polymer; SR: stiff rod polymer.

ically, flexible biopolymers, such as dextran, are less adsorptive than stiffer polymers with comparable chemical composition (e.g., linear β-glucans). Likewise, longer molecules are more adsorptive than shorter ones.

Many bacteria tend to attach in larger numbers to hydrophobic than to hydrophilic surfaces (Fletcher and Loeb, 1979; Pringle and Fletcher, 1983); for others the opposite is true (Dexter, 1979; Dexter et al., 1975). The molecular basis for hydrophobicity has not been established. On one hand, electron microscopic investigations have suggested that acidic polymers, resembling polysaccharides, are involved in the microbial adsorption process; on the other hand, polysaccharides are usually regarded as hydrophilic. Proteins, however, may contain amino acids with typically hydrophobic side groups, and may serve as the "polymeric adhesives" according to purely physicochemical considerations as well as experimental evidence (Paul and Jeffrey, 1985). However, polysaccharides may exhibit amphophilic properties (Symes, 1982), as typified by a polysaccharide isolated from a marine *Pseudomonas* species NCMB 2021, which exhibited unusual solution properties (Christensen et al., 1985). Despite the abundance of hydroxyl groups found in carbohydrates, hydrophobic properties may be present, either due to hydrophobic groups such as methyl groups (in 6-deoxyhexoses such as rhamnoses or fucose) and acetyl groups, or due to a "polarized" arrangement of the hydroxyl groups (hydrophilic) and the ring protons (hydrophobic). The conformation of the carbohydrate ring, i.e., a ${}^4C_1$-form versus a ${}^1C_4$-form, as well as the type of sugars, will influence the extent of hydrophobicity from this phenomenon.

4.5 PHYSICAL PROPERTIES OF BIOFILMS

The chemical and physical properties of biofilms are related to those of their main components, i.e., the microbial cells and the extracellular macromolecules. The extracellular polymer matrix can account for as much as 50 to 90% of the biofilm organic carbon (Bakke et al., 1984). Microbial extracellular and cell surface polymers are important in biofilm phenomena for at least two reasons. First, the interaction between a bacterial cell and the substratum or another bacterial is largely determined by the physical properties of the macromolecules at the cell surface. Secondly, the presence of extracellular polymers is responsible for the integrity of the biofilms. Consequently, a biofilm can be viewed as polymer gel with organisms dispersed within it. In addition, substrate, electron acceptor, and other soluble components can be found diffusing in the aqueous interstices of the gel.

Noncellular particulate material (e.g., extracellular polymeric materials) in biofilm has rarely been considered in the past. As a result, cellular growth rates in the biofilm have generally been overestimated. Generally the biofilm is *spatially heterogeneous*, i.e., the components within the gel are not necessarily uniformly distributed.

The macroscopic properties of the biofilm frequently control the rate and extent of the processes occurring within it. For example, metabolic activity within the biofilm may be rate-controlled by diffusion of substrate or electron acceptor to individual cells, so biofilm diffusivity is a relevant property. The physical, chemical, and biological properties of the biofilm are also dependent on the environment in which the biofilm has accumulated. The physical and chemical components of the aquatic and substratum environment influence the selection of the prevalent biofilm organisms. The predominant organisms, in turn, modify the microenvironment of the substratum in a manner specific to their metabolic activity.

A biofilm is largely composed of water. Reported biofilm water content ranges from 87 to 99% (Kornegay and Andrews, 1967; Characklis et al., 1981; Zelver, 1979). The quantitative estimation of biofilm water content is dependent, to some extent, on the measurement technique as discussed in Chapter 3. Characklis et al. (1981) report biofilm water content to range from 98.1 to 99%, based on 63 determinations from nine experiments using a mixed population biofilm in a tubular reactor. Glucose loading rate was the only variable in these experiments (0.06–1.14 mg m^{-2} s^{-1}) and had no significant effect on water content.

Biofilms are generally very hydrophilic. EPS are mainly responsible for this fact, since polysaccharides, the major EPS component, contain many hydrophilic sugar residues (Section 4.4.3). Certainly, the quantity and composition of EPS and inorganic material (e.g., Ca^{+2}) in the biofilm influences water content. In biological systems and living cells, water is considered to be more highly ordered than in simple solutions (Drost-Hansen, 1969), and ions that promote water structure (e.g., Na^+, Li^+, Ca^{+2}, Mg^{+2}) tend to be excluded, whereas ions that inhibit water structure (e.g., K^+, Cl^-) are not.

4.5.1 Thermodynamic Properties

Thermodynamic properties of the biofilm include its volume and mass which can be used to calculate biofilm density. Practically, biofilm density is determined by measuring dry biofilm mass on a known substratum area and dividing by a measured biofilm thickness:

$$\rho_f = \frac{X_f''}{L_f} \qquad [4.3]$$

where ρ_f = biofilm mass density ($M\ L^{-3}$)
X_f'' = areal biofilm concentration ($M\ L^{-2}$)
L_f = biofilm thickness (L)

Analytical methods for biofilm thickness are rather system-specific, and many techniques have been used (Chapter 3). Recently, an *in situ*, nondestructive method for determining biofilm thickness has been reported (Bakke and Olsson, 1986). L_f must be measured while the biofilm is wet to eliminate shrinkage from dehydration. Biofilm mass, on the other hand, refers to biomass remaining after drying, generally for 1 h at 103°C. Because L measurement methods vary considerably, useful comparisons of ρ_f between experimenters are sometimes difficult (Table 4.2).

4.5.1.1 *Biofilm Thickness* Biofilm thickness is often a very important characteristic in the analysis of biofilm processes, since it determines the diffusional length and is necessary for calculating fluid frictional resistance and heat transfer resistance. However, accurate measurement of biofilm thickness is difficult (Chapter 3). The biofilm thickness may vary considerably over a given substratum due to morphological features of the biofilm. Bakke (1986) has reported the spatial variation in biofilm thickness over the width of a rectangular duct under laminar flow conditions (Chapter 9). The variation in thickness was also a function of the biofilm "age." Others have observed a very nonuniform thickness, which certainly increases fluid frictional resistance as well as advective mass transfer at the interface (Picologlou et al., 1980). In slow-moving, nutrient-rich water, the biofilm thickness can exceed 30 mm (e.g., algal/bacterial mats). Biofilm thickness also seems to be influenced by the microbial species diversity. For example, a pure culture *Pseudomonas aeruginosa* biofilm rarely exceeded a thickness of 50 μm in several laboratory studies (Trulear, 1983; Bakke, 1986), whereas a mixed culture biofilm under virtually the same experimental conditions frequently attained a thickness greater than 120 μm (Characklis, 1980).

A biofilm system is characterized by several characteristic lengths which further delineate its nature:

$$\begin{aligned} \text{Cell size} &= 1\text{–}10\ \mu\text{m} \\ \text{Mass transfer boundary layer} &= 10\text{–}100\ \mu \\ \text{Diffusional length} &= 10\text{–}1000\ \mu\text{m} \\ \text{Biofilm thickness} &= 10\text{–}1000\ \mu\text{m} \\ \text{Reactor media size} &= 1000\text{–}100{,}000\ \mu\text{m} \end{aligned}$$

4.5.1.2 *Biofilm Density* Accurate measurement of biofilm mass density is directly related to accuracy of the thickness measurement (Table 4.2). Thus, reported biofilm densities must be considered in relation to the techniques used to measure biofilm thickness. Biofilm mass densities have been reported as high as 105 kg m^{-3} (Hoehn and Ray, 1973) and as low as 10 kg m^{-3} (Characklis, 1980). Biofilm density can vary with depth in the biofilm (Table 4.3). Watanabe (unpublished data) used a "microslicer" to separate the layers in a biofilm over 500 μm thick. An increase in density with depth was observed. Many other factors influence biofilm thickness and density, including microbial species present in the biofilm, fluid shear stress, and substrate flow.

Selective attachment and/or selective growth of certain microbial species from the available population may influence biofilm density. For example, sev-

TABLE 4.2 Reported Biofilm Thickness and Biofilm Density (Dry Mass per Unit Wet Volume)

Reference	Thickness, L_f (μm)	Density (kg m^{-3})	Type Biofilm[a]
Kornegay and Andrews, 1967	160–210	66–130	A
Hoehn and Ray, 1973	30–1300	20–105	B
Williamson and McCarty, 1976	150–580	42–109	C
Rittman and McCarty, 1978	100	50[b]	A
Rittman and McCarty, 1980	119–126	5[c]	B
Rittman and McCarty, 1981	0–125	5[c]	B
Trulear and Characklis, 1982	10–124	10–65	B
Trulear, 1983	36–47	17–47[d]	D
Bakke, 1986	0–60	27	D

[a]A, steady state, heterotrophic, mixed population; B, heterotrophic, mixed population; C, steady state, nitrifying, mixed population; D, steady state, *Pseudomonas aeruginosa*.

[b]Calculated assuming biofilm is 80% volatile solids.

[c]Calculated assuming biofilm is 50% carbon.

[d]Calculated from measured thickness corrected for refractive index of biofilm.

TABLE 4.3 Variation of Biofilm Density with Biofilm Depth[a]

	Thickness of Biofilm Section (μm)	Depth from Water-Biofilm Interface (μm)	Density (kg m^{-3})
Surface film	400	0–400	37
Intermediate	200	400–600	98
Base film	130	600–730	102

[a]Watanabe (unpublished results). Biofilm was approximately 730 μm thick.

eral investigators (Picologlou et al., 1980; Trulear and Characklis, 1982; McCoy and Costerton, 1982) have observed the predominance of filamentous organisms in the biofilm leading to a low biofilm density (<40 kg m^{-3}). The filamentous morphology may be elicited by a microbial metabolic response to environmental stress. For example, McCoy and Costerton (1982) observed peculiar filamentous pseudomonads in biofilms subjected to turbulent flows. Pseudomonads are not generally considered to be filamentous, but Jensen and Woolfolk (1985) have observed complete filamentation of *Ps. putida* 40 under oxygen limitation in the latter stages of batch growth. In a biofilm, the cells in the lower layers may frequently be limited by oxygen, lack of which may trigger filamentous growth. Wardell et al. (1980) also observed filamentation in biofilms.

Increasing initial shear stress increases ρ_f (Figure 4.7) according to experi-

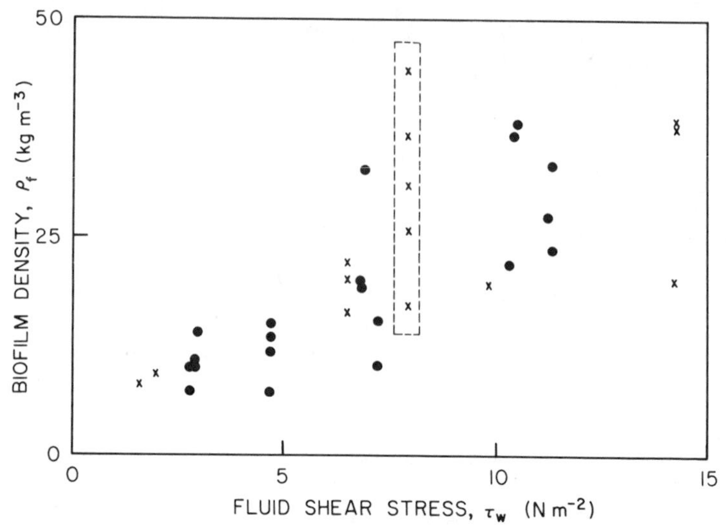

Figure 4.7 Influence of fluid shear stress on biofilm density as reported by Characklis (1980). Experiments were conducted with mixed microbial populations in a tubular reactor at constant velocity at 30–40°C (●) or constant shear stress at 30°C (×). The loading rate of the substrate [glucose (×) or glucose plus trypticase soy broth (●)] varied from 0.008 mg C m^{-2} s^{-1} upward. Data in box are plotted in Figure 4.8.

ments reported by Characklis (1980). In these experiments, the volumetric flow remained constant while the pressure drop and hence the shear stress increased due to biofilm accumulation (fluid dynamics is discussed in Chapter 9). The substrate (glucose) flux J_S was low (<0.16 mg m^{-2} s^{-1} as organic carbon). In another set of experiments, the pressure drop and hence the shear stress remained constant while the volumetric flow rate decreased due to biofilm accumulation in the tube. Increasing shear stress resulted in a proportional increase in biofilm density (Figure 4.7). Kornegay and Andrews (1967) conducted experiments in a RotoTorque reactor (Chapter 3) with the inner cylinder rotating at different speeds. Increasing shear stress had no significant effect on the biofilm density. However, the substrate flux in these experiments was relatively high (1.5 mg m^{-2} s^{-1}).

Results of experiments in the RotoTorque and a tubular reactor (Characklis, 1980) indicate that there is a significant effect of substrate loading and flux on biofilm density up to a limiting flux (Figure 4.8). Trulear (1983) reported increasing cell concentration in the biofilm with increasing substrate loading rate (Table 4.4). There may be a significant interactive effect between shear stress and substrate flux on biofilm density with shear stress not having a significant effect at high J_S. The *type* of substrate may also influence biofilm density, since different microbial communities may result from different substrate compositions. In addition, Turakhia (1986) has observed that increasing calcium concentration increases the biofilm cellular density (Table 4.4).

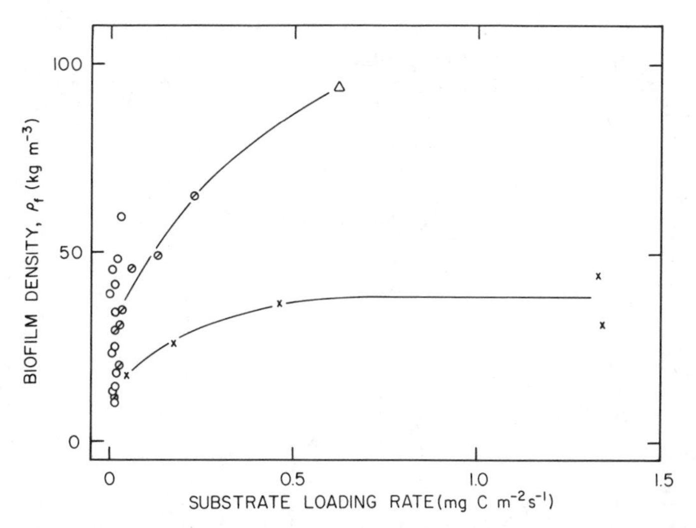

Figure 4.8 Influence of substrate loading rate on biofilm density. Experiments were conducted with mixed microbial populations: ∅, RotoTorque, $\tau_w = 2.5$ N m^{-2}, T = 30°C, glucose (data from Trulear, 1980); △, RotoTorque $\tau_w = 1.1$ N m^{-2}, T = 25°C, glucose (data from Kornegay and Andrews, 1967); ○, RotoTorque, $\tau_w = 1.9$–2.9 N m^{-2}, T = 30–40°C, glucose plus trypticase soy broth (data from Characklis, 1980); ×, tubular reactor, $\tau_w = 7.9$ N m^{-2}, T = 30°C, glucose (data from Characklis, 1980). data points × are plotted in box in Figure 4.7.

The biofilm density can be significantly higher in environments containing suspended particulates of inorganic salts with low solubility that become integrated in the biofilm. Simplistically, the density of a biofilm consisting of a mixture of two components, one organic and one inorganic, can be expressed as follows:

$$\rho = \frac{\rho_o V_o + \rho_i V_i}{V_o + V_i} \qquad [4.4]$$

where ρ_o, ρ_i = density of organic and inorganic components, respectively (ML^{-3})

V_o, V_i = volume of each component (L^3)

Since ρ_i may easily exceed $2\rho_o$, the volume of the organic component may have to be twice the inorganic component volume to be of equal weight in the biofilm. The water content may also be affected, since the inorganic fraction generally immobilizes less water than the organic fraction.

Some published information suggests that biofilm density increases with age. Hoehn and Ray (1973) observed a maximum (105 kg m^{-3}) in biofilm density during their experiments with a mixed population biofilm containing macroor-

TABLE 4.4 Influence of Substrate Loading Rate and Calcium Concentration on Biofilm Cellular Composition[a]

Variable	Areal Density $g\ m^{-2}$	Cell Density: Areal (10^{12} cells m^{-2})	Cell Density: Volumetric (10^{16} cells m^{-3})
	Turakhia, 1986: $S_i^b = 15\ (g\ C)\ m^{-3}$		
$Ca^{+2} = 0.4\ g\ m^{-3}$	1.5	1.6	4.0[c]
25	1.9	3.3	8.2[c]
50	2.3	8.6	21.5[c]
	Trulear, 1983: $Ca^{+2} = 25\ g\ m^{-3}$		
$S_i^b = 3\ (g\ C)\ m^{-3}$	0.5[d]	1.7	4.7
10	1.4[d]	7.6	33.0
17	2.4[d]	19	40.4

[a] Substrate was glucose, and all experiments were conducted in a RotoTorque reactor.
[b] Loading rate for $S_i = 1\ g\ m^{-3}$ is $5 \times 10^{-6}\ g\ m^{-2}\ s^{-1}$.
[c] Calculated assuming a biofilm thickness of 40 μm.
[d] Calculated assuming that the biofilm is 50% carbon.

ganisms as well as microorganisms. Trulear (1983), experimenting with *Ps. aeruginosa*, observed increasing biofilm density with time; this was attributed to EPS accumulation, since the cell mass remained constant (Figure 4.9). Increasing biofilm density may also reflect the continued accumulation of inorganic components within the biofilm.

4.5.2 Transport Properties

Transport properties of biofilm are of critical importance in quantifying the momentum, heat, and mass transfer processes in biofilm systems (Chapter 9). Transport properties, however, are sometimes difficult to quantify accurately and precisely. In some cases, the definition of a transport coefficient is subject to argument.

4.5.2.1 *Biofilm Rheology* A biofilm is not a rigid substance. The EPS matrix that holds the biofilm together behaves as a gel. The rheological properties of the biofilm influence the transfer of fluid momentum at the biofilm-water interface. Thus, the biofilm has rheological properties that influence its relation to the immediate aquatic environment. Rheological measurements conducted with a Weissenburg rheogoniometer on an *in situ,* mixed population biofilm (Characklis, 1980) indicate that the biofilm is viscoelastic. The results, at an

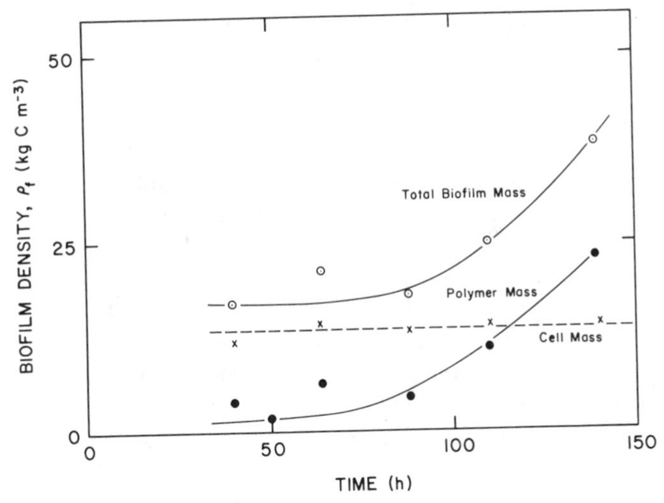

Figure 4.9 Change in biofilm density during the accumulation period. Experiment conducted in a RotoTorque with *Ps. aeruginosa* (data from Trulear, 1983).

excitation frequency of 6 Hz, indicate (Table 4.5) that the elastic modulus of the biofilm is of the same order of magnitude as that of a cross-linked protein gel having the same water content (e.g., a fibrinogen network).

Thus, the biofilm–water interface is a compliant surface that tends to snap back when deformed by fluid shear forces, i.e., mechanical energy in the fluid is absorbed by the biofilm. The effect is to significantly increase frictional drag when water is flowing past the surface. In addition, the deformation may lead to advective mass transport at the bulk water–biofilm interface or within the biofilm. Thus, the viscous properties of the biofilm contribute to increased fluid frictional resistance in flow conduits. Although the immobilized EPS in the biofilm contributes to increases in fluid drag, the same EPS in solution may decrease drag (Section 4.4.1)

TABLE 4.5 A Comparison of the Viscoelastic Properties of Biofilm with Coagulated Fibrinogen[a]

Material	Elastic (Storage) Modulus (N m^{-2})	Viscous (Loss) Modulus (N m^{-2})
Biofilm[b]	59.5	118
Coagulated fibrinogen (ID)[c]	85	7

[a] Characklis, 1980. Measurements conducted on a Weissenburg rheogoniometer at an excitation frequency of 6 Hz by Dr. L.V. McIntire, Rice University, Houston, TX.

[b] Environmental conditions: $T_b = 40°C$, $\tau_w = 3.3$ N m^{-2}, $J_S = 0.1$ mg m^{-2} s^{-1}.

[c] 2300 g m^{-3} plasma.

4.5.2.2 Thermal Conductivity The thermal conductivity reflects the rate at which heat is conducted through a material in which a temperature gradient exists. A high thermal conductivity indicates that the material is a good heat conductor. The thermal conductivity of a mixed culture biofilm has been reported by Characklis et al. (1981) to be the same as that of water—not surprising, considering the high water content of biofilms. The glucose loading rate was a variable in these experiments but had no significant effect on thermal conductivity. The thermal conductivity of biofilms, as well as other materials relevant to industrial and laboratory heat transfer equipment, is reported in Table 4.6. A comparison of thermal conductivity of biofilm (≈ 0.6 W m^{-1} K^{-1}) indicates that a biofilm provides approximately 27 times as much resistance to heat transfer as stainless steel for equal thickness. Thus, a very thin biofilm can significantly restrict heat transfer through a stainless steel tube. More detailed heat transfer calculations related to biofouling of heat exchangers are presented in Chapters 9 and 14.

TABLE 4.6 Thermal Conductivity of Biofilm and Other Selected Materials

Material	Thermal Conductivity (W m^{-1} K^{-1})	Temperature (°C)	References
Biofilm	0.68 ± 0.27	28.3 ± 0.3	Characklis et al. (1981)
	0.71 ± 0.39	26.7 ± 0.3	
	0.57 ± 0.10	28.3 ± 0.3	
Water	0.61	26.7	Weast, 1973
	0.62	28.3	
Carbon steel	51.92	0–100	Perry and Chilton, 1973
Steel	46.86	18	AEC[a] 1955
Stainless steel (type 316)	16.30	0–100	Pery and Chilton, 1973
Aluminum 5052	138.46	20	Perry and Chilton, 1973
	205.85	100	AEC[a] 1955
Cupronickel 10% 706	44.71	0–100	Perry and Chilton, 1973
Copper	384	18	AEC[a] 1955
Titanium (commercial pure)	16.44	0–100	Perry and Chilton, 1973
Glass	0.6–0.9	—	Weast, 1973

[a] Atomic Energy Commission.

4.5.2.3 *Diffusion Coefficient* The diffusion coefficient, by analogy with the thermal conductivity, reflects the rate at which molecules diffuse through a material as a result of a concentration gradient. In an active biofilm, microorganisms are consuming substrate (electron donor), electron acceptor, and other nutrients that diffuse through the biofilm. Consequently, portions of the biofilm may be deficient in one or another component, depending on their relative diffusion rate, their reaction rate, and the stoichiometry between the reacting components (Chapter 2). Hence, sulfate-reducing bacteria, which are obligate anaerobes, can proliferate deep within a biofilm in contact with a flowing stream of oxygenated water (Figure 4.10). The diffusion coefficient is characteristic of the material as well as the diffusing molecule. Measured diffusion coefficients for various molecules through biofilms and bioflocs are reported in Table 4.7. Bioflocs are microbial aggregates held together by EPS and are suspended in the growth fluid rather than attached to a substratum. Hence, some analogies between bioflocs and biofilms are to be expected. For the most part, diffusion coefficients for biofilms and bioflocs are within 80% of the values determined for water. Experimental techniques for measurement of diffusion coefficients in biofilm and in bioflocs vary considerably, so that differences in results between investigators may be significant.

Diffusivity within the biofilm may change with biofilm age. For example, Siegrist and Gujer (1985) report a slight decrease in diffusion with experimental increasing run time. Matson and Characklis (1976) also observed a decrease in oxygen diffusion coefficient in bioflocs with increasing sludge age. The relationship between age and density, however, has not been quantified.

In situ observations by Siegrist and Gujer (1985) indicate that a decrease in diffusion coefficient occurs with increasing biofilm thickness (mixed culture).

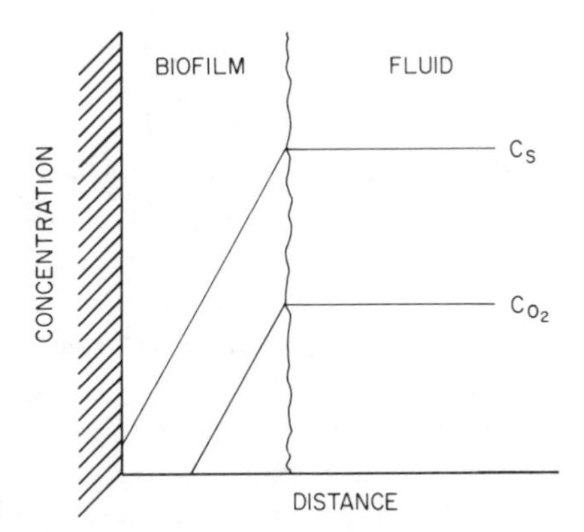

Figure 4.10 A hypothetical biofilm with linear concentration gradients of substrate (S) and oxygen (O_2) schematically indicated. This biofilm contains aerobic and anaerobic environments.

TABLE 4.7 Experimentally Determined Diffusivities in Biofilms and Bioflocs

Diffusing Species	Diffusivity D_f (10^{-10} m^2 s^{-1})	D_f/D_{H_2O}	Microbial Aggregate	Reference
Oxygen	15	0.70	Film	Tomlinson and Snaddon, 1966
Oxygen	21	0.08	Floc	Mueller et al., 1968
Glucose	0.48	0.08	Floc	Baillod and Boyle, 1970
Glucose	0.6–6.0	0.1–1.0	Floc	Pipes, 1974
Oxygen	22	0.9	Film	Williamson and McCarty, 1976
Ammonia	13	0.8		
Nitrate	14	0.9		
Oxygen	4–20	0.2–1.0	Floc	Matson and Characklis, 1976
Glucose	0.6–2.1	0.1–0.3		
Oxygen		0.95	Film	Bungay et al., 1969
Glucose		≈0.5	Film	Siegrist and Gujer, 1985
Sodium (Na^+)		≈0.6	Film	
Bromide (Br^-)		≈0.6	Film	

The diffusion coefficient varied from 40 to 140% of the corresponding value in water. They explained the apparent higher diffusion rate in water on eddy diffusion in the vicinity of the biofilm-water interface caused by increasing biofilm roughness as the film thickness increased. Bakke (1986) has observed increasing biofilm roughness with age (see Chapter 9). Biofilm roughness will be discussed in more detail in Chapter 9.

Extracellular polymeric substances, irrespective of their charge density or their acidic nature, can be expected to act in some manner as diffusion barriers, molecular sieves, and adsorbents. External layers of polysaccharides around cells, whether they are discrete capsules, irregular but continuous slime layers, or the matrix of a biofilm, influence physicochemical processes such as diffusion and fluid frictional resistance. Synthetic gels composed of polysaccharides or other synthetic materials may provide a useful physical model for capsules and biofilms. For example, diffusivity through polyacrylamide, a synthetic polymer gel, decreases with increasing gel density (Figure 4.11), consistent with observations of Siegrist and Gujer (1985) for decreasing biofilm diffusivity with increasing biofilm density. If synthetic polymer gels are good physical models of biofilms, there is probably an influence of biofilm density on biofilm transport coefficients as well, although the biofilm density may not always reach a high enough level to cause a significant effect. Diffusion in (agar) gels has been related to diffusion in solution by the following expression (Dudman, 1977):

$$D' = \frac{D}{1 + (a_D - 1)\phi_D} \qquad [4.5]$$

where D', D = diffusion coefficients in solution and in the gel, respectively ($L^2\, t^{-1}$)
a_D = coefficient = 5/3 for randomly oriented rods
ϕ_D = coefficient

ϕ_D is dependent on the ratio of the volume occupied by the gel substance to the total volume of the gel (Lauffer, 1961; Schantz and Lauffer, 1962). In the simplest case, ϕ_D would be the volume fraction occupied by the gel substance, but it is possible that ϕ_D could be larger because of hydration or because of exclusion of solute particles from the neighborhood of the gel particles for mechanical, electrostatic, or other reasons. ϕ_D is important also because it is involved in a correction for obstruction to diffusion. The value for ϕ_D in 1.5% agar gel is approximately 0.05 (Schantz and Lauffer, 1962). Thus, for gels of 1.5% agar, $D' = 0.0986D$, indicating a very small effect on diffusion rates. Dudman (1977) used data of Duguid and Wilkinson (1953) to estimate that the capsule of *En-*

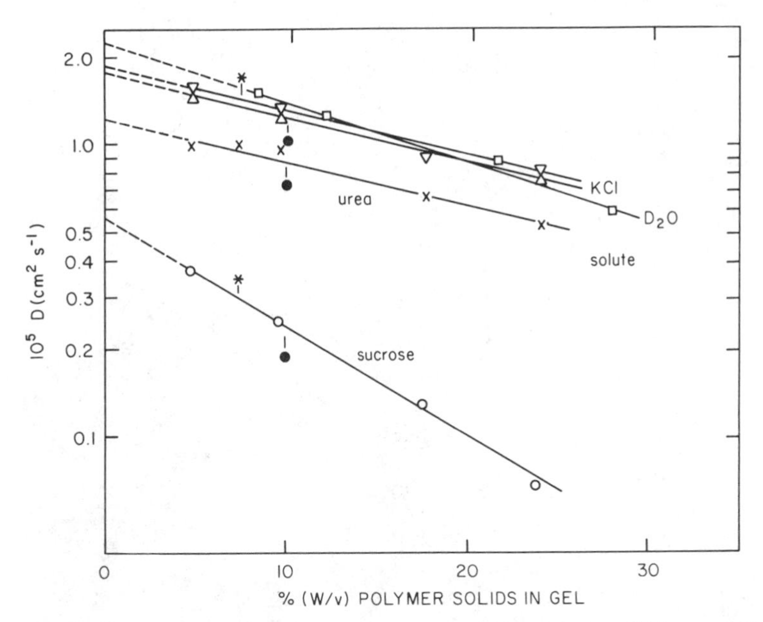

Figure 4.11 Diffusion coefficients in polyacrylamide gel (after White and Dorion, 1961). Sucrose (○), urea (×), KCl unstirred (△), KCl stirred (▽), D_2O (□) in 95% acrylamide : 5% methylene-*bis*-acrylamide. Two other polymer gels were tested—90% acrylamide : 10% methylene-*bis*-acrylamide (●), and 99% acrylamide : 1% methylene-*bis*-acrylamide (*).

terobacter aerogenes contains 1–2% polysaccharide, the same as the agar concentration used in ordinary laboratory media. Thus, the capsule layer might not be expected to present a significant diffusional resistance, at least for small molecules.

However, gels are three dimensional networks with pores and channels through which solutes must diffuse. The effective radii of the pores in agar gels have been estimated to be 95 and 44 nm in gels of 1 and 2% agar, respectively (Ackers and Steers, 1962). Figure 4.12 is a scanning electron micrograph of an *Acetobacter acetii* cell on 2% agar (bar is 0.2 μm). The photograph indicates a pore size on the order of the estimates of Acker and Steers. This imposes an

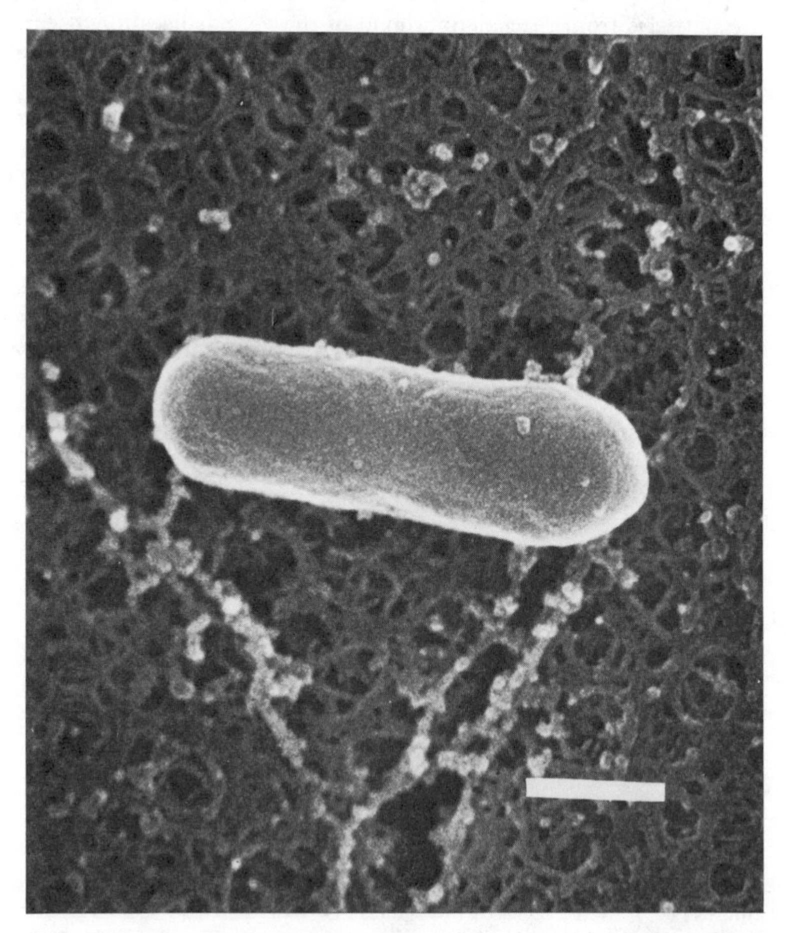

Figure 4.12 Scanning electron micrograph of an *Acetobacter acetii* cell on a 2% agar gel. The bar is 0.2 μm, so the pore size is estimated as approximately 40–70 μm (courtesy W. Lee).

upper limit on the size of particles that may diffuse through 1-2% agar gels. Because of the pore size limit, EPS can be expected to act as a molecular sieve. The equivalent pore size may become progressively smaller as one proceeds from the outer slime layer toward the cell wall.

Measurements of diffusion coefficients for oxygen and nitrogen through gels of 2, 4, and 6% agarose containing 0.35% hyaluronic acid showed little deviation from the diffusion of these gases in water (McCabe and Laurent, 1975). Diffusion coefficients for oxygen in biofilms, however, range from 8 to 95% of the value in water (Table 4.7). The differences in these results could be attributed to the differences in experimental methods. Another factor which contributes to the discrepancy between biofilm results and gel experiments is the heterogeneous nature of the biofilm "gel," which may vary considerably with biofilm growth conditions. Irregular morphology of the biofilm–liquid interface may increase advective mass transfer, which may be overlooked in diffusion measurements (Characklis et al., 1981; Picologlou et al., 1980; Siegrist and Gujer, 1985). Finally, cells represent approximately 20–50% of the biofilm volume and may reduce diffusion rates considerably.

Itamunoala (1987) observed that the diffusion coefficient for glucose in calcium alginate gel (2–8% alginate) was at least 24% lower than in water. The diffusion coefficient decreased with decreasing glucose concentration for glucose concentrations less than 18 kg m^{-3}. Increasing the alginate concentration substantially decreased the diffusion coefficient.

Tanaka et al. (1984) also measured diffusion rates of several substrates of varying molecular size into and from calcium alginate gel beads. In contrast to Itamunoala, he found that substrates such as glucose, L-tryptophan, and α-lactoalbumin (molecular weights less than 20,000) diffused freely into and out of the gels without disturbance by the pores in the gel beads. Increasing the concentration of the Ca alginate in the beads did not influence diffusion rates. It would appear that the molecular size of the molecules was sufficiently small to permit free diffusion. In the case of higher molecular weight substances such as albumin (MW = 69,000), γ-globulin (MW = 154,000), and fibrinogen (MW = 341,000), no diffusion occurred from the bulk solution into the beads, but diffusion of these molecules out of the beads was observed. However, diffusion out of the beads was considerably slower than calculated theoretically. The observed lower diffusion rate out of the gel may have been caused by lower solubility of the molecules in the gel. Why were the large molecules unable to penetrate the gel from the outside, but able to diffuse outward from the gel? The authors suggest that the structure of the Ca alginate gel may have been influenced by the presence of the large molecules in the gel. Other factors that influence diffusion in gels—and probably biofilms—include electrostatic forces (Donnan potential), pH, ion composition, and ionic strength. Siegrist and Gujer (1985) report no effect of ionic strength above 1.5×10^{-3} M for diffusion of small molecules in biofilms. Theoretical aspects of electrolyte diffusion in biofilms are presented in Chapter 11.

4.5.3 Geometric Properties

Biofilms are not necessarily uniform spatially, as is shown by observations of patchy distributions of biofilms on substrata. In addition to spatial variation in composition, surface morphology may also influence properties and interaction with the immediate environment. The necessity for segregated biofilm models in the base film (i.e., no significant biofilm roughness) hinges on the spatial distribution of the cells in the biofilm. There is ample electron micrograph evidence (Figure 4.13) for aggregation or colonies of specific cells in biofilms to make segregation an important concept, especially when two species depend on each other for growth (e.g., Nitrosomonas and Nitrobacter species). Gujer (1987) has estimated that performance of nitrifying biofilm reactors could conceivably be affected if distances between these species, or *microcolonies*, are greater than 10 μm.

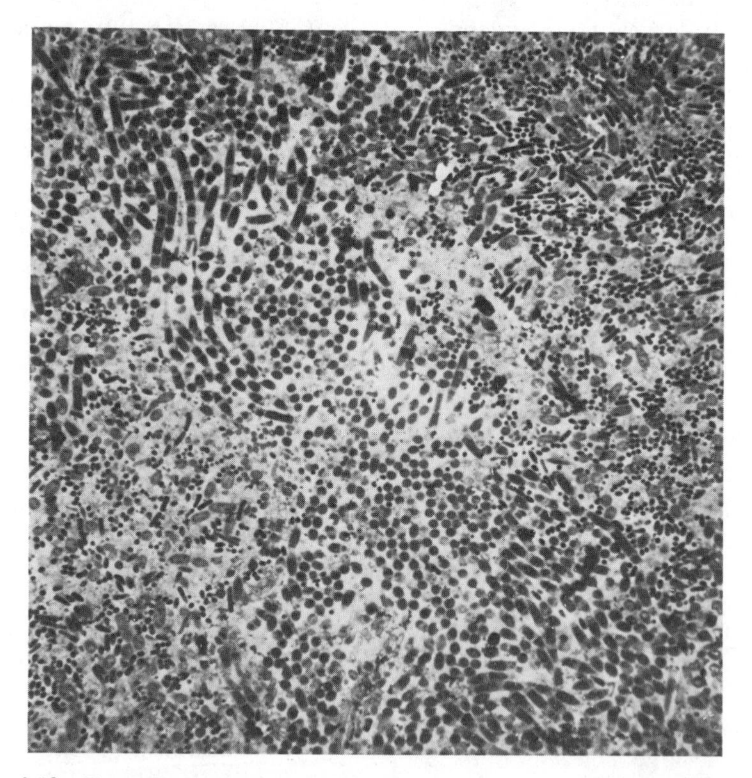

Figure 4.13 Transmission electron micrograph of a granular sludge from a laboratory fluidized reactor, indicating segregation of various microbial species within a biofilm. The diagonal from the top left to the bottom right is a colony of Methanothrix-like bacteria. Various other species may be distinguished by their morphology and electron density in the micrograph (courtesy J.T.C. Grotenhuis, Wageningen, The Netherlands).

The surface film may possess a very distinctive morphology. Its roughness may, in some cases, be characterized by an amplitude and period, which can change with time (Chapter 9). In other cases, the morphology may be due to filamentous microbial growth, as indicated in Figure 4.14. The influence of cell physiology/ecology on filamentous habit of microorganisms is not clear, but experimental observations indicate that filaments have a tendency to form in oligotrophic environments or niches that are oxygen transfer-limited (McCoy and Costerton, 1982). The effect of filaments on hydraulic resistance has been documented (Picologlou et al., 1980) and is discussed further in Chapter 9.

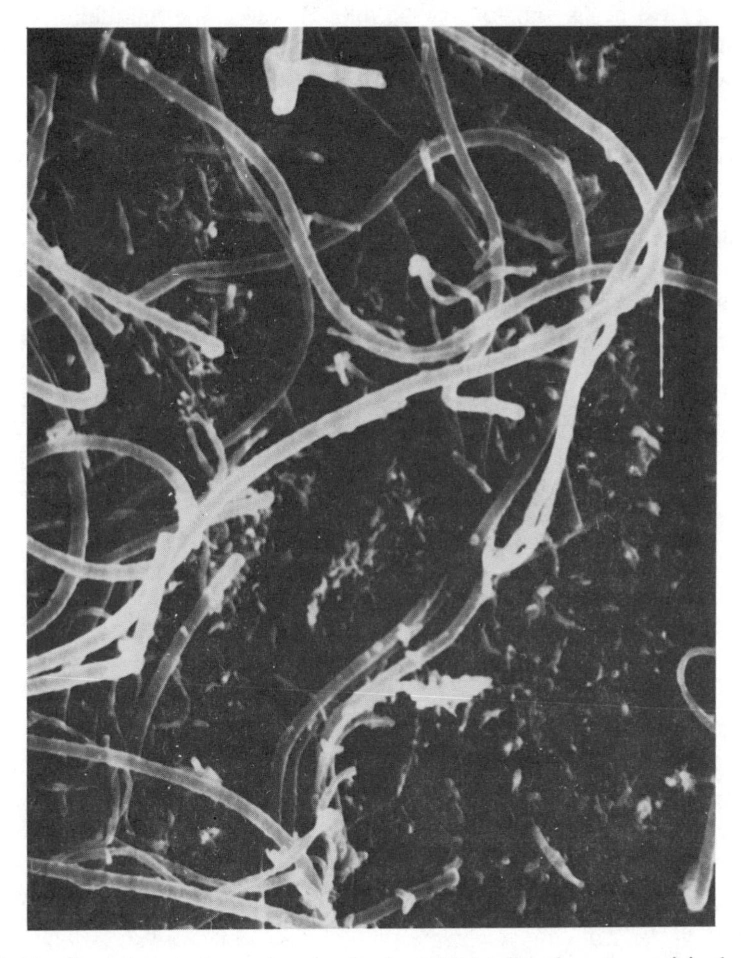

Figure 4.14 Scanning electron micrograph of a RBC biofilm from a municipal wastewater treatment plant. Filaments are clearly evident in association with a "mat" of rod shaped bacteria. (Reprinted with permission from Alleman et al., 1982.)

4.6 ELEMENTAL COMPOSITION

The elemental composition of various biofilms is presented in Table 4.8. More information on elemental composition is presented in Chapter 6. The relative amounts of inorganic and organic biofilm components can be determined by combusting a dried (103°C for 1 h) biofilm sample at 550°C. Volatile and fixed solids generally reflect the organic and inorganic fraction, respectively. Bound water and carbonate salts resulting from precipitation or corrosion products in the biofilm may also volatilize at the high temperatures and cause significant errors in these tests. The volatile fraction has been reported as high as 80% of the biofilm dry weight in laboratory experiments where biotic components dominate (e.g., Kornegay and Andrews, 1967). However, the volatile (organic) fraction of a biofilm can be considerably less than the volatile fraction of a suspended microbial cell population (>90% volatile solids), primarily because of the inorganic constituents adsorbed, entrapped, or precipitated within the biofilm matrix. The inorganic fraction is especially high in biofilms accumulating in natural aquatic ecosystems where silt, clay, and sediments are entrapped in the matrix. The carbon to nitrogen ratios (C/N) in some of the biofilms are considerably higher than in microbial cells (approximately 5). The high C/N may reflect a large proportion of EPS (generally low in N) or a preponderance of carbonate salts if total carbon rather than organic carbon is measured.

The inorganic composition of the biofilm undoubtedly influences the physical properties of the biofilm. For example, calcium, and probably other multivalent cations (except magnesium) increase intermolecular bonding between EPS. Turakhia et al. (1983) have demonstrated the relative importance of Ca in biofilm mechanical strength by effectively disrupting biofilm by applying a cal-

TABLE 4.8 Reported Composition for Biofilms as Compared with Suspended Biomass

	% Dry Weight		% Volatile Solids			
Reference	Ash	Volatile	C	H	N	Type[a]
	Biofilms					
Turakhia, 1986	13.3	86.7	53.1	6.1	12.4	a
Kornegay and Andrews, 1967	8	92	29.2	6.1	9.8	b
Characklis, 1980	20	80	42.8		10	c
Characklis, 1980			6–14		0.5–3	c
	Suspended Biomass					
Wang et al., 1976	3	97	45.6	6.7	12.7	d
Herbert, 1976	3.6	96.4	48.9	7.1	12.3	e

[a] a, laboratory *Pseudomonas aeruginosa* biofilm; b, laboratory heterotrophic, mixed population biofilm; c, natural, heterotrophic, mixed population biofilm; d, *Pseudomonas* $C_{12}B$; e, *Klebsiella aerogenes*.

cium-specific chelant. Characklis et al. (1983) have demonstrated that corrosion products from the substratum become entrapped in the biofilm and influence the macroscopic biofilm properties such as density and hydraulic (or relative) roughness. High levels of manganese and iron have been found in some biofilms, especially those that receive frequent chlorination (Chapter 14). The waters containing high biofilm manganese and iron levels sometimes contain low concentration of these elements in the water, but frequently contain significant organic carbon concentrations. Corrosion in pumps or piping systems is sometimes the source of these metals, but their presence in the water supply, combined with the redox chemistry of the biofilm and/or chlorine, is a contributing factor to their accumulation in biofilms. Silica is also found in high concentrations in biofilms accumulating in natural waters.

4.7 ECOLOGICAL ASPECTS OF EXOPOLYMER FORMATION

4.7.1 Variation in EPS Chemical Composition

Chemical compositions of extracellular polysaccharides reportedly change during the bacterial growth cycle. Uhlinger and White (1983) noted that the amount of galactose in the exopolymer fraction recovered from a batch culture of *Pseudomonas atlantica* decreased during the growth cycle relative to the other monosaccharides. The high relative amount of galactose observed occurred only very early in the growth cycle, when the absolute amounts of other sugars were low. Later, the ratio between the different sugars was essentially constant. The excess galactose observed initially may have been due to incomplete separation of the polymers from the medium, which contained galactose as the major carbon source.

Christensen et al. (1985) report that *Pseudomonas* sp. NCMB 2021 produces more than one polymer. The different polymers were produced in different stages of the growth cycle (batch system). Thus, the relative amounts of the polysaccharide constituents vary significantly for *Pseudomonas* sp. NCMB 2021. With very few exceptions (e.g., alginate), major structural modifications of polysaccharide chains (e.g., incorporation or removal of certain monosaccharides) are not to be expected after the polymerization step (Sutherland, 1977). Therefore, variation in sugar composition during the growth cycle, if observed, strongly suggests production of more than one polymer. Structural modifications and, consequently, altered physical properties of extracellular polysaccharides may occur under certain growth conditions. By using different nutrient limitations, both the composition and amount of xanthan could be varied (Sutherland, 1985).

4.7.2 EPS in Planktonic Cells versus Biofilms

Will an organism produce the same extracellular polymers in the laboratory as in the natural environment? Costerton et al. (1978) suggest that many bacteria lose the ability to secrete copious amounts of EPS observed in environmental samples when the bacteria are cultivated in laboratory media. An alternative explanation, in addition to possibilities of spontaneous mutation, might be that the large observed EPS accumulation around the cells over a long period in nature may occur even when the specific polymer production rate is low. When grown in rich media, bacteria produce polymer at a high rate but release their polymers into the medium more rapidly as a slime, rather than retaining polymer as a distinct capsule. For example, bacterial strains used in the industrial production of polysaccharides frequently have a polysaccharide production rate too low to be detected by increased broth viscosity (Christensen et al., 1985), a method commonly used in industry for monitoring extracellular polysaccharide production. Thus, visual observations unaccompanied by polymer analyses may wrongly indicate that bacteria have lost their ability to produce EPS. More detailed chemical analyses and better defined growth conditions for the microorganism will settle this controversy. The EPS matrix in biofilms appears to contain the same polymers in liquid cultures for the same organisms, in view of the following observations:

1. Without identifying the polymers, it was found that the production rate of EPS did not *quantitatively* differ between biofilms and suspended cultures (Bakke et al., 1984) of *Pseudomonas aeruginosa* when the growth conditions were otherwise the same.
2. Antibodies made against polymers produced in liquid cultures have been shown to react with biofilm matrices *in situ* (Costerton et al., 1981).

The production of different polysaccharides during exponential growth and in the stationary phase, as observed in laboratory batch cultures (Christensen et al., 1985), may have a parallel in natural environments. It has been observed that induced starvation of exponentially growing cells leads to extensive release of soluble, viscous polysaccharide, whereas no production of the same polysaccharide was observed in the growing cells (Wrangstadt et al., 1986) suggesting that starvation triggers the production of a different polymer.

4.7.3 EPS as a Nutrient Trap

The anionic character of most EPS gives biofilms cation exchange properties. Thus, bacteria may presumably concentrate and utilize cationic nutrients (e.g., amines) by means of this process, especially under oligotrophic conditions (Costerton et al., 1981). However, for the bacteria to utilize adsorbed nutrients, several requirements must be fulfilled. First, organic cations must be present in the environment. Secondly, these organic cations must bind to the biofilm more strongly than some of the inorganic cations naturally present. Finally, the organisms must transport these relatively strongly bound nutrients from the biofilm EPS matrix through the cell wall. The process will depend largely upon the cation-binding characteristics of the EPS and the concentration and nature of the salts in the medium. It is unlikely—especially in a marine environment, where salts and especially divalent cations are abundant—that the biofilm can function effectively as a trap for soluble nutrients. On the other hand, particulate nutrients may be effectively immobilized in a biofilm.

ACKNOWLEDGMENT

Dr. O. Smidsrød, Institute for Biotechnology, University of Trondheim, Trondheim, Norway, is gratefully thanked for reviewing this chapter.

REFERENCES

Abbot, A., R.C.W. Berkeley, and P.R. Rutter, "Sucrose and deposition of *Streptococcus mutans* at solid-liquid interfaces," in R.C.W. Berkeley, J.M. Lynch, J. Melling, P.R. Rutter, and D. Vincent (eds.), *Microbial Adhesion to Surfaces,* Ellis Horwood, Chichester, 1980, pp. 117–142.

Ackers, G.K. and R.L. Steers, *Biochim. Biophys. Acta,* **59**, 137 (1962).

Albertsson, P.-A., *Adv. Protein Chem.,* **24**, 309 (1970).

Alleman, J.E., J.A. Veil, and J.T. Canady, *Wat. Res.,* **16**, 543 (1982).

Aspinall, G.O. *The Polysaccharides,* Vol. 1, Academic Press, New York, 1982.

Aspinall, G.O. *The Polysaccharides,* Vol. 2, Academic Press, New York, 1983.

Atomic Energy Commission, *Reactor Handbook,* Vol. 2, AECD-3646, U.S. GPO, Washington, 1955.

Baillod, R.C. and W.C. Boyle, *J. San. Eng. Div. ASCE,* **96**(SA4), 525 (1970).

Bakke, R., "Biofilm detachment," unpublished doctoral thesis, Montana State University, Bozeman, MT, 1986.

Bakke, R. and P.Q. Ollson, *J. Microbiol. Meth.,* **5**, 93–98 (1986).

Bakke, R., M.G. Trulear, J.A. Robinson, and W.G. Characklis, *Biotech. Bioeng.,* **26**, 1418–1424 (1984).

Bhattacharjee, A.K., H.J. Jennings, and C.P. Kenney, *Biochem. Biophys. Res. Commun.,* **61**, 489 (1974).

Brant, D. and B.A. Burton, in D. Brant (ed.), *Solution Properties of Polysaccharides,* ACS Symp. Ser. **150**, 1981, p. 81.

Bungay, H.R., W.J. Whalen, and W.M. Saunders, *Biotech. Bioeng.,* **11**, 765 (1969).

Characklis, W.G., *Biofilm Development and Destruction,* Final Report, EPRI CS-1554, Project RP902-1, Electric Power Research Institute, Palo Alto, CA, 1980.

Characklis, W.G., M.J. Nimmons, and B.F. Picologlou, *J. Heat Transfer Eng.,* **3**, 23 (1981).

Characklis, W.G., N. Zelver, C.H. Nelson, R.O Lewis, D.E. Dobb, and G.K. Pagenkopf, Paper No. 250, in *Corrosion 83,* Natl. Assoc. Corr. Engrs., 1983.

Christensen, B.E., J. Kjosbakken, and O. Smidsrød, *Appl. Environ. Microbiol.,* **50**, 837 (1985).

Corpe, W.A., *Dev. Ind. Microbiol.,* **11**, 402 (1970).

Costerton, J.W., G.G. Geesey, and K.-J. Cheng, *Sci. Am.*, **238**, 86 (1978).

Costerton, J.W., R.T. Irvin, and K.-J. Cheng, *Ann. Rev. Microbiol.*, **35**, 299 (1981).

Dexter, S.C., *J. Coll. Int. Sci.*, **70**, 346 (1979).

Dexter, S.C., J.D. Sullivan, Jr., J. Williams, III, and S.W. Watson, *Appl. Microbiol.*, **30**, 298 (1975).

DiMarzio, E.A. and M. Bishop, *Biopolymers*, **13**, 2331 (1974).

Drost-Hansen, W. *Ind. Eng. Chem.*, **61**, 10 (1969).

Dudman, W.F., "The role of surface polysaccharides in natural environments," in I.W. Sutherland (ed.), *Surface Carbohydrates of the Procaryotic Cell*, Academic Press, New York, 1977, p. 359.

Duguid, J.P. and J.F. Wilkinson, *J. Gen. Microbiol.*, **9**, 174 (1953).

Fletcher, M. and G.D. Floodgate, *J. Gen.Microbiol.*, **74**, 365 (1973).

Fletcher, M. and G.I. Loeb, *Appl. Env. Microbiol.*, **36**, 67 (1979).

Gorin, P.A.J. and J.F.I. Spencer, *Can. J. Chem.*, **44**, 993 (1966).

Gujer, W., *Wat. Sci. Tech.*, **19**, 495 (1987).

Hacking, A.J., I.W.F. Taylor, T.R. Jarman, and J.R.W. Govan, *J. Gen. Microbiol.*, **129**, 3473 (1983).

Harada, T., A. Misaki, and H. Saito, *Arch. Biochem. Biophys.*, **124**, 292 (1968).

Haug, A. and B. Larsen, *Carbohydr. Res.*, **32**, 317 (1971).

Herbert, D., "Stoichiometric aspects of microbial growth," in A.C.R. Dean, D.C. Ellwood, C.G.T. Evans, and J. Melling (eds)., *Continuous Culture 6: Applications and New Fields*, Ellis Horwood, Chichester, 1976, p. 1.

Hoehn, R.C. and A.D. Ray, *J. Wat. Poll. Contr. Fed.*, **46**, 2302 (1973).

Holzwarth, G.M., "Is xanthan gum a wormlike chain or a rigid rod?," in D.A. Brant (ed.), *Solution Properties of Polysaccharides*, ACS Symp. Ser. **150**, 1981, p. 15.

Itamunoala, G.F., *Biotechnology Prog.*, **3**, 115–120 (1987).

Jann, B., P. Hofmann, and K. Jann, *Carbohydr. Res.*, **120**, 131 (1983).

Jann, K. and B. Jann, "Bacterial polysaccharide antigens," in I.W. Sutherland (ed.), *Surface Carbohydrates of the Procaryotic Cell*, Academic Press, New York, 1977, p. 247.

Jansson, P.E., L. Kenne, and B. Lindberg, *Carbohydr. Res.*, **50**, 275 (1975).

Jensen, R.H. and C.A. Woolfolk, *Appl. Env. Microbiol.*, **50**, 364 (1985).

Kenne, L. and B. Lindberg, "Bacterial polysaccharides," in G.O. Aspinall (ed.), *The Polysaccharides*, Vol. 2, Academic Press, New York, 1983, p. 287.

Kornegay, B.H. and J.F. Andrews, *Characteristics and Kinetics of Fixed-Film Biological Reactors*, Final report, Grant WP-01181, Federal Water Pollution Control Administration, U.S. GPO, Washington, 1967.

Lamblin, G., M. Lhermitte, P. Degand, P. Roussel, and H. Slayter, *Biochimie*, **61**, 23 (1979).

Lauffer, M.A., *Biophys. J.*, **1**, 205 (1961).

Lindberg, B., J. Løngren, and J.L. Thompson, *Acta Chem. Scand.*, **27**, 1819, (1973).

Linker, A. and R.S. Jones, *J. Biol. Chem.*, **241**, 3845 (1966).

Marshall, K.C., "Bacterial adhesion in natural environments," in R.W.C. Berkeley,

J.M. Lynch, J. Melling, P.R. Rutter, and B. Vincent, (eds.), *Microbial Adhesion to Surfaces,* Ellis Horwood, Chichester, 1980, p. 187.

Marshall, K.C. and R.H. Cruickshank, *Arch. Mikrobiol.*, **91**, 29 (1973).

Matson, J.V. and W.G. Characklis, *Wat. Res.*, **10**, 877 (1976).

McCabe, N. and T.C. Laurent, *Biochem. Biophys. Acta,* **399**, 131 (1975).

McCoy, W.F. and J.W. Costerton, *Dev. Ind. Microbiol.*, **23**, 441 (1982).

Mikkelsen, A., B.T. Stokke, B.E. Christensen, and A. Elgsaeter, *Biopolymers,* **24**, 1683 (1985).

Morris, E.R., D.A. Rees, G. Young, M.D. Walkinshaw, and A. Darke, *J. Mol. Biol.*, **110**, 1 (1977).

Mueller, J.A., W.C. Boyle, and E.N. Lightfoot, *Biotech. Bioeng.*, **10**, 331 (1968).

Norton, I.T., D.M. Goodall, S.A. Frangou, E.R. Morris, and D.A. Rees, *J. Mol. Biol.*, **175**, 371 (1984).

Paul, J.H. and W.H. Jeffrey, *Appl. Environ. Microbiol.*, **50**, 431 (1985).

Perry, R.H. and C.H. Chilton, *Chemical Engineers' Handbook,* 5th ed., McGraw-Hill, New York, 1973.

Picologlou, B.F., N. Zelver, and W.G. Characklis, *J. Hyd. Div. ASCE,* **106**, 733–746 (1980).

Pipes, D.M. "Variations in glucose diffusion coefficients through biological floc," unpublished Master of Science thesis, Rice University, Houston, 1974.

Pringle, J.H. and M. Fletcher, *Appl. Env. Microbiol.*, **45**, 811 (1983).

Rees, D.A., E.R. Morris, D. Thom, and J.K. Madden, "Shapes and interactions of carbohydrate chains," in G.O. Aspinall, (ed.), *The Polysaccharides,* Vol. 1, Academic Press, New York, 1982, p. 195.

Rittmann, B.E. and P.L. McCarty, *J. Env. Eng. Div. ASCE,* **104**, 889 (1978).

Rittmann, B.E. and P.L. McCarty, *Biotech. Bioeng.*, **22**, 2343 (1980).

Rittmann, B.E. and P.L. McCarty, *J. Env. Eng. Div. ASCE,* **107**, 831 (1981).

Robb, I.D., "Stereo-biochemistry and function of polymers," in K.C. Marshall (ed.), *Microbial Adhesion and Aggregation* (Dahlem Konferenzen), Springer-Verlag, Berlin, 1984, p. 39.

Rutter, P.R. and B. Vincent, "The adhesion of microorganisms to surfaces: Physicochemical aspects," in R.W.C. Berkeley, J.M. Lynch, J. Melling, P.R. Rutter, and B. Vincent (eds.), *Microbial Adhesion to Surfaces,* Ellis Horwood, Chichester, 1980, pp. 79–92.

Saito, H., "Conformation, dynamics, and gelatin mechanism of gel-state (1-3)-β-D-glucans revealed by C^{13} NMR," in D.A. Brant (ed.), *Solution Properties of Polysaccharides,* ACS Symp. Ser. **150**, 1981, p. 125.

Sandford, P.A. and J. Baird, "Industrial utilization of polysaccharides," in G.O. Aspinall (ed.), *The Polysaccharides,* Vol. 2, Academic Press, New York, 1983, p. 411.

Sato, T., T. Norisuye, and H. Fujita, *Polym. J.* (Tokyo), **16**, 423 (1984).

Schantz, E.J. and M.A. Lauffer, *Biochemistry,* **1**, 658 (1962).

Siegrist, H. and W. Gujer, *Wat. Res.*, **19**, 1369 (1985).

Sigler, K., *Biopolymers,* **13**, 2553 (1974).

Skjåk-Bræk, G., H. Grasdalen, and B. Larsen, *Carbohydr. Res.* **154**, 239 (1986).

Smidsrød, O., *Faraday Discuss. Chem. Soc.*, **57**, 263 (1974).

Smidsrød, O., in *27th International Congress on Pure and Applied Chemistry (Proc.)*, 1980, p. 317.

Smidsrød, O. and A. Haug, *Biopolymers*, **10**, 1213 (1971).

Stokke, B.T., A. Elgsaeter, and O. Smidsrød., *Int. J. Biol. Macromol.*, **8**, 217-225 (1986).

Sutherland, I.W., "Bacterial exopolysaccharides—their nature and production," in I.W. Sutherland (ed.), *Surface Carbohydrates of the Procaryotic Cell*, Academic Press, New York, 1977, p. 27.

Sutherland, I.W. *Ann. Rev. Microbiol.*, **39**, 243 (1985).

Symes, K.C., *Carbohydr. Polymers*, **2**, 276 (1982).

Tanaka, T., *Sci. Am.*, **244**(1), 124 (1981).

Tanaka, H., M. Matsumura, and I.A. Veliky, *Biotech. Bioeng.*, **26**, 53 (1984).

Tanford, C., *Physical Chemistry of Macromolecules*, Wiley, New York, 1961.

Thom, D., G.T. Grant, E.R. Morris, and D.A. Rees, *Carbohydr. Res.*, **100**, 29 (1982).

Tomlinson, T.G. and D.M. Snaddon, *Int. J. Air Wat. Poll.*, **10**, 865 (1966).

Trulear, M.G. "Dynamics of biofilm processes in an annular rotating reactor," unpublished Master of Science thesis, Rice University, Houston, TX, 1980.

Trulear, M.G., "Cellular reproduction and extracellular polymer formation in the development of biofilms," unpublished doctoral thesis, Montana State University, Bozeman, MT, 1983.

Trulear, M.G. and W.G. Characklis, *J. Wat. Poll. Control Fed.*, **54**, 1288 (1982).

Turakhia, M.H. "The influence of calcium on biofilm processes," unpublished doctoral thesis, Montana State University, Bozeman, MT, 1986.

Turakhia, M.H., K.E. Cooksey, and W.G. Characklis, *Appl. Env. Microbiol.*, **46**, 1236 (1983).

Uhlinger, D.J. and D.C. White, *Appl. Env. Microbiol.*, **45**, 64 (1983).

Wang, H.Y., D.G. Mou, and J.R. Swartz, *Biotech. Bioeng.*, **18**, 1811 (1976).

Ward, J.B., and R.W.C. Berkeley, "The microbial cell surface and adhesion," in R.C.W. Berkeley, J.M. Lynch, J. Melling, P.R. Rutter, and B. Vincent (eds.), *Microbial Adhesion to Surfaces*, Ellis Horwood, Chichester, 1980, p. 47.

Wardell, J.N., C.M. Brown, and D.C. Ellwood, "A continuous culture study of the attachment of bacteria to surfaces," in R.C.W. Berkeley, J.M. Lynch, J. Melling, P.R. Rutter, and B. Vincent, (eds.), *Microbial Adhesion to Surfaces*, Ellis Horwood, Chichester, 1980, p. 221.

Weast, R.D. (ed.), *Handbook of Chemistry and Physics*, 50th ed., CRC Press, Cleveland, OH, 1973.

White, M.L. and G.H. Dorion, *J. Polymer Sci.*, **55**, 731 (1961).

Wiggins, P.M., *J. Theor. Biol.*, **32**, 131 (1971).

Williamson, K. and P.L. McCarty, *J. Wat. Pollut. Control Fed.*, **48**, 281 (1976).

Wrangstad, M., P.L. Conway, and S. Kjelleberg, *Arch. Microbiol.*, **145**, 220 (1986).

Zelver, N., "Biofilm development and associated energy losses in water conduits," unpublished Master of Science thesis, Rice University, Houston, TX, 1979.

5

THE MICROBIAL CELL

WILLIAM G. CHARACKLIS

Montana State University, Bozeman, Montana

KEVIN C. MARSHALL

University of New South Wales, Kensington, New South Wales, Australia

GORDON A. McFETERS

Montana State University, Bozeman, Montana

SUMMARY

A wide variety of microorganisms are found in biofilms. In this chapter, the diversity and major functions of the predominant microbial groups are presented. Algae are important components of biofilms exposed to light, such as those found in open channels and on rocks in streams. Predatory protozoa play a role in mature biofilms where their continual grazing of bacteria at the surface of the biofilms may keep the bacteria in an active physiological state. Aquatic fungi are important in the colonization and deterioration of wooden structures immersed in water. However, bacteria are the primary colonizing organisms at solid-liquid interfaces and normally constitute the predominant organisms in most biofilms. Extracellular polymeric substances (EPS) produced by bacteria provide the biofilm matrix within which the organisms are embedded.

The types of bacteria found in biofilms depend on the rate and extent of their growth. The extent of growth can be deduced from the stoichiometry of the growth process, which includes the energetics of the overall growth reaction (Chapter 6). The rate of growth is characterized from a kinetic equation in which the limiting nutrient concentration is the extensive variable (Chapter 8). However, the rate and extent of microbial growth in a biofilm also depends on the immediate biofilm environment. Factors characterizing the environment include the nutritional status of the bulk water compartment, the properties of the substratum, position in the biofilm (where nutrient and O_2 transport may be limiting, as described in Chapter 9), and other environmental variables (temperature, etc.).

Heterotrophic bacteria are the most important types found in biofilms, a fact that suggests these organisms are able to obtain sufficient energy substrate from the water flowing past surfaces. The efficiency with which bacteria colonize and grow at surfaces seems to be related to their small size (they occupy the relatively quiescent water in the viscous sublayer), rapid growth rates, adaptability to changing environments (as a result of derepression of genes coding for specific enzymes, mutations, and various modes of genetic exchange, especially by plasmid transfer), and ability to produce extracellular structures or substances (e.g., fimbriae and EPS) that are important in adsorption and aggregation phenomena.

Understanding the behavior of a complex ecosystem such as a biofilm necessitates consideration of a microbial cell and microbial populations as model systems. To do so, certain biological principles have been reviewed. Various control elements exist in these microbial systems and are characterized by significantly different response times; this fact permits simplifications in model equations. For these reasons, bacteria provide ideal model components for the more complex modeling of biofilm processes, as can be seen in Chapters 11 through 13.

5.1 INTRODUCTION

Biofilms consist of two major components: (1) microorganisms that are embedded in (2) a matrix of extracellular polymer substances (EPS) of microbial origin. Such biofilms result from the initial adsorption of microorganisms to solid substrata followed by growth of the microorganisms, EPS production, and the capture or entrapment of other microbial cells from the aqueous phase. The properties of biofilms are presented in Chapter 4. In this chapter, we present an introduction to the types of microorganisms, their structure, and their growth and reproduction as related to biofilm formation.

In order to emphasize the importance of microorganisms in biofilms it should be noted that mature biofilms may contain as many as 10^{16} cells/m^{-3} (10^{13} cells/liter), a cell concentration considerably higher than normally achieved in suspended culture (Chapter 4, Table 4.4). Organisms found in biofilms range from viruses to complex multicellular organisms with specialized organelles. However, bacteria are normally the dominant microorganisms in most biofilms, and consequently the properties of bacteria are emphasized in this chapter.

5.2 DIVERSITY OF MICROORGANISMS

Microorganisms, by definition, are those living organisms whose detailed form can only be seen under a light microscope. Under this definition, we can include the microalgae, fungi, protozoa, bacteria and, with reservations, viruses. This group of organisms spans the range from complex cellular structures (eucaryotes) to simple cellular structures (procaryotes), and to noncellular entities (viruses). Ignoring the viruses for a moment, both the eucaryotic and procaryotic microorganisms possess most of the complex biochemical systems for energy conservation, cell synthesis, and substrate decomposition found in higher organisms. Eucaryotic and procaryotic microorganisms are schematically compared in Figure 5.1.

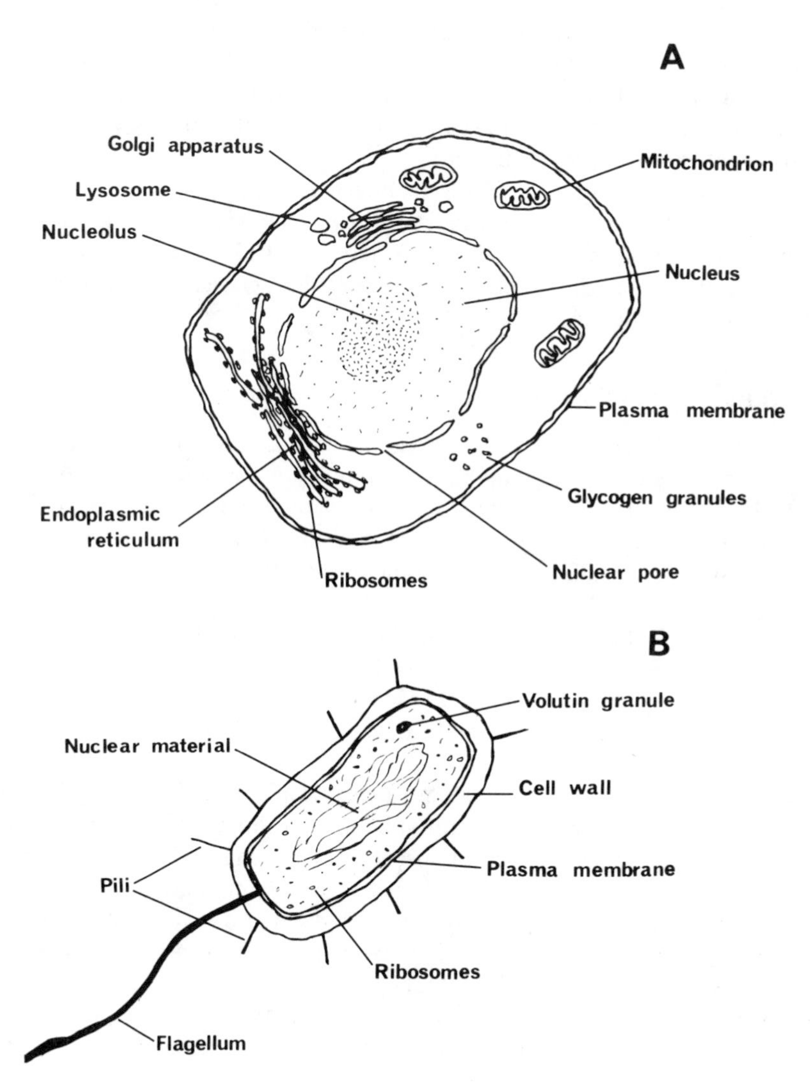

Figure 5.1 Diagrammatic representations of (a) a eucaryotic cell and (b) a procaryotic cell.

5.2.1 Viruses

Viruses are noncellular infectious agents containing genetic material in the form of *either* deoxyribonucleic acid (DNA) *or* ribonucleic acid (RNA). In the extracellular (infectious) state, viruses are so small that they cannot be seen under a light microscope; hence they are termed *submicroscopic*. These virus particles (virions) consist of nucleic acid surrounded by a protein coat (the capsid) and, sometimes, other materials. The virions are metabolically inert, lacking the ability to generate energy or to synthesize new virions. Whether viruses are regarded as living organisms depends on the definition of life itself. The virion functions to carry the viral nucleic acid from an infected host cell to another similar host. The sequence of virus replication in a host cell consists of attachment to the susceptible cell, penetration of the cell by the virus or the nucleic acid, replication of the virus nucleic acid, production of capsid proteins, assembly of new virions, and release of mature virions from the cell.

Viruses of different types are able to infect animals, plants, and bacteria (where they are termed *bacteriophages*, or just phages). They cause a variety of diseases by disrupting the functions of infected cells. In some instances, viruses cause minor changes in the genetic (hereditable) characteristics of the host cells themselves by introducing new genes.

5.2.2 Eucaryotic Microorganisms

The cells of all higher plants and animals, as well as those of the microalgae, fungi, and protozoa frequently found in biofilms, are *eucaryotic* (i.e., possess a true nucleus). Eucaryotic cells possess a well-defined nuclear membrane and chromosomes, and they exhibit mitotic cell division. In addition, certain metabolic functions are localized in membrane-enclosed organelles such as mitochondria (respiratory energy production) and chloroplasts (photosynthesis). The plasma membrane (Figure 5.1) of eucaryotes contains sterols, and there are extensive internal membrane systems (endoplasmic reticulum, Golgi apparatus). Many other differences between eucaryotic and the simpler procaryotic cells exist, and these are presented in detail in Brock et al. (1984).

5.2.2.1 *Microalgae* Some algae are macroscopic (such as seaweeds), but the majority are microscopic, the simplest being unicellular. Algae contain chloroplasts and hence are capable of photosynthesis (the conversion of light energy into chemical energy in the form of reduced organic compounds). The microalgae are responsible for most of the primary production in aquatic habitats and form a major component, although not the primary colonizers, of biofilms in waters exposed to light (Jordan and Staley, 1976).

The major groups of algae are listed in Table 5.1. Microalgae are found in all groups except the Phaeophyta (brown algae, such as the seaweeds).

5.2.2.2 *Fungi* The fungi are a group of eucaryotes that have rigid cell walls and lack chlorophyll. Most fungi have a characteristic filamentous morphology,

TABLE 5.1 Groups of Algae

Division	Common Name	Morphology	Cell Wall	Chlorophylls and Other Pigments
Chlorophyta	Green algae	Unicellular, filamentous, multicellular (plantlike)	Cellulose	a, b
Euglenophyta	Euglenids	Unicellular	No wall	a, b
Pyrrophyta	Dinoflagellates	Unicellular, a few filamentous	Cellulose	a,c, and carotenoids
Chrysophyta	Diatoms, chrysophytes	Unicellular, filamentous	Two overlapping halves, often contain silica	a or c and carotenoids
Phaeophyta	Brown algae	Plantlike multicellular	Cellulose and alginate	a, c, and caretenoids
Rhodophyta	Red algae	Unicellular, plantlike multicellular	Cellulose	a and phycobilins

which is usually branched. Each individual filament is termed a hypha, and a bundle of hyphae is called a mycelium. Yeasts are an exception in that they are unicellular, but under certain growth conditions many yeasts are capable of filament formation. Fungi have been related to corrosion in kerosene fuel tanks, and fungal biofilms are known to attack wooden structures, including cooling tower packing.

The fungi are classified into four main groups as follows:

1. *Phycomycetes* are considered to be primitive fungi, lacking septa (cross walls) in the hyphae. This group consists of terrestrial types that lack motile reproductive cells and aquatic types that have motile reproductive cells. The so-called *water molds*, or aquatic Phycomycetes, are found on the surface of decaying plant or animal debris in waters, and are regarded as nuisance fouling organisms on wooden piles in man-made structures in aquatic habitats.
2. *Ascomycetes* are septate and are characterized by forming spores in a structure termed an ascus. Common ascomycetes include bread molds and yeasts.
3. *Basidiomyctes* are septate fungi forming complex fruiting bodies (as in mushrooms) with spores formed on a differentiated tip of hyphae termed a *basidia*.

4. *Deuteromycetes* (or fungi imperfecti) are septate fungi which lack a recognizable sexual stage in their life cycle. Where a sexual stage of such fungi has been identified, the organism is usually classified into either the ascomycete or the basidiomycete group.

5.2.2.3 *Protozoa* The protozoa are a very diverse group of unicellular, nonphotosynthetic organisms that lack a true cell wall. Protozoa feed either by the uptake of macromolecules by a process termed pinocytosis, whereby a portion of the cell membrane invaginates and pinches off to form an internal membranous vacuole containing macromolecules derived from the external environment, or by phagocytosis, whereby bacteria, algae, and other particles are ingested and subsequently digested as a source of nutrients. Protozoa are frequently observed grazing in biofilms within wastewater treatment facilities (e.g., trickling filters) and, by doing so, influence biofilm accumulation and performance. At the microscopic level, motile protozoa may even influence mass transfer to deeper portions of the biofilm by creating turbulence through their motion within the biofilm.

There are four main groups of protozoa, namely:

1. *Mastigophora:* These are motile by means of one or more flagella, and cell division is by longitudinal binary fission.
2. *Sarcodina:* These are motile by means of pseudopodia (ameboid movement), and cell division is by binary fission. Most are naked, but some produce tests (shells) of calcium carbonate (foraminifera) or of silica (radiolaria).
3. *Sporozoa:* These are a diverse group that are often parasitic and form spores. Usually nonmotile.
4. *Ciliophora:* These are motile by means of numerous cilia providing a coordinated locomotory system. They possess both a macronucleus and a micronucleus, and divide by transverse binary fission.

5.2.3 Procaryotic Microorganisms

Bacterial cells are procaryotic: they do not possess a true nucleus. In these organisms, the DNA is present as a single circular molecule that is neither complexed with histones nor surrounded by a nuclear membrane. Cell division is preceeded by a simple replication of the DNA molecule. The respiratory system is associated with the plasma membrane (see Figure 5.1), and the photosynthetic system, where present, is associated with internal membranous lamellae or vesicles. Again, a detailed comparison of procaryotic and eucaryotic cells may be found in Brock et al. (1984).

Bacteria are now regarded as belonging to an entirely separate kingdom, the Procaryotae, which is divided into four divisions (Bergey, 1984):

Division I—Gracilicutes: Bacteria with a Gram-negative cell wall structure (i.e., stable cell walls with a thin peptidoglycan layer—see Section 5.4.3)

CLASS I—SCOTOBACTERIA: Nonphotosynthetic, Gram-negative bacteria (i.e., most Gram-negative bacteria).

CLASS II—ANOXYPHOTOBACTERIA: Photosynthetic bacteria not evolving oxygen that carry out photosynthesis under anaerobic conditions.

CLASS III—OXYPHOTOBACTERIA: Photosynthetic bacteria (cyanobacteria and *Prochloron*) that evolve oxygen and are able to carry out photosynthesis under aerobic conditions.

Division II—Firmicutes: Bacteria with a Gram-positive cell wall structure (i.e., stable cell walls with a thick peptidoglycan layer).

CLASS I—FIRMIBACTERIA: Simple, nonbranched Gram-positive bacteria.

CLASS II—THALLOBACTERIA: Gram-positive bacteria that show branching at some stage in their life cycle.

Division III—Tenericutes: Highly pleomorphic bacteria without a rigid cell wall. Require a high osmotic pressure for maintenance of cell structure. Commonly known as mycoplasmas. Only one class.

CLASS I—MOLLICUTES: The mycoplasmas.

Division IV—Mendosicutes: Cells lack peptidoglycan, but do possess a rigid cell wall. Gram-positive or Gram-negative. Found in a variety of extreme environments. Have unusual membrane lipids, ribosomes, and RNA sequences (apparently primitive bacterial types). Only one class.

CLASS I—ARCHAEOBACTERIA: Includes methanogenic bacteria, acidothermophiles (*Sulfolobus*), and halophiles (*Halobacterium*).

5.3 PHYSIOLOGICAL GROUPS OF BACTERIA

The procaryotes are also grouped according to the methods by which they capture energy for their metabolism and growth. The main sources of energy utilized by procaryotes include light, reduced inorganic compounds, and reduced organic compounds. Chapter 6 deals with the energetics and stoichiometry of microbial metabolism.

5.3.1 Photosynthetic Bacteria

All photosynthetic bacteria utilize sunlight as a source of energy, but the type of hydrogen donor employed varies according to the physiological type of organism involved.

5.3.1.1 *Anoxyphotobacteria* These bacteria carry out photosynthesis under anaerobic conditions and do not evolve oxygen. Their photosynthetic pigments are known as bacteriochlorophylls, and the hydrogen donor may be H_2S or a reduced low molecular weight organic compound:

1. H_2S, as in the green (e.g., *Chlorobium*) or purple (e.g., *Chromatium*) bacteria. Elemental sulfur is deposited either outside (green sulfur) or inside the cells (purple bacteria). The overall photosynthetic reaction may be represented by

$$2H_2S + CO_2 \xrightarrow{\text{light}} (CH_2O) + 2S + H_2O \qquad [5.1]$$

2. Reduced low molecular weight organic compounds such as ethanol, butyric acid, butanol; etc. In this case, the organic compound is oxidized according to the following reaction (where H_2A represents the reduced organic compound):

$$2H_2A + CO_2 \xrightarrow{\text{light}} (CH_2O) + 2A + H_2O \qquad [5.2]$$

This reaction is carried out particularly by certain purple bacteria (e.g., *Rhodospirillum*).

5.3.1.2 *Oxyphotobacteria* These bacteria use water as the hydrogen donor and evolve oxygen. They possess true chlorophyll *a (cyanobacteria)* or chlorophylls *a* and *b* (*Prochloron*). The generalized reaction for oxygenic photosynthesis is:

$$2H_2O + CO_2 \xrightarrow{\text{light}} (CH_2O) + O_2 + H_2O \qquad [5.3]$$

Certain cyanobacteria (e.g. *Oscillatoria limnetica*) are able to utilize H_2S as the hydrogen donor during photosynthesis under anaerobic conditions (Cohen et al., 1975), and thus provide an evolutionary bridge between the anoxygenic and oxygenic types of photosynthetic procaryotes.

5.3.2 Chemolithotrophic Bacteria

Chemolithotrophic bacteria obtain their energy from the oxidation of inorganic compounds and obtain their carbon for synthesis of organic compounds from CO_2.

Typical *energy* reactions of these organisms follow, with a reference to their biofilm habitat:

Hydrogenomonas (in methanogenic biofilms)

$$H_2 + 0.5O_2 \rightarrow H_2O \qquad [5.4]$$

Nitrosomonas (in nitrifying biofilms)

$$NH_4^+ + 1.5O_2 \rightarrow NO_2^- + 2H^+ + H_2O \quad [5.5]$$

Nitrobacter (in nitrifying biofilms)

$$NO_2^- + 0.5O_2 \rightarrow NO_3^- \quad [5.6]$$

Thiobacillus thiooxidans (biofilms on pyrite ores)

$$HS^- + 2O_2 \rightarrow SO_4^{-2} + H^+ \quad [5.7]$$

Thiobacillus ferrooxidans (biofilms on iron or mild steel pipe)

$$2Fe^{++} + 2H^+ + 0.5O_2 \rightarrow 2Fe^{+++} + H_2O \quad [5.8]$$

In most instances, the activities of chemolithotrophic bacteria are important in altering the availability of the above inorganic compounds in various biofilm ecosystems.

5.3.3 Heterotrophic Bacteria

The majority of bacteria require reduced organic carbon substrates as their energy source, and these are termed heterotrophic (or, more correctly, chemoheterotrophic) bacteria. Within the heterotrophs, we can distinguish physiological groups with respect to their response to oxygen.

5.3.3.1 Aerobic Heterotrophs Many bacteria are able to grow only in the presence of oxygen. These organisms break down complex organic molecules in a series of steps and gain most of their energy during respiration by passing electrons along a series of electron acceptors (the electron transport system, see Stanier et al., 1976) to oxygen, which is reduced to water. In this way chemical energy is converted to the form of adenosine triphosphate (ATP) for use in biosynthetic processes. The yield of ATP from respiration is approximately 38 ATP molecules per molecule of glucose (Brock et al., 1984). Respiration is the most efficient process in energy metabolism, and (apart from the carbon used in cell synthesis) most of the substrate carbon is released as CO_2.

5.3.3.2 Anaerobic Respiration Some respiratory bacteria are able to substitute alternative electron acceptors for oxygen under anaerobic conditions. Nitrate is one of the most common alternative electron acceptors: electrons are passed along the electron transport system to nitrate which is reduced to nitrogen gas. This particular anaerobic process is also termed denitrification. The yield of ATP per glucose molecule is less than in the case where O_2 is the terminal electron acceptor, but anaerobic respiration remains a very efficient energy-yielding process.

5.3.3.3 *Anaerobic Heterotrophs* Some bacteria only grow in the absence of oxygen. These organisms gain their energy from fermentation reactions, which result in the partial breakdown of reduced organic compounds to yield low molecular weight organic products such as ethanol, lactate, acetate, succinate, etc., along with some CO_2 evolution. By comparison with aerobic respiration, fermentation reactions are inefficient, involving no ATP production via the electron transport system. Only two ATP molecules are produced per molecule of glucose (Brock et al., 1984).

Other anaerobes are capable of further decomposing the low molecular weight organic products of fermentation. These are the sulfate-reducing bacteria (SRBs), which at the same time use sulfate as a terminal electron acceptor (but cannot use oxygen at all) to generate energy via the electron transport chain. At lower redox potentials and in the absence of sulfate, the methanogenic bacteria (MBs) also utilize the low molecular weight products of fermentation (including CO_2 and H_2) to obtain energy, producing methane as an end product.

5.3.3.4 *Facultative Anaerobes* There are a number of bacteria that have the best of both worlds. The facultative anaerobes are capable of fermentative processes under anaerobic conditions, yet in the presence of oxygen they are able to utilize an electron transport chain to pass electrons to oxygen and thus behave as respiratory heterotrophs.

5.4 THE BACTERIAL CELL

The majority of bacteria are unicellular and possess rigid cell walls (but see Section 5.2.3), which protect the plasma membrane from sudden changes in osmotic pressure in the external environment. Although most metabolic functions of cells are contained within the plasma membrane, it is the cell wall and extracellular components (those outside the cell wall) that are in direct contact with the external environment and hence play a vital role in biofilm development. Table 5.2 indicates the molecular composition and biosynthetic capabilities of a bacterial cell.

5.4.1 Plasma Membrane

The plasma (or cell) membrane defines the boundary of the metabolically reactive portion of the cell. Although some reactions do occur outside the plasma membrane, none release energy for use by the cell. The membrane thickness is about 8.0 nm (10% of cell dry weight), and it consists mainly of phospholipids and protein. Phospholipids are *amphiphilic* molecules, that is, they possess both hydrophobic and hydrophilic regions in their structure. These molecules disperse in water in such a way that the hydrophobic groups associate together to form a double-layered membrane with hydrophilic outer surfaces masking the hydrophobic groups from the water phase. The membrane lipids in the eubacteria (Gracilicutes, Firmicutes, Tenericutes; see Section 5.2.3) are ester-linked, whereas those of the archaebacteria (Mendosicutes; see Section 5.2.3) are ether-linked. Bacterial membranes lack sterols (found in eucaryote membranes), but often contain the functional equivalents termed hopanoids. The major membrane proteins are hydrophobic and are embedded in the phospholipid matrix. Other hydrophilic proteins are associated with the ionic (polar) groups of the phospholipids. Under the electron microscope, thin sections stained with heavy metals reveal two electron-dense outer layers separated by an electron-light layer—a structure termed a *unit membrane*.

The plasma membrane functions as a selectively permeable barrier preventing the passive movement of polar solutes through the membrane. Active transport mechanisms, sometimes involving transmembrane proteins (permeases), are required for the transport of nutrients and products in either direction in a highly specific manner. Fat-soluble substances may penetrate the membrane by becoming dissolved in the lipid phase of the membrane. The enzyme systems responsible for electron transport to oxygen and oxidative phosphorylation (ATP production) are located on the inner surface of the plasma membrane. Damage to the plasma membrane results in the leakage of internal cell contents and usually causes death of the cell.

5.4.2 Inside the Cell

The internal content of the cell, the *cytoplasm,* contains the ribosomes (composed of RNA and protein), DNA, and the enzymes necessary for the metabolic processes of the cell. The ribosomes are the site of protein (including enzyme) synthesis. They are densely packed granules (10–20 nm) containing 90% of the cell RNA and constitute approximately 40% of the cell dry weight.

5.4.3 Cell Walls

Most bacterial cells are bounded by a rigid cell wall (but see Section 5.2.3) that is responsible for the characteristic morphology of the cell and for its structural integrity in variable osmotic conditions or in the presence of high local shear stress. The rigid component of the walls of most eubacteria is peptidoglycan (Figure 5.2). The results of the most common staining procedure used in bacteriology, the Gram stain, depend on the relative thickness of the peptidoglycan layer in the bacterial cell walls. The Gram stain, in essence, consists of staining the cells with crystal violet, mordanting with iodine, decolorizing with alcohol (or acetone), and counterstaining with dilute carbol fuchsin. Cells that appear blue (i.e. retain the first stain even after the alcohol treatment) under the microscope are termed Gram-positive, and those that appear red (i.e. lose the blue stain and stain with carbol fuchsin) are termed Gram-negative.

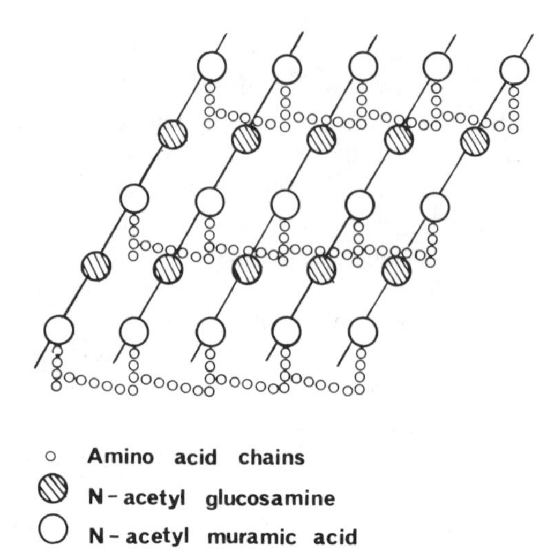

Figure 5.2 Diagrammatic representation of the peptidoglycan layer of a Gram-positive bacterium, showing the amino acid cross links between chains of N-acetylglucosamine and N-acetylmuramic acid.

Peptidoglycan accounts for 30 to 50% of the dry weight of isolated cell walls of Gram-positive bacteria. Other secondary cell wall polymers in these bacteria include teichoic acids, teichuronic acids, neutral polysaccharides and sometimes protein. The Gram-positive cell wall, as seen under the electron microscope, is relatively thick and simple in its structure, whereas that of the Gram-negative cell is more complex (Figure 5.3). The peptidoglycan layer in Gram-negative bacteria represents only 2 to 10% of the dry weight of the wall and varies from 3 to 10 nm in thickness. Between the plasma membrane and the peptidoglycan layer is a space, the periplasmic space, that contains some hydrolytic enzymes. These enzymes play a role in substrate utilization and in the postprocessing of secreted materials. Outside the peptidoglycan layer of Gram-negative bacteria is the outer membrane, also a unit membrane, which consists of phospholipid (25%), lipopolysaccharide (30%), and protein (45%). This outer membrane is an atypical membrane in that it acts as a permeability barrier to hydrophobic substances. This property results from the asymmetry of the outer membrane in that all of the lipopolysaccharide is found on the outer side of the bilayer while the phospholipid is found on the inner side. A detailed account of Gram-positive and Gram-negative cell wall structures has been presented by Wicken (1985).

5.4.4 Extracellular Components

Many bacteria are able to produce extracellular polymeric substances (EPS), and these appear in a variety of forms (Chapter 4). Consequently, various terms are used to describe these polymers, such as slime layers and capsules. Costerton

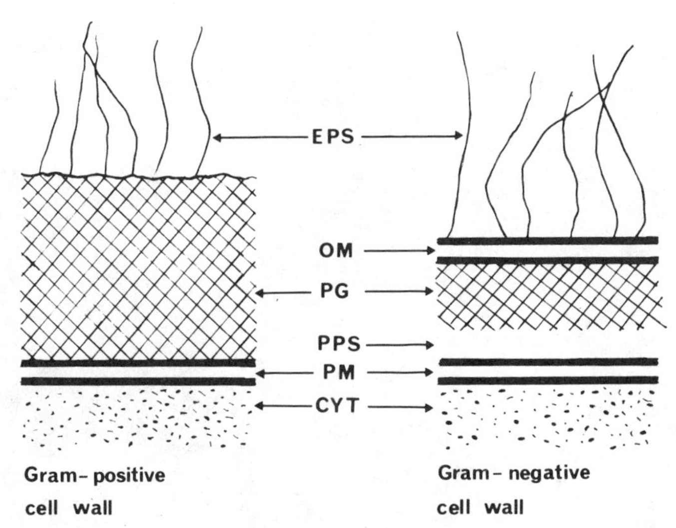

Figure 5.3 Diagrammatic representations of Gram-positive and Gram-negative cell wall structures. EPS, extracellular polymeric substances; OM, outer membrane; PG, peptidoglycan; PPS, periplasmic space; PM, plasma membrane; CYT, cytoplasm.

et al. (1978) have used the term *bacterial glycocalyx* to describe such polymers. In most cases, the polymers are homo- or heteropolysaccharides, and then the term *glycocalyx* is satisfactory. In some bacteria, however, the extracellular polymer is protein or glycoprotein.

The extracellular polymer may remain firmly attached as a discrete covering layer of each cell, or it may part freely from the cell as it is formed. In the former case, it is called a capsule; in the latter, free slime or the slime layer. Capsules and slime may be distinct from the morphological and biochemical standpoint (Chapter 4). The capsule is part of the cell; the slime is a secretion. Capsules are of definite shape and of more or less definite density throughout, and the outer capsule interface is quite distinct. The capsule may be several times thicker than the cell; it supplements the protection of the cell wall. The slime layers, in contrast, are rather amorphous; they are concentrated in the vicinity of the cells and decrease in density with increasing distance from a cell.

Bacteria adsorb to many surfaces, including the human tooth and lung, as well as ship hulls and pipe surfaces. The adsorption may be mediated by the mass of tangled polymeric fibers that extend from the cell surface surrounding an individual cell (Fletcher and Floodgate, 1973; Marshall and Cruickshank, 1983). The capsule and slime material also play a major role in the aggregation of bacteria in flocs and films.

Leppard (1986) and Bakke have presented electron microscopic evidence (Figure 5.4) for fibrils in biofilms as thin as 5 nm and claim that as much as 90% of the organic extracellular matrix is composed of colloidal, electron-opaque fibrils. Robinson et al. (1985) indicate that the protoplast of *Methano-*

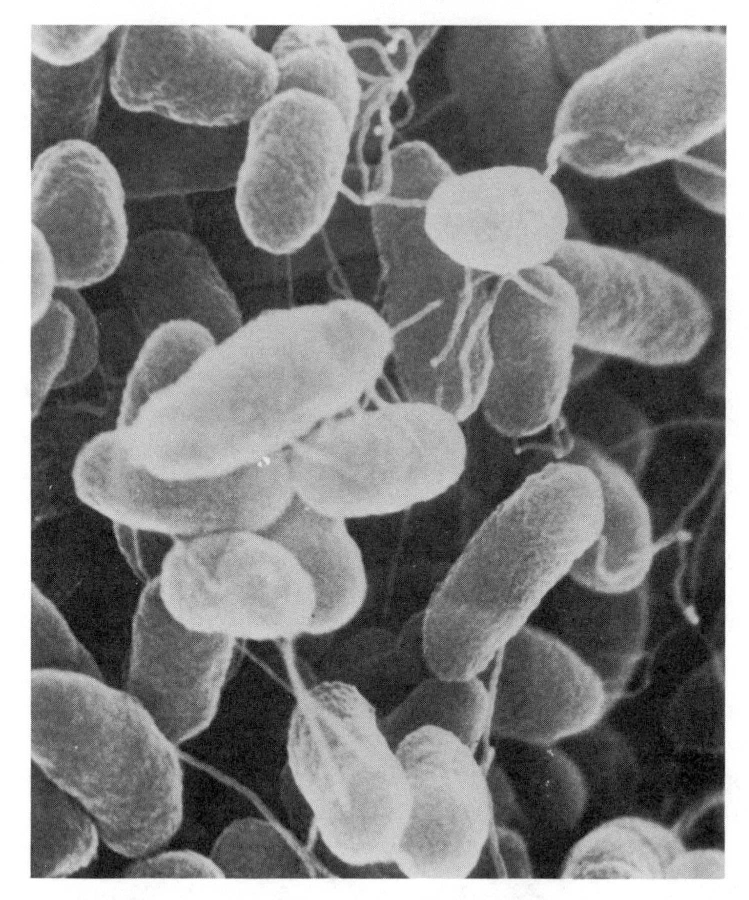

Figure 5.4 A scanning electron micrograph of a *Pseudomonas aeruginosa* biofilm displaying the fibrillar nature of the extracellular material (courtesy of R. Bakke). The fibrils are approximately 30 nm in diameter.

sarcina mazei is bounded by a typical plasma membrane outside of which is a matrix of loose fibrils. The matrix is 30–60 nm thick and is 95% polysaccharide. The presence and compactness of the matrix are responsible for aggregation of these cells, and colony disaggregation seems to be the result of matrix shedding and degradation.

Cell surface hydrophobicity appears as an important factor in the adsorption and aggregation of bacteria (Rosenberg and Kjelleberg, 1986). It is likely that proteins and amphipathic molecules in surface components contribute to the relative surface hydrophobicity of various bacteria.

5.4.5 Surface Appendages

Many bacteria produce a variety of surface appendages that often play a vital role in motility, buoyancy, adsorption, aggregation, and other functions.

5.4.5.1 *Flagella* Many bacteria (motile bacteria) are capable of rapid motion in a liquid, primarily with flagella that are anchored in the plasma membrane. In quiescent or very slow flow conditions, motility is the major mechanism for cell transport to and from a substratum, and therefore is important in the cell adsorption process. Indeed, the flagellum may be the anchor structure for adsorption in some cases. The flagella, consisting of protein subunits, are only 10–20 nm in diameter, but may be as long as 15–20 μm. The bacteria may move at speeds as high as 50 μm s^{-1}, approximately 20 times their own length every second. There may be more than one flagellum per cell; a cell with many flagella (located evenly around the cell) is termed *peritrichous*. Motility can be guided in a semisolid medium if the motile bacteria move in a gradient of limiting nutrient concentration to a region of optimum concentration. This movement in response to an external influence is termed *taxis* or, if the influence is nutritional or chemical, *chemotaxis*. Such a response to air is termed *aerotaxis* and to light, *phototaxis*.

Flagella are not the only means for motility. Movement along a surface by gliding bacteria, analogous to Stefan adhesion, has been reported in which the organisms move on the substratum while remaining adsorbed by EPS (Humphrey et al., 1979).

5.4.5.2 *Sex Pili and Fimbriae* There are two recognized groups of nonflagellar proteinaceous surface structures: (1) fimbriae, which are adhesive factors, and (2) sex pili, which are involved in cell to cell contacts that result in conjugation. These surface structures do not confer motility to the cell and are not essential to growth under normal conditions. Both pili and fimbriae are generally shorter and thinner than flagella. These appendages are 10 μm or more in length.

Fimbriae have been implicated in bacterial attachment to specific organism surfaces and thus may perform an important ecological function in natural habitats. Pili play a major role in genetic exchange between bacteria. In a biofilm, there is ample opportunity for genetic exchange between immobile cells in close proximity. Thus, pili may be instrumental in determining genetic activity within a biofilm.

5.4.5.3 *Prosthecae* The stalk like structures called *prosthecae* are extensions of the cell wall and plasma membrane. Prosthecae appear to support a variety of functions, including the budding of daughter cells (*Hyphomicrobium*), adsorption to surfaces (*Caulobacter*), increased surface area for nutrient absorption, and increased cell buoyancy.

5.5 OTHER CHARACTERISTICS OF BACTERIAL CELLS

5.5.1 Cell Composition

The general composition can be approximated by the ratio of carbon to nitrogen to phosphorus (C:N:P), which is approximately 100:20:5 for carbon-limited cells (Bratbak, 1985). The carbon:hydrogen:nitrogen ratios (C:H:N) for *Pseudomonas* $C_{12}B$ and *Klebsiella aerogenes* grown under one set of environmental conditions (carbon-limited) are 100:15:28 and 100:15:25 (Chapter 4, Table 4.4). However, the presence of capsules or a slime layer may significantly influence the cellular composition determined by traditional techniques. Since both capsules and extracellular polymers are mainly polysaccharide (Fletcher and Marshall, 1982; Stanier et al., 1976), they may lead to an overestimate of the carbon per cell even though the amount of nitrogen and phosphorus per cell is unaffected. Bratbak (1985) has estimated that the carbon content of one bacterial cell is approximately 1×10^{-13} g.

5.5.2 Morphology

Size and shape are important as identifying characteristics of bacteria, and since these parameters determine the surface area to volume ratio, they influence metabolic rates and sensitivity to environmental conditions. Bacteria exhibit three general fundamental shapes:

1. spheres, or *cocci* (singular, coccus)
2. rods, or *bacilli* (singular, bacillus)
3. rod-shaped spirals, or *spirilla* (singular, spirillum)

The spherical bacteria divide in one (e.g., streptococci), two (e.g., staphylococci), or three dimensions or planes (e.g., sarcinae). Division in one plane results in single cells, pairs of cells, or chains. Division in irregular planes results in clusters, and division in three planes results in packets. The rod is usually a cylinder with the ends more or less rounded. Filamentous forms are also frequently observed, consisting of linear multicellular extensions.

The shape depends to a large extent upon certain environmental factors such as temperature, culture age, concentration of limiting nutrient, and composition of the medium. Rod-shaped and filamentous organisms are hypothesized to increase efficiency of nutrient uptake, since most bacteria found in lakes and oceans, where the concentration of organic material is low (oligotrophic environments), are filamentous or rod-shaped (Brock, 1966). Conventional rod-shaped pseudomonads have been found to produce long filaments under certain conditions, including biofilm environments (McCoy and Costerton, 1982). Thus, cellular size and shape are dependent on physiological state. In a biofilm, size and/or shape may also vary with biofilm depth, since concentration gradients and thus physiological state may vary.

5.5.3 Size and Specific Gravity

Bacteria show great variation in size, but none can be seen without the aid of a microscope. The majority of the commonly encountered bacteria measure about 0.5 μm in diameter for spherical cells and 0.5 by 2 to 3 μm for the rod forms.

Assuming a coccus form with a 1 μm diameter, a cubic centimeter (a milliliter) contains 1×10^{12} cells. The surface area of one of these cells is 4.2 μm^2. [Borzani (1961) reports that the surface area of a *Sarcina lutea* cell is approximately 4.5 μm^2.] The surface area of 1×10^{12} cells is then 3.1 m^2, and the surface area to volume ratio is 3,100,000 m^{-1}. A rod 1 $\mu m \times 3$ μm has a volume of 2.36 μm^3, and 4×10^{11} cells would occupy a cubic centimeter. The total surface area of all these cells is 3.8 m^2, and the surface area to volume ratio is 3,900,000 m^{-1}.

Growth conditions can have a profound effect on cell properties. For example, a starved cell may have a diameter as small as 0.2 μm.

The specific gravity of bacterial cells is about 1.07, not very different from water. The actual value may depend on the proportion of substances present in the cell and their specific gravities. Such cellular substances include proteins (1.5), carbohydrates (1.4–1.6), nucleic acids (2.0), and mineral salts (2.5). Kubitschek et al., (1984) determined that the specific gravity of *E. coli* is independent of growth rate and remains unchanged over extended periods. The measured specific gravity was approximately 1.09 with an error much less than 1%. The presence of lipid material or gas vacuoles (cyanobacteria) will reduce the specific gravity. The dry weight of an individual cell is approximately $1.7–68 \times 10^{-13}$ g (Gray et al., 1974; Holme, 1957). Thus, a cell should settle through a water column, since it is more dense than water. However, the rate of fall is a function of the frictional force resisting the fall as well as the cell density and fluid viscosity (Stokes's law), and it may be negligibly small. Objects more dense than water may even float if they are supported by surface tension. Thus, some aerobic bacteria grow on the surface of a liquid medium by virtue of their surface properties and form a film or pellicle.

5.5.4 Mechanical Strength of Microorganisms

Microorganisms are frequently subjected to high osmotic or hydrostatic pressure in biofilm environments as a result of concentration grȧdients in the film. Microbial cells are also subjected to shear force when suspensions are pumped and when adsorbed to a substratum in a turbulent flow. Such forces are often regarded as a possible cause of protein denaturation and loss of microbial activity.

In a microbial cell, the proteins, lipids, and nucleic acid are contained within a fragile semipermeable membrane, which is protected from rupture by a strong, rigid outer cell wall. The cell wall is responsible for structural integrity even in solutions of high or low osmotic pressure or in the presence of high local shear stress. Adsorbed and biofilm microorganisms frequently experience high osmotic pressures and high fluid shear stress. Changes in their physiology under these stresses may cause unusual behavior and deserve further research effort.

5.5.4.1 *Osmotic Pressure* The osmotic pressure within a biofilm may change dramatically with depth, although measurements have not been reported. Certainly the osmotic pressure changes in the vicinity of some interfaces, such as clay–water interfaces, where microbial adsorption occurs. Most microbial environments have solute concentrations considerably lower than those within the cell. Therefore, the cell maintains a high internal osmotic pressure. Water passes from a region of low solute concentration to one of high solute concentration (i.e., the cell) by a process called osmosis, and the increased water content in the cell results in higher pressure within the cell. The microorganisms protect themselves by a rigid cell wall that permits high internal osmotic pressure to build up without causing lysis. Gram-positive cells can have internal osmotic pressure of 20 atm or higher, and Gram-negative cells have internal osmotic pressure of about 10–15 atm. For example, *Staphylococcus aureus* has been found to have an osmotic pressure of 20–25 atm (Slayter, 1967).

In contrast, cells that grow in high solute concentration (e.g., *Halobacterium*) must perform more work to extract water from a solution, which usually results in a lower growth yield or a lower growth rate (Brock, 1979).

5.5.4.2 *Hydrostatic Pressure* Bacteria and fungi can live under very high pressure, due to water (the ocean bottom) or other loads (deep oil or sulfur wells and coal deposits). Viable bacteria are recovered from environments where the water depth exceeds 10,000 m and the pressure is up to 1140 atm.

Growth of microorganisms at high pressure is as much as 1000 times slower than at optimal conditions (1 atm, 25°C). For example, *Pseudomonas bathycetes* at 1000 atm and 3°C has a lag phase of 33 days and a generation time of 33 days. The mechanisms by which an organism can withstand such high pressure are not clearly understood.

5.6 BIOENERGETICS

Bioenergetics describes the transformations of energy that occur via cellular metabolism. Metabolism consists of catabolism and anabolism. *Catabolism* is the manipulation and partial conservation of energy within cells, while in *anabolism* energy is used for synthetic processes required for cellular growth. Since bacteria, as well as other organisms, obey the laws of thermodynamics that relate to the flow of energy, living bacteria can be defined in bioenergetic terms as follows:

Bacteria are living *if they exist at a lower state of entropy than their surroundings. They have a capacity to use external energy to maintain their low entropic state.*

Bacteria must constantly maintain the process of energy metabolism to conserve their structure and viability. In this sense, bacteria are open systems and depend on a continuous supply of an adequate level of a usable energy source for their viability.

Because of their metabolic diversity bacteria are able to use a variety of different energy sources: organic compounds, reduced inorganic compounds, and visible light can be used by different types of bacteria. Because of this diversity, bacteria are found in a variety of environments, including some so extreme that few other organisms can reproduce there. Biofilms can also tolerate a range of various physical and chemical characteristics that might restrict the energy metabolism of higher organisms. The metabolic capabilities of biofilm bacteria allow them to grow in these environments.

5.6.1 Bioenergetic Mechanisms

Bacterial energy metabolism depends largely upon oxidation–reduction reactions. Such metabolic schemes require the input of reduced energy donors or electron donors, which are oxidized, whereupon some of the available energy is stored by the bacterium in a useful form. Broadly speaking, the energy metabolism of bacteria is classified as aerobic or anaerobic depending on whether oxygen is used as the terminal electron acceptor. However, some bacteria, termed *facultative,* can metabolize both aerobically and anaerobically. Aerobic pathways require oxygen, while anaerobic ones can use a wide variety of inorganic electron acceptors. In addition, some bacteria utilize organic electron acceptors in a process that is termed *fermentation*. (See Section 5.3.3)

All of these bioenergetic schemes result in the conservation of a portion of the available energy from the electron source in the form of ATP. The electron transport pathway (ETP), located within the cytoplasmic membrane, couples anabolism and the synthesis of ATP. In the ETP, a series of oxidoreductase enzymes remove energy from the electrons taken from the electron donors by oxidation and transform it into a transmembrane form of energy, called the *membrane potential*. The membrane potential has two components, (1) due to a charge separation and (2) a differential concentration of protons across the membrane. The membrane potential is used to fuel a number of energy-requiring bacterial processes, such as motility and active transport of nutrients, in addition to the synthesis of ATP. This description is termed the *chemiosmotic model* of energy conservation and has been generally accepted only recently.

The amount of energy that is conserved by bacteria per mole of electron donor oxidized is highly variable. Bacteria utilizing the aerobic catabolism of glucose to carbon dioxide are relatively efficient, while fermentation and some forms of anaerobic energy metabolism have a much lower yield of useful energy per mole of substrate catabolized (Chapter 6). Thus, anaerobic and fermenting bacteria must metabolize greater quantities of energy sources to sustain growth, and generally grow more slowly. On the other hand, such bacteria can often successfully occupy habitats where there is little competition.

5.6.2 Energy Flow through Biofilms

5.6.2.1 ***Product Formation*** Bacteria within biofilm communities employ the same range of energy-related metabolic processes as their planktonic counterparts. However, the physical and chemical microenvironments within biofilms impose constraints on the metabolism of the bacteria in those niches.

For instance, the deepest regions of some biofilms are anaerobic and provide a suitable habitat for bacteria that are facultative, fermentative, or anaerobic. Those zones may also be enriched in inorganic ions when the biofilm is formed on metal surfaces. The water–biofilm interface of the same biofilm, on the other hand, may represent an aerobic niche subject to significant fluid shear forces. There is good evidence that products of fermentation or other anaerobic metabolism excreted deep in the biofilm serve as substrate for the cells in the aerobic layer (Schaftel, 1982). Thus, the bioenergetics of a biofilm is of major concern and must be distinguished clearly from bioenergetics of the individual cells within it.

5.6.2.2 *Electromagnetic Effects* Bakke (1983), experimenting with *Pseudomonas aeruginosa* and mixed populations, observed some interesting biofilm phenomena possibly related to energy transfer mechanisms. The substrate flux to the biofilm prior to any substrate transient (the turnover rate of substrate by the biofilm organisms) was on the order of 2.2×10^5 lactate molecules per cell per second (compare with Table 5.2). The substrate flux is coupled to a proton flux across the cell membrane, since each lactate molecule requires one or more protons (Otto et al, 1980). An active cell membrane is characterized by a strong electric field ($\approx 10^7$ V m^{-1}) due to charge separation, as described by the chemiosmotic theory (Mitchell, 1961, 1966). As a consequence, variations in electric and magnetic fields can be expected around cells when proton flux changes are imposed across the membrane. Proton flux changes in yeast cells have been observed with nuclear magnetic resonance spectra of ^{31}P (den Hollander et al., 1981). The increased substrate flux to the yeast cells was accompanied by an initial drop in membrane potential, followed by a strong increase 3 min later. A dramatic consequence of such stimuli is "substrate-accelerated

TABLE 5.2 The Biosynthetic Capabilities of a Bacterial Cell[a]

Chemical component	Percent of dry weight	Approximate molecular weight	Number of molecules per cell	Number of molecules synthesized per second	Number of molecules of ATP required to synthesize per second	Percent of total synthetic energy required
DNA	5	2,000,000,000	1	0.00083	60,000	2.5
RNA	10	1,000,000	15,000	12.5	75,000	3.1
Protein	70	60,000	1,700,000	1,400	2,120,000	88.0
Lipids	10	1,000	15,000,000	12,500	87,500	3.7
Polysaccharides	5	200,000	39,000	32.5	65,000	2.7

[a] Reprinted with permission from Lehninger (1971). *Escherichia coli* is about $1 \times 1 \times 3$ μm in size; it has a volume of 2.25 μm^3, a total weight of 10×10^{13} g, and a dry weight of 2.5×10^{-13} g. The rates of biosynthesis were averaged over a 20 minute cell division cycle.

death," where the membrane potential collapses as a result of increased substrate flux (Dykhuizen and Hartl, 1978).

Bakke stimulated variations in proton flux across the membrane of the biofilm cells by imposing step increases in substrate loading. The cells were able to step up their substrate turnover rate almost instantaneously to accommodate the increase in substrate loading. Massive sloughing of biofilm material was observed immediately after the substrate transient. Interestingly, the biofilm cell numbers did not change as a result of the sloughing, which suggests that most of the sloughed biofilm material was EPS. The sloughing may have occurred as a result of biofilm density changes amplified by the shear stress on the biofilm (3 N m^{-2}). Bakke estimates that the electric field strength due to a 40% increase in substrate flux could change by as much as 10^4 V m^{-1} within 5 ms. The rapid change in electric field strength may have a dramatic effect on the EPS of the biofilm. For example, Tanaka et al. (1982) have observed a hundredfold increase in piezoelectric polymer gel density due to an increase of 10^2 V m^{-1} in electric field strength. The change in polymer gel density, if it occurs in the biofilm, may influence many biofilm properties, including diffusivity.

The electrochemical characterization of the microenvironment experienced by a cell in the biofilm may lead to the explanation of many phenomena observed in biofilms. However, methods for such observations are still in a primitive state.

5.7 MICROBIAL GROWTH

The term *growth,* as commonly applied to bacteria and other microorganisms, usually refers to changes in the entire crop of cells rather than to change in the individual organism. More frequently than not, the inoculum contains thousands of organisms; thus growth denotes the increase in number or mass beyond that present in the original inoculum. When bacteria are inoculated into a medium containing all requirements for growth and incubated under appropriate conditions, a tremendous increase in numbers occurs within a short time. With some species the maximum population is reached within 24 hours, whereas others require a much longer period of incubation.

5.7.1 Reproductive Processes

The most common process for reproduction in bacteria is that known as *binary fission,* or *transverse fission*. The terminal events of this reproductive process find the single cell dividing into two, after the development of a transverse cell wall to separate the intracellular contents, as shown schematically in Figure 5.5. This is an asexual reproductive process. The exact morphological transitions leading up to the event of binary fission are not clearly understood. However, if we start with a single viable bacterium in a growing culture, we can postulate the

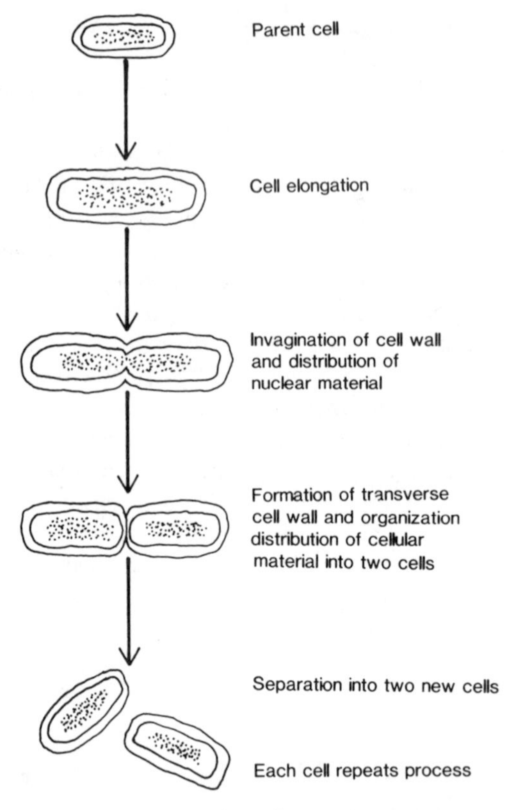

Figure 5.5 Diagrammatic representation of bacterial multiplication by binary fission.

following developments: The nutrients from the medium are taken into the cell by a selective process. The enzyme systems of the bacterium then convert the assimilated chemicals (nutrients) into macromolecules characteristic of the particular organism. A doubling of DNA occurs, and cell elongation may follow (this is more evident in bacilli than in cocci). The contents of the cell undergo reorganization to distribute the material for two cells, which are formed by a transverse wall, or septum, that subsequently develops by an invagination of the cytoplasmic membrane.

Another means of reproduction observed in some bacteria (e.g., Actinomycetales) is the formation of a filamentous growth followed by fragmentation into small units, which then develop into cells of normal size. Again, some bacteria are capable of reproducing by budding. An outgrowth, or bud, develops from the parent cell and after a period of enlargement separates from the parent as a new cell.

5.7.2 Specific Growth Rate

The prevailing means of bacterial reproduction, as already indicated, is binary fission: one cell divides, producing two new cells. Thus, if we start with a single bacterium, the increase in population is by geometric progression: 1, 2, 4, 8, 16, 32, etc. The time interval required for the cell to divide, or for the population to double, is known as the *mean generation time*. Not all bacteria have the same generation time; for some, such as *Escherichia coli,* it may be as little as 15 to 20 min; for others it may be several hours (Table 5.3). Similarly, the generation time is not the same for a particular bacterium under all conditions. The amount and kind of nutrient available in the medium and the specific environmental conditions influence the generation time.

5.7.3 Starvation Survival and Growth at Surfaces

Bacteria growing in biofilms may be copiotrophs, that is, they require high levels of energy substrate for growth (Poindexter, 1981). In very low nutrient waters (oligotrophic environments), copiotrophic bacteria tend to starve. In order to survive such conditions, these bacteria resort to size reduction (0.2 to 0.5 μm diameter) and significantly lowered endogenous metabolism (utilization of internal organic substrates). This process has been termed *starvation survival* (Morita, 1982). ZoBell (1943) postulated that the advantage to bacteria of colonizing surfaces was the concentration at the surfaces of macromolecules and other molecules that could serve as nutrient sources for the bacteria. Starved bacteria have been reported to be more adhesive than well-fed bacteria (Dawson et al., 1981), and it has been demonstrated that starved bacteria in oligotrophic conditions are able to utilize energy substrates bound at surfaces for growth and reproduction (Kjelleberg et al., 1982; Power and Marshall, 1988). In the study by Kjelleberg et al. (1982), the marine *Vibrio* DW1 had a generation time of 37 min in a nutrient-rich medium. It did not grow in the aqueous phase of an oligo-

TABLE 5.3 Generation Times of Several Microbial Species

Species	Medium	Temp. (°C)	Reactor	Generation Time (h)
Bacillus mycoides	Synthetic	37	Batch	0.5
Bacillus thermophylus	Synthetic	55	Batch	0.3
Escherichia coli	Synthetic	37	Batch	0.3
Lactobacillus acidophilus	Milk	37	Batch	1-1.5
Mycobacterium tuberculosis	Synthetic	37	Batch	13-15
Staphylococcus aureus	Synthetic	37	Batch	0.5
Streptococcus lactis	Milk	37	Batch	0.5
Pseudomonas aeruginosa	Glucose	25	Chemostat	1.5

trophic medium, but showed a mean generation time of about 57 min at a surface exposed to the oligotrophic medium.

5.8 MICROORGANISMS AS MODEL SYSTEMS

A major focus of this text is modeling of biofilm processes. A fundamental prerequisite to model development is definition of purpose. The model structure (i.e., mathematical format and degree of abstraction, level of temporal and spatial detail, and requirement for supporting physical rationale) should be tailored to the intended use. Modeling biofilm processes has as its focus understanding the *integrated* functioning of the cellular populations within the biofilm. As a prelude, the microorganism, the most essential component of the biofilm, will be considered as a model system.

5.8.1 The Cell

Very roughly, a functional cell can be regarded simply as a population of interacting chemical species. The production and consumption of the various chemical species are controlled by the familiar rate laws of chemistry, sometimes referred to as the *law of mass action*. Although this view of the microbe is useful, a better approximation is obtained by regarding the cell as a population of interacting chemical species in which the reactions are mediated by specific catalysts or enzymes. A suitable family of such enzymes converts a rather chaotic collection of interacting species into a *reaction network*.

An even better approximation arises from the realization that some of the enzymes in the biochemical network are also *control elements*. Their functioning rate is dependent on the concentration of various chemical species in the cell milieu. The presence of these control elements converts the reaction network into an "information-processing" network that permits the integration of various specific reaction pathways into an adaptive biological system.

Finally, we must recognize that a cell is not simply a reaction network controlled by enzymes, but also involves a crucial genetic constituent. The genes represent another kind of reaction network superimposed on the enzymatic one, and subject to similar kinds of control. Thus, the repression of a gene and the inhibition of an enzyme can be treated in a formally identical manner. In the same way, mutation in an axenic culture and genetic transfer in a mixed population represent analogies to adaptive (as opposed to constitutive) enzymes.

5.8.2 The Cell Population

Rarely in the modeling process is the individual cell considered. Rather, the population is the focus of the modeling. A population is a collection of interacting cells that, in some sense, constitutes a system. The interacting cells are considered subsystems. In this way, nature can be described as a hierarchy of systems. A biofilm cell population is a collection of cells, usually consisting of many different species, frequently segregated in colonies within the biofilm. Then, each species may be considered a subsystem of the biofilm, and each individual cell a subsystem of its species within the biofilm. Carrying the illustration a step further, a specific reaction pathway is a subsystem of a single cell. An important line of questions arise regarding the control of this hierarchy of systems, starting with intracellular control.

5.8.3 Intracellular Control Systems

We have implicitly described three types of control exerted by the microbial cell or population(s) of microbial cells:

1. *The law of mass action* refers to the familiar rate laws, such as the Michaelis–Menten equation (Monod equation for microbial growth), which describes the rate of an enzyme reaction as a function of substrate concentration. If concentration drops significantly, the rate of reaction decreases. The response time for this type of control is very short. For example, a biofilm responds instantaneously to a step increase in substrate concentration by increasing its specific activity and sometimes by sloughing major portions into the bulk liquid compartment (Bakke, 1983).
2. *Control of enzyme activity* is accomplished by feedback or feedforward mechanisms through the mediation of inhibitors or activators. The response time is longer than that due to mass action, but it is generally less than a generation time. Thus, an effective inhibitor added to control biofouling exerts its effect rapidly.

Control of enzyme synthesis is accomplished genetically by the action of repressors or derepressors. The time necessary for action is on the order of a several generation time. As a result, attaining true steady state in a biofilm, even a monospecies biofilm (Bakke, 1986), may require weeks of operation at constant environmental conditions.

One of the challenges of modeling microbial biofilms is the development of sufficiently simple mathematical expressions that describe the behavior of cells with such sophisticated control systems. The cell *populations* are also subjected to ecological controls, which influence the population distributions in biofilms. Complex models are frequently not useful. For the most part, model validation derives more from long-term accumulation of phenomenological evidence than from mechanistic rigor. However, a physical rationale can be used in support of mathematical functions and to define (or extend) appropriate models in instances where the current state of knowledge or available data are insufficient for model definition.

5.8.4 Open versus Closed Systems

One of the characteristic features of biological systems (e.g., cells), in general, is that they are *open*, i.e., they are characterized by a high degree of interaction with their environment. In a thermodynamic sense, they exchange mass and energy with their environment. As a result, classical thermodynamic principles do not rigorously apply although they have been used with good success in restricted but important cases (Roels, 1983; this book, Chapter 6). Thus, closed experimental systems are often inappropriate for modeling biofilm processes, whether in technology or in ecology.

5.9 BIOLOGICAL PRINCIPLES

There are at least four biological principles that are of general importance in model building according to Frederickson et al. (1970).

5.9.1 Phenotype and Genotype

The first principle is physiological and ecological; it states that the activities of an organism (phenotype) and rates at which these activities are conducted are dependent not only on the hereditary characterization of the organism (genotype), but also on its environment. It is rare that an organism fully expresses its genotype at any one time. A corollary of this principle is that the constitution or state of the environment depends upon the activities of organisms contained in it. This principle and its corollary are explicitly incorporated in most biofilm models. For example, the concept of *limiting nutrient* describes an influence of the environment on the organisms (the limiting nutrient controls metabolism) and an influence of the organism on the environment (the organism consumes the limiting nutrient, thus altering the state of the environment).

5.9.2 Past History

The second principle is drawn from ecology and genetics. The current phenotype (constitution or state) of an organism depends not only on its genotype but also on the past history of environments experienced by the organism. Thus, cultivating an organism in a changing environment (e.g., batch culture), one may observe differing physiological function and morphologic structure within the same genotype. Although this principle is recognized, it is seldom incorporated satisfactorily in microbial process models. If this principle is incorporated, the model is said to be *structured;* otherwise, it is unstructured (Chapter 2).

5.9.3 Taxonomic Classification

The third principle states that organisms can be classified on the basis of their morphology, growth form, and mode of reproduction. Thus, pseudomonads are unicellular and generally reproduce by binary fission; they may be distinguished from *Sphaerotilus* sp., which may be filamentous or multicellular, exhibit apical growth, and reproduce in several different ways. The implications of this principle are frequently not realized when modeling microbial processes. For example, the same model is frequently used for describing growth of *Pseudomonas* sp. and *Sphaerotilus* sp.

In biofilm process analysis, the *function* of the organism in a specific environment is of much greater interest than its taxonomic classification. The classification of strains into various categories is a matter of personal judgement as well as scientific method.

5.9.4 Mutation and Genetic Transfer

The fourth principle is the ability of organisms to alter their genotype. Microorganisms mutate on the order of every 10^8 divisions. Indeed, the first chemostats were developed for genetic studies to capitalize on this rate of mutation in rapidly growing populations (Monod, 1950; Novick and Szilard, 1950). For over thirty years, genetic characteristics of bacteria were believed to be relatively static except for infrequent "beneficial" mutations. We now accept that bacteria and bacterial viruses readily exchange genetic material, which leads to a greater variety of potential genetic characteristics and/or genotypes. In addition, environmental conditions, including those imposed by man, select for traits that are most favorable for survival of the bacterium.

5.9.4.1 *Mechanism of Gene Transfer* Bacteria and viruses exchange genetic material naturally by three mechanisms. These processes usually involve two cell types (donor and recipient) that are closely related but genetically distinct. The first involves the uptake of naked DNA by competent bacteria and is termed *transformation*. Secondly, *transduction* involves the transfer of genetic material from a donor to a recipient bacterium by way of a bacterial virus that does not destroy the latter. Finally, *conjugation* is the transfer of genetic material between bacteria by way of a tubelike bridge (pilus) that forms between the cells. The chemical incorporation of the genetic material into the structure of the recipient cell chromosome frequently occurs, but expression of the donor characteristic(s) does not depend on incorporation.

Smaller DNA structures called *plasmids* are involved in both natural and modulated genetic exchange. Plasmids naturally encode important traits such as resistance to antibiotics and toxin production. These genetic elements are transferred between bacteria by conjugation under natural circumstances. However, plasmids can also be deliberately inserted into recipient cells by transformation. Plasmids are attractive to practitioners of genetic engineering as vehicles for inserting genetic traits from dissimilar organisms into bacteria, where they are expressed as foreign proteins. The production of human insulin by common intestinal bacteria is an example that has been commercially exploited recently.

For some time, gene transfer has been an accepted laboratory methodology for mapping bacterial genes and for studying their function by complementation. In addition to the classical genetic transfer mechanisms of conjugation, transduction, and transformation, plasmids are determinants of many specialized phenotypic characters in natural populations. All of these processes, plus that of transposition, lead to an analogy of genes to enzymes in that both may be classified as constitutive or adaptive components of the cell. The "adaptive" genes, by analogy, are responsive to the environment of the cell.

5.9.4.2 *Genetic Transfer in Biofilm* The high volumetric cell concentrations observed in biofilms (Table 4.4) and electron micrograph evidence indicate the close proximity of bacteria within biofilms and suggest that genetic exchange might be facilitated in that setting. Suzuki et al. (1984) report that the transfer rate of sex pheromones in a Streptococcus population is increased by 100 times when the cells are on a surface. In addition, bacteria that are stationary within the biofilm are exposed to all of the genetically different bacteria and viruses that pass in the water column. Thus, biofilm bacteria are also exposed to large diversity of genetic material for potential exchange.

REFERENCES

Bakke, R., "Dynamics of biofilm processes," unpublished Master of Science thesis, Montana State University, Bozeman, MT, 1983.

Bakke, R., "Biofilm detachment," unpublished doctoral thesis, Montana State University, Bozeman, MT, 1986.

Bergey's Manual of Systematic Bacteriology, N.R. Krieg (ed.), **1**, Williams and Wilkins, Baltimore, 1984.

Borzani, W., *J. Biochem. Microbiol. Tech. Engrg.*, **3**, 235–240 (1961).

Bratbak, G., *Appl. Env. Microbiol.*, **49**, 1488–1493 (1985).

Brock, T.D., "Ecology of saline lakes", in M. Shilo (ed.), *Strategies of Microbial Life in Extreme Environments*, Verlag Chemie, Weinheim, 1979. pp. 29–48.

Brock, T.D., *Principles of Microbial Ecology*, Prentice-Hall, Englewood Cliffs, NJ, 1966.

Brock, T.D., D.W. Smith, and M.T. Madigan, *Biology of Microorganisms*, 4th Edition, Prentice-Hall, Englewood Cliffs, NJ, 1984.

Cohen, Y., B.B. Jorgensen, E. Padan, and M. Shilo, *Nature*, **257**, 489–492 (1975).

Costerton, J.W., G.G. Geesey, and K.-J. Cheng, *Sci. Am.*, **238**, 86 (1978).

Dawson, M.P., B.A. Humphrey, and K.C. Marshall, *Curr. Microbiology*, **6**, 195–198 (1981).

den Hollander, J.A., K. Ugurbil, T.R. Brown, and R.G. Shulman, *J. Am. Chem. Soc.*, **20**, 5871–5880 (1981).

Dykhuizen, D. and D. Hartl, *J. Bact.*, **135**, 876–882 (1978).

Fletcher, M.M., and G.D. Floodgate, *J. Gen. Microbiol.*, **74**, 325–344 (1973).

Fletcher, M. and K.C. Marshall, in K.C. Marshall (ed.), *Adv. Microbial Ecol.*, **6**, 1982, pp. 199–236.

Frederickson, A.G., R.D. Megee, and H.M. Tsuchiya, *Adv. Appl. Microbiol.*, **13**, 419-465 (1970).

Gray, T.R.G., R. Hisset, and T. Duxbury, *Rev. Ecol. Biol. Sol.*, **11**, 15-26 (1974).

Holme, T., *Acta Chem. Scand.*, **11**, 763-765 (1957).

Humphrey, B.A., M.R. Dickson, and K.C. Marshall, *Arch. Microbiol.*, **120**, 231-238 (1979).

Jordan, T.L., and J.T. Staley, *Microb. Ecol.*, **2**, 241-251 (1976).

Kjelleberg, S. and M. Hermansson, *Appl. Env. Microbiol.*, **48**, 497-503 (1984).

Kjelleberg, S., B.A. Humphrey, and K.C. Marshall, *Appl. Env. Microbiol.*, **43**, 1166-1172 (1982).

Kubitschek, H.E., W.W. Baldwin, S.J. Schroeter, and R. Graetzer, *J. Bact.*, **158**(1), 296-299 (1984).

Lehninger, A.L., *Bioenergetics*, 2nd ed., W.A. Benjamin, Menlo Park, CA, 1971.

Leppard, G.G., *Wat. Res.*, **20**, 697-702 (1986).

Marshall, K.C. and R.H. Cruickshank, *Arch. Mikrobiol.*, **91**, 29-40 (1973).

McCoy, W.F. and J.W. Costerton, *Dev. Ind. Microbiol.*, **23**, 441 (1982).

Mitchell, P., *Nature*, **191**, 144-148 (1961).

Mitchell, P., *Chemiosmotic Coupling and Energy Transduction*, Glynn Research Ltd., Bodmin, UK, 1966.

Monod, J., *Ann. Inst. Pasteur*, **79**, 390-410 (1950).

Morita, R.Y., *Adv. Microbial Ecol.*, **6**, 171-198 (1982).

Novick, A. and I. Szilard, *Science*, **112**, 715-716 (1950).

Otto, R., J. Hugenholtz, W.N. Konings, and A. Veldkamp, *FEMS Microbiol. Ltrs.*, **9**, 85-88 (1980).

Poindexter, J.S., *Adv. Microbial Ecol.*, **5**, 63-89 (1981).

Power, K. and K.C. Marshall, *Biofouling*, **1**, 163-174 (1988).

Robinson, R.W., H.C. Aldrich, S.F. Hurst, and A.S. Bleiweis, *Appl. Env. Microbiol.*, **49**, 321-327 (1985).

Roels, J.A., *Energetics and Kinetics in Biotechnology*, Elsevier Biomedical Press, Amsterdam, 1983.

Rosenberg, M. and S. Kjelleberg, *Adv. Microbial Ecol.*, **9**, 353-393 (1986).

Schaftel, S.O., "Processes of aerobic/anaerobic biofilm development," unpublished Master of Science thesis, Montana State University, Bozeman, MT, 1982.

Slayter, J.H., "Genetic interactions in microbial communities" in M.G. Klug and C.A. Reddy (eds.), *Current Perspectives in Microbial Ecology*, American Society for Microbiology, Washington, DC, 1984, pp. 87-93.

Stanier, R.Y., A. Adelberg, and J.L. Ingraham, *The Microbial World*, Prentice-Hall, Englewood Cliffs, NJ, 1976.

Suzuki, A., M. Mori, Y. Sakagani, A. Isogai, M. Fujino, C. Kitada, R. Craig, and D. Clewell, *Science*, **226**, 849-850 (1984).

Tanaka, T., I. Nishio, S.T. Sun, and S. Veno-Nishio, *Science*, **218**, 467-469 (1982).

Wicken, A.J., "Bacterial cell walls and surfaces", in D.C. Savage and M. Fletcher (eds.), *Bacterial Adhesion, Mechanisms and Physiological Significance*, Plenum, New York, NY, 1985, pp. 45-70.

Zobell, C.E., *J. Bact.*, **46**, 39-56 (1943).

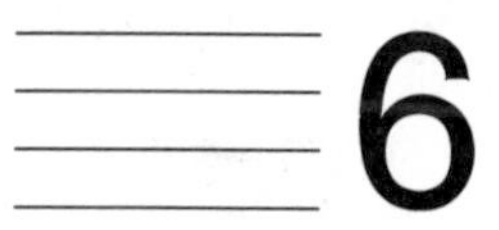

ENERGETICS AND STOICHIOMETRY

WILLIAM G. CHARACKLIS
Montana State University, Bozeman, Montana

SUMMARY

Two process characteristics are essential to the analysis of biofilm processes: (1) stoichiometry and (2) kinetics. Stoichiometry describes the extent of a process and provides information relating quantities of reactants consumed to quantities of products formed. Necessarily, energy transformations occur during a reaction, and these energy changes are related to reaction stoichiometry. This chapter focuses on the energetics and stoichiometry of microbial processes through the use of conservation principles presented in Chapter 2. Factors influencing the macroscopic energetics of microbial processes are discussed, and measures of energy "efficiency" and yields are presented. Biomass composition and the concept of formula weight for biomass are introduced as a convenience for discussing the stoichiometry of a microbial process.

The accurate determination of the microbial stoichiometry permits the calculation of all reactant and product concentrations from experimental measurements of one product (or reactant). Thus, the reactant or product that is easiest to measure is determined and all other reactants and products can be calculated. In some cases, the easiest quantity to analyze is not suitable because of other competing reactions. For example, in sulfate reducing systems, sulfide appears to be a convenient and useful quantity to measure. However, sulfide can easily and rapidly precipitate or react in other ways, so that the results are not easy to interpret. In such cases, sulfate may be a more appropriate quantity to monitor. In aerobic systems, oxygen remains a key variable for monitoring biofilm processes.

Stoichiometric calculations may also be useful in estimating the nutrient limitation (e.g., nitrogen or phosphorus) in a biofilm system. Industrial and utility experience indicates that, in some cases, undesirable biofilm accumulation has been unwittingly enhanced by addition of rate-limiting nutrients in the form of corrosion (nitrogen or phosphorus compounds) or scale (phosphorus compounds) inhibitors.

The importance of extracellular polymeric substances and other metabolic products indicates the need for further efforts in elucidating the stoichiometry of their production. The physical and electrochemical properties of these components are extremely important for the structure and function of the biofilm (Chapter 4).

The usefulness of the fundamental stoichiometric principles for biofilm microorganisms is assessed. As long as the microbial cell constituents can be distinguished from the extracellular constituents of the biofilm, the stoichiometry of microbial biofilm processes can be addressed with conventional stoichiometry calculations.

6.1 INTRODUCTION

Two process quantities are essential to design, analysis, or control of biofilm processes: (1) stoichiometry and (2) kinetics. Stoichiometry is generally associated with equilibrium in abiotic or chemical systems and describes the extent of a process. Stoichiometry is information relating quantities of reactants consumed to quantities of products formed. Necessarily, energy transformations occur during a reaction, and in abiotic systems these energy changes have been related to reaction stoichiometry. The same principles developed in abiotic systems relate to biofilms. This chapter deals with the energetics and stoichiometry of microbial processes through the use of conservation principles that were presented in Chapter 2.

The factors influencing the macroscopic energetics of microbial processes are discussed, and measures of energy efficiency and yields are presented. Biomass composition and the concept of formula weight for biomass are introduced as a convenience for discussing the stoichiometry of a microbial process. On this basis, the stoichiometries of the fundamental macroscopic microbial processes are developed and related to the overall or observed stoichiometry in batch and continuous microbial reactors.

6.2 ENERGY CONSERVATION AND UTILIZATION

The conservation principles discussed in Chapter 2 are necessary, but not very useful for a thorough process analysis until additional information concerning the stoichiometry of the chemical transformation is introduced as follows (previously presented as Eq. 2.24):

$$C_6H_{12}O_6 + 2.5O_2 + 0.7NH_3 \rightarrow 3.5CH_{1.4}O_{0.4}N_{0.2} + 2.5CO_2 + 4.6H_2O \quad [6.1]$$

Equation 6.1 describes the overall equation for microbial growth on glucose which results from numerous reactions occurring in sequence and parallel within the cell. $CH_{1.4}O_{0.4}N_{0.2}$ is an empirical formula for the biomass produced. So Eq. 6.1 is the net sum of all the reactions termed metabolism. As a result, if a cycle in metabolism does not result in the output of new cellular material or products, then it does not contribute to the net reaction. Examples of such cycles are the ADP-ATP system and the NAD-$NADH^+$ system (Lehninger, 1971):

$$ADP + P_i \rightleftarrows ATP, \qquad NAD + H^+ \rightleftarrows NADH^+ \quad [6.2]$$

Hence, detailed knowledge about these cycles is unnecessary in a macroscopic treatment of microbial processes using net stoichiometric equations. Complexity due to metabolic patterns and cycles, such as those above, arises in *structured* microbial cell models.

Energy flow through a microbial system will be presented in this section from a macroscopic standpoint, using the laws of thermodynamics to describe energy flows resulting from microbial reactions. Microbial reactions are basically oxidation-reduction reactions, and the influence of electron donor-electron acceptor combination on reaction stoichiometry is discussed.

6.2.1 Thermodynamics, Open Systems, and Microbial Growth

Microbial growth occurs at the expense of energy released by the flow of electrons from donors to acceptors within the cell. The free energy thus released is a thermodynamic state variable, which represents the maximum work that can be accomplished at constant temperature and pressure. But this maximum work can only be attained under reversible or equilibrium conditions. Since bacterial cells are open systems in which irreversible processes are occurring, only a portion of the free energy released can be captured for useful work. The remainder escapes as heat.

The extent to which microbial growth and other microbial processes occur is a function of the energy released by the electron transfer and the efficiency of energy utilization by the organism.

Organisms cannot carry out macroscopic reactions that are thermodynamically impossible. From the standpoint of macroscopic chemical transformations, microorganisms act only as catalysts of oxidation-reduction. Consequently, microorganisms do not oxidize glucose or reduce oxygen, but rather mediate the oxidation of glucose by oxygen (i.e., electron transfer from glucose to oxygen). Thus, thermodynamics predicts that glucose in an aerobic environment will be oxidized to CO_2 and H_2O either chemically (over a long period) or by a path which includes microbial oxidation with the production of biomass. However, the biomass is a metastable product, which will also be oxidized in a closed system with an excess of oxygen (Figure 6.1).

6.2.1.1 The Laws of Thermodynamics The *first law of thermodynamics* is a statement of the conservation equations for mass and energy, i.e., mass and energy are conserved in an isolated system. All microbial reaction systems are consistent with the first law.

The *second law of thermodynamics* states that systems in isolation spontaneously tend toward states of greater disorganization. Microbial reaction systems obey the second law also, but the reasons are not so clear without more precise statement of the second law. The degree of disorganization, or entropy, of a system is not the only property necessary to describe spontaneous processes. The free energy is also an important parameter; it can be defined as the energy capable of performing work. A process can only proceed spontaneously when there is a loss of free energy. In an isolated system at constant temperature, the change in free energy and the change in entropy are related to each other by the following equation:

$$\Delta G = \Delta H - T\,\Delta S \qquad [6.3]$$

where ΔG = change in free energy
ΔH = change in heat content (enthalpy)
T = absolute temperature
ΔS = change in entropy

Although the second law predicts that a system, when left to itself, will tend to decrease its state of organization, it nevertheless allows for an increase in organization when free energy is supplied to the system. And this is what occurs in the case of living systems. The high free energy content and low state of disorganization of living systems are maintained and, at times, even extended by a continuous supply of free energy. As soon as this supply is cut off, living systems proceed spontaneously to a great state of disorganization (death). This type of labile system, maintained at a certain level of organization by a continuous supply of free energy, is described as a steady state. A steady state is not, in general, an equilibrium state in which the system has achieved the lowest possible free energy and the highest possible disorganization.

A steady state is a state of a system away from equilibrium, which is maintained in apparent constancy by the continuous supply of free energy.

A physical model for such a situation is a thermoregulated water bath maintaining a constant temperature higher than that of its environment.

A supply of free energy is only one requirement for the maintenance of a

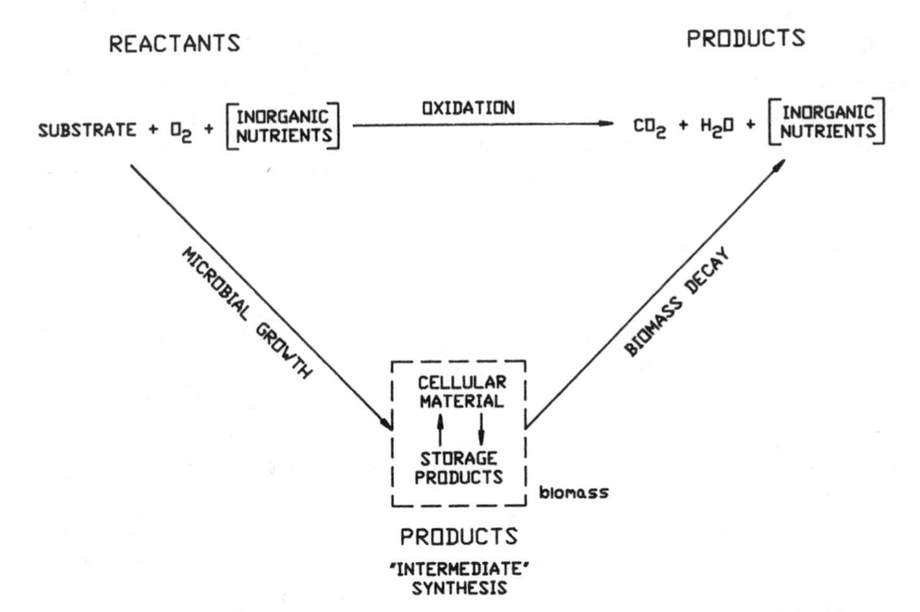

Figure 6.1 The complete oxidation of an organic substrate can be accomplished by two methods: (1) chemical oxidation and (2) microbial oxidation followed by chemical oxidation of the microbial cells. The microbial cells are a metastable product in the process.

steady state. There must also be a minimal organization capable of absorbing and channeling the energy in a usable manner. A biological reaction system as a whole does not violate the second law, since it is powered by a continuous and generous supply of free energy which, in nature, has its origins in the nuclear reactions of the sun.

6.2.1.2 Thermochemistry Thermodynamics is a set of rules that govern the transition of a material system (such as a chemical reaction) from one state to another. When considering energy changes in a chemical system, it is not immediately obvious that they should be looked at strictly from the point of view of the substances taking part. For instance, if carbon is burned in oxygen, the energy produced is regarded as a positive quantity. But in terms of the reaction itself,

$$C_6H_{12}O_6 + O_2 \rightarrow 6CO_2 + 6H_2O \qquad [6.4]$$

$$\text{energy change} = -2880 \text{ KJ mole}^{-1}$$

the negative sign expresses what is happening to the reactants from their own point of view: They are losing energy. Another way of expressing the reaction is as follows (see Section 2.6):

$$6CO_2 + 6H_2O - 6O_2 - C_6H_{12}O_6 + 2880 \text{ KJ} = 0 \qquad [6.5]$$

In a reversible reaction system, e.g., $A + B \rightleftarrows Y + Z$, if the energy change is negative, the reaction, once started, tends to go spontaneously from left to right. When equilibrium has been attained, the energy change is zero, and no further useful chemical energy can be extracted from the system, because it no longer has a tendency to move in either direction. If we want to synthesize $A + B$ from $Y + Z$, energy must in some way be supplied, and the energy change from this reverse reaction is positive. In brief, under constant conditions of temperature and pressure:

$\Delta G < 0$ Reaction can occur spontaneously to equilibrium point
$\Delta G = 0$ Equilibrium has been attained
$\Delta G > 0$ Reaction has to be forced away from the equilibrium

where ΔG is the free energy of a reversible chemical reaction and is defined by Eq. 6.3.

If a reaction gives off heat, the enthalpy change, ΔH, will be negative; the system is losing heat to its surroundings. Hence, if the term $T\ \Delta S$ (ΔS is the entropy change) is not too large, ΔG will be negative also, indicating that the reaction, once started, can proceed spontaneously to equilibrium. Although the term $T\ \Delta S$ should not be overlooked, in the majority of biochemical transformations it is small compared with ΔG or ΔH because many of the reactant and product molecules are of similar shape and size and are already in solution. Hence, for many practical purposes, $\Delta G = \Delta H$. However, it is because of the $T\ \Delta S$ term that ΔG is considered more useful than ΔH as a measure of the work available from a chemical reaction.

The more dilute a solution, the greater the entropy, and therefore the entropy change of a reaction in solution varies with the concentration of the reactants and products taking part. ΔH, however, is much less sensitive to changes in concentration, so ΔG changes with concentration primarily because of the alteration in ΔS. The change in ΔG with concentration can be derived from the equation

$$\Delta G = 2.3\ \mathrm{RT} \log \frac{\text{Product of molar conc. of products}}{\text{Product of molar conc. of reactants}} = 2.3\ \mathrm{RT} \log K \qquad [6.6]$$

where R = universal gas constant
K = equilibrium constant

In Eq. 6.6, the first term on the right hand side represents the concentrations of products and reactants actually taking part in the reaction; K represents the equilibrium concentrations (the equation properly applies to activities not concentrations, but in the case of dilute solutions these terms are almost equivalent). At equilibrium, $\Delta G = O$ because the two terms cancel.

Conventionally, ΔG° is defined by conditions at which all products and reactants are at unit (molar) concentration and Eq. 6.6 becomes

$$\Delta G^\circ = -2.3 \text{ RT} \log K \qquad [6.7]$$

where ΔG° = the "standard free energy" change.

ΔG° is a constant characteristic of any particular reaction, because it is the value of ΔG when the reactants and products are all present under defined conditions. Microbial reactions, which occur in aqueous systems, can more conveniently be expressed in terms of $\Delta G^\circ(\mathrm{W})$, which is analogous to ΔG° except that $[H^+]$ and $[OH^-]$ are assigned their concentrations (more formally, their activities) in neutral water. Values of $\Delta G^\circ(\mathrm{W})$ for 25°C thus apply to unit concentrations (activities) of reactants and products at pH 7.0.

6.2.2 Oxidation-Reduction (Redox) Reactions

Microbial reactions occur at the expense of energy released by the flow of electrons from donors to acceptors mediated by the microorganism. In this section, equilibrium relations of relevant redox (oxidation-reduction) components in a microbial reaction system are emphasized. It is important to know that the concentrations of oxidizable or reducible species may be quite different from their equilibrium concentrations, i.e., those predicted thermodynamically, because redox reactions are generally slow. Biological mediation increases reaction rates and consequently the rate of approach to equilibrium frequently depends on the activity of the organisms.

A redox reaction consists of two parts: (1) an oxidation reaction in which a substance loses or donates electrons and (2) a reduction reaction in which the substance gains or accepts electrons. These electrons must be conserved. The coupling of these two "half reactions" is by the electrons generated or consumed. This concept is used in balancing redox reactions, which consists in balancing the mass of each constituent and then balancing the charge. An illustration of the procedure for balancing a redox reaction is presented in the Appendix at the end of this chapter.

6.2.2.1 *Direction of Redox Reactions (after McCarty, 1972)* Redox stoichiometry can be expressed by the methods in Chapter 2, similarly to any other reaction. Therefore, a redox reaction can be expressed in the following general way:

$$\sum_i \nu_i A_i = 0 \qquad [6.8]$$

where ν_i = stoichiometric coefficient for ith component (mol) ($\nu_i < 0$ for reactants, $\nu_i > 0$ for products)

M_i = molecular weight of the ith reaction component (M mol^{-1})

The Gibbs free energy change ΔG for the reaction is expressed as

$$\Delta G = \Delta G^\circ + \mathrm{RT} \sum_i \nu_i \ln a_i \qquad [6.9]$$

where a_i = the activity of component i (dimensionless)

Electrode or redox potentials are related to the Gibbs free energy change per mole of electrons associated with a given reaction as follows:

$$E^\circ = \frac{\Delta G^\circ}{n_e F_A} \qquad [6.10]$$

where F_A = the faraday, the charge per mole of electrons
n_e = number of electrons involved in the reaction

Potentials are usually written with respect to the standard hydrogen electrode (E_H). Values of $E_\mathrm{H}^\circ(\mathrm{W})$ determined in this way are listed in Table 6.1 and are for all constituents at unit activity except H^+ and OH^-, which are assumed to have the activities associated with neutral water. The redox potential for other than unit activity is determined from the Peters–Nernst equation:

$$E_\mathrm{H} = E_\mathrm{H}^\circ + \frac{\mathrm{RT}}{n_e F_A} \sum_i \nu_i \ln a_i \qquad [6.11]$$

The pH (pH $= -\log H^+$) of a solution measures the relative tendency of a solution to accept protons. In an acid solution, this tendency (and the pH) is low. In alkaline solution, it is high. The electron activity is an equally convenient parameter for measuring the relative tendency of a solution to accept or transfer electrons at equilibrium:

$$p\epsilon = -\log e^- \qquad [6.12]$$

In a highly reducing environment, the relative tendency for accepting electrons is very low, as is $p\epsilon$. Thus, a high $p\epsilon$ indicates a high tendency to oxidation. In equilibrium equations, H^+ and e^- are treated in an analogous way (Stumm and Morgan, 1981). $p\epsilon$ is a convenient measure of the oxidizing intensity of a system at equilibrium and is related to the reversible redox potential (E_H) and Gibbs free energy as follows:

$$p\epsilon = E_\mathrm{H} \frac{F_A}{2.3\mathrm{RT}} = \frac{\Delta G}{2.3 n_e \mathrm{RT}} \qquad [6.13]$$

TABLE 6.1 Free Energies for Various Half Reactions Where Reactants and Products Are at Unit Activity except $[H^+] = 10^{-7}$[a]

	Half Reaction	$\Delta G^0(W)$ [kJ/(mol e$^-$)$^{-1}$]	$E_H^0(W)$ (V)	$p\epsilon^0(W)$
a.	$\frac{1}{3}NO_2^- + \frac{4}{3}H^+ + e^- = \frac{1}{6}N_2 + \frac{2}{3}H_2O$	−93.15	0.97	16.4
b.	$\frac{1}{4}O_2 + H^+ + e^- = \frac{1}{2}H_2O$	−78.14	0.81	13.7
c.	$Fe^{+3} + e^- = Fe^{+2}$	−74.39	0.77	13.0
d.	$\frac{1}{5}NO_3^- + \frac{6}{5}H^+ + e^- = \frac{1}{10}N_2 + \frac{3}{5}H_2O$	−71.66	0.74	12.0
e.	$\frac{1}{2}NO_3^- + H^+ + e^- = \frac{1}{2}NO_2^- + \frac{1}{2}H_2O$	−39.43	0.41	6.9
f.	$\frac{1}{8}NO_3^- + \frac{5}{4}H^+ + e^- = \frac{1}{8}NH_4^+ + \frac{3}{8}H_2O$	−39.50	0.36	6.0
g.	$\frac{1}{8}SO_4^{+2} + \frac{19}{16}H^+ + e^- = \frac{1}{16}H_2S + \frac{1}{16}HS^- + \frac{1}{2}H_2O$	21.28	−0.22	−3.8
h.	$\frac{1}{8}CO_2 + H^+ + e^- = \frac{1}{8}CH_4 + \frac{1}{4}H_2O$	24.11	−0.25	−4.2
i.	$\frac{1}{8}CO_2 + \frac{1}{8}CHO_3^- + H^+ + e^- = \frac{1}{8}CH_3COO^- + \frac{3}{8}H_2O$	27.65	−0.29	−4.8
j.	$\frac{15}{92}CO_2 + \frac{1}{92}HCO_3^- + H^+ + e^- = \frac{1}{92}CH_3(CH_2)_{14}COO^- + \frac{31}{92}H_2O$	27.85	−0.29	−4.8
k.	$\frac{1}{7}CO_2 + \frac{1}{14}HCO_3^- + H^+ + e^- = \frac{1}{14}CH_3CH_2COO^- + \frac{5}{14}H_2O$	27.88	−0.29	−4.9
l.	$\frac{1}{8}CO_2 + H^+ + e^- = \frac{1}{12}CH_3CH_2OH + \frac{1}{4}H_2O$	31.76	−0.33	−5.6
m.	$\frac{1}{6}CO_2 + \frac{1}{12}HCO_3^-$ $+ \frac{1}{12}NH_4^+ + H^+ + e^- = \frac{1}{12}CH_3CHNH_2COOH + \frac{5}{12}H_2O$	31.69	−0.33	−5.6
n.	$\frac{1}{6}CO_2 + \frac{1}{12}HCO_3^- + H^+ + e^- = \frac{1}{12}CH_2CHOHCOO^- + \frac{1}{3}H_2O$	32.94	−0.34	−5.8
o.	$\frac{1}{5}CO_2 + \frac{1}{10}HCO_3^- + H^+ + e^- = \frac{1}{10}CH_3COCOO^- + \frac{2}{5}H_2O$	35.75	−0.37	−6.3
p.	$H^+ + e^- = \frac{1}{2}H_2$	40.46	−0.42	−7.1
q.	$\frac{1}{4}CO_2 + H^+ + e^- = \frac{1}{24}C_6H_{12}O_6 + \frac{1}{4}H_2O$	41.92	−0.43	−7.4
r.	$\frac{1}{2}HCO_3^- + H^+ + e^- = \frac{1}{2}HCOO^- + \frac{1}{2}H_2O$	48.03	−0.50	−8.4

[a] After McCarty (1972).

At 25°C, $2.3RT/F_A = 0.059$ V equivalent^{-1} and $2.3RT = 5.699$ kJ equivalent^{-1}, so

$$p\epsilon = \frac{E_H}{0.059} = \frac{-\Delta G}{5.699 n_e} \qquad [6.14]$$

$p\epsilon$ is an intensity factor and represents the electron free energy level per mole of electrons. It has a computational advantage over other free energy parameters:

$$p\epsilon = p\epsilon^\circ + \sum_i \nu_i \log a_i \qquad [6.15]$$

Values for $p\epsilon^\circ(W)$ are listed in Table 6.1.

6.2.2.2 *Energy Relations* Free energy values for various relevant half reactions are presented in Table 6.1. Each reaction has been normalized to a single electron. Free energy values for microbial reactions of interest can be obtained by subtracting one half reaction from another. For example, the free energy value for the oxidation of lactic acid by oxygen is obtained by subtracting Eq. o from Eq. b in Table 6.1:

$$\tfrac{1}{12}CH_3CHOHCOO^- + \tfrac{1}{4}O_2 \rightleftarrows \tfrac{1}{6}CO_2 + \tfrac{1}{12}HCO_3^- + \tfrac{1}{6}H_2O$$
$$\Delta G^\circ(W) = -111.08 \text{ kJ (mol e}^-)^{-1} \qquad [6.16]$$

This equation can be rewritten by adding a hydrogen ion (H^+) to each side:

$$\tfrac{1}{12}CH_3CHOHCOOH + \tfrac{1}{4}O_2 \rightleftarrows \tfrac{1}{4}CO_2 + \tfrac{1}{4}H_2O$$
$$\Delta G^\circ(W) = -111.09 \text{ kJ (mol e}^-)^{-1} \qquad [6.17]$$

The oxidation of glucose by sulfate is obtained by subtracting Eq. q from Eq. g in Table 6.1:

$$\tfrac{1}{24}C_6H_{12}O_6 + \tfrac{1}{8}SO_4^{-2} + \tfrac{1}{16}H^+ \rightleftarrows \tfrac{1}{4}CO_2 + \tfrac{1}{16}HS^- + \tfrac{1}{4}H_2O$$
$$\Delta G^\circ(W) = -20.65 \text{ kJ (mol e}^-)^{-1} \qquad [6.18]$$

Equation 6.18 is worthy of further consideration because it cannot be accomplished by any known single microbial species. Only a limited number of organisms, the sulfate-reducing bacteria (SRB), can utilize sulfate (SO_4^{-2}) as a final electron acceptor. However, the SRB do not have the metabolic capability to break down (catabolize) glucose. Therefore, Eq. 6.18 describes a thermodynamically favored reaction that cannot be catalyzed by a single microbial species. Breakdown by a heterogeneous population or *consortium* containing SRB, however, is possible (D'Alessandro et al., 1974).

The energy released by various oxidation–reduction reactions, calculated using data from Table 6.1, is illustrated in Figure 6.2. Formate contains the most available energy per electron mole, and methane the least. When oxygen is used as the electron acceptor, the difference in available energy between these two compounds is only about 20% ($\approx$100–120 kJ/mol). As a result, biomass yields per electron equivalent of the electron donor are nearly constant when oxygen is the electron acceptor (Table 6.2 and Figure 6.2). When carbon dioxide is used as the electron acceptor, the energy yield from the two compounds ranges from zero (for methane) to -23.93 kJ per electron mole (for formate). This is quite a large relative difference and partially explains the relatively low and widely varying biomass yields observed in methane-forming systems where carbon dioxide is the electron acceptor.

6.2.2.3 Sequence of Microbially Mediated Redox Reactions (after Stumm and Morgan, 1981) In a closed aqueous system containing organic material (e.g., $C_6H_{12}O_6$), microbial oxidation of organic matter is observed to occur first by reduction of O_2, $p\epsilon(W) = 13.8$. When oxygen is essentially depleted and organic material remains, NO_3^- will be reduced. The progression from one electron acceptor to another occurs as $p\epsilon$ decreases. Reduction of MnO_2, if present (not indicated in Table 6.1), will occur at the same $p\epsilon$ level as NO_3^- reduction followed by reduction of Fe^{2+}. At lower $p\epsilon$ levels, fermentation and SO_4^{-2} and CO_2 reduction may occur almost simultaneously.

The reaction sequence described above is consistent with the ordering of the thermodynamic driving forces ΔG. The electron donor ($C_6H_{12}O_6$) will supply electrons to the lowest unoccupied electron level (O_2). If that electron level becomes saturated, successive levels (NO_3^-, SO_4^{-2}, etc.) will be filled up. This succession of reactions can be observed in the vertical distribution of nutrient-

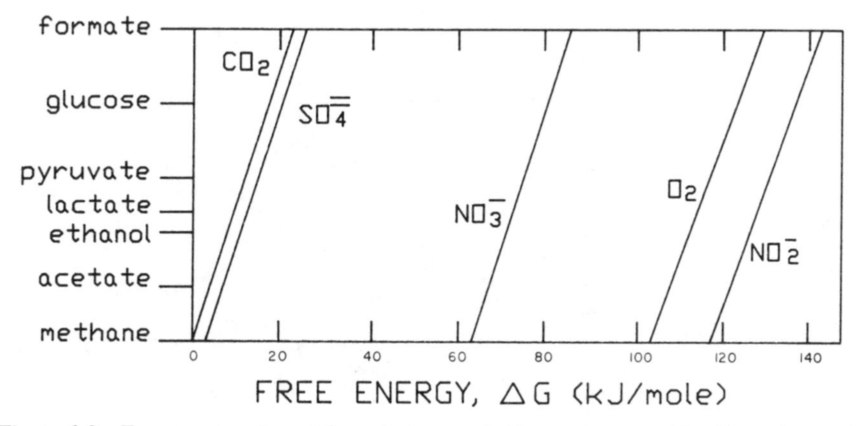

Figure 6.2 Free energy released (per electron mole) for various combinations of organic electron donors and inorganic electron acceptors (after McCarty, 1972).

TABLE 6.2 Theoretical and Observed Yields for Various Substrates with Oxygen as an Electron Acceptor[a]

Substrate	Molecular Weight ($g\ mol^{-1}$)		Available Electrons (mol^{-1})	Mass of Substrate per Mole of Available Electrons [$g\ (mol\ e^{-})^{-1}$]	Theoretical Biomass Yield ($g\ g^{-1}$)	Observed Biomass Yield ($g\ g^{-1}$)	Reference
Glucose	180	24	7.5	3.15	0.42	0.44	Busch, 1971
Glycerol	92	14	6.6	2.76	0.46	0.50	Payne, 1970
Acetate	60	8	7.5	2.70	0.36	0.36	Payne, 1970
Palmitic acid	256	92	2.8	2.81	1.01	0.98	Busch, 1971
Glutamic acid	147	16	9.2	3.22	0.35	0.35	Busch, 1971

[a] After Sykes (1975).

rich lakes, in temporal changes in a batch reactor such as a batch anaerobic digester, and in the spatial distribution of microbial metabolism in groundwater systems (Chapter 18).

The sequence of chemical reactions is necessarily paralleled by an ecological succession of microorganisms (aerobic heterotrophs, denitrifiers, fermenters, sulfate reducers, and methane-forming bacteria). There appears to be a tendency for more energy-yielding, microbially mediated reactions to take precedence over processes that are less energy-yielding.

6.3 BIOMASS AND PRODUCT FORMATION

It would be useful and convenient if theoretical equations could be developed relating biomass production and/or oxygen requirements to the free energy of oxidation ΔG of the limiting nutrient (i.e., the substrate), as is done in abiotic chemical reactions. Then, ΔG° values, determined with the help of Table 6.1, could provide estimates of yields. Roels (1983) has provided the background and methodology for these *a priori* yield estimates. Experimentally determined yield values for organisms aerobically grown on various substrates as the sole carbon and energy source are presented in Table 6.2 as well as predictions based on thermodynamic calculations.

Where metabolic pathways are known, the production of ATP per unit substrate ($Y_{ATP/S}$) can be calculated and theoretical biomass yields determined (Servizi and Bogan, 1963). Bauchop (1958) and Bauchop and Elsden (1960) provided the basis for these calculations by estimating Y_{ATP} from their experiments with a limited number of organism–substrate systems, which resulted in a biomass yield per unit ATP consumed given by

$$Y_{X/ATP} = 10.5 \text{ (g biomass) (mol ATP)}^{-1}$$

This average value of $Y_{X/ATP}$ ($\pm 20\%$) has been verified by many researchers using batch, nutrient-saturated systems (Payne, 1970).

6.3.1 Stoichiometry in a Batch Microbial Reactor

The stoichiometry of microbial reactions has frequently been determined in a batch reactor system in which the initial state is characterized by a relatively large amount of nutrients and a small number of viable cells, i.e., the reaction system is nutrient-saturated. Such systems are frequently referred to as "growth-limited," meaning that the growth rate of cells is presumably limited by their internal metabolic machinery. The subsequent events of substrate uptake and microbial growth are illustrated in Figure 6.3 and are frequently referred to collectively as the "bacterial growth cycle" (see Chapter 2). Microbial growth and reproduction proceed exponentially in this system until the substrate reaches such a low concentration level that it is not equally accessible to all the organ-

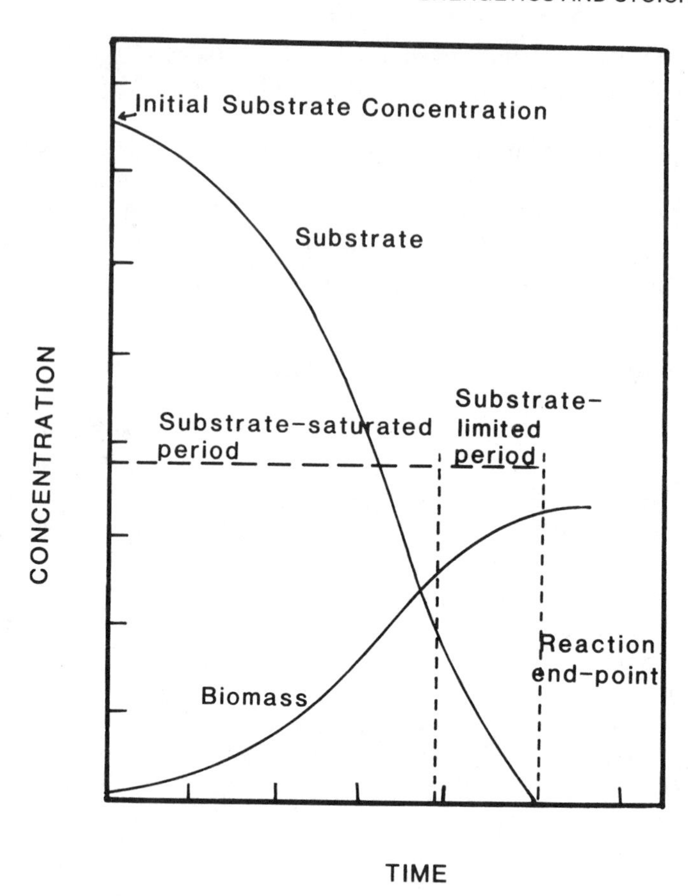

Figure 6.3 The progression of microbial growth in a closed (batch) system illustrating the increase in biomass and cumulative oxygen consumption as substrate is depleted by microbial metabolism for energy and synthesis processes. In the early period, substrate is in large supply for a very few organisms and the system is substrate-saturated (or growth-controlled). As growth proceeds, substrate is depleted and the increasing number of organisms must share a reduced food supply; the system is substrate-limited.

isms. At that point, the reactor system becomes "substrate-limited" as opposed to "growth-limited." Shortly thereafter, the reaction (i.e., microbial growth) comes to a temporary halt due to the complete disappearance of the limiting nutrient. The "plateau" in the growth cycle is analogous to a reaction endpoint because any further activity in the reactor is attributable to biomass decay and not microbial growth. *Biomass decay* is a general term that comprises biomass reduction resulting from many processes, including endogenous metabolism, maintenance energy requirements, cell lysis, and (in undefined, mixed populations) even predation. The biomass decay reaction in an aerobic environment can be expressed as follows:

$$\text{biomass} + O_2 \rightarrow CO_2 + H_2O + \text{inorganic nutrients} \qquad [6.19]$$

Since the equilibrium state for a closed system consisting of an organic compound in an aerobic environment results in the production of CO_2 and H_2O, the cell population resulting from growth (at the stationary state in the "growth cycle") is a metastable state on the pathway to equilibrium. The relationship between chemical oxidation and microbial oxidation of an organic nutrient, including the influence of growth and decay, is illustrated in Figure 6.1.

An important physiological concept is the law of the limiting factor attributed to Liebig:

The rate and/or extent of a microbial process is controlled by the environmental factor that is limiting.

Although the limiting factor can be temperature, pH, or even fluid shear stress, for most applications considered here it will be a nutrient. There are two important distinctions to be considered:

1. There are stoichiometric and kinetic limiting nutrients or factors, and they need not be the same nutrient.
2. More than a single factor may present a kinetic or stoichiometric limitation for the same microbial process.

For example, the stoichiometric limiting factor is the first to become exhausted during the batch growth of a microbial population. The kinetic limiting factor limits the rate but not necessarily the extent of the reaction. The stoichiometric limiting factor provides a convenient, reproducible endpoint, or final state, for determining the stoichiometry of microbial reaction processes in closed or batch reactors.

6.3.2 Yield Expressions

The synthesis and oxidation products of microbial growth reactions are generally expressed as yields, i.e., ratios of amount of products formed to amount of reactants consumed. The reactant of interest is usually, but not necessarily, the limiting factor, i.e., the substrate.

Consider the following stoichiometry, which describes a microbial growth system in which glucose is the substrate—a system that has been precisely defined by experiments of Busch (1971):

Energy:

$$C_6H_{12}O_6 + 6O_2 \rightarrow 6CO_2 + 6H_2O \qquad [-120.11\ \text{kJ (mol e}^-)^{-1}] \qquad [6.20a]$$

Synthesis:

$$C_6H_{12}O_6 + 1.2NH_3 \rightarrow 6CH_{1.4}O_{0.4}N_{0.2} + 3.6H_2O \qquad [6.20b]$$

Overall Stoichiometry:

$$C_6H_{12}O_6 + 2.5O_2 + 0.7NH_3 \rightarrow 3.5CH_{1.4}O_{0.4}N_{0.2} + 2.5CO_2 + 4.6H_2O \qquad [6.1]$$

Equation 6.1 represents the stoichiometric equation for removal of glucose by a mixed microbial population in a batch reactor which is initially substrate-saturated. The reaction is complete when all of the substrate (glucose in this case) is exhausted. The empirical formula $CH_{1.4}O_{0.4}N_{0.2}$ represents the biomass composition for the particular reaction environment. Reported cell compositions from other reaction environments are listed in Table 6.3.

The overall stoichiometry (Eq. 6.1) is the net result of the energy reaction (Eq. 6.20b) and the synthesis reaction (Eq. 6.20a). The energy reaction indicates the amount of useful work that can be accomplished by the cell. The synthesis reaction indicates the transformation that will consume the energy produced. Equation 6.1 can be balanced in many ways, because experimental data are needed to evaluate the coefficients in a given reaction environment. To balance the overall stoichiometric equation, the biomass yield coefficient must be determined experimentally. The reaction is complete when all the glucose has been converted to oxidation (respiration) or synthesis products.

Yield can be expressed in many ways, and in view of the many methods used to estimate biomass and nutrient concentration, one must take care to define it precisely. Therefore, the units for yield will be reported in terms of the method used for the measurement of biomass and nutrient. The importance of consistent units for material balance purposes cannot be overemphasized; here mass units will be used most often:

$$Y_{X/S} = \frac{\text{g cells formed}}{\text{g nutrient used}} \qquad [6.21]$$

Experimentally determined biomass yields for different microbial populations on various substrates are presented in Table 6.4. The table also contains data for biomass yields on various substrates when endogenous electron accep-

TABLE 6.3 Elemental Composition of Microorganisms.

Organism	Composition (%)				Empirical Formula	Reference
	C	H	N	O		
C. utilis	50.4	7.7	5.8	36.1	$CH_{1.83}N_{0.10}O_{0.54}$	Herbert, 1976
C. utilis	46.9	7.3	10.8	35.1	$CH_{1.87}N_{0.20}O_{0.56}$	Herbert, 1976
C. utilis	50.4	7.7	11.0	31.0	$CH_{1.83}N_{0.19}O_{0.46}$	Herbert, 1976
C. utilis	46.9	7.3	11.0	34.9	$Ch_{1.87}N_{0.20}O_{0.56}$	Herbert, 1976
K. aerogenes	50.6	7.4	13.0	29.0	$CH_{1.75}N_{0.22}O_{0.43}$	Herbert, 1976
K. aerogenes	50.0	7.2	14.1	28.7	$CH_{1.73}N_{0.24}O_{0.43}$	Herbert, 1976
K. aerogenes	50.7	7.4	10.1	31.8	$CH_{1.75}N_{0.17}O_{0.47}$	Herbert, 1976
K. aerogenes	50.0	7.2	14.1	28.7	$CH_{1.73}N_{0.24}O_{0.43}$	Herbert, 1976
S. cerevisiae	49.3	6.7	9.8	34.2	$CH_{1.64}N_{0.16}O_{0.52}$	Harrison, 1967
S. cerevisiae	47.7	7.3	9.4	35.6	$CH_{1.83}N_{0.17}O_{0.56}$	Kok and Roels, 1980
S. cerevisiae	49.3	6.7	9.8	33.5	$CH_{1.81}N_{0.17}O_{0.51}$	Wang et al., 1976
Pseudomonas denitrificans	48.4	7.3	11.4	32.9	$CH_{1.81}N_{0.20}O_{0.51}$	Stouthamer, 1977
Pseudomonas denitrificans	51.0	6.4	11.3	31.3	$CH_{1.51}N_{0.19}O_{0.46}$	Shimizu et al., 1978
E. coli	48.1	7.1	13.4	31.4	$CH_{1.77}N_{0.24}O_{0.49}$	Bauer and Ziv, 1976
Pseudomonas $C_{12}B$	47.0	7.8	12.6	32.6	$CH_{2.00}N_{0.23}O_{0.52}$	Mayberry et al., 1968
A. aerogenes	45.9	7.0	13.4	33.7	$CH_{1.83}N_{0.25}O_{0.55}$	Mayberry et al., 1968
P. aeruginosa	55.3	7.9	14.6	22.2	$CH_{1.73}N_{0.23}O_{0.30}$	Turakhia, 1986

tors (i.e., fermentation) are used. These data can be compared with the $\Delta G°$ values in Figure 6.2. All experimental yield values were determined in batch reactors except for those reported by Robinson et al. (1984) for *Pseudomonas aeruginosa* in a chemostat, and by Bakke et al. (1984) and Turakhia (1986) for *Ps. aeruginosa* biofilms.

The method for determining the stoichiometric equation can be used for many microbial growth systems. For example, the sulfate-reducing bacteria can utilize lactate as substrate while using sulfate as the electron acceptor:

Energy with lactate:

$$CH_3CHOHCOOH + 1.5H_2SO_4 \rightarrow 1.5H_2S + 3H_2CO_3 \quad [-11.67 \text{ kJ (mol e}^-)^{-1}] \quad [6.22a]$$

TABLE 6.4 Observed Biomass Yields from Glucose When Oxygen (Aerobic) or an Endogenous (Anaerobic) Electron Acceptor is Used

		$Y_{X/S}$ (g g^{-1})[a]		
Organism	Reactor	Oxygen[b]	Endogeneous[b]	Reference
Aerobacter aerogenes	Batch	0.42	0.14	Hadjipetrou and Stouthamer, 1965
Escherichia coli	Batch	0.47	0.13	Hernandez and Johnson, 1967
Ps. aeruginosa	Biofilm	0.33	—	Turakhia, 1986
Ps. aeruginosa:	Chemostat			Robinson et al., 1984
total biomass		0.54	—	
cell mass		0.30	—	
Ps. fluorescens	Batch	0.38	—	Nagai, 1979
Rhodopseudomonas spheroides		0.45		Nagai, 1979
Propionibacterium	Batch	—	0.21	Bauchop and Elsden, 1960
Candida utilis	Batch	0.51	—	Hernandez and Johnson, 1967
Saccharomyces cerevisiae	Batch	0.42	0.12	Bauchop and Elsden, 1960
Saccharomyces cerevisiae	Batch	0.50		Nagai, 1979
Mixed populations	Batch	0.44	—	Busch, 1971
	Batch	0.40	—	Servisi and Bogan, 1963

[a] Multiply values by 1.125 to convert to grams of biomass carbon per gram of glucose carbon.
[b] Electron acceptor.

Synthesis with lactate:

$$CH_3CHOHCOOH + 0.6NH_3 \rightarrow 3CH_{1.4}O_{0.4}N_{0.2} + 1.8H_2O \qquad [6.22b]$$

Overall stoichiometry lactate:

$$CH_3CHOHCOOH + 2.16H_2SO_4 + 0.0036NH_3$$
$$\rightarrow 0.18CH_{1.4}O_{0.4}N_{0.2} + 2.16H_2S + 2.82CO_2 + 2.93H_2O \quad [6.22c]$$

The overall stoichiometric equation was balanced using the experimentally determined biomass yield determined by Traore et al (1982). Each mole of lactate will transfer 12 electrons. The oxidation of lactate with sulfate rather than oxygen (Eq. 6.16) results in considerably less energy production for synthesis [-11.67 versus -111.08 kJ (mol $e^-)^{-1}$]. The effect of the difference in available energy on biomass synthesis (yield) is significant. Equation 6.22c indicates that if one measures the sulfate depletion, the production of cell mass and depletion of lactate can be calculated.

If the overall stoichiometry is known, then all reactant and product concentrations can be calculated from measurements of one product (or reactant).

The sulfate reducers can also use acetate as an electron donor:

Energy with acetate:

$$CH_3COOH + H_2SO_4 \rightarrow H_2S + 2H_2CO_3$$
$$[-6.36 \text{ kJ (mol e}^-)^{-1}] \quad [6.23a]$$

Synthesis with acetate:

$$5Ch_3COOH + 2NH_3 \rightarrow 10CH_{1.4}O_{0.4}N_{0.2} + 6H_2O \qquad [6.23b]$$

Overall stoichiometry with acetate:

$$CH_3COOH + 0.915H_2SO_4 + 0.034NH_3$$
$$\rightarrow 0.17CH_{1.4}O_{0.4}N_{0.2} + 0.915H_2S + 1.83CO_2 + 1.93H_2O \quad [6.23c]$$

In the case of acetate, the experimental yield value obtained by Middleton and Lawrence (1977) was used to balance the overall stoichiometric equation. Each mole of acetate will transfer 8 moles of electrons. Thus, oxidation of a mole of lactate will result in release of approximately three times as much energy as the oxidation of a mole of acetate. This disparity in "energy yield" is not reflected in biomass yields. Thus, the experimental measurement of biomass yield or the cellular metabolism must be examined more closely to explain the apparent anomaly.

6.3.3 Elementary Composition of Biomass (after Herbert, 1976)

Attempts have been made to express cell composition by simple molecular formulae such as $C_4H_8O_2N$ (Mayberry et al., 1968) or $C_5H_7O_2N$ (Porges et al., 1956). In such complicated mixtures of substances as bacterial cells, the elements are unlikely to occur in simple whole number ratios. The elemental formulas reported in Table 6.3 contain one gram-atom of carbon. Herbert (1976) advocates the use of this *C-mole* for the following reasons: (1) the carbon content of microorganisms varies little with growth conditions and limiting nutrient (see Table 6.5), (2) carbon is the most abundant element in the cell, so that changes in other elements have little effect on the formula weight, and (3) the carbon balance is usually of greater importance.

The chemical composition of microorganisms can be considerably affected by the growth rate and the nature of the growth-limiting nutrient (Herbert, 1958, 1961). Some illustrative data are shown in Table 6.5, which gives analyses for protein, total carbohydrate, RNA, and DNA of *Klebsiella aerogenes* and *Candida utilis* (a yeast) grown in a chemostat at high and low dilution rates (μ_{max} values for these strains were 0.95 and 0.55 h^{-1}, respectively). In both organisms,

TABLE 6.5 Molecular Composition of *Klebsiella aerogenes* and *Candida utilis* Cells[a]

	Klebsiella aerogenes NCTC 418 (37°C, pH 7.0)			
Limiting nutrient:	Glycerol		Ammonia	
Dilution rate (h^{-1})	0.1	0.85	0.1	0.85
Protein (%)	74.1	72.0	65.6	73.4
Carbohydrate (%)	1.4	1.9	18.6	1.8
RNA (%)	11.2	21.6	8.2	19.9
DNA (%)	4.4	2.5	3.1	3.2
	Candida utilis NCYC 321 (30°C, pH 5.5)			
Limiting nutrient:	Glucose		Ammonia	
Dilution rate (h^{-1})	0.05	0.45	0.05	0.45
Protein (%)	58.8	55.8	43.2	56.7
Carbohydrate (%)	20.9	19.5	36.2	19.0

[a]Grown in continuous culture at different growth rates. All data expressed as percentage of dry weight of cells on an ash-free basis. Mean ash content was 3.6% for *Klebsiella aerogenes* NCTC 418 and 7.0% for *Candida utilis* NCYC 321 (Evans et al., 1970).

there is a large increase in total carbohydrate (mainly glycogen) when grown at low dilution rates in N-limiting conditions; they differ, however, in that the yeast has a considerable carbohydrate content (mainly in the cell wall) even under C-limiting conditions, while C-limited *Kl. aerogenes* has a low carbohydrate content. Characklis and Dydek (1976) observed increases in capsule and slime material in biofilm bacteria with increasing C:N ratio in the water. The increase in capsular material also was reflected in a higher chlorine demand of the cells. There is a large increase in RNA content with dilution rate in *Kl. aerogenes*, whatever the limiting nutrient; similar, though smaller, changes are found in *C. utilis*. Thus, chemical composition of cells is not necessarily a constant.

One of the difficulties in assessing the composition of microbial cellular mass is the definition of biomass. In most cases, biomass includes the cellular mass plus any products that are intimately associated with the cells, such as the extracellular polymeric substances (EPS) that make up the capsule and slime layer (see Section 5.4.4). This most certainly is the case with *Kl. aerogenes*, which is known to produce a substantial extracellular capsule (Table 6.5). Since the oxygen equivalents of cell mass and polysaccharides are similar, changes in relative fractions of EPS and cells in suspended biomass or biofilms may be difficult to detect with such insensitive analyses as chemical oxygen demand (COD) (Grady et al., 1975). Methods have been developed (Chapter 3) that permit the structured modeling of biofilms in which the biofilm mass is separated into two compartments: (1) cellular carbon and (2) EPS carbon (Robinson et al., 1984; Bakke et al., 1984). The significance of this carbon distribution on process modeling is discussed in Chapter 13.

Other elements may also influence biological processes. Turakhia and Characklis (1989) studied the influence of Ca on biofilm processes. Increasing Ca in the water significantly increased Ca levels in the biofilm (Figure 6.4). Although there was no influence on microbial growth rate between Ca^{+2} at 25 and 50 g m^{-3}, the rate of cellular detachment from the biofilm decreased significantly, as indicated by increased cell carbon in the biofilm (Figure 6.5). However, Ca biofilm content had little influence on the biofilm composition (Table 6.6).

Turakhia (unpublished results) also conducted an experiment with cadmium in a biofilm reactor. After the mature biofilm had achieved steady state, cadmium (0.05 g m^{-3}) was added to the reactor continuously for 24 h at 0.18 (mg Cd) h^{-1}. After 24 h, the biofilm contained 3.0 mg Cd per gram of biofilm (dry weight). Approximately 10% of the applied Cd had been immobilized in the biofilm. Thus, other minor elements may be found in biofilm as a result of metabolism or resulting from physicochemical interactions with the organic EPS matrix.

6.3.4 Product Formation

Extracellular polymeric substances (EPS) are extracellular products formed by bacteria, which are of major importance in biofilm systems. Product (EPS) for-

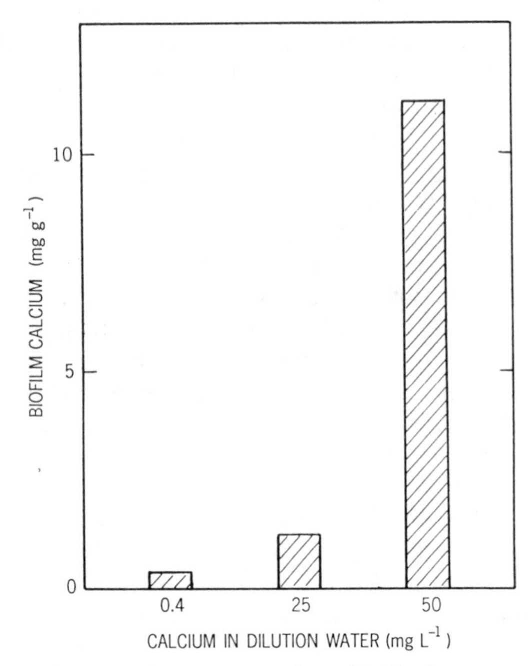

Figure 6.4 Calcium in a *Pseudomonas aeruginosa* biofilm increases with increasing calcium concentration in the nutrient solution (Turakhia and Characklis, 1988).

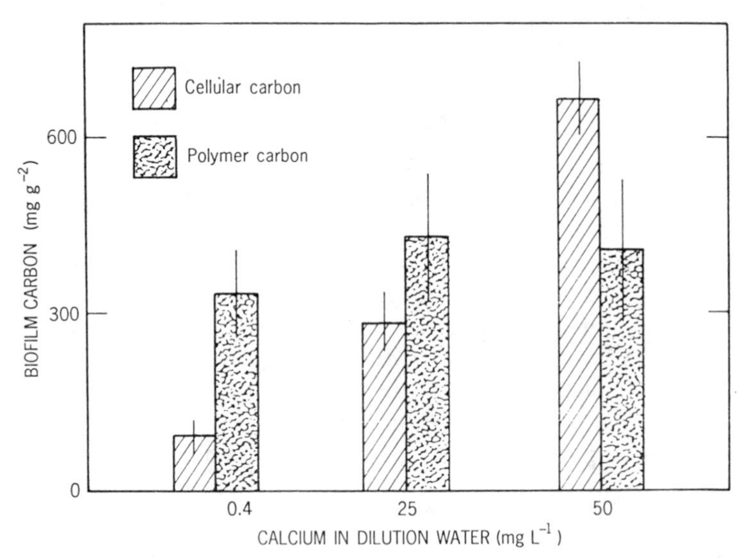

Figure 6.5 Biofilm cellular carbon increased with increasing calcium concentration in the nutrient solution, while EPS carbon did not change significantly (Turakhia and Characklis, 1988).

TABLE 6.6 Elemental Composition of Biofilm (Ash-Free Basis) as a Function of Calcium Concentration in the Nutrient Solution

Calcium	Composition (% Dry Weight)				
(g m^{-3})	C	H	N	O[a]	Elemental Formula
0.4	55.0	8.0	14.3	22.7	$CH_{1.74}N_{0.22}O_{0.3}$
25	55.5	8.1	14.2	22.2	$CH_{1.75}N_{0.22}O_{0.30}$
50	55.3	7.8	14.9	22.9	$CH_{1.70}N_{0.23}O_{0.30}$

[a] By difference.

mation rate has frequently been related to specific growth rate by the Leudeking–Piret equation:

$$r_p = k\mu + k' \tag{6.24}$$

where r_p = specific product formation rate ($M_P M_X{}^{-1} t^{-1}$)
k = growth-associated rate coefficient ($M_P M_X{}^{-1}$)
k' = nongrowth-associated rate coefficient ($M_P M_X{}^{-1} t^{-1}$)

Several sets of product formation rate coefficients are listed in Table 6.7 for comparison.

Product formation, especially EPS, is critically important in interpreting biomass yield determinations. Neglecting product formation by *Ps. aeruginosa* leads to significant overestimates in cellular yields. By lumping the biomass into one term, Robinson et al. (1984) calculated a biomass yield for *Ps. aeruginosa* in a chemostat equal to 0.6 g of biomass-carbon formed per gram of glucose carbon. If polymer carbon is considered as a separate entity, $Y_{X/S}$ = 0.3 g of cell carbon per gram of glucose carbon consumed. The difficulty with this yield is that some of the substrate carbon consumed was channeled to synthesis of EPS. Thus, the substrate carbon (S-carbon) is partitioned into the following processes: (1) S-carbon is oxidized to provide energy for cell synthesis, (2) S-carbon is used for cell carbon, (3) S-carbon is oxidized to provide energy for EPS synthesis, and (4) S-carbon is used for EPS carbon (Figure 6.6). Quantitative data for distinguishing between these various sinks for substrate carbon are not available. However, Robinson et al. (1984) suggest that more substrate is converted by *Ps. aeruginosa* to EPS at low growth rates. The observation is consistent with the findings of others (Jarman, 1979) that the EPS per cell is inversely related to the specific growth rate. The results of Robinson and Jarman were obtained in a chemostat and may reflect the residence time of the cells in the chemostat. For example, the EPS formation rate was slow at low growth rate. However at low growth rate (i.e., high cell residence times in the chemostat), the cells have more time to *accumulate* EPS, which remains in intimate contact with the cell. No experimental data to test this hypothesis are available.

Turakhia and Characklis (1988), working with a *Ps. aeruginosa* biofilm and

TABLE 6.7 Rate Coefficients for Product Formation Rate[a]

Organism	Product	k [(g P) (g X)$^{-1}$]	k' [(g P) (g X)$^{-1}$ s^{-1}]	Reference
Ps. aeruginosa	EPS	0.3	0.04	Robinson et al., 1984
Ps. aeruginosa	EPS	2.1	0	Turakhia, 1986
Ps. aeruginosa[b]	EPS	3.3	0.1	Mian et al., 1978; Williams and Wimpenny, 1977; Moraine and Rogovin, 1971
Xanthomonas campestris	Xanthan gum	1.8	0.2	Ollis, 1983

[a] Units for coefficients are in terms of product carbon and cell carbon. Experiments of Turakhia (1986) were with *Pseudomonas aeruginosa* biofilms.
[b] Nitrogen-limited system.

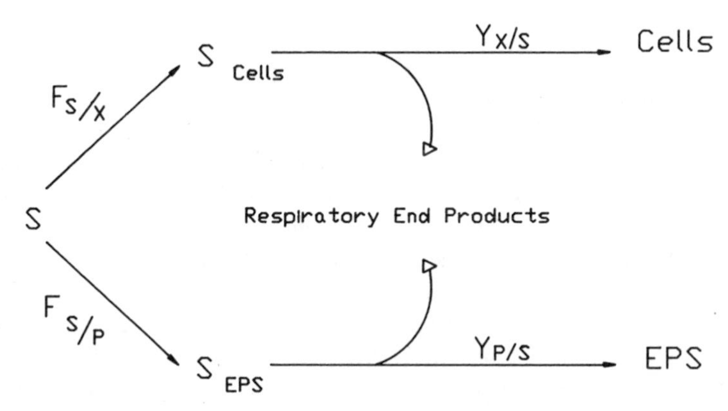

Figure 6.6 Substrate carbon is partitioned into metabolic energy, cells, and EPS within a *Pseudomonas aeruginosa* biofilm (after Robinson et al., 1984). The substrate carbon is partitioned into the following processes: (1) substrate is oxidized to provide energy for cell synthesis, (2) substrate is used for cell carbon, (3) substrate is oxidized to provide energy for EPS synthesis, and (4) substrate is used for EPS carbon.

using the same media as Robinson et al. (1984), determined coefficients for EPS formation that were significantly larger than reported by Robinson and his co-workers. At least two possible explanations can be considered: (1) the biofilm organisms were generally growing slower and/or (2) biofilm organisms produce more EPS than their planktonic counterparts. The results clearly indicate the need for more definitive studies that can distinguish between biofilm cell carbon and biofilm EPS carbon.

Namkung and Rittmann (1986) examined soluble product formation by an aerobic, mixed population biofilm, utilizing phenol as the primary substrate at low inlet concentrations [0.1–1.0 (g C) m^{-3}]. The soluble products were mostly high molecular weight, ranging from below 500 to over 10,000 daltons, while the substrate molecular weight was 94 daltons. The growth-associated product formation was much greater than the non-growth-associated product formation, as was reported for EPS formation by *P. aeruginosa* above. Trulear (1983), on the other hand, observed no measurable soluble organic products in experiments with *P. aeruginosa* for inlet glucose concentrations ranging from 2 to 16 (g C) m^{-3}. Differences between the experiments of Trulear and those of Namkung and Rittmann, which account for the divergent observations, may include the microbial populations and the substrate composition.

6.4 NUTRIENTS

A number of metallic and other inorganic substances are necessary in microbial metabolism and are generally termed *essential nutrients*. Nitrogen, phosphorus, and sulfur are in this category. Other elements such as Mg, Ca, K, Fe, Mn, Zn,

Cu, and Co are also essential and are usually termed *trace elements* because of the relatively small amount needed. In most applications, nitrogen and phosphorus are the most critical.

Busch and Myrick (1960) used a closed, aerobic, mixed population, growth-limited system (the BOD bottle system) to determine the NH_4^+-N requirements for *maximum biomass yield* from glucose carbon (Eq. 6.1): C:N = 7.3. Although this results in maximum biomass yield, C:N in the biomass is 4.3 (Eq. 6.1). The difference between the two ratios represents the carbon oxidized to CO_2 for energy. The conclusion from these and other results is that C:N > 7.3 results in nitrogen-limited growth in an aerobic system. Nitrogen-limited growth generally results in the formation of higher quantities of extracellular polymeric substances. Escher (unpublished results) indicates that adsorbed cells grown on nitrogen-limited media do not generate the adhesive force to the substratum observed with carbon-limited cells. This suggests that proteins, whose synthesis is retarded in nitrogen-limited media, play a major role in adsorption processes.

Lewis and Busch (1964) determined the NO_3^--N requirements for a similar system and found C:N = 9.0. The C:N is higher for NO_3^--N because more substrate energy is required to reduce the NO_3^--N to NH_4^+-N for synthesis. Thus, more carbon must be oxidized to CO_2.

Schaezler et al. (1970) established the phosphorus requirement in a closed, mixed culture, growth-limited system (the BOD bottle system): C:P = 49.4. They also conducted experiments with *E. coli* and mixed cultures in shake flasks to determine phosphorus requirements for maximum synthesis, with the following results: C:P = 81.6 for *E. coli*; C:P = 65.6 for mixed culture. Higher substrate concentration in the shake flask system may have increased the amount of nutrient cycling in the cell population, thus resulting in higher C:P in the shake flasks. However, mixing may also have an influence. The shake flask is mixed while the BOD bottle is not. The mixing may also result in more rapid cycling of nutrients within the population. Nutrient cycling may occur to an even greater extent in anaerobic systems. C:N and C:P will be much higher in anaerobic systems, presumably because cell yields are low.

6.5 OXYGEN

Oxygen consumption is stoichiometrically related to CO_2 and biomass production as indicated in Eq. 6.1. Therefore, it is reasonable to use oxygen consumption to follow the progress of a microbial reaction in an aerobic environment. Cumulative oxygen consumption, like biomass production, is a logistic function of time. The progression of cumulative oxygen consumption by heterotrophic mixed cultures of bacteria utilizing a single organic carbon nutrient exhibits a plateau corresponding to the utilization of all primary nutrient (Busch, 1958, 1965). A plateau can also result from consumption of all the oxygen if initial dissolved oxygen is stoichiometrically limiting. In addition, several plateaus may result in multicomponent substrate mixtures, although acclimated mixed cultures may subsequently produce a single plateau for the same mixture.

The relationship of oxygen consumption to microbial stoichiometry in a batch, substrate-saturated system is illustrated in Figure 6.7. A material balance approach, using oxygen equivalents, is convenient for analysis of this system. As an example, consider the oxidation of glucose in abiotic and biotic environments. The oxygen demand for complete oxidation of glucose in a abiotic environment can be determined from the following equation:

$$C_6H_{12}O_6 + 6O_2 \rightarrow 6CO_2 + 6H_2O \quad [6.25]$$

which indicates a theoretical oxygen demand (ThOD) of 1.07 (g oxygen) (g glucose)$^{-1}$ [6 (mol O_2) (mol glucose)$^{-1}$]. Glucose can also be oxidized biochemically in a batch, substrate-saturated microbial reactor. A plateau in oxygen consumption will result when all substrate is exhausted and may persist for as long as 8 h (Figure 6.8). An oxygen material balance at plateau indicates that only 41% of the glucose ThOD [0.44 (g oxygen) (g glucose)$^{-1}$] has been supplied by consumption of all the glucose. This fraction of the ThOD corresponds to the oxidation of glucose to CO_2, which is the primary source of metabolic energy. The remaining substrate oxygen demand resides in the synthesis products, i.e., the biomass. The substrate oxygen demand residing in the biomass can be determined from the biomass ThOD:

$$3.5CH_{1.4}O_{0.4}N_{0.2} + 3.5O_2 \rightarrow 3.5CO_2 + 1.4H_2O + 0.7NH_3 \quad [6.26]$$

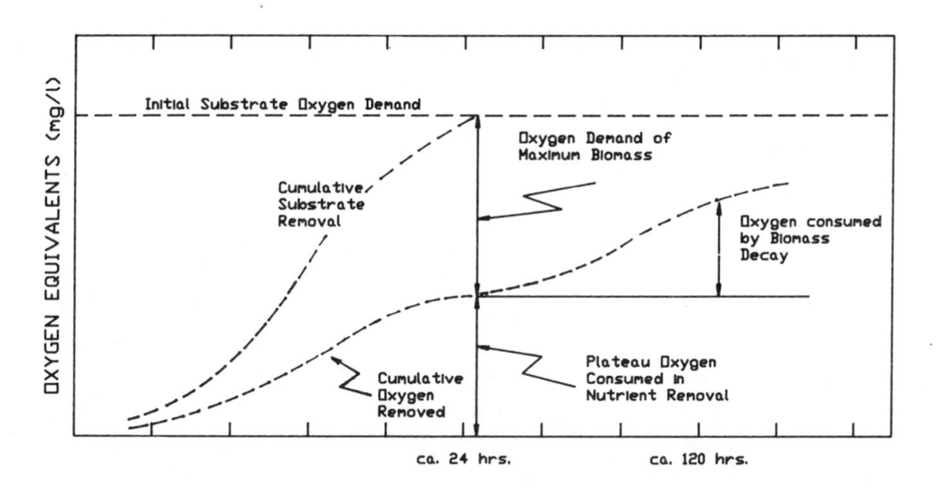

Figure 6.7 Progression of oxygen consumption by microbial cells metabolizing substrate in a batch, aerobic, initially substrate-saturated system. The earlier part of the progression represents oxygen demand of the substrate for growth. The later part represents oxygen demand for biomass decay after the primary substrate is depleted. (After Busch, 1971.)

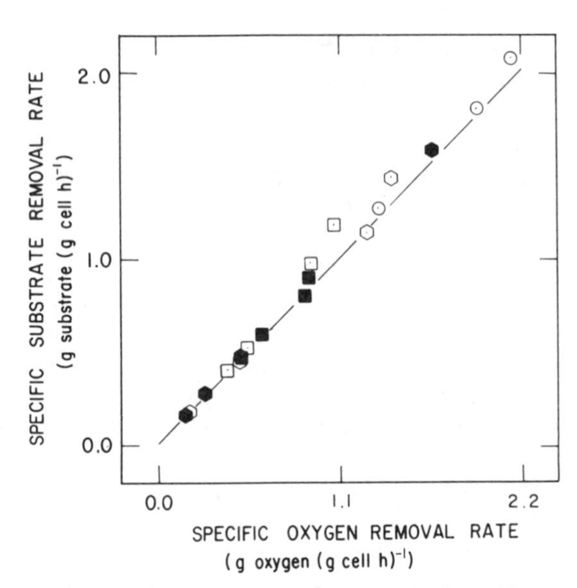

Figure 6.8 Specific substrate removal rate as a function of specific oxygen removal rate in a *Ps. aeruginosa* biofilm. The line represents the best linear fit, and the various symbols refer to different calcium concentrations in the nutrient solution (Turakhia and Characklis, 1988). The results are consistent with Eq. 6.1. Various symbols refer to different experimental runs.

Then, the sum of Eqs. 6.26 and 6.1 describes the total oxidation of glucose in a biotic environment:

$$C_6H_{12}O_6 + 2.5O_2 + 0.7NH_3 \rightarrow 3.5CH_{1.4}O_{0.4}N_{0.2} + 2.5CO_2 + 4.6H_2O \quad [6.1]$$

$$+\ 3.5CH_{1.4}O_{0.4}N_{0.2} + 3.5O_2 \rightarrow 3.5CO_2 + 0.7NH_3 + 1.4\ H_2O \quad [6.26]$$

$$C_6H_{12}O_6 + 6O_2 \rightarrow 6CO_2 + 6H_2O \quad [6.25]$$

The oxygen material balance is complete, since all of the oxygen (6 moles of molecular oxygen from Eq. 6.25) has been recovered. By using the plateau as the reaction endpoint, the oxygen consumed by the biomass in removing soluble nutrient can be separated from the oxygen required for the subsequent oxidation of biomass (i.e., biomass decay in Eq. 6.26). The plateau also provides other reaction information important to the process. It is the minimum oxygen required to degrade the soluble organic substrate aerobically. In addition, the time required to reach the plateau represents the maximum time necessary for the aerobic degradation of the substrate to CO_2, biomass, and water.

The plateau is characteristic of a bacterial growth process that is limited by organic carbon concentration. A deficiency of an essential nutrient, the degradation of particulate or colloidal nutrients, or the presence of a large fraction of microbial predators can obscure the plateau (Busch, 1971).

Turakhia (1986) conducted experiments with *Ps. aeruginosa* in a batch respirometer, in a chemostat, and in a RotoTorque (biofilm) to monitor substrate (glucose), dissolved oxygen, soluble organic carbon, and cellular carbon (biofilm and suspended). The average value for substrate (glucose) carbon consumed per unit oxygen consumed in the batch respirometer was $Y_{S/O} = 0.89 \pm 0.06$ g of substrate carbon per gram of oxygen. This value compares very well with Eq. 6.1 (calculated $Y_{S/O} = 0.90$), validated with mixed populations. Turakhia and Characklis (1988) also determined $Y_{S/O}$ for a *Ps. aeruginosa* biofilm by plotting specific substrate (glucose) removal rate versus oxygen removal rate (Figure 6.8). The slope of the line is $Y_{S/O} = 0.92 \pm 0.02$ g of substrate carbon per gram of oxygen. No significant diffusional resistance was observed in the experiments of Turakhia and Characklis. We may conclude that $Y_{S/O}$ is the same in a biofilm as in a batch reactor as long as the microenvironment of the cell is the same in terms of dissolved oxygen and substrate.

There appears to be no major influence of the biofilm environment on microbial stoichiometry.

APPENDIX: BALANCING A REDOX REACTION

Balance the reaction in which ferrous iron (Fe^{+2}) is oxidized to ferric iron (Fe^{+3}) by oxygen (O_2) in aqueous solution. The oxygen is reduced to water (H_2O). The reaction takes place in alkaline solution.

Solution

Working with half reactions, the following procedure is used for balancing a redox reaction:

1. Identify the principal reactants and products (i.e., other than H^+ and OH^-, and write each half reaction without stoichiometric coefficients).
2. Balance the atoms other than H and O in each half reaction.
3. Balance the oxygen (O) by using water (H_2O).
4. Balance the hydrogen (H) by using hydrogen ion (H^+).
5. Balance the electrical charge by using electrons (e^-).
6. Multiply each half reaction by an appropriate integer so that both contain the same number of electrons.
7. Add the two balanced half-reactions.

This procedure sometimes produces an equation containing H^+ as a reactant or product. If you know that such a reaction takes place in alkaline solution, add the reaction for dissociation of water to the balanced equation to eliminate H^+ and form H_2O.

1. Reactants and products are

$$Fe^{+2} \rightleftharpoons Fe^{+3}$$
$$O_2 \rightleftharpoons H_2O$$

2. Atoms other than H and O are already balanced.
3. Balance oxygen with water:

$$Fe^{+2} \rightleftharpoons Fe^{+3}$$
$$O_2 \rightleftharpoons 2H_2O$$

4. Balance hydrogen with H^+:

$$Fe2^{+} \rightleftharpoons Fe^{+3}$$
$$O_2 + 4H^+ \rightleftharpoons 2H_2O$$

5. Balance the charge with electrons (e^-):

$$Fe^{+2} \rightleftharpoons Fe^{+3} + e^-$$
$$O_2 + 4H^+ + 4e^- \rightleftharpoons 2H_2O$$

6. Multiple the Fe half-reaction by 4 and add the half-reactions.

$$4Fe^{+2} \rightleftharpoons 4Fe^{+4} + 4e^-$$
$$\underline{O_2 + 4H^+ + 4e^- \rightleftharpoons 2H_2O}$$
$$4Fe^{+2} + O_2 + 4H^+ \rightleftharpoons 4Fe^{+3} + 2H_2O$$

7. The equation is balanced, but you were told that the reaction takes place in alkaline solution. Add the equation for dissociation of water to eliminate H^+ as a reactant:

$$4Fe^{+2} + O_2 + 4H^+ \rightleftharpoons 4Fe^{+3} + 2H_2O$$
$$\underline{4H_2O \rightleftharpoons 4H^+ + 4OH^-}$$
$$4Fe^{+2} + O_2 + 2H_2O \rightleftharpoons 4Fe^{+3} + 4OH^-$$

In this example, ferrous iron (Fe^{+2}) is the reducing agent and oxygen is the oxidizing agent. Fe^{+2} donates its electrons, which are accepted by the oxygen.

REFERENCES

Bakke, R., M.G. Trulear, J.A. Robinson, and W.G. Characklis, *Biotech. Bioeng.*, **26**, 1418-1424 (1984).

Bauchop, T., *J. Gen. Microbiol.*, **18**, 7 (1958).

Bauchop, T. and S.R. Elsden, *J. Gen. Microbiol.*, **23**, 457-469 (1960).

Bauer, S. and E. Ziv, *Biotech. Bioeng.*, **18**, 81-94 (1976).

Busch, A.W., "BOD progression in soluble substrates," in *Proc., 13th Industrial Waste Conference,* Purdue University, West Lafayette, IN, (1958), pp. 1336-1342.

Busch, A.W., *Chem. Engrg.*, **72**, 83 (1965).

Busch, A.W., *Aerobic Biological Treatment of Wastewaters,* Gulf Publishing Co., Houston, TX, 1971.

Busch, A.W. and N. Myrick, *J. Wat. Poll. Contr. Fed.*, **32**, 741 (1960).

Characklis, W.G. and S.T. Dydek, *Wat. Res.*, **10**, 515-522 (1976).

D'Alessandro, P.A., W.G. Characklis, M.A. Kessick, and C.H. Ward, *Dev. Ind. Microbiol.*, **15**, 294-302 (1974).

Evans, C.G.T., D. Herbert, and D.W. Tempest, "The continuous cultivation of microorganisms," in J.R. Norris and D.W. Ribbons, (eds.), *Methods in Microbiology,* Vol. 2, Academic Press, London, 1970, p. 277.

Grady, C.P.L., P.L. Findley, and R.E. Muck, *Biotech. Bioeng.*, **17**, 859-872 (1975).

Hadjipetrou, L.P. and A.H. Stouthamer, *J. Gen. Microbiol.*, **38**, 29 (1965).

Harrison, J.S., *Process Biochem.*, **2**, 41 (1967).

Herbert, D., "Some principles of continuous culture," in G. Tunevall (ed.), *Recent Progress in Microbiology,* Blackwell Scientific Publishers, Oxford, (1958), pp. 381-396.

Herbert, D., *Symp. Soc. Gen. Microbiol.*, **11**, 391 (1961).

Herbert, D., "Stoichiometric aspects of microbial growth" in A.C.R. Dean, D.C. Ellwood, C.G.T. Evans, and J. Melling (eds.), *Continuous Culture 6: Applications and New Fields,* Ellis Horwood Ltd., Chichester, 1976, pp. 1-30.

Hernandez, E. and M.J. Johnson, *J. Bacteriol.*, **94**, 991-995 (1967).

Jarman, T.R., in R.C.W. Berkeley, G.W. Gooday, and D.C. Ellwood (eds.), *Microbial Polysaccharides and Polysaccharidases,* Academic Press, New York, 1979, pp. 35.

Kok, H.E. and J.A. Roels, *Biotech. Bioeng.*, **22**, 1097-1104 (1980).

Lehninger, A.L., *Bioenergetics,* 2nd ed., W.A. Benjamin, Menlo Park, CA, 1971.

Lewis, J.W. and A.W. Busch, "BOD progression in soluble substrates VII: The quantitative error due to nitrate as a nitrogen source" in *Proc. 19th Industrial Waste Conference,* Purdue University, West Lafayette, IN, 1964.

Mayberry, W.R., G.L. Prochazka, and W.J. Payne, *J. Bact.*, **96**, 1424 (1968).

McCarty, P.L., "Energetics of organic matter degradation," in R. Mitchell (ed.), *Water Pollution Microbiology,* Wiley-Interscience, New York, 1972, pp. 91-118.

McWhorter, T.R. and H. Heukelekian, "Growth and endogenous phases in the oxidation of glucose," in *Advances in Water Pollution Research,* Water Pollution Control Federation, Washington, 1964.

Mian, F.A., T.R. Jarman, and R.C. Righelato, *J. Bact.*, **134**, 418 (1978).

Middleton, A.C. and A.W. Lawrence, *J. Wat. Poll. Contr. Fed.*, **49**, 1659-1670 (1977).

Morain, R.A. and P. Rogovin, *Biotech. Bioeng.*, **13**, 381 (1971).

Nagai, S. "Mass and energy balances for microbial growth kinetics," in *Advances in Biochemical Engineering,* T.K. Ghose, A. Fiechter, and N. Blakebrough (eds.), Springer-Verlag, New York, 1979, pp. 53.

Namkung, E. and B.E. Rittmann, *Water Res.*, **20**, 795-806 (1986).

Ollis, D. F., *Ann. N.Y. Acad. Sci.*, **413**, 144 (1983).

Payne, W.J., *Ann. Rev. Microbiol.*, **24**, 17 (1970).

Porges, N., L. Jasewicz, and S.R. Hoover, "Principles of biological oxidation," in J. McCabe and W.W. Eckenfelder (eds.), *Biological Treatment of Sewage and Industrial Wastes,* Rheinhold Publishing Corp., New York, NY (1956), pp. 35-48.

Robinson, J.A., M.G. Trulear, and W.G. Characklis, *Biotech. Bioeng.*, **26**, 1409-1417 (1984).

Roels, J.A. *Energetics and Kinetics in Biotechnology,* Elsevier Biomedical Press, Amsterdam, 1983.

Schaezler, D.J., A.W. Busch, and C.H. Ward, *Dev. Ind. Microbiol.*, **11**, 255-277 (1970).

Servizi, J.A. and R.H. Bogan, *J. Sanit. Engrg. Div. ASCE,* **89**, 17-40 (1963).

Shimizu, T., T. Furuki, T. Waki, and K. Ichikawa, *J. Ferment. Tech.*, **56**, 207 (1978).

Stouthamer, A.H., *Antonie van Leeuwenhoek,* **43**, 351-367 (1977).

Stumm, W. and J.J. Morgan, *Aquatic Chemistry,* 2nd Edition, John Wiley and Sons, New York, NY (1981).

Sykes, R.M., *J. Wat. Poll. Contr. Fed.*, **47**, 591-600 (1975).

Traore, A.S., C.E. Hatchikian, J. Legall, and J.P. Belaich, *J. Bact.*, **149**, 606-611 (1982).

Trulear, M.G., "Cellular reproduction and extracellular polymer formation in the development of biofilms," unpublished doctoral thesis, Montana State University, Bozeman, 1983.

Turakhia, M.H., "The influence of calcium on biofilm processes," unpublished doctoral thesis, Montana State University, Bozeman, MT, 1986.

Turakhia, M.H. and W.G. Characklis, *Biotech. Bioeng.*, **33**, 406-414 (1988).

Wang, H.Y., D.G. Mou, and J.R. Swartz, *Biotech. Bioeng.*, **18**, 1811-1814 (1976).

Williams, A.G. and J.W.T. Wimpenny, *J. Gen. Microbiol.*, **102**, 13 (1977).

PART III

PROCESS RATES

BIOFILM PROCESSES

WILLIAM G. CHARACKLIS
Montana State University, Bozeman, Montana

SUMMARY

The progression of biofilm accumulation frequently takes the form of a sigmoidal curve, which can be arbitrarily divided into three sequential phases: (1) initial events, (2) exponential or log accumulation, and (3) plateau or steady state. The individual biofilm processes can be grouped into various transport, interfacial transfer, and transformation processes that contribute to biomass accumulation at a substratum. Thus, biofilm accumulation is the net result of the following physical, chemical, and biological processes: (1) First, organic molecules accumulate at the substratum, resulting in a "conditioned" substratum, (2) planktonic microbial cells are transported from the bulk water to the conditioned substratum, (3) a fraction of the cells that reach the substratum reversibly adsorb to the substratum, (4) some reversibly adsorbed cells desorb, (5) a fraction of the reversibly adsorbed cells remain immobilized and become irreversibly adsorbed, and (6) the irreversibly adsorbed cells grow at the expense of substrate and nutrients in the bulk water, increasing biofilm cell numbers and forming other metabolic products. The extracellular polymeric substances (EPS) that hold the biofilm together are especially important to the integrity of the biofilm. Then (7) cells and other particulate matter attach to the biofilm, increasing biofilm accumulation, and, finally, (8) portions of the biofilm detach and are reentrained in the bulk water.

This progression and the processes contributing to accumulation provide a framework for further analysis of the rate and extent of biofilm accumulation.

7.1 INTRODUCTION

In most, if not all, reported research on biofilms, certain observed or measured quantities are reported: net biofilm accumulation and/or substrate (or oxygen) removal. A difficulty with these observed quantities is that they reflect the contribution of several processes of more fundamental significance. For example, net biofilm accumulation results from the combination of the following processes: (1) transport of cells to the substratum, (2) adsorption of cells to the substratum, (3) growth and other metabolic processes within the biofilm, and (4) detachment of portions of the biofilm. If all of the processes occur in series, the slowest step of the sequence may exert the greatest influence and limit the overall process rate. This step is called the *rate-determining step* or *rate-limiting step*. If the overall process consists of a number of parallel processes (or processes in series and parallel), the slowest process becomes the *rate-controlling step*. Identifying the rate-controlling and/or rate-limiting step is critical to successful scaleup procedures, and its determination contributes significantly to the insight gained from experimental results. Process analysis permits the determination of the rate-limiting or rate-controlling step in the overall process at different environmental, operating, or physiological conditions.

Biofilm accumulation is the net result of the following physical, chemical, and biological processes:

1. Organic molecules are transported from the bulk fluid to the substratum, where some of them adsorb, resulting in a *conditioned* substratum (Figure 7.1a).
2. A fraction of the planktonic microbial cells are *transported* from the bulk water to the conditioned substratum (Figure 7.1b).
3. A fraction of the cells that strike the substratum adsorb to the substratum for some finite time, and then desorb. This process is termed *reversible adsorption* (Figure 7.1b).
4. *Desorption* may result from fluid shear forces, but other physical, chemical, or biological factors may also influence the process (Figure 7.1b).
5. A fraction of the reversibly adsorbed cells remain immobilized beyond a "critical" residence time and become *irreversibly adsorbed* (Figure 7.1b).
6. The irreversibly adsorbed cells grow at the expense of substrate and nutrients in the bulk water, increasing biofilm cell numbers (Figure 7.1c). The cells may also form significant amounts of products, some of which may be excreted. One class of products is the extracellular polymeric substances (EPS), which hold the biofilm together. Thus, biofilm accumulation increases through *microbial metabolism* at the expense of substrate energy in the bulk water.
7. Cells and other particulate matter attach to the biofilm, increasing biofilm accumulation (Figure 7.1d). *Attachment* is the immobilization of cells and

other particulate matter in the biofilm, while adsorption refers to the same processes occurring on the substratum.

8. Portions of the biofilm detach and are reentrained in the bulk water (Figure 7.1d). *Detachment* refers to the loss of material from the biofilm, while desorption is loss of cells and other material from the substratum. Detachment may be termed erosion or sloughing, depending on the nature of the biofilm loss. Cell multiplication can also lead to release of daughter cells into the bulk water.

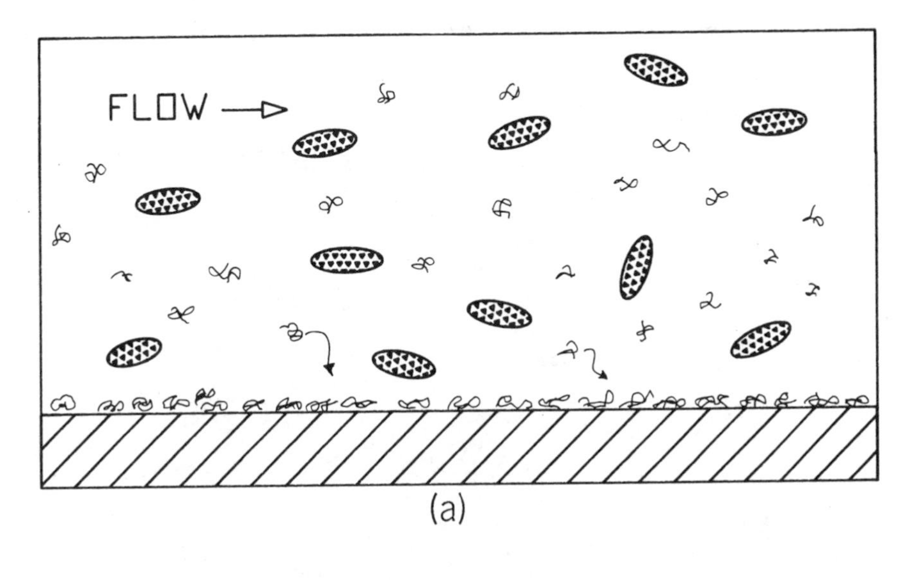

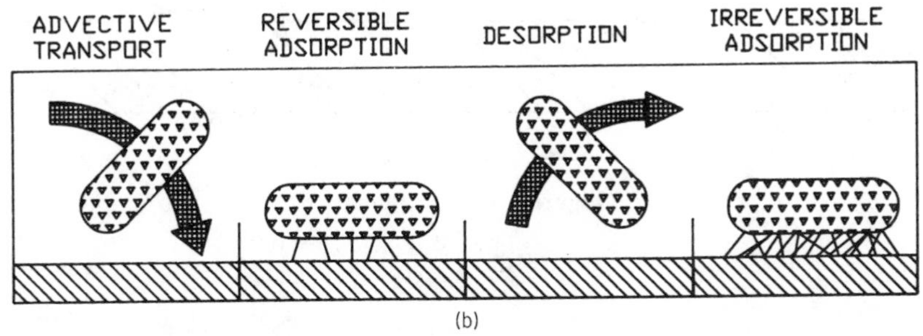

Figure 7.1 (a) Transport and adsorption of organic molecules on a clean substratum, forming a conditioning film. (b) Transport of microbial cells to the conditioned substratum, and adsorption, desorption, and irreversible adsorption of cells at the substratum. (c) Growth and multiplication of microbial cells at the substratum at the expense of substrate in the water. (d) Attachment and detachment of cells and other particles to and from the biofilm.

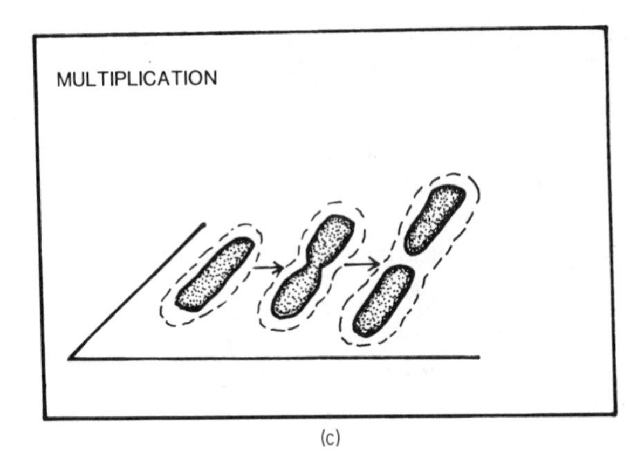

(c)

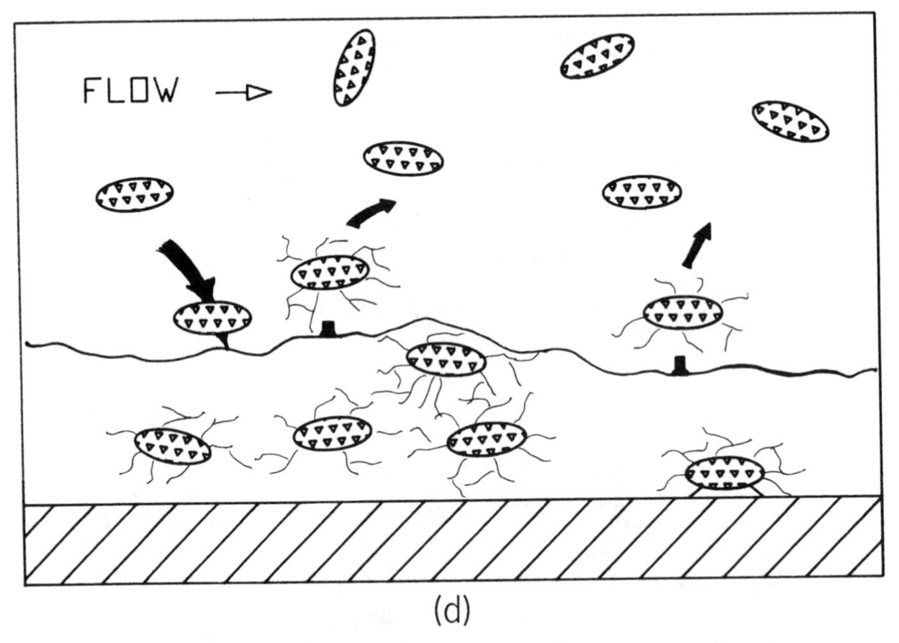

(d)

Figure 7.1 Continued

The progression of biofilm accumulation frequently takes the form of a sigmoidal curve (Figure 7.2), which can be divided arbitrarily into three phases: initial events, exponential or log accumulation, and plateau or steady state. The individual processes can be grouped into transport, interfacial transfer, and transformation processes.

Interfacial transfer processes describe the transfer of energy or material between the compartments or phases (interphase transport), while *transport* de-

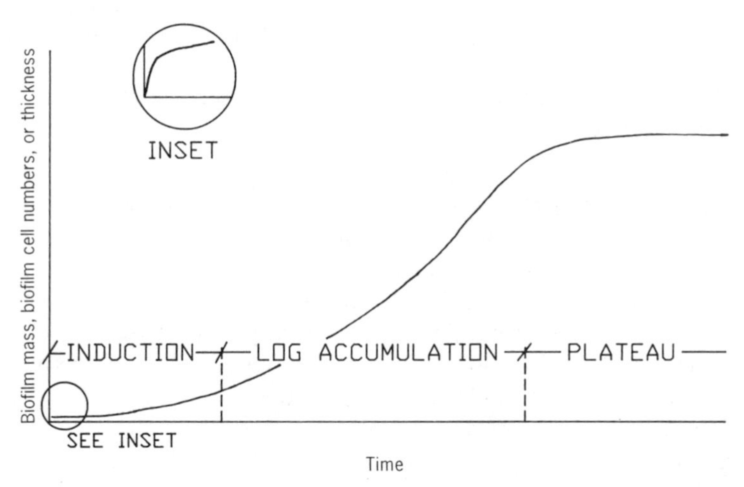

Figure 7.2 The progression of biofilm accumulation frequently takes the form of a sigmoid curve. The progression can be arbitrarily divided into three phases: (1) initial events, (2) exponential or log accumulation, and (3) plateau or steady state. The inset describes the initial accumulation of the conditioning film, which is essentially negligible in terms of thickness or mass deposition.

scribes the transport of energy or material within the system compartments or phases (intraphase transport). Of the fundamental processes listed above, all can be classified as interfacial transfer or transport processes except for metabolic processes of the immobilized cells, which are transformation processes. More detail on interfacial transfer and transport processes and the rate expressions describing them is provided in Chapter 9.

Transformation processes are those resulting in molecular transformations of matter, i.e., chemical or biochemical reactions. Transformation processes are described by rate equations or, more specifically, constitutive or kinetic expressions of the following form:

$$r = r(C_1, \ldots, C_n) \qquad [7.1]$$

where $C_1, \ldots, C_n$ are concentrations of the various components. Kinetic expressions are the focus of Chapter 8. The microbial transformation processes can be classified further: (1) microbial growth and reproduction, (2) product formation (e.g., EPS), (3) maintenance, and (4) decay. These are unstructured processes (see Chapter 2) in that specific intracellular transformations are not considered.

The above processes are considered fundamental because accepted mathematical descriptions for each process exist. In addition, further subdivision of these processes is difficult if dependence on observable quantities is required for validation of the accuracy of the mathematical expressions.

7.2 TRANSPORT AND INTERFACIAL TRANSFER PROCESSES

7.2.1 Substratum Conditioning: Transport and Adsorption of Organic Molecules

7.2.1.1 *Transport of Molecules* Transport of molecules and small particles (<0.01–0.1 μm) in quiescent or laminar flow is described satisfactorily in terms of molecular diffusion, i.e., Fick's law (Chapter 9). In turbulent flow, the diffusion equation must be modified to include turbulent eddy transport (an eddy is a current or bundle of fluid moving contrary to the main current). Compared to larger particles such as bacterial cells, transport of molecules and small particles is quite rapid. Consequently, adsorption of an organic conditioning film is frequently reported to occur "instantaneously." In further discussions regarding the cell–substratum interaction, the substratum is generally presumed to be conditioned.

7.2.1.2 *Adsorption of the Conditioning Film* Adsorption of an organic film is an interfacial transfer process (i.e., the molecule is transferred from the bulk liquid compartment to the substratum compartment) and occurs within minutes of exposure, causing changes in the properties of the wetted surface. Little and Zsolnay (1985) measured as much as 0.8 mg m^{-2} organic matter on stainless steel after 15 min exposure in seawater. Bryers (1980) observed 15 mg m^{-2} of organic matter (measured as chemical oxygen demand) within minutes on glass in a laboratory system. The conditioning film on platinum in seawater may reach a thickness of 0.03–0.08 μm in 10 h (Loeb and Neihof, 1975). Rates of adsorption are discussed further in Chapter 9.

Adsorption of an organic conditioning film is very rapid compared to the other biofilm processes.

Investigators have shown that materials with diverse surface properties (e.g., wettability, surface tension, electrophoretic mobility) are rapidly conditioned by adsorbing organics when exposed to natural waters with low organic concentrations. These organic molecules frequently appear to be polysaccharides or glycoproteins. The layer is not static, as evidenced by the results of Brash and Samak (1978), indicating significant turnover in molecular (proteinaceous) films on polyethylene. Protein molecules in the bulk fluid (laminar flow) were continuously exchanging with adsorbed proteins. Little and Zsolnay (1985) indicate that some of the adsorbed molecules on metal surfaces in seawater desorb or disappear with exposure time even though the total accumulated material increases or remains unchanged. With increasing molecular weight, polymers adsorb more strongly due to multiple binding sites and they may displace molecules of lower molecular weight (Cohen-Stuart et al., 1980).

The conditioning film may be dynamic, i.e., turnover of molecules may occur.

Since polymers may have as many as 10^5 units in their chain, each macromolecule may have many bonds to the substratum. The probability of having all its segments on the substratum at one time is small, but a polymer may have 30 to 60% of its segments in contact with the substratum (Barnett et al., 1981; Botham and Theis, 1970; Cafe and Robb, 1982). Thus, once approximately 10^4 segments are bonded to the surface, the polymer is unlikely to desorb. Consequently, the rate at which *segments* adsorb may control the period for molecular desorption and thus turnover at the substratum.

The conditioning film is generally observed or presumed to be uniform in both composition and coverage. But there appears to be little conclusive evidence that the spatial distribution of the conditioning film is uniform so that a "patchy" distribution is possible (see Section 7.2.1.5). The film may be heterogeneously distributed over the substratum and may not cover the entire surface, especially when viewed at the scale of a microorganism or at the scale of the appendages and polymers that first interact with the substratum.

7.2.1.3 Composition of the Conditioning Film Different substrata may accumulate conditioning films of different composition because of their differing surface properties such as potential, charge, and critical surface tension. The influence of the substratum on composition of the conditioning film, however, is negligible at the beginning, and thereafter, the influence only lasts for about one hour of exposure in seawater as determined by Little and Zsolnay (1985). After 4 h of exposure, the composition of the conditioning film was the about the same for similarly high surface energy substrata (titanium, aluminum, copper–nickel, and stainless steel). Baier and various coworkers have characterized conditioning films as largely glycoprotein (Baier, 1980; Baier and Weiss, 1975; Marshall, 1979), using internal total reflectance infrared spectrophotometric analysis on the earliest films adsorbed on pure germanium prisms. Loeb and Neihof (1975) concluded that decreased wetting of a platinum surface was due to an organic film, possibly rich in humic acids. The contact angle (Zisman, 1964) of the platinum–water or platinum–methylene iodide interface increased considerably when the platinum was exposed to natural seawater. The phenomenon did not occur when the organic fraction of the seawater had been photooxidized with ultraviolet radiation.

7.2.1.4 Influence of Conditioning Film on Substratum Properties Substratum alterations as a result of the conditioning film include decrease in hydrophobicity (Baier, 1975); both positively and negatively charged surfaces acquire net negative charges (Loeb and Neihof, 1975), and zeta potentials, contact potentials, and critical surface tensions are increased or decreased (Baier, 1975), depending on the initial surface energy. Adsorption of a conditioning film decreases the surface energy of clean, high energy surfaces (70 dyn

cm^{-1}), but has little effect on low energy surfaces (20 dyn cm^{-1}) (Baier, 1980). One would expect that surfaces of initially differing energies, even after conditioning with similar adsorbed layers of protein, would continue to influence bacterial adsorption. This appears to be the case. Baier (1980) has shown that the apparent strength of adhesion, as measured by spread areas of mammalian cells, is related to the substratum surface energy prior to the conditioning process. Thus, siliconized surfaces promoted desorption of adsorbed cells, even after protein conditioning of those surfaces. Baier (1980) concludes that the configurations of even similarly composed molecules within the conditioning films must be influenced by the initial surface state of the substratum.

Surface modification by organic macromolecules usually does change the surface charge. Loeb and Neihof (1975) showed (by means of microelectrophoresis experiments) the convergence of surface charge on various types of particles when exposed to natural seawater.

7.2.1.5 Influence of Conditioning Film on Bacterial Adsorption A conditioning film probably forms prior to microbial adsorption in all aquatic environments because of the universal presence of organic macromolecules in natural waters and differential transport rates of the two components (soluble organics and microbial cells). For example, Marshall (1979) reported protein and polysaccharide infrared adsorption signals on a germanium prism before the onset of bacterial adsorption, suggesting that conditioning was required for microbial adsorption. However, there is no absolute evidence that microorganisms *require* a conditioned substratum to adsorb.

A conditioning film forms prior to microbial adsorption in all aquatic environments.

Some possible conditioning films, primarily adsorbed protein, may reduce or inhibit bacterial adsorption. Fletcher (1976) reports that adsorption of a marine Pseudomonad to polystyrene was inhibited by deliberate precoating with albumin, gelatin, fibrinogen, and pepsin. Meadows (1971) found that albumin inhibited adsorption in his system, while casein and gelatin enhanced bacterial adsorption. Fletcher and Marshall (1982) measured contact angles (see below) of experimental surfaces in both the "clean" and the conditioned state in an aqueous system using an air bubble contact method. They found that the relative adsorption of bacteria to plastic substrata could be correlated with preadsorption of various proteins and these modifications were reflected in a change in measured contact angles.

Chamberlain (personal communication) has concentrated organics from seawater and conducted adsorption experiments with the organics at varying concentrations up to 100 g L^{-1}. As expected, more organics adsorbed at the higher organic concentrations. The conditioned substrata were then exposed to microbial populations to observe cell adsorption. The extent of microbial adsorption was inversely proportional to the concentration of adsorbed organics. Since the substratum may not be completely covered with adsorbed organics at lower organic concentrations, Chamberlain's results may indicate that microorganisms are adsorbing to "clean" patches of the substratum and not to the conditioned surface. Thus, inhibition of adsorption by conditioning films may reflect the degree of surface coverage.

Baier (personal communication) indicates that if a substratum is precoated with protein to equilibrium in the absence of particles (e.g., cells), bacterial adsorption is not as strong, i.e., the surface has been "passivated." On the other hand, during the natural accumulation process when cells are being transported to the interface, the substratum is accumulating a *fuzzy* coating that may interact efficiently with the cell surface, its polymers, or its organelles (e.g., pili). The most effective fuzzy coating, as assessed by the sticking efficiency of the cells arriving at the substratum, is generally about 15-20 nm thick.

7.2.2 Transport of Microbial Cells

When a clean surface is immersed in natural water containing dispersed microorganisms, nutrients, and organic macromolecules, transport of these components to the substratum controls the initial rate of accumulation. In very dilute dispersions of microbial cells and nutrients, such as open ocean waters or distilled water storage tanks, transport of microbial cells to the substratum may be the rate-controlling step in biofilm accumulation for long periods of time. So mass transport of cells and/or nutrients is critical in a rate analysis of biofilm accumulation.

Mass transport processes are influenced strongly by the mixing in the bulk fluid, which is generally related to the fluid flow regime (laminar or turbulent flow). Transport of molecules and small particles (<10 μm) in quiescent or laminar flow is described satisfactorily in terms of gravity and diffusion. In turbulent flow, the diffusion equation must be modified to include turbulent eddy transport. In addition, larger particles develop a "sluggishness" with respect to the surrounding fluid in turbulent flow. As the particle approaches the substratum, eddy transport diminishes and the viscous sublayer (Chapter 8) exerts a greater influence. Transport in the sublayer region depends on the molecule or particle characteristics.

7.2.2.1 Quiescent Conditions Under quiescent conditions, transport of bacteria in the fluid is by gravitational forces (i.e., sedimentation), by Brownian diffusion, or by motility for those organisms capable of it. Sedimentation rates are low for bacteria because of their size and specific gravity (approximately 1.05-1.10). Microorganisms of size 1-4 μm^3 undergo little Brownian motion (have a small Brownian diffusivity; see below). Therefore, motility may be a more important transport process in quiescent systems. Motility (motion under the organism's own power) is frequently related to some form of *taxis*—i.e., response to external stimuli such as a concentration gradient.

7.2.2.2 Motility Jang and Yen (1985) determined the Brownian diffusivity of different microorganisms to be approximately 50×10^{-6} mm^2 s^{-1}, as compared to the calculated non-Brownian diffusivity (due to motility) of approximately 0.4×10^{-3} to 5.6×10^{-3} mm^2 s^{-1}. The following equation was used to calculate non-Brownian diffusion:

$$D_c = \frac{v_r d_r}{3(1 - \cos \alpha)} \qquad [7.2]$$

where D_c = diffusivity (L^2t^{-1})
v_r = velocity of motility (Lt^{-1})
d_r = free length of random run (L)
α = angle of turn (—)

α is 90° if chemotaxis is not occurring. Thus, by ignoring motility as a contributing factor in the process of transport to the substratum, the transport rate could be underestimated by as much as 20 to 50 times. The non-Brownian diffusivity can be calculated from the velocity and the mean free path. Vaituzi and Doetsch (1969) measured speeds up to 55.8 μm s^{-1} for *Pseudomonas aeruginosa* with track photography. Their results suggest a mean free path of random run in the range of 50 to 85 μm. These values yield a non-Brownian diffusivity (Eq. 7.2) of 10^{-3} mm^2 s^{-1} for *P. aeruginosa*.

7.2.2.3 Laminar Flow For laminar flow, the mechanism for mass transport of cells in the liquid phase is diffusion, as described by Fick's first law (Chapter 9). Fick's law states that the diffusive flux is proportional to the concentration gradient. Fick's law is mainly used for describing diffusion of soluble components, but can also be useful for large molecules or small particles, such as microbial cells diffusing in water. In the case of particles, the Brownian (or non-Brownian) diffusion coefficient is used in Fick's law. If the cells are motile, their transport rate is increased significantly.

7.2.2.4 Turbulent Flow Within a turbulent flow regime, larger particles suspended within the fluid are transported to the solid surface, primarily by fluid dynamic forces. The particle flux to the surface increases with increasing particle concentration. However, the particle flux is also strongly dependent on the physical properties of the particles (e.g., size, shape, density) and is influenced by many other forces as the particles near the substratum.

Larger particles develop a sluggishness with respect to the surrounding fluid as they approach the wetted surface; eddy transport diminishes and the viscous sublayer exerts a greater influence. For soluble matter and small particles, diffusion can adequately describe transport in the viscous layer (Lister, 1981; Lin et al., 1953; Wells and Chamberlain, 1967). For larger particles, other mechanisms must be considered to explain experimental observations. Microbial cells

(0.5–10.0 μm effective diameter) can be transported from the bulk fluid to the wetted surface by several mechanisms, including: diffusion (Brownian and non-Brownian); gravity; thermophoresis; and fluid dynamic forces (inertia, lift, drag, drainage, and downsweeps).

Eddy diffusion may be instrumental in dispersing particles in the turbulent core region, thus maintaining a relatively uniform concentration in that region. Particles in turbulent flow are transported to within short distances of the surface by eddy diffusion and are propelled into the viscous (or laminar) sublayer under their acquired momentum. Turbulent eddies supply the initial impetus, and viscous drag slows down the particle as it penetrates the sublayer (Friedlander and Johnstone, 1957; Beal, 1970). For microbial cells, the inertial forces are very small because of their small diameter and density (in relation to water).

If the particle is traveling along the wall but faster than the fluid in the region of the wall, the *lift force* directs the particle toward the wall (Rouhiainen and Stachiewicz, 1970). This would normally be the case if the particle density is greater than fluid density and the particle is settling toward the wall. *Frictional drag forces* can be significant, especially in the viscous sublayer region. The drag force slows down the particle as it approaches the surface; it is proportional to the difference between particle velocity and fluid velocity.

If the mass density of the particle differs substantially from the fluid density, the *gravity force* may be significant. For microbial cells in turbulent flow, the gravity force is generally negligible. *Thermophoresis* is only relevant when particles are being transported in a temperature gradient (Lister, 1981). If the surface is hot and the bulk fluid is cold (e.g., a power plant condenser), the thermophoretic force will repel the particle from the surface. *Brownian diffusion* contributes little to the transport of microbial cells (>1.0 μm diameter) in turbulent flow. For starved cells, which are on the order of 0.2–0.4 μm in diameter, Brownian diffusion will be more important. The *fluid drainage force* is also significant in aqueous systems (Lister, 1981). The drainage force supplies the resistance encountered by the particle near the wall due to the pressure in the draining fluid film between the two approaching surfaces. This force is quite large for a microbial cell as it approaches the wall.

Recent evidence indicates that *advective downsweeps* or *turbulent bursts* of fluid, resembling microscopic tornadoes, originate in the turbulent core and penetrate all the way to the wall (Cleaver and Yates, 1975, 1976). Particles in the bulk fluid can be transported to the wall in this manner. Aside from the lift force, this is the only fluid dynamic force directing the bacterial cell to the wall. Thus, downsweeps appear to be quite important in the transport of cells in turbulent flow. For a Reynolds number = 30,000 in a circular tube, the bursts resulting from the downsweeps have the following characteristics: burst diameter = 1.1 mm; average axial distance between bursts = 5.0 mm; mean time between bursts = 0.006 s.

The minimum transport rate of particles will be observed when the particle diameter is approximately 0.1 μm. For smaller particles, Brownian diffusion becomes important. The calculated particle flux to the pipe wall for a bulk fluid particle concentration of 10^{10} particles m^{-3} is approximately 1000 particles m^{-2} s^{-1}.

7.2.2.5 Geometric Factors Transport is also influenced by the macroscopic geometry of the experimental system. Figure 7.3 is a composite of several microenvironments a cell encounters in a power plant condenser system. The fluid forces in the various environments are quite different and influence transport rates significantly. The resulting microenvironments (e.g., crevices), will also determine the type (e.g., aerobic or anaerobic) of microbial activity in the biofilm. In addition, surface roughness influences transport rate by increasing advective mass transport near the substratum. If the surface roughness elements are larger than the viscous sublayer, the roughness can be measured by hydraulic methods. If they are smaller (microroughness), quantitative methods are more difficult and the results are not easily interpretable. Nevertheless, particle deposition may be very sensitive to roughness too small to be detected by fluid dynamic methods (Browne, 1974).

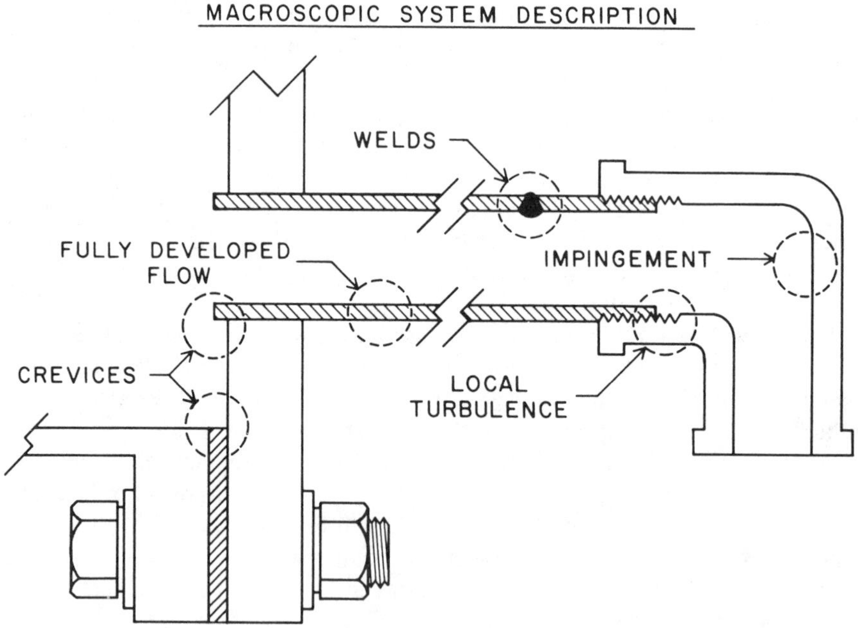

Figure 7.3 A composite representation of various geometries existing within a heat exchanger.

7.3 INTERFACIAL TRANSFER PROCESSES

7.3.1 Adsorption of Cells to the Substratum

7.3.1.1 Reversible and Irreversible Adsorption Adsorption of cells and other components to the substratum is an interfacial transfer process, since the components leave the bulk liquid compartment and become part of the substratum compartment. Previous research and *in situ* observations in quiescent (Zobell, 1943; Marshall et al., 1971) as well as laminar flow systems (Powell and Slater, 1983) suggest the existence of a two stage adsorption process: (1) reversible adsorption followed by (2) irreversible adsorption.

Reversible adsorption refers to an initially weak interaction of the cell with the substratum such that the cell can sometimes even exhibit Brownian motion. Reversible adsorption involves primarily long range interaction forces between the cell and the substratum, including the following: London–van der Waals forces, double layer (electrostatic) interactions, steric interactions, and possibly polymer bridging. These interactions are generally characterized by a low heat of adsorption per chemical bond (20–50 kJ mol^{-1}). Reversible adsorption, in some cases, may reflect low specificity between the cell and the substratum.

Irreversible adsorption is generally considered a permanent bonding to the substratum, frequently mediated by extracellular polymers.Irreversible adsorption is characterized by a high heat of adsorption per chemical bond (40–400 kJ mol^{-1}) and a one-way chemical interaction (thus, irreversible adsorption is sometimes referred to as chemisorption) involving dipole–dipole (Keesom) interactions, dipole-induced dipole (Debye) interactions, ion–dipole interactions, hydrogen bonds, hydrophobic interactions, or significant polymeric bridging.

Discrimination between reversible and irreversible adsorption in a quiescent experimental system is dependent on the assay technique, which generally includes a rinse step to remove "reversibly" adsorbed cells. The rinse step should provide calibrated or calculable shear force if the results are to be meaningful and comparable with others. In a flow system, a cell is transported to the substratum and reversibly adsorbs to the substratum.The reversibly adsorbed cell remains at the substratum for a finite time, t_{rc} (cell residence time at the substratum), then desorbs, possibly due to fluid shear forces, and is reentrained in the bulk water. If the cell residence time exceeds a critical time, t'_{rc}, the cell is irreversibly adsorbed to the substratum. Observations by Escher (1986) in laminar flow indicate that t'_{rc} is on the order of minutes (Figure 7.4).

The "stoichiometry" of the cell population balance can be written in the following way:

$$\underset{\substack{\text{suspended}\\\text{cells}}}{X_{\text{bulk}}} \rightleftharpoons \underset{\substack{\text{reversibly}\\\text{adsorbed cells}}}{X_{\text{rev}}} \rightleftharpoons \underset{\substack{\text{irreversibly}\\\text{adsorbed cells}}}{X_{\text{irrev}}} \qquad [7.3]$$

where

$$\underset{\text{total adsorbed cells}}{[X_{tot}]} = \underset{\text{reversibly adsorbed cells}}{[X_{rev}]} + \underset{\text{irreversibly adsorbed cells}}{[X_{irrev}]} \quad [7.4]$$

Using these "stoichiometric" equations, the kinetics and surface concentrations can be determined (Chapters 9 and 12).

7.3.1.2 Reversible Adsorption Forces that reversibly bind a cell to a surface have been reviewed at various levels of mathematical complexity (Pethica, 1961, 1980; Baier, 1980; Daniels, 1980; Dolowy, 1980; Fletcher and Beachey, 1980; Gingell and Vince, 1980; Rutter, 1980; Rutter and Vincent, 1980). Despite the large number of reviews and a considerable amount of work, theory does not explain the natural phenomena very well. There are basically two theories of the initial interactions of cells and substrata. In the first (DLVO theory),

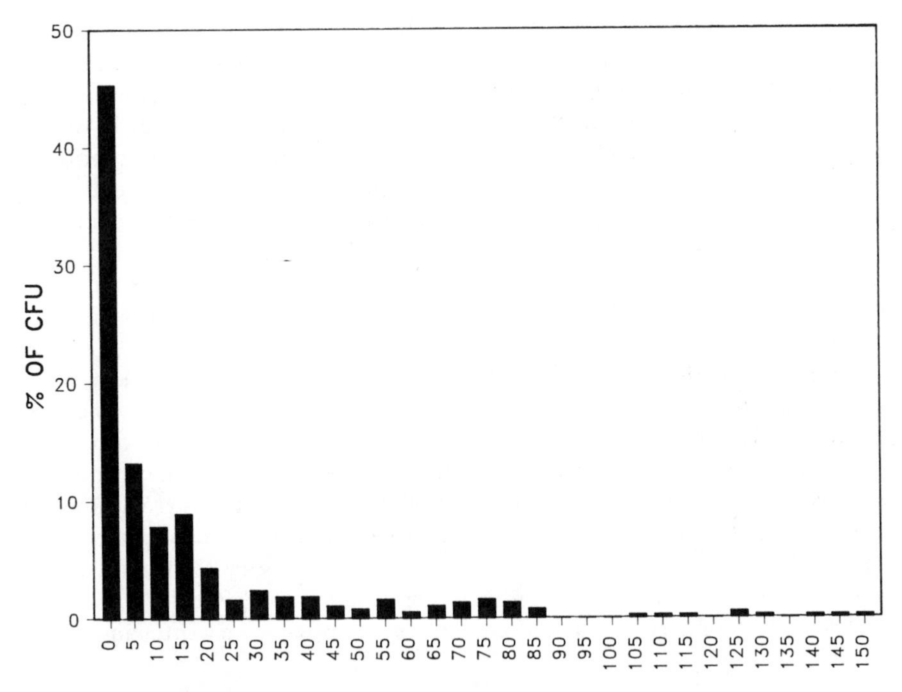

Figure 7.4 Finite residence time of adsorbed *Ps. aeruginosa* colony-forming units (CFU) at a glass substratum. Shear stress was 0.5 N m^{-2}, and mean cell residence time of the adsorbed cells at the substratum was 16 minutes. (Reprinted with permission from Escher, 1986.)

the electrostatic properties of the system are considered, whereas the second ("wettability" theory) considers the interfacial free energy of the system.

Double Layer (Electrostatic) Interactions. The double layer, or DLVO, theory (Derjaguin and Landau, 1941; Verwey and Overbeek, 1948) involves the electrostatic forces and the London–van der Waals forces (Rutter and Vincent, 1984):

$$\text{energy of interaction} = \text{London–van der Waals energy} + \text{electrostatic energy} \quad [7.5]$$

The theory predicts two possible positions for attraction (Figure 7.5): (1) the primary minimum at small separations and (2) the secondary minimum at larger distances of separation. At some point between these two minima, repulsive forces are maximal. Problems with this approach reside in the values used for the charges on the surfaces, the different geometry at the attachment site, and the varying dielectric constant of the liquid as the two surfaces approach. In addition, Hamaker's constant cannot be measured in these types of systems (Rutter and Vincent, 1980).

The theory suggests that reversible adsorption of microbial cells can take place at the secondary minimum (about 5–10 nm from the substratum), as has been described by Marshall et al. (1971). The time spent at this distance may be sufficient for other adhesive forces to become effective, e.g., polymer bridging.

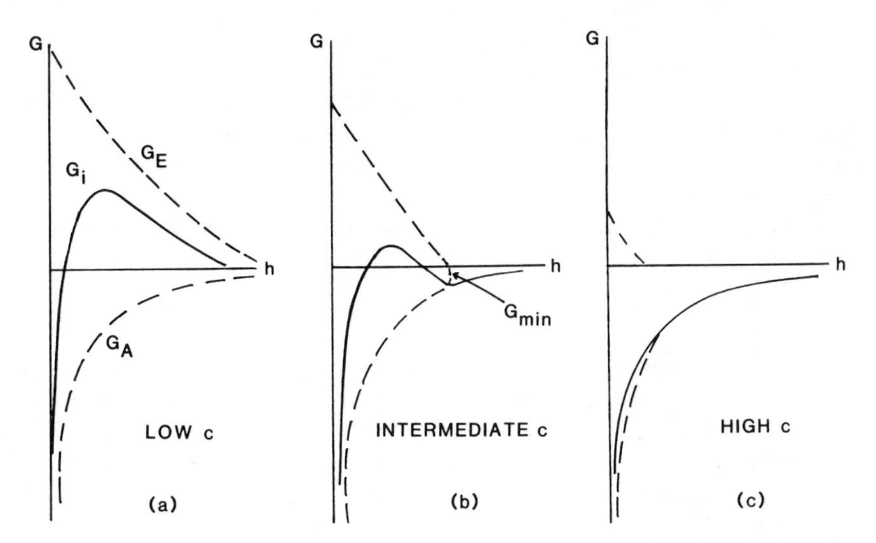

Figure 7.5 Interaction free energy G_i versus separation h as a function of electrolyte concentration c for a charged particle approaching a macroscopic surface of the same sign charge. G_A is the free energy of attraction, and G_E is the free energy of repulsion.

However, it is unlikely that cells are able to approach a substratum sufficiently closely (say, less than 1 nm) to overcome the repulsive peak between the primary and secondary minima. For instance, it has been calculated that the energy developed by a pseudomonad swimming at 33 $\mu m\ s^{-1}$ is insufficient to overcome this barrier (Marshall et al., 1971). Abbott et al. (1980) observed oscillation of *S. mutans* adsorbed on plastic (Perspex) and hypothesized that the movement reflected reversible adsorption at the secondary minimum. They tested this hypothesis by transferring adsorbed cells to a medium of lower ionic strength, which theoretically should increase the thickness of the double layer. When the cells did not desorb, they concluded that it was unlikely that the oscillation was due to deposition at the secondary minimum.

The mathematical expression of the DLVO theory of particle interaction includes the radius of the particles. As the radius decreases, the repulsive energy barrier decreases. Thus, when cells are able to reduce their effective radius, as in the production of filopodia (e.g., mammalian cells) or fimbriae (bacteria), they may overcome the repulsive maximum and adhere at the primary attraction minimum (Weiss and Harlos, 1977).

All of the results mentioned above have been obtained in systems with little or no fluid shear stress—a situation that rarely obtains in the natural environment.

The DLVO theory was developed for abiotic colloidal particles, which differ from viable microbial cells in that the latter have a continuous energy flux through their boundaries, resulting in a pH gradient and proton motive force across the cell wall (Chapter 5). Thus, for this and other reasons, considerable modifications and extensions are necessary to describe the microbial adsorption processes with DLVO theory. For example, substratum and cell surface geometry and roughness must be considered as well as potential microbial cell deformations as the cell approaches the substratum. The application of the DLVO theory to microbial adsorption has been successful for interactions occurring at medium to long range, i.e., >3 nm in aquatic environments (Pethica, 1980). More detail on DLVO theory as applied to microbial adsorption can be found elsewhere (Pethica, 1980; Rutter and Vincent, 1980; Tadros, 1980). Thus although long range forces may control the time the cell spends in close proximity to the substratum (reversible adsorption), other forces determine whether irreversible adsorption (adhesion) will occur.

Interfacial Free Energy. Theoretically, if the total free energy of a system containing a cell and an adjacent substratum is reduced by contact, then adsorption of the cell to the substratum will result. Thus, the "wetting" theory of cell adsorption predicts that a reduction in free energy will result from the lowering of the total interfacial free energy. The work of adsorption is

$$W = \sigma_{ab} - \sigma_a - \sigma_b \qquad [7.6]$$

where W = work of adsorption per unit area (M t^{-2})
σ_{ab} = total free energy per unit area of contact of two surfaces a and b at the contact separation distance h, made up of contributions σ_{ah} and σ_{bh} from the opposing surfaces (M t^{-2})
$= \sigma_{ah} + \sigma_{bh}$
σ_a, σ_b = interfacial free energies (surface tensions) for the two surfaces at large separations (M t^{-2})

The wetting approach does not depend on chemical detail, but rather relies on an appropriate method for measuring changes in surface tension. The most commonly used procedure has been the comparison of contact angles between model liquids and the surfaces of microbial cells. The contact angles can be used to determine the critical surface tension σ_c for the surface before and after conditioning. The critical surface tension is the surface tension of a hypothetical liquid that would just wet the test surface with a zero contact angle (Zisman, 1964; Baier et al., 1968). The technique has been used for describing microbial adsorption to oral surfaces (Baier and Glantz, 1978) and microbial fouling (DePalma et al., 1979). The results are useful in a qualitative manner and permit the ranking of contact angles for both cells and the substrata. Minimum adsorption of bacteria to a variety of substrata in seawater occurs for σ_c in the range 20–25 dyn cm^{-1} (Dexter, 1979). *In vitro* experiments with a *Pseudomonas* species indicate, conversely, that primary adsorption is best at the lowest σ_c materials, which are usually the most hydrophobic (Fletcher and Loeb, 1979).

Pethica (1980) finds the relationship between critical wetting tension and adsorption of particles (cells) to be qualitative at best, since the theory (the Young equation) assumes that particles are homogeneous and the surface is insoluble in the wetting liquid used to measure the contact angle.

Flagella, Fimbriae, and Pili. Many microbial cells have surface appendages. The presence of such microfilaments or cell protuberances will influence processes occurring at the first contact of cell with the substratum. Present evidence supports the importance of projections in the adsorption process and suggests that cells may "probe" the substratum in some cases, searching for a local inhomogeneity suitable for adsorption (Pethica, 1980). The projections can be classified as flagella, fimbriae, or pili (Chapter 5).

The flagellum is 10–20 nm wide and can be as long as 15–20 μm. Bacterial flagella are the source of motility and, thus, can play an important role in transport of cells to the substratum, especially if a chemotactic response is evoked. The adsorption of a marine pseudomonad was inhibited through removal of the flagella, and the efficiency of adsorption was reversed with flagella synthesis (Fletcher and Beachey, 1980). Stanley (1983) decreased the extent of *Ps. aeruginosa* adsorption by 90% by mechanically removing the flagella. Meadows (1971) concluded from his experimental observations that cells that attached at one end appeared to do so by their polar flagella. However, Marshall et al.

(1971) indicate that the attachment at the poles could be due to other surface components. Wood (1980) found no difference between the adsorption of complete and mechanically deflagellated *E. coli*. Thus, there is no unequivocal evidence for the general involvement of flagella in adsorption, although it is certain that they are significantly involved in transport.

Pili and fimbriae are generally 5–25 nm wide and 1–2 μm in length. They occur, as many as several hundred per cell, on many Gram-negative bacteria and an occasional Gram-positive bacterium (Ward and Berkeley, 1980). Pili and fimbriae are composed primarily of protein, frequently with significant hydrocarbon character in the side chains. Thus, the pili are hydrophobic, which may be the cause of their general adhesive properties (Ward and Berkeley, 1980). Heckels et al. (1976) observed that pili conferred an advantage on cells in the adsorption process. They suggest, as does Pethica (1961), that pili facilitate adsorption by providing a cell extension of very small relative radius, which can more easily penetrate the electrostatic barrier (according to the DLVO theory). Rosenberg et al. (1982) describe a hydrocarbon-degrading *Acinetobacter calcoaceticus* which adsorbs to hydrocarbons and other hydrophobic substrata. The organism possesses numerous thin (≈ 3.5 nm) fimbriae, which the authors believe are a major factor in cell adsorption to polystyrene and hydrocarbon and enable the cells to grow on hydrocarbons.

Bridging of a cell to the substratum could also possibly occur through mediation of the pili. Thus, gonococcal adsorption to epithelial cells is hypothesized to occur as follows (Watt, 1980): pili penetrate the electrostatic barrier between the gonococcus and the mucosal cell surface. The pilus then bridges the gonococcus to the epithelial cell membrane in a very specific manner. Most of the interest in adsorption mediated by pili has been focused on enteric bacteria and their influence on intestinal disease in which adsorption occurs at very specific substratum sites. The role of pili (or fimbriae) in more nonspecific adsorption is not clear. EPS may also bridge cells to the substratum.

7.3.1.3 Irreversible Adsorption Irreversible adsorption is sometimes referred to as chemisorption involving short range forces such as dipole–dipole (Keesom) interactions, dipole-induced dipole (Debye) interactions, ion–dipole interactions, hydrogen bonds, hydrophobic interactions, or polymeric bridging.

Irreversible adsorption can also be considered as *adhesion* when mediated by an adhesin (e.g., EPS). Polymeric bridging may be the most often cited mechanism for irreversible adsorption. Extracellular polymeric substances (EPS) are the adhesins most often considered and appear to play a major role in irreversible adsorption, although EPS are probably not necessary for reversible adsorption. A microbial cell may form more than one type of EPS, and polysaccharides, lipids, and proteins have all been implicated in microbial adsorption; this may lead to some of the conflicting results regarding the role of EPS in microbial adsorption.

Robb (1984) has summarized several important points regarding the polymer–cell–substratum interaction (Figure 7.6):

1. Polymers irreversibly adsorb to surfaces as a result of weak polymer surface bonds. Even though the bonds are weak, the large number of bonds results in firm adsorption. This applies to "free" polymers, which may form the conditioning film, as well as polymers immobilized in a capsule of a bacterial cells.
2. Bacteria, which generally are negatively charged, can adsorb (via polymers) to negatively charged substrata without the need for any specific biological mechanism. Rates of adsorption (Chapter 9) may be reduced considerably, however, when both surface and bacteria are negatively charged.
3. Extracellular proteins may play an important role in cellular adsorption because the positive groups can form strong bonds to negative adsorption sites on the substratum even though the protein may have a net negative charge. Paul and Jeffrey (1985) suggest the existence of separate mechanisms for the irreversible adsorption of *V. proteolytica* to hydrophilic and hydrophobic substrata. Proteins are suspected of playing a major role in the adsorption of ions to the hydrophobic substratum.
4. When cell adsorption occurs primarily via charged polymers, the extent of adsorption will be strongly influenced by their ionic strength.
5. Both small molecules and other polymers may prevent the adsorption of macromolecules on the substratum. Thus, these interfering molecules may alter the ability of a cell to adsorb via its bound EPS.
6. The net charge on a bacterium with a capsule or other bound EPS arises from groups on the polymer itself.

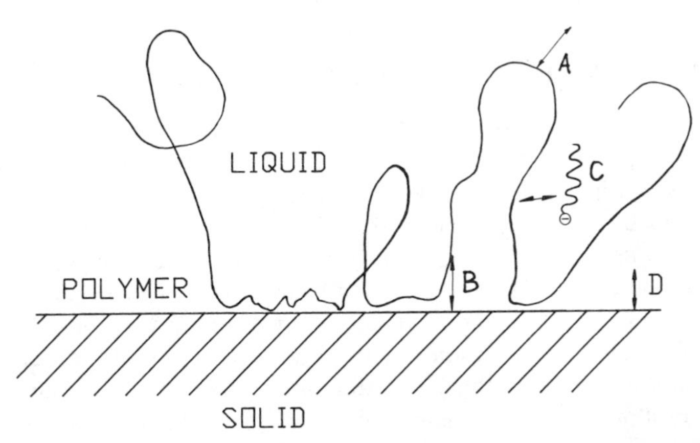

Figure 7.6 Representation of a polymer adsorbed at a solid substratum. Important interactions are between (A) polymer segment and solvent, (B) polymer segment and substratum, (C) polymer segment and other soluble components, and (D) substratum and soluble components. (After Robb, 1984.)

7. The strength of adhesion increases with time. This effect may be due partially to viscous flow of the EPS between the cell and the substratum.

EPS (not necessarily of the same type implicated in adsorption) may also protect the cell, provide a means of intracellular and intercellular communication, help to conserve or store cellular energy, and provide resistance to chemical toxins (Costerton et al., 1981). A better characterization of the role of EPS would be desirable.

7.3.1.4 Adsorption to Substrata That Also Serve as Substrates Microorganisms sometimes can degrade the substratum (e.g., hydrocarbons, cellulosic materials) to obtain energy and/or carbon. The adsorption of cells in these circumstances may cause more specific cell–substratum interactions. For example, Miura et al. (1977) have shown that rate of hydrocarbon degradation increases with increasing strength of organism adsorption to the hydrocarbon droplets. Certain conditions are apparently necessary for microbial growth on particulate substrates: (1) the cell must be intimately associated with the substrate surface (in this case, the substrate is the substratum), (2) some cellular product (e.g., an enzyme or a lipid) must be excreted which partially hydrolyzes (or emulsifies) the substrate. Energetically, it may be important for the cell to be attached to the substrate by a polymer matrix so that diffusion of the enzyme (or other product) to the substrate and the hydrolyzate to the cell are largely restricted to the region between the cell and substrate. Clearly, adsorption on surfaces may permit organisms to retain desirable extracellular activity in its immediate environment since the extracellular polymers (e.g., enzymes, lipids) are essentially immobilized at the substratum or within the EPS matrix. The surface adsorption may even elicit the enhanced formation of required extracellular compounds (e.g., enzymes) for degradation of particulate substrates.

7.3.2 Desorption of Cells and Net Deposition at the Substratum

The existence of reversibly as well as irreversibly adsorbed cells requires that desorption of cells from the substratum occur. Desorption is an interfacial transfer process consisting in the transfer of cells and other adsorbed components from the substratum compartment to the bulk liquid compartment. Powell and Slater (1983) and Escher (1986) clearly show that if one assumes that all cells contacting the surface become irreversibly adsorbed, one will grossly overestimate the surface cell population. Thus, many of the data reported reflect only deposition of cells during initial events (Figure 7.2) in biofilm accumulation. *Deposition* should refer to the net accumulation of irreversibly adsorbed cells at the substratum arising from the combined effect of adsorption and desorption.

Escher (1986) has observed the individual processes of adsorption and desorption *in situ* and nondestructively with an image analysis system (Chapter 3). A series of experiments were conducted with *Ps. aeruginosa* and a rectangular glass capillary serving as the substratum. All experiments were conducted in laminar flow. A typical progression of substratum colonization is presented in Figure 12.10 (Chapter 12). Adsorption rates were found to increase with cell concentration in the bulk fluid and increase with fluid velocity because of the resulting increase in cell transport rate to the substratum. The adsorption rate was independent of the number of cells at the substratum. *Ps. aeruginosa* clearly demonstrated a reversibly adsorbed and an irreversibly adsorbed state, indicating that cells have a distinct probability of desorbing within a short time after adsorbing. However, the longer a cell remains reversibly adsorbed to the substratum, the smaller its probability of desorbing. Hence, the adsorbing cell requires some time to irreversibly adsorb to the substratum. As a result, desorption rates are closely linked to adsorption rates and not to the concentration of adsorbed cells.

Desorption did not occur to any significant extent in Escher's experiments for some 50 min into the run. After 50 min, the desorption rate was proportional to the adsorption rate with no apparent dependence on adsorbed cell numbers. The time lag prior to observation of any desorption events was termed the *offset time* (Chapter 12) and was hypothesized to be due to modification of the substratum properties by the conditioning film (Escher, 1986).

7.3.2.1 Hydrodynamics and Desorption Hydrodynamic forces often control desorption (Powell and Slater, 1982; Timperley, 1981). The desorption rate generally increases with increasing fluid velocity. Both macroscopic geometry and microroughness may also significantly influence desorption, since cells adsorbed in cavities will be sheltered from severe hydrodynamic shear.

Cell desorption primarily results from two fluid dynamic forces: shear, and lift or "upsweeps." Shear forces along a conduit are exerted parallel to the substratum. Hence, a cell desorbed from the substratum by shear in laminar or turbulent flow may be transported close to the substratum (in the viscous or laminar sublayer) for a significant distance, i.e., it may bounce or roll. This behavior will result in numerous collisions with the substratum and hence more opportunities for repeated adsorption/desorption events. In turbulent flow, a significant lift force, or *upsweep,* is exerted; it may transport the desorbed cell quickly into the bulk liquid. Upsweeps are analogous to the downsweeps discussed in relation to transport. Both downsweeps and upsweeps result from turbulent bursts that move to and away from the substratum into the bulk flow. Upsweeps generate a lift force normal to the substratum that can influence desorption. Drag (viscous shear) forces, which act in the direction of flow on adsorbed cells, are approximately 1000 times greater than the lift forces acting on them. Thus although viscous shear may dislodge a cell, unless a lift force is present, the cell will presumably roll along the substratum until another surface adsorption site is found.

7.3.2.2 Motility Kefford and Marshall (1984) conducted adsorption experiments with *Leptospira biflexa* serovar *patoc* in which reversible and irreversible adsorption were observed. Reversible adsorption was unchanged in the presence of a chemical additive that inhibits cell motility. However, desorption was significantly reduced when the cells were treated with the uncoupler. Thus, motility appears to significantly influence desorption in motile species.

7.3.2.3 Polarization Gordon et al., (1983) report that anodic polarization reduced the interfacial pH at metal surfaces while cathodic polarization increased it. The polarization influenced the deposition of two *Vibrio* species in seawater.

7.3.2.4 Sticking Efficiency The quantity that is inevitably measured in microbial adsorption studies is the net rate of accumulation, i.e., the number of cells that are transported to the substratum and that irreversibly adsorb. The number of cells transported to the substratum is unknown, since direct measurement of it is difficult. Certainly, the dominant mechanisms for cell transport in various liquid flow regimes are different. Nevertheless, transport rates can be calculated with confidence if the fluid flow regime is defined and a collision rate can be estimated. Thus, a sticking efficiency can be defined:

$$\text{sticking efficiency} = \frac{\text{number of cells adsorbing to the substratum}}{\text{number of cells transported to the substratum}} \quad [7.7]$$

Fletcher (1977) measured the net accumulation rate under quiescent conditions in which sedimentation was probably the dominant transport mechanism although the cell was motile. Under these conditions, desorption is presumed negligible. Powell and Slater (1983) conducted experiments in laminar flow where cell transport is controlled by diffusion. Experimental data on the net accumulation rate from these two reports are compared with calculated transport rates, assuming the mechanism for transport was known, in Table 7.1. Several observations are striking:

1. The rate of cell transport (calculated), mainly by sedimentation, is significantly greater in the quiescent system. In laminar flow, diffusive transport dominates and the cell transport rates increase with increasing shear stress. Thus, the difference in transport rates between quiescent and laminar flow is attributed to a change in the dominant transport mechanism.
2. The net rate of accumulation is significantly greater in the quiescent system. It is greater in the laminar flow experiment with lower shear stress than in the one with higher shear stress. One apparent explanation for these results is that the desorption rate increases with increasing shear stress (Powell and Slater, 1983).
3. The calculated sticking efficiency in the various experiments is proportional to the shear stress (see Chapter 11).

TABLE 7.1 A Comparison between the Measured Net Cell Accumulation Rate at a Surface and the Calculated Cell Transport Rate to the Surface

Reference	Fluid Shear Stress ($N\ m^{-2}$)	Rate of Transport (10^4 cells $m^{-2}\ s^{-1}$)	Rate of Accumulation (10^4 cells $m^{-2}\ s^{-1}$)	Cell Concentration (cells m^{-3})	Presumed Transport Mechanism
Fletcher (1977)	0	5000	4170	10.0×10^{13}	Sedimentation
Powell and	0.80[a]	472	3	1.4×10^{13}	Diffusion
Slater (1983)	0.09[b]	167	31	1.4×10^{13}	Diffusion

[a] Reynolds number = 11.
[b] Reynolds number = 1.2.

Cell coverage of the substratum at saturation appears to be strongly influenced by the flow regime, with higher shear stresses resulting in lower adsorbed cell densities.

7.3.2.5 Extent of Adsorption The extent of adsorption can be determined at the saturation point, sometimes referred to erroneously as the equilibrium state. True equilibrium cannot exist in a system containing viable cells. In most experiments conducted in closed, quiescent systems, growth is discouraged by washing the cells repeatedly and then resuspending in water with no energy source. The experiments are conducted and the results analyzed as if abiotic adsorption were occurring. Fletcher (1977) observed approximately 45% coverage of the substratum surface at the saturation point in a quiescent system. Absolom et al. (1983) observed on the order of 12–18% coverage.

Other investigators have used open flow systems for observing adsorption processes. In one experimental system, the microbial suspension flows through a rectangular glass capillary and the processes are observed *in situ* and nondestructively through the capillary with a microscope (Chapter 3). With this system, Powell and Slater (1983) observed 1–5% coverage at saturation in laminar flow, and Leech and Hefford (1980) observed as much as 30% coverage in the capillary experimental system. Escher (1986), using the same system, observed 1-5% coverage as long as adsorption was the predominant mechanism of cell accumulation at the substratum. A close-packed monolayer of adsorbed cells has not been observed in a flow system until a significant amount of *growth* has occurred on the substratum. Cell replication was observed by Escher as well as Leech and Hefford. Escher also observed that the spatial distribution of the adsorbed cells tended toward uniformity as the experiment progressed (Chapter 11).

Nelson et al. (1985) conducted experiments in turbulent flow which resulted in approximately 0.01% coverage at saturation and uniform distribution of cells across the surface. The uniform distribution and low saturation coverage, as also observed by Escher (1986) in laminar flow, suggest that there exists a "zone of inhibition" around the adsorbed cells, possibly defined by EPS or other chemosensory mechanisms. Others have observed a similar spatial distribution with abiotic particles adsorbing in a flow system and have attributed the behavior to particle oscillations on the microscale, which discourage deposition in the vicinity of an adsorbed, oscillating particle (Dabros and van de Ven, 1983). Abbott et al. (1980) have observed oscillations of adsorbed *S. mutans* in a rotating disk apparatus at laminar flow. The adhesion was mediated by a polymer of undetermined composition.

7.3.2.6 Influence of Physiological State The physiological state of the organism influences the rate and, possibly, the net accumulation during initial events (Fletcher, 1977). Bryers and Characklis (1982) observed that accumulation during initial events was directly proportional to growth rate in a mixed culture system when feeding the biofilm reactor from a chemostat (Figure 7.7). Nelson et al. (1985) have observed a decrease in adsorption rate with increasing specific growth rate for a *Pseudomonas* species (Figure 7.8) in a turbulent flow environment.

Leech and Hefford (1980) observed an increase in deposition rate with increasing flow velocity in a capillary tube (laminar flow) when cells were growing slowly ($\mu = 0.04\ h^{-1}$) in a chemostat. At high growth rates (0.2 and 0.5 h^{-1}), deposition rates decreased with increasing flow velocity. Increasing flow velocity will increase transport, and thus the collision rate of the cells with the sub-

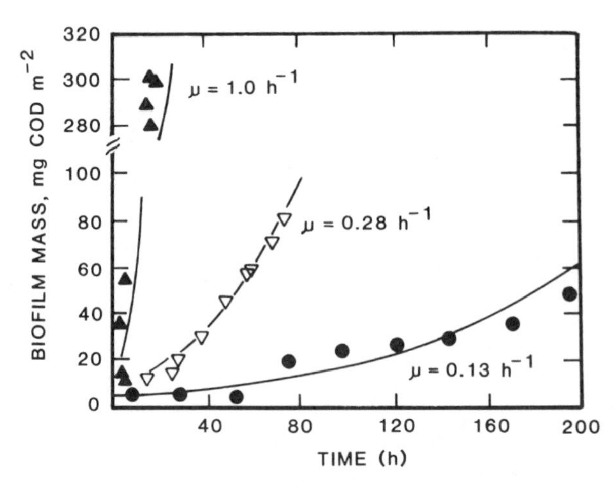

Figure 7.7 Accumulation of biofilm during the initial phase as determined by Chemical Oxygen Demand. A mixed population was cultivated in a chemostat prior to flow past the substratum (glass). The rate of biofilm accumulation (all chemically oxidizable components) increases with increasing growth rate in the chemostat (Bryers and Characklis, 1982).

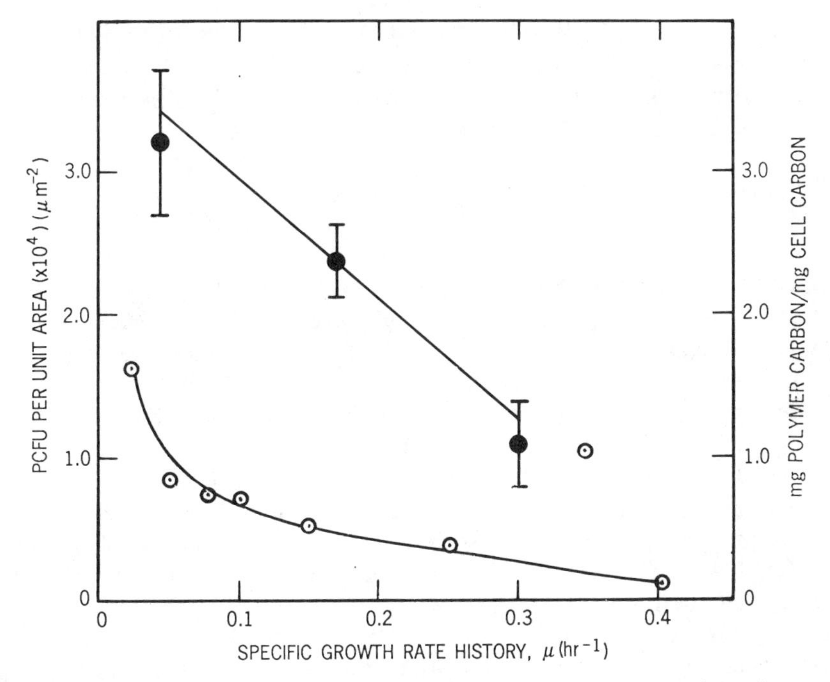

Figure 7.8 Saturation concentration of adsorbed *Ps. aeruginosa* on glass. *Ps. aeruginosa* was cultivated in a chemostat prior to its flow past the substratum. Adsorbed cell concentration (●) at saturation decreases with increasing growth rate of the cells in the chemostat (Nelson et al., 1985). The EPS (polymer) content (⊙) of the cells in the chemostat was reported by Robinson et al. (1984).

stratum. Hence, increased deposition for the slow-growing cells with increasing flow velocity is understandable. For the fast-growing cells, it appears the sticking efficiency is decreasing, possibly as a result of EPS-related mechanisms. *Ps. aeruginosa*, cultured in a chemostat, accumulated more EPS at lower growth rates (Robinson et al., 1984). At higher growth rates, the cells may contain less EPS or EPS of different composition and thus require more residence time at the substratum in order to adsorb irreversibly.

7.3.2.7 Influence of Substratum Microroughness The surface roughness of the substratum may be a significant factor that has been overlooked in experimental studies purporting to simulate a specific environment. The microroughness of various roll finishes in stainless steel tubes is shown schematically in Figure 7.9, which makes clear the probable role of microroughness in both transport and adsorption of macromolecules and microbial cells. Convective transport rates near the microrough surface will be greater than at a smooth surface. Once the cell arrives at the surface, adsorption rates will probably be higher for at least two reasons:

1. Microroughness shelters the adsorbed cell from shear forces, thereby reducing the desorption rate.
2. Microroughness increases the surface area available for cell–substratum contact and hence adsorption.

Crude observations indicate that net cell accumulation rate is greater on rougher surfaces, but direct quantitative evidence is not available.

7.3.2.8 Purposeful Desorption and Detachment: Cleaning Procedures Desorption may also result from chemical treatment (e.g., chelants, surfactants, oxidants) and/or physical treatment (e.g., increased fluid velocity, ultrasound). Little fundamental information is available on the kinetics of desorption, and as a result, no well-accepted comparative criteria for testing chemical control treatments are available.

Both Powell and Slater (1982) and Timperley (1981) conducted studies to determine the influence of fluid dynamics on desorption. Both investigators observed an increase in desorption with an increase in Reynolds number. Timperley (1981) suggests that the shear force near the wall, even in turbulent flow, is small. Shear stress increases as the square of the mean velocity and, as with the viscous sublayer, is hardly affected by differences in pipe diameter if the mean velocity of flow is the same in the different pipes. Timperley conducted experi-

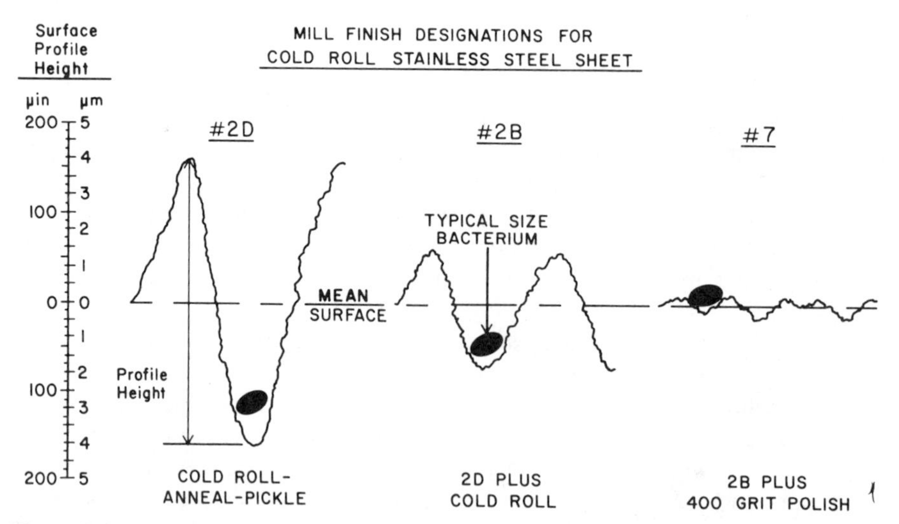

Figure 7.9 Diagrammatic comparison of the size of microroughness on stainless steel tubing with the size of a microbial cell.

ments in turbulent flow with two different pipe diameters (38 and 76 mm OD) "coated with a tenacious soil containing indicator organisms." The cell numbers remaining after hydrodynamic cleaning decrease dramatically with increasing mean fluid velocity (Figure 7.10). Within experimental error, both pipes were cleaned to the same extent at the same fluid velocity. The shear stress in the two pipes is different at the same mean fluid velocity. Thus, the performance appears to be dependent on the mean fluid velocity.

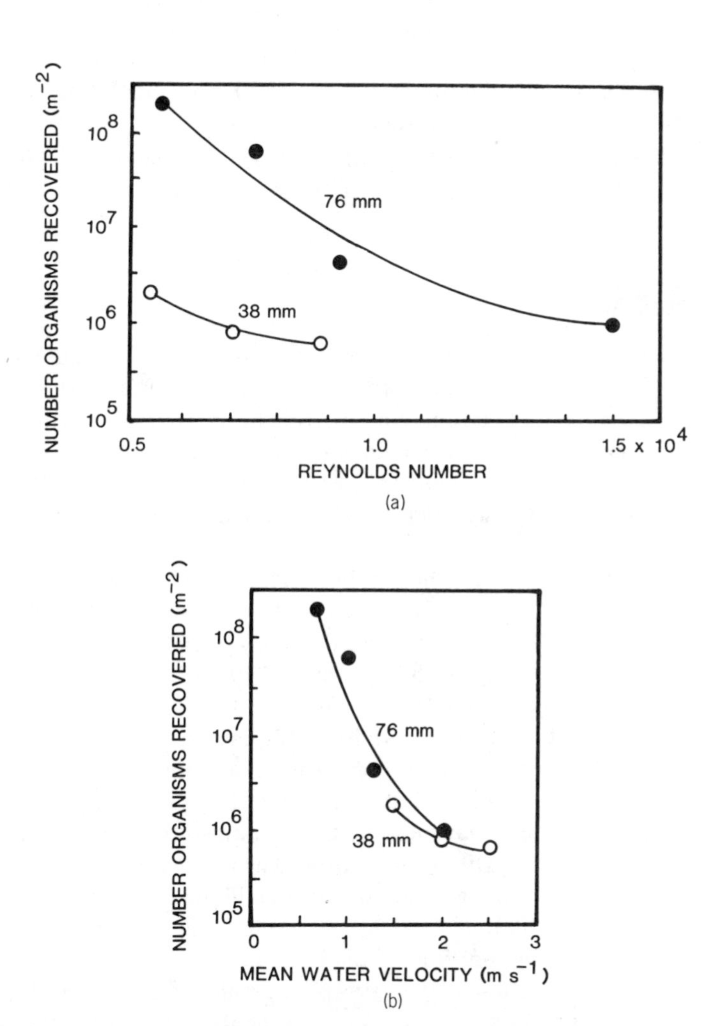

Figure 7.10 Effect of Reynolds number (a) and mean fluid velocity (b) on purposeful detachment of cells from tubing by hydrodynamic forces. The same number of organisms are removed in the different tubes when they are operated at the same fluid velocity. Tubes were 38 and 76 mm OD (Timperley, 1981.)

7.3.3 Attachment

Attachment is capture of particles (cells or abiotic particles) from the fluid by the biofilm, i.e., the immobilized biotic component. Thus, attachment is an interfacial transfer process describing the transfer of cells and other components from the bulk liquid compartment to the biofilm compartment. Attachment and adsorption differ only in that the capture surface is the biofilm and the substratum, respectively. Thus, the transport and attachment mechanisms are the same as described in Section 7.2.3. The biofilm, however, is different from most substrata in several ways, including the following:

1. The biofilm is generally rougher (Chapter 4), and so advective transport mechanisms are more important.
2. The sticking efficiency of the biofilm is probably influenced by its roughness and its EPS content, which can be substantial.
3. The biofilm is a compliant surface.

7.3.4 Detachment

Detachment is an interfacial transfer process which transfers cells and other components from the biofilm compartment to the bulk liquid. Desorption of microbial cells and related biofilm material from the substratum occurs from the moment of initial adsorption. Detachment, however, is material loss from the biofilm matrix as opposed to material loss from the substratum. Detachment phenomena can be arbitrarily categorized as erosion, sloughing, or abrasion.

7.3.4.1 Erosion Erosion is the continuous loss of small portions of biofilm. It is highly dependent on fluid dynamic conditions. For example, the rate of erosion increases with increasing biofilm thickness and fluid shear stress at the biofilm–fluid interface (Trulear and Characklis, 1982).

Trulear and Characklis (1982) have observed that biofilm detachment increases with both fluid shear stress and biofilm mass. As the fluid velocity increases, the viscous sublayer thickness decreases. Consequently, the region near the tube wall subject to relatively low shear forces (i.e., the viscous sublayer) is reduced. As a result, there may be some upper limit to the effectiveness of any cleaning operation based on fluid shear stress. The viscous sublayer may provide a valuable *a priori* criterion for predicting the maximum effectiveness (the minimum thickness attainable) of cleaning techniques which depend on fluid dynamic forces for removing portions of the biofilm.

Attachment and detachment processes play a major role in the ecology of the biofilm. Attachment and detachment of microorganisms to and from the biofilm provide the means for interaction between dispersed (planktonic) organisms and the biofilm organisms, i.e., cell turnover (individual cell residence time) in the biofilm. As a biofilm develops, succession in species is observed. Trulear (1983) developed a biofilm of *Ps. aeruginosa* in conditions of high shear stress and then challenged it with *Sphaerotilus natans*. *S. natans*, which grows significantly faster than *Pseudomonas aeruginosa* in the medium, became the dominant species in the biofilm within 24 h (Figure 7.11). Detachment, influenced strongly by fluid shear stress, combined with growth processes in the biofilm may serve to "wash out" the slower-growing organisms from the biofilm. Thus, an individual cell may have a finite residence time in the biofilm within which to reproduce. If reproduction does not occur in that time period, the species will be diluted out of the biofilm (see Chapter 11).

7.3.4.2 Sloughing Sloughing refers to a rapid, massive loss of biofilm (Howell and Atkinson, 1976). Sloughing is more frequently witnessed with thicker biofilms developed in nutrient-rich environments. Erosion may also be occurring simultaneously with sloughing.

Sloughing may occur as a result of an artificial stimulus. Bakke (1983) ob-

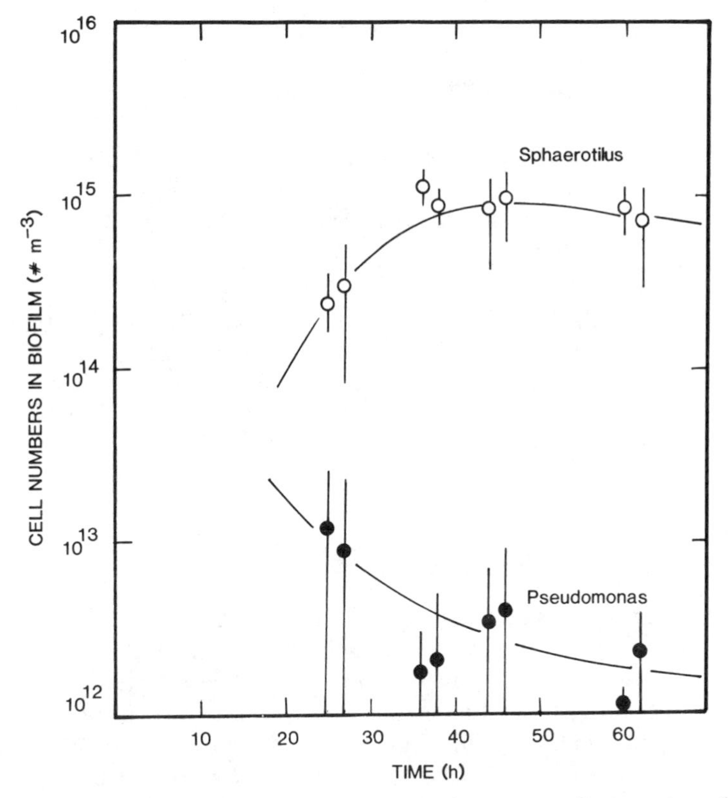

Figure 7.11 The succession of *Sphaerotilus natans* in a mature *Ps. aeruginosa* biofilm. The monopopulation *Ps. aeruginosa* biofilm was challenged with *Sphaerotilus natans* at zero time. Inlet cell numbers for both species were on the order of 10^{10} m^{-3} during the challenge period.

served massive detachment when substrate loading to the biofilm was instantaneously doubled. One hypothesis is that the cell membrane potential plays a key role in the phenomena (Chapter 5). Turakhia et al. (1983) have observed sloughing upon the addition of a calcium-specific chelant (EGTA), suggesting the importance of calcium for the cohesiveness or adhesiveness of the biofilm.

Detachment of biofilm is the major objective of many antifouling additives used in industry. Very little is known regarding the kinetics of detachment and the factors affecting the removal. Such kinetic expressions would be useful for modeling purposes and would provide useful comparative criteria for testing of antifouling treatments. Many chemical treatments have been used to detach biofilm material, with varying success, including: (1) oxidizing biocides (e.g., chlorine), (2) surfactants, and (3) nonoxidizing biocides (see Chapter 15).

7.3.4.3 Abrasion Abrasion is the loss of biofilm due to repeated collisions between substratum particles. Biofilm in fluidized bed is subject to abrasion detachment, as is the biofilm in a sand filter during the backwash cycle.

7.4 TRANSFORMATION PROCESSES

Transformation processes are those in which molecular rearrangements occur, i.e., reactions. After a cell adsorbs to the substratum, microbial transformations (metabolic processes) become significant and *colonization* will ensue (Escher, 1986).

Studies of transformation processes within biofilms have generally relied on an unstructured approach to analyze the biomass component. Unstructured models (Chapter 2) characterize the biotic component only in terms of biomass, with little attention given to the various biofilm components, physiological state of the organisms, microbial species, or EPS content. Processes in unstructured models are limited to those that can be evaluated without considering activities within the cell, i.e., a more macroscopic approach.

7.4.1 Fundamental Transformations

Four fundamental rate processes can be identified: (1) growth, (2) product formation, (3) maintenance and/or endogenous decay, and (4) death and/or lysis. Any or all of these processes may be occurring in a biofilm at any time. Growth refers to cell growth and multiplication. The cells also form products, some of which are retained in the biofilm (e.g., EPS) and some of which diffuse out into the bulk fluid. The cells also have to sustain their internal structure, and the process of maintenance consumes energy. If nutrients are depleted or toxic substance are present, death and/or lysis occurs.

The microbial processes occurring in a biofilm are more complex than suggested by the four fundamental processes defined above. However, this classification has been useful in determining, to some extent, the flow of substrate energy through the biofilm. Mathematical description of the kinetic expressions has also been accomplished and is discussed further in Chapter 9. Further structuring of biofilm processes within the biofilm (as opposed to the influence of the processes on the overlying liquid phase) and more specific identification of the products being formed are required in order to better understand the behavior of biofilm systems.

7.4.2 Observed Transformations

The rates of the microbial transformation processes are difficult to measure directly and are generally inferred from more easily observed rate processes. The more familiar observed rate processes include (1) substrate removal rate, (2) electron acceptor (e.g., oxygen) removal rate, (3) biomass production rate, and (4) product formation rate. The relationship between transformation and observed process rates is presented in Table 8.3. The stoichiometry of the process is qualitatively represented by each row in the matrix (− refers to reactants and + to products). The columns of the matrix indicate the fundamental rate processes that may contribute to the observed rates (last row in the matrix). For example, substrate removal (column 1) is the net result of growth, maintenance, and product formation.

Recent studies (Robinson et al., 1984; Bakke et al., 1984) have provided more information on biofilm metabolic processes by using pure cultures (*Ps. aeruginosa*) and measuring EPS content. The results suggest the following:

1. The growth rate of cells (*Ps. aeruginosa*) in the biofilm can be estimated from their growth rate in chemostats when the substrate concentration in the microenvironment of the cell is equal in the two conditions. Results of experiments in a chemostat and in a biofilm reactor suggest the same energy metabolism in both microbial environments. The maximum specific growth rate, saturation coefficient, and yield are approximately the same in a chemostat and in the biofilm. No significant diffusional resistances existed in the biofilm during this study. Another concern is that the substratum surface should not release any components that influence metabolism. In these studies, acrylic plastic was the substratum.
2. The EPS formation rate and EPS yield by biofilm cells (*Ps. aeruginosa*) were essentially the same for dispersed cells. However, the EPS accumulation rate in the biofilm was quite high. Therefore, EPS may be the dominant component in the biofilm (on the order of 50% of the biofilm organic carbon in this study).
3. Maintenance requirements are essentially negligible until the biofilm becomes very thick. Even then, the results of EPS formation or anaerobic metabolism deep within the biofilm may be mistaken for maintenance energy requirements if resolute measurements are lacking.

One of the important conclusions from these results is that rate and stoichiometric coefficients derived from chemostat experiments may be used for modeling biofilm processes.

Various methods (e.g., heterotrophic potential) have been used to determine biofilm activity which require removing the biofilm from its environment and incubating after substrate addition (e.g., a radiolabeled substrate). Removing the cells from the substratum, or removing the substratum with the biofilm from its environment, may lead to dramatic changes in biofilm activity because the attached cells may be subjected to environmental changes that will influence the activity. These methods should be avoided.

7.4.3 Influence of Mass Transfer and Diffusion on Transformations in a Biofilm

Analyses of biofilm process rates and stoichiometries are frequently complicated by significant mass transfer resistances in the liquid or diffusional resistances within the biofilm. Trulear and Characklis (1982) have observed that the substrate removal rate increases in proportion to biofilm thickness up to a critical thickness, beyond which the removal rate remained constant. The critical, or *active,* thickness was observed to increase with increasing substrate concentration (Figure 7.12). This behavior can be attributed to diffusional resistance within the biofilm. Once the biofilm thickness exceeds the depth of substrate penetration into the biofilm, the substrate removal rate is unaffected by further biofilm accumulation.

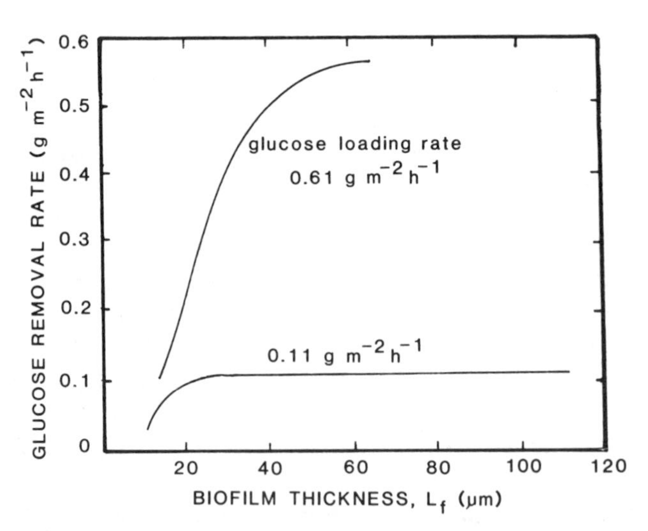

Figure 7.12 The influence of biofilm thickness and glucose loading rate on glucose removal rate by a mixed population biofilm. The glucose removal rate increases with thickness up to a critical thickness sometimes termed the "active" thickness. (After Characklis, 1980.)

The biofilm rate processes may also be controlled by mass transfer limitation in the bulk fluid phase. For example, substrate removal rate is dependent on fluid velocity past the biofilm. At low fluid velocities, a thick enough mass transfer boundary layer can cause a fluid phase mass transfer resistance that decreases the substrate concentration at the fluid–biofilm interface and thereby decreases the substrate removal rate (see Figure 9.32, Chapter 9). Two factors may result in low mass transfer rates from the bulk fluid to the biofilm: (1) low fluid velocities, and (2) low liquid phase concentration of the material being transported. Much biofilm research has been conducted at low flows or in quiescent conditions. Mass transfer may be the rate-controlling step in the overall process in these studies and, without further analysis, may be confused with the rates of more fundamental processes such as growth and adsorption. In highly turbulent systems, mass transfer in the liquid phase is rarely a significant factor. Mass transfer limitations in biofilm systems are discussed in more detail in Chapter 9.

7.5 NET ACCUMULATION

The progression of net biofilm accumulation has been described in terms of the various contributing processes. The progression typically follows a sigmoidal function. The overall process of accumulation can be visualized in the composite diagram presented as Figure 7.13.

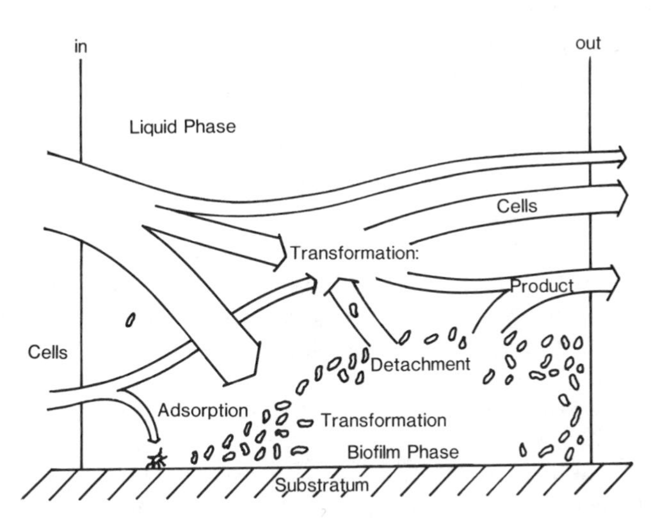

Figure 7.13 A composite of all processes contributing to biofilm accumulation: (1) transport and adsorption of a molecular conditioning film, (2) transport of cells to the substratum, (3) adsorption/desorption of cells at the substratum, (4) growth, product formation, endogenous respiration, and death within the biofilm, and (5) attachment/detachment at the biofilm–bulk water interface.

REFERENCES

Abbot, A., R.C.W. Berkeley, and P.R. Rutter, "Sucrose and deposition of *Streptococcus mutans* at solid-liquid interfaces," in R.C.W. Berkeley, J.M. Lynch, J. Melling, P.R. Rutter, and D. Vincent (eds.), *Microbial Adhesion to Surfaces*. Ellis Horwood, Chichester, 1980, pp. 117–142.

Absolom, D.R., F.V. Lamberti, Z. Polikova, W. Zingg, C.J. van Oss, and W. Neumann, *Appl. Env. Microbiol.*, **46**, 90–97 (1983).

Baier, R. E., "Applied chemistry at protein interfaces," in R.E. Baier (ed.), *Applied Chemistry at Protein Interfaces,* Advances in Chemistry Series, **145**, Amer. Chem. Soc., Washington, 1975, pp. 1–25.

Baier, R. E., "Substrata influences on adhesion of microorganisms and their resultant new surface properties," in G. Bitton and K. C. Marshall (eds.), *Adsorption of Microorganisms to Surfaces,* Wiley, New York, 1980, pp. 59–104.

Baier, R.E. and P.O. Glantz, *Acta Odontologica Scandanavica,* **36**, 289–301 (1978).

Baier, R.E., E.G. Shafrin, and W.A. Zisman, *Science,* **162**, 1360–1368 (1968).

Baier, R.E. and L. Weiss, "Demonstrations of the involvement of adsorbed proteins in all adhesion and cell growth on solid surfaces," in R.E. Baier (ed.), *Applied Chemistry at Protein Interfaces,* Advances in Chemistry Series, **145**, Amer. Chem. Soc., Washington, 1975, pp. 300–307.

Bakke, R., "Dynamics of biofilm processes: Substrate load variations," unpublished Master of Science thesis, Montana State University, Bozeman, MT, 1983.

Bakke, R., M.G. Trulear, J.A. Robinson, and W.G. Characklis, *Biotech. Bioeng.*, **26**, 1418–1424 (1984).

Barnett, K.G., T. Cosgrove, B. Vincent, D.S. Sissons, and M. Cohen-Stuart, *Macromolecules,* **14**, 1018–1020 (1981).

Beal, S.K., *Nucl. Sci. Eng.*, **40**, 1–11 (1970).

Botham, R.A. and C. Theis, *J. Poly. Sci.* (Part C), **30**, 369–380 (1970).

Brash, J.L. and Q.M. Samak, *J. Colloid Interface Sci.*, **65**, 495–504 (1978).

Browne, L.W.B., *Atmos. Environ.*, **8**, 801 (1974).

Bryers, J.D., "Dynamics of early biofilm formation in a turbulent flow system," Ph.D. thesis, Rice University, Houston, TX, 1980.

Bryers, J.D. and W.G. Characklis, *Biotech. Bioeng.*, **24**, 2451–2476 (1982).

Cafe, M.C. and I.D. Robb, *J. Coll. Interface Sci.*, **86**, 411–421 (1982).

Characklis, W.G., *Biofilm Development and Destruction*, Final Report, EPRI CS-1554, Project RP902-1, Electric Power Research Institute, Palo Alto, CA, 1980.

Characklis, W.G., "Biofilm development: A process analysis," in K.C. Marshall (ed.), *Microbial Adhesion and Aggregation* (Dahlem Konferenzen), Springer-Verlag, Berlin, 1984, pp. 137–157.

Cleaver, J.W. and B. Yates, *Chem. Eng. Sci.*, **30**, 983–992 (1975).

Cleaver. J.W. and B. Yates, *Chem. Eng. Sci.*, **31**, 147–151 (1976).

Cohen-Stuart, M.A., J.M.H.M. Scheutjens, and G.J. Fleer, *J. Polymer Sci.* (Part B), **18**, 559 (1980).

Costerton, J.W., R.T. Irvin, and K.J. Cheng, *Ann. Rev. Microbiol.*, **35**, 299–324 (1981).

Dabros, T. and T.G.M. van de Ven, *Colloid. Polym. Sci.*, **261**, 694 (1983).

Daniels, S.L., "Mechanisms involved in sorption of microorganisms to solid surfaces," in G. Bitton and K.C. Marshall (eds.), *Adsorption of Microorganisms to Surfaces,* Wiley, New York, 1980, pp. 8-58.

DePalma, V.A., D.W. Goupil, and C.K. Akers, presented at 6th Offshore Thermal Energy Conversion Conference, Washington, 1979.

Derjaguin, B.V. and I. Landau, *Acta Physicochim. URSS,* **14**, 633-662 (1941).

Dexter, S.C., *J. Colloid Interface Sci.*, **70**, 346-354 (1979).

Dolowy, K., in A.S.G. Curtis and J.D. Pitts (eds.), *Cell Adhesion and Motility,* Cambridge University Press, London and New York, 1980, pp. 39-63.

Escher, A.R., "Bacterial colonization of a smooth surface: An analysis with image analyzer," unpublished doctoral thesis, Montana State University, Bozeman, MT, 1986.

Fletcher, M., *J. Gen. Microbiol.*, **94**, 400-404 (1976).

Fletcher, M., *Can. J. Microbiol.*, **23**, 1-6 (1977).

Fletcher, M., in E. H. Beachey, (ed.), *Bacterial Adherence,* Chapman & Hall, London, 1980, pp. 347-374.

Fletcher, M. and G.I. Loeb, *Appl. Environ. Microbiol.*, **37**, 67-72 (1979).

Fletcher, M. and K.C. Marshall, *Appl. Environ. Microbiol.*, **44**, 184-192 (1982).

Friedlander, S.K. and H.F. Johnstone, *Ind. Eng. Chem.*, **49**, 1151-1156 (1957).

Gingell, D. and S. Vince, in A.S.G. Curtis and L.D. Pitts, (eds.), *Cell Adhesion and Motility,* Cambridge University Press, London and New York, 1980, pp. 1-37.

Gordon, A.S., S.M. Gerchakov, and F.J. Milliero, *Appl. Environ. Microbiol.*, **45**, 411-417 (1983).

Heckels, J.E., B. Blackett, S.J. Everson, and M.E. Ward, *J. Gen. Microbiol.*, **96**, 359-364 (1976).

Howell, J.A. and B. Atkinson, *Wat. Res.*, **10**, 307-315 (1976).

Jang, L.K. and T.F. Yen, "A theoretical model of diffusion of motile and nonmotile bacteria toward solid surfaces," in J.E. Zajic and E.C. Donaldson, (eds.), *Microbes and Oil Recovery,* **1**, International Bioresources Journal, 1985, pp. 226-246.

Kefford, B. and K.C. Marshall, *Arch. Microbiol.*, **138**, 84-88 (1984).

Leech, R. and R.J.W. Hefford, "The observation of bacterial deposition from a flowing suspension," in R.C.W. Berkeley, J.M. Lynch, J. Melling, P.R. Rutter, and D. Vincent, (eds.), *Microbial Adhesion to Surfaces,* Ellis Horwood, Chichester, 1980, pp. 544-545.

Lin, C.S., R.W. Moulton, and G.L. Putnam, *Ind. Eng. Chem.*, **45**, 636-640 (1953).

Lister, D. H. "Corrosion products in power generating systems," in E.F.S. Somerscales and J.G. Knudsen (eds.), *Fouling of Heat Transfer Equipment,* Hemisphere, Washington, 1981, pp. 135-200.

Little, B.J. and O.J. Jacobus, *Org. Geochem.*, **8**, 27-33 (1985).

Little, B.J. and A. Zsolnay, *J. Colloid. Int. Sci.*, **104**, 79-86 (1985).

Loeb, G.I. and R.A. Neihof, "Marine conditioning films," in R.E. Baier (ed.), *Applied Chemistry at Protein Interfaces,* Advances in Chemistry Series, **145**, Amer. Chem. Soc., Washington, 1975, pp. 319-335.

Marshall, K.C., in M. Shilo (ed.), *Strategies of Microbial Life in Extreme Environments,*

Dahlem Konferenzen Life Sciences Research Report 13, Verlag Chemie, Weinheim, 1979, pp. 281-290.

Marshall, K.C., R. Stout, and R. Mitchell, *J. Gen. Microbiol.*, **68**, 337-348 (1971).

Meadows, P.S., *Arch. Mikrobiol.*, **75**, 374-381 (1971).

Miura, Y., M. Okazaki, S. Hamada, S. Murakawa, and R. Yugen, *Biotech. Bioeng.*, **19**, 701-714 (1977).

Neihof, R.A. and G.I. Loeb, *Limnol. Oceanogr.*, **17**, 7-16 (1972).

Nelson, C.H., J.A. Robinson, and W.G. Characklis, *Biotech. Bioeng.*, **27**, 1662-1667 (1985).

Paul, J.H. and W.H. Jeffrey, *Appl. Environ. Microbiol.*, **50**, 431-437 (1985).

Pethica, B.A., *Exp. Cell Res. Suppl.*, **8**, 123-140 (1961).

Pethica, B.A., "Microbial and cell adhesion," in R.C.W. Berkeley, J.M. Lynch, J. Melling, P.R. Rutter, and D. Vincent (eds)., *Microbial Adhesion to Surfaces*, Ellis Horwood, Chichester, 1980, pp. 19-45.

Powell, M.S. and N.K.H. Slater, *Biotech. Bioeng.*, **24**, 2527-2537 (1982).

Powell, M.S. and N.K.H. Slater, *Biotech. Bioeng.*, **25**, 891-900 (1983).

Robb, I.D., "Stereo-biochemistry and function of polymers," in K.C. Marshall (ed.), *Microbial Adhesion and Aggregation* (Dahlem Konferenzen), Springer-Verlag, Berlin, 1984, p. 39.

Robinson, J.A., M.G. Trulear, and W.G. Characklis, *Biotech. Bioeng.*, **26**, 1409-1417 (1984).

Rosenberg, M., E.A. Bayer, J. Delarea, and E. Rosenberg, *Appl. Environ. Microbiol.*, **44**, 929-937 (1982).

Rouhiainen, P.O. and J.W. Stachiewicz, *J. Heat Transfer* (Trans. ASME), **92**, 169-177 (1970).

Rutter, P.R. in A.S.G. Curtis and L.D. Pitts (eds.), *Cell Adhesion and Motility*, Cambridge University Press, London and New York, 1980, pp. 103-135.

Rutter, P.R. and B. Vincent, "The adhesion of microorganisms to surfaces: Physicochemical aspects," in R.C.W. Berkeley, J.M. Lynch, J. Melling, P.R. Rutter, and D. Vincent (eds.), *Microbial Adhesion to Surfaces*, Ellis Horwood, Chichester, 1980, pp. 79-92.

Rutter, P.R. and B. Vincent, "Physicochemical interactions of the substratum, microorganisms, and the fluid phase," in K.C. Marshall (ed.), *Microbial Adhesion and Aggregation* (Dahlem Konferenzen), Springer-Verlag, Berlin, 1984, pp. 21-38.

Stanley, P.M., *Can. J. Microbiol.*, **29**, 1493-1499 (1983).

Tadros, T.F., "Particle-surface adhesion," in R.W.C. Berkeley, J.M. Lynch, J. Melling, P.R. Rutter, and B. Vincent (eds.), *Microbial Adhesion to Surfaces*, Ellis Horwood, Chichester, 1980, pp. 93-116.

Timperley, D., "Effect of Reynolds number and mean velocity on the cleaning-in-place of pipelines," in B. Hallstrom, D.B. Lund, and C. Tragardh (eds.), *Fundamentals and Applications of Surface Phenomena Associated with Fouling and Cleaning in Food Processing*, Lund University, Tylosand, Sweden, 1981, pp. 402-412.

Trulear, M.G. and W.G. Characklis, *J. Wat. Poll. Control Fed.*, **54**, 1288-1310 (1982).

Trulear, M.G., "Cellular reproduction and extracellular polymer formation in the development of biofilms," unpublished doctoral thesis, Montana State University, Bozeman, MT, 1983.

Turakhia, M.H., K.E. Cooksey, and W.G. Characklis, *Appl. Env. Microbiol.*, **46**, 1236–1238 (1983).

Vaituzi, Z. and R.N. Doetsch, *Appl. Environ. Microbiol.*, **17**, 584–588 (1969).

Verwey, E.J.W. and J.T.G. Overbeek, *Theory of the Stability of Lyophobic Colloids*, Elsevier, Amersterdam, 1948.

Ward, J.B. and R.C.W. Berkeley, "The microbial cell surface and adhesion," in R.C.W. Berkeley, J.M. Lynch, J. Melling, P.R. Rutter, and D. Vincent (eds.), *Microbial Adhesion to Surfaces*, Ellis Horwood, Chichester, 1980, pp. 47–66.

Watt, P.J., "The adhesive properties of *Neisseria gonorrhoeae*," in R.W.C. Berkeley, J.M. Lynch, J. Melling, P.R. Rutter, and B. Vincent (eds.), *Microbial Adhesion to Surfaces*, Ellis Horwood, Chichester, 1980, pp. 455–472.

Weiss, L. and J.P. Harlos, in W.C. deMello (ed.), *Intercellular Communications*, Plenum, New York, 1977, pp. 33–59.

Wells, A.C. and A.C. Chamberlain, *Br. J. Appl. Physiol.*, **18**, 1793–1799 (1967).

Wood, J.M., "The interaction of microorganisms with ion exchange resins," in R.W.C. Berkeley, J.M. Lynch, J. Melling, P.R. Rutter, and B. Vincent (eds.), *Microbial Adhesion to Surfaces*, Ellis Horwood, Chichester, 1980, pp. 163–185.

Zisman, W.A., in F.M. Fowkes (ed.), *Contact Angle Wettability and Adhesion*, Advances in Chemistry Series, **43**, Am. Chem. Soc., Washington, 1964, pp. 1-51.

Zobell, C.E., *J. Bacteriol.*, **46**, 39–59 (1943).

KINETICS OF MICROBIAL TRANSFORMATIONS

WILLIAM G. CHARACKLIS
Montana State University, Bozeman, Montana

SUMMARY

The kinetic expressions for the relevant microbial transformations are presented. In Chapter 6, the stoichiometric principles of the transformations, including yield concepts, are discussed. In this chapter, the rate expressions and stoichiometric ratios (i.e., the yields) are combined in material balances through the matrix method described in Chapter 2. Thus, the reaction rate of different components can be related through stoichiometry. The relevant reactions include growth, maintenance or decay, product formation, and death. The rates of these fundamental processes are difficult to measure, so the data must be extracted from *observed processes*. Observed rates include the specific substrate removal rate, specific oxygen removal rate, and specific biomass production rate. The stoichiometry of each fundamental process are expressed in terms of its reactants and products.

8.1 INTRODUCTION

Transformations or reactions have two features of importance to the scientist and engineer: (1) *stoichiometry,* which determines the extent of the reaction and the composition of the products, and (2) *rate,* which determines how fast changes will occur.

In this chapter, kinetic expressions that describe the rate of transformations mediated by microbial metabolism are presented. Transformation or reaction processes refer to those processes in which molecular rearrangements occur. Unstructured kinetic models (Section 2.2.3, Chapter 2) for microbial transformations are emphasized in this chapter. They characterize the biotic component only in terms of its mass, with little attention given to physiological state or microbial species diversity. Another way of considering unstructured models is that they reflect only the changes observed external to the cell. Unstructured models are only useful under balanced growth conditions.

8.1.1 The Rate Equation

A rate equation characterizes the rate of a reaction. The form of the equation may be suggested by a theoretical analysis based upon a given mechanistic model or may simply be the result of an empirical curve fit procedure. In either case, the values of the rate coefficients must be determined by experiment, since predictive methods are unlikely to be of universal value. The formulation of a rate equation traditionally involves the study of the influence of concentration on the rate, assuming all other conditions are constant. Subsequently, the influence of environmental factors, such as temperature and pH, on the rate coefficients may be investigated. Changes in environmental factors are reflected in the rate coefficient and are not considered to influence the functional dependence of the rate on concentration unless the reaction mechanism changes.

Knowledge of the true reaction mechanism is valuable for extrapolating the experimental data to regions other than those investigated. Difficulties arise, however, because the reaction mechanism frequently changes when reaction environment changes. Describing reaction kinetics in terms of the reaction mechanism results in a large number of parameters, requiring considerable experimental effort, numerous equations, and statistical uncertainty in estimates of the rate coefficients.

For these and other reasons, the *nth order rate equation* (Churchill, 1974; Denbigh and Turner, 1971; Levenspiel, 1972) has been the most frequently used kinetic expression in chemical engineering reactor theory:

$$r_V = k\,C^n \qquad [8.1]$$

where r_V = rate of reaction ($ML^{-3}t^{-1}$)
k = reaction rate coefficient ($M^{1-n}L^{3n-3}t^{-1}$)
C = concentration (ML^{-3})
n = reaction order

The equation has two parameters: the rate coefficient k and the reaction order n. The equation generally provides a good empirical curve fit over restricted ranges of experimental data. One of the objectives of microbiological or biochemical reactor theory has been to develop an analogous equation or equations for describing behavior in fermentation, wastewater treatment, and microbial ecosystems in general. The classical rate equation used in microbial engineering is the saturation or Monod expression for microbial growth (Section 8.2.3):

$$\mu = \frac{\mu_{\max} S}{K_S + S} \tag{8.2}$$

where $\mu_{\max}$ = maximum specific growth rate (t^{-1})
K_S = rate (saturation) coefficient (ML^{-3})
S = substrate concentration (ML^{-3})

The equation also has two parameters, and the growth rate μ becomes saturated (approaches $\mu_{\max}$) as S increases to high values.

8.1.2 Experimental Data

Kinetic models for microbial reactions are empirical models. Hence, evaluation of kinetic models requires experimentation and analysis of the data to determine the best estimates of the rate coefficients. Therefore, this chapter includes examples using actual experimental data to illustrate the determination of reaction coefficients. The processes and coefficients emphasized are those for which biofilm data are also known and will be referred to in other chapters. The data of Trulear (1983) have been presented elsewhere (Robinson et al., 1984; Bakke et al., 1984; this book, Chapter 13) in the context of biofilm processes and hence are convenient for this development. Trulear conducted chemostat experiments with *Pseudomonas aeruginosa* growing in a glucose and mineral salts medium at 23°C. Glucose was the only carbon source and was the growth-limiting nutrient. The steady state data from the chemostat experiments are presented in Table 8.1. All data are in units of carbon mass. Suspended solids were measured gravimetrically (0.2 μm filtration), but the concentration was converted to carbon units by multiplying by 0.45 g C per gram of suspended solids.

8.1.3 Some Simplifications

The kinetic developments require some simplification, since the details of microbial metabolism are not described in their entirety:

1. Unstructured models represent the biotic component only by mass and/or cell numbers. The extracellular polymeric substances (EPS), however, will be introduced as a nonviable component separate from the cell mass so that the biomass can be segregated into two components.
2. Balanced growth (e.g., steady state in an open system or log growth in a closed system) is assumed and is a requirement for unstructured models. Steady state chemostat data are used primarily throughout this chapter.
3. A single species population, *Ps. aeruginosa,* is presumed throughout the developments.

TABLE 8.1 Steady State Experimental Data from a Chemostat Containing *Ps. aeruginosa* Growing in Glucose and Mineral Salts[a]

Dilution Rate D (h^{-1})	Substrate Feed S_i [(g C) m^{-3}]	Suspended Solids[b] X [(g C) m^{-3}]	Cells X_c [(g C) m^{-3}]	Substrate S [(g C) m^{-3}]	Product P_E [(g C) m^{-3}]
0.025	37	13.5 ± 1.3	4.9 ± 1.2	0.1 ± 0	7.8 ± 4.6
0.05	32	14.2 ± 3.5	8.7 ± 2.1	0.2 ± 0	7.2 ± 0.2
0.075	38	15.8 ± 3.8	8.6 ± 0.3	0.8 ± 0.3	6.3 ± 1.5
0.1	39	14.4 ± 0	9.6 ± 0.1	0.4 ± 0	6.9 ± 0.3
0.15	39	15.6 ± 1.6	10.9 ± 1.7	0.3 ± 0	6.0 ± 0.4
0.25	37	13.4 ± 1.7	7.8 ± 0.2	7.5 ± 0.8	3.0 ± 0.2
0.35	36	4.0 ± 1.2	2.0 ± 0.1	28.6 ± 0.8	2.1 ± 0.8
0.4	37	0	0.9 ± 0.3	33.2 ± 0.3	0.1 ± 0.1

[a] Glucose was the sole carbon source and the growth-limiting nutrient (Trulear, 1983).

[b] Calculated in units of carbon by multiplying suspended solids mass by 0.45 (g C) $(g\ X)^{-1}$.

8.2 GROWTH

The equations presented in this section describe rates of growth for an unstructured system. The cell composition is assumed constant throughout the experiment, i.e., we assume balanced growth. Balanced growth can be realized during the exponential growth phase in a batch reactor and at steady state in an open or continuous reactor.

8.2.1 Exponential Growth

The requisite conditions for microbial growth include a viable inoculum, an energy source, nutrients for synthesis, and a suitable physicochemical environment. The rate of microbial growth is proportional to the biomass present, which results in the following *rate equation:*

$$r_X = \mu X \qquad [8.3]$$

where μ = specific growth rate (t^{-1})
X = biomass concentration (ML^{-3})
r_X = reaction rate ($ML^{-3}t^{-1}$)

This equation describes exponential or logarithmic growth and is a useful description of microbial growth as long as environmental conditions remain constant and the constitution of the biomass remains constant (balanced growth). In a batch reactor, the rate of biomass accumulation in the logarithmic growth phase is described by the following *balance equation* (Chapter 2):

$$\frac{dX}{dt} = \mu X \qquad [8.4]$$

The equation can be integrated to yield

$$X = X_0 2^{t/g} \qquad [8.5]$$

or

$$g = \frac{\ln 2}{\mu} \qquad [8.6]$$

where g = generation time or doubling time of the population (t)
X_0 = initial biomass concentration (the inoculum) (ML^{-3})

But the specific growth rate μ is influenced by many environmental variables. The most important variable is the concentration of the limiting nutrient, the substrate. So we set

$$\mu = f(S) \qquad [8.7]$$

Other variables of interest include temperature, pH, ionic strength, and concentration of inhibiting substances. Therefore,

$$\mu = f(S, T, \text{pH}, S_{N1}, S_{N2}, \ldots S_{Nn}) \qquad [8.8]$$

where S_{Ni} = concentration of ith component in growth medium (ML^{-3})

8.2.2 The Logistic Equation

The logistic equation describes a familiar S-shaped curve frequently observed in batch growth of microorganisms (Fig. 8.1):

$$r_X = \mu_{\max} X\left(1 - \frac{X}{X_m}\right) \qquad [8.9]$$

where $\mu_{\max}$ = maximum specific growth rate (t^{-1})
X_m = maximum biomass population attained (ML^{-3})

This equation requires the growth rate to be fixed if the biomass concentration is specified. Hence, the growth rate cannot depend on the substrate concentration, and the equation loses its usefulness as a growth rate equation. Despite its limitation with respect to growth processes, the logistic equation is useful for describing the *accumulation rate* of various components in a reactor. For example, in a batch reactor the growth curve is indicative of the rate of accumulation of cells, not necessarily their growth rate. In addition, the progression of biofilm accumulation is frequently described by a logistic function.

For purposes of describing accumulation rate, the logistic equation can be expressed as follows:

$$\frac{dX}{dt} = kX(1 - k'X) \qquad [8.10]$$

where k = rate coefficient (t^{-1})
k' = rate (saturation) coefficient (L^3M^{-1})

The relationship of the rate coefficients to the shape of the function is presented in Figure 8.1. The integrated form of the equation is

$$X = \frac{X_0 e^{kt}}{1 - k'X_0(1 - e^{kt})} \qquad [8.11]$$

The logistic equation will be used in subsequent chapters to describe the progression of biofilm accumulation and its influence on fluid frictional resistance and heat transfer resistance.

8.2.3 The Saturation Rate Equation

The most widely used expression for describing the rate of microbial growth as a function of nutrient (substrate) concentration is that attributed to Monod (1942, 1949), which describes a rectangular hyperbola (Figure 8.2):

$$\mu = \frac{\mu_{max} S}{K_S + S} \qquad [8.2]$$

where K_S = rate (saturation) coefficient (ML^{-3})
S = substrate concentration (ML^{-3})

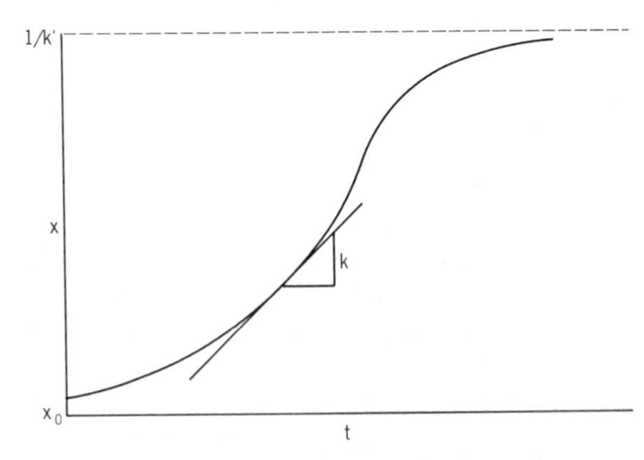

Figure 8.1. The progression of biofilm accumulation is described by a logistic equation. The geometric representation of the rate coefficients from Eq. 8.10 is indicated on the graph.

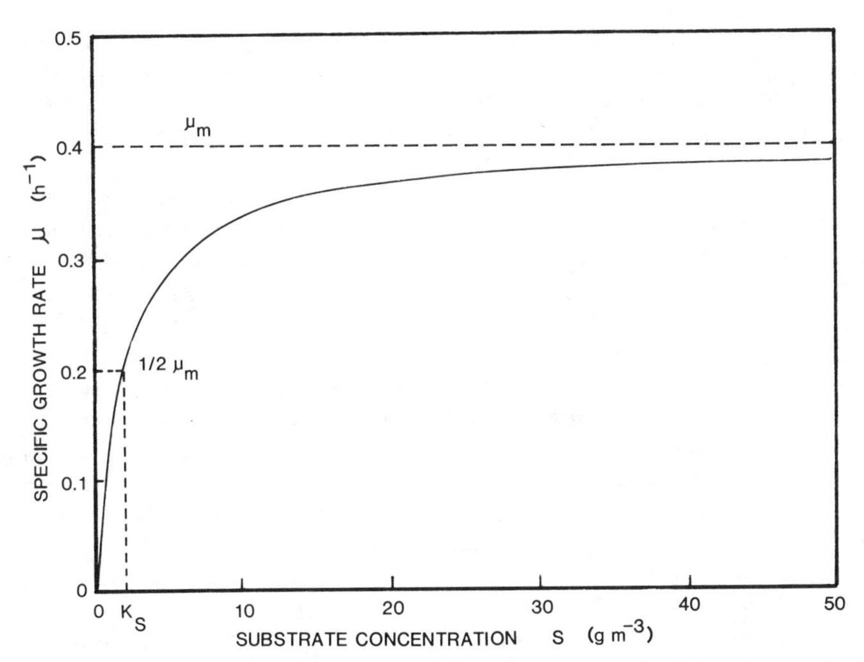

Figure 8.2. A rectangular hyperbola describes the saturation rate (Monod expression). The geometric representations of the rate coefficients from Eq. 8.2 are presented. The numerical values of the coefficients are for glucose-limited *Pseudomonas aeruginosa* in a chemostat. (Reprinted with permission from Robinson et al., 1984.)

The expression can be considered as the combination of two expressions that describe the behavior at very low and very high concentrations:

$$\mu = \mu_{\max} \qquad \text{for} \quad S \gg K_S \qquad [8.12]$$

$$\mu = \frac{\mu_{\max}}{K_S} S \qquad \text{for} \quad S \ll K_S \qquad [8.13]$$

At low concentrations, the rate is first order with respect to substrate concentration (Eq. 8.13). In many systems of relevance to biofilm formation, the most important data are those at low substrate concentration (oligotrophic conditions). Unfortunately, there are sometimes analytical limitations on the evaluation of data in this region. At low concentrations, it is also likely that diffusion or mass transport will limit the overall rate of growth (Chapter 9). At high concentrations, the rate is independent of concentration, i.e., zero order with respect to substrate (Eq. 8.12). The basis for the rate expression can be visualized by the way in which it relates to saturation of the organism with substrate. At low concentration, the organism still has a significant reaction potential available, and an increase in substrate supply will increase its growth rate. As the concentration is increased, however, the cell can no longer assimilate the substrate that is being provided: it becomes saturated. Because of this behavior, Eq. 8.2 is frequently referred to as the *saturation rate equation*. The same equation is used to describe enzyme kinetics (Michaelis and Menten, 1913) and adsorption phenomena (Langmuir, 1918). All of these processes are characterized as having active sites that become saturated at high concentration.

The coefficients in the saturation rate equation are frequently used to categorize microorganisms as *oligotrophic* (grow in a very dilute culture medium) or *copiotrophic* (grow in a concentrated culture medium). According to this classification, an oligotrophic organism has a characteristically low K_S-value (<1 g m^{-3}). The difficulty with this classification is that the same organism may have a high K_S in one culture medium and a low K_S in another. Martin and MacLeod (1984) present strong evidence for this phenomenon and conclude that two such broad classes such as oligotrophs and copiotrophs probably do not exist (see Chapter 10). Thus, the proliferation of organisms in a particular environment will depend on the specific nutrients available and their concentration. This behavior is not totally surprising in light of the similarity between the kinetic expressions describing microbial growth and enzymes within the cell. Depending on the limiting nutrient, different enzymes within the cell may limit growth, and they probably exhibit different saturation coefficients.

The similarity between the equations describing microbial growth and enzyme kinetics suggests a mechanistic relationship between the two. Monod (as well as Kessick, 1974) indicates that the models are of the same form but suggested that the saturation coefficient K_s should be expected to be related to the apparent dissociation coefficient of the enzyme involved in the first step of the breakdown of the substrate.

Thus, Monod found that relatively simple empirical laws conveniently express the relation between exponential growth rate and concentration of an essential nutrient. He also pointed out that several mathematically different formulations could be made to fit the data. However, he found it both convenient and logical to adopt a hyperbolic equation similar to an adsorption isotherm or to the Michaelis–Menten equation. In summary, Monod was the first to use a hyperbolic equation to describe a microbial growth rate as a function of limiting nutrient concentration. The curve fit was good, and the method has received wide acceptance.

Example 8A. A chemostat with microbial growth. Example 8A and subsequent examples use the matrix method of notation described in Chapter 2. Consider a continuous flow stirred tank reactor with a feed containing sterile substrate at a concentration S_i, i.e., a chemostat (Figure 8.3). The chemostat liquid volume is V, and the volumetric flow rate is Q. The reactor substrate concentration is S, and the reactor biomass concentration is X. The only fundamental process occurring is *growth,* and the growth rate depends on substrate concentration according to the saturation equation. The inlet substrate concentration is $S_i = 40$ (g C) m^{-3} (glucose), the temperature is 23°C, and the pH is 7.0. Dissolved oxygen is in excess.

The chemostat is operated at one dilution rate until steady state is reached at

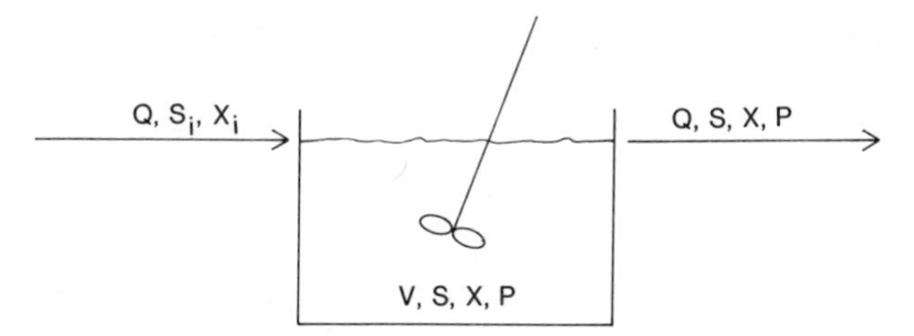

Figure 8.3. Schematic diagram of a chemostat indicating the appropriate variables from Examples 8A, 8B, 8C, and 8D.

which time S and X are determined. At each dilution rate, a different steady state S and X are measured (Figure 8.5)

The chemical state vector for this system is

$$\mathbf{S} = \begin{bmatrix} S \\ X \end{bmatrix} \qquad [8A.1]$$

The transport vector for this CFSTR is

$$\boldsymbol{\phi} = \begin{bmatrix} D(S_i - S) \\ D(X_i - X) \end{bmatrix} \qquad [8A.2]$$

where D = dilution rate, Q/V (t^{-1})

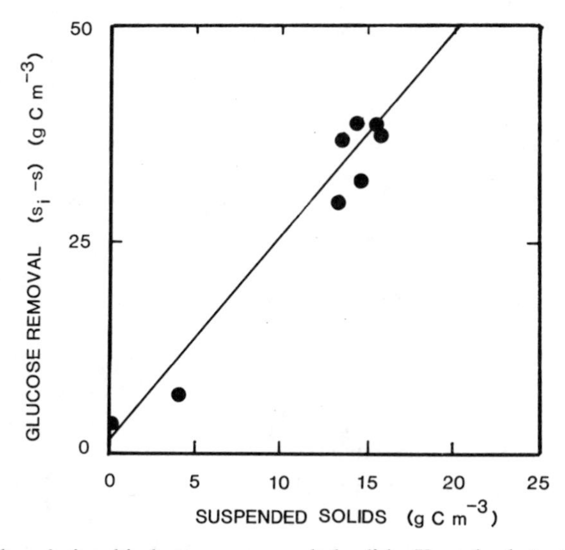

Figure 8.4. The relationship between suspended solids, X, and substrate removal, $S_i - S$, to determine yield. Data are for glucose-limited *Pseudomonas aeruginosa* in a chemostat (Robinson et al., 1984).

The stoichiometry of the system can be expressed as follows:

$$a_1\, C_6H_{12}O_6 + a_2\, O_2 + a_3\, NH_3 \rightarrow a_4\, CH_{1.4}O_{0.4}N_{0.2} + a_5\, CO_2 + a_6\, H_2O \quad [8A.3]$$

or, since only substrate and biomass are considered in this model,

$$a_1 S \rightarrow a_4 X \quad [8A.4]$$

Therefore, the stoichiometry matrix is

$$\mathbf{a} = \begin{bmatrix} -\frac{a_1}{a_4} & 1 \end{bmatrix} \quad [8A.5]$$

The rate of reaction is

$$r_X = \mu X \quad [8A.6]$$

So the resulting reaction rate vector is

$$\mathbf{r} = \mu X \quad [8A.7]$$

Then

$$\mathbf{r} \cdot \mathbf{a} = \begin{bmatrix} \dfrac{-\mu X}{a_1/a_4} \\ \mu X \end{bmatrix} \quad [8A.8]$$

But a_1/a_4 is usually denoted by $Y_{X/S}$, the yield of biomass from substrate. So

$$\mathbf{r} \cdot \mathbf{a} = \begin{bmatrix} -\mu X/Y_{X/S} \\ \mu X \end{bmatrix} \quad [8A.9]$$

In general,

$$\frac{d\mathbf{S}}{dt} = \boldsymbol{\phi} + \mathbf{r} \cdot \mathbf{a} \quad [8A.10]$$

Then from Eqs. 8A.2, 8A.9, and 8A.10,

$$\frac{dS}{dt} = D(S_i - S) - \frac{\mu X}{Y_{X/S}} \quad [8A.11]$$

$$\underset{\substack{\text{rate of}\\ \text{accumulation}\\ \text{in reactor}}}{\frac{dX}{dt}} = \underset{\substack{\text{net rate of}\\ \text{transport into}\\ \text{reactor}}}{D(X_i - X)} + \underset{\substack{\text{rate of}\\ \text{production}}}{\mu X} \qquad [8A.12]$$

which are the material balances for substrate and biomass, respectively. The derivatives with respect to time are rates of accumulation, whereas the last terms on the right hand side are process rates. For sterile feed, $X_i = 0$, and at a steady state the derivatives vanish. Then Eq. 8A.11 and 8A.12 become

$$\frac{D(S_i - S)}{X} = \frac{\mu}{Y_{X/S}} \qquad [8A.13]$$

$$D = \mu \qquad [8A.14]$$

All quantities on the left hand side of Eqs. 8A.13 and 8A.14 are measurable. In fact, the left hand side of Eq. 8A.13 is termed the specific substrate removal rate or specific activity, q_S:

$$q_S = \frac{D(S_i - S)}{X} \qquad [8A.15]$$

Equation 8A.14 indicates that the specific growth rate can be controlled by the experimenter by varying the flow rate, Q since $D = Q/V$. The conclusion is that at steady state, the rate of cell production equals the rate of cell washout.

The resulting two equations, 8A.13 and 8A.14, contain two unknowns, μ and $Y_{X/S}$, which can be determined from experimental data. By substituting Eq. 8A.14 into Eq. 8A.13, we can eliminate D with the following result:

$$Y_{X/S} = \frac{X}{S_i - S} \qquad [8A.16]$$

The data from Table 8.1 for suspended solids and substrate are plotted in this way in Figure 8.4. $Y_{X/S} = 0.69$ (g C X)/(g C S) by this analysis.

These experimental chemostat data behave consistently with the model that has been developed. By inserting the saturation rate equation in Eq. 8A.14, μ_{max} and K_S can be determined by methods described in Section 8.2.3.1. The dependence of S and X on the dilution rate D (or the specific growth rate) can be assessed with these coefficients (Figure 8.5). S is very low at low dilution rate (i.e., at high liquid residence time), where sufficient time results in almost complete substrate depletion. The biomass concentration is relatively high, since the large amount of substrate consumption goes to biomass production. At high dilution rate, S increases (eventually reaching S_i at $D > D_c$) because of short residence times, which also provide greater dilution of cells from the reactor.

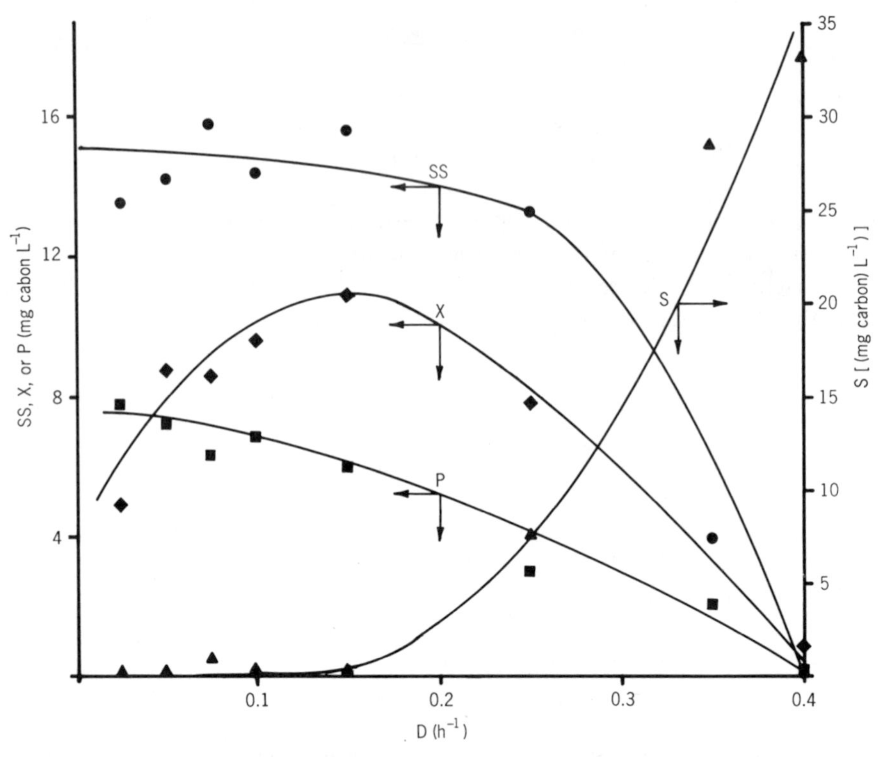

Figure 8.5. The relationship between the substrate concentration S, the EPS concentration P, the biomass concentration X, and the dilution rate D in a chemostat for glucose-limited *Pseudomonas aeruginosa*. SS refers to measured suspended solids ($X + P$). (Reprinted with permission from Robinson et al., 1984).

8.2.3.1 Evaluation of Rate Coefficients

Linear Regression. Evaluation of parameters in the saturation expression traditionally has required linearization of Eq. 8.2. Each linearization yields different values for the parameters and each has its limitations:

1. *Lineweaver–Burk.*

$$\frac{1}{\mu} = \frac{1}{\mu_{\max}} + \frac{K_S}{\mu_{\max}}\left(\frac{1}{S}\right) \qquad [8.14]$$

The most accurately determined experimental points are clustered about the origin, where specific growth rate and substrate concentration are both large and easy to measure accurately. The least accurate data will be away from the origin, where specific growth rate and substrate concentration are low. These points will dominate the determination of the slope, K_S/μ_{max}. The value of μ_{max}, however, can be determined with accuracy from a linear regression of this form.

2. *Eadie–Hofstee.*

$$\mu = \mu_{max} - K_S \frac{\mu}{S} \qquad [8.15]$$

Both coordinates contain the measured variable, μ, which may be subject to the large errors, especially when μ is small.

3. *Unnamed.*

$$\frac{S}{\mu} = \frac{K_S}{\mu_{max}} + \frac{S}{\mu_{max}} \qquad [8.16]$$

The intercept in this equation generally occurs close to the origin, so that determination of K_s is subject to large errors. However, μ_{max} can be determined relatively accurately with this equation.

Linear regression, although useful because of its simplicity, does not necessarily lead to the calculation of the best estimates of the parameters from the data. In fact, the linearized version of each equation yields different estimates for the coefficients. The experimental data from the chemostat (Table 8.1) have been used to obtain the rate coefficients using the three linearized forms of the saturation equation. The resulting Monod or saturation rate equations are presented graphically with the data in Figure 8.6. The coefficients are presented in Table 8.2 along with the standard deviation. In addition, the correlation between μ_{max} and K_S is indicated. While the model development presumes that μ_{max} and K_S are uncorrelated, the results indicate relatively strong correlation between μ_{max} and K_S in the linearized equations.

Nonlinear Regression. Use of the linearized form of the saturation equation violates an important assumption of least squares analysis, namely that the independent variable (either μ or S) is free of error. Generally, the confidence limits are not reported (Table 8.1). Thus in most cases, it is better to fit nonlinear data directly to the nonlinear form of the model using nonlinear regression techniques (Robinson and Characklis, 1984; Dabes et al., 1973). The results of the nonlinear regression of the data are presented in Table 8.2 and Figure 8.6. By most statistical criteria, the results of the nonlinear regression are better estimates of the rate coefficients.

Approximation. Bailey and Ollis (1986) recommend the following strategy for approximating the saturation equation parameters:

1. Determine μ_{max} from Eq. 8.14 (intercept) or Eq. 8.16 (slope).
2. Return to a graph similar to Figure 8.6. S at $\mu = \frac{1}{2}\mu_{max}$ is K_S. By this method, $\mu_{max} = 0.3\ h^{-1}$ and $K_S \approx 1.5\ g\ m^{-3}$.

8.2.3.2 *Double Saturation Kinetics* It is conceivable that two or more reaction components are simultaneously limiting or influencing growth. For example, substrate and oxygen may be limiting at some depth in a biofilm. Several equations have been proposed to describe this phenomenon. All of the equations reduce to the saturation equation as one of the two components increases to saturation level. Tsao and Hansen (1975) proposed the most general approach with the following equation:

$$r_X = \left(1 + \sum_i \frac{k_i S_{Ei}}{K_{Ei} + S_{Ei}}\right)\left(\prod_i \frac{\mu_{m,i} S_{si}}{K_{si} + S_{si}}\right)X \qquad [8.17]$$

where S_{si} = concentration of ith essential substrate or nutrient
S_{Ei} = concentration of ith nonessential but growth-enhancing substance

A special case of Eq. 8.17 for two substrates has been proposed by Megee (1970):

$$r_X = \left(\frac{\mu_{m1}\mu_{m2} S_1 S_2}{(K_{S1} + S_1)(K_{S2} + S_2)}\right)X \qquad [8.18]$$

Another equation for dual limitation has been described by Howell (1983):

$$r_X = \left(\frac{\mu_m S_1 S_2}{(K_{S1} + S_1)(K_{S2} + S_2)}\right)X \qquad [8.19]$$

More experimental observations are necessary in different growth environments before any one multiple substrate rate equation receives wide acceptance.

8.2.3.3 *Particulate Substrates* In many microbial ecosystems, the substrate is not dissolved. Examples of such systems include natural waters (Kirchman, 1983; Schink and Zeikus, 1982), municipal wastewaters (Gujer, 1980; Hrudey, 1982), and industrial fermentations utilizing starch (Rollings et al., 1983), cellulose, hydrocarbons (Mallee and Blanch, 1977; Miura et al., 1977a, 1977b), and other insoluble substrates (Suga et al., 1975; Watson and Jones, 1977). The equations presented thus far have only been rigorously tested with dissolved substrates. Undoubtedly, the temporal and spatial interaction between the particulate substrate and the microorganism will play a major role in deter-

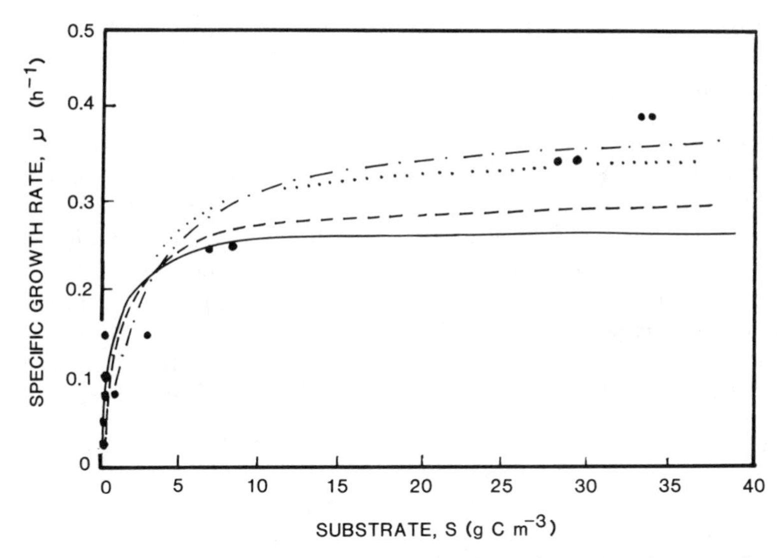

Figure 8.6. The relationship between μ and S for the various regression equations presented in Table 8.2: Eq. 8.14 (dashed curve), Eq. 8.15 (solid curve), Eq. 8.16 (dot–dash curve), and non-linear regression (dotted curve). Data were obtained by Trulear (1983).

TABLE 8.2 Statistical Comparison of Rate Coefficients for Glucose-Limited *Pseudomonas aeruginosa* Calculated by Different Regression Equations[a]

Equation	μ_{max} Mean (h^{-1})	μ_{max} S.D. (%)	K_S Mean ($g\ m^{-3}$)	K_S S.D. (%)	RSS[b]	Correlation Coefficient[c]
Nonlinear regression	0.37	7	1.6	32	0.03	0.59
Eq. 8.14	0.30	32	1.1	37	151	0.99
Eq. 8.15	0.40	4	2.5	25	292	0.67
Eq. 8.16	0.27	14	0.7	29	0.15	0.73

[a] Data from Trulear, 1983; analysis by M.A. Siebel.

[b] Residual sum of squares.

[c] Correlation coefficient between μ_m and K_s.

mining the rate and extent of degradation and hence the resulting growth rate of the organism on the substrate. This particle–microorganism interaction is discussed in more detail in Section 7.3.1.4 (Chapter 7).

A saturation growth expression may well result even with particulate substrates, but the saturation level may reflect a surface area limitation as opposed to a nutrient limitation.

8.3 MAINTENANCE AND ENDOGENOUS METABOLISM

8.3.1 Maintenance

Microorganisms require energy to maintain existing structures and for processes such as motility—the "cost of life." Maintenance processes may significantly influence intracellular processes during starvation survival within a biofilm (see Chapter 10). Thus, maintenance may result in reduced size of biofilm cells in substrate-depleted regions of the biofilm. The maintenance rate, therefore, reflects a diversion of substrate away from synthesis or growth. Consequently, maintenance decreases the overall (or observed) yield of cells from substrate. The maintenance rate term is found in the substrate material balance as part of the substrate depletion term. The kinetic expression for maintenance rate is

$$r_m = k_m X \tag{8.20}$$

where k_m = maintenance coefficient ($\mathrm{MM^{-1}t^{-1}}$)

Maintenance may be represented quantitatively as the substrate mass oxidized per unit cell mass per unit time to provide energy for maintenance. The rate expression could also be expressed as follows:

$$r_m = \frac{k'_m X}{Y_{m/S}} \tag{8.21}$$

where k'_m = specific maintenance coefficient ($\mathrm{t^{-1}}$)
$Y_{m/S}$ = yield of maintenance energy per unit substrate consumed for the maintenance processes ($\mathrm{M_X M_S^{-1}}$).

More complicated rate expressions for the maintenance rate have been proposed. Their usefulness, however, is questionable for unstructured systems. In fact, the maintenance rate is rarely determined directly, since no easily discernible product results from the process other than carbon dioxide.

Example 8B. A chemostat with growth and maintenance. This example is very similar to Example 8A except that two fundamental process are occurring: *growth* and *maintenance*. The chemical state vector for this system is the same except for the addition of respiration products, P_R:

$$\mathbf{S} = \begin{bmatrix} S \\ X \\ P_R \end{bmatrix} \tag{8B.1}$$

The transport vector is

$$\phi = \begin{bmatrix} D(S_i - S) \\ D(X_i - X) \\ D(P_{Ri} - P_R) \end{bmatrix} \qquad [8B.2]$$

The stoichiometry of the system is like the one in Eq. 8A.3 for growth but must also include a reaction for maintenance energy use. Therefore, the stoichiometry can be simply expressed as follows:

Growth

$$a_1\,C_6H_{12}O_6 + a_2\,O_2 + a_3\,NH_3 \rightarrow a_4\,CH_{1.4}O_{0.4}N_{0.2} + a_5\,CO_2 + a_6\,H_2O \qquad [8B.3]$$

and maintenance

$$a_7\,C_6H_{12}O_6 + a_8\,O_2 \rightarrow a_9\,CO_2 + a_{10}\,H_2O \qquad [8B.4]$$

or

$$a_1\,S \rightarrow a_4\,X + a_5\,P_R \qquad [8B.5]$$

$$a_7\,S \rightarrow \qquad a_9\,P_R \qquad [8B.6]$$

Then, the stoichiometric matrix is

$$\mathbf{a} = \begin{bmatrix} -a_1/a_4 & 1 & a_5/a_4 \\ -a_7/a_9 & 0 & 1 \end{bmatrix} \qquad [8B.7]$$

The reaction rate vector is

$$\mathbf{r} = [\mu X \quad k_m' X] \qquad [8B.8]$$

Then

$$\mathbf{r}\cdot\mathbf{a} = \begin{bmatrix} \dfrac{-\mu X}{Y_{X/S}} - \dfrac{k_m' X}{Y_{m/S}} \\ \mu X \\ \dfrac{\mu X}{a_4/a_5} + k_m' X \end{bmatrix} \qquad [8B.9]$$

where $Y_{m/S}$ = yield of maintenance energy from substrate ($M_m M_S^{-1}$)

Then

$$\frac{dS}{dt} = D(S_i - S) - \frac{\mu X}{Y_{X/S}} - \frac{k_m' X}{Y_{m/S}} \quad [8B.10]$$

$$\frac{dX}{dt} = D(X_i - X) - \mu X \quad [8B.11]$$

which are the material balances for substrate and biomass, respectively. The material balance for respiration products is omitted, since generally the direct measurements necessary for validating the balance are not made, i.e., respiration products are not often measured. There is no difference in the biomass balance equation from Eq. 8A.12. Only the substrate balance has changed.

For a sterile feed, $X_i = 0$, and at steady state the derivatives vanish. Then Eqs. 8B.10 and 8B.11 become

$$\frac{D(S_i - S)}{X} = \frac{\mu}{Y_{X/S}} + \frac{k_m'}{Y_{m/S}} \quad [8B.12]$$

$$D = \mu \quad [8B.13]$$

All quantities on the left hand side of Eqs. 8B.12 and 8B.13 are measurable. However, there are four parameters to calculate and only two equations. Therefore, graphical methods are used if several data points at various dilution rates are available, since every experiment at a different dilution rate results in another degree of freedom. Equation 8B.12 can be expressed in the following way to eliminate one parameter:

$$q_S = \frac{D}{Y_{X/S}} + k_m \quad [8B.14]$$

A plot of q_s versus D (Figure 8.7) is linear, with

$$\text{slope} = \frac{1}{Y_{X/S}} \quad [8B.15]$$

$$\text{intercept} = k_m \quad [8B.16]$$

The data for suspended solids carbon (X) and cellular carbon (X_c) have been plotted in Figure 8.7. The data for X do not indicate a maintenance energy requirement, while the data for X_c do indicate a nonzero intercept (approximately 0.1 g S carbon per g X), but the intercept is not significantly different from zero (95% confidence).

Note that growth yield and observed yield are not the same in this analysis. The observed or effective yield is

$$Y_0 = \frac{X}{S_i - S}, \qquad [8B.17]$$

or by combining Eqs. 8B.17, 8B.14, and 8B.12

$$\frac{1}{Y_0} = \frac{1}{Y_{X/S}} + k_m \qquad [8B.18]$$

Thus, the observed yield decreases as a result of maintenance processes.

8.3.2 Endogenous Metabolism

Another way of rationalizing a decreased observed cellular yield is by considering endogenous metabolism: the degradation of cellular (i.e., endogenous) components. Endogenous metabolism may also play an important role in starvation survival behavior of organisms in a substrate-depleted biofilm (see Chapter 10). The endogenous metabolism term is found in the biomass material balance and

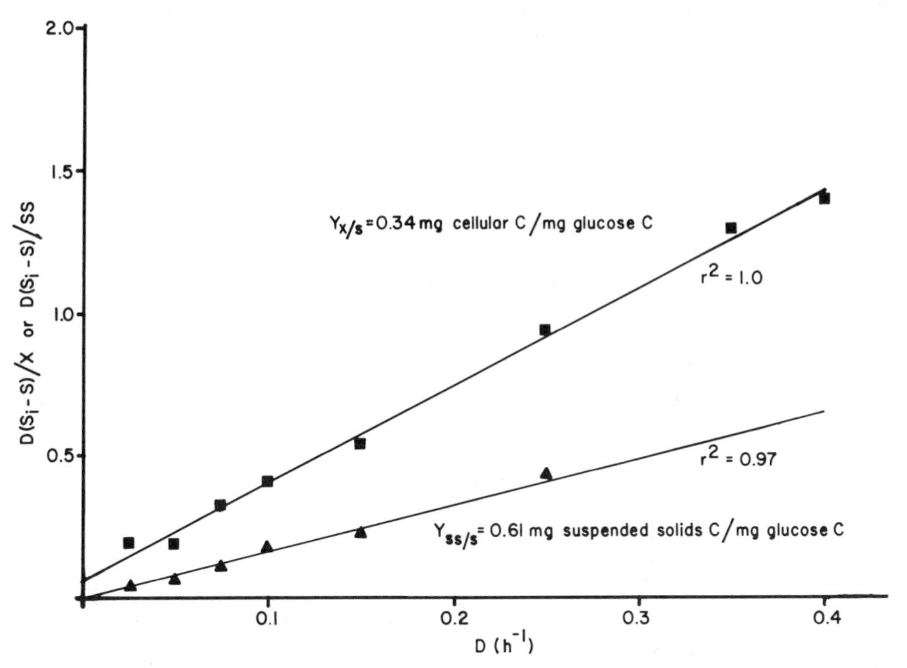

Figure 8.7. The relationship between specific substrate removal rate q_S (Eq. 8B.14) and dilution rate D. The maintenance coefficient k_m is the specific substrate removal rate q_S at $D = 0$ (the intercept). ■, cell carbon; ▲, suspended solids. (Reprinted with permission from Robinson et al., 1984.)

essentially reflects the decay of biomass. The rate expression for endogenous metabolism is first order in the biomass concentration:

$$r_E = k_e X \tag{8.22}$$

where k_e = endogenous decay coefficient (t^{-1})

Example 8C. A chemostat with microbial growth and endogenous metabolism. This example is very similar to Example 8A except that *growth* and *endogenous respiration* are occurring. The chemical state vector for this system is as follows:

$$\mathbf{S} = \begin{bmatrix} S \\ X \\ P_R \end{bmatrix} \tag{8C.1}$$

The transport vector is

$$\phi = \begin{bmatrix} D(S_i - S) \\ D(X_i - X) \\ D(P_{Ri} - P_R) \end{bmatrix} \tag{8C.2}$$

The stoichiometry of the system is the same as expressed in Eq. 8A.3 for growth, but must also include a reaction for endogenous respiration. Endogenous respiration may occur at the expense of various intracellular components, including ribosomes or storage products such as poly-β-hydroxybutyrate. If for simplicity, the overall degradation of biomass is considered, the stoichiometry can be simply expressed as follows:

Growth

$$a_1\, C_6H_{12}O_6 + a_2\, O_2 + a_3\, NH_3 \rightarrow a_4\, CH_{1.4}O_{0.4}N_{0.2} + a_5\, CO_2 + a_6\, H_2O \tag{8C.3}$$

and

Endogenous respiration

$$a_7\, CH_{1.4}O_{0.4}N_{0.2} + a_8\, O_2 \rightarrow a_9\, CO_2 + a_{10}\, H_2O \tag{8C.4}$$

or

$$a_1S \rightarrow a_4X + a_5P_R \quad [8C.5]$$

$$a_7S \rightarrow \qquad a_9P_R \quad [8C.6]$$

Then, the stoichiometric matrix is

$$\mathbf{a} = \begin{bmatrix} -a_1/a_4 & 1 & a_5/a_4 \\ 0 & -a_9/a_7 & 1 \end{bmatrix} \quad [8C.7]$$

The reaction rate vector is

$$\mathbf{r} = [\mu X \quad k_e X] \quad [8C.8]$$

Then

$$\mathbf{r} \cdot \mathbf{a} = \begin{bmatrix} \dfrac{-\mu X}{Y_{X/S}} \\ \mu X - \dfrac{a_9}{a_7} k_e X \\ \dfrac{a_5}{a_4} \mu X + k_e X \end{bmatrix} \quad [8C.9]$$

Then

$$\frac{dS}{dt} = D(S_i - S) - \frac{\mu X}{Y_{X/S}} \quad [8C.10]$$

$$\frac{dX}{dt} = D(X_i - X) + \mu X - k_e X \quad [8C.11]$$

which are the material balances for substrate and biomass, respectively. The material balance for respiration products is omitted, since the measurements necessary for this balance are not generally made. There is no difference in the substrate balance equation from Eq. 8A.11. Only the biomass balance has changed.

For a sterile feed, $X_i = 0$, and at steady state, the derivatives vanish. Then Eq. 8C.10 and 8C.11 become

$$\frac{D(S_i - S)}{X} = \frac{\mu}{Y_{X/S}} \quad [8C.12]$$

$$D = \mu - k_e \quad [8C.13]$$

All quantities on the left hand side of Eqs. 8C.12 and 8C.13 are measurable. In Eq. 8C.13, D is sometimes called the *net growth rate,* μ_{net}, since it reflects the difference between growth and endogenous respiration.

The two equations 8C.12 and 8C.13 contain three unknowns, μ, k_e, and $Y_{X/S}$. If Eq. 8C.13 is solved for μ and substituted in Eq. 8C.12, the following equation with two unknowns results:

$$q_S = \frac{D + k_e}{Y_{X/S}} \quad [8C.14]$$

If q_s is plotted versus D for a set of steady state experiments at various dilution rates, a straight line results with

$$\text{slope} = \frac{1}{Y_{X/S}} \quad [8C.15]$$

$$\text{intercept} = \frac{k_e}{Y_{X/S}} \quad [8C.16]$$

Thus, k_e and $Y_{X/S}$ can be determined. Growth yield and observed yield are not the same in this model either. By substituting Eq. 8C.13 into Eq. 8C.12 and considering the definition of Y_o (Eq. 8B.18), we can eliminate D, with the following result:

$$\frac{1}{Y_o} = \frac{1}{Y_{X/S}} + \frac{k_e}{DY_{X/S}} \quad [8C.17]$$

This result should be compared with that in Eq. 8B.18. By inspection,

$$k_m = \frac{k_e}{DY_{X/S}} \quad [8C.18]$$

and the models for maintenance and endogenous respiration are seen to be related. k_e for *Ps. aeruginosa* can be determined from $k_m = 0.1$ g S per g X_C and $Y_{X/S} = 0.69$ g X_C per g S (Example 8A).

The ribosomes, which are produced rapidly and to a great extent during rapid growth, are frequently degraded as exogenous substrate energy is depleted and starvation ensues. Using RNA or ribosome degradation as an illustration of endogenous metabolism, a more structured model for endogenous metabolism would include the intracellular accumulation of product (ribosomes) during rapid growth and its metabolism during starvation.

A clear distinction between endogenous metabolism and maintenance may not be possible on an operational level. However, endogenous metabolism only occurs in the absence of significant exogenous substrate. Maintenance energy, on the other hand, is a requirement of the living cell which may be obtained from an exogenous or endogenous substrate. The effects of k_m or k_e are generally observed at low specific growth rates.

8.4 PRODUCT FORMATION

Generally, biomass, CO_2, and water are all microbial products. However, for our purposes, the term "product" will be reserved for other materials, not necessarily related to respiration or cell synthesis, that accumulate in the cell or in the culture medium. Gaden (1959) proposed a classification scheme for a broad range of product formation in fermentations, and Bailey and Ollis (1986) have elaborated on the scheme. This section only addresses two classes of product formation.

8.4.1 Energy Production-Associated Products

The kinetics of product formation processes associated with microbial energy production can be described by the Leudeking–Piret equation (1959):

$$r_P = k_P \mu X + k_P' X \qquad [8.23]$$

where k_P = growth-associated product formation coefficient ($M_P M_X^{-1}$)
k_P' = non growth-associated product formation coefficient ($M_P M_X^{-1} t^{-1}$)
r_P = product formation rate ($M_P L^{-3} t^{-1}$)

The main product appears as a result of primary energy metabolism and may even result from direct oxidation of a carbohydrate substrate, e.g.,

glucose → extracellular polysaccharides

For this case, metabolic routes are primarily sequential and each rate process (e.g., cell synthesis, product formation, respiration) exhibits a maximum close to that of others. This can explain the proportionality between product formation rate, r_P, and μ. Example 8D presents an analysis of the rate of EPS formation by *Pseudomonas aeruginosa* in a chemostat that obeys this rate law.

The cell may also have a constitutive enzyme that controls the product formation rate. Then k' may represent units of product-forming activity per mass of cells. The main product may also appear as an indirect result of primary energy metabolism if reaction patterns are complex. The maxima for the various rate processes may appear at quite different times or conditions.

Example 8D. A chemostat with growth and product formation In this example, growth and product formation are occurring in a chemostat. Thus, it is similar to Example 8A except that product formation is added. In this case, the product is extracellular polymeric substances immobilized on the outside of the individual cells (P_E). The chemical state vector is as follows:

$$\mathbf{S} = \begin{bmatrix} S \\ X \\ P_E \end{bmatrix} \qquad [8D.1]$$

The transport vector is

$$\boldsymbol{\phi} = \begin{bmatrix} D(S_i - S) \\ D(X_i - X) \\ D(P_{Ei} - P_E) \end{bmatrix} \qquad [8D.2]$$

The stoichiometry of the system is the same as expressed in Eq. 8A.3 for growth, but must also include a reaction for product formation. Therefore, the stoichiometry can be simply expressed as follows:

Growth

$$a_1\, C_6H_{12}O_6 + a_2\, O_2 + a_3\, NH_3 \rightarrow a_4\, CH_{1.4}O_{0.4}N_{0.2} + a_5\, CO_2 + a_6\, H_2O \qquad [8D.3]$$

and product formation

$$a_7\, C_6H_{12}O_6 \rightarrow a_8\, CH_2O + a_9\, H_2O \qquad [8D.4]$$

or

$$a_1\, S \rightarrow a_4\, X \qquad [8D.5]$$

$$a_7\, S \rightarrow \qquad a_8\, P_E \qquad [8D.6]$$

where CH_2O is an approximate composition for P_E. Then, the stoichiometric matrix is

$$\mathbf{a} = \begin{bmatrix} -a_1/a_4 & 1 & 0 \\ -a_7/a_8 & 0 & 1 \end{bmatrix} \qquad [8D.7]$$

The reaction rate vector is

$$\mathbf{r} = [\mu X \quad k\mu X + k'X] \tag{8D.8}$$

Then

$$\mathbf{r} \cdot \mathbf{a} = \begin{bmatrix} -\mu X/Y_{X/S} - r_P X/Y_{P/S} \\ \mu X \\ r_P X \end{bmatrix} \tag{8D.9}$$

where $Y_{P/S}$ = yield of product from substrate ($M_P M_S^{-1}$)
r_P = specific product formation rate ($M_P M_X^{-1} t^{-1}$)
$= k\mu + k'$

Then

$$\frac{dS}{dt} = D(S_i - S) - \frac{\mu X}{Y_{X/S}} - \frac{r_P X}{Y_{P/S}} \tag{8D.10}$$

$$\frac{dX}{dt} = D(X_i - X) - \mu X \tag{8D.11}$$

$$\frac{dP}{dt} = D(P_{Ei} - P_E) + r_P X \tag{8D.12}$$

which are the material balances for substrate, biomass, and product, respectively. The material balance for respiration products is omitted, since the measurements necessary for this balance are not generally made. There is no difference in the biomass balance equation from Eq. 8A.12. However, the substrate balance is different from Eq. 8A.11, and a product balance has been added.

For a sterile feed, $X_i = 0$, $P_{Ei} = 0$, and steady state, the derivatives vanish and Eqs. 8D.10, 8D.11, and 8D.12 become

$$\frac{D(S_i - S)}{X} = \frac{\mu}{Y_{X/S}} + \frac{r_P}{Y_{P/S}} \tag{8D.13}$$

$$D = \mu \tag{8D.14}$$

$$\frac{DP_E}{X} = r_P \tag{8D.15}$$

All quantities on the left hand side of Eqs. 8D.13, 8D.14, and 8D.15 are measurable. μ is directly determined from the dilution rate. However, there are three more parameters ($Y_{X/S}$, $Y_{P/S}$, r_P) to calculate and only two more equations. r_P is determined directly from experimental measurements (Eq. 8D.15). Then, graphical methods are used if several data points at various dilution rates are available. Equation 8D.13 can be expressed as follows:

$$q_S = D\left(\frac{1}{Y_{X/S}} + \frac{r_P}{DY_{P/S}}\right) \quad [8D.16]$$

A plot of q_s versus D (Figure 8.7) is linear, with

$$\text{slope} = \frac{1}{Y_{x/s}} + \frac{k_P}{Y_{p/s}} = 3.4 \text{ (g S) (g}^{-1} \text{ X}_C) \quad [8D.17]$$

and

$$\text{intercept} = \frac{k_P'}{Y_{p/s}} = 0.063 \text{ (g S) (g}^{-1} \text{ X}_C) \text{ h}^{-1} \quad [8D.18]$$

Note that growth yield and observed yield are not the same in this analysis. The observed or effective yield is (Figure 8.4)

$$Y_o = \frac{X}{S_i - S} = 0.61 \quad [8D.19]$$

k_P and k_P' can be obtained by plotting DP/X versus D (Figure 8.8), which results in the following:

$$\text{slope} = k_P = 0.27 \text{ (g P) (g X}_C)^{-1} \quad [8D.20]$$

$$\text{intercept} = k_P' = 0.035 \text{ (g P) (g}^{-1} \text{ X}_C)^{-1} \text{ h}^{-1} \quad [8D.21]$$

From Eqs. 8D.18 and 8D.21,

$$Y_{P/S} = 0.56 \text{ (g P) (g S)}^{-1} \quad [8D.22]$$

From Eqs. 8D.17, 8D.20, and 8D.22,

$$Y_{X/S} = 0.34 \text{ (g X}_c) \text{ (g S)}^{-1} \quad [8D.23]$$

The growth yield is $Y_{X/S} = 0.34$ and is not equal to the observed yield based on suspended solids ($Y_o = 0.69$; Figure 8.4) or based on cell carbon alone ($Y_o = 0.29$; Figure 8.7).

The ratio of glucose conversion to cells ($F_{S/X}$) and EPS ($F_{S/P}$) may be estimated using the following expressions:

$$F_{S/X} = \frac{X_c}{Y_{X/S}(S_i - S)} \quad [8D.24]$$

$$F_{P/X} = \frac{P}{Y_{S/P}(S_i - S)} \quad [8D.25]$$

Substrate carbon oxidized for energy, yielding respiratory products as well, is incorporated into $F_{S/X}$ and $F_{S/P}$ (Figure 6.6, Chapter 6). The ratio $F_{S/X}/F_{S/P}$ generally decreases with increasing growth rate, indicating that more glucose was shunted to EPS formation at low growth rates (Figure 8.9).

8.4.2 Product Formation Not Associated with Energy Metabolism

Biosynthesis of complex molecules not directly resulting from energy metabolism is illustrated by the penicillin fermentation. Cell and metabolic activities reach their maximum rates quickly, followed later by the formation of the desired product. This type of product formation will generally require a structured model in order to predict the process rates.

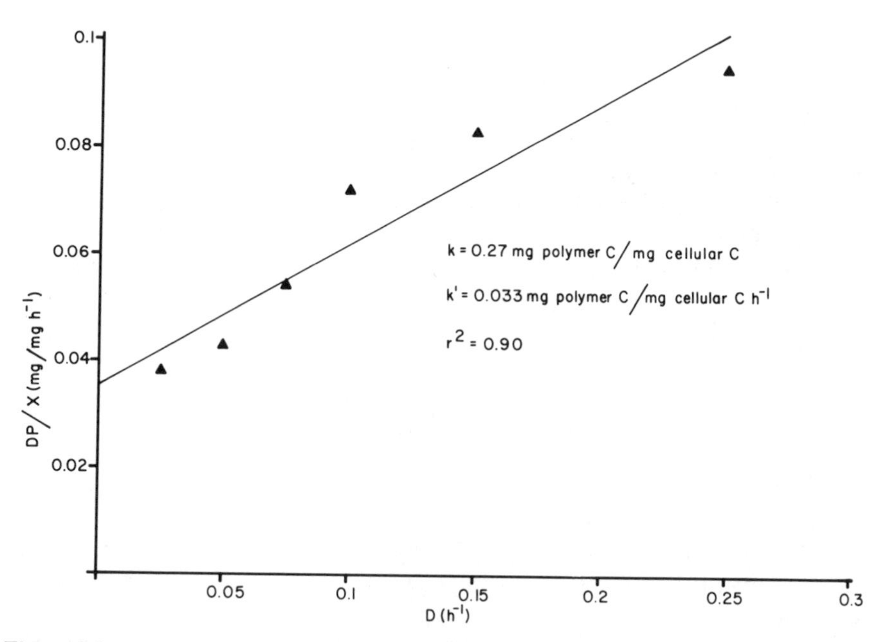

Figure 8.8. The relationship between the specific product formation rate $r_P = DP/X$ and the dilution rate D. The non growth-associated product formation rate coefficient, k, is the slope of the line. The growth-associated product formation rate coefficient, k', is the intercept when $D = 0$. (Reprinted with permission from Robinson et al, 1984.)

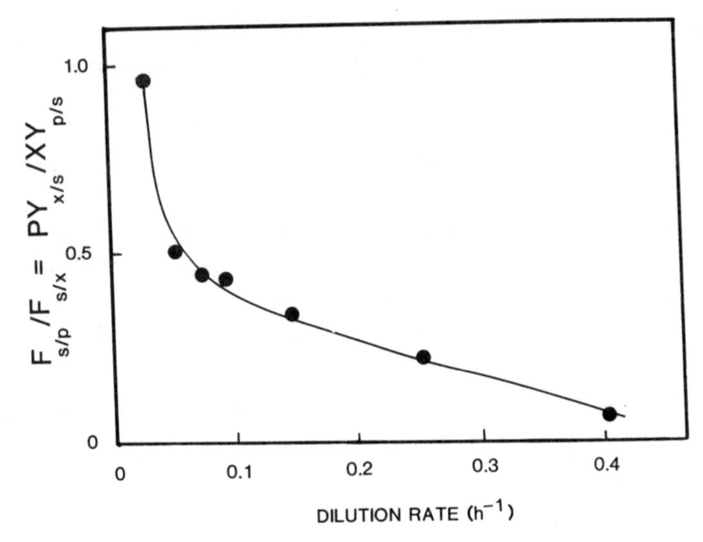

Figure 8.9. EPS carbon per unit cell carbon for glucose-limited *Pseudomonas aeruginosa* as a function of dilution rate in a chemostat. The EPS was immobilized on the suspended cells in the chemostat (Trulear, 1983; Robinson et al., 1984).

8.5 DEATH

Death is difficult to define in bacteria. Inability to form colonies on a selected agar medium does not imply death. Often, cells are merely injured and can recover given the appropriate growth conditions.

Death may result from substrate depletion, toxic metabolite production, or introduction of some toxic material (e.g., a biocide) into the cell environment.

Cell death resulting from substrate depletion is generally considered to be first order with respect to cell concentration:

$$r_D = k_D X_c \tag{8.24}$$

where k_D = specific death coefficient (t^{-1})

Prior to death and subsequent lysis, starved cells may shrink in size but still remain viable. If a food source becomes available, these cells begin to grow and reproduce again.

Cell death due to a biocide is generally considered to follow a variant of Eq. 8.24 termed Chick's Law (Chick, 1908), which explicitly indicates a dependence on biocide concentration and cell concentration:

$$r_D = -\alpha C_B^n X_C^m \tag{8.25}$$

where α = the specific "lethality" coefficient ($L^3M^{-1}t^{-1}$ if $m = 0$)
C_B = biocide concentration (ML^{-3})
m, n = coefficient (—)

The extent of death, or percentage reduction in viable cells, is an important performance criterion in many systems receiving biocide treatment. The extent of death can be calculated from rate parameters if the treatment duration is known. Consider a viable cell balance in a batch reactor:

$$\frac{dX_c}{dt} = r_D \qquad [8.26]$$

where r_D = rate of cell death (t^{-1})

Substituting Eq. 8.25 into Eq. 8.26, assuming $n = 1$ and $m = 0$, and rearranging,

$$\int_{X_{c0}}^{X_{ct}} dX_c = -\alpha \int_{t_0}^{t} C_B \, dt \qquad [8.27]$$

where X_{c0} = viable cell number concentration at $t = 0$ (L^{-3})
X_{ct} = viable cell number concentration at $t = t$ (L^{-3})

Integrating Eq. 8.27 assuming $C_B \approx$ constant,

$$X_{ct} - X_{c0} = -\alpha C_B D_U \qquad [8.28]$$

where D_U = treatment duration (t)

Then,

$$\frac{X_{ct}}{X_{c0}} = 1 - \frac{\alpha C_B D_U}{X_{c0}} \qquad [8.29]$$

where

$$\frac{X_{ct}}{X_{c0}} \times 100 = \%\text{ cells ``killed''} \qquad [8.30]$$

Also note that the integral on the right hand side of Eq. 8.27 is the biocide dose, F_B:

$$F_B = \int_{t_0}^{t} C_B \, dt \qquad [8.31]$$

TABLE 8.3 A Matrix Representation for the Fundamental Microbial Transformation Processes[a]

		Stoichiometry							
		Reactants			= Products				
					Biomass		Product		
Fundamental Process	Rate Coefficient	Substrate S	Nutrient S_N	Electron Acceptor S_E	X	X_C	P_E	P_I	Metabolite P_m
Growth	μ	−	−	−	+	+	+	(+)	+
Maintenance:									
Exogenous	m	−		−	−		+		+
Endogenous	k_e		+	−	−	−	+	−	+
Product formation	k_p	−	−	−	(+)		+	+	+
Death	k_d	(+)	(+)		−	−	+		
Observed rate		q_S	q_N	q_E	μ_{net}	μ	q_{Pe}	q_{Pi}	q_M

q_i = specific production or removal rate of component i (t^{-1})
μ = specific *cellular* growth rate (t^{-1})
μ_{net} = net specific *biomass* production rate (t^{-1})
X = biomass concentration ($M_X L^{-3}$)
X_C = cellular mass concentration ($M_{XC} L^{-1}$)
P_E = extracellular microbial product concentration ($M_P L^{-3}$)
P_I = intracellular microbial product concentration ($M_P L^{-3}$)
S = substrate concentration (ML^{-3})
S_N = nutrient concentration (ML^{-3})
S_E = electron acceptor concentration (ML^{-3})

[a] Characklis (1984).

If C_B is constant during the duration of treatment ($D_U = t - t_0$), then

$$F_B = C_B D_U \qquad [8.32]$$

Many versions of the kinetic expression (Eq. 8.25) are possible, depending on the values of m and n. The values $m = 0$ and $n = 1$ are frequently assumed.

8.6 CONCLUDING STATEMENT

The rate expressions and stoichiometric ratios (i.e., the yields) have been combined in material balances through the matrix method described in Chapter 2. Thus, the reaction rate of different components can be related through stoichiometry. The rate and stoichiometry for the various microbial transformations can be expressed in a summary matrix (Table 8.3). The relevant reactions include growth, maintenance or decay, product formation, and death. The rates of these fundamental processes are difficult to measure so the data must be extracted from *observed processes*. Observed rates include specific substrate removal rate (q_S), specific oxygen removal rate (q_E), and specific biomass production rate (μ_{net}). The stoichiometry of each fundamental process can be expressed in terms of its reactants and products.

REFERENCES

Bailey, J.E. and D.F. Ollis, *Biochemical Engineering Fundamentals,* 2nd ed., McGraw-Hill, New York, 1986.

Bakke, R., M.G. Trulear, J.A. Robinson, and W.G. Characklis, *Biotech. Bioeng.,* **26**, 1418–1424 (1984).

Characklis, W.G., "Biofilm development: A process analysis," in K.C. Marshall (ed.), *Microbial Adhesion and Aggregation,* Springer-Verlag, Berlin, 1984, pp. 137–158.

Chick, H., *J. Hyg.,* **8**, 698 (1908).

Churchill, S.W., *The Interpretation and Use of Rate Data: The Rate Concept,* McGraw-Hill, New York, 1974.

Dabes, J.N., R.K. Finn, and C.R. Wilkie, *Biotech. Bioeng.,* **15**, 1159–1177 (1973).

Denbigh, K.G. and J.C.R. Turner, *Chemical Reactor Theory,* 2nd ed., Cambridge University Press, Cambridge, England, 1971.

Gaden, E.L., *J. Biochem. Microbiol. Tech. Eng.,* **1**, 63 (1959).

Gujer, W., *Prog. Wat. Tech.,* **12**, 79–95 (1980).

Howell, J.A., "Mathematical models in microbiology: Mathematical tool kit," in M. Bazin (ed.), *Mathematics in Microbiology,* Academic Press, London, 1983, pp. 37–76.

Hrudey, S.E., *J. Wat. Poll. Contr. Fed.,* **54**, 1207–1214 (1982).

Kessick, M.A., *Biotech. Bioeng.,* **16**, 1545–1547 (1974).

Kirchman, D., *Limnol. Oceanog.,* **28**, 858–872 (1983).

Langmuir, I., *J. Am. Chem. Soc.*, **40**, 1361 (1918).

Leudeking, R. and E.L. Piret, *J. Biochem. Microbiol. Technol. Eng.*, **1**, 393 (1959).

Levenspiel, O., *Chemical Reaction Engineering*, 2nd ed., Wiley, New York, 1972.

Mallee, F.M. and H.W. Blanch, *Biotech. Bioeng.*, **19**, 1793–1816 (1977).

Martin, P. and R.A. MacLeod, *Appl. Env. Microbiol.*, **47**, 1017–1022 (1984).

Megee, R.D., unpublished doctoral dissertation, University of Minnesota, Minneapolis, MN, 1970.

Michaelis, L. and M.L. Menten, *Biochemie*, **49**, 333–369 (1913).

Miura, Y., M. Okazaki, S.I. Hamada, S.I. Murakawa, and R. Yugen, *Biotech. Bioeng.*, **19**, 701–714 (1977a).

Miura, Y., M. Okazaki, S.I. Hamada, S.I. Murakawa, and R. Yugen, *Biotech. Bioeng.*, **19**, 715–726 (1977b).

Monod, J., *Recherches sur la Croissance des Cultures Bacteriennes*, Hermann, Paris, 1942.

Monod, J., *Ann. Rev. Microbiol.*, **3**, 371–394 (1949).

Robinson, J.A. and W.G. Characklis, *Microbial Ecol.*, **10**, 165–178 (1984).

Robinson, J.A., M.G. Trulear, and W.G. Characklis, *Biotech. Bioeng.*, **26**, 1409–1417 (1984).

Rollings, J.E., M.R. Okos, and G.T. Tsao, *ACS Symp. Ser.*, **207**, 443–461 (1983).

Schink, B. and J.G. Zeikus, *J. Bact.*, **128**, 393–404 (1982).

Suga, K., G. van Dedem, and M. Moo-Young, *Biotech. Bioeng.*, **17**, 185–201 (1975).

Trulear, M.G., "Cellular reproduction and extracellular polymer formation in the development of biofilms," unpublished doctoral thesis, Montana State University, Bozeman, MT, 1983.

Tsao, G.T. and T.P. Hansen, *Biotech. Bioeng.*, **17**, 1591–1598 (1975).

Watson, S.W. and N. Jones, *Wat. Res.*, **11**, 95–100 (1977).

TRANSPORT AND INTERFACIAL TRANSFER PHENOMENA

WILLIAM G. CHARACKLIS
Montana State University, Bozeman, Montana

MUKESH H. TURAKHIA
Houston, Texas

NICHOLAS ZELVER
Boston, Massachusetts

SUMMARY

Momentum, energy, and mass transport are compared, and close analogies between the three quantities are presented. The analogies hold both in laminar and in turbulent flow. The equations of motion, heat, and mass transport along with the flux laws permit the use of differential balances in a solid, in a biofilm, or in water in laminar flow. The differential balances can only be used for examining intraphase transport. Obviously, methods for describing interphase transport between the water and the biofilm are necessary. Because of the complexity of interphase transport in turbulent flow, simpler expressions describing transfer between a turbulent flow and a surface (e.g., a biofilm) are presented, based on Newton's law of cooling. However, differential balances in a turbulent liquid phase can be used by incorporating time-smoothed or empirical turbulent velocity profiles.

Biofilm accumulation and metabolic activity are generally controlled by one of the transport processes. However, biofilms, as they accumulate, have a significant influence on the transport processes. Biofilms increase hydrodynamic frictional resistance and, in so doing, increase advective heat and mass transfer. For predictive purpose, it is essential that a relationship (deterministic or empirical) between biofilm properties and hydrodynamic roughness be established. For example, roughness may be a simple function of biofilm thickness. If so, then fluid frictional resistance can be predicted as a function of biofilm thickness. If fluid frictional resistance can be predicted from biofilm thickness, then so can advective heat and mass transfer at the biofilm–bulk water interface. Conductive heat transfer within the biofilm is a function of biofilm thickness and biofilm thermal conductivity. Thus, predicting the biofilm accumulation can lead to prediction of both advective and conductive heat transfer. The metabolic activity within the biofilm is frequently controlled by the rate of diffusion of essential nutrients or electron acceptor through the biofilm.

The interfacial transfer processes such as adsorption and attachment require more experimental study. The behavior of the biofilm as a filter or trap for particles may be critical to biofilm accumulation in many important systems.

The important, if not critical, role of transport and interfacial transfer processes in biofilm systems is emphasized and the importance of biofilm thickness measurements underscored. More attention must be turned to measuring interfacial concentrations and concentrations within biofilms (e.g., using microelectrodes) if further progress is to be made in predicting and manipulating biofilm systems.

9.1 INTRODUCTION

Three types of rate processes are important for material balance equations (Chap. 3): accumulation, transport, and transformation. Microbial transformation processes and their contribution to biofilm processes were the focus of Chapter 8. However, transport processes such as fluid flow, heat transport, and mass transport (diffusion) influence the progression of biofilm development as well as biofilm properties. Interfacial transfer phenomena such as adsorption and attachment are, more or less, a subset of transport phenomena and are necessary in the discussion of biofilm processes. Attention to these transport and transfer processes is justified in that the important role of transport and transfer processes is the major feature distinguishing biofilm systems from more classical suspended microbial systems.

Not only do transport processes influence biofilm processes, but the accumulation of biofilm, in turn, influences bulk liquid compartment processes such as hydrodynamics, heat transfer, and mass transfer. Thus, we shall examine the contribution of biofilms to increasing fluid frictional resistance and heat transfer resistance, phenomena frequently termed *fouling* (Chapter 14). The three transport processes are closely related, and a change in one will almost always affect the others. A typical scenario is presented in Figure 9.1. The flow of water through a clean tube is associated with a pressure drop as schematically presented in Figure 9.1a. As biofilm accumulates on the inside wall of the tube, the pressure drop increases and other changes in the transport processes are observed (Figure 9.1b):

1. The biofilm accumulation changes the surface texture, resulting in a "rougher" surface. Biofilm development may result in unusually high fluid frictional resistance losses in water conduits (Chapter 14) because the rheology and morphology (e.g., roughness) of the biofilm influence the *fluid frictional resistance* in a conduit. The biofilm accumulation may constrict the cross-sectional area of the tube, so that more energy is required to transport the same quantity of fluid through the tube at the same rate. Alternatively, the biofilm roughness causes eddy currents, which dissipate some of the energy available for flow through the tube.
2. The added biofilm coating thermally insulates the surface, so that heat passes through the tube wall at a reduced rate. Biofilms accumulating on surfaces of heat transfer tubes significantly reduce the heat transfer rate (Characklis et al., 1981) because the thermal conductivity of the biofilm is

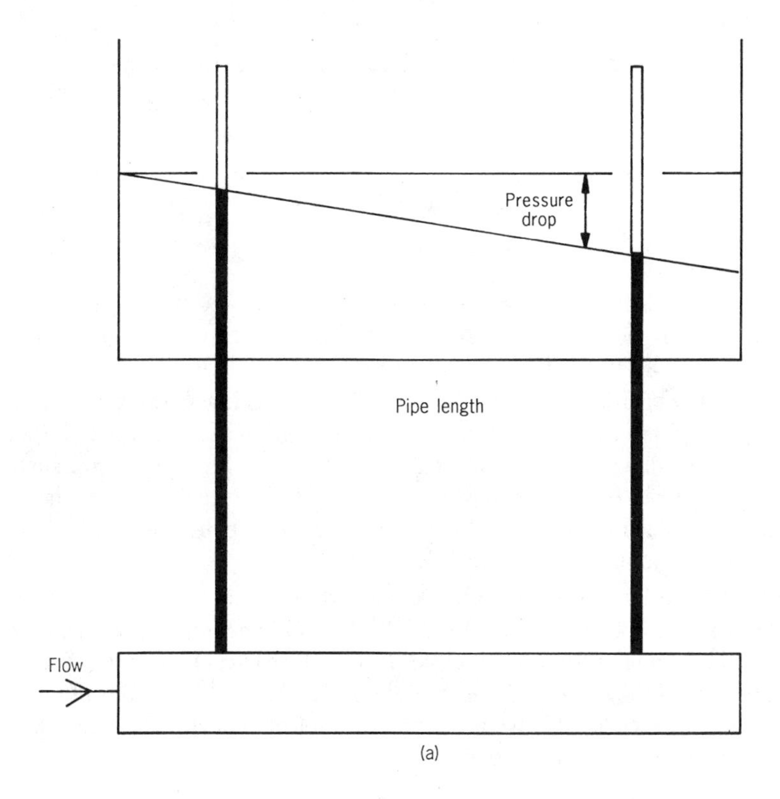

Figure 9.1 The energy loss in a rough pipe (b) is significantly greater than in a smooth tube (a) as reflected by the pressure differences along the two tubes.

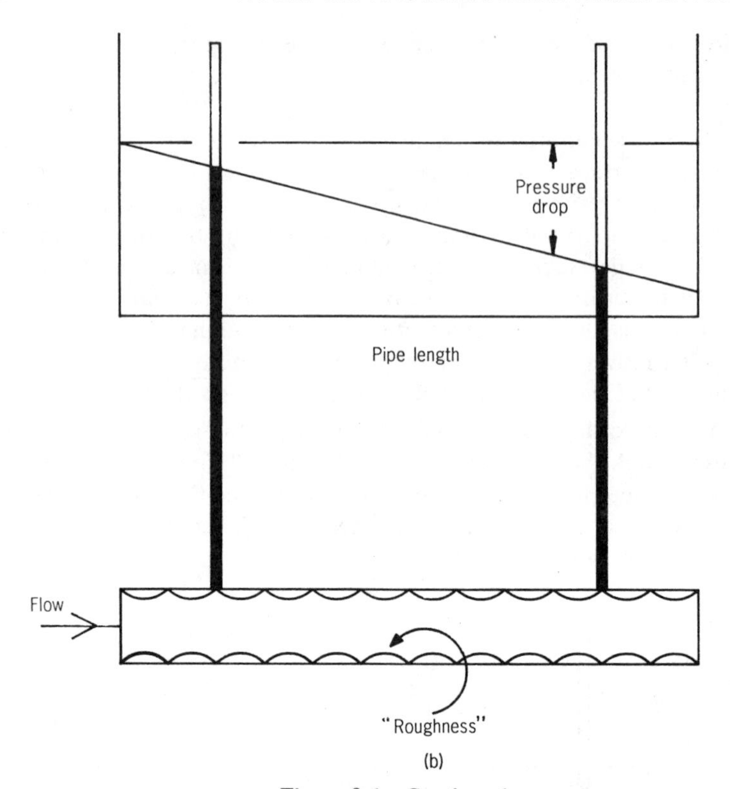

Figure 9.1 Continued

significantly less than that of the metal heat transfer surface. Thus, the biofilm reduces *conductive heat transfer*.

3. The eddy currents increase the rate of heat and mass transport from the bulk liquid to the biofilm by *advective heat and mass transport*. The eddy currents enhance the transport of mass to and from the biofilm surface. Thus, nutrients are transported to the biofilm surface faster. At the same time, the accumulated biofilm is subjected to a greater shear force, resulting in more biofilm detachment.
4. As the biofilm increases in thickness, components (soluble and particulate) entering the biofilm from the bulk liquid encounter a *diffusional resistance,* which ultimately may control the rate of metabolism in the lower portions of the biofilm. Thus, metabolic activity within the biofilm may be rate-controlled by diffusion of substrate, electron acceptor, nutrients, or inhibitors to individual cells.

Transport processes in the bulk liquid and within the biofilm may significantly influence biofilm accumulation, which in turn influences transport processes in the bulk liquid. Therefore, transport properties of the biofilm can significantly influence the rate and extent of the processes occurring within it.

On the other hand, the physical, chemical, and biological properties of the resultant biofilm are dependent on the particular environment in which it has accumulated. Environmental factors (e.g., fluid velocity, temperature, nutrient concentration) influence the selection of the prevalent biofilm organisms. The predominant organisms, in turn, modify the microenvironment of the substratum in a manner that reflects their metabolic activity.

9.2 TRANSPORT PHENOMENA

The physical laws, as well as the "biological laws," governing biofilm processes must be understood to obtain a thorough analysis of a specific biofilm ecosystem, the performance of engineering equipment, or the design of process equipment in which biofilm processes are occurring. Thus, we wish to predict the rate of momentum, heat, and mass transport in the equipment. The design may be accomplished by expensive and time-consuming experimental measurements to determine coefficients, which are correlated through the use of dimensionless empirical equations. These empirical equations, however, only fit the data over a certain range of operating conditions. A frequently less expensive and more reliable approach is to predict the coefficients from equations based on the laws of nature.

We focus on the systematic and integrated study of momentum, energy, and mass transport. Understanding biofilms requires familiarity with specialized areas of these classical topics in engineering science. *Fluid dynamics* is the subtopic of momentum transport that concerns us because of its influence on heat and mass transport. *Conductive and advective heat transport* is the subset of energy transport most relevant to fouling biofilms. Both *diffusive and advective mass transport* frequently influence the overall rate of a biofilm process and the observed activity of organisms within the biofilm The study of transport phenomena also makes use of the similarities between the equations used to describe the processes of momentum, heat, and mass transport. These analogies are often related to similarities in the physical mechanisms by which the transport takes place. This makes concepts and mathematical solutions easier to formulate and comprehend.

9.2.1 The Balance Equations

The study of transport phenomena is primarily concerned with the prediction of velocity, temperature, and concentration variations within a single phase. These *profiles,* or *distributions,* are obtained with the help of the following sets of equations: (1) the balance equations (or conservation laws), (2) the flux laws, and (3) the rate equations describing production or depletion (transformation terms). A general review of overall balance equations was presented in Chapter 2. The general expression for the balance of a conservative quantity is expressed mathematically (Eq. 2.1, Chapter 2) as follows:

$$\frac{d}{dt}\int_V S\,dV = \int_V r_{VS}dV + \int_A \mathbf{n}\cdot\mathbf{J}_S\,dA \qquad [9.1]$$

accumulation rate of S in control volume V = volumetric transformation rate of S in control volume V + transport rate into control volume V through the boundary surface A

where r_{VS} = volumetric reaction rate of S in volume V
$\mathbf{J}_S$ = surface flux vector for S across the control volume boundary
$\mathbf{n}$ = inward normal unit vector on the control volume surface

Equation 9.1 applies to the whole system when macroscopic balances are being conducted, but can also be used for microscopic or differential balances on parts of the system. Some special cases of Eq. 9.1 will be presented that are useful for aquatic biofilm systems. The equations are useful for laminar and turbulent flow. For laminar flow, intensive variables such as velocity, temperature, and concentration and the deterministic variables such as position and transport coefficients are solely dependent on material properties. For a turbulent flow, time-averaged intensive variables are assumed, and the transport coefficients are dependent on local turbulence and not on the materials (Schlicting, 1968; Davies, 1972; Bird et al., 1960).

For a differential volume, the balance *for momentum* in water, assuming constant density and viscosity, results in the equation of motion (Bird et al., 1960):

$$\rho\frac{\partial v}{\partial t} + \mathbf{v}\cdot\nabla v = -\nabla p + \eta\nabla^2\mathbf{v} + \rho\mathbf{g} \qquad [9.2]$$

where ρ = fluid density (ML^{-3})
η = fluid viscosity ($ML^{-1}t^{-1}$)
$\mathbf{v}$ = velocity vector (Lt^{-1})
p = pressure ($ML^{-1}t^{-2}$)
$\mathbf{g}$ = acceleration of gravity (Lt^{-2})

The balance *for heat* in a solid (e.g., a biofilm) of constant thermal conductivity is

$$\rho C_p\frac{\partial T}{\partial t} = k_T\nabla^2 T \qquad [9.3]$$

where T = temperature (T)
k_T = thermal conductivity ($MLt^{-3}T^{-1}$)
C_p = heat capacity of the fluid at constant pressure ($L^2t^{-2}T^{-1}$)

If we apply the conservation principles to a binary mixture of reactive species A in species B, then the balance for *mass of component A* is

$$\frac{\partial S_A}{\partial t} + (\mathbf{v} \cdot \nabla S_A) = D_{AB} \nabla^2 S_A + r_{VA} \qquad [9.4]$$

where S_A = mass concentration of component A (ML^{-3})
D_{AB} = diffusion coefficient of A in B (L^2t^{-1})
r_{VA} = volumetric reaction rate of A ($ML^{-3}t^{-1}$)

This equation is useful for diffusion in dilute aqueous solutions (suspensions) at constant temperature and pressure. For solids, such as biofilms, or stagnant water, $\mathbf{v} = 0$:

$$\frac{\partial S_A}{\partial t} = D_{AB} \nabla^2 S_A \qquad [9.5]$$

Equation 9.5 is Fick's second law of diffusion or simply the *diffusion equation*.

Overall or macroscopic balances, as illustrated in Chapter 2, are useful especially when the velocity, temperature, and concentration are uniform in the system. However, the nature of biofilm processes usually results in *gradients*, which significantly influence the microenvironment of the biofilm organisms. In such cases, *differential* models or *lumped parameter* models must be employed rather than macroscopic balances. Some of the more important concepts relating biofilm processes to fluid dynamics, heat transport, and mass transport are addressed in the following sections. More details on these topics and the methods of transport phenomena can be found elsewhere (Bird et al., 1960; Fahien, 1983). Bailey and Ollis (1986) present many transport phenomena applications of more general interest in microbial and biochemical systems.

9.2.2 Dimensional Analysis

Nature knows no units, so it is not surprising that a dimensional analysis can produce considerable insight into the workings of the system. Use of dimensionless variables has proven convenient in engineering analysis and has helped reduce the number of variables and parameters that need be varied in an experimental program. For example, the Reynolds number, *Re* (see Section 9.2.3), is a result of dimensional analysis that incorporates four parameters (fluid velocity, tube diameter, fluid viscosity, and fluid density) into one. Dimensional analysis is also a useful and convenient method for solving differential equations and determining the variables to be measured in an experimental program (Fahien, 1983; Bird et al., 1960; Churchill, 1974).

9.2.3 Laminar Versus Turbulent Flow

Hydrodynamics can be conveniently subdivided into *laminar* and *turbulent* flow. Although turbulent flow is the rule in engineering equipment, laminar flow is observed in many natural systems and is also an effective tool in experimental apparatus.

The distinction between laminar and turbulent flow can be made by describing an experiment similar to the one conducted by O. Reynolds over 100 years ago. A fluid is being pumped through a long tube. A stream of dye is injected through a small syringe at any point in the fluid (Figure 9.2). If the flow is laminar, the dye will move with the fluid in a coherent, straight path for a great distance. If injected at a given radial and angular position, the dye remains at that position as it moves downstream with the fluid. The fluid and thus the dye move in *laminae,* or sheets, in laminar flow. If the flow is turbulent, the dye will be mixed in the eddy currents as it travels downstream and will rapidly be dispersed throughout the cross section of the tube.

The criterion for turbulent flow is the dimensionless Reynolds number, *Re*:

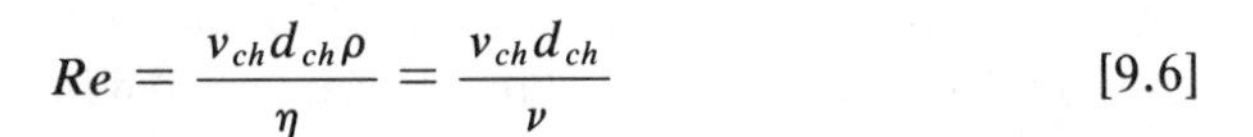

$$Re = \frac{v_{ch} d_{ch} \rho}{\eta} = \frac{v_{ch} d_{ch}}{\nu} \qquad [9.6]$$

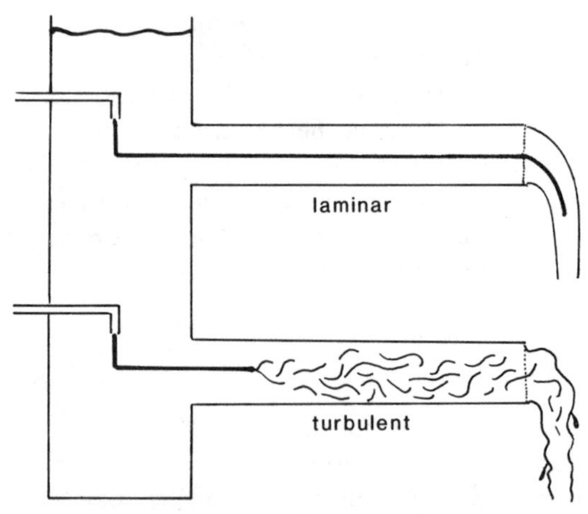

Figure 9.2 A stream of dye is injected into a laminar flow (top) and a turbulent flow (bottom) to demonstrate the differences between the two flow regimes. The dye in the laminar flow remains within a contiguous stream because there is little intermixing between the layers (or laminae). The turbulent flow is characterized by random fluctuations that mix the dye in the fluid thoroughly.

where v_{ch} = characteristic fluid velocity
d_{ch} = characteristic length (e.g., tube diameter)

The Reynolds number is the ratio of *inertial forces,* as described by Newton's second law of motion, to *viscous forces*. If the Reynolds number is high, inertial forces dominate and turbulent flow exists. If it is low, viscous forces prevail and we have laminar flow. Typical values for η and ν in water are listed in Table 9.1.

For the tube flow example, the characteristic velocity is the average linear velocity in the axial direction, and the characteristic length is the tube diameter. If $Re < 2100$, then the flow is laminar and for $Re > 2100$ turbulent flow is occurring. The exact transition Re, Re_{tr}, between laminar and turbulent flow may vary somewhat depending on entrance conditions and the roughness of the tube surface. Re_{tr} also depends on the system geometry (Schlicting, 1968). So, for example, Re_{tr} for flow past a sphere ranges from 1 to 1000.

9.3 INTRAPHASE TRANSPORT: LAMINAR FLOW AND SOLIDS

9.3.1 Flux Laws

The three intensive variables of most concern in transport phenomena within biofilm systems are fluid velocity (momentum transport), temperature (heat transport), and concentration (mass transport). These variables are sometimes referred to as *potentials*. Velocity is a vector quantity (has magnitude and direction), while temperature and concentration are scalars (have only magnitudes). The flux laws relate the flux of a potential to the potential gradient. Heat flux and mass flux are vector quantities, while momentum flux is a tensor quantity

TABLE 9.1 Viscosity and Density of Water as a Function of Temperature.

Temperature (°C)	Viscosity, η (10^{-3} kg m^{-1} s^{-1})	Density (kg m^{-3})	Kinematic Viscosity, ν (10^{-6} m^2 s^{-1})
0	1.792	999.87	1.792
5	1.519	999.99	1.519
10	1.308	999.73	1.308
15	1.140	999.13	1.140
20	1.005	998.23	1.007
25	0.894	997.07	0.897
30	0.801	995.67	0.804
40	0.656	992.25	0.661
60	0.469	983.24	0.477
80	0.357	971.83	0.367

(has a magnitude and *two* directions). The flux laws or equations take the following form:

$$\text{flux} = \frac{\text{flow rate}}{\text{area}}$$

$$= \text{transport property} \times \text{potential gradient} \qquad [9.7]$$

where "flow rate" refers to the flow of momentum, heat, or mass. The flux laws in one dimension (in this case, the x-direction) are as follows:

Newton's Law of Viscosity.

$$\tau_{xy} = -\eta \frac{dv_y}{dx} \qquad [9.8]$$

Fourier's Law of Heat Conduction.

$$q_{Hx} = -k_T \frac{dT}{dx} \qquad [9.9]$$

Fick's Law of Diffusion.

$$J_{Sx} = -D_S \frac{dS}{dx} \qquad [9.10]$$

where τ_{xy} = flux of y-momentum in the x-direction (shear stress) ($ML^{-1}t^{-2}$)
v_y = fluid velocity in the y-direction (Lt^{-1})
q_{Hx} = heat flux in x direction (Mt^{-3})
J_{Sx} = mass flux in the x-direction ($ML^{-2}t^{-1}$)

η, k_T, and D_S are the transport coefficients, and the derivatives represent the potential gradient. The potential gradient is the *driving force* for the flux. In heat conduction, for example, heat flows in the direction of decreasing temperature, and the rate of heat flow, q_{Hx}, is proportional to the temperature gradient. τ_{xy} is the momentum flux, frequently termed shear stress, and has units of force per unit area. The force is transmitted in the x-direction due to a fluid moving at a velocity v_y in the y-direction (perpendicular to the x-direction). Hence, shear stress is a tensor with two directions specified (in this case, x and y). Shear stress at a surface may significantly influence microbial cell adsorption and desorption. q_{Hx} is the conductive heat flux in the x-direction. J_{Sx} is the diffusive mass (not molar) flux of component S (e.g., glucose) through component W (e.g., water) in the x-direction.

The analogies between the various transport processes are more clearly presented in Table 9.2 where the transport properties have been combined with several thermodynamic properties into "diffusivities" that have consistent units. Finally, the potentials are transformed into "concentrations": momentum per unit volume, energy per unit volume, and mass per unit volume. From Table 9.2 the flux laws can be written in the following form:

$$\text{flux} = -\text{"diffusivity"} \times \text{"concentration" gradient} \qquad [9.11]$$

The flux "laws" are not laws at all, but are only definitions for the transport property in the equation. The transport property frequently cannot be measured with great accuracy, and in some cases even its definition is open to debate. The flux laws, as stated above, are only valid in solids, fluids in laminar flow, or quiescent fluids. Thus, the transport properties can only be measured in apparatus in laminar flow, in solids, or in quiescent fluids.

9.3.2 Transport in Laminar Flow and Solids: Steady State Differential Balances

9.3.2.1 *Laminar Flow in a Circular Tube* While engineers are almost exclusively concerned with turbulent flows, laminar flow is of microbiological interest, as illustrated by the movement of motile bacteria through slow-moving or

TABLE 9.2 The One-Dimensional Flux Equations

	Momentum	Energy (heat)	Mass
Flux	τ_{xy}	q_{Hx}	J_{Sx}
Transport property	η [$ML^{-1}t^{-1}$] (viscosity)	k_T [$MLt^{-3}T^{-1}$] (thermal conductivity)	D_S [L^2t^{-1}] (diffusivity)
Thermodynamic property	ρ [ML^{-3}] (specific gravity)	C_p [$L^2t^{-2}T^{-1}$] (heat capacity at const. pressure)	
"Diffusivity"	$\nu = \dfrac{\eta}{\rho}$ [L^2t^{-1}] (kinematic viscosity or momentum diffusivity)	$a = \dfrac{k_T}{\rho C_p}$ [L^2t^{-1}] (thermal diffusivity)	D_s [L^2t^{-1}] (mass diffusivity)
Potential	v_y [Lt^{-1}]	T [T]	S [ML^{-3}]
Potential gradient	$\dfrac{dv_y}{dx}$	$\dfrac{dT}{dx}$	$\dfrac{dS}{dx}$
Potential "concentration"	ρv_y	$\rho C_p T$	S
Potential "concentration" gradient	$\dfrac{d\rho v_y}{dx}$	$\dfrac{d\rho C_p T}{dx}$	$\dfrac{dS}{dx}$

quiescent water. Laminar flow is also useful in experimental systems because the hydrodynamics can be described analytically.

Laminar flow of water in a circular tube may be analyzed by means of a differential momentum balance. For momentum transport, the general conservation law for transport in one direction through an arbitrary control volume is

net rate of accumulation of momentum

= net rate of input by transport + sum of forces [9.12]

This expression is similar to Eq. 2.2 except that the force term replaces the process rate or production term. In this case, certain forces can be interpreted as production sources for momentum. For example, a pump produces a pressure (force per unit area), which is a "source" of momentum. Similarly, fluid flow from an elevated reservoir is "driven" by gravity forces. To illustrate these concepts, laminar flow in a tube is described analytically in Example 9B. The flow in the circular tube geometry has been studied extensively and is the most widely used conduit geometry in technology.

9.3.2.2 *Conductive Heat Transfer* Biofilms can significantly reduce heat flux at a surface, and this poses one of the major problems of fouling (Chapter 14). Hence, it is important to be able to find the rate of heat flow at a surface in a piece of equipment (e.g., a power plant condenser). The procedure for solving this transport problem is as follows:

1. Use the differential balance (in this case, an energy balance) to derive a differential equation for the heat flux as a function of position in the biofilm.
2. Use the flux law (in this case, Fourier's law of heat conduction) to relate the heat flux to the temperature. Use the boundary conditions to solve (i.e., integrate) the differential equation(s) to obtain a temperature distribution.
3. Use the flux law and the equation for the temperature distribution to obtain the flux at each boundary.
4. Obtain the total transport rate or flow at the boundary by multiplying the flux at the boundary by the area perpendicular to the flux.

The problem of heat transfer through a rectangular solid slab will be solved using rectangular coordinates and neglecting any convective heat transfer and any possible generation of heat. The general procedure can be used for more complicated problems in both momentum, heat, or mass transfer.

Example 9A. Steady State Heat Transfer Through a Solid. Consider a large slab of thin dimension L_x (Figure 9.3). The other dimensions, L_y and L_z, are

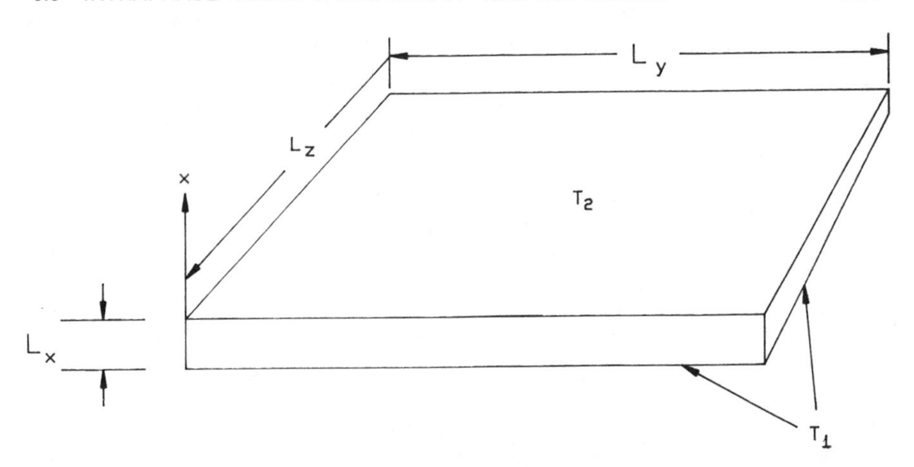

Figure 9.3 The bottom of a rectangular slab having dimensions L_x, L_y, and L_z is heated and maintained at a temperature T_1. The top of the slab is maintained at a lower temperature, T_2.

large compared to L_x, so that heat is only transported in the x-direction when there is a temperature difference between the top and bottom faces of the slab. The slab bottom is at $x=0$, and the top is at $x=L_x$. The lower face of the slab is maintained at temperature T_1 while the top face is maintained at a lower temperature T_2. Heat will flow from the bottom face to the top face. T_2 may be maintained constant by a uniform flow of cooling water. The constant heat flow at the bottom face may be the result of steam condensing. The heat flux q_x through the slab is

$$q_{Hx} = Q_H/A_x \qquad [9A.1]$$

where Q_H is the heat flow rate and A_x is the area of the cross section perpendicular to the x-direction.

1. *Steady state differential heat balance*
Perform the heat balance over the differential volume $\Delta x L_z L_y$

$$\underset{\text{input}}{A_x q_{Hx}|_x} - \underset{\text{output}}{A_x q_{Hx}|_{x+\Delta x}} = 0 \qquad [9A.2]$$

Dividing by the differential volume $A_x\ \Delta x$ and taking the limit as $\Delta x \to 0$,

$$\lim_{\Delta x \to 0} = \frac{q_{Hx}|_x - q_{Hx}|_{x+\Delta x}}{\Delta x} = 0 \qquad [9A.3]$$

or, by the definition of a derivative,

$$\frac{dq_{Hx}}{dx} = 0 \qquad [9A.4]$$

So q_{Hx} is a constant, say

$$q_{Hx} = q_{H0} \qquad [9A.5]$$

2. *Insert the flux law and state the boundary conditions*
Insert Fourier's law into Eq. 9A.5 for q_{Hx}:

$$-k_T \frac{dT}{dx} = q_{H0} \qquad [9A.6]$$

The boundary conditions for this problem are

B.C. 1: at $x = 0, \quad T = T_1$

B.C. 2: at $x = L_x, \quad T = T_2$

Integrating Eq. 9A.6,

$$T = -\frac{q_{H0}}{k_T} x + C_1 \qquad [9A.7]$$

By inserting B.C. 1 into Eq. 9A.7, we see that

$$C_1 = T_1 \qquad [9A.8]$$

So

$$T = -\frac{q_{H0}}{k_T} x + T_1 \qquad [9A.9]$$

By inserting B.C.2 into Eq. 9A.9, we find q_{H0}:

$$q_{H0} = -k_T \frac{T_2 - T_1}{L_x} \qquad [9A.10]$$

By substituting Eq. 9A.10 into Eq. 9A.9, the temperature distribution is obtained:

$$\frac{T - T_1}{T_2 - T_1} = \frac{x}{L_x} \qquad [9A.11]$$

2. *Find the heat flux at the boundaries*

The heat flux at the either face is defined by

$$q_{Hx}|_{x=0} = -k_T \left.\frac{dT}{dx}\right|_{x=0} \qquad [9A.12]$$

or

$$q_{Hx}|_{x=0} = k_T \frac{T_1 - T_2}{L_x} \qquad [9A.13]$$

Remember that q_{Hx} is independent of x (Eq. 9A.5).

4. *Total transport rate at the boundary*

The total transport is determined by multiplying the flux at the boundary (Eq. 9A.13) by the area normal to the flux. So

$$Q_H = Q_H|_{x=0} = q_{Hx}A_x|_{x=0} = k_T A_x \frac{T_1 - T_2}{L_x} \qquad [9A.14]$$

The heat flow rate is the same at $x = L_x$, because q_{Hx} is independent of x.

9.3.3 Summary

Differential material balances can be used to formulate rigorous models describing transport and reaction phenomena in laminar flow systems or in solids. The simplification in these systems is the negligible role of advective transport processes, which are difficult to describe in turbulent flow and in some geometries.

9.4 INTERPHASE TRANSPORT: TURBULENT TRANSPORT TO A SURFACE

Turbulent flow is extremely complex, and at present no deterministic theory exists for prediction of turbulent velocity profiles with the same degree of certainty as with laminar flows. For example, the balance equations (Eqs. 9.1-9.5) and flux laws (Eqs. 9.8-9.10) pertain to laminar and turbulent flow regimes. However, in turbulent flow, the variables ($\mathbf{v}$, T, and S) must be time-averaged and the transport coefficients are functions of the turbulent intensity and not solely of the properties of the fluid.

Our objective for interphase transport is to develop models that will predict transfer coefficients describing transport from the bulk water to the biofilm as functions of other variables (e.g., mean velocity, water properties, tube diameter). With these transfer coefficients, an overall rate of mass, heat, or momentum transport in the bulk water can be calculated for design and analysis of operating equipment or environmental systems in which biofilm processes are significant.

9.4.1 Differential Models: Three Region Model for Turbulent Flow in a Pipe

It is customary in engineering analysis of turbulent flow to divide the flow in a pipe into three regions: (1) the wall region, or *viscous sublayer,* in which viscous forces dominate, (2) the *turbulent core,* where inertial forces prevail, and (3) and a transition region, termed the *buffer region,* where viscous and inertial forces are comparable (Figure 9.4).

The viscous sublayer (or momentum transfer boundary layer), δ_v, is defined as the region where viscous forces are dominant and is generally very thin. The viscous sublayer is where almost all of the resistance to flow resides in turbulent flow regimes. δ_v decreases with increasing *Re* (Figure 9.5). Thus, for a pipe of a given diameter, increasing velocity will decrease δ_v. The viscous sublayer *is not* completely in laminar flow. There are turbulent fluctuations, or bursts, which are constantly penetrating the sublayer, causing local velocity perturbations in all directions, even opposite to the main direction of flow. These turbulent bursts

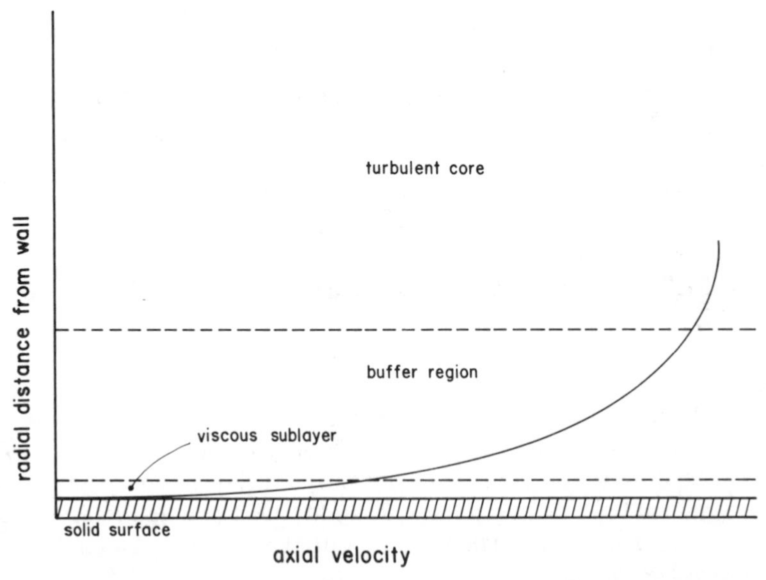

Figure 9.4 The turbulent velocity profile in a circular tube can be divided into three regions: The core region is extremely well mixed, and the velocity is essentially uniform throughout the region; inertial forces dominate. The buffer region is influenced by both inertial and viscous forces. The viscous sublayer is dominated by viscous forces, and mixing is poor within this region.

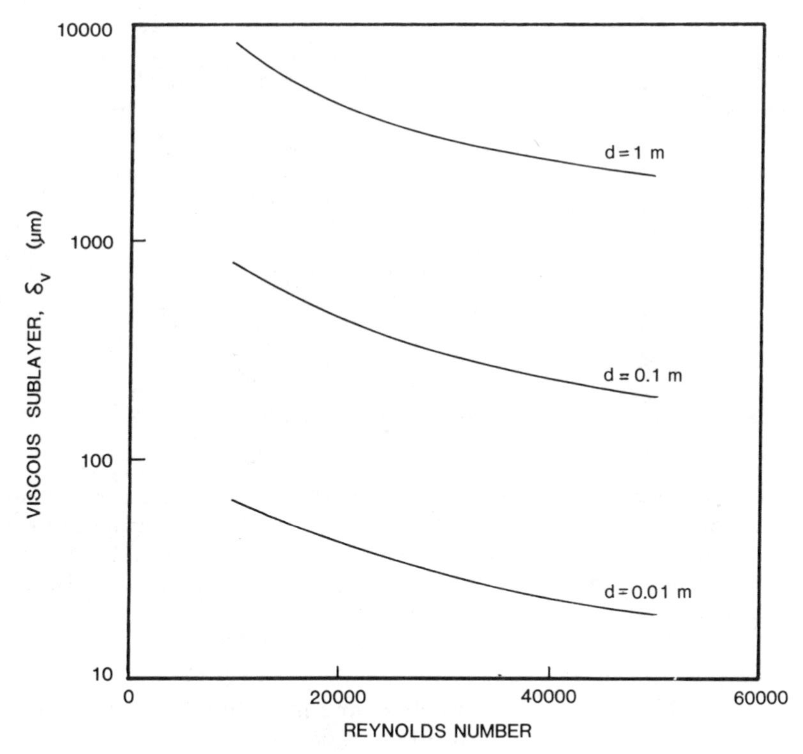

Figure 9.5 The viscous sublayer thickness δ_v decreases exponentially with increasing Reynolds number in a circular tube (diameter d) at 30°C.

may be largely responsible for the transport of colloidal particles like bacteria to pipe walls in turbulent flow (Cleaver and Yates, 1975, 1976).

A velocity profile has been determined for each of these regions (Bird et al., 1960; Fahien, 1983), and transport rates can be determined from these profiles in a manner similar to that in laminar flow. In the following section, our objective is to obtain transfer coefficients as functions of other macroscopic variables without using the velocity profiles.

9.4.2 Lumped Parameter Models

9.4.2.1 Film Theory Model Observations in turbulent flow systems indicate that temperature and concentration profiles change rapidly near the wall of the vessel and are very flat away from the wall (e.g., heat exchange tube wall or biofilm reactor surface) (Figure 9.6). Therefore, since the major resistance to transfer is near the wall, it is convenient to postulate that *all* the resistance to heat or mass transport occurs in the thin film of fluid near the wall. Outside the film, the temperature and concentration have their bulk values. Within the film, laminar flow is *assumed,* even though the flow outside the film is turbulent. The

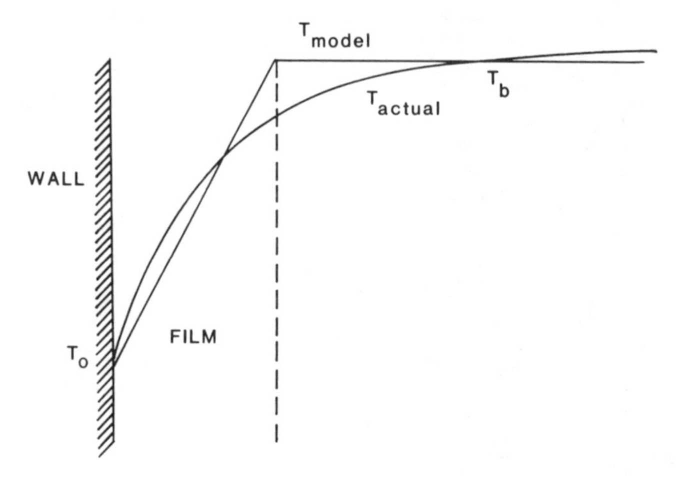

Figure 9.6 The film theory model (Eq. 9.13) for turbulent heat transfer presumes that all temperature change from the wall (T_0) to the bulk water (T_b) occurs within a thin film, the thermal boundary layer. The thermal boundary layer does not have the same thickness as the viscous sublayer. The model gradient is presumed linear. The actual temperature gradient is shown for comparison.

effective film thickness is δ, and the heat and mass transfer coefficients that describe advective transport are sometimes called *film coefficients*.

The actual variation in T is illustrated in Figure 9.6. The effective film thickness is *not* the thickness of the viscous sublayer, but that of a fictitious film that would be required to account for the entire heat transfer resistance if it were only due to molecular transport as expressed in the flux law.

9.4.2.2 *Turbulent Heat Transfer Coefficient* Consider a heat exchanger consisting of a pipe through which a fluid is flowing turbulently. If the temperature of the tube wall, T_0, is greater than that of the fluid, the fluid will be heated as it passes through the pipe. The purpose of the heat exchanger is to carry away heat, and the heat flux is expressed in terms of a heat transfer coefficient h:

$$q_{H0} = h(T_b - T_0) \tag{9.13}$$

where q_{H0} is the heat flux at the substratum, T_b is the temperature of the bulk water, and T_0 is the temperature at the substratum. Equation 9.13, *Newton's law of cooling,* is the defining equation for h. In the actual heat exchanger, q_{H0}, T_b, and T_0 may vary with axial distance.

9.4.2.3 *Turbulent Mass Transfer Coefficient* Now consider a substratum with a uniform reactive coating. Component S is being transported from the bulk liquid to the substratum (solid phase). If S_0 is the concentration at the substratum and S_b is the concentration in the bulk water, then

$$J_S = k_D(S_b - S_0) \qquad [9.14]$$

where J_S is the flux of substrate to the biofilm–water interface, and k_D (D for diffusion) is the mass transfer coefficient.

9.4.2.4 *Turbulent Momentum Transfer Coefficient: Friction Factor*

The general definition of friction factor is expressed as follows:

$$f = \frac{\text{drag force/characteristic area}}{\text{kinetic energy/volume}} \qquad [9.15]$$

The turbulent flow drag force at the substratum–bulk water interface on a pipe wall equals the shear stress at the interface, τ_0, resulting from the viscous forces and the gravity forces that are overcome by the pressure drop. The characteristic area is the substratum surface area. The kinetic energy per unit volume is $\frac{1}{2}\rho v_b^2$, where v_b is the mean bulk water velocity. So, for a circular tube,

$$f = \frac{\tau_0}{0.5\, \rho v_b^2} \qquad [9.16]$$

A momentum transfer coefficient could be defined as $\frac{1}{2} f v_b$:

$$\tau_0 = 0.5 f v_b(\rho v_b - \rho v_0) \qquad [9.17]$$

or, since the velocity at the wall (v_0) is zero (no slip condition),

$$\tau_0 = 0.5 f v_b(\rho v_b - 0) \qquad [9.18]$$

or

$$\tau_0 = \frac{f}{2} \rho v_b^2 \qquad [9.19]$$

However, friction factor rather than "momentum transfer coefficient" is generally used. A comparison of Eqs. 9.13, 9.14, and 9.18 demonstrates the analogy between heat, mass, and momentum transport (Table 9.3).

9.4.2.5 *Nusselt Numbers*

The friction factor is dimensionless, while the heat and mass transfer coefficients are not. Since it is desirable to use dimensionless variables, it is customary to define dimensionless heat and mass transfer coefficients as follows:

$$Nu = \frac{h d_{ch}}{k_T} \qquad [9.20]$$

TABLE 9.3 A Comparison of the Defining Equations for Momentum, Heat, and Mass Transfer Coefficient.[a]

General Expression

flux = transfer coefficient × driving force

Momentum Transfer Coefficient (Friction Factor)

$$\tau_0 = 0.5 f v_b (\rho v_b - 0) \quad [9.18]$$

Heat Transfer Coefficient

$$q_{H0} = h(T_b - T_0) \quad [9.13]$$

Mass Transfer Coefficient

$$J_S = k_D(S_b - S_0) \quad [9.14]$$

[a]The defining equation is based on Newton's law of cooling.

and

$$Sh = Nu_S = \frac{k_D d_{ch}}{D_S} \quad [9.21]$$

where Nu is the heat transfer Nusselt number, Nu_S is the Nusselt number for mass transfer, and Sh is the Sherwood number for mass transfer. The Nusselt numbers are ratios of advective transport rates to diffusive transport rates.

9.4.3 Summary

Interphase transport is extremely important for biofilm processes because it brings a variety of nutrients, cells, and other materials from the bulk liquid compartment to the biofilm–bulk water interface. For example, growth of biofilm cells may be rate-limited by the rate of transport of nutrients from the bulk liquid to the biofilm interface. In addition, the extent of biofilm accumulation or the maximum biofilm thickness may be controlled by detachment of materials from the biofilm. These topics will be dealt with in more detail later in the chapter.

9.5 HYDRODYNAMICS

9.5.1 Intraphase Transport: Laminar Flow

A differential momentum balance in the bulk liquid of a circular tube in laminar flow results in an analytical solution for velocity as a function of distance from the conduit surface. Example 9B is a derivation of the laminar velocity profile in a tube.

Example 9B. Laminar Flow in a Circular Tube. Water, an incompressible Newtonian fluid (a fluid that obeys Newton's law of viscosity, Eq. 9.8) is flowing at steady state through a long circular tube of length L and radius R (Figure 9.7). The section of tube being considered is far removed from the entrance (approximately 50 diameters), so that *fully developed* flow exists, i.e., the velocity profile does not change with distance along the tube. The axial coordinate z is in the direction of flow. The distance from the center of the tube is r. The velocity in the $+z$ direction, v_z, depends on the radial position r. At $z = 0$, the uniform pressure over the cross section, πR^2, is p_0, At $z = L$, the pressure is p_L.

The differential momentum balance includes three forces in the tube system: (1) viscous forces due to fluid friction, (2) pressure forces, and (3) gravity forces. The pressure forces must be greater than the combined viscous and gravity forces, or there will be no flow. The resulting velocity profile is the Hagen–Poiseuille equation (Figure 9.7):

$$v_z = \frac{(P_0 - P_L)R^2}{4\eta L}\left[1 - \left(\frac{r}{R}\right)^2\right] \qquad [9B.1]$$

where P is the total pressure ($p - \rho gz$), p is the pressure, and L is the distance between pressure measurement points. It can be seen that $v_z = 0$ at $r = R$, i.e.,

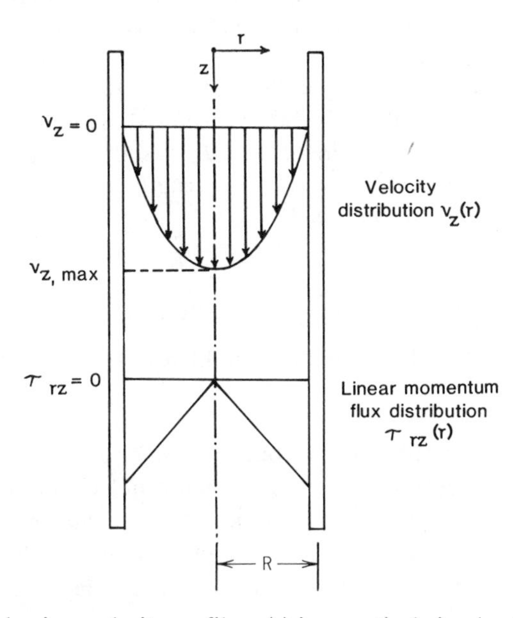

Figure 9.7 The laminar velocity profile $v_z(r)$ in a vertical circular tube of radius R is parabolic and is known the Hagen–Poiseuille equation (Eq. 9B.1).

at the wall. This is the result of the *no slip* condition, i.e., the fluid immediately adjacent to the wall has the same velocity as the wall.

By integrating v_z (Eq. 9A.1) over the cross section, the average velocity v_b is obtained:

$$v_b = \frac{(P_0 - P_L)R^2}{8\eta L} \tag{9B.2}$$

The average velocity is important because it permits calculation of the volumetric flow rate, $Q = v_b A$:

$$Q = \frac{\pi(P_0 - P_L)R^4}{8\eta L} \tag{9B.3}$$

The momentum flux or shear stress at the wall is

$$\tau_0 = P_0 - P_L = (p_0 - p_L) + \rho g L \tag{9B.4}$$

The force at the tube wall is the product of the shear stress at the wall and the wall surface area:

$$F_0 = \pi R^2[(p_0 - p_L) + \rho g L] \tag{9B.5}$$

Thus, the net force acting downstream on the cylinder of water due to the pressure difference and gravity forces is balanced by the viscous force, which tends to resist the motion.

The Hagen–Poiseuille law is only useful for $Re < 2100$. End effects have been neglected. An entrance length to the tube is required to attain a fully developed flow and, for laminar flow, is on the order of $L_e = 0.058d \cdot Re$. If the section of pipe being considered includes the entrance length, a correction to Eq. 9B.3 must be applied (Perry and Chilton, 1973).

9.5.2 Interphase Transport: Friction Factors in Turbulent Flow

There are several reasons for addressing problems associated with the friction factor.

1. Friction factor describes the turbulent state of a hydrodynamic system. For example, a typical engineering design problem is to obtain the pressure drop, ΔP, required to pump fluid through a pipe of given radius and length L at an average velocity. The result is then used to select a pump.
2. Advective heat transfer and mass transfer are influenced by the friction factor.

3. Biofilms influence friction factor in many technologically important systems. Much of the interactive effect between friction factor and biofilms results from the surface film (Figure 1.2, Chapter 1), which can be quite rough and frequently extends into the bulk flow.

The pressure drop ΔP in laminar flow can be calculated from Eq. 9B.3. For turbulent flow, there is no generally accepted theory for predicting the pressure drop. Consequently, engineers have relied heavily on empirical correlations of experimental data.

The general definition of the friction factor (Eq. 9.15) is the basis for designing pipe systems. In pipe flow, the drag force equals the shear stress at the wall resulting from the viscous forces and the gravity forces that are overcome by the pressure drop. In Eq. 9.15, the characteristic area is the tube surface area, and the kinetic energy per unit volume is $\frac{1}{2}\rho v_b^2$. So, for a circular tube,

$$f = \frac{1}{2}\frac{d}{L}\frac{P_0 - P_L}{\rho v_b^2} \qquad [9.22]$$

where d is the tube diameter, and L is the tube length. If the Hagen–Poiseuille equation for average velocity (Eq. 9B.2) is substituted in Eq. 9.22,

$$f = \frac{16}{Re} \qquad [9.23]$$

This result is only useful for laminar flow ($Re < 2100$). Friction factors in laminar flow are of little use, however, because the pressure drop can be calculated directly.

The Fanning friction factor diagram describes the influence of Re on f (Figure 9.8). The friction factor is a strong function of Re (flow) in laminar flow, as evidenced by Eq. 9.23 and Figure 9.8. The friction factor in turbulent flow ($Re > 2100$), however, is a weak function of Re (Figure 9.8). For smooth tubes, the relationship between friction factor and Re is described by the Blasius equation for $2100 < Re < 100{,}000$:

$$f = \frac{0.0791}{Re^{0.25}} \qquad [9.24]$$

The Fanning friction factor is used primarily in the chemical engineering literature. Civil engineers use a friction factor which is larger by a factor of four ($f = 0.316/Re^{0.25}$). Experimental results in this chapter use the civil engineering friction factor.

In turbulent flow, f is not influenced a great deal by Re. However, the roughness of the pipe exerts a large effect on the friction factor. If the pipe is rough, then higher pressure drops are needed for a given flow rate than in a smooth tube, and higher pumping costs result. Thus, the relative roughness, e/d (where e is the height of the roughness elements), enters the f-Re relationship (Figure 9.8). The relationship of friction factor to roughness can be described by the Colebrook equation as follows:

$$\frac{1}{\sqrt{f}} = -4.0 \log\left[\frac{e}{d} + \frac{4.67}{Re\sqrt{f}}\right] + 2.28 \qquad [9.25]$$

where e = height of roughness elements (L).

As biofilms accumulate, hydrodynamic roughness and friction factor increase, resulting in costly fouling problems. The relationship between hydrodynamic roughness and friction factor can be explained in terms of the viscous sublayer. If the roughness height, $e < \delta_v$, then the friction factor will fall on the line for smooth tubes. If $e > \delta_v$, then the tube is determined to be "rough," since an increased f is observed (Figure 9.8). Thus, a tube can go from a smooth condition to a rough condition as a biofilm accumulates (see Section 9.5.3.1).

The relation between friction factor and Reynolds number also implicitly includes the influence of temperature because the fluid properties (ρ and η) vary with temperature.

The Fanning friction factor diagram is only useful for flow through conduits. Thus, a different Re and f must be defined for flows in other geometries. For example, a different relationship exists between friction factor and Re in porous media (Chapter 18). The friction factor for a RotoTorque reactor is derived in Example 9C using Eq. 9.15 as a starting point. The derivation illustrates the procedure for arriving at a friction factor relationship for unique hydrodynamic systems.

Example 9C. Definition of a Friction Factor for the Annular Reactor (RotoTorque™). The RotoTorque (Section 3.3.1, Chapter 3) consists of two concentric cylinders: a stationary outer cylinder and a rotating inner cylinder (Figure 9.9). The rotational velocity is controlled by a variable speed motor. A tachometer–torque transducer, mounted on the shaft between the motor and the annular reactor, continuously monitors rotational speed Ω and torque T_q. The RotoTorque is operated with a continuous flow of liquid through the system, resulting in an axial flow component, which must be included in the analysis.

Re for the rotational or tangential velocity component is

$$Re_R = \frac{\Omega \alpha R_0^2 \rho}{\eta} \qquad [9C.1]$$

where R_0 is the inner radius of the outer cylinder and α is the ratio of the radius of the inner cylinder to the outer cylinder (i.e., αR_0 is the radius of the inner

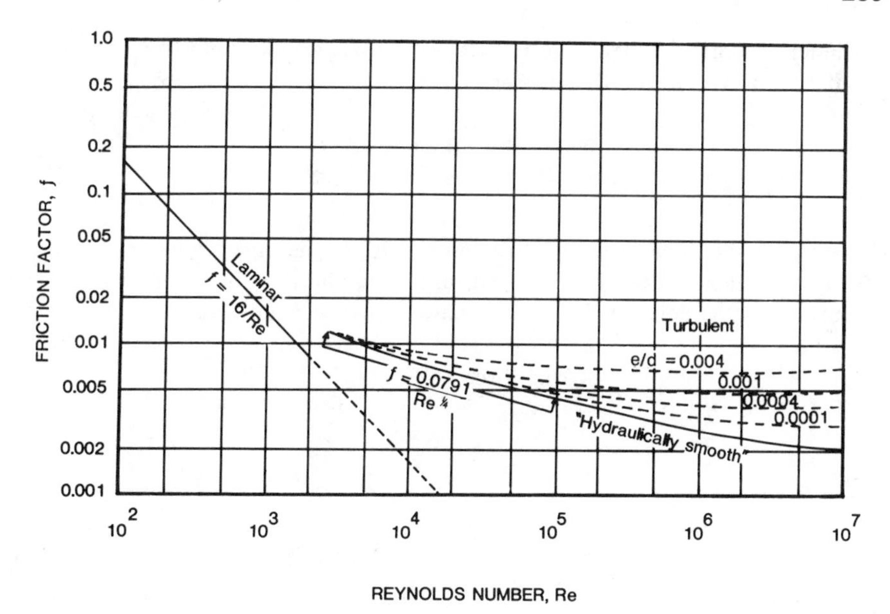

Figure 9.8 Fanning friction factors in a tube flow vary with Reynolds number and tube roughness.

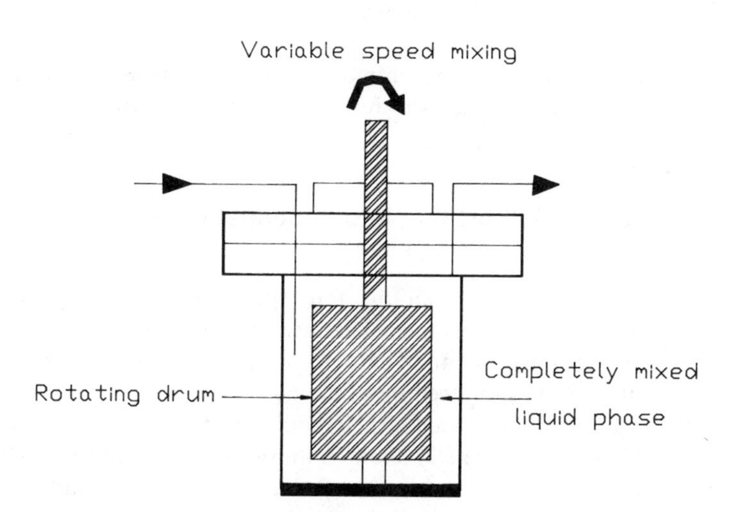

Figure 9.9 A schematic diagram of the RotoTorque reactor (also see Chapter 3).

cylinder). Another dimensionless number, the Taylor number (Taylor, 1923), is analogous to the Reynolds number:

$$Ta = \frac{v_{in} R_0 (1 - \alpha)^{1.5}}{\nu \sqrt{\alpha}} \qquad [9C.2]$$

where v_{in} is the linear velocity of the outer edge of the inner cylinder. To define the friction factor, the drag force per unit area and the kinetic energy per unit volume must be defined. Using the characteristics of the system,

$$\frac{\text{drag force}}{\text{area}} = \frac{T_q}{2\pi^2 \alpha R_0^2 H} \qquad [9C.3]$$

and

$$\frac{\text{kinetic energy}}{\text{volume}} = 0.5 \rho \alpha^2 R^2 \Omega^2 \qquad [9C.4]$$

where H is the height of the cylinder. So the friction factor (Eq. 9.15) for the RotoTorque is

$$f_R = \frac{T_q}{0.5 \pi \rho \alpha^2 R_0^2 \Omega H} \qquad [9C.5]$$

The relationship of f_R to Taylor number is presented in Figure 9.10. f_R, a dimensionless variable, includes the influence of five variables on frictional resistance.

9.5.3 Influence of Biofilms on Fluid Frictional Resistance in Conduits

Biofilm accumulation in water conduits and on ship hulls causes pronounced increases in fluid frictional resistance. The resulting energy losses are of major concern in water transmission and in ship propulsion. Deterioration of pipeline capacity attributed to biofilm accumulation can be substantial. For example, Seifert and Krueger (1950) reported a 55% reduction of capacity in a 80 km long water supply pipeline (600 mm ID) due to a thin slimy layer approximately 650 μm thick. A summary of case histories of deterioration of flow in conduits resulting from biofilm accumulation is presented in Table 9.4. Characklis (1972) described these cases in more detail.

9.5.3.1 *Influence on Friction Factor in Conduits* Characklis and coworkers (Zelver, 1979; Picologlou et al., 1980) have documented the effects of biofilm accumulation on friction factor in a tube (12.7 mm ID) with a hydrauli-

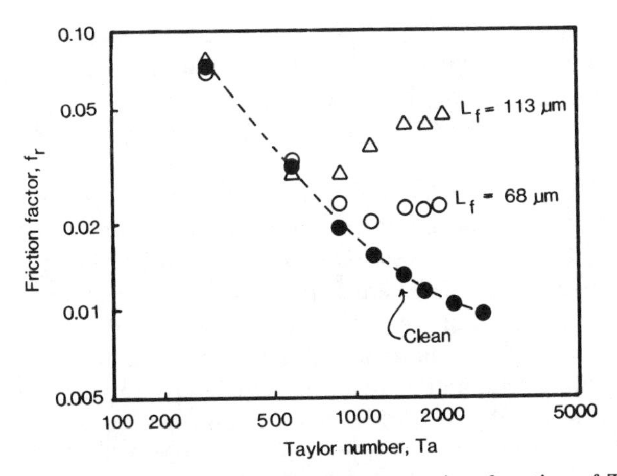

Figure 9.10 Friction factor in the RotoTorque system is a function of Taylor number (Eq. 9C.2). The solid symbols refer to a clean reactor; the open symbols show the influence of biofilm with thickness L_f. (Trulear, 1980.)

TABLE 9.4 Case Histories of Conduits Experiencing Extraordinary Energy Losses due to Biofilms[a]

Diameter (m)	Length (km)	Surface	Biofilm Thickness (mm)	Frictional Head (m)	Loss in Capacity (%)	Reference
1.07	11.2	Cement	0.8–1.6		12	Minkus, 1954
0.91	11.2	Concrete			23	Minkus, 1954
3.35	7.2	Concrete	12.7	4.9[b]		Pollard and House, 1959
0.91	35.2	Steel	3.2–6.4		16	Arnold, 1963
0.61	80	Steel	0.6		55	Wiederhold, 1949
0.36	2	Steel			35	Derby, 1947

[a] Characklis, 1972.
[b] At 25.5 m^3/s.

cally smooth surface. The tubular reactor system (Section 3.3.2, Chapter 3) incorporated a high recycle rate, so that the system could be analyzed as a continuous flow stirred tank reactor (CFSTR). The microbial inoculum was an undefined, mixed bacterial population (from sewage), and the medium was glucose and mineral salts for most of the experiments. Two measurements indicated the hydraulic condition in the conduit:

1. *Fluid Flow Velocity* (v_b). The flow velocity in conduits is directly related to the flow capacity (volume rate of flow), which is the product of v_b and the cross-sectional area of the conduit.

2. *Pressure Drop* (ΔP). The pressure drop across a section of conduit with flowing water is related to the loss of fluid energy. Increased surface roughness, reduced conduit diameter, and other factors can increase the pressure drop as turbulence dissipates the hydraulic energy.

If the water velocity is maintained constant in the tube, then the pressure drop will increase as biofilm accumulates. The increase in pressure differential as biofilm accumulates in a 12.7 mm ID tube is indicated in Figure 9.11. The increased pressure drop will increase pumping costs. If the biofilm forms on a ship hull, the power consumption for propulsion at constant velocity will increase.

If the pressure drop is maintained constant, then the fluid velocity will decrease with biofilm accumulation. The decrease in flow capacity (water velocity) when a biofilm accumulates in a 12.7 mm ID tube operated at a constant pressure differential or pressure head is presented in Figure 9.12. This situation occurs in gravity flow pipelines. The biofilm thickness was 180 μm after 100 h in this experiment.

The friction factor, described in Eq. 9.22, incorporates both v_b and ΔP and describes the energy loss per unit volume of fluid in a conduit. As biofilm accumulates on a tube or pipe surface in turbulent flow, the friction factor increases

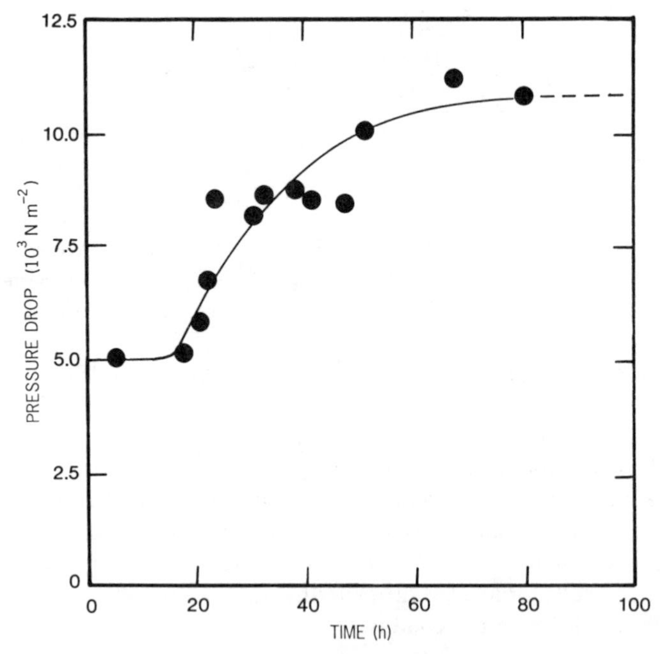

Figure 9.11 Progression of pressure drop in a circular tube (12.7 mm ID) due to biofilm accumulation for constant water velocity 1.5 m s^{-1}. The glucose loading rate was 0.8 g m^{-2} s^{-1}, and the temperature was 40°C. (Reprinted with permission from Picologlou et al., 1980.)

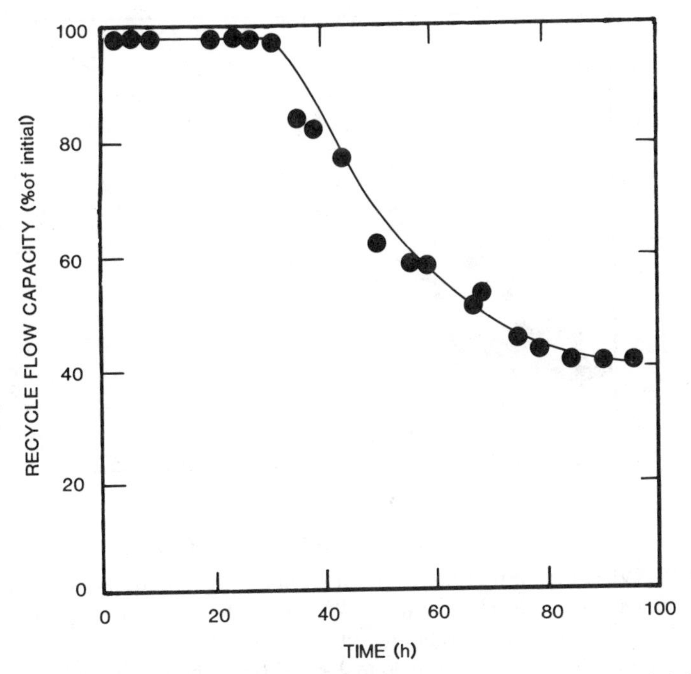

Figure 9.12 Progression of flow capacity (water velocity) in a circular tube (12.7 mm ID) at constant pressure drop (shear stress was 7.9 N m^{-2}). Initial water velocity was 1.85 m s^{-1}, glucose loading rate was 4.3 g m^{-2} s^{-1}, and temperature was 30°C. (Reprinted with permission from Picologlou et al., 1980.)

regardless of whether velocity or pressure drop is maintained constant. Thus, the friction factor is generally the variable considered. After an initial lag, it increases with increasing biofilm thickness if the velocity is maintained constant (Figure 9.13). The lag in frictional resistance increase, frequently observed in biofilm systems, ends when a characteristic critical thickness (in this case approximately 30 μm) is reached. The critical thickness is dependent on flow conditions (see discussion on rigid roughness below).

Tube Constriction. One mechanism that increases frictional losses is the constriction of the tube cross sectional area by the biofilm. The increase in pressure differential and biofilm thickness with time for one experiment is presented in Figure 9.14. The *calculated* increase in pressure drop for a decrease in radius equal to the measured biofilm thickness is also presented *assuming* that the biofilm was smooth and rigid. Pressure drop attributed to a decrease in tube radius resulting from biofilm accumulation is calculated by the Blasius equation for a smooth tube (Eq. 9.24). Constriction of the tube accounted for no more than a 10% increase in calculated pressure drop, while the measured pressure

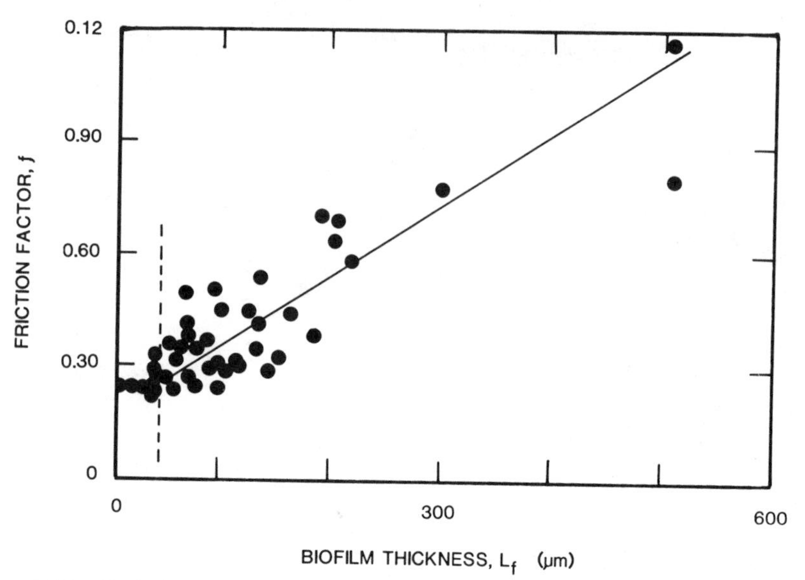

Figure 9.13 Variation of friction factor (at constant pressure differential) with biofilm thickness in a circular tube (12.7 mm ID). The shear stress for all experiments was 6.5-7.9 N m^{-2}. The glucose loading rate varied from 0.42 to 33.6 g m^{-2} s^{-1}, while the experimental run temperature varied from 30 to 35°C. (Reprinted with permission from Picologlou et al., 1980.)

drop due to biofilm accumulation increased approximately 110%. Clearly, the effect of a reduction in tube diameter by biofilm accumulation was minimal.

Rigid Roughness. When the roughness of an inside tube wall surface is sufficiently coarse, eddy currents develop that result in energy losses. Extensive tests on flow within rough pipes were conducted by Nikuradse (1933), using known size sand grains fixed to the inside of tubes. Nikuradse's work describes the relationship of friction factor to Reynolds number (Re) for a range of sand roughness diameters. The friction factor is related to Re and the equivalent sand roughness e_s through the empirical Colebrook-White equation (Schlichting, 1968). This equation leads to a correlation between friction factor and Re for various "commercially rough" tubes:

$$e_s = \frac{d}{2}\left[10^{(0.87-0.50\sqrt{f})} - \frac{18.7}{Re\sqrt{f}}\right] \qquad [9.26]$$

where e_s = equivalent sand roughness (L).

This equation is useful in biofouled tubes, since an equivalent sand roughness for biofilm can be calculated from a measurement of the flow rate and pressure drop. The progression of e_s during which biofilm is accumulating is presented in

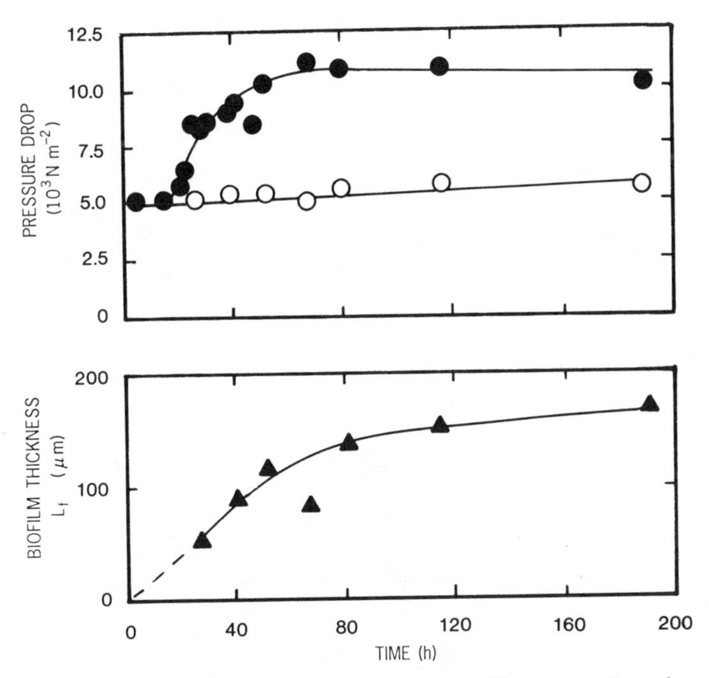

Figure 9.14 Progression of measured pressure drop (●) compared to the calculated pressure drop (○) due to a decrease in tube radius resulting from increasing biofilm thickness (▲). (Reprinted with permission from Picologlou et al., 1980.)

Figure 9.15. e_s generally increases with biofilm thickness (Figure 9.16), although the equivalent sand roughness of the biofilm can be greater than the actual film thickness. The biofilm thickness measurement was an average thickness measurement and did not take account of height of roughness peaks. The average biofilm thickness will be less than the height of some of the peaks. The equivalent sand roughness depends on the roughness peaks, but it is not numerically equal to their size. It is not unusual for the equivalent sand roughness to be greater than the height of the roughness peaks (Schlichting, 1968). If the biofilm indeed increases the effective roughness of the tube wall, a dependence of calculated equivalent sand roughness on biofilm surface morphology as well as biofilm thickness or roughness peaks is expected. It has been well established that different surface roughness configurations having identical sizes of roughness peaks can have different equivalent sand roughnesses (Schlichting, 1968).

The flow regime (smooth, transitional, or fully rough) depends on the magnitude of e_s relative to the size of the viscous sublayer (δ_v) as given by Schlichting (1968):

$$\delta_v = \frac{10\text{d}}{Re}\left(\frac{f}{2}\right)^{-0.5} \qquad [9.27]$$

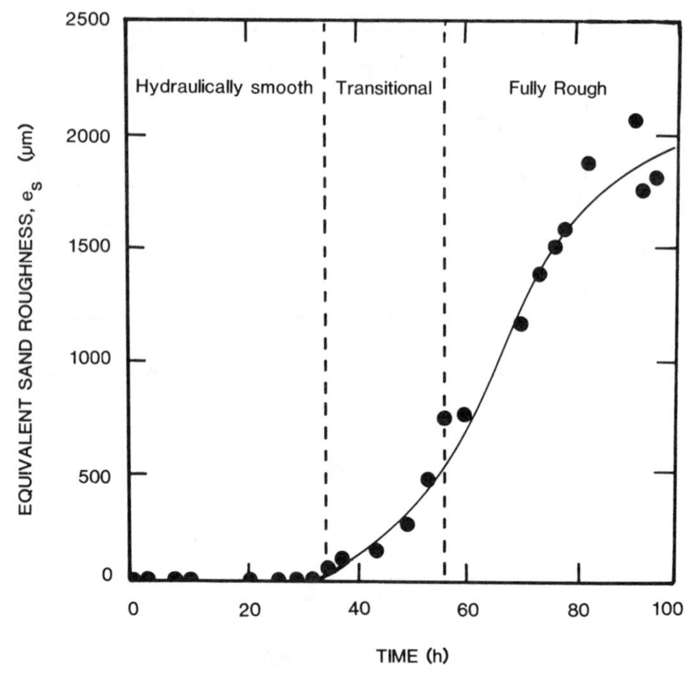

Figure 9.15 Progression of equivalent sand roughness in a circular tube (12.7 mm ID) resulting from biofilm accumulation (shear stress 7.9 N m^{-2}). Initial water velocity was 1.85 m s^{-1}, glucose loading rate was 4.3 g m^{-2} s^{-1}, and temperature was 30°C. (Reprinted with permission from Picologlou et al., 1980.)

More specifically, when $e_s < \delta_v$, the pipe is considered hydraulically smooth; when $14\delta_v > e_s > \delta_v$, the flow is in the transitional regime; when $e_s > 14\delta_v$, the flow is in the fully rough regime (Figure 9.15).

The increased friction factor resulting from biofilm accumulation has been compared with the classical relationships developed for friction factor due to *rigid roughness* in tubes by Zelver (1979) and Picologlou et al. (1980). The relationship between friction factor and *Re* for a range of biofilm thicknesses in a tube is presented in Figure 9.17. The friction factor in a tube with biofilm varies with *Re* in the same way as for rigid roughness over the wide range of *Re* investigated (5000–48,000), as indicated by a comparison of Figures 9.8 and 9.17. The data were obtained by periodically reducing the flow (shear stress) in discrete steps from its initial value in a given experiment and then calculating friction factor and *Re* at each step. Reduction, rather than increase, of the flow (shear stress) from the initial condition minimized biofilm detachment during the flow excursions.

The friction factor is dependent on biofilm thickness only after a critical biofilm thickness is attained. This thickness is approximately equal to the thickness of the viscous sublayer. It corresponds to the protrusion of surface film ir-

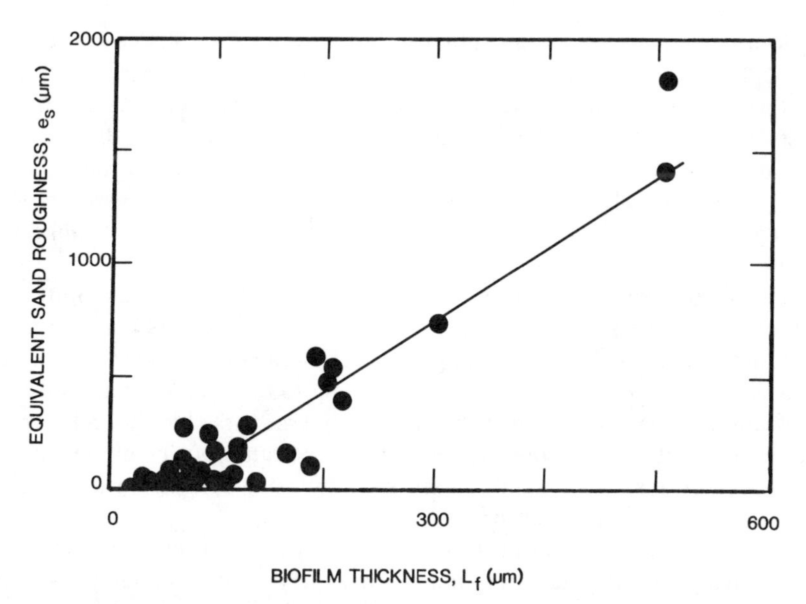

Figure 9.16 Variation in calculated equivalent sand roughness with biofilm thickness at a fluid shear stress of 6.5–7.9 N m^{-2}. The glucose loading rate varied from 0.42 to 33.6 g m^{-2} s^{-1}, while the experimental run temperature varied from 30 to 35°C. (Reprinted with permission from Picologlou et al., 1980.)

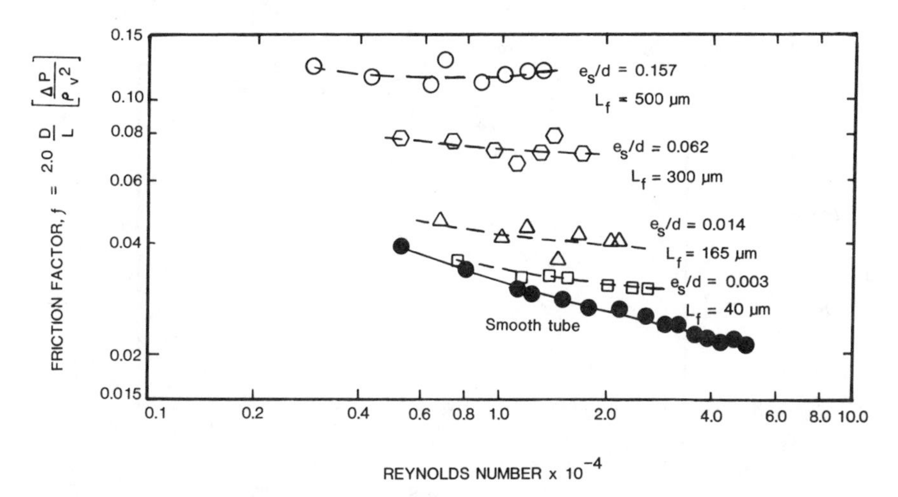

Figure 9.17 Friction factor in a circular tube (12.7 mm ID) as a function of relative roughness e_s/d, measured biofilm thickness, and *Re*. Initial water velocity was 1.85 m s^{-1}, shear stress was 7.9 N m^{-2}, glucose loading rate was 4.3 g m^{-2} s^{-1}, and temperature was 30°C. (Reprinted with permission from Picologlou et al., 1980.)

regularities through the viscous sublayer. Until the critical thickness is reached, the roughness peaks are smaller than the viscous sublayer thickness and the friction factor does not increase (the tube is hydraulically smooth). At a wall shear stress of 6.5–7.9 N m^{-2}, the viscous sublayer is approximately 40 μm, which compares well with the observed critical thickness (30–35 μm) for the same wall shear stress range (Figure 9.18). Good correlation has been observed between the viscous sublayer thickness and the critical biofilm thickness for a range of *Re* (Figure 9.19).

In turbulent flow in conduits with compliant boundaries, the possibility exists that when *Re* exceeds a certain value, rippling of the compliant boundary takes place, accompanied by drastic changes in friction factor (Schuster, 1971). Such a phenomenon would manifest itself, according to the preceding analysis, as a significant change in the equivalent sand roughness for the biofilm. Such transitions were not observed in our experiments for the range of *Re* numbers investigated. A single equivalent sand roughness at any one time was sufficient to correlate the friction factor and the *Re*.

The roughness of the clean tube may influence the progression of energy losses during biofilm accumulation. Two experiments conducted by Zelver

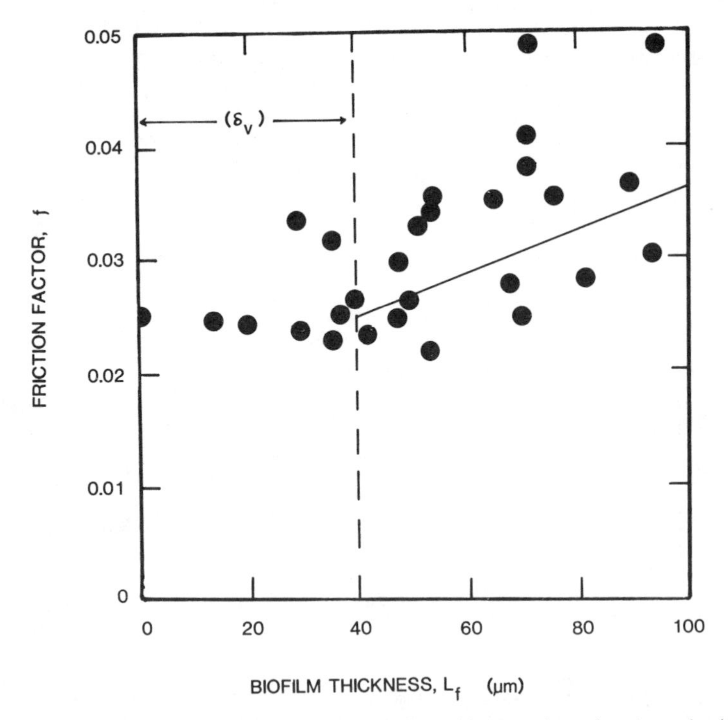

Figure 9.18 Change in friction factor with biofilm thickness in a circular tube (12.7 mm ID) with v_b = 1.8 m s^{-1} and T = 30°C. The critical biofilm thickness where friction factor begins to increase is approximately 30–35 μm. (Zelver, 1979.)

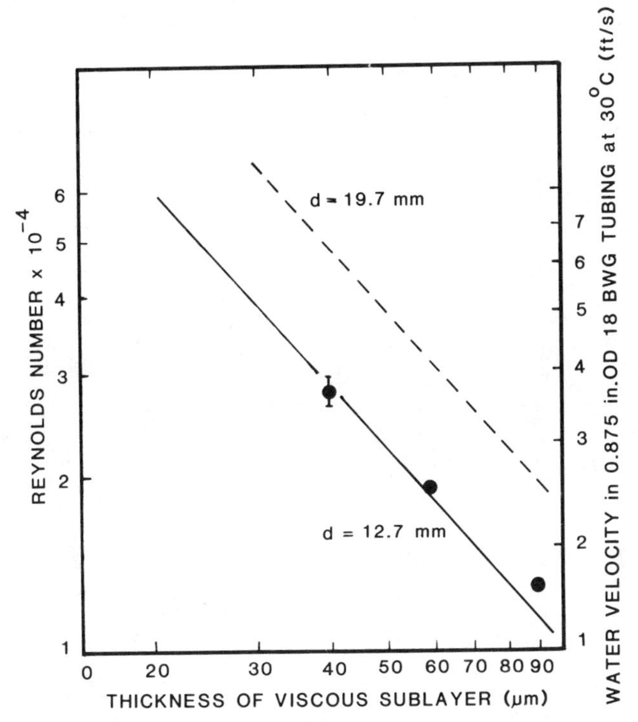

Figure 9.19 Relationship between the viscous sublayer and biofilm thickness at the start of increase in friction factor for experiments run over a range of Reynolds numbers. The lines represent the theoretical viscous sublayer thickness. The data points are observed critical biofilm thicknesses for onset of increased frictional resistance. d is the tube diameter.

(1979) were identical except that one surface was initially hydraulically smooth ($e_s/\delta_v = 0.20$) while the other was initially fully rough ($e_s/\delta_v = 36$). The fully rough condition was due to sand grains (average diameter 220 μm) immobilized on the inside surface. The following results were evident (Figure 9.20):

1. The initial (clean condition) friction factor was greater in the rough tube, and the frictional resistance remains greater there at all times.
2. The frictional resistance is reduced slightly during the first 30 hours in the rough tube.

The initial decrease in frictional resistance in the rough tube suggests that the biofilm developed between the sand grains and "smoothed" the rough surface for approximately 30 hours. Thus, erosion of biofilm from the rough surface may not be significant until the biofilm accumulates beyond the roughness elements.

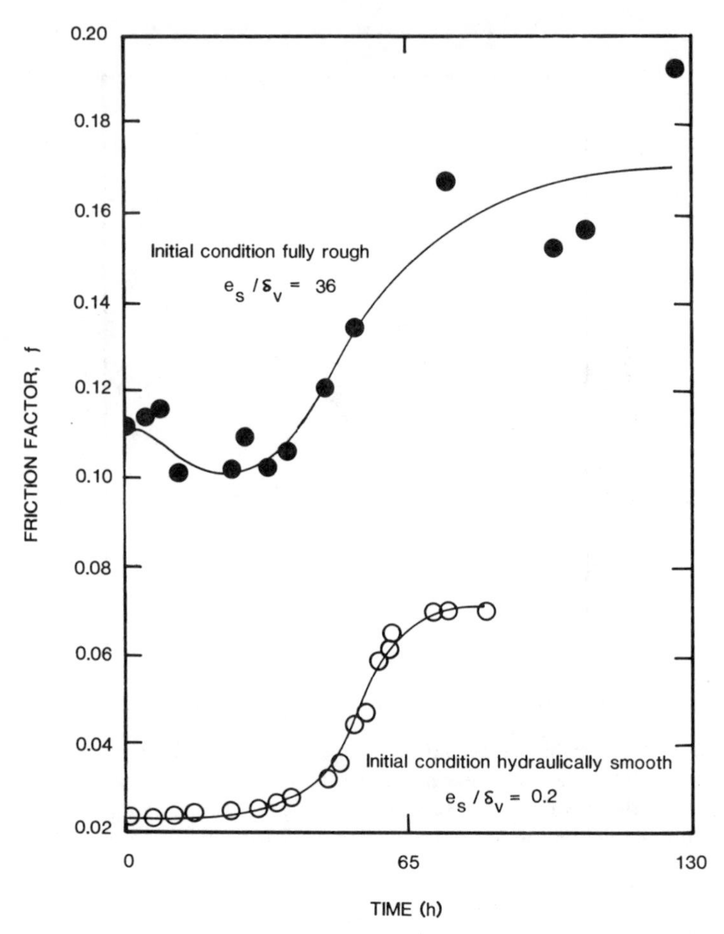

Figure 9.20 Comparison of friction factor progression in a preroughened tube (initial $e_s/\delta_\nu = 36$) and in a smooth tube (initial $e_s/\delta_\nu = 0.2$). Initial water velocity was 1.8 m s^{-1} in both tubes, shear stress was 7.9 N m^{-2}, glucose loading rate was 0.86 g m^{-2} s^{-1}, and temperature was 30°C. (Zelver, 1979.)

Although the frictional resistance effects of biofilm can be adequately described by formulae suitable for rigid rough surfaces, the conclusion should not be made that, indeed, the biofilm presents a rigid rough surface to the flow. Such a notion is an oversimplification and cannot account for all experimental observations.

9.5.3.2 Other Possible Mechanisms Contributing to Energy Losses

Biofilm Creep. Brauer (1963) performed experiments on form stability of the interior of asphalt-lined pipes as a function of the parameters of the flowing water. At higher temperatures, the asphalt coating assumed a rippled surface structure that was accompanied by an unusual increase in frictional resistance. Brauer explained the phenomenon as an actual flow of the coating under the action of fluid shear stresses. Energy is dissipated by the asphalt being dragged along the pipe surface. Transport of biofilm in tubes by creep seems an unlikely explanation for the high frictional resistance observed by Picologlou et al (1980), because the biofilm coating always appeared uniform throughout the tubing (biofilm transport would require a steady supply of film, or the wall coating would disappear).

Biofilm Oscillation. The possibility exists for a viscoelastic surface to absorb energy from the flowing fluid, such energy being eventually dissipated through viscous action. Rheological measurements performed on biofilm grown on platens of a Weissenberg rheogoniometer established the viscoelastic nature of the biofilm (Table 4.5, Chapter 4). As a result of the large viscous modulus (the viscous modulus was larger than the elastic modulus at all frequencies tested between 7 and 12 Hz), the possibility exists that the biofilm draws energy from the flow, such energy being eventually dissipated through viscous action. The situation is quite complex and defies analysis, particularly since there is a nonlinear coupling between the structure of the turbulent flow and the biofilm response. Nevertheless, oscillation or resonance of the biofilm in the turbulent flow could be a contributor to the energy losses observed.

Fluid Viscosity. The fluid viscosity in described experiments varied no more than 2.0% from water for any experiment, so it did not contribute significantly to the energy losses observed.

Filamentous Organisms. McCoy and Costerton (1982) measured filaments of attached *Sphaerotilus natans* and correlated filament development and size with increase in friction factor in an admiralty brass tube (14.2 mm ID). The results indicate that growth of the filamentous biofilm was responsible for an increase in f. Picologlou et al. (1980) also observed filaments fluttering with a frequency related to the fluid velocity, and the frictional resistance increased with increasing filament length. Such increases are analogous to increased drag in streams due to bottom vegetation, and similar phenomena occur in the study of atmospheric boundary layers due the presence of grassy vegetation. There can be a significant contribution of filamentous organisms to increases in frictional resistance.

9.5.3.3 Influence of Biofilms on Fluid Frictional Resistance in the RotoTorque The hydrodynamics of the RotoTorque reactor were described in Example 9C. Trulear (1980), experimenting with undefined mixed population biofilms, observed the increase in friction factor f_R as a function of biofilm thickness in several experiments with varying substrate (glucose) loading rates (Figure 9.21). The results clearly indicate a viscous sublayer (based on the critical

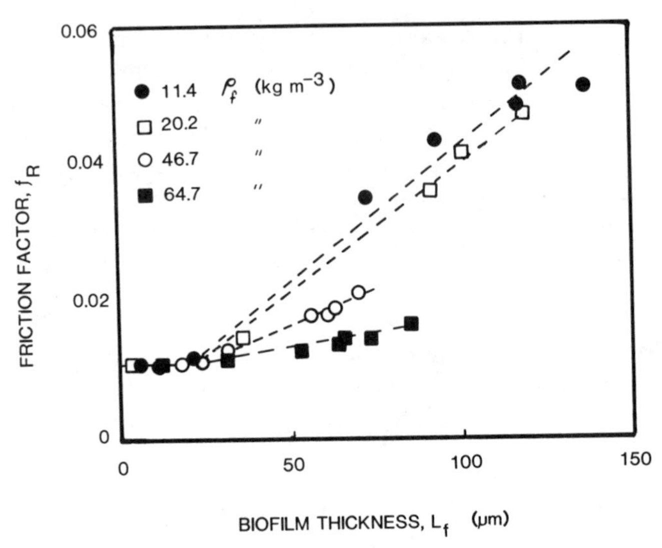

Figure 9.21 Relationship between friction factor and biofilm thickness in a RotoTorque. The critical biofilm thickness for onset of frictional resistance is approximately 30 μm. The increasing slopes of the lines are inversely proportional to the measured biofilm density [11.4, 20.2, 46.7, 64.7 kg m^{-3}]. (Trulear, 1980.)

biofilm thickness) on the order of 30–40 μm thick. The results are analogous to those for the tube presented in Figure 9.18. The data also indicate that the biofilm density significantly influenced the roughness of the biofilm. At the lower substrate loading rates, filamentous organisms were easily detected in the biofilm community, and increased frictional resistance was observed for a given biofilm thickness. The increased f_R was possibly due to the filamentous organisms. However, since the measured biofilm thickness (by the volumetric displacement method—Chapter 3) did not take full account of the filaments, the increased f_R may be due to underestimating the biofilm thickness.

In subsequent experiments with *Ps. aeruginosa* biofilms in the RotoTorque, Trulear (1983) observed no increase in fluid frictional resistance. Apparently, the effective roughness of the *Ps. aeruginosa* biofilm was not sufficient to influence hydrodynamics at the biofilm–bulk water interface.

9.5.4 Summary

A minimal understanding of hydrodynamics is essential to understanding and appreciating biofilm behavior. For example, fluid shear forces may control the ultimate accumulation by influencing biofilm detachment. Are there ways to control biofilm at an optimal thickness for a wastewater treatment process? On the other hand, biofilm accumulation in a pipe can lead to dramatic increases in fluid frictional resistance and its costly consequences (Chapter 14). Can fouling biofilm thickness be minimized by fluid shear forces? In addition, hydrodynamics influences both mass and heat transfer rate, significant industrial concerns related to biofilm accumulation. Thus, hydrodynamics is an essential component of any biofilm analysis, and the reader should be familiar with the subject.

9.6 HEAT TRANSPORT

Heat is only one form of energy and, according to the first law of thermodynamics, it is energy and not heat that is conserved. Nevertheless, in biofilm systems, the other forms of energy are generally negligible. Heat transport occurs frequently in nature and in process plants, e.g., the cooling of water as it exits a hot spring in Yellowstone Park or the heating of river water as it passes through a power plant condenser. There are two mechanisms for heat transport that are relevant to biofilm systems: (1) *conduction,* which is the transport of heat from a high temperature to a low temperature within a phase such as a solid or fluid by the motion of molecules or electrons, and (2) *advection,* which is the transport of heat that results from bulk fluid motion, usually in connection with interphase transport.

9.6.1 Intraphase Transport: Conduction

Conduction is simplest to study in solids, because no advection is possible. We have already considered such a problem in Example 9A. The heat transfer rate can be expressed as follows:

$$\text{heat transfer rate} = \frac{\text{driving force}}{\text{resistance}} \qquad [9.28]$$

The driving force is the temperature difference across the solid in Example 9A. The resistance quotient is defined from Fourier's law. Consider the rectangular slab in Example 9A:

$$Q_H = k_T A_x \frac{T_1 - T_2}{L_x} \qquad [9.29]$$

By comparing Eqs. 9.28 and 9.29 and noting that the driving force is $T_1 - T_2$, the resistance to heat transfer can be deduced:

$$R_t = \text{resistance to heat transfer} = L_x / k_T A_x \qquad [9.30]$$

So

$$Q_H = \frac{T_1 - T_2}{R_t} \qquad [9.31]$$

There can be more than one resistance to heat transfer in series. Since the tubular geometry is extremely important in heat transfer problems, an example of heat *conduction* in a tube follows (Example 9D). It considers a tube consisting of two composite materials. The example illustrates the influence of biofilms on heat transfer (Chapter 14).

Example 9D. Conductive Heat Transfer in Composite Tubular Walls. A composite cylindrical tube wall consists of two different materials: a metal alloy and a biofilm (Figure 9.22). Each material has a characteristic thermal conductivity. Fluid is flowing inside the tube and over the outside of the tube. The system is at steady state with the heat flux at $r_0 q_{H0}$. A differential thermal energy balance on a shell of volume $2\pi r L \Delta r$ for the tube (t) region results in

$$r_0 q_{H0} = -k_{Tt} r \frac{dT^{(t)}}{dr} \qquad [9D.1]$$

The thermal conductivity, k_{Tt}, is the proportionality constant in Eq. 9D.1 and is a physical characteristic or property of the material. k_{Tt} does not vary a great deal with temperature and is usually considered constant within small temperature ranges. Values of k_{Tt} for various materials are presented in Table 4.6 (Chapter 4). Confidence limits on the values can be as much as $\pm$15–20%.

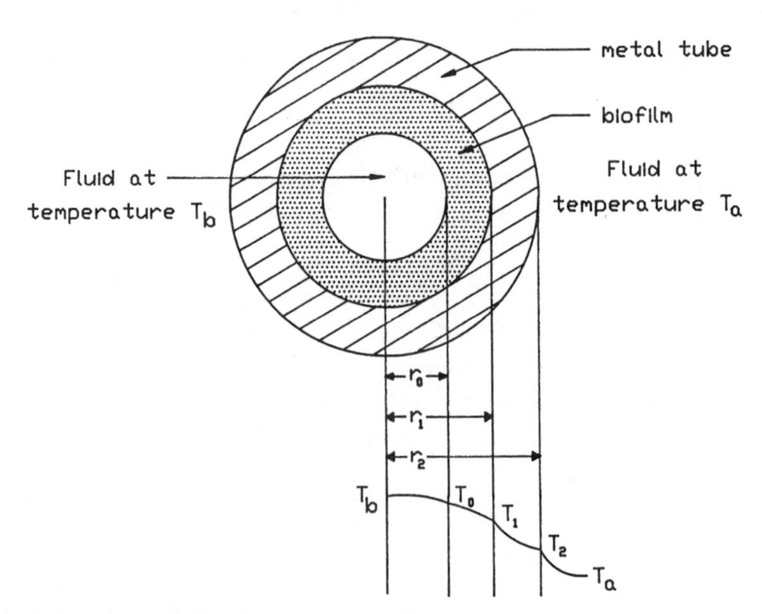

Figure 9.22 Heat transfer through a composite tube composed of a metal tube coated with a biofilm on the inside surface. The fluid on the inside of the tube is at temperature T_a, and the fluid on the outside is at temperature T_b.

A similar balance over the biofilm (f) region results in the following:

$$r_0 q_{H0} = -k_{Tf} r \frac{dT^{(f)}}{dr} \qquad [9D.2]$$

Integration of these equations for constant thermal conductivity gives

$$T_0 - T_1 = r_0 q_{H0} \frac{\ln(r_1/r_0)}{k_{Tf}} \qquad [9D.3]$$

$$T_1 - T_2 = r_0 q_{H0} \frac{\ln(r_2/r_1)}{k_{Tt}} \qquad [9D.4]$$

At the two solid–fluid interfaces (see Section 9.6.2)

$$T_b - T_0 = \frac{q_{H0}}{h_0} \qquad [9D.5]$$

$$T_2 - T_a = \frac{q_2}{h_2} = \frac{q_{H0} r_0}{h_2 r_2} \qquad [9D.6]$$

Addition of Eqs. 9D.3–9D.6 gives an expression for $T_b - T_a$, which is solved for q_{H0}. Multiplication of q_{H0} by the heat transfer surface area gives Q_{H0}:

$$Q_{H0} = 2\pi r_0 q_{H0} = \frac{2\pi L(T_b - T_a)}{\dfrac{1}{r_0 q_{h0}} + \dfrac{\ln(r_1/r_0)}{k_{Tf}} + \dfrac{\ln(r_2/r_1)}{k_{Tt}} + \dfrac{1}{r_2 h_2}} \qquad [9D.7]$$

An overall heat transfer coefficient, U, can be defined as follows:

$$Q_{H0} = U(2\pi r_0 L)(T_b - T_a) \qquad [9D.8]$$

By combining Eqs. 9D.7 and 9D.8,

$$U = r_0^{-1}\left[\frac{1}{r_0 h_0} + \frac{\ln(r_1/r_0)}{k_{Tf}} + \frac{\ln(r_2/r_1)}{k_{Tt}} + \frac{1}{r_2 h_2}\right]^{-1} \qquad [9D.9]$$

where U is calculated based on r_0. Thus, there are four resistances to heat transfer in this system, represented by the four terms in brackets in Eq. 9D.9: (1) advective heat transfer in the tube, (2) conductive resistance in the biofilm, (3) conductive resistance in the tube, and (4) advective resistance on the outside of the tube, respectively.

The conductive resistance of the biofilm and the tube wall are represented by the second and third terms on the right hand side of Eq. 9D.9 and can be compared by considering the relative thicknesses of the two materials necessary to provide the same resistance to heat transfer. Suppose the tube is titanium 1 mm thick and has an inside radius of 12 mm. From Table 4.6 (Chapter 4), the thermal conductivities for biofilm and a titanium tube are

$$k_{Tf} = 0.7 \text{ W m}^{-1} \text{ K}^{-1}, \qquad k_{Tt} = 16.44 \text{ W m}^{-1} \text{ K}^{-1}$$

Then, a biofilm approximately 41 μm thick offers the same resistance to heat transfer as a titanium tube 1000 μm thick. The result clearly illustrates the detrimental effect of even a thin biofilm on heat transfer.

9.6.2 Advection

Not only can heat be transported by temperature gradients, but it can also be transported by bulk fluid motion. Heat transfer in a tubular geometry with advective transport will be considered in Example 9E. In this case, the total heat flux has contributions from two sources:

$$\underset{\text{total flux}}{q_{Hx}^{\text{tot}}} = \underset{\text{conductive flux}}{q_H} + \underset{\text{advective flux}}{(\rho\, C_p T)\, v_x} \qquad [9.32]$$

Generally, the advective heat flux is important in dealing with heat transfer at interfaces.

9.6.2.1 *Heat Transfer Coefficients in Smooth Tubes* In engineering design problems the transport of heat from a solid–liquid interface, i.e., a boundary between two phases, must frequently be determined. This is in contrast to the transport of heat by conduction, which is within one phase. Experimentally, the temperature profile in turbulent flow has been found to be very flat away from the wall (Figure 9.6). Since the major resistance to transfer is near the wall, engineers generally postulate that all the resistance occurs in the thin film of fluid near the wall. Outside of this film, the temperature is the uniform bulk temperature. The film thickness, δ_T, is called the effective heat transfer film thickness or the *thermal* or *heat transfer* boundary layer, and the heat transfer coefficient is frequently termed the film coefficient. The heat transfer coefficient is analogous to the friction factor in that it describes the interphase transport of heat in turbulent flow (Section 9.4.2.2). However, the heat transfer coefficient, unlike the friction factor, is dimensional. Consequently, it is customary to use a dimensionless version called the Nusselt number, Nu (Eq. 9.20), for calculations.

Consider the flow of a cooling water through a heat exchange tube, a very important practical geometry (Figure 9.23). Assume the velocity distribution at

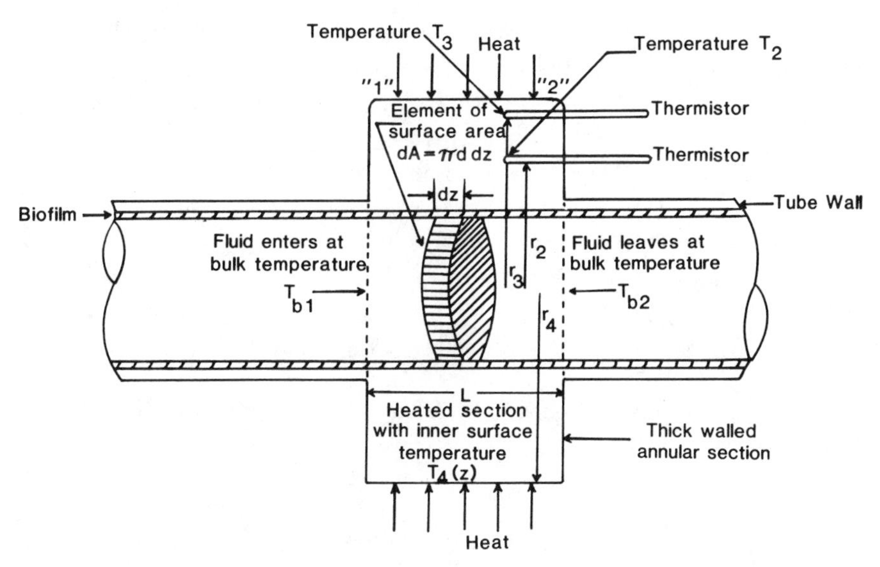

Figure 9.23 Schematic representation of a thick-walled heat exchanger with biofilm on the inner tube (radius R) surface. Fluid enters from the left. The heated section begins at location 1 and ends at 2 with length L. Two thermistors are placed at r_2 and r_3.

location 1 is known and the wall temperature T_1 in the heated section from $z = 0$ to $z = L$ is constant. For a tube of radius R and length L, the total heat flow into the water at the pipe wall is

$$Q_H = \int_0^L \int_0^{2\pi} + k_T \left.\frac{\partial T}{\partial r}\right|_{r=R} R \, d\Theta \, dz \qquad [9.33]$$

which is valid for laminar or turbulent flow. The flux sign is positive because heat is being added to the system in the $-r$ direction. The heat transfer coefficient can be defined as follows:

$$Q_H = h_1(2\pi RL)(T_1 - T_{b1}) \qquad [9.34]$$

or

$$Q_H = h_{\ln}(2\pi RL)\left(\frac{(T_{11} - T_{b1}) - (T_{12} - T_{b2})}{\ln[(T_{11} - T_{b1})/(T_{12} - T_{b2})]}\right) \qquad [9.35]$$

where h_1 is based on the initial temperature difference $(T_1 - T_b)_1$, and $h_{\ln}$ is based on the logarithmic mean temperature difference $(T_1 - T_b)_{\ln}$. $h_{\ln}$ is frequently preferable because it is less dependent on L/d than h_1. Thus, if we equate Q_H from Eqs. 9.33 and 9.34 (or Eq. 9.35), we get

$$h_1 = \frac{1}{2\pi RL(T_1 - T_{b1})} \int_0^L \int_0^{2\pi} + k_T \frac{\partial T}{\partial r}\bigg|_{r=R} R\, d\Theta\, dz \qquad [9.36]$$

Then we define dimensionless variables as follows:

$$r^* = \frac{r}{R}, \qquad z^* = \frac{z}{L}, \qquad T^* = \frac{T - T_1}{T_{b1} - T_1}$$

Multiplying both sides of Eq. 9.36 by $2R/k_T$ after inserting the dimensionless variables gives

$$Nu_1 = \frac{1}{2\pi L/d} \int_0^{L/d} \int_0^{2\pi} - \frac{\partial T^*}{\partial r^*}\bigg|_{r^*=0.5} d\Theta\, dz^* \qquad [9.37]$$

Thus, the Nusselt number is a dimensionless temperature gradient averaged over the heat transfer surface. The heat transfer coefficient is a function of the geometry, the flow rate, the fluid properties, and even the roughness of the heat transfer surface. Consequently, Nu has been related to Re, which incorporates flow rate and geometry, while the fluid properties (C_p, η, and k_T) are combined into the dimensionless Prandtl number Pr:

$$Nu_1 = Nu_1(Re, Pr, L/d) \qquad [9.38]$$

where $Pr = C_p\eta/k_T$

The effects of temperature on viscosity can be significant in some cases, and then there is an additional parameter in the Nu relationship.

Several correlations between Nu and Re have been developed which are analogous to the friction factor-Re correlations (Figure 9.8). The Sieder–Tate (1936) relation is presented in Figure 9.24. For highly turbulent flows, the curves for $L/d > 10$ converge to a single line which, for $Re_b > 20{,}000$, is described as follows:

$$\frac{hd}{k_{Tb}} = 0.026\left(\frac{dF}{\eta}\right)^{0.8}\left(\frac{C_p\eta}{k_T}\right)^{0.33}\left(\frac{\eta}{\eta_0}\right)^{0.14} \qquad [9.39a]$$

$$Nu = 0.026 Re_b^{0.8} Pr^{0.33}\left(\frac{\eta}{\eta_0}\right)^{0.14} \qquad [9.39b]$$

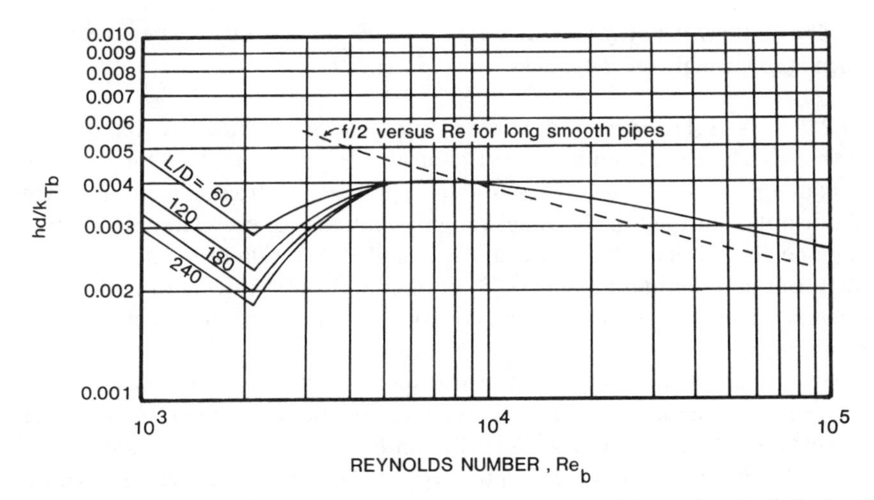

Figure 9.24 Heat transfer coefficients as a function of Reynolds number for fully developed flow in smooth circular pipes. L/D is the ratio between the pipe length and diameter. (Reprinted with permission from Sieder and Tate, 1936.)

where Re_b = Reynolds number (DG/η) based on bulk water properties (—)
F = mass flow rate of water (Mt^{-1})
η_0 = water viscosity at the heated wall ($ML^{-1}t^{-1}$)
k_{Tb} = thermal conductivity of bulk water ($MLt^{-3}T^{-1}$)

Equation 9.39 is accurate ±20% in the range $Re_b = 10^4$–10^5, $Pr = 0.6$–100, and $L/d > 10$. For $Re_b > 10{,}000$, the heat transfer ordinate is approximately equal to $f/2$ for long, smooth pipes. Thus, Colburn proposed the following empirical analogy in this range of operation:

$$j_H = \frac{f}{2} \qquad [9.40]$$

and

$$j_H = \frac{h}{C_p G}\left(\frac{C_p \eta}{k_T}\right)^{0.67}_{\text{film}} \qquad [9.41]$$

or, equating Eqs. 9.40 and 9.41,

$$h = 0.031 f C_p^{0.33} \eta^{-0.67} k_T^{0.67} \rho_f v \qquad [9.42]$$

where "film" refers to properties evaluated at the "film temperature" $(T_b + T_0)/2$, in which T_b and T_0 are averages of the terminal values; C_p is evaluated at

T_b. The analogy is not valid below $Re = 10{,}000$ or for rough tubes, because f depends so strongly on roughness.

9.6.2.2 *Heat Transfer Coefficients in Rough Tubes* Advective heat transfer can be calculated with the Colburn or Reynolds analogy for turbulent flow through *smooth* tubes. When biofilm is present, the friction factor increases significantly and the tube becomes rough. A number of correlations have been proposed for advective heat transfer in rough tubes (Dipprey and Sabersky, 1963; Bott and Gudmundsson, 1978; Walker and Bott, 1973),and each gives a different dependence on friction factor, the primary parameter in the calculation.

All correlations for advective heat transport in tubes with significant roughness require some knowledge of the surface roughness height ϵ. A roughness Reynolds number, Re_r, is a useful criterion for determining the roughness regime (Walker and Bott, 1973; Dipprey and Sabersky, 1963; Picologlou et al., 1980):

$$Re_r = Re \cdot \frac{e}{d}\left(\frac{f}{8}\right)^{1/2} \qquad [9.43]$$

where e is the height of the roughness elements. Re_r divides the flow into three regimes (Nikuradse, 1933):

$$\text{smooth:} \qquad Re_r < 5 \qquad [9.44]$$

$$\text{transition:} \qquad 5 < Re_r < 70 \qquad [9.45]$$

$$\text{fully rough:} \qquad Re_r > 70 \qquad [9.46]$$

In the smooth regime, roughness elements protrude only slightly into the viscous sublayer. In the transition regime, the roughness elements are approximately the same height as the viscous sublayer, and in the rough regime, some roughness elements penetrate beyond the viscous sublayer into the fully turbulent core.

In a series of experiments with rough tubes, Walker and Bott (1973) developed a correlation for Nusselt number and friction factor:

$$Nu = 0.0114(f/8)^{1/2} Re^{1.174} Pr^{1/2} \qquad [9.47]$$

The above correlation was obtained for a range of Reynolds and Prandtl numbers using drawn stainless steel tubes of various sizes and manufactured by different methods. Nearly all of these data fall in the transition regime for roughness, with a roughness height from 2.5–18 μm ($e/d = 0.0002$–0.0008). Since bacterial cells are the order of 0.5–3.0 μm in diameter, a bacterial colony 8–30 cells high can produce an equivalent range of e/d-values.

Dipprey and Sabersky (1963) developed the following similarity law for correlating the effects of surface roughness with advectional heat transfer and frictional losses:

$$\frac{f/2\,St - 1}{\sqrt{f}} + A = g(Re_r, Pr) \qquad [9.48]$$

where $A = 8.48$, a constant, and g is a similarity function of Re_r and Pr. For fully developed roughness,

$$g(Re_r, Pr) = K_f(Re_r)^p(Pr)^m \qquad [9.49]$$

where p, m = constants universally valid for fully rough hydraulic behavior ($p = 0.2$, $m = 0.44$)

$$K_f = \begin{cases} 5.19 & \text{for close-packed roughness} \\ 6.37 & \text{for two dimensional roughness} \end{cases}$$

Thus, the heat transfer similarity law for "fully rough" flow is

$$h = vC_p \frac{f/8}{1 + (f/8)^{1/2}[K_f[(f/8)^{1/2} \cdot Re \cdot e/d]^{0.2} Pr^{0.44} - 8.48]} \qquad [9.50]$$

The correlation requires knowledge of the roughness height e for effective prediction. A number of additional correlations have been developed which relate the roughness height to the friction factor (f):

1. Dipprey and Sabersky (1963):

$$\frac{e}{d} = \exp\left(\frac{3.0 - (f/8)^{-1/2}}{2.5}\right) \qquad [9.51]$$

2. Colebrook and White (1939) (adapted from Picologlou et al., 1980):

$$\frac{e}{d} = \frac{1}{2}\left(10^{(0.87 - 4f^{-1/2})} - \frac{18.70}{Re\,f^{1/2}}\right) \qquad [9.52]$$

3. Bott and Gudmundsson (1978):

$$\frac{e}{d} = \frac{1}{2}\exp\left(-\frac{0.75 + (f/8)^{-1/2}}{2.5}\right) \qquad [9.53]$$

The Dipprey-Sabersky roughness height correlation is for sand grain roughness and fully rough flow. The Colebrook-White equation is suitable for the smooth, transition, and fully rough regimes, and yields an equivalent sand grain roughness. While the Colbrook–White equation covers a wide range of flow regimes, it can result in roughness heights less than zero at some conditions. For $f = 0.04$, the Dipprey–Sabersky correlation is in agreement with the Colebrook–White equation. The correlation by Bott and Gudmundsson is used for rippled silica deposits of different geometry (i.e., transverse ripples, rather than random or evenly spaced protuberances) and larger e than the other correlations. The biofilm may have elastic properties (Picologlou et al., 1980), resulting in calculated equivalent sand grain roughness approximately three times the biofilm thickness, a plausible result if the film were to flex and deform under the action of turbulent bursts.

The correlation of Walker and Bott (1973) appears suitable for roughness induced by biofilm accumulation. Their equation is based on a number of observations of heat transfer resistance in "naturally" rough tubes of stainless steel and titanium:

$$Nu = 0.0114(f/8)^{0.5} Re^{1.174} Pr^{0.5} \quad [9.54]$$

The Walker–Bott correlation was determined for surface roughness ranging from 3 to 20 μm (microbial cell diameters are approximately 1 μm). Thus, even sparsely distributed cell colonies are within the transition regime for heat transfer at the wall.

9.6.3 Influence of Biofilms on Heat Transport

Conductive heat transfer resistance results from the insulating layer formed by the biofilm and generally increases as the biofilm accumulates. It is the difference between overall and advective heat transfer resistance and can be calculated *a priori* if the biofilm thickness and thermal conductivity are known. The conductive heat transfer resistance is dependent on the thermal conductivity and thickness of the deposit. Advective heat transfer resistance results from fluid motion or turbulence and generally decreases as biofilm accumulates, since the roughness of the biofilm increases turbulence in the interfacial region. Advective heat transfer resistance can be calculated from the friction factor and the properties of water. Thus, the advective heat transfer resistance depends on the roughness characteristics of the biofilm and the shear stress at the heat transfer surface.

The overall heat transfer resistance determines the influence of biofilm on heat exchanger performance. However, advective and conductive heat transfer resistance may be important because differences in deposit properties (apparent roughness and thermal conductivity) can result in significant differences in the contribution of conductive and advective processes to overall heat transfer resistance.

Measurements related to biofilm accumulation and general fouling processes include monitoring overall heat transfer resistance and characterizing the biofilm or fouling deposit. A variety of different monitoring devices have been developed to obtain accurate fouling data under various experimental conditions. A detailed description of some of the devices can be found elsewhere (Knudsen, 1981). One type of instrument is the thick-walled heat exchanger (Characklis et al., 1981; Turakhia and Characklis, 1984; Zelver et al., 1985). Constant heat is supplied to the thick-walled heat exchanger (TWHE), and the overall heat transfer resistance is measured. In addition, the flow rate, pressure drop, and biofilm thickness can be measured. Figure 9.23 is a schematic cross-sectional diagram of the TWHE, which consists of a thick metal annular block clamped firmly to a heat exchange tube. The heat transferred at the outside annular surface is determined by measuring the temperature at two radial positions (r_2 and r_3) within the metal block:

$$q = \frac{k_{T\text{block}}\, 2\pi L(T_3 - T_2)}{\ln(r_3/r_2)} \qquad [9.55]$$

The contact resistance between the metal block and the tube alloy can be determined using a graphical technique developed by Wilson (1915). The Wilson plot calibration will give the total resistance between the inner temperature thermistor and the inner surface of the tube. The overall heat transfer resistance is determined from measurements as follows:

$$U^{-1} = \left[(T_2 - T_b) - \frac{q \ln(r_2/R_0)}{2\pi L k_{T\text{block}}}\right] \frac{2\pi R_0 L}{q} \qquad [9.56]$$

where R_0 is the outside diameter of the tube.

In terms of processes, the overall heat transfer resistance (U^{-1}) is the sum of conductive and advective resistances. For a tube with biofilm,

$$U^{-1} = \left[\underbrace{\frac{R_0}{(R_1 - L_f)h}}_{\text{advective resistance}} + \underbrace{\frac{R_0 \ln[R/(R - L_f)]}{k_{Tf}}}_{\text{conductive resistance of the biofilm}} + \underbrace{\frac{R_0 \ln(R_0/R)}{k_{T\text{tube}}}}_{\text{conductive resistance of the tube}}\right] \qquad [9.57]$$

If we ignore the tube resistance because it is constant,

$$U^{-1} = \left[\frac{R_0}{(R - L_f)h} + \frac{R_0 \ln[R/(R - L_f)]}{k_{Tf}}\right] \qquad [9.58]$$

and

$$U^{-1} = U_{\text{adv}}^{-1} + U_{\text{cond}}^{-1} \qquad [9.59]$$

Therefore, the overall heat transfer resistance is represented by U^{-1}, while the advective heat transfer resistance is determined by the reciprocal of the heat transfer coefficient:

$$U_{\text{adv}}^{-1} = \frac{R_0}{(R - L_f)h} \qquad [9.60]$$

The conductive heat transfer resistance *due to the biofilm* is

$$U_{\text{cond}}^{-1} = U^{-1} - U_{\text{adv}}^{-1} - \frac{R_0 \ln(R_0/R)}{k_{T\text{tube}}} \qquad [9.61]$$

In a clean, smooth tube, the second term in Eq. 9.57 is zero and h is predicted from the Colburn analogy (Eq. 9.42). As biofilm accumulates, the overall heat transfer resistance (U^{-1}) for this tube changes due to the increased biofilm conductive resistance and decreased advective heat transfer resistance resulting from increasing biofilm roughness. Characklis et al. (1981) have determined the relative changes in conductive and advective heat transfer resistance in a tube during biofilm accumulation (Figure 9.25). At the beginning of the experiment, the conductive resistance due to the biofilm is zero and only advective resistance exists. As biofilm accumulates, the conductive resistance increases in proportion to the biofilm thickness. Advective resistance decreases as a result of increased roughness, reflected by an increase in friction factor. The Colburn analogy was

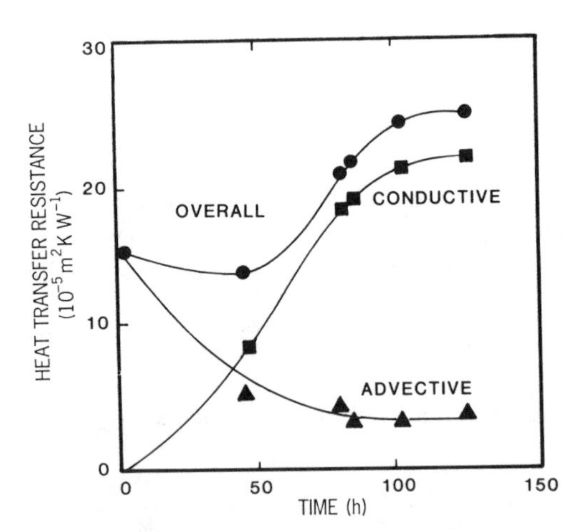

Figure 9.25 Progression of overall, conductive, and advective heat transfer resistance as a biofilm accumulates in a heated circular tube. Conductive heat transfer resistance (■), increases and dominates the overall heat transfer resistance (●), despite a decrease in advective heat transfer resistance (▲), resulting from increased roughness of the biofilm. Reprinted with permission from Turakhia and Characklis, 1984.)

used to calculate the advective resistance from friction factor measurements, the result is only an approximation to the actual advective resistance. The biofilm thermal conductivity was measured in separate experiments.

The thermal conductivity of the biofilm, k_{Tf}, can be determined by measuring the overall heat transfer coefficient U at a constant heat flux if k_{tube} and h have been determined. If $T_2 - T_b$ varies insignificantly along the longitudinal distance L, then

$$k_{Tf} = \frac{R_0 \ln[R/(R - L_f)]}{U_{\text{cond}}^{-1}} \qquad [9.62]$$

If the biofilm contains precipitates (e.g., scale) and/or corrosion products, a different rate and extent of heat transfer resistance progression will be observed. For example, in case of a pure scale, the increase in overall heat transfer resistance is largely due to increase in conductive heat transfer resistance (Figure 9.26) because scale generally exhibits a low relative roughness and the advective resistance remains essentially constant. In addition, the thermal conductivity of calcium carbonate scale is considerably higher than that for biofilm. Measured values for the thermal conductivity and relative roughness of biofilm and $CaCO_3$ scale are compared in Table 9.5.

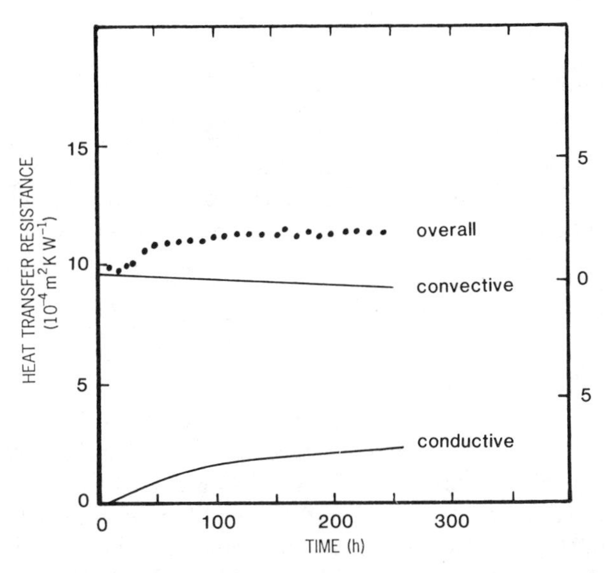

Figure 9.26 Progression of overall, conductive, and advective heat transfer resistance as a calcium carbonate scale accumulates in a heated circular tube. The influence of the scale on advective heat transfer resistance is essentially negligible. (Reprinted with permission from Turakhia and Characklis, 1984.)

TABLE 9.5 A Comparison of Measured Values for Thermal Conductivity and Relative Roughness of Biofilm and $CaCO_3$ Deposits[a]

Deposit	Thickness (μm)	Relative Roughness (dimensionless)	Thermal Conductivity ($W\ m^{-1}\ K^{-1}$)	Reference
Biofilm	40	0.003		Characklis, 1981a
	165	0.014		Characklis, 1981a
	300	0.062		Characklis, 1981a
	500	0.157		Characklis, 1981a
$CaCO_3$	165[b]	0.0001		
	224[b]	0.0002		
	262[b]	0.0006		
$CaCO_3$			2.26–2.93	Sherwood et al., 1975
$CaSO_4$			2.31	Sherwood et al., 1975
$Ca_3(PO_4)_2$			2.60	Sherwood et al., 1975
$Mg_3(PO_4)_2$			2.16	Sherwood et al., 1975
Fe_2O_3 (magnetic)			2.88	Sherwood et al., 1975
Analcite			1.27	Sherwood et al., 1975
Biofilm	≈100		0.63	Characklis et al., 1981

[a] Turakhia and Characklis (1984).
[b] Calculated from overall heat transfer resistance assuming a thermal conductivity for $CaCO_3$ of 2.6 $W\ m^{-1}\ K^{-1}$.

9.6.4 Summary

The major concern relating heat transfer and biofilms is fouling of heat exchange equipment surfaces leading to costly reduction in performance. The topic of heat exchange fouling is considered in more detail in Chapter 14. Suffice it to say that biofilms can readily form the largest resistance to heat transfer in systems cooled by water within the temperature ranges conducive to microbial growth.

9.7 MASS TRANSPORT

Mass transport can influence the observed rate of various transformations in biofilm systems. For example, fixed film wastewater treatment processes are generally believed to be rate-limited by diffusion of oxygen or substrate through the biofilm. In rivers or streams, where substrate (e.g., organic carbon) concentrations are generally very low (i.e., oligotrophic), mass transport in the water phase undoubtedly influences the rate of transformations and hence controls the rate of the observed microbial activity. Transport of both soluble and particulate components are of importance in biofilm systems.

Mass is transported by two mechanisms (Eq. 9.4): (1) *diffusion*, which is the transport of mass within a phase from a high concentration to a low concentration in a solid or fluid, and (2) *advection* (sometimes termed convection), which results from bulk fluid motion and influences both intraphase and interphase transport. Most intraphase transport in biofilm systems will be related to the *base film*, while the interphase transport will generally be related to the *surface film* (Figure 1.2, Chapter 1).

9.7.1 Intraphase Transport: Diffusion (Water and Base Film)

9.7.1.1 *Bulk Water Compartment (Zero Flow)*

Molecular Diffusion in Water. Molecular diffusion in water (constant density) is described by Fick's first law of diffusion, which, for one dimensional transport, is described as follows:

$$J_{Sz} = -D_S \frac{dS}{dz} \qquad [9.63]$$

where J_{S_z} is the flux of S in the z-direction and D_S is the diffusion coefficient for S. Molecular diffusion coefficients in water for several important compounds are presented in Table 9.6.

Diffusion of Cells (Particles) in Water. Under laminar flow conditions, particles are transported by diffusion perpendicular to the flow. Nonmotile microorganisms with size 1–4 μm^3 display little Brownian motion and hence have

TABLE 9.6 Molecular Diffusion Coefficients in Water for Some Important Compounds[a]

Diffusing Species	Counter-ion	Diffusion Coefficient (10^{-10} m^2 s^{-1})	Temperature (°C)	Reference
Ammonia	Cl^-	17	18	Washburn (1929)
	SO_4^{-2}	9	20	
Carbon dioxide		19.5	25	Landoldt-Börnstein (1969)
Glucose		6.7	25	Weast (1973)
Methane		18.8	25	Witherspoon and Saraf (1965)
Methanol		15	25	Perry and Chilton (1973)
Nitrate	Na^+	15	25	Gray (1972)
Nitrogen		22	24	Landoldt-Börnstein (1969)
Oxygen		23	23.5	Landoldt-Börnstein (1969)
Raffinose		4.3	25	Gray (1972)
Sucrose		5.3	25	Gray (1972)

[a] After Harremöes (1978).

small Brownian diffusivity. The Stokes–Einstein equation describes the diffusion coefficient for spherical cells for which the no slip condition applies at the cell–water interface:

$$D_{XB} = \frac{k_B T}{3\pi\eta d_C} \qquad [9.64]$$

where D_{XB} = Brownian diffusion coefficient for cells (L^2t^{-1})
k_B = Boltzmann constant, 1.3805×10^{-23} J K^{-1}
d_C = cell diameter (L)
T = absolute temperature (T)

For a nonmotile, spherical microbial cell with a diameter of 1 μm, calculations yield $D_{XB} = 5.5 \times 30^{-13}$ m^2 s^{-1} at 30°C.

Cell motility has often been ignored, but may significantly affect the transport rate. The non-Brownian diffusivity incorporates cell motility and can be calculated from the velocity and the mean free path (Jang and Yen, 1985):

$$D_c = \frac{v_r d_r}{3(1 - \cos\alpha)} \qquad [9.65]$$

where D_c = non-Brownian diffusion coefficient for cells (L^2t^{-1})
v_r = velocity of motility (Lt^{-1})
d_r = free length of a random run (L)
α = turn angle (—)

$\cos\alpha$ can be assumed to be zero as long as chemotaxis is not occurring. Jang and Yen (1985) calculated the non-Brownian diffusivity for different microorganisms to be in the range of 0.4×10^{-9} to 5.6×10^{-9} m^2 s^{-1}. Vaituzis and Doetsch (1969) measured speeds up to 55.8 μm s^{-1} for *Pseudomonas aeruginosa* with "track photography." Their results suggest a mean free path of random run in the range of 50–85 μm. These values yield a non-Brownian diffusivity of approximately 1×10^{-9} m^2 s^{-1} for *Pseudomonas aeruginosa.* Comparing the calculated Brownian diffusivity ($\approx 5.5 \times 10^{-13}$ m^2 s^{-1}) from Eq. 9.64 with the calculated (Eq. 9.65) and measured (Vaituzis and Doetsch) non-Brownian diffusivity, we see that diffusive transport rate of cells may be underestimated by as much as 100 times if motility is ignored.

9.7.1.2 Base Film Compartment

Molecular Diffusion in the Base Film. Diffusion through the base film is generally expressed as diffusion through a porous solid where no advection is occurring. Molecular diffusion through the base film is also expressed by Eq. 9.10, except that the diffusion coefficient D in the base film replaces that in water. In most cases, we shall be interested in simultaneous diffusion and reaction in the biofilm, so that case is addressed in Example 9F. Diffusion and reaction in biofilms has been the subject of significant modeling efforts by a number of research groups in the past 25 years, including Busch (1971), Atkinson (1974), Harremöes (1978), and McCarty and coworkers (Williamson and McCarty, 1975, 1976; Rittmann and McCarty, 1980a, 1980b). Diffusion coefficients for various components in biofilm are reported in Table 4.7 (Chapter 4).

Example 9E. Molecular Diffusion and Reaction in a Biofilm (Half-Order Kinetics). Harremöes (1978) determined that the specific growth rate can be assumed zero order with respect to S in biofilms, since the intrinsic saturation coefficient is generally very small. Therefore, the biofilm volume in which non-zero order kinetics is valid is very small. The analysis also assumes that substrate is the only limiting factor for growth in the biofilm. Oxygen is presumed in excess throughout the biofilm volume. Then

$$\mu = \mu_{\max} \qquad [9E.1]$$

For one dimensional diffusion (x-direction) and steady state (Figure 9.27), Eq. 9.4 becomes

$$D_{Sf} \frac{d^2S}{dx^2} = r_{VS} \qquad [9E.2]$$

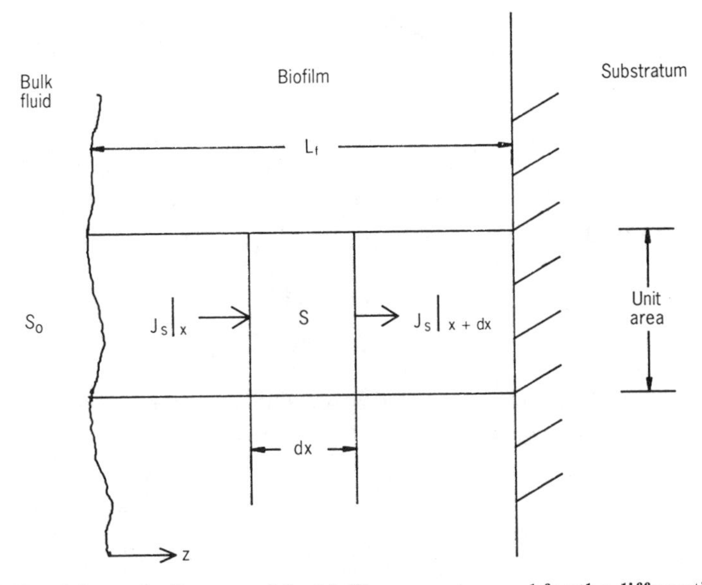

Figure 9.27 Schematic diagram of the biofilm geometry used for the differential material balance in Example 9E.

where r_{VS} is the volumetric rate of substrate removal in the biofilm. Define the following dimensionless variables:

$$S^* = \frac{S}{S_1}, \qquad x^* = \frac{x}{L_f} \tag{9E.3}$$

where S_1 = substrate concentration at the biofilm–bulk water interface (ML^{-3}). Since no mass transfer resistance in the water phase is indicated, S_1 also equals the bulk water concentration. Then

$$\frac{d^2S^*}{dx^{*2}} = \frac{r_{VS}L_f^2}{D_{Sf}S_1} \tag{9E.4}$$

and, since zero order kinetics are assumed,

$$r_{VS} = \frac{\mu_{\max} X_f}{Y_{X/S}} \tag{9E.5}$$

or

$$r_{VS} = q_{Sf,\max} X_f \tag{9E.6}$$

where X_f = biofilm density ($M_X L^{-3}$)
$q_{Sf,\max}$ = maximum specific substrate removal rate ($M_S M_X^{-1} t^{-1}$)

Then

$$\frac{d^2S^*}{dx^{*2}} = \frac{2}{\Omega_p^2} \tag{9E.7}$$

where Ω_p is the "penetration ratio":

$$\Omega_p = \left[\frac{q_{Sf,\max} X_f L_f^2}{D_{Sf}S_1}\right]^{-1/2} \tag{9E.8}$$

The boundary conditions for Eq. 9E.7 depend on whether there is complete ($\Omega_p > 1$) or incomplete ($\Omega_p < 1$) penetration of substrate. For $\Omega_p > 1$,

$$S^* = 1 \quad \text{at} \quad x^* = 0 \tag{9E.9a}$$

$$\frac{dS^*}{dx^*} = 0 \quad \text{at} \quad x^* = 1 \tag{9E.9b}$$

Then,

$$S^* = \frac{x^{*2}}{\Omega_p^2} - \frac{2x^{*2}}{\Omega_p^2} + 1 \qquad [9E.10]$$

The substrate concentration distribution in the biofilm for incomplete penetration of substrate is presented in Figure 9.28. For $\Omega_p < 1$,

$$S^* = 1 \quad \text{at} \quad x^* = 0 \qquad [9E.11a]$$

$$S^* = 0 \quad \text{at} \quad x^* = x_A^* \qquad [9E.11b]$$

where $x_A^* = L_{fA}/L_f$

L_{fA} = biofilm depth to which substrate "penetrates" (L)

Then,

$$S^* = \frac{x^{*2}}{x_A^{*2}} - \frac{2x^*}{x_A^*} + 1 \qquad [9E.12]$$

The substrate concentration distribution in the biofilm for complete penetration of substrate is shown in Figure 9.29.

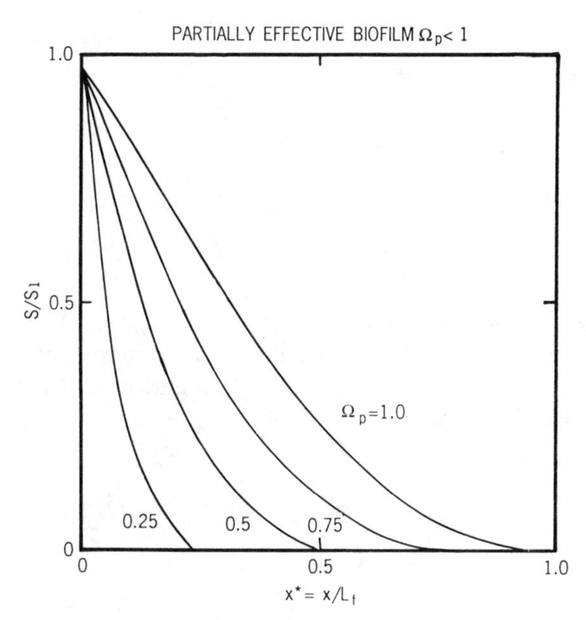

Figure 9.28 Substrate concentration distribution in a partially penetrated biofilm (after Harremöes, 1978).

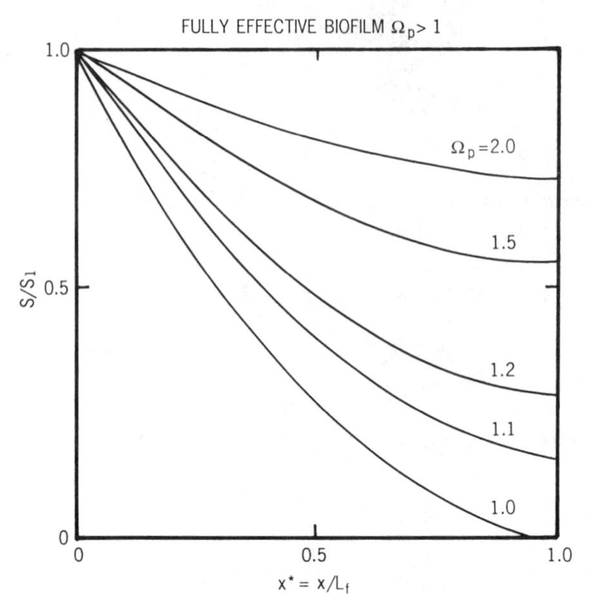

Figure 9.29 Substrate concentration distribution in a completely penetrated biofilm (after Harremöes, 1978).

The rate of substrate removal by the biofilm (flux) is expressed as follows:

$$J_{Sx} = -D_{Sf} \left.\frac{dS}{dx}\right|_{x=0} \tag{9E.13}$$

Then, for $\Omega_p > 1$,

$$J_{Sx} = q_{Sf,\max} X_f L_f \tag{9E.14}$$

which indicates zero order (with respect to bulk liquid S) substrate removal rate. Harremöes (1978), LaMotta (1976), and Trulear and Characklis (1982) have observed zero order kinetics in biofilm reactors with glucose (substrate) being consumed by an undefined multispecies biofilm.

For $\Omega < 1$,

$$J_{Sx} = (2D_{Sf} q_{Sf,\max} X_f S_1)^{1/2} \tag{9E.15}$$

or

$$r_S'' = k_{1/2} S_1^{1/2} \tag{9E.15a}$$

where $k_{1/2} = (2D_{Sf}q_{Sf,\max}X_f)^{1/2}$, which indicates half-order (with respect to bulk liquid S) substrate removal rate when diffusion limitations restrict substrate penetration in the biofilm. The derivation is an approximation of saturation kinetics, which are presumed to be valid in the biofilm. Nevertheless, since few biofilm kinetic studies have been accomplished with a defined microbial population with known kinetic coefficients (i.e., $\mu_{\max}$ and K_S), the half-order kinetic expression is most convenient for analyzing undefined, mixed population biofilm results.

Measurements of the glucose removal rate in a RotoTorque reactor (Trulear and Characklis, 1982) with a mixed population biofilm compare very well with the theory. Trulear's data are compared with calculated values from Harremöes's theory in Figure 9.30. The active biofilm thickness is 25-30 μm for the two glucose loading rates depicted. Trulear calculates that for loading rates B''_{AS} above 2.8 mg/m^2 min, oxygen was probably limiting in the biofilm.

The effect of glucose loading rate on substrate flux into the biofilm traces a saturation function (Figure 9.31). The glucose reaction rate increases linearly with glucose loading rate B''_{AS} up to approximately 2.8 mg m^2 min. At that loading rate, oxygen becomes limiting and the reaction rate increases more slowly with loading rate.

In most biofilm systems, the electron acceptor (e.g., oxygen) as well as the substrate (electron donor) can be rate-limiting or stoichiometrically limiting. In

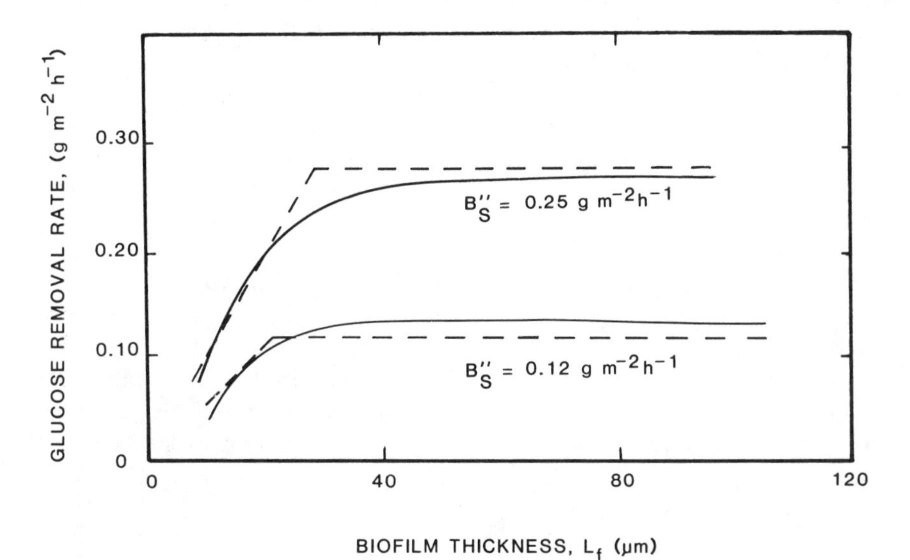

Figure 9.30 Comparison of observed (—) and calculated (---) (after Harremöes, 1978) glucose removal rate as a function of biofilm thickness in a RotoTorque reactor (Trulear, 1980).

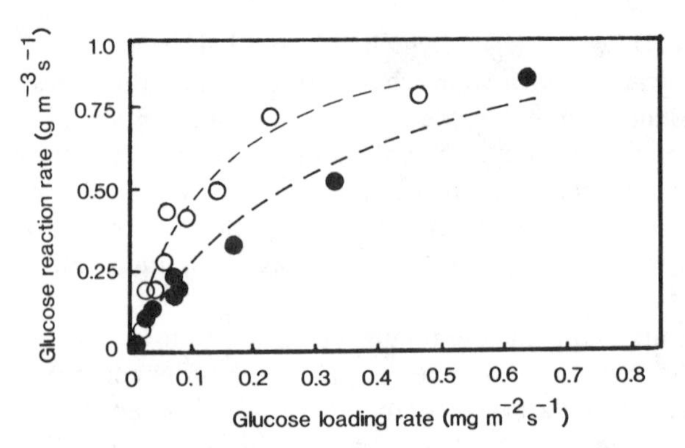

Figure 9.31 The relationship between substrate (glucose) reaction rate and substrate loading rate in mixed population biofilms. Data are from (○) Trulear and Characklis (1982) and (●) LaMotta (1976).

some biofilms, oxygen may be limiting biofilm processes in some layers while substrate limits them in other layers. Williamson and McCarty (1975, 1976) present a thorough analysis of a biofilm with dual substrate (ammonia nitrogen and oxygen) limitation.

Advection in Laminar Flow and Biofilms. Mass can be transported by advection or bulk flow as well as by concentration gradients. When mass transport by advective means is significant in the z-direction, Eq. 9.4 can be reduced to the following form for steady state flow:

$$\underset{\text{total flux}}{J_{Sx}^{\text{tot}}} = \underset{\text{diffusive flux}}{J_{Sz}} + \underset{\text{advective flux}}{Sv_z} \qquad [9.66]$$

For laminar flow, the Hagen–Poiseuille equation (Eq. 9B.1) can be substituted for v_z. In the biofilm, $v_z = 0$ and Eq. 9.66 reverts to Eq. 9.10.

Advection can also be important within the biofilm with regard to the movement of particulate components, such as microbial cells. Modeling efforts by Wanner and Gujer (1986) and Kissel et al. (1985) have focused on distribution of microbial species within multispecies biofilms. Wanner and Gujer (1985a, 1985b, 1986) present mathematical models describing microbial competition in biofilms (Chapter 11). They distinguish three processes in a biofilm that affect the microbial species composition:

1. Substrate diffusion from the bulk liquid into the biofilm
2. Microbial conversion of substrate into microbial mass
3. Volume expansion of microbial mass

The spatial distribution of microorganisms in the biofilm (perpendicular to the substratum) may determine the performance of a biofilm reactor in terms of substrate conversion. The Wanner–Gujer model describing population dynamics within a biofilm is based on two general assumptions (Chapter 11):

1. Biofilms expand because each of the microbial species expands in proportion to its growth rate and its volume fraction.
2. The volume fraction of any one microbial species changes with time because its growth rate differs from the mean growth rate in the biofilm.

The volume expansion of each microbial species is considered as an advective mechanism for transport of microbial cells and other particles within the biofilm. The transport property, in this case, is related to the volumetric growth rate of the microbial species. One of the *a priori* assumptions of the Wanner–Gujer (1986) model is that biofilm properties vary only in the direction perpendicular to the substratum. Kissel et al. (1984) assume that differences in composition, such as accumulation of inert material, may have a bearing on the location at which detachment occurs and hence emphasize the need to model spatial variations of biofilm composition in three dimensions. Gujer (1987) has commented on three dimensional spatial variation in biofilms.

9.7.2 Interphase Transport (Surface Film)

9.7.2.1 *Mass Transfer Coefficients for Smooth Substrata*

The transport of dissolved substances across an interface, especially in turbulent flow, is described by Newton's law of cooling in the following manner:

$$J_S = k_D(S_b - S_1) \qquad [9.67]$$

where J_S = interfacial mass flux of component S ($ML^{-2}t^{-1}$)
k_D = mass transfer coefficient (Lt^{-1})
S_b = bulk water concentration (ML^{-3})
S_1 = interfacial concentration (ML^{-3})

In turbulent flow over a flat plate, most of the resistance to mass transport occurs near the interface, in the *mass transfer boundary layer* (Section 9.4.2.1). Obviously, the mixing intensity near the interface will influence the rate of mass transfer at the interface. The Sherwood number, Sh, is a dimensionless mass transfer coefficient (sometimes called the mass transfer Nusselt number, Nu_S) that has been used to correlate interfacial mass transfer rates to f, an indicator of mixing (turbulent intensity) at the interface. Using analogous methods to those used in correlating heat transfer coefficients (Eq. 9.38), the following result is obtained (also see Eq. 9.21):

$$Sh = \frac{k_D d_{ch}}{D} = Sh(Re, Sc, L/D) \quad [9.68]$$

where Sc = Schmidt number
$= \eta/\rho D = \nu/D \approx 1000$ for water

In addition, in analogy to Eq. 9.40,

$$j_H = j_D = \frac{f}{2} \quad [9.69]$$

where

$$j_D = \frac{k_D}{S v_\infty} \left(\frac{\eta}{\rho D_S} \right)^{0.67} \quad [9.70]$$

or, equating Eqs. 9.69 and 9.70,

$$k_D = 0.5 f S v_\infty \eta^{-0.67} \rho^{0.67} D^{0.67} \quad [9.71]$$

where v_∞ = water velocity far from the plate (Lt^{-1}). The mass transport of *particles* over a flat plate can be described as follows (Friedlander, 1977):

$$k_D = 0.678 \frac{d}{L_p} Re^{0.5} Sc^{0.33} \quad [9.72]$$

where L_p = length of flat plate tangential to flow (L). Bouwer (1987) has attempted to relate transport of particles in a biofilm system to particle capture by the biofilm.

9.7.2.2 *Influence of Biofilms on Mass Transfer Coefficients* In turbulent flow, most of the resistance to mass transport occurs near the interface, in the *mass transfer boundary layer*. As a consequence, the roughness of the interface strongly influences the mass transfer coefficient. The hydrodynamic roughness attributed to biofilms results in increased f (Section 9.5.3) and strongly influences the mass transfer coefficient (Eq. 9.71). Thus, like heat transfer coefficients, mass transfer coefficients are influenced by biofilms because of the "roughness" created by an accumulating biofilm. As indicated in Figure 1.2 (Chapter 1), the biofilm roughness is a characteristic of the surface film.

Mass transfer at a rough surface (e.g., a biofilm) may be as much as three times higher than at a smooth surface, as suggested by the following relationship for Sh in a rough tube (Davies, 1972):

$$Sh = a\, Re\, Sc^{0.5} (e/d)^{0.15} \quad [9.73]$$

where a = a dimensionless coefficient. For example, for a sand-rough rotating disk electrodes of radius R_D (Cornet et al., 1969),

$$Sh = a\ Re\ Sc^{0.33}(e/R_D)^{0.27} \quad [9.74]$$

Several investigators have concluded that mass transfer resistance, external to the biofilm, may limit the rate of substrate removal by the biofilm (Tomlinson and Snaddon, 1966; Kornegay and Andrews, 1967; Maier et al., 1967; LaMotta, 1976; Trulear, 1980; Siegrist and Gujer, 1985).

LaMotta (1976) reported increased substrate removal rate with increasing rotational speed in a RotoTorque type system. The biofilm in his experiments was only 5–8 μm thick, so the influence of biofilm morphology was negligible. The relationship of mass transfer coefficient to rotational velocity was obtained by regression:

$$k_D = 5.33 \times 10^{-6}\ \Omega^{0.66}\ \mathrm{m\ s^{-1}} \quad [9.75]$$

where Ω = rotational speed in s^{-1}. Trulear (1980) also observed increased glucose removal rate with increasing rotational speed in a RotoTorque system (Figure 9.32). The biofilm was approximately 110 μm thick but had a surface film with some filamentous structures.

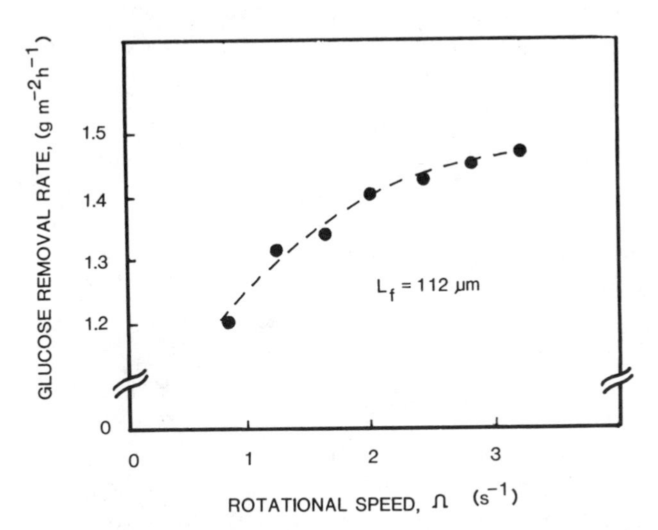

Figure 9.32 Influence of rotational speed in a RotoTorque reactor on glucose removal rate (reprinted with permission from Trulear and Characklis, 1982). The biofilm (112 μm) accumulated at a rotational speed $\Omega = 3.7\ s^{-1}$ and at a glucose loading rate B''_{AS} = 0.45 mg $m^{-2}\ s^{-1}$. The data indicate a mass transfer resistance in the bulk liquid compartment up to a rotational speed of approximately 150 rpm ($3s^{-1}$).

Example 9F. Mass Transfer at the Biofilm–Bulk Liquid Interface and Reaction within the Biofilm. Example 9E for zero order biofilm kinetics can be modified to incorporate the effect of a mass transfer resistance in the bulk water phase by changing the boundary conditions (Harremöes, 1978). For one dimensional diffusion (x-direction) and steady state, Eq. 9.4 becomes

$$D_{Sf}\frac{d^2S}{dx^2} = r_{VS} \tag{9F.1}$$

Define the following dimensionless variables:

$$S^* = \frac{S}{S_b}, \qquad x^* = \frac{x}{L_f} \tag{9F.2}$$

where S_b = substrate concentration in the bulk water (ML^{-3})
r_{VS} = volumetric substrate removal rate in the biofilm ($ML^{-3}t^{-1}$)
L_f = biofilm thickness (L)

The solution to Eq. 9F.1 is

$$S^* = \left(\frac{r_{VS}L_f^2}{S_bD_{Sf}}\right)x^{*2} + C_1x^* + C_2 \tag{9F.3}$$

The boundary conditions for this system are as follows:

No flux at the substratum:

$$\text{At} \quad x^* = x_0^*, \qquad S^* = 0 \text{ and } \frac{dS^*}{dx^*} = 0 \tag{9F.4a}$$

Mass transfer at the biofilm–bulk water interface:

$$\text{At} \quad x^* = 1, \qquad S^* = S_1^* \text{ and } \frac{dS^*}{dx^*} = \frac{k_DL_f}{D_{Sf}}(1 - S_1^*) \tag{9F.4b}$$

where $S_1^* = S_1/S_b$
k_D = mass transfer coefficient (Lt^{-1})

Incorporating the boundary conditions, Eq. 9F.3 becomes

$$\frac{r_S''}{k_DS_b} = \frac{1}{2\,\Delta^2}[(1 + 4\,\Delta^2)^{1/2} - 1] \tag{9F.5}$$

where Δ = ratio of mass transfer rate to zero order reaction rate (—)
$= k_D S_b / k_{1/2} S_b^{1/2}$
r_S'' = substrate flux at biofilm–bulk fluid interface ($ML^{-2}t^{-1}$)

As $\Delta \to 0, r_S'' \to hS_b$. As $\Delta \to \infty, r_S'' \to k_{1/2} S_b^{1/2}$. Figure 9.33a indicates the effect of mass transfer resistance in the water compartment on overall rate of substrate by the biofilm. Figure 9.33b presents the results of microelectrode measurements of dissolved oxygen in a biofilm with the bulk water flowing over it. The mass transfer coefficient will depend on the roughness of the biofilm and the water velocity flowing past the biofilm.

9.7.3 Summary

The major distinction between the behavior of microbial biofilms and sessile microbial populations is that generally mass transport and diffusion limit the microbial activity in biofilms. Mass transport in the bulk liquid may sometimes control the rate of growth of biofilm organisms because the transport rate of nutrients to the biofilm from the bulk liquid is slower than the nutrient uptake by the organisms in the biofilm. Diffusion of nutrients within the biofilm may control the growth of microorganisms deep within the film. Concentration gradients resulting from diffusional resistance probably determines the population distribution of organisms within the biofilm. For example, the depletion of oxy-

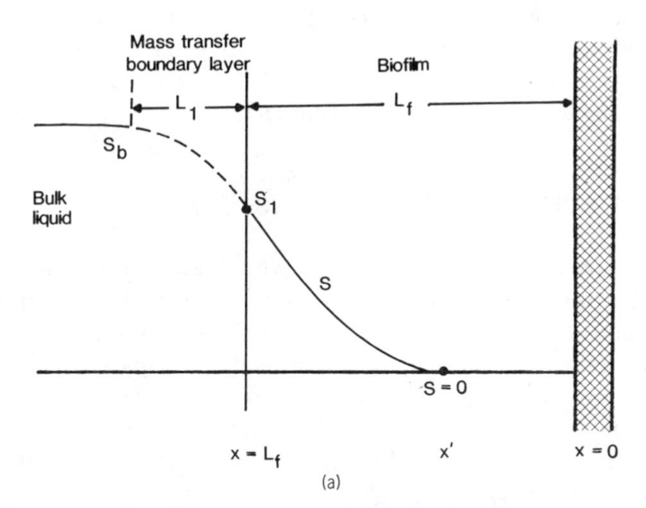

Figure 9.33a Influence of mass transfer resistance on the substrate concentration gradient in the bulk water compartment. The concentration continues to decrease in the biofilm as a result of substrate removal by microorganisms and diffusion in the biofilm. (After Harremöes, 1978.)

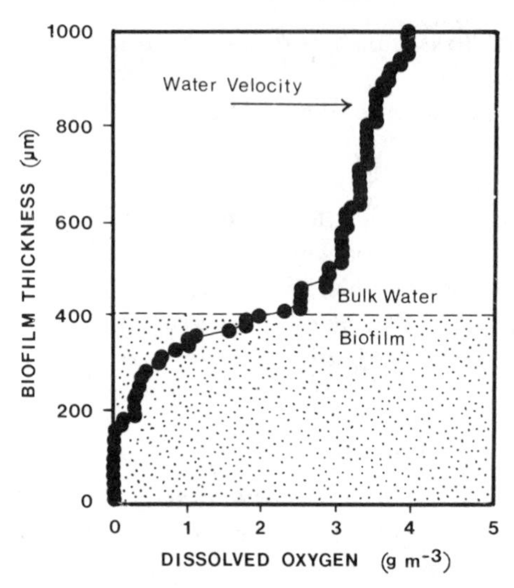

Figure 9.33b The dissolved oxygen (DO) concentration profile in a biofilm system was determined with a microelectrode (tip diameter approximately 5 μm). The mass transfer boundary layer in the bulk water is pronounced within approximately 100 μm of the biofilm. The DO gradient is higher just below the bulk water-biofilm interface where microorganism oxygen uptake is probably the highest. At about 180 μm from the substratum, the biofilm is anaerobic. (Courtesy Z. Lewandowski).

gen with depth in the biofilm permits the proliferation of anaerobic organisms within the film despite the presence of dissolved oxygen in the bulk liquid. Thus, mass transport in the bulk liquid phase and diffusion within the biofilm are essential processes to consider in any biofilm analysis.

9.8 INTERFACIAL TRANSFER PHENOMENA

Interfacial transfer processes can be considered as transport processes in that no molecular rearrangements occur because of them. In fact, transport is essential to the interfacial transfer processes, since transport moves the component to the interface, where it is transferred to another biofilm compartment. It is only because of the movement of a component from one compartment to another that interfacial transfer processes have been distinguished from transport processes.

In many cases of relevance to biofilm systems, interfacial transfer processes involve particulate materials.

9.8.1 Molecular Adsorption and Desorption

Loeb and Neihof (1975) and DePalma et al. (1979) have measured adsorption rates of organic molecules to solid substrata in seawater, and Bryers (1980) has observed analogous adsorption rates in a laboratory system. The *net* rate of adsorption in these studies can be described as follows:

$$r_A'' = k_a S_i \left[1 - \frac{S_i''}{k_s} \right] \qquad [9.76]$$

where r_A'' = rate of net adsorption ($ML^{-2}t^{-1}$)
k_a = adsorption rate coefficient (Lt^{-1})
k_s = saturation coefficient (ML^{-2})
S_i'' = areal concentration of i (ML^{-2})

Since

$$S_i'' = \rho_i L_f \qquad [9.77]$$

adsorption can be described in terms of adsorbed film thickness. Adsorption rates for organic matter in seawater as high as 0.027 μm h^{-1} were observed by Loeb and Neihof and by Bryers, but the maximum extent of accumulation from molecular fouling was always less than 0.1 μm (Table 9.7). Thickness measurements show that molecular fouling can have no significant effect on fluid flow or heat transfer. Nevertheless, the surface properties resulting from adsorption of

TABLE 9.7 The Rate and Extent of Adsorption of Organic Matter on Clean Substrata[a]

Maximum Rate (nm/min)	Maximum Accumulation (nm)	Maximum Accumulation [(μg COD) cm^{-2}]	Surface	Reference
0.15–0.45	30–80		Pt[b]	Loeb and Neihof, 1975
0.004	7.1		Ge[c]	DePalma et al., 1979
0.044	77.3		Ti[c]	
0.01[d]	13.5[d]	1.5	Glass[e]	Bryers, 1980
0.22[d]	22.5[d]	2.5	Glass[f]	

[a] After Characklis (1981b).

[b] Immersed in quiescent Chesapeake Bay water (3–4°C) containing 2.3 mg carbon, salinity 9–16‰, and pH 7.9–8.2.

[c] Gulf of Mexico water (22°C) flowing past the surface at a fluid shear stress of 7.1 N m^{-2}. Salinity was 34‰. Carbon concentration not reported.

[d] Estimated from measurements of chemical oxygen demand (COD) adsorbed per unit area. Assumed COD of protein is 0.855 (mg COD)/(mg protein), and protein density is 1.3 (g protein) cm^{-3}.

[e] Medium consisted of a sterile 1 : 1 w/w of trypticase soy broth–glucose mixture (34°C; pH 8). The glass surfaces were immersed in tubes placed in a mechanical shaker. Carbon concentration was approximately 80 mg L^{-1}.

[f] Medium was effluent (30°C; pH 8) from a chemostat (10–20 mg L^{-1} COD, 3 mg L^{-1} polysaccharide) with no primary substrate remaining. Microorganisms were present (approximately 10^6 cells ml^{-1}, but no cells attached during the period of interest. Fluid shear stress was 3.8 N m^{-2}.

an organic film may affect the sequence of microbial events that follow. Thus, the rate of molecular fouling can be considered instantaneous since its rate is much greater than that of microbial fouling.

Brash and Samak (1978) present experimental evidence that significant turnover occurs in molecular (proteinaceous) fouling films on polyethylene. Protein molecules in the bulk fluid are continuously exchanging with adsorbed proteins. Thus, the composition of the molecular conditioning film may be changing continuously, and clearly the kinetic expressions presented must describe *net* adsorption. This also suggests that planktonic microbial cells and their associated extracellular material may be continually exchanging with biofilm material at the wall.

9.8.2 Cellular Adsorption and Desorption

Cellular adsorption is the immobilization of a cell at the substratum for a time. Desorption is the mobilization of the cell, that is, its complete removal from the substratum. Desorption is the reverse of adsorption. These two processes have been observed independently by image analysis methods (Chapters 3 and 12).

9.8.2.1 Stoichiometry A two stage cellular adsorption process can frequently be observed: (1) reversible adsorption followed by (2) irreversible adsorption. Reversible adsorption is an initially weak adsorption: the cell can still exhibit Brownian motion and is readily removed by mild (i.e., low shear force) rinsing. In contrast, irreversible adsorption is a permanent bonding to the substratum, perhaps mediated by EPS. Irreversibly adsorbed cells can only be removed by rather severe mechanical or chemical treatment. Thus, initial colonization of a substratum can be described by the following "stoichiometry":

$$\underset{\text{suspended cells}}{X} \rightleftarrows \underset{\text{reversibly adsorbed cells}}{X_r''} \rightarrow \underset{\text{irreversibly adsorbed cells}}{X_e''} \qquad [9.78]$$

$$\underset{\text{Total adsorbed cells}}{X_{\text{tot}}''} = \underset{\text{reversibly adsorbed cells}}{X_r''} + \underset{\text{irreversibly adsorbed cells}}{X_e''} \qquad [9.79]$$

Bowen et al. (1976) proposed an analysis for transport and adsorption of particles in a rectangular channel, assuming a first order reaction approximation for the substratum–particle capture rate, which leads to an expanded Graetz solution. The solution converges well for large Peclet numbers and proved to be accurate for inert particles adsorbing to charged surfaces. Escher (1986) found that the theory compared well with results for adsorption of *Pseudomonas aeruginosa* in a rectangular glass capillary as well (Chapter 12). The derivations are presented in detail in Chapter 12. At this point, suffice it to say that adsorption rate is first order with respect to suspended cell concentration:

$$r''_C = \frac{X_C D_{Cb}}{L_d} \tag{9.80a}$$

or

$$r''_C = k_C X_C \tag{9.80b}$$

where r''_C = adsorption rate of cells to the substratum (cells $L^{-2}t^{-1}$)
L_d = diffusion distance (L)
X_C = bulk liquid cell concentration (cells L^{-3})
D_{Cb} = diffusion coefficient for cells in the bulk liquid (L^2t^{-1})
k_C = cell adsorption rate coefficient (Lt^{-1})

In the special case where all cells transported to the substratum adsorb, the substratum acts as a total sink for suspended particles, so that the total flux of cells to the substratum is effectively determined. Then Eq. 9.80 becomes

$$B''_{AC} = \frac{X_C D_{Cb}}{L_d} \tag{9.81a}$$

or

$$B''_{AC} = k_{AC} X_C \tag{9.81b}$$

where B''_{AC} = flux of cells to substratum (cells $L^{-2}t^{-1}$)

In the case where all cells transported to the substratum adsorb, $k_{AC} = k_C$. As we shall see, only a small fraction (less than 1%) of the calculated number of cells reaching the substratum actually adsorb.

9.8.2.2 *Sticking Efficiency* The sticking efficiency is the ratio of the adsorption rate to the flux of cells to the substratum. Thus, it is an overall probability of adsorption. The sticking efficiency can be described by the ratio of Equations 9.80 to 9.81 as follows:

$$\alpha_e = \frac{\text{number of cells adsorbing to substratum}}{\text{number of cells transported to substratum}} \tag{9.82}$$

Thus, the sticking efficiency α_e has a range of zero to one. It is the probability that a cell being transported to the substratum will adsorb.

Thus, Eq. 9.80b can be expressed as

$$r''_C = \alpha_e k_{AC} X_C \tag{9.83}$$

where $k_{AC}X_C$ describes the rate of transport of cells to the substratum. Sticking efficiencies calculated from experiments conducted under different flow conditions are compared in Table 7.1 (Chapter 7). The experimentally observed sticking efficiencies (Escher, 1986; Powell and Slater, 1983; Hermanowicz et al., 1989) are discussed in more detail in Chapter 12 and demonstrate the important role of hydrodynamics in initial events of biofilm accumulation.

9.8.3 Attachment and Detachment

9.8.3.1 Attachment Biofilm accumulation may be strongly influenced by attachment of particles from the bulk water compartment. Transport of the particles through the bulk water to the biofilm is described by equations for mass transfer (Section 9.7.2.1):

$$J_X = k_{DX}X \quad [9.84]$$

where J_X = flux of particles to the biofilm-water interface ($ML^{-2}t^{-1}$)
k_{DX} = particle mass transfer coefficient (Lt^{-1})
X = particle concentration in water (ML^{-3})

The particle concentration at the biofilm-bulk water interface is presumed negligible. Using this equation, Bouwer (1987) has estimated that biofilm systems in porous media (substrata) with small diameter pores and/or long hydraulic residence times (e.g., fixed bed reactors) can be effective in removing submicron particles through interception, a mechanism critical to acceptable filtration performance. In systems with slow flows and horizontal substrata, large particles ($>$10–50 μm) can be removed by sedimentation. Once transported to the biofilm, many factors influence the attachment of the particles to the biofilm. Unfortunately, little is known about the attachment efficiency, so the rate equation (Eq. 9.84) for transport is simply modified by a sticking efficiency factor to describe the rate of attachment:

$$r_X'' = \alpha_e k_{DX}X \quad [9.85]$$

where r_X'' = rate of attachment of particles to the biofilm ($ML^{-2}t^{-1}$).

The sticking efficiency has been estimated for several geometries and environmental systems and is generally very small ($\ll 1\%$) for circumstances relevant to biofilms (O'Melia, 1986; Beal, 1970).

9.8.3.2 Detachment Detachment can be classified as erosion, sloughing, or abrasion. The erosion rate from a biofilm is generally presumed to be a function of the particulate material concentration in the biofilm (Trulear and Characklis, 1982). The cellular erosion rate from a *Pseudomonas aeruginosa* biofilm has been modelled by Bakke et al. (1984) as follows:

$$r''_{EC} = -k_{EC}X_{fC} \qquad [9.86]$$

where r''_{EC} = rate of cellular erosion ($ML^{-2}t^{-1}$)
k_{EC} = specific cellular erosion rate coefficient (Lt^{-1})
X_{fC} = cellular concentration in the biofilm (ML^{-3})

k_{EC} depends on the local hydraulic shear stress and probably on properties of the biofilm contributing to its cohesive strength. For example, Trulear and Characklis (1982) report that increasing the shear stress increases the rate of erosion from a mixed population biofilm in a RotoTorque system (Figure 9.34). Rittmann (1982) has used this first order (in biofilm concentration) kinetic expression to extrapolate the data of Trulear and Characklis to other geometries and a wider range of hydraulic shear stress.

A more rational expression for the erosion rate includes the possibility that k_{EC} depends on biofilm concentration—in particular, that it varies as the square of the biofilm concentration. This is consistent with observations on the progression of biofilm accumulation, which generally occurs according to a logistic

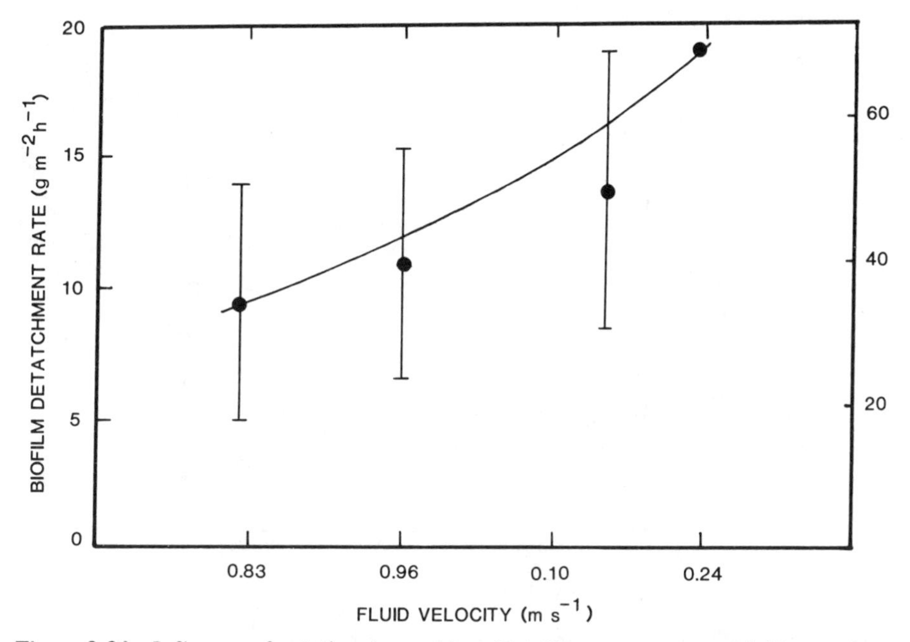

Figure 9.34 Influence of rotational speed in a RotoTorque reactor on biofilm erosion rate. The data have been fitted by regression to Eq. 13.30. The biofilm mass was between 0.75 and 0.80 g m^{-2} at 30°C. (Reprinted with permission from Trulear and Characklis, 1982.)

function (Section 8.2.2, Chapter 8). The logistic function, in terms of biofilm cell concentration, is as follows:

$$\underset{\substack{\text{accumulation}\\ \text{rate of cells}\\ \text{in the biofilm}}}{\frac{dX_{fC}}{dt}} = \underset{\substack{\text{accretion}\\ \text{rate of cells}\\ \text{in the biofilm}}}{k_1 X_{fC}} - \underset{\substack{\text{erosion}\\ \text{rate of cells}\\ \text{from the biofilm}}}{k_2 X_{fC}^2} \qquad [9.87]$$

Accretion of biofilm cells may be the net result of adsorption, growth, etc. Nevertheless, we may deduce that erosion is proportional to the biofilm concentration squared:

$$r''_{EC} = -k'_{EC} X_C^2 \qquad [9.88]$$

where k''_{EC} = specific cellular erosion rate coefficient (L^{-4} $cells^{-1}$ t^{-1})
X_C = cell concentration in the biofilm (cells L^{-3})

Erosion rate data from Trulear and Characklis (1982) are compared with a second order polynomial fit in Figure 9.35.

Wanner and Gujer (1986), in a theoretical modeling effort, hypothesized a similar general expression for the detachment rate. In this case, the erosion rate

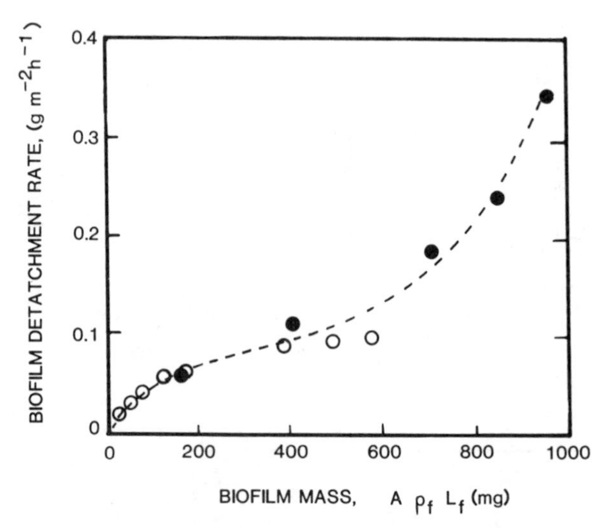

Figure 9.35 Influence of biofilm mass (on 0.2 m^2 reactor surface) on erosion rate (reprinted with permission from Trulear and Characklis, 1982). The line represents the best fit for second order dependence on biofilm mass for biofilm mass greater than 200 mg. Circles denote two different experiments.

is proportional to the biofilm mass concentration and also to the square of the biofilm thickness:

$$r_E = -k_E L_f^2 X \qquad [9.89]$$

where r_E = rate of biofilm erosion ($\mathrm{Mt^{-1}}$)
k_E = specific biofilm erosion rate ($\mathrm{L^{-1}t^{-1}}$)
X = mass per unit biofilm volume ($\mathrm{ML^{-3}}$)

Eq. 9.89 has some advantages in incorporating two properties of the biofilm:

1. The hydrodynamic effects are probably related to the biofilm thickness.
2. Biofilm strength may be related to biofilm density.

9.8.4 Summary

Interfacial transfer processes are essentially transport processes, since no molecular rearrangements result. Intraphase transport is essential to the interfacial transfer processes, since it moves the component to the interface, where it is transferred to another biofilm system compartment. In many cases of relevance to biofilm systems, interfacial transfer processes involve particulate materials. Sticking efficiency is a crucial parameter for distinguishing particle transport from particle transfer. It is because of the movement of a component from one compartment to another and the low probability of transfer (i.e., low sticking efficiency) that interfacial transfer processes have been distinguished from transport processes. The transfer processes are obviously essential to any comprehensive understanding of biofilms, since a substratum cannot be colonized until at least one cell adsorbs. The interfacial transfer processes are the least understood of all the biofilm processes and require much more experimental scrutiny before accurate predictions can be made regarding their relative importance in biofilm reactors.

REFERENCES

Arnold, G.E., *Eng. News Rec.*, **116**, 774–775 (1963).

Atkinson, B., *Biochemical Reactors,* Pion, London, 1974.

Bailey, J.E. and D.F. Ollis, *Biochemical Engineering Fundamentals,* 2nd ed., McGraw-Hill, New York, 1986.

Bakke, R., M.G. Trulear, J.A. Robinson, and W.G. Characklis, *Biotech. Bioeng.*, **26**, 1418–1424 (1984).

Beal, S.K., *Nucl. Sci. Eng.*, **40**, 1–11 (1970).

Bird, R.B., W.E. Stewart, and E.N. Lightfoot, *Transport Phenomena*, Wiley, New York, 1960.

Bott, T.R. and J.S. Gudmundsson, in *Proc. 76th International Heat Transfer Conference*, Toronto, 1978, pp. 373–378.

Bouwer, E.J., *Wat. Res.*, **21**, 1489–1498 (1987).

Bowen, B.D., S. Levine, and N. Epstein, *J. Colloid. Interface Sci.*, **54**, 375–390 (1976).

Brash, J.L. and Q.M. Samak, *J. Colloid. Interface Sci.*, **65**, 495–504 (1978).

Brauer, H., *Chem. Ztg.*, **87**, 199 (1963).

Bryers, J.D.,"Dynamics of early biofilm formation in a turbulent flow system," unpublished doctoral dissertation, Rice University, Houston, TX, 1980.

Busch, A.W., *Aerobic Biological Treatment of Waste Waters*, Gulf Publishing Co., Houston, TX, 1971.

Characklis, W.G., *Wat. Res.*, **7**, 1249–1258, 1972.

Characklis, W.G., *Biotech. Bioeng.*, **23**, 1923–1960, 1981a.

Characklis, W.G., "Microbial fouling: A process analysis", in E.F.C. Somerscales and J.G. Knudsen (eds.), *Fouling of Heat Transfer Equipment*, Hemisphere, Washington, 1981b, pp. 251–291.

Characklis, W.G., M.J. Nimmons, and B.F. Picologlou, *Heat Transfer Eng.*, **3**, 23 (1981).

Churchill, S.W., *The Interpretation and Use of Rate Data*, McGraw-Hill, New York, 1974.

Cleaver, J.W. and B. Yates, *Chem. Eng. Sci.*, **30**, 983–992 (1975).

Cleaver, J.W. and B. Yates, *Chem. Eng. Sci.*, **31**, 147–151 (1976).

Colebrook, C.F., *J. Inst. Civil Engrs.*, **11**, 133 (1939).

Cornet, I., W.N. Lewis, and R. Kappesser, *Trans. Inst. Chem. Eng.*, **47**, T222 (1969).

Davies, J.T., *Turbulence Phenomena*, Academic Press, New York, 1972.

DePalma, V.A., D.W. Goupil, and C.K. Akers, "Field demonstration of rapid microfouling in model heat exchangers: Gulf of Mexico," in *Proc. 6th OTEC Conference*, Washington, 1979.

Derby, R.L., *J. Am. Wat. Wks. Assoc.*, **39**, 1107–1114 (1947).

Dipprey, D.F. and R.H. Sabersky, *Int. J. Heat Mass Transfer*, **6**, 329–353 (1963).

Escher, A. "Bacterial colonization of a smooth surface: An analysis with image analyzer," unpublished doctoral thesis, Montana State University, Bozeman, MT, 1986.

Fahien, R.W., *Fundamentals of Transport Phenomena*, McGraw-Hill, 1983.

Fletcher, M., *Can. J. Microbiol.*, **23**, 1–6 (1977).

Friedlander, S.K., *Smoke, Dust, and Haze*, Wiley, New York, 1977.

Gray, D.D., *American Institute of Physics Handbook*, 3rd ed., McGraw-Hill, New York, 1972.

Gujer, W., *Wat. Sci. Tech.*, **19**, 495 (1987).

Harremöes, P., "Biofilm kinetics," in R. Mitchell (ed.), *Water Pollution Microbiology*, Vol. 2, Wiley, New York, 1978. pp. 71–110.

Hermanowicz, S.W., R.E. Danielson, and R.C. Cooper, *Biotech. Bioeng.*, **33**, 157–163 (1989).

Jang, L.K. and T.F. Yen, "A theoretical model of diffusion of motile and nonmotile bacteria toward solid surfaces," in Zajic and Donaldson (eds.), *Microbes and Oil Recovery,* Vol. 1, Int. Bioresources J., 1985, pp. 226–246.

Kissel, J.C., P.L. McCarthy, and R.L. Street, *J. Env. Eng.,* **111**(4), 549–551 (1985).

Knudsen, J.G., "Apparatus and techniques for measurement of fouling of heat transfer surfaces," in E.F.C. Somerscales and J.G. Knudsen (eds.), *Fouling of Heat Transfer Equipment,* Hemisphere, Washington, 1981, pp. 57–81.

Kornegay, B.H. and J.F. Andrews, "Characteristics and kinetics of fixed-film biological reactors," Final Report, Grant WP-01181, Fed. Wat. Poll. Contr. Admin., U.S. GPO, Washington, 1967.

LaMotta, E.J., *Biotech. Bioeng.,* **18**, 1359–1370 (1976).

Loeb, G.I. and R.A. Neihof, "Marine conditioning films," in R.E. Baier (ed.), *Applied Chemistry at Interfaces,* Adv. Chem. Series 145, Am. Chem. Soc., Washington, 1975, pp. 319–335.

Maier, W.J., V.C. Behn, and C.D. Gates, *J. San. Eng. Div. ASCE,* **93**, 91 (1967).

McCoy, W.F. and J.W. Costerton, *Dev. Ind. Microbiol.,* **23**, 441 (1982).

Minkus, A.J., *J. New Engl. Wat. Wks. Assoc.,* **68**, 1–10 (1954).

Nikuradse, J., *Forsch. Arb. Ing.-Wes.,* No. 361 (1933).

O'Melia, C. R., "Particle-particle interactions," in W. Stumm (ed.), *Aquatic Surface Chemistry,* Wiley, New York, 1986.

Perry, R.H. and C.H. Chilton, *Chemical Engineers' Handbook,* 5th ed., McGraw-Hill, New York, 1973.

Picologlou, B.F., N. Zelver, and W.G. Characklis, *J. Hyd. Div. ASCE,* **106**, 733–746 (1980).

Pollard, A.L. and H.E. House, *J. Pwr. Div. ASCE,* **85**, PO6, 163–171 (1959).

Powell, M.S. and N.K.H. Slater, *Biotech. Bioeng.,* **25**, 891–900 (1983).

Rittmann, B.E., *Biotech. Bioeng.,* **24**, 501–506 (1982).

Rittmann, B.E. and P.L. McCarty, *Biotech. Bioeng.,* **22**, 2343 (1980a).

Rittmann, B.E. and P.L. McCarty, *Biotech. Bioeng.,* **22**, 2359 (1980b).

Schlichting, H., *Boundary Layer Theory,* 6th ed., McGraw-Hill, New York, 1968.

Schuster, H., Fluid friction in the presence of non-rigid boundaries, unpublished doctoral thesis, Johns Hopkins University, Baltimore, 1971.

Seifert, L. and W. Krueger, *VDI Z.,* **92**, 189–191 (1950).

Sherwood, T.K., R.L. Pigford, and C.R. Wilkie, *Mass Transfer,* McGraw-Hill, New York, 1975.

Sieder, E.N. and G.E. Tate, *Ind. Eng. Chem.,* **28**, 1429–1435 (1936).

Siegrist, H. and W. Gujer, *Wat. Res.,* **19**, 1369–1378 (1985).

Taylor, G.I., *Phil. Trans.,* **223**, 289–293 (1923).

Tomlinson, T.G. and D.M. Snaddon, *Air Water Pollut.,* **10**, 865 (1966).

Trulear, M.G., "Dynamics of biofilm processes in an annular reactor," unpublished Master of Science thesis, Rice University, Houston, TX, 1980.

Trulear, M.G., "Cellular reproduction and extracellular polymer formation in the development of biofilms," unpublished doctoral thesis, Montana State University, Bozeman, MT, 1983.

Trulear, M.G. and W.G. Characklis, *J. Wat. Poll. Contr. Fed.*, **54**, 1288 (1982).

Turakhia, M.H. "The influence of calcium on biofilm processes," unpublished doctoral thesis, Montana State University, Bozeman, MT, 1986.

Turakhia, M.H. and W.G. Characklis, *Heat Transf. Eng.*, **5**, 93–101 (1984).

Vaituzis, Z. and R.N. Doetsch, *Appl. Env. Microbiol.*, **17**, 584–588 (1969).

Walker, R.A. and T.R. Bott, *The Chemical Engineer*, **271**, 151–156 (1973).

Wanner, O. and W. Gujer, *Wat. Sci. Tech.*, **17**, 27–44 (1985a).

Wanner, O. and W. Gujer, *EAWAG News 18/19*, **1**, 4 (1985b).

Wanner, O. and W. Gujer, *Biotech. Bioeng.*, **28**, 314–328 (1986).

Washburn, E.N., *International Critical Tables of Numerical Data*, McGraw-Hill, New York, 1929.

Weast, R.C., *Handbook of Chemistry and Physics*, 50th ed., CRC Press, Cleveland, OH, 1973.

Wiederhold, W., *Gas WassFach*, **90**, 634–641 (1949).

Williamson, K.F. and P.L. McCarty, *Biotech. Bioeng.*, **17**, 915 (1975).

Williamson, K.F. and P.L. McCarty, *J. Wat. Poll. Contr. Fed.*, **48**, 9–24 (1976).

Wilson, E.E., *Trans. ASME*, **37**, 47 (1915).

Witherspoon, P.A. and D.N. Saraf, *J. Phys. Chem.*, **69**, 3752 (1965).

Zelver, N., "Biofilm development and associated energy losses in water conduits," unpublished Master of Science thesis, Rice University, Houston, TX, 1979.

Zelver, N., F.L. Roe, and W.G. Characklis, "Potential for monitoring fouling in the food industry", in D. Lund, E. Plett, and C. Sandu (eds.), *Fouling and Cleaning in the Food Industry*, Department of Food Science, University of Wisconsin, Madison, WI, 1985.

10

PHYSIOLOGICAL ECOLOGY IN BIOFILM SYSTEMS

WILLIAM G. CHARACKLIS
Montana State University, Bozeman, Montana

GORDON A. MCFETERS
Montana State University, Bozeman, Montana

KEVIN C. MARSHALL
University of New South Wales, Kensington, New South Wales, Australia

SUMMARY

Three factors differentiate the ecology of biofilms from that of planktonic microbial populations: 1) the proximate substratum, 2) the relatively high volumetric concentration of cells, and 3) the long residence time of the neighboring cells in the biofilm. In addition, microbial cells in biofilms are generally growing at rates much below their maximum, largely because of the trophic state of the bulk water compartment. Thus, a transport or transfer process is usually rate controlling the overall "performance" of the reactor.

The microbial ecology of biofilms concerns the interactions of the microbial cells in the biofilm with each other and the microenvironment within the biofilm including the substratum. Most microorganisms adsorb to many surfaces. The substratum properties and those of the conditioning film may influence the rate and extent of adsorption, but for virtually all substrata in any natural aquatic environment, microbial colonization will proceed. After a short period (on the order of hours), the dominant processes adding cells to the biofilm will be growth and reproduction. The stoichiometry and kinetics of the adsorbed cells in the biofilms do not appear to change from those observed for the same organisms in the planktonic (suspended) state as long as the microenvironment of the cells is the same. In other words, the immobilized condition does not directly or significantly influence cell physiology. Indirectly, the biofilm may significantly influence microenvironmental conditions, which in turn will influence physiology. For example, a microbial species may not colonize a copper substratum because it is inhibited by copper ions leaching from the substratum to the cells. However, another species with a substantial EPS content may colonize because the EPS immobilizes the copper ions. Then, the first species may colonize in intimate contact with the second species.

Oligotrophic–copiotrophic classification may be unnecessary for discussions of steady state biofilms, since the characteristics are implicit in the growth kinetic expressions. However, the classification may be useful in transient environments for predicting the persistence of species under changing conditions (e.g., feast–famine regimes).

The high cell densities and the (electro)chemical properties of the EPS influence transport of soluble and particulate components, possibly including genetic material. Conceptually, the diffusion process can be described with presently available models. However, the factors influencing detachment (erosion and sloughing) processes are not so obvious. Since detachment influences several processes, including the population distribution, more attention to causes and effects of detachment is necessary. In addition, the integration of the biofilm with the inorganic chemistry of the bulk liquid has not been addressed sufficiently. How do heavy metals interact with the EPS, and how do they influence cellular activity? In a similar manner, the relationship of substratum chemistry (especially in relation to microbial corrosion) to the biofilm needs clarification.

Various environmental variables (e.g., pH, temperature, salinity) influence microbial stoichiometry and kinetics and, thus, physiological ecology, in a biofilm. In this chapter, temperature is used to illustrate the effects of an environmental variable on biofilm structure and function. The influence of temperature is incorporated in the coefficients of quantitative expressions for the kinetics and stoichiometry of fundamental processes (e.g., growth). The effect of temperature on overall structure and function of the biofilm can then be deduced from mathematical models incorporating these expressions.

The various physical, chemical and biological processes occurring in the biofilm and their interaction are complex. Thus, the integration of these processes into a thorough understanding of biofilm structure and function requires systematic mathematical modeling in conjunction with well-defined experiments.

10.1 INTRODUCTION

Microbial ecology is concerned with the interactions between different microorganisms and between organisms and their environment. Because of the diversity of microbial types in natural habitats (Chapter 5), a wide range of physiological activities are attributable to microorganisms within any particular ecosystem.

Biofilm systems are of particular interest to microbial ecologists because of the following:

1. The initial presence of two compartments, water and the substratum, leads to unique physical forces being operative at the substratum–water interface. Microbial activity at such interfaces is modified by microenvironmental changes such as the concentration or depletion of nutrients or inhibitors, altered extracellular enzyme activity, altered gas availability, pH changes, and macromolecular adsorption leading to possible modifications of substratum properties (Marshall, 1976).
2. The subsequent development of the biofilm is the result of adsorption, growth, and extracellular polymeric substance (EPS) production by microorganisms at the substratum surface.

The main microorganisms involved in biofilm formation are heterotrophic bacteria. The bacteria are more or less uniformly distributed throughout the EPS matrix of the biofilm (see Section 10.5). Because of the gel-like structure of the biofilm, transport processes generally control the microbial behavior within the biofilm (Chapter 9). Hence, diffusive transport of electron donors, nutrients, oxygen, and/or other electron acceptors (Chapter 5) will determine the physiology of the populations at various depths in a thick biofilm. Thus, while pure culture biofilms (particularly where aerobic bacteria are employed) provide an accurate representation of processes occurring at depth in biofilms, mixed culture biofilms (particularly those incorporating various types of aerobes, microaerophiles, anaerobes and facultative anaerobes) should provide a more relevant model for the study of varied physiology resulting from transport limitations. The comments by Brock (1987) on the need for studies on natural microbial communities are particularly relevant in the context of biofilm ecology.

10.2 WHY DO BACTERIA ADSORB TO SURFACES?

Many bacteria are inherently sticky and tend to rapidly adsorb to a range of surfaces. Other bacteria appear to respond gradually to the presence of a surface and become more firmly adsorbed to the surface as time passes. These differences between bacteria and between different locations may result from the nutritional condition of the bacteria in the aqueous phase, the nature of the substratum surface, and the prior macromolecular conditioning of the substratum surface.

10.2.1 Trophic State of Bacteria in Aqueous Habitats

Microbial communities in freshwater and marine environments are influenced by biotic and abiotic factors. Among the abiotic factors, the nutrient status of the water is usually of paramount importance, since most natural bodies of water are oligotrophic (nutrient-limited). However, the microbial nutrients present in such systems are heterogeneously distributed in both time and space. As a result, native (autochthonous) bacteria may exist alternatively under feast and famine conditions, with the latter predominating. These microorganisms must adapt to persist in such adverse environments. Their adaptability partially explains their survival under limiting and highly variable nutrient conditions. Poindexter (1981) has defined an oligotrophic habitat as one with a flux of less than 1 (g C) m^{-3} day^{-1}. This definition may require modification for a biofilm system by reference to a unit area, so that the oligotrophic criterion is expressed as a flux (units of mass per unit area per unit time). Biofilm formation has been observed on surfaces immersed in oligotrophic waters, as well as in waters with higher fluxes of organic nutrients (mesotrophic and eutrophic waters).

10.2.1.1 Oligotrophic Environments Oligotrophic bacteria are generally defined as those capable of growth and reproduction, albeit slowly, in oligotrophic habitats. This appears to be a simplistic definition, as recent studies (Martin and MacLeod, 1984) have shown that certain marine bacteria grow at oligotrophic levels of certain energy substrates, but require very much higher concentrations of other energy substrates. Thus, the classification of such bacteria as oligotrophic in natural aquatic habitats may depend not only on the quantity but also the quality of the assimilable organic carbon.

The term oligotrophic *is better suited for describing aquatic environments than aquatic microorganisms.*

Bacteria that grow in oligotophic environments represent a group of adapted autochthonous aquatic organisms (Hirsch, 1979; Poindexter, 1981) that are efficient, versatile, and capable of proliferation in very dilute media. The observed maximal growth rate of these bacteria in the planktonic state, however, is rather low, quite possibly because their growth is limited by the transport of nutrients from the bulk water to the cell. Such organisms are uniquely suited to low nutrient environments without competitors and may be readily isolated from such habitats on immersed glass slides (Poindexter, 1981). Even so, some of these bacteria further adapt morphologically to fluctuations in nutrient levels. *Caulobacter* is a genus that grows in oligotrophic environments and produces a cellular stalk, termed the prostheca, as an attachment structure with an adhesive distal end (Poindexter, 1981). Prosthecae are synthesized specifically in response to ambient nutrient depletion. In this way, the bacteria can benefit from being attached to the nutritionally enriched substratum as well as being stationary in a moving supply of potential nutrients. Thus, bacteria in oligotrophic environments not only are well suited to conditions in natural waters but are capable of further advantageous morphological adaptation. These bacteria are capable of primary colonization of surfaces immersed in oligotrophic waters (Kjelleberg et al., 1985). Adsorbed bacteria can grow in conditions where the nutrient concentration is below the level required by most other organisms. These bacteria can also occur in a succession following colonization and partial nutrient removal by copiotrophic bacteria (Fletcher and Marshall, 1982a). However, these slow-growing organisms never become a significant component of mature biofilms, wherein the dominant organisms are fast-growing, copiotrophic bacteria (see Section 10.3.2). Chapter 11 presents a model that may be helpful in describing the competition between these two apparently disparate groups of bacteria in biofilms.

10.2.1.2 Eutrophic Environments Many bacteria depend for growth on the exploitation of a nutrient flux at least 50-fold higher than that found in oligotrophic waters and where the level of available carbon does not fall to near zero for prolonged periods (Poindexter, 1981), conditions termed *eutrophic* or *copiotrophic*. Such bacteria, sometimes termed *copiotrophs,* grow well in eutrophic waters but are unable to grow in oligotrophic waters. In what seems to be a paradoxical situation, relatively large numbers of copiotrophic bacteria are found in oligotrophic waters and are the dominant organisms in biofilms developing on substrata in such waters.

Copiotrophic bacteria confronted with typical oligotrophic nutrient conditions are faced with a subminimal supply of nutrients. Under these conditions of starvation, physiological and morphological adaptations are evoked to maintain viability. The formation of miniaturized cells is one of the first manifestations of adaptive starvation survival processes (Morita, 1982). These cells, in turn, display the altered metabolic properties of significantly reduced endogenous respiration and an increased surface to volume ratio that provide such cells an advantage under conditions of nutrient limitation (Novitsky and Morita, 1977). This state of reduced metabolic activity supports the concept of dormancy proposed by Stevenson (1978) to account for the prolonged survival of copiotrophs in oligotrophic environments for extended periods of time. Qualitative and quantitative changes in macromolecular components also accompany this transition to the survival contingency, such as increased levels of certain catabolic enzymes and decreased concentrations of some anabolic enzymes, as reviewed by Matin (1979). Results of a recent study by Amy et al. (1983) with a marine *Vibrio* under oligotrophic conditions support this notion. Starvation conditions may also activate chemotactic competency in some bacteria (Morita, 1982)—a valuable phenotype for starving bacteria in search of nutrient-rich areas such as interfaces. Nutrient transport mechanisms may also change in starving copiotrophs (Morita, 1982) as reflected by increasing substrate affinities (a lower K_S-value in the saturation rate equation, Eq. 8.2). Uhlinger and White (1983) have shown that starvation conditions and exposure to greater surface area activated a marine pseudomonad to accumulate EPS in the biofilm. They also demonstrated that EPS synthesis was coincident with cellular stress, as measured by adenylate pools, and proposed that the quantity of EPS produced may be used as an estimate of the nutritional status of these microbes. Since increased EPS production rates were not clearly demonstrated and the portion of EPS detached was unknown, more measurements are necessary for validation.

It is somewhat surprising that the adenylate energy charge and cellular ATP pools remain high in starving copiotrophs (Uhlinger and White, 1983). On the other hand, maintenance of activated membranes for rapid nutrient uptake and a timely recovery, once adequate substrates are available, is dependent on processes that also maintain the adequacy of high energy adenylates. Once the starved cells locate an area of nutrient enrichment, rapid nutrient uptake may take place as a prerequisite to growth. In fact, dwarf cells immediately resumed rapid respiration without any lag, enlarged to normal size and subsequently commenced rapid growth on a surface when nutrients became available (Kjelleberg et al., 1982; Amy et al., 1983). That no macromolecular synthesis is necessary and starved dwarf cells are ready to mount this response was shown by similar results when chloramphenicol was present. The subject of survival strategies of bacteria in natural habitats has been reviewed recently by Roszak and Colwell (1987).

Sufficient nutrients for the growth of copiotrophs in oligotrophic waters are found when (1) there is an intermittent input of energy substrate into the water (death of a large organism, upwelling of nutrients, nutrient input from terrestrial sources), or (2) the bacteria approach an interface (solid–liquid or gas–liquid) where suitable nutrients may accumulate (see Section 10.2.2). Rubio and Wilderer (1986) have observed significantly different accumulation of biofilm in a laboratory rotating biological contactor by imposing a varying substrate load. The biofilm exposed to feast–famine conditions accumulated to a much higher extent than the biofilm fed continuously at the same daily rate.

10.2.1.3 Oligotrophy versus Copiotrophy in Biofilms In too many cases, the basis for classifying a cell as oligotrophic or copiotrophic is evidence from batch experiments. If the cells are adsorbed or in biofilms, this may lead to misconceptions because the adsorbed cells are exposed to a continuous flux of nutrients. Thus, the flow rate may compensate for a relatively low nutrient concentration. For example, using Poindexter's criterion for oligotrophic conditions [1 (g C) m^{-3} day^{-1}] in a RotoTorque, which has a surface area to volume ratio of approximately 35,000 m^{-1}, as many as 3×10^7 cells m^{-2} may accumulate on the RotoTorque walls in one day. Is this an oligotrophic environment or not?

10.2.2 Conditioning Films

The adsorption of macromolecules and other hydrophobic molecules occurs almost instantaneously upon exposure of a clean substratum surface to a natural aqueous environment. Such adsorbed molecular layers are termed conditioning films, because they alter both the surface charge (Neihof and Loeb, 1974) and surface free energy (Baier, 1980; Fletcher and Marshall, 1982b) of the substratum. Some adsorbed macromolecules inhibit bacterial adhesion to the substratum, whereas others have little effect on bacterial adhesion (Fletcher, 1976; Fletcher and Marshall, 1982b). Presumably, conditioning films in nature play a role in modifying the extent of bacterial adhesion to immersed surfaces.

The suggestion by ZoBell (1943) that bacteria grow preferentially on surfaces when exposed to nutrient limitation provided the impetus for microbiologists to further study the importance of interfaces in many marine and aquatic systems. However, it was not until the past decade that experimental approaches were made to some of the more important questions implicit in the original concept. In that time, conflicting evidence has accumulated, some reports stating that growth or activity at interfaces is less than the comparable freely suspended bacteria, and others stating the opposite (see discussion of this topic by Fletcher, 1984). The explanation of this discrepancy was one of the underlying objectives of a study by Bright and Fletcher (1983), who concluded that there was no generalized "surface effect" that is either stimulatory or inhibitory, relative to freely suspended cells. Rather, the effect(s) of surfaces on microbial activity depended heavily on environmental conditions, such as the substrate and its concentration, as well as the substratum properties. However, most recent studies on comparisons of behavior of bacteria in the aqueous phase and at surfaces have been based on the use of low molecular weight, soluble substrates (see Section 10.3.2). In oligotrophic waters such substrates would be rapidly utilized by bacteria in the aqueous phase and would not be expected to accumulate at surfaces. On the other hand, macromolecules and smaller hydrophobic molecules should partition at surfaces to form conditioning films. It is these substrates that may provide an advantage for starving copiotrophic bacteria when they encounter a surface.

Starved bacteria have been used to demonstrate the rapid metabolism of a model surface-bound energy substrate. Kefford et al. (1982) and Kjelleberg et al. (1983) found that a number of bacteria were able to utilize ^{14}C-labeled stearic acid bound to surfaces. In the case of Leptospira (Kefford et al., 1982), many of the bacteria scavenged stearic acid from the surface and then returned to the aqueous phase. Hermansson and Marshall (1985) deliberately isolated a marine bacterium, *Vibrio* MH3, that was unable to firmly adhere to surfaces; they demonstrated that this organism successfully scavenged bound stearic acid from a substratum.

Despite some reports to the contrary, it seems certain that molecular adsorption to substratum surfaces, particularly in flowing systems where nutrients are continually replenished, provides a source of nutrients for bacterial growth in the immediate vicinity of the solid–liquid interface.

By means of studies with a dialysis microchamber combined with time lapse video recording techniques, Kjelleberg et al. (1982) and Power and Marshall (1988) have observed cellular growth and reproduction of starved bacteria adsorbing to a nutrient-enriched substratum. Starved cells of *Vibrio* DW1 did not grow in the aqueous phase of a nutrient medium, where the level of energy substrate was reduced one thousandfold, but the bacteria did grow and reproduce on a dialysis membrane surface, where a continuous flow of the medium allowed nutrients to accumulate (Kjelleberg et al., 1982). What limited the growth rate of the bacteria? Transport and adsorption of nutrients to the substratum or cell physiology?

Individual cells of *Vibrio* DW1 attached to the membrane in a perpendicular manner and then produced a succession of daughter cells that were released directly into the aqueous phase (Kjelleberg et al. 1982). Using stearic acid bound to the membrane surface as the only available energy source, Power and Marshall (1988) observed growth and cell division of previously starved cells of *Pseudomonas* JD8 following attachment to the membrane. After division, the daughter cells slowly migrated over the membrane surface, a phenomenon that may be related to changes in the substratum surface free energy following utilization of the stearic acid substrate in the vicinity of individual cells. The reversibly adsorbed *Vibrio* MH3 was found by Power and Marshall (1988) to grow from the starved minicell form to normal size and to begin the process of binary fission prior to returning to the aqueous phase, where the division process was completed. Similar observations have been made, using image analysis techniques (Chapter 12), on the behavior of bacteria during initial colonization of surfaces (Lawrence and Caldwell, 1987; Lawrence et al., 1987; Escher, 1986), although no attempt was made in these studies to control the nature or location of the available energy substrate.

10.2.3 Adsorption of Bacteria

10.2.3.1 Mechanisms of Adsorption Bacteria possess a net negative charge at pH levels normally encountered in natural habitats, and this creates an apparent problem in the adsorption process, since substratum surfaces in nature either are negatively charged or rapidly acquire a negatively charged conditioning film (Marshall, 1976). It is generally conceded that, while the main body of the bacterial cell does not make direct contact with the substratum surface, adsorption is mediated by a process of bridging to the substratum by fine extracellular structures capable of overcoming the repulsion effects by a combination of Brownian displacement, chemical bonding, dipole interactions, and hydrophobic interactions. The extracellular structures involved in polymer bridging vary with different bacterial types and include pili, fimbriae, flagella, and extracellular polymeric substances (EPS) of many different chemical constitutions (see Chapter 5, Sections 5.4.4 and 5.4.5).

The precise mechanisms for adsorption of bacteria to surfaces are still the subject of considerable debate. In general, two distinct schools of thought are prominent:

1. The *interfacial forces approach,* wherein the adsorption of bacteria to the substratum is considered in terms of (a) *long range forces* that involve the interaction between electrical double layer repulsion forces and van der Waals attraction forces and may be important in *reversible* adsorption, and (b) *short-range forces* that involve chemical bonding, dipole interac-

tions, and hydrophobic interactions between the extracellular components of the bacteria and the substratum, leading to *irreversible* adsorption.

2. The *surface free energy approach,* whereby adsorption is considered in terms of the interaction of the surface free energy of the bacterium, the surface free energy of the substratum, and the surface tension of the liquid phase.

For more detail on the mechanisms of adsorption/adhesion see the books by Marshall (1984) and Savage and Fletcher (1985).

From an ecological viewpoint, it is important to note that different bacteria adsorb in different ways to a range of substrata.

The types and numbers of bacteria colonizing different surfaces, consequently, will vary because of the selective adsorption processes operating in the different situations (Mills and Maubrey, 1981; Baker 1984).

Mechanisms of adsorption must be considered in the context of results in flow systems that indicate that only 0.01–1.0% of the cells contacting the substratum actually adsorb to it. These observations have generally been made in experimental systems with monopopulations where the homogeneity of the population is high. Why do such a small fraction of the collisions result in adsorption? Valeur et al. (1988) suggest that a small percentage of the population have unique properties more conducive to adsorption. If the substratum is rough, are molecular mechanisms of adsorption important to the accumulation of cells at the substratum?

10.2.3.2 Active and Passive Adsorption In their original description of the concepts of reversible and irreversible adsorption, Marshall et al. (1971a) observed a time-dependent, firm adsorption of a marine bacterium and suggested that the time delay may involve a triggering of the production of bridging polymer following the close association of the cell with the substratum surface. Subsequent studies (Fletcher, 1980) revealed that some bacteria exhibit an almost instantaneous adsorption to surfaces. In this case, it was presumed that EPS had been preformed by the bacteria and the firm adhesion was a physicochemical (passive), rather than a physiological (active), process. The suggested differences in process rates, however, may also be dependent on other environmental factors such as the fluid shear stress at the substratum.

The response of bacteria to the presence of a surface is sometimes more complicated than indicated above. For instance, Fletcher and Marshall (1982b) reported that the marine *Pseudomonas* NCMB2021 exhibited a time-dependent irreversible adsorption (adhesion) to a hydrophobic substratum but an essentially instantaneous adhesion to a hydrophilic surface (Figure 10.1). The reason for this difference in behavior is uncertain, but could result from (1) prior production of polymer capable of bridging to a hydrophilic surface but requiring

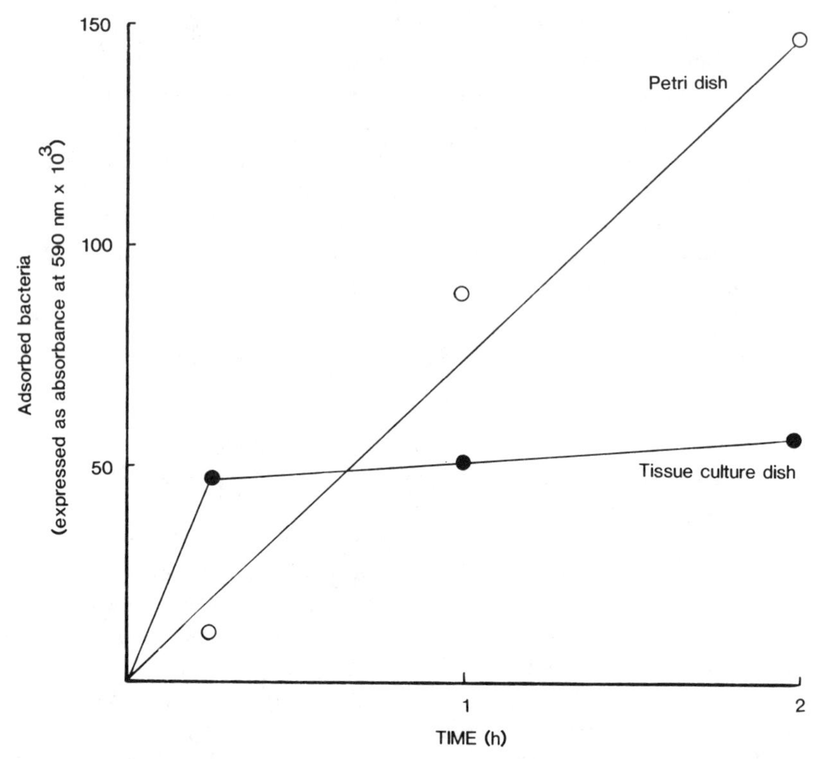

Figure 10.1 Adhesion of *Pseudomonas* NCMB2021 to a hydrophilic tissue culture dish (●) and a hydrophobic petri dish (○) (both polystyrene) at various time intervals (adapted from Fletcher and Marshall, 1982b).

some surface-induced modification for binding to a hydrophobic surface, (2) production of two different polymers by the same cells with only one of the polymers stimulated by surface contact, or (3) the presence of a heterogeneous population of cells, some of which are constitutive for the production of polymer capable of binding to a hydrophilic surface, whereas others respond to the presence of a surface by producing a polymer capable of binding to a hydrophobic surface.

If certain bacteria exhibit a physiological response following a close association with a surface, how do the bacteria sense the presence of the surface and how does this trigger the physiological response? Some suggestions include (1) some form of elastic deformation in the bacterial cell envelope when near a surface (Fletcher, 1984), (2) the possibility of proton accumulation between the bacterium and the substratum, leading to an alteration in the proton motive force across the plasma membrane on the side of the bacterium adjacent to the substratum (Ellwood et al., 1982; Fletcher, 1984), (3) a response to microbially produced surfactants that may be adsorbed to the substratum (Humphrey and Marshall, 1984), and (4) a response to nutrients adsorbed at the solid–liquid interface (Kjelleberg et al., 1982; Power and Marshall, 1988; see Section 10.2.2).

10.2.4 Substratum Roughness

The extent of microbial colonization at a substratum appears to increase with increasing roughness of the substratum (Geesey and Costerton, 1979; Baker, 1984; Characklis, 1984). The potential role of microroughness in adsorption of microorganisms is illustrated in Figure 7.9 (Chapter 7). Advective transport rates near a rough surface will be greater than for a smooth surface, and once a cell arrives at a surface, its sticking efficiency (Chapters 7, 9, 12) will probably be higher as a result of (1) reduced desorption due to lower shear forces experienced by cells shielded from the main fluid flow and (2) more substratum surface area available for adsorption (Characklis, 1984). Although Weise and Rheinheimer (1978), Geesey and Costerton (1979), and Beeftink and Staugaard (1986) found (Figure 10.2a) that attached microorganisms on particles were mostly concentrated in crevices rather than on smooth, flat surfaces, Baker (1984) reported that bacteria did not selectively colonize cavities and grooves despite the fact that roughening the substratum surface greatly increased the rate of bacterial colonization (Figure 10.2b). Although initial colonization may be in crevices and grooves, subsequent spreading of the colonies due to growth may occur over the entire surface. Certainly geometry and fluid dynamics will influence the spatial distribution of colonization.

Although substratum roughness probably plays an important role in the rate and extent of bacterial colonization, little conclusive evidence is available to quantitatively evaluate the effects.

Do surface irregularities serve as anchoring points for bacterial bridging polymer (Baker, 1984)? Does surface roughness encourage the adsorption of an entirely different suite of bacteria than that found on a smooth surface? Does surface roughness reduce the degree of cell desorption (Characklis, 1984)?

10.3 MICROBIAL ACTIVITY WITHIN BIOFILMS

The rate and extent of biofilm accumulation depends on a variety of biological, chemical, and physical factors that are often characteristic of a particular habitat. For example, transport and interfacial transfer processes probably control the ultimate thickness of a biofilm in high hydrodynamic shear stress environments. Likewise, microbial activity within the biofilm can influence many important properties of it, including the microbial population distribution and product formation.

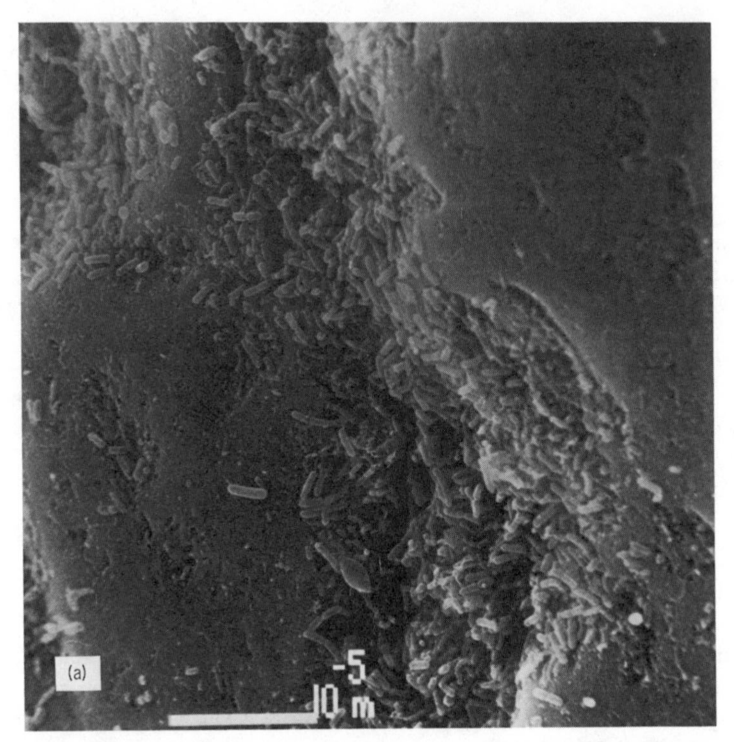

Figure 10.2 (a) Scanning electron micrograph of mixed population bacterial colonization of a sand grain in an experimental fluidized bed reactor (line = 10 μm). Colonization during the early stages of operation is limited to the crack and crevices of the particle. (Reprinted with permission from Beeftink and Staugaard, 1986.) (b) The average population densities of bacteria adhering to untreated (○) and scratched (●) glass and to untreated (□) and scratched (■) polystyrene. (Adapted from Baker, 1984.)

10.3.1 Microbial Succession

The fate of bacteria responsible for initial colonization of a freshly immersed solid substratum depends on the flow rate, the trophic state of the bulk water, and the type and physiological state of the bacteria adsorbing to the surface. In flowing systems, many apparently adsorbed bacteria readily desorb from the surface and play no role in biofilm formation (Escher, 1986; Lawrence and Caldwell, 1987). Other primary colonizing bacteria remain immobilized on the surface but do not grow or even lyse, whereas many of the biofilm bacteria actually grow and reproduce (Bott and Brock, 1970; Kjelleberg et al., 1982; Escher, 1986; Lawrence and Caldwell, 1987; Power and Marshall, 1988).

Growth and reproduction of the primary colonizing bacteria is not the only contribution to biofilm accumulation. These primary organisms may modify the surface characteristics of the substratum (including formation of corrosion products), rendering it suitable for subsequent colonization by secondary micro-

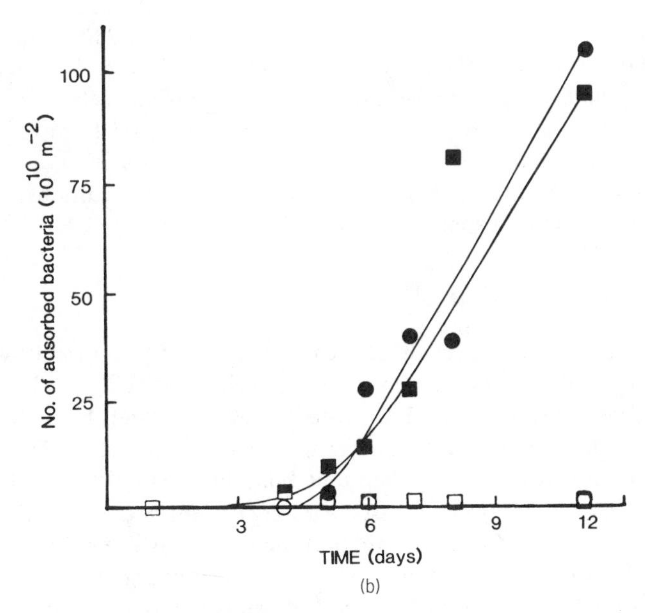

Figure 10.2 Continued

organisms. Marshall et al. (1971b) and Corpe (1973) observed that initial colonization of substrata immersed in seawater by rod-shaped bacteria was followed by other bacteria such as *Caulobacter, Hyphomicrobium,* and *Saprospira*. These studies have been confirmed by the use of scanning electron microscopy, which indicated increasing complexity in the microbial community of a biofilm with time and a composition that varies markedly with the substratum (Mack et al., 1975; Gerchakov et al., 1977; Marszalek et al., 1979; Dempsey, 1981). For example, microalgae are generally primary colonizers of surfaces exposed to light (Jordan and Staley, 1976), but diatoms, other algae, fungi, and protozoa are found in more mature biofilms (Mack et al., 1975; Gerchakov et al., 1977; Marszalek et al., 1979).

10.3.2 Specific Growth Rates

The specific growth rate of microorganisms within a biofilm will be a significant factor influencing the population distribution in the biofilm. Those organisms that can grow fastest in the various layers or niches (microenvironments) encountered within the biofilm will probably dominate within the film, at least according to the mathematical model of biofilm population dynamics described in Chapter 11. Can the specific growth rate in the biofilm state be predicted by measurements conducted in the planktonic state? Alternatively stated, does the mere presence of the organism in a biofilm matrix alter its physiology?

Experimental results with *Pseudomonas aeruginosa* (Bakke et al., 1984) suggest that the growth rate of a microorganism in the biofilm and in the planktonic state are the same so long as the microenvironment of the cell is the same. The results are based on *in situ* measurements of microbial growth rates in a biofilm (RotoTorque) as compared with chemostat experiments using the same growth medium (glucose and mineral salts) and environmental conditions. A more detailed analysis of these experiments is presented in Chapter 13.

10.3.3 Product Formation

A comparison of available results regarding the rate of EPS formation in planktonic and in biofilm states does not produce such a clear picture. Trulear (1983) and Turakhia (1986) measured the rate of EPS formation by a *Pseudomonas aeruginosa* biofilm in a RotoTorque system. The kinetics of the EPS formation rate were suitably described by the Leudeking–Piret equation (Eq. 8.23, Chapter 8) in which the non-growth-associated coefficient was essentially zero. Jarman and Pace (1984), however, state that the maximum theoretical yield of EPS from carbohydrates is not strongly dependent on the energy efficiency of the cell. Energy can come both from catabolism and from oxidation during the formation of certain polysaccharide components.

Robinson et al. (1984) determined the rate coefficients for this same process in a chemostat (Table 6.7, Chapter 6). Figures 8.8 and 8.9 are from the chemostat experiments with *Pseudomonas aeruginosa* and indicate that the EPS formation rate increases with increasing specific growth rate ($D = \mu$) while the extent of EPS accumulated per cell decreases with increasing specific growth rate. Could it be that at low dilution rates (high residence times) the cell has more time to accumulate EPS even though EPS is being produced at a lower rate? At low growth rates, the cell carbon was approximately equal to the EPS carbon (Figure 8.9), and carbon dioxide accounted for a very small portion of the glucose carbon utilized. In a sulfur-limited *Xanthamonas campestris* chemostat, EPS accounted for 62% of the glucose carbon utilized by the culture (Jarman and Pace, 1984). Cell carbon accounted for 11%, and carbon dioxide for 30%.

The kinetic coefficients obtained in the chemostat (planktonic state) and in the RotoTorque (biofilm state) do not compare well in the case of Turakhia's results (Table 6.7). On the other hand, kinetic coefficients determined from Trulear's biofilm experiments are not significantly different from those in the chemostat. Results from other experimenters in a nitrogen-limited system yield coefficients similar to Turakhia's (Table 6.7).

The role of EPS in biofilm activity is not clear. However, EPS is known to concentrate ions from the aqueous phase. This concentration process is being explored at present to determine the efficacy of biofilms for concentration of heavy (generally commercially or strategically valuable) metals from dilute aqueous solution. Many observations of metal concentration by microorganisms have been made in the marine environment (Cowen and Silver, 1984), in wastewater treatment (Brown and Lester, 1979, 1980; Rudd et al., 1984), and in research studies (Hsieh et al., 1985; Norrberg and Persson, 1984). The metal immobilization has generally been attributed to the EPS surrounding the cell. In some cases, specificity of microbial cell adsorption to metal substrata has been attributed to the specificity of the EPS–metal interaction (Zaidi et al., 1984).

Namkung and Rittmann (1986) have determined the rate of soluble product formation by mixed population biofilms using phenol at very low substrate loading rates as the sole energy and carbon source. Their results are consistent with *Pseudomonas aeruginosa* EPS formation in that growth-associated product formation is more important than non-growth-associated. The products were rather high molecular weight, more than 70% of the effluent SOC having a molecular weight greater than 1000 daltons.

Other environmental factors may influence product formation. For example, Adamkiewicz and Pilon (1983) have even observed the stimulation of polysaccharide production by magnetic fields. *Streptococcus mutans* accumulated up to 66% more polysaccharide on surfaces facing magnetic north than on surfaces facing south. Reversal of the magnetic field by 180° caused a similar reversal of accumulation. Quinlan (1983) has discussed the potential effect of temperature on metabolic excretion rates.

10.3.4 Influence of Substrate Loading Rate

The biofilm–substrate reaction rate, also described as the substrate flux into the biofilm, increases with increasing substrate loading rate (Figure 9.31, Chapter 9).

Generally, increasing the substrate loading rate increases the substrate flux into the biofilm.

At a high enough substrate loading rate, the substrate flux into the biofilm will reach a constant value as a result of one of the following: (1) the growth rate of the microbial population in the biofilm reaches its maximum, (2) the biofilm thickness exceeds the penetration depth of the substrate into the biofilm, or (3) the electron acceptor (e.g., oxygen) or another nutrient becomes limiting.

The influence of substrate loading on substrate flux may be due to the increased density of cells observed at higher substrate loading rates (Table 4.4 and Figure 4.8, Chapter 4). The cell accumulation rate at the substratum, even in the early stages of colonization, is dominated by growth processes (Chapter 12). Thus, high substrate loading leads to higher cell densities in the biofilm (Figure 10.3) and results in increased substrate flux into the biofilm.

10.3.5 Gradients

The metabolic activity of a biofilm is rate-controlled by transport processes. This is the major fact distinguishing biofilm activity from planktonic microbial activ-

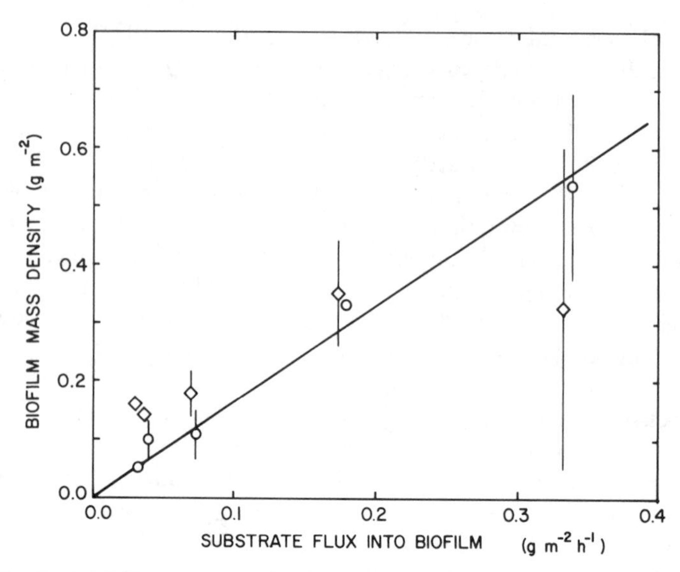

Figure 10.3 Both biofilm cell density (○) and biofilm EPS density (◇) increase with increasing substrate flux into the biofilm. The biofilm cell density is essentially directly proportional to the substrate flux (Trulear, 1983).

ity (Chapter 9). A good example of the influence of transport processes on microbial activity is the depletion of oxygen within a biofilm, which results in an aerobic and an anaerobic layer. Schaftel (1982) has demonstrated dissimilatory sulfate reduction in a RotoTorque reactor in which the effluent dissolved oxygen concentration exceeded 2 g m^{-3}. Since dissimilatory sulfate reduction cannot occur unless the redox potential is at -100 mV or lower, an anaerobic microniche exists in the lower layers of the biofilm.

Alldredge and Cohen (1987) used microelectrodes to demonstrate the existence of persistent oxygen and pH gradients around flocculent, macroscopic marine particles. Oxygen could be completely depleted in large fecal pellets. Mass transfer boundary layers hundreds of micrometers thick were observed despite advective processes around the particle. They conclude that existence of chemical microniches on the scale of millimeters around macroscopic particles may significantly influence the distribution and activity of marine microorganisms.

Concentration gradients can influence microbial population diversity and spatial distribution as well as microbial metabolic activity within biofilms.

Transport limitations not only control process rates but can sometimes alter the stoichiometry of processes within the biofilm. For example, Schaftel (1982) presents results indicating that anaerobic layers within biofilms form organic products that can be used by the overlying aerobic biofilm layers as substrate.

One interesting observation from numerous biofilm studies is that monopopulation biofilms generally attain a smaller steady state thickness than mixed population biofilms. For example, experiments with *Pseudomonas aeruginosa* biofilms attain a maximum thickness of approximately 30–40 μm, while mixed populations under the same conditions can be as thick as 130 μm (Trulear, 1980, 1983). These observations suggest that various microniches may result in a more diverse microbial population that is more efficient at either producing biomass (cells and EPS) or immobilizing it at the substratum.

10.3.6 Changes in Cell Morphology

Their presence in a biofilm may influence the morphology of certain microbial species. For example, several investigators (Picologlou et al., 1980; Trulear and Characklis, 1982; McCoy and Costerton, 1982) have observed the predominance of filamentous organisms in biofilms leading to a relatively low biofilm density (Section 4.5.1.2, Chapter 4). The filamentous characteristic of the biofilm can significantly increase the fluid frictional resistance resulting from biofilm accumulation (Section 9.5.3.2, Chapter 9). The filamentous morphology may be elicited by a microbial metabolic response to environmental stress. For example, McCoy and Costerton (1982) observed characteristic filamentous pseudomonads in biofilms subjected to turbulent flows. Pseudomonads are not generally considered to be filamentous, but Jensen and Woolfolk (1985) have observed complete filamentation of *Ps. putida* 40 under oxygen limitation in the latter stages of batch growth. In a biofilm, the cells in the lower layers may frequently be limited by oxygen, lack of which may trigger filamentous growth.

Ou and Alexander (1974) observed that *Bacillus megaterium* growing in the presence of glass beads (29 and 53 μm average diameter) were frequently filamentous, sometimes reaching lengths of 600 μm. The filaments were nonseptate. The formation of filaments was prevented by addition of magnesium but not by several other cations.

10.3.7 Effects of Light

Photosynthesis is the major source of energy in many natural aquatic habitats as well as some technological environments. For example, algal-bacterial biofilms can foul the distribution deck (plenum) of an open recirculating cooling tower, causing blockage of water flow; they can increase corrosion rates, and they form dissolved organic carbon compounds, which serve as nutrients for bacteria in the closed part of the cooling water recirculation system (Grade and Thomas, 1981). The algal biofilms may also harbor various troublesome bacteria such as Legionella species (Soracco et al., 1983).

Obviously, the photosynthetic organisms will grow in direct proportion to the light and nutrients available, but chemosynthetic organisms (autotrophs and heterotrophs) also grow in most of these mixed biofilms (Figure 10.4). The interac-

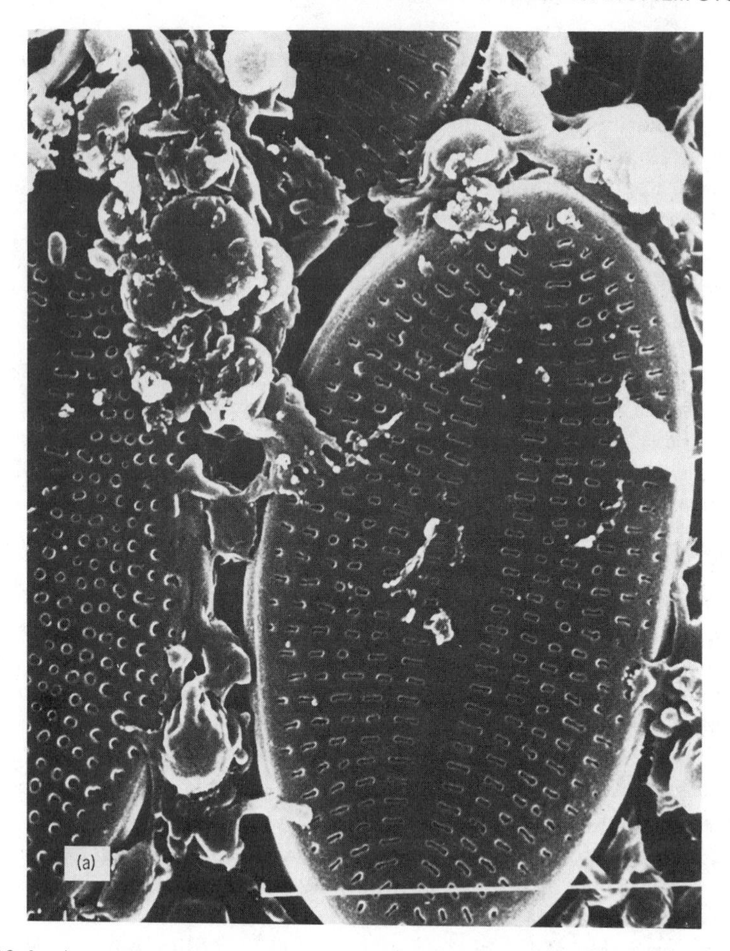

Figure 10.4 A scanning electron micrograph of a diatom and bacterial biofilm from a clean mountain stream. (a) A bacterial biofilm surrounds a diatom cell but does not colonize it. Line = 5 μm. (b) The community includes diatoms and bacteria in close association. (Escher and Characklis, 1982.)

tions between bacteria and algae in aquatic systems has been reviewed by Cole (1982), and a schematic illustration of the cycling of oxygen and carbon within an algal–bacterial biofilm is presented in Figure 10.5. Photosynthetic biofilms, or *mats,* receive their energy from light. Haack and McFeters (1982) observed that heterotrophic bacteria were responding to the metabolic activities of the phototrophic population in an alpine stream. Their measurements indicated that there was a direct flux of dissolved organic algal products into the heterotrophic bacteria. The cycling of carbon and oxygen in algal bacterial aggregates and biofilms has been observed in the marine environment as well (Azam and Ammerman, 1984). A conceptual model explaining some of the unexpected observations resulting from nutrient cycling within photosynthetic biofilms has

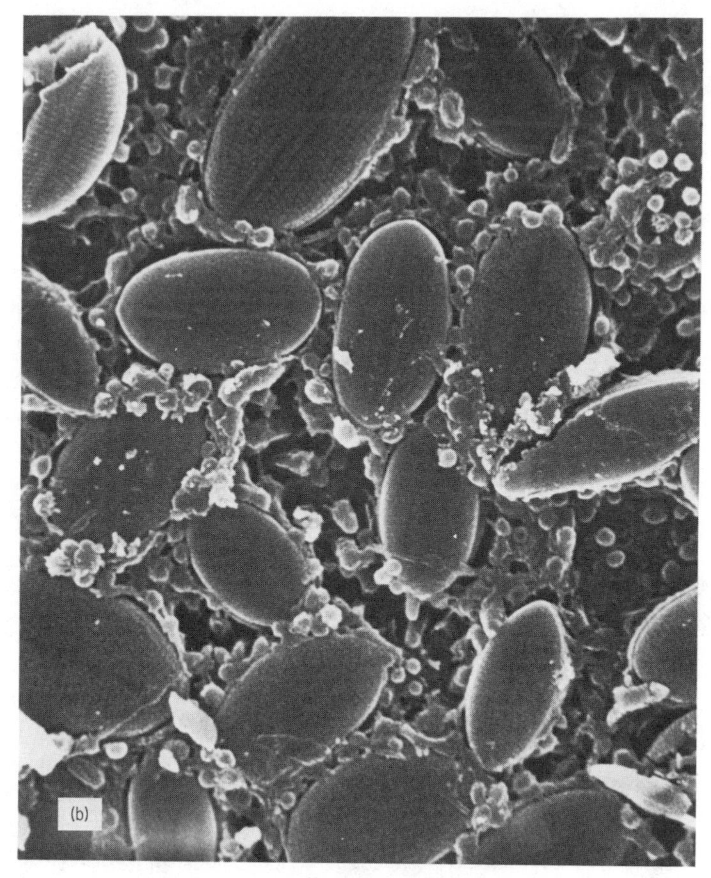

Figure 10.4 Continued

been proposed by Escher and Characklis (1982). Chemical gradients in these films dictate the species diversity and the spatial distribution of the populations, as evidenced by the microelectrode investigations of Jorgensen (1980). Although the biofilm–bulk water interface in photosynthetic systems is generally aerobic, methanogenic and sulfate-reducing bacterial activity have been measured in the deeper layers (Ward, 1978).

10.3.8 Microbial Interactions

The most important characteristic of natural microbial communities, in contrast with pure cultures of specific microorganisms, is the wide range of interactions occurring between the various microorganisms. The extent of such interactions is usually large in biofilms because of the high population densities and long cell

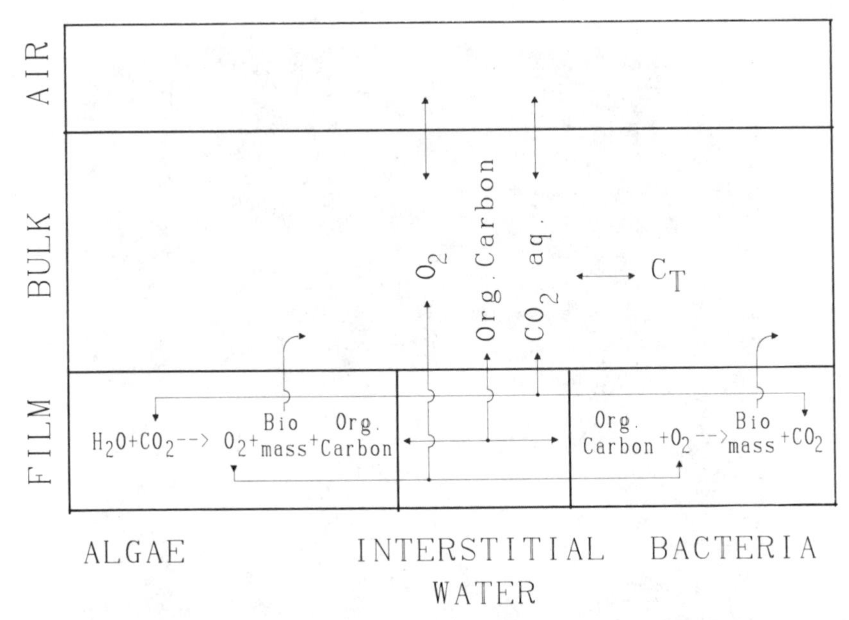

Figure 10.5 A schematic illustration of the cycling of carbon and oxygen in an algal-bacterial biofilm. Algae or diatoms obtain energy from sunlight, consume carbon dioxide (produced by bacteria), and produce organic carbon and oxygen. The bacteria consume the oxygen and organic carbon produced by the algae and produce carbon dioxide. Some of the products escape into the bulk water compartment. (Reprinted with permission from Escher and Characklis, 1982.)

residence times found in such films. The major types of interactions (Odum, 1971) expected in biofilms are described in the following sections.

10.3.8.1 No Interaction A lack of interaction between two organisms is termed *neutralism* and, in most circumstances, is only observed at very low population densities. It is unlikely that neutralism will occur in biofilm communities.

10.3.8.2 Beneficial Interactions When one population within a community benefits from the activity of the second population and the latter population is unaffected, the interaction is termed *protocooperation*. For example, the second population may excrete vitamins or other growth factors, or it may degrade complex energy substrates to yield simple substrates that stimulate the growth of the first population. In many situations, the activity of each population results in a stimulation of the other population (cross feeding effects). If the association of the populations is not obligatory, it is termed *commensalism;* if it is obligatory, it is termed *mutualism*. There is no doubt that beneficial interactions occur between microbial populations in biofilms, as in other microbial communities, but little or no work on this important aspect of microbial activity in biofilms has been reported.

10.3.8.3 Detrimental Interactions

Competition, in the strict ecological sense, refers to the detrimental effect on both interacting populations resulting from the common requirement for some limiting factor, such as space, nutrients, or oxygen. This is almost certainly the most common form of detrimental interaction occurring in a high population density biofilm and is the major interaction considered in modeling biofilms at present (Chapter 11). A direct antagonism towards another population is termed *amensalism,* and may consist in the production of organic acids, antibiotics, or even lytic enzymes (enzymes that break down cell wall components) by one or other of the interacting populations. Again, such interactions must occur in biofilms, but have not been studied in any detail.

10.3.8.4 Mixed Interactions

Mixed interactions occur where one population (the host) is destroyed for the benefit of the second population (the parasite or predator). *Parasitism* refers to the situation where the aggressor invades the host and normally multiples within the host. Bacterial viruses (phages) are regarded as bacterial parasites, as are the small parasitic bacteria of the genus *Bdellovibrio* that penetrate the cell walls of other bacteria and multiply in the space between the wall and the plasma membrane. *Predation* refers to the situation where the aggressor is larger than the host, which it engulfs and digests internally. Many types of protozoa are predators on bacteria and microalgae. Amebae engulf smaller microorganisms by means of highly fluid pseudopodia, whereas flagellates and ciliates sweep the host into some form of opening in their cells. In all cases, the host cells become enveloped in a membrane (the food vacuole) and are digested by the predator.

Probably the best descriptions of the biological communities and their interactions in biofilms are of those in trickling filters (Cooke, 1959; Curds, 1965; Güde, 1979; Mack et al., 1975). Following bacterial colonization of the substratum, algal growth (particularly diatoms) is observed along with active protozoan predation on the biofilm community (Curds, 1965; Mack et al., 1975). Although protozoan predation is usually regarded as a detrimental effect on the host cells, grazing of biofilms by protozoa may maintain the bacteria there in a more active physiological state (Fenchel, 1986). Protozoan grazing of bacteria in activated sludge systems can be a highly selective process, leading to an altered dominance of certain bacterial types in the presence of the protozoa (Güde, 1979). Motile protozoa may also provide increased transport rates for nutrients into the depths of the biofilm by virtue of turbulence generated by their movement through the biofilm.

10.3.9 Genetic Exchange in Biofilms

Genetic exchange between planktonic bacteria in aquatic habitats is normally limited by their low population densities, which lead to low collision frequencies and short contact times. The substantially higher population densities in biofilms, combined with greater physiological activity where nutrients are continuously replenished in flowing systems, may lead to greater opportunities for various forms of genetic exchange within populations in the biofilms (see Chapter 5, Sections 5.9.4.1 and 5.9.4.2). On the other hand, the EPS matrix may impede transport of the genetic material from one cell to another unless intimate cell contact exists. Increased stability of a recombinant plasmid of *Escherichia coli* in continuous culture has been demonstrated in artificially immobilized cells as compared to planktonic forms (Nasri et al., 1987). However, the increased stability was due neither to increased plasmid transfer nor to an increase in plasmid copy number. More investigations are necessary to comprehend genetic processes within biofilms and their importance in ecological and technological processes.

10.4 DETACHMENT

Detachment is defined as the transfer of biomass from the biofilm to the bulk liquid compartment. Biofilm detachment can be the rate-limiting process that determines the metabolic state (average specific cellular growth rate) of steady state biofilms (Bakke et al., 1984). Biofilm detachment may often control biofilm behavior in nature and may also serve as a means of artificial control. The central role of detachment processes in determining the composition and performance of mixed population biofilm reactors, and the need for further investigation of detachment processes, have been demonstrated through theoretical analysis of biofilm reactors (Howell and Atkinson, 1976; Wanner and Gujer, 1985). For example, in fixed film wastewater treatment systems detachment transfers the immobilized biomass into suspended solids, which must be removed prior to discharge of the treated wastewater. The performance of a wastewater treatment plant is assessed in part by the suspended solids in the effluent; excessive suspended solids degrade process performance.

Detachment from the biofilm is a major influence on biofilm system performance.

In drinking water distribution systems, biofilm cells on the pipe walls detach, leading to increased planktonic cell numbers, which degrade the quality of the drinking water.

10.4.1 Erosion

Erosion is the continuous removal of small particles from the biofilm at the biofilm–liquid interface, at least partially due to momentum transfer from the bulk liquid. The friction imposed by the biofilm on the liquid is, in other words, experienced as a shear force on the biofilm. The kinetic energy of the liquid is dissipated by breaking bonds in the biofilm, causing detachment. Erosion, therefore, is a function of a momentum transfer rate or hydrodynamic shear stress at the biofilm–water interface.

Erosion is the predominant means of detachment in biofilms receiving low substrate loadings under turbulent conditions. Some data on erosion rate as a function of biofilm thickness and fluid shear stress have been collected (Chapters 7 and 9). Environmental factors may influence erosion rates. For example, Turakhia and Characklis (1989) have observed that free calcium in the bulk water decreases the erosion rate in a *Pseudomonas aeruginosa* biofilm (Chapter 13). Oakley (unpublished data) has observed differences in erosion rate with various substrata (plastic surfaces). Chang and Rittmann (1988) indicate that erosion may not be evident in the initial stages of biofilm accumulation when the substratum is rough. During the early stages of colonization, the biofilm accumulates in the roughness crevices, where it is protected from the fluid shear forces.

Detachment (erosion) has been directly related to microbial growth rate in the biofilm in pure culture studies (Bakke et al., 1984).

10.4.2 Sloughing

Sloughing refers to sporadic detachment of large fragments of biofilm resulting from changing conditions within the biofilm. These conditions may evolve slowly and cause sloughing at random, or they may be triggered by transitions in the environment (internal or external to the film). Sloughing is frequently observed at high substrate loadings and laminar flow or low shear stresses. No quantitative phenomenological models for sloughing exist, although Howell and Atkinson (1976) prepared a model in which high organic loadings in trickling filters (laminar flow) was the driving force for oxygen depletion in the lower depths of the biofilm, which in turn caused massive sloughing. Under anaerobic conditions, gas formation may lead to sloughing. Jansen and Kristensen (1980) have observed the formation of N_2 bubbles under denitrifying biofilms and reason that sloughing may even result from pressure changes causing the bubbles to expand. Siegrist (1985) suggests electrochemical phenomena may contribute to sloughing because charge accumulation in the biofilm from diffusing and reacting species increases the osmotic pressure within the film. Matson (unpublished results) has observed a similar phenomenon on rapidly changing the biofilm environment from aerobic to anaerobic, but this may due as much to the transient as to the anaerobic environment.

Howell and Atkinson (1976) suggest that sloughing has a major influence on the performance of (trickling) filters and must be investigated more thoroughly, for without proper understanding of the sloughing process, optimal design and operation of trickling filters is unlikely.

10.4.2.1 Transients: Substrate Loading Shock or pulse loading has been suggested as a probable cause of sloughing. Bakke (1983) observed a remarkable biofilm phenomenon. He observed massive sloughing immediately upon increasing the substrate loading rate on a mixed population or *Ps. aeruginosa* biofilm in a RotoTorque. Bakke used step increases in inlet substrate concentration (10–20 g m^{-3}) on a RotoTorque to observe the response of mature biofilms. Using lactate or lactose as the substrate and *Ps. aeruginosa* or a mixed population, he increased the supply of growth substrate to the biofilm stepwise, with the following results: (1) biofilm material was immediately sloughed, (2) biofilm cell numbers remained constant, and (3) specific substrate removal rate and product formation rate increased immediately. These observations suggest that the biofilm organisms may slough their exopolymers in response to the shock. In addition, the immobilized cells appear to possess a *reaction potential* (Busch, 1971), which is expressed in response to an instantaneous increase in substrate loading.

Bakke suggests that electrochemical mechanisms may be important in his observations of substrate shock loading. Transport of lactate and lactose through cell membranes requires H^+ (a proton) to be transported in symport (i.e., in parallel) with the substrate (Chapter 5). Glucose, on the other hand, does not require proton transport in symport. The sloughing was considerably less when glucose was the substrate. Thus, Bakke hypothesized that rapid changes in electric field occurred within the biofilm in response to a step increase in substrate, especially in the immediate proximity of the cells. The effect was not observed to the same degree with glucose as the substrate, presumably because glucose does not require a proton for transport through the cell membrane. Thus, the step increase in substrate presumably causes changes in the electrochemical properties of the biofilm EPS, since EPS are essentially polyelectrolytes and may interact significantly with other charged species in the film. There is some indication that microbial EPS form gels that are piezoelectric (Bakke, unpublished observations), causing volumetric expansion or contraction in response to changes in potential. The EPS are largely responsible for the cohesive nature of the biofilm, and a change in EPS volume, especially in the presence of hydrodynamic shear forces, may result in sloughing.

Lee et al. (1975) modeled the aggregation–disaggregation phenomenon in activated sludge bacterial flocs in a manner consistent with Bakke's hypothesis for biofilm sloughing. The authors suggest that the rate of aggregation of cells is proportional to the numbers of single cells present. However, the disaggregation rate is proportional to substrate concentration as well as aggregate number concentration, even during pulse or step increases in substrate concentration. Thus, step increases in substrate concentration will cause disaggregation according to the model. The hypothesis is consistent with superficial observations, in operating activated sludge facilities, that effluent suspended solids increase when shock substrate loadings are encountered.

10.4.2.2 Other Transients Other factors may also influence sloughing. Polymerases (e.g., proteases, polysaccharases) produced by biofilm organisms may lead to weakening of the EPS matrix, making the film more susceptible to fluid shear forces. The polymer degradation may occur in nutrient-poor regions in the biofilm, i.e., near the substratum in a thick biofilm. Predation may also influence activated sludge disaggregation, a process analogous to biofilm sloughing (Curds, 1965, 1973).

Even changes in temperature in the bulk water or at the substratum (e.g., as illustrated by heat exchangers) may cause a sloughing. Sloughing has also been observed in conjunction with flow velocity variations (Chapters 14 and 15).

10.4.3 Relationship between Sloughing and Erosion

Erosion and sloughing are not necessarily exclusive phenomena. Erosion shears particles from the surface film as a function of hydrodynamic shear forces. The "peaks" of a rough surface film are presumably more subject to erosion and, when removed, result in a smoother surface film, which leads to decreased mass and momentum transfer. If substrate limitations in the film cause sloughing, as Howell and Atkinson (1976) suggested, then decreased substrate flux into the film due to smoothing by erosion may lead to sloughing. Removal of biofilm fragments during sloughing will lead to increased biofilm roughness, and thereby to increased mass and momentum transfer. Increased momentum transfer may in turn lead to increased erosion. Since erosion may enhance sloughing and vice versa, it is probable that most biofilms will reach some balance between the two processes over a long time period. A biofilm will display a rougher morphology when sloughing is the dominating process than when erosion is. However, a rough morphology does not necessarily imply that sloughing is the dominant detachment process.

10.4.4 Influence of Detachment on Biofilm System Performance

Howell and Atkinson concluded that sloughing may have major impact on biofilm reactor performance and cause widely fluctuating output even if influent concentration and operating conditions are constant. By comparing extreme cases of sloughing and erosion, Wanner and Gujer (1985) demonstrated that detachment plays an important role in multispecies biofilm progressions, composition, and behavior (Chapter 11). Theoretical examination, as well as experimentation, is required on detachment processes in mixed population biofilms to quantify the effect of detachment on biofilm system performance.

Drinking water distribution systems offer an illustration of the potential critical role of detachment on system performance. Mixed population biofilms accumulate on the pipe surfaces and, for the most part, do not result in any detectable problems. However, a sloughing episode may result in rapid release of undesirably high numbers of coliforms, requiring intervention to maintain health standards. It is probable that Legionella is sometimes released in this manner from a variety of water systems, resulting in serious health hazards.

10.5 GEOMETRIC AND SPATIAL PROPERTIES OF BIOFILMS

A biofilm frequently can be thought of as being composed of a base film and a surface film (Chapter 1). The surface film represents the roughness of the biofilm and the manner in which it interacts with the bulk liquid. Hydrodynamic measurements (see Chapter 9) clearly indicate an increasing "equivalent sand roughness" as a mixed population biofilm accumulates (Figure 9.15, Chapter 9). For example, biofilm roughness may result from filamentous organisms proliferating in the biofilm (Section 10.3.6). However, Bakke (unpublished results) has reported increasing biofilm roughness in a monopopulation *Pseudomonas aeruginosa* biofilm as well. After 50 h accumulation, the biofilm thickness measured *in situ* (Bakke and Olsson, 1986) is uniform across the width of the flow channel (Figure 10.6a and b). After 300 h accumulation, the average thickness has not changed significantly but the variation in thickness resulting from channelization of the film is obvious (Figure 10.7a and b). The variance in the biofilm thickness measurement increases throughout the experiment, further emphasizing the changing roughness of the biofilm (Figure 10.8). The influence of changing roughness on metabolic activity in a monopopulation biofilm is not clear. One hypothesis suggests that roughness decreases the diffusional length, permitting greater respiratory activity in the lower depths of the film. Oxygen flux to the obligate aerobe *Pseudomonas aeruginosa* in the lower depths is critical to the activity in these regions, and so may also limit the average thickness of these films. Comparative observations indicate that monopopulation *Pseudomonas aeruginosa* biofilms attain a maximum thickness of approximately 40 μm, while mixed population biofilms under the same nutrient and oxygen loading reach thicknesses of as much as 140 μm. Perhaps facultative organisms can accumulate and proliferate in the lower layers of the mixed population biofilm, and their cell volume contributes to the biofilm thickness.

Heterogeneous patchy spatial distributions characterize biofilm processes and may influence the population spatial distribution within the film. Nelson et al. (1985) and Escher (1986) have observed that *Pseudomonas aeruginosa* tends to adsorb to glass in a uniform (as opposed to random or aggregated) pattern. Siebel (1987) has observed similar behavior of *Pseudomonas aeruginosa* on polycarbonate. However, Siebel also investigated the adsorption of *Klebsiella pneumoniae* on polycarbonate and observed a very aggregated distribution of colonies. When *Pseudomonas aeruginosa* and *Klebsiella pneumoniae* were mixed, the colonization pattern appeared additive, i.e., *Pseudomonas aeruginosa* uniformly colonized the substratum in the area between *Klebsiella pneumoniae* aggregates. Siebel suggests that certain known characteristics of the species dictate the observed pattern. *Pseudomonas aeruginosa* is motile, while *Klebsiella pneumoniae* is not. Thus, *Pseudomonas aeruginosa* is transported (in laminar flow) 50 to 100 times faster to the substratum. *Klebsiella pneumoniae* grows approximately 5 times faster than *Pseudomonas aeruginosa*, so that when a *Klebsiella pneumoniae* adsorbs, a colony (i.e., an aggregate) forms very rapidly. *Klebsiella pneumoniae* also produces more EPS, so that its

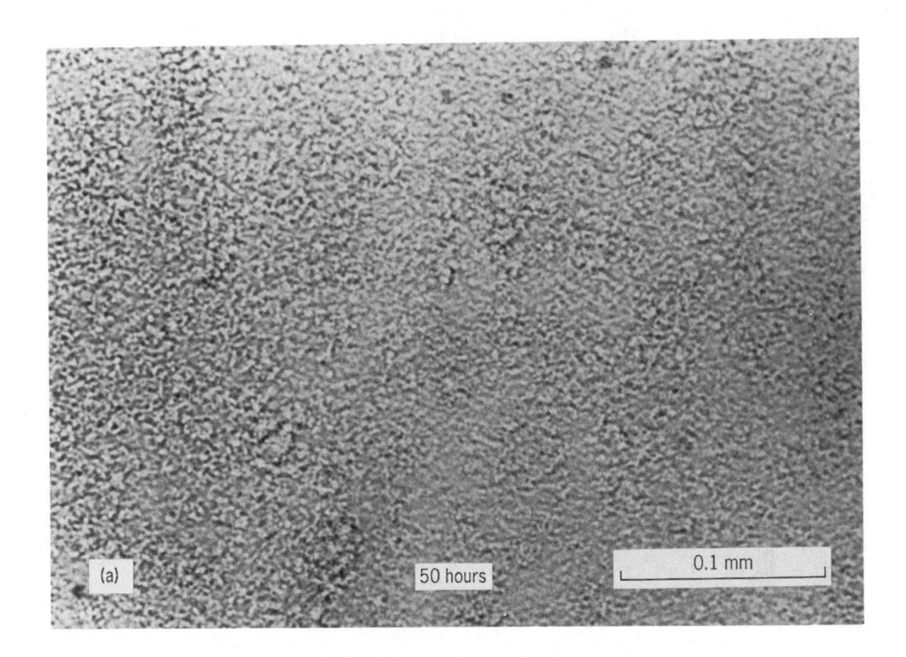

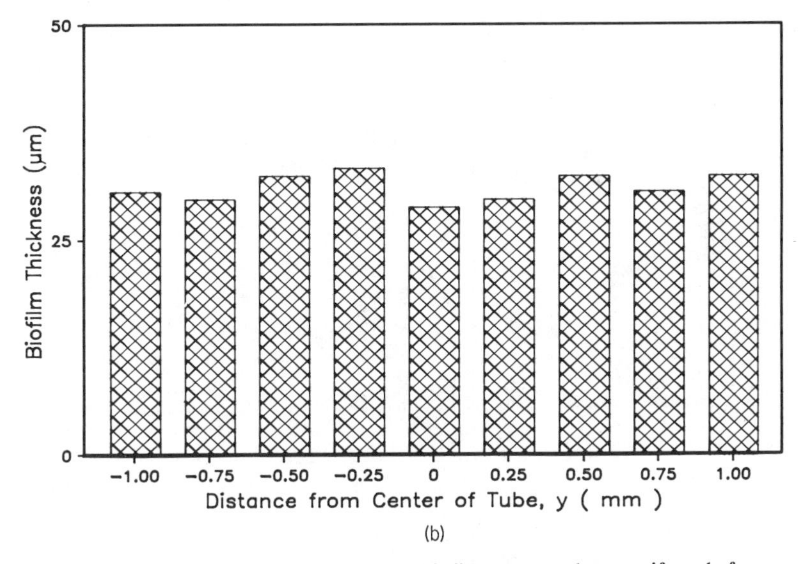

Figure 10.6 (a) A *Pseudomonas aeruginosa* biofilm accumulates uniformly for as much as 50 h. (b) The biofilm thickness is fairly uniform after 50 h. (Courtesy of R. Bakke.)

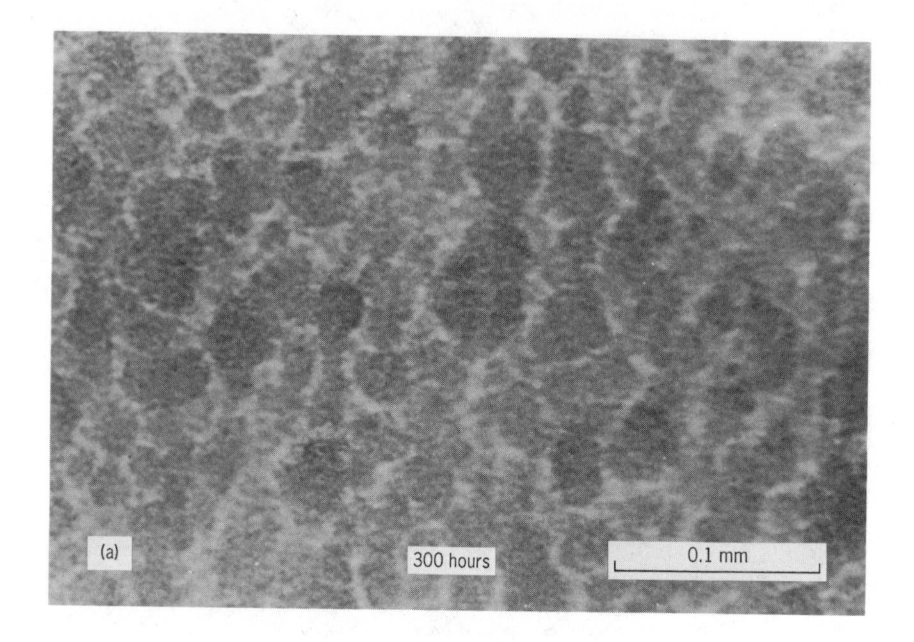

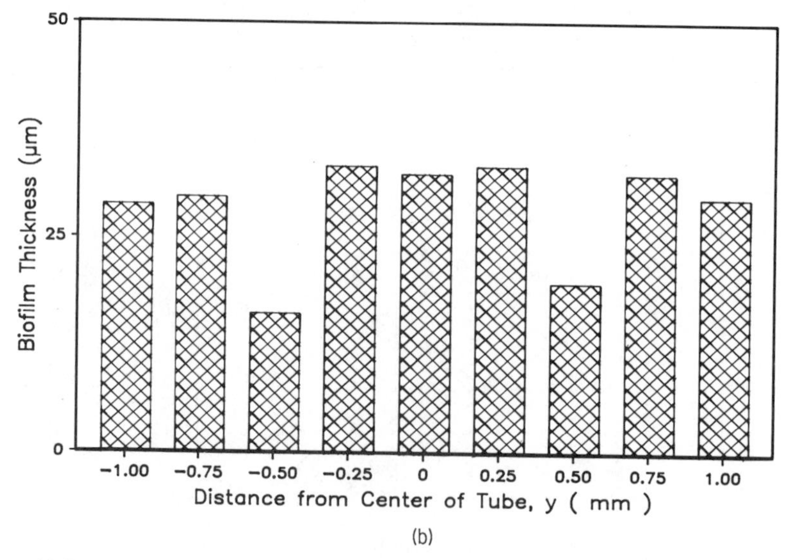

Figure 10.7 After 300 h accumulation, (a) channels can be seen in the *Pseudomonas aeruginosa* biofilm; (b) the average thickness has not changed significantly, but the variation in thickness resulting from channelization of the film is obvious. (Courtesy of R. Bakke.)

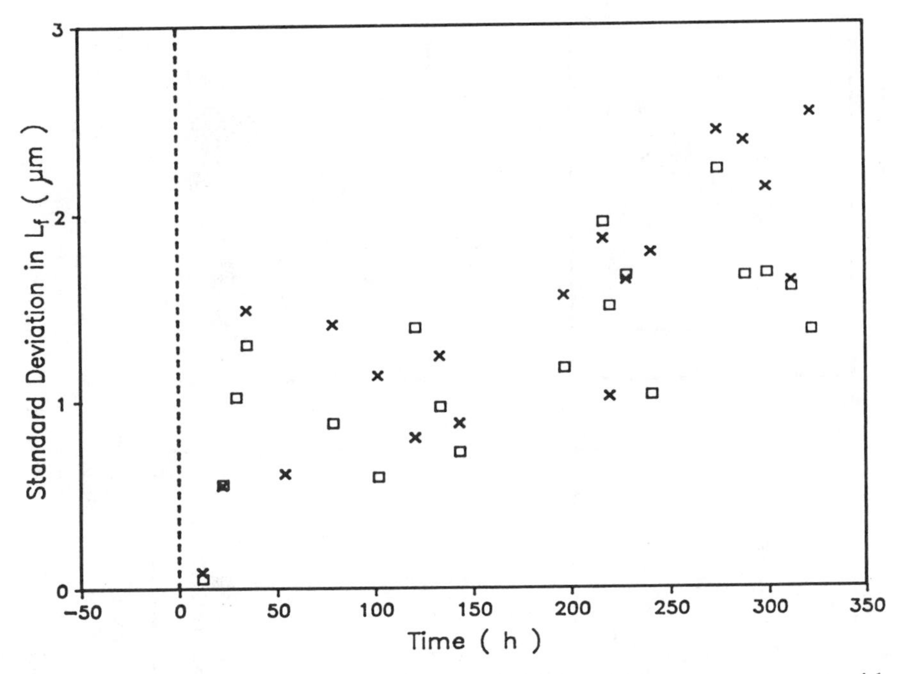

Figure 10.8 The increasing standard deviation in biofilm thickness measurements with time reflects the increasing roughness of the biofilm. The different symbols represent results from two different experiments.

daughter cells may resist detachment to a greater degree. A schematic diagram of the individual and mixed accumulation patterns is presented in Figure 10.9. The spatial pattern remains fairly patchy even after the film accumulates for approximately approximately 5 days. The biofilm thickness is quite irregular, containing *Klebsiella pneumoniae* colonies as much as 100 μm high and a relatively uniform *Pseudomonas aeruginosa* film of about 30–40 μm. Since *Klebsiella pneumoniae* is a facultative organism, metabolism continues in the deeper regions of the *Klebsiella pneumoniae* colonies despite an apparent lack of oxygen (Siebel, 1987). The relative numbers of the two species, equal at the beginning of the experiment, change with time, so that *Klebsiella pneumoniae* comprises approximately 25% of the cells in the biofilm after 220 h (pseudosteady state).

Siebel's observations suggest that areal population spatial distributions, in addition to the spatial distribution with depth, can be significant. Gujer (1987) has theoretically considered the areal distribution of populations and suggests that as little as 10 μm separation of interacting species may influence biofilm reactor performance. There is ample microscopic evidence that segregation of species occurs in some biofilms, while others appear to be homogeneous (Figure 10.10). In some cases, the degree of homogeneity depends on the magnification used to view the biofilm.

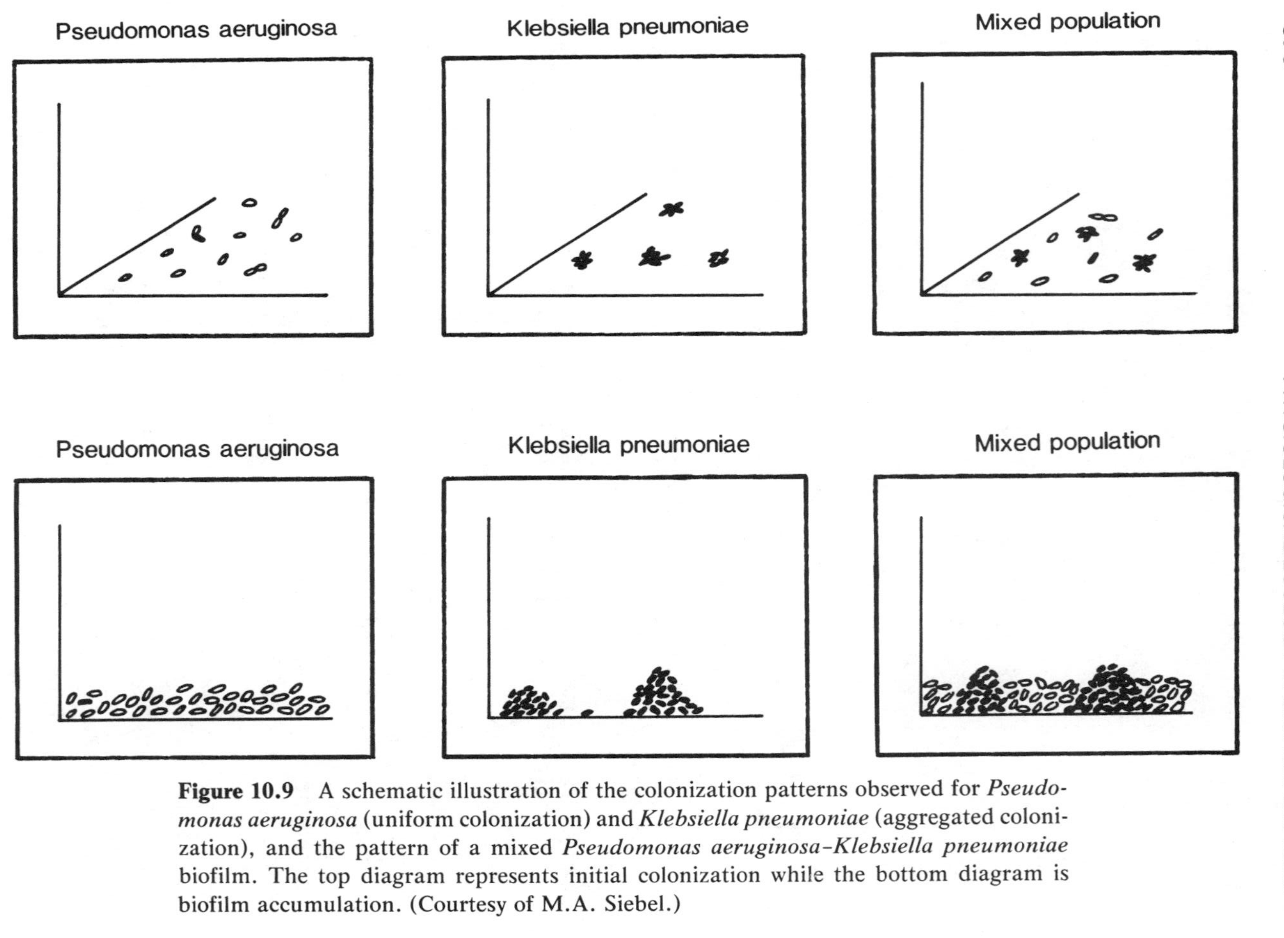

Figure 10.9 A schematic illustration of the colonization patterns observed for *Pseudomonas aeruginosa* (uniform colonization) and *Klebsiella pneumoniae* (aggregated colonization), and the pattern of a mixed *Pseudomonas aeruginosa–Klebsiella pneumoniae* biofilm. The top diagram represents initial colonization while the bottom diagram is biofilm accumulation. (Courtesy of M.A. Siebel.)

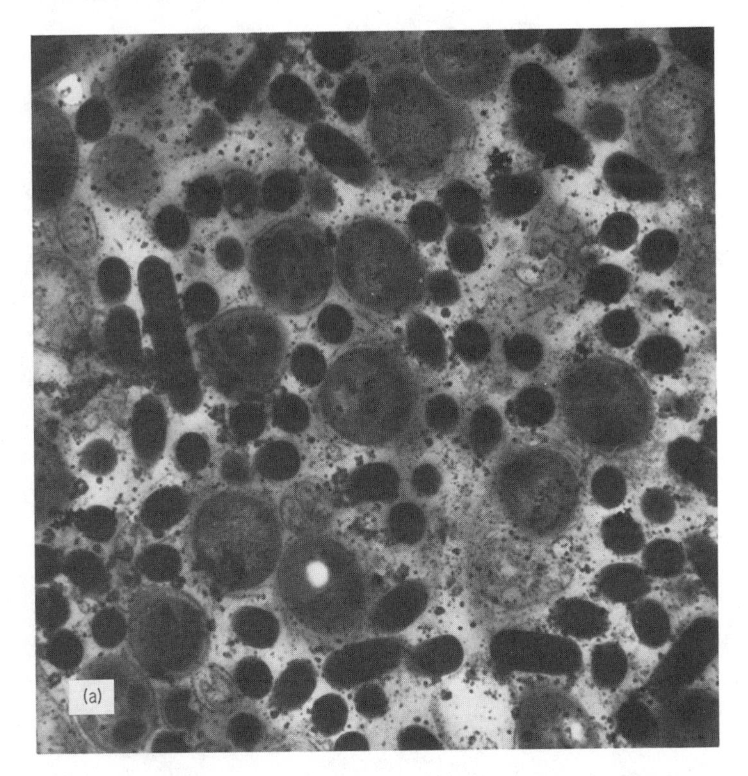

Figure 10.10 (a) Transmission electron micrograph of a granular, microbial aggregate accumulated in an anaerobic laboratory upflow sludge blanket (UASB) reactor. Large light-colored bacteria are surrounded by small dark ones. (b) The segregation is also observed at lower magnification in the same type of aggregates. (Courtesy J.T.C. Grotenhuis, Wageningen, The Netherlands.)

Geometric properties and spatial distributions of cells within biofilms probably are important influences on the functioning of the biofilm.

How important is mixing of species in a biofilm for the overall biofilm activity? If important, can it be influenced by external stimuli? How do detachment processes influence the spatial distribution of populations? Wanner and Gujer (1985) suggest that erosion only influences the variation with depth. However, sloughing influences the three dimensional distribution. Considerable experimentation, in conjunction with systematic modeling, is needed to answer to these questions.

10.6 TEMPERATURE

There are several parameters that frequently influence processes occurring in microbial ecosystems, including temperature, pH, salinity, inhibitor concentra-

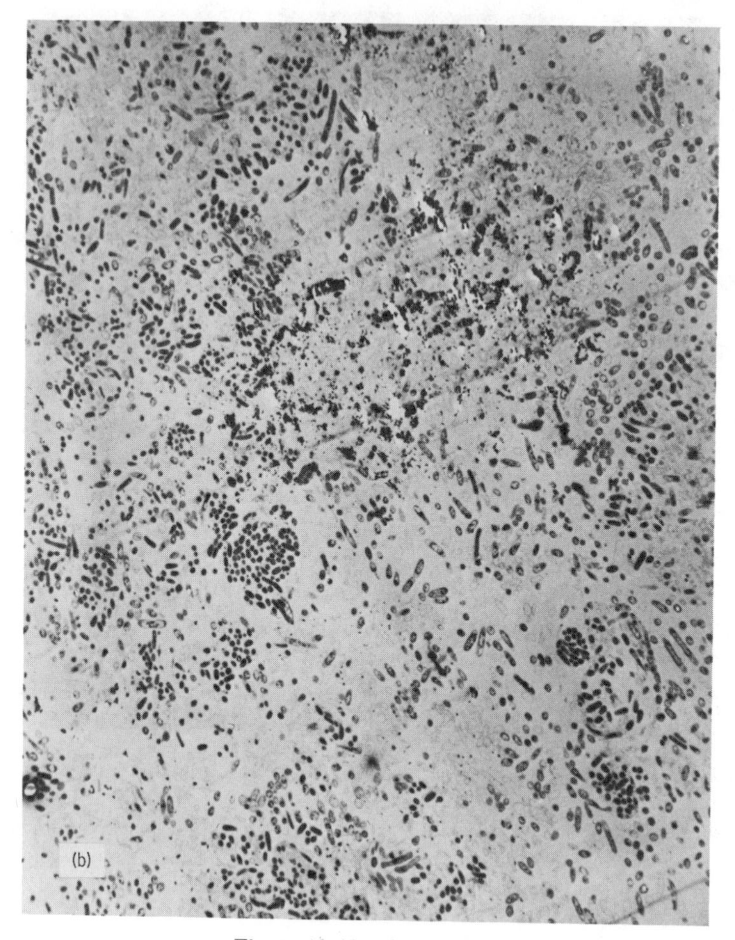

Figure 10.10 Continued

tion, and solar radiation. These parameters influence one or more transformation, transport, or interfacial transfer processes in a biofilm system. The parameters regulate a process by influencing one or both of its important characteristics:

1. *Stoichiometry,* which determines what changes will occur and to what extent.
2. *Rate:* how fast the changes will occur.

The factors that influence process stoichiometry and rate must be identified in order to satisfactorily design and control technical scale processes, analyze aquatic ecosystems, and design laboratory experiments. In this section, we shall illustrate by examining the *effect of temperature* on microbial processes in general and on biofilm processes more specifically (Characklis and Gujer, 1979).

10.6.1 Kinetics

10.6.1.1 The Arrhenius Equation The effect of temperature on rate is generally described by the Arrhenius equation, which implies that rate is an ever increasing function of temperature:

$$k = A_f \exp(-E_a/RT) \qquad [10.1]$$

where k = rate coefficient (t^{-1})
A_f = Arrhenius frequency factor (t^{-1})
R = universal gas constant ($ML^2t^{-2}mol^{-1}T^{-1}$)
= 8.314 J mol^{-1} K^{-1}
T = temperature (T)
E_a = activation energy (ML^2t^{-2} mol^{-1})

Rigorously, the Arrhenius equation only applies to single reactions or multiple reactions in which the first step is rate-limiting. For this and other reasons relating to its derivation, the concept of *activation energy* in microbial systems is not clear. Consequently, the reported activation energies are considered as empirical coefficients that relate process rate to temperature.

When the process rate is evaluated at two temperatures, the Arrhenius expressions can be combined to give

$$\ln \frac{k_1}{k_2} = E_a \frac{T_1 - T_2}{RT_1T_2} \qquad [10.2]$$

When temperature differences are small,

$$\ln \frac{k_1}{k_2} = \Theta(T_1 - T_2) \qquad [10.3]$$

and

$$\Theta = E_a/RT_1T_2 = 0.0015\, E_a \qquad [10.4]$$

where E_a is in kJ/mol. Θ and $10^{0.434\Theta}$ are frequently used for relating wastewater microbial process performance to temperature. Q_{10} is sometimes used to indicate the fractional change in the rate (i.e., k_1/k_2) for a 10°C rise in temperature. Then

$$Q_{10} = \exp\left(E_a \frac{10}{RT_1(T_1 - 10)}\right) \qquad [10.5]$$

or

$$Q_{10} = \exp(0.014\, E_a) \tag{10.6}$$

Note that the Q_{10}-value for a process may vary according to the temperature interval used in the measurement. The maximum specific growth rate varies with temperature according to an Arrhenius function (Figure 10.11).

10.6.1.2 Enzyme Reaction (in Vitro) The rate of enzyme reactions as a function of substrate concentration is generally expressed as a saturation function, which Michaelis and Menten (1913) developed from reaction mechanism considerations:

$$r_S = \frac{k_1 S}{k_2 + S} \tag{10.7}$$

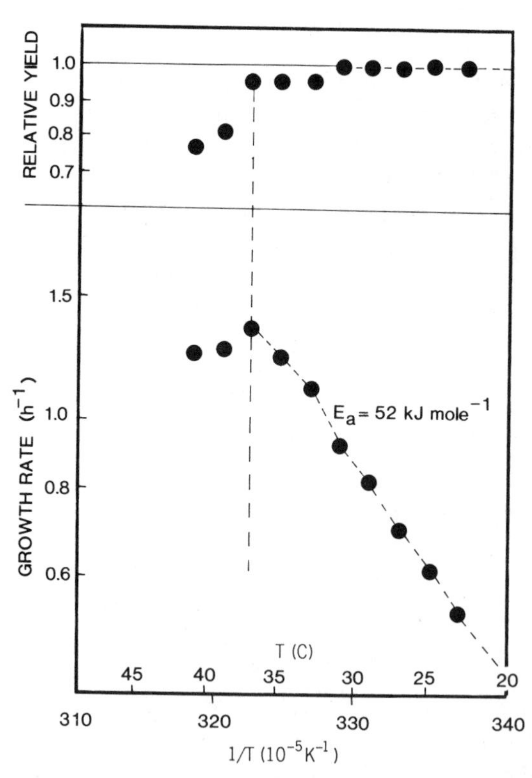

Figure 10.11 Temperature significantly influences specific growth rate but not biomass yield for *Escherichia coli* grown in a batch reactor on a glucose-mineral salts medium. The specific growth rate obeys an Arrhenius function (Eq. 10.1). The yield remains nearly constant as long as the specific growth rate is continuously increasing (Monod, 1942).

where r_S = reaction rate (t^{-1})
k_1 = maximum reaction rate (t^{-1})
k_2 = saturation coefficient (ML^{-3})
S = substrate concentration (ML^{-3})

The rate of the enzymatic decomposition of hydrogen peroxide increases from 1°C to approximately 53°C with E_a = 17.6 kJ mol^{-1} (Figure 10.12). Above 53°C, the reaction rate decreases rapidly, presumably due to enzyme denaturation. The data refer to the effect of temperature on k_1, since $S \gg k_2$ in these experiments. There are some data reported on the effect of temperature on k_2.

At constant temperature, a number of enzyme systems have been observed to denature irreversibly with time according to a first order rate expression:

$$\frac{dE}{dt} = k_e'E \tag{10.8}$$

where E = concentration of active enzyme (ML^{-3})
k_e'' = decay coefficient (t^{-1})

The decay coefficient k_e' depends on temperature, often obeying an Arrhenius function (Bailey and Ollis, 1986).

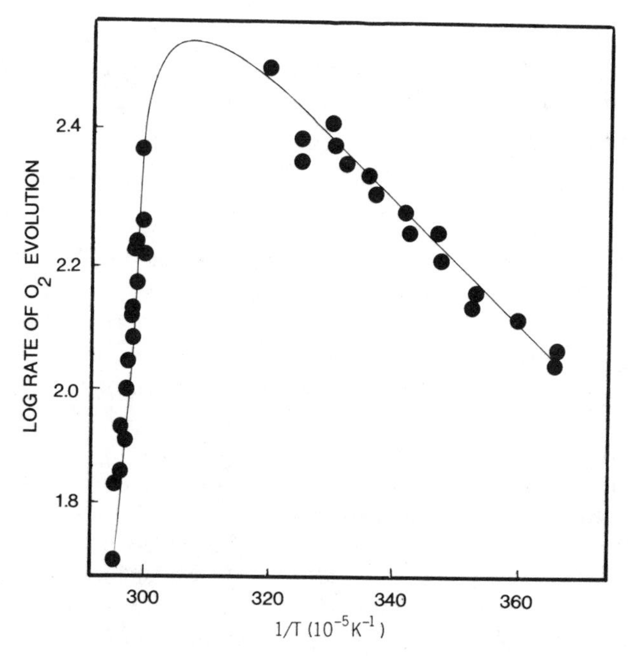

Figure 10.12 Effect of temperature on the rate of decomposition of hydrogen peroxide catalyzed by crystalline catalase (after Sizer, 1944).

10.6.1.3 Mass Transfer versus Reaction Rate Limitations There are two complicating factors that must be considered in microbial (and especially biofilm) systems beyond those considered in homogeneous chemical reactions:

1. Since more than one phase is present, the transport of components from one phase to another must be considered in the rate equation. Consequently, the rate expression will contain mass transfer terms as well as transformation (reaction) coefficients.
2. Because of the different phases present, many different contacting patterns for the two phases are possible. Each contacting pattern will generate a different mass transfer rate term, so the rate equation in each system may be somewhat different.

Mass transfer and diffusion limitations can be significant in biofilm systems, as described in many references (Atkinson, 1974; Characklis, 1978; Kornegay and Andrews, 1968; LaMotta, 1976a, 1976b). Research and plant performance studies frequently report rates in terms of the saturation kinetics model but fail to report the characteristic length of the biofilm, i.e., the biofilm thickness. Consequently, reaction rates and mass transfer rates cannot be separated, and the effect of temperature on the process rate is somewhat difficult to analyze. Some information can be inferred from activation energy values, since microbial growth rates generally have an activation energy of 50 kJ or more (Table 10.1).

Example 10A. The Influence of Temperature on Substrate Removal in a Biofilm. LaMotta (1976b) has indicated that the following expression describes substrate flux to a microbial film if zero order reaction kinetics within the biofilm is assumed:

$$r_S'' = (2D_e r_S)^{0.5} S_1^{0.5} \qquad [10A.1]$$

where r_S'' = substrate flux at the biofilm surface ($ML^{-2}t^{-1}$)
D_e = effective diffusivity of substrate in the biofilm (L^2t^{-1})
r_S = "true" rate of substrate removal by the cells ($ML^{-3}t^{-1}$)
S_1 = substrate concentration at the bulk water–biofilm interface (assumes no liquid phase mass transfer resistance) (ML^{-3})

If the biofilm is a nitrifying biofilm with very little organic carbon present, then very little heterotrophic activity will be present in the biofilm. Knowles et al. (1965) report that a temperature increase from 10 to 20°C in a nitrifying system will result in r_S increasing by a factor of 2.6. The molecular diffusivity will increase by a factor of approximately 1.3 for a temperature increase of 10°C (Bird et al., 1960). Then according to Eq. 10A.1, substrate (ammonia) removal rate will increase by a factor of approximately 1.9 as a result of a 10°C temperature increase.

TABLE 10.1 Activation Energies and Temperature Range for Maximum Specific Growth Rate Measured in Closed (Batch) Systems[a]

Activation Energy ($kJ\ mol^{-1}$)	Temperature (°C)	Organism	Substrate	Reference
52.3	23–33	*E. coli*	Glucose, mannitol, sorbitol, maltose	Monod, 1942
47.1	10–37	Enrichment	Glutamic acid	Flegal and Schroeder, 1976
69.4	20–30	*E. coli*	Glucose	Ng, 1969
30.2	24.8–37	*D. desulfuricans*	N_2	Senez, 1962
29.1	24.8–37	*D. desulfuricans*	NH_4^+	Senez, 1962
42.6	23–37	*A. aerogenes*	Glucose	Senez, 1962
63.2	18–39.4	*E. coli*	Glucose	Johnson and Lewis, 1946
118.4	10–30	*P. aeruginosa*	Glucose	Brown, 1957
71.1	0–20	A psychrophile	Glucose	Brown, 1957
41.8	20–30	A psychrophile	Glucose	Brown, 1957
145.6	0–15	*Pseudomonas*	Lactose	Green and Jezeski, 1954
85.8	0–30	*Pseudomonas*	Lactose	Green and Jezeski, 1954
45.6	8–20	*G. fujikuroi*	Glucose	Borrow, 1964
21.3	20–28			
76.0	8.3–30.8	*Nitrosomonas*	NH_4^+	Knowles et al., 1965
44.0	8.3–30.8	*Nitrosomonas*	NO_2^-	Knowles et al., 1965
77.9	6–14	*Nitrosomonas*	NH_4^+	Gujer, 1977
68.9	6–14	*Nitrosomonas*	NH_4^+	Gujer, 1977

[a] After Characklis and Gujer, 1979.

10.6.1.4 Batch Reactors In batch reactors, the organisms are substrate-saturated and microbial growth rate is characterized by μ_m, since $S \gg K_S$. Experimental results from such systems (Flegal and Schroeder, 1976; Ingraham, 1958; Monod, 1942; Senez, 1962) indicate that μ_m increases with temperature as described by an Arrhenius expression (Figure 10.11). Tables 10.1 and 10.2 list activation energies for μ_m and also thermal inactivation as reported in the literature. The variation in K_S with temperature for batch reactors is presented in Figure 10.13.

10.6.1.5 Continuous Reactors In continuous reactors, the specific growth rate is generally less than μ_m and the concentration of the limiting nutrient is quite low. Consequently, the equations describing biomass production and effluent substrate concentration in a chemostat contain three rate coefficents (μ_m, K_S, k_e), which depend on temperature (Example 8C, Chapter 8): From Eq. 8C.13 (Chapter 8),

$$S = \frac{K_S}{[\mu_m/(D + k_e)] - 1} \qquad [10.9]$$

From Eq. 8C.12 (Chapter 8).

$$X = \frac{DY_{X/S}(S_i - S)}{D + k_e} \qquad [10.10]$$

where X = biomass concentration (ML^{-3})
D = dilution rate (t^{-1})
μ_m = maximum specific growth rate (t^{-1})
K_S = saturation coefficient (ML^{-3})
k_e = endogenous decay coefficient (t^{-1})

The effect of temperature on μ_m, as measured in continuous reactors, is similar to that observed in batch reactors (Table 10.3). Activation energies for μ_m range from 35 to 83 kJ mol^{-1} with a typical value of approximately 60 kJ mol^{-1}.

In continuous reactors, especially at lower dilution rates, the growth rate is considerably less than its maximum and endogenous decay becomes significant. *Endogenous decay* is a general term including processes such as endogenous metabolism, maintenance requirements, and cell lysis that tend to degrade biomass. The biomass decay coefficient k_e approximately doubles with a 10°C increase (Table 10.3). Endogenous decay can obviously contribute to reduced levels of effluent biomass concentration from continuous reactors. The observed yield Y_o will be influenced by endogenous decay throughout wide temperature ranges, but the intrinsic biomass yield $Y_{X/S}$ will not (Section 10.6.2).

The temperature effect on the saturation coefficent K_s is not clear. Although K_S does not have units of rate, it has been related to the mass transfer rate

TABLE 10.2 Activation Energies and Temperature Range for Thermal Inactivation of Microorganisms Measured in Closed (Batch) Systems[a]

Activation Energy ($kJ\ mol^{-1}$)	Temperature (°C)	Organism	Substrate	Reference
−252	37–42	*A. aerogenes*	Glucose	Senez, 1962
−167.5	37–42	*D. desulfuricans*	N_2	Senez, 1962
−173.9	37–42.2	*D. desulfuricans*	NH_4^+	Senez, 1962
−225.9	40–50[b]	*H. polymorpha*	Methanol	Cooney and Makiguchi, 1976

[a] After Characklis and Gujer, 1979.

[b] In a continuous reactor.

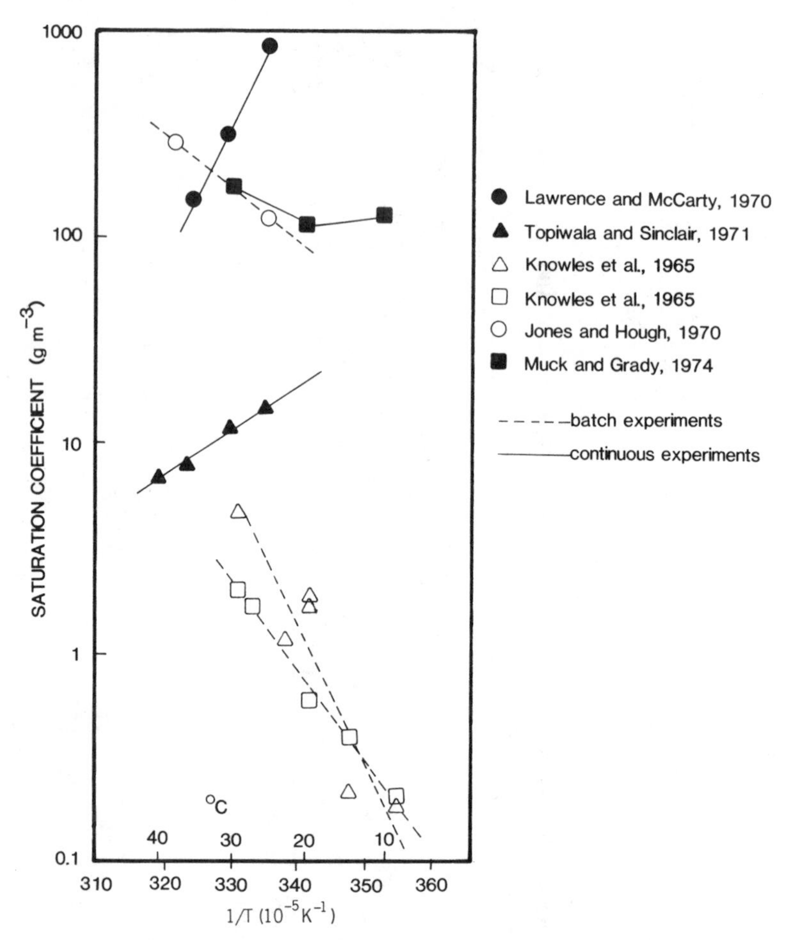

Figure 10.13 Temperature influences the saturation coefficient. The data obey an Arrhenius function (after Characklis and Gujer, 1979). The dashed line represents batch experiments and indicates an increasing K_S with increasing temperature. The solid line represents continuous reactor results and indicates that K_S decreases with increasing temperature.

(Characklis, 1978), and so data from several investigations are presented in Arrhenius form in Figure 10.13. The activation energy for K_S is positive for batch data and negative for continuous reactor data, which may indicate some inadequacy of the saturation kinetics form for the description of both reaction systems.

10.6.1.6 Substrate Removal Rate The substrate removal rate is a continuous function of S and X. At high S, it is independent of S and the reaction can be considered *substrate-saturated*. Under these conditions, the process is rate-limited by the microbial growth. This is the case for batch growth processes (Ta-

TABLE 10.3 Activation Energies for Maximum Specific Growth Rates and Endogenous Decay in Continuous Reactors

Activation Energy ($kJ\ mol^{-1}$)				
μ_m	k_e	Organism	Substrate	Reference
35.1	37.7	*A. aerogenes*	Glucose	Topiwala and Sinclair, 1971
55.6	—	*E. coli*	NH_4^+	Ryu and Mateles, 1968
69.5	76.1	*P. fluorescens*	Glucose	Mennet and Nakayama, 1971
—	35.3	*P. fluorescens*	Glucose	Palumbo and Witter, 1969
52.3	52.3	Enrichment	Glucose	Muck and Grady, 1974
—	55.0	Activated sludge	Endogenous	Benedek and Farkas, 1971
70.3[a]	—	Activated sludge	NH_4^+[b]	Downing et al., 1964
57.9[c]	—	Activated sludge	NH_4^+[b]	Downing et al., 1964
83.2[d]	—	Activated sludge	NH_4^+[b]	Downing et al., 1964
75.8	—	Activated sludge	NH_4^+[b]	Gujer, 1977

[a]Fill and draw system containing 50 g m^{-3} biomass.

[b]Nitrification.

[c]Fill and draw system containing 500 g m^{-3} biomass.

[d]Continuous reactor containing 2000–6500 g m^{-3} biomass.

bles 10.1 and 10.2). The effect of temperature on substrate removal rate in such cases is similar to that of the maximum microbial growth rate, i.e., $E_a = 60$ kJ mol^{-1}.

When the removal rate is dependent on S, the process is *substrate-limited*. This is the case of continuous reactors except for those operated at very high growth rates (near "washout"). Under substrate-limited conditions, the substrate removal rate is not so sensitive to temperature. Nitrification provides a convenient example for the effect of temperature on substrate removal. The data of Knowles et al. (1965), obtained from batch reactors, have been used in Figure 10.14 to indicate the effect of temperature on residual substrate (NH_4^+) concentration in steady state continuous reactors at various growth rates. The results clearly indicate that the residual substrate is independent of temperature at low growth rates and becomes increasingly more dependent on temperature as the growth rate increases. In general, the residual substrate will be higher at lower temperatures. Similar behavior is evident in the removal of soluble organic substrates (Figure 10.15).

The relative insensitivity to temperature at low growth rates demonstrates the importance of comparing the *reaction time* with the *residence time* in a system (Busch, 1971). Consider the performance of a plug flow reactor in removing substrate as a function of temperature (Figure 10.16). Using the effluent substrate

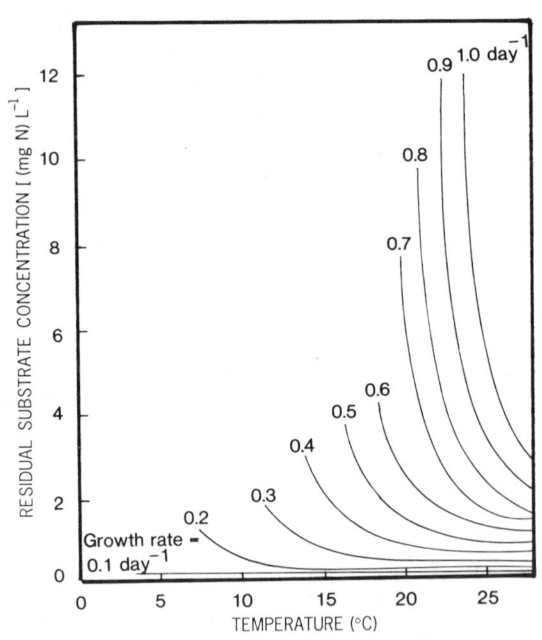

Figure 10.14 Residual or effluent substrate (NH_4-N) concentration from a chemostat depends on temperature. The calculations are based on data from Knowles et al. (1965) obtained in batch experiments. (After Characklis and Gujer, 1979.)

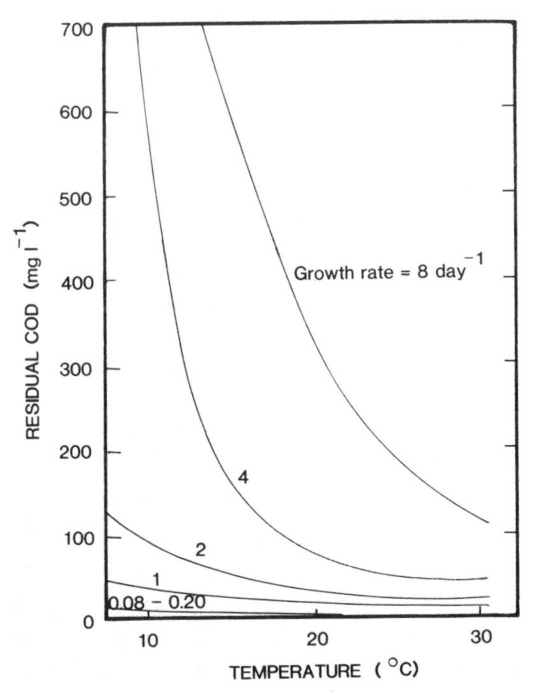

Figure 10.15 Residual or effluent substrate (glucose) concentration from a continuous reactor for a mixed population with glucose, measured as chemical oxygen demand, depends on temperature (after Characklis and Gujer, 1979). Calculations are based on data from Muck and Grady (1974) and Friedman and Schroeder (1971) for continuous reactors.

concentration as a performance criterion, increasing the temperature from 10 to 20°C improves performance. However, increasing the temperature from 20 to 30°C does not improve performance. Is the process insensitive to temperature above 20°C? As indicated in Figure 10.16, the reactor system may be insensitive to temperature above 20°C, but the microbial process is not. The longitudinal gradients in the plug flow reactor clearly indicate an influence of temperature on the substrate removal above 20°C, but it cannot be measured by analysis of the effluent. Thus, the reaction time decreases with temperature and remains less than the residence time, so that the effluent concentration remains constant above 20°C.

Continuous flow microbial reactors are not as sensitive to temperature at long residence times, which usually coincide with low net microbial growth rate.

The contrast between reaction time and residence time has also been applied to inhibition by Lewandowski (1987), who has constructed a rather fundamental theory that predicts chemical inhibition of microbial processes as a function of

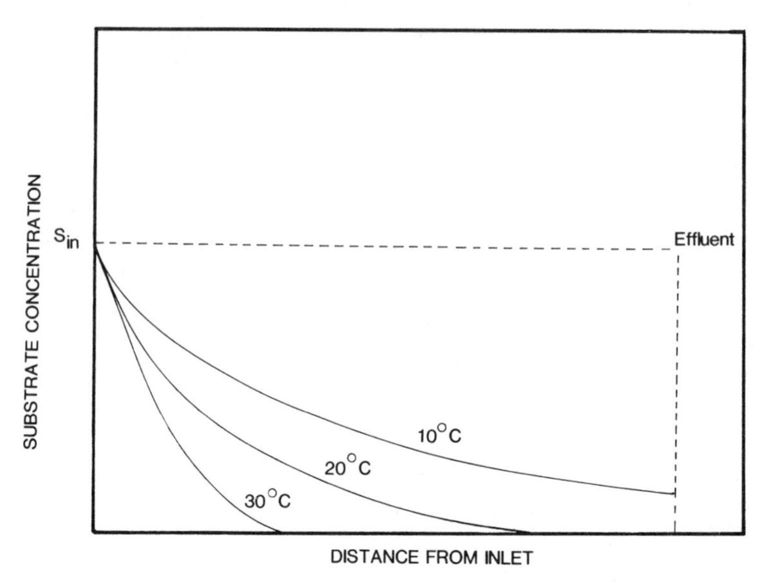

Figure 10.16 Hypothetical substrate concentration profile in a plug flow reactor at various temperatures. At 10°C, the effluent substrate concentration is detectable. At 20 and 30°C, the effluent substrate concentration is not detectable and no effect of temperature on performance is indicated. Yet, the concentration profile indicates that temperature was still influencing the reaction process.

reactor operating parameters. Other variables such as pH can also be addressed with a similar phenomenological theory.

10.6.2 Stoichiometry

The literature (Flegal and Schroeder, 1976; Monod, 1942, 1949; Ng, 1969; Senez, 1962) indicates that the biomass yield $Y_{X/S}$ is constant over a wide temperature range for many organisms and substrates. Generally, the yield is constant in the temperature range where the *growth rate* behaves according to an Arrhenius function. Outside of this temperature range, the yield may change dramatically. Thus, the yield for *Escherichia coli* grown in a closed (batch) reactor on a glucose–mineral salts medium is nearly constant from 20 to 37°C but decreases significantly thereafter (Figure 10.11). The specific growth rate also increases continuously up to 37°C and then decreases.

Microbial growth stoichiometry is unaffected by temperature over wide temperature ranges.

The good correlation of the specific growth rate with the Arrhenius model over a wide temperature range seems to suggest the temperature dependence of a single reaction, i.e., the rate-limiting step. Outside of the characteristic temperature range, a change in rate-limiting step or a change in metabolic sequence may occur. A concurrent change in yield, especially in the latter case, would be expected. Monod (1949) has suggested that the reasonably good fit of microbial growth data to the saturation kinetics model occurs because the overall rate is limited by one reaction, i.e., the rate-limiting step.

The temperature range in which the yield remains constant may be greater than evidenced by much of the microbiological data. Lower yields, frequently attributed to higher temperatures, may result from oxygen limitation due to a lower saturation oxygen concentration. Higher oxygen solubilities and lower reaction rates at lower temperatures increase oxygen availability and consequently yield.

In closed systems, biomass composition does not change as a function of temperature (Schaechter et al., 1958). Biomass composition in continuous reactors, however, does change with temperature. Tempest and Hunter (1965) presented an explanation based on intracellular activities, which rationalized these discrepancies between composition changes in closed and open systems. Regardless of the intracellular mechanism, the behavior appears to be described well by presuming a biomass decay process whose rate is dependent on temperature:

$$\text{biomass} + O_2 \rightarrow CO_2 + H_2O + \text{inorganic nutrients} \qquad [10.11]$$

Because of the rapid growth of the organisms in substrate-saturated closed systems, biomass decay is insignificant. However in continuous reactors, biomass decay is a function of the imposed growth rate (i.e., dilution rate) and temperature (Tempest and Hunter, 1965), which suggests that multiple reactions (biomass production and decay) must be considered if the temperature effects are to be determined in a systematic manner.

$$\text{substrate} + O_2 + \text{nutrients} \rightarrow \text{biomass} + \text{respiration product} \qquad [10.12]$$

$$\text{biomass} + O_2 \rightarrow \text{respiration products} + \text{nutrients} \qquad [10.13]$$

The relationship between the two microbial reactions and the abiotic oxidation of the substrate has been schematically diagrammed in Figure 6.1 (Chapter 6).

10.6.3 Population Dynamics

Competition between microorganisms may result in the dominance of one species because of its short generation time and high population density. In fact, the coexistence of several species at steady state in continuous culture is observed much more readily when more than one substrate is available. For example, the diversity of nutrient sources in wastewater treatment must be a key factor in supporting a heterogeneous population of organisms. Yoon et al. (1977) have demonstrated these principles in laboratory experiments with a two substrate, two organism system.

Goldman and Ryther (1976) have effectively illustrated the effect of temperature on the population dynamics of algae in enriched continuous cultures in the laboratory. Five species were cultured axenically and in combination at temperatures varying from 10 to 30°C. Figure 10.17a indicates the change in steady state biomass carbon with varying temperature for the various species grown axenically. The effect of temperature on competitive advantage was tested in chemostat experiments (Figure 10.17b). Axenic cultures of *S. costatum* were grown in a chemostat at 10 and 15°C. In each case, *S. costatum* was contaminated with *P. tricornutum*. At 10°C, *S. costatum* maintained its dominant position. However, at 15°C, *P. tricornutum* gained the dominant position after 6–8 days. The results presented in Figure 10.17b can be predicted from the axenic culture results in Figure 10.17a.

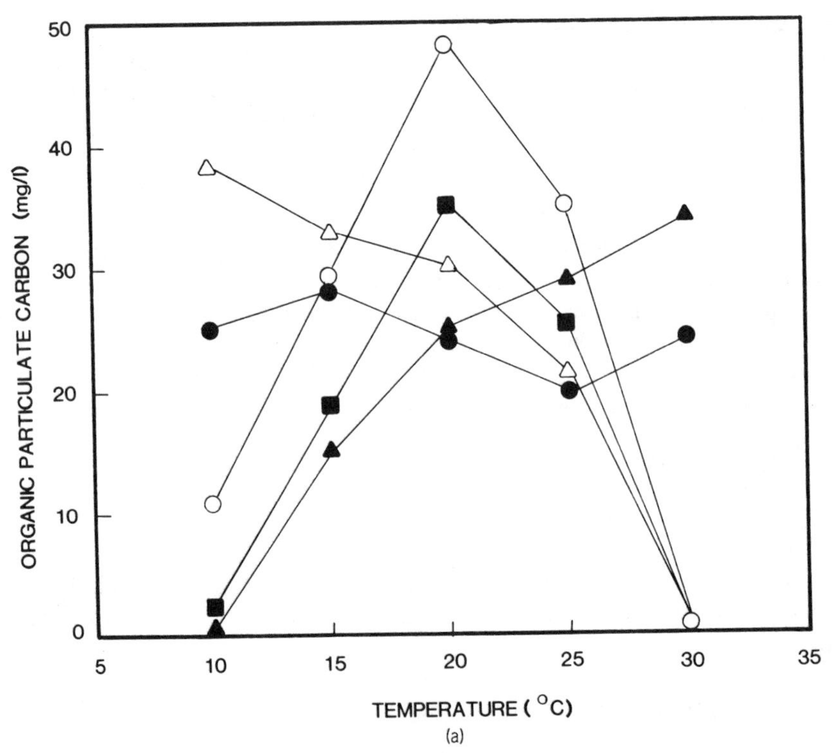

Figure 10.17 (a) Steady state biomass concentration (as particulate organic carbon) varies with temperature for five different algal species in laboratory continuous reactors (Goldman and Ryther, 1976). *S. costatum* (△), *T. pseudonana* (●), *P. tricornutum* (○), *M. lutheri* (■), *D. tertiolecta* (▲). (b) Experiments were conducted at 10 and 15°C in continuous reactors with two algal species, *P. tricornutum* (○) and *S. costatum* (●). In each case, monopopulations of *S. costatum* were challenged by an inoculum of *P. tricornutum* (Goldman and Ryther, 1976). At 10°C, *P. tricornutum* could not be maintained in the reactor; at 15°C, *P. tricornutum* became the dominant organism.

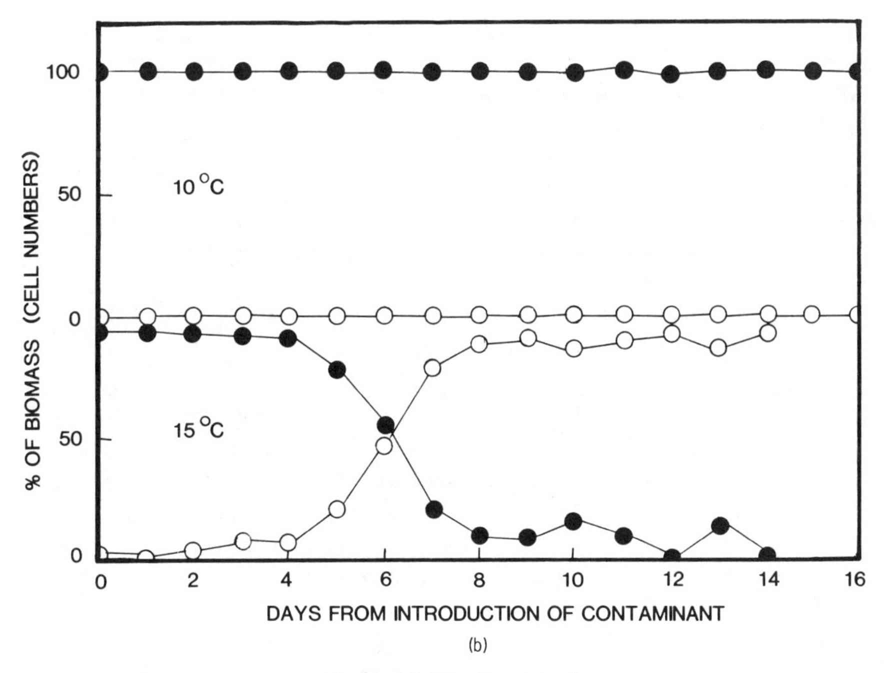

Figure 10.17 Continued

10.6.4 Nitrification in a Trickling Filter

Trickling filters have successfully been used to nitrify effluents with negligible organic carbon (i.e., secondary effluents). The detailed modeling of the behavior of a nitrifying biofilm is complex (Chapter 11), but a few simple calculations will illustrate how empirical nitrification data from trickling filters can be interpreted.

Frank-Kamenetzkii (1969) developed a relationship that may be used to predict when the electron donor (NH_4^+) or the electron acceptor (O_2) is limiting the flux into the biofilm:

$$K = \frac{D_O \nu_N M_N S_O}{D_N \nu_O M_O S_N} \qquad [10.14]$$

where K = dimensionless number

D_O, D_N = effective diffusivity of oxygen and ammonium in the biofilm, respectively (L^2t^{-1})

ν_O, ν_N = stoichiometric coefficients for O and N, respectively (mol)

M_O, M_N = molecular weight of O and N, respectively ($M\ mol^{-1}$)

S_O, S_N = oxygen and ammonium concentrations at the bulk water-biofilm interface (ML^{-3})

$K > 1$ indicates that NH_4^+ limits the nitrification rate of the biofilm; otherwise O_2 is limiting. Haug and McCarty (1971) theoretically predict $D_O/D_N = 1.43$, and Wezernak and Gannon (1967) report $\nu_O M_O/\nu_N M_N = 4.33$ (g O_2)/(g NH_4–N). Therefore, $K = 1$ if $S_N = 0.33 S_O$.

In a nitrifying trickling filter receiving secondary effluent (i.e., heterotrophic activity is low), it may be assumed that the flowing water is saturated with oxygen. For this situation, the Frank-Kamenetskii relationship predicts the boundary between oxygen and ammonium limitation in a trickling filter surprisingly well (broken line in Figure 10.18).

At high ammonium concentration, the ammonium flux into the biofilm becomes independent of NH_4^+ concentration and is oxygen-limited. The oxygen-limited ammonium flux is characterized by a $Q_{10} = 1.6$, which is comparable to the value of 1.86 calculated from Eq. 10A.1. The transition from NH_4^+ limitation to oxygen limitation is a function of temperature because oxygen saturation depends on temperature. The dashed line in Figure 10.18 represents the transition from NH_4^+ to oxygen limitation and was predicted from the oxygen saturation temperature dependence and the Frank-Kamenetskii relationship. Thus, fundamental data provide useful information even in calculations of mass-transfer-limited processes.

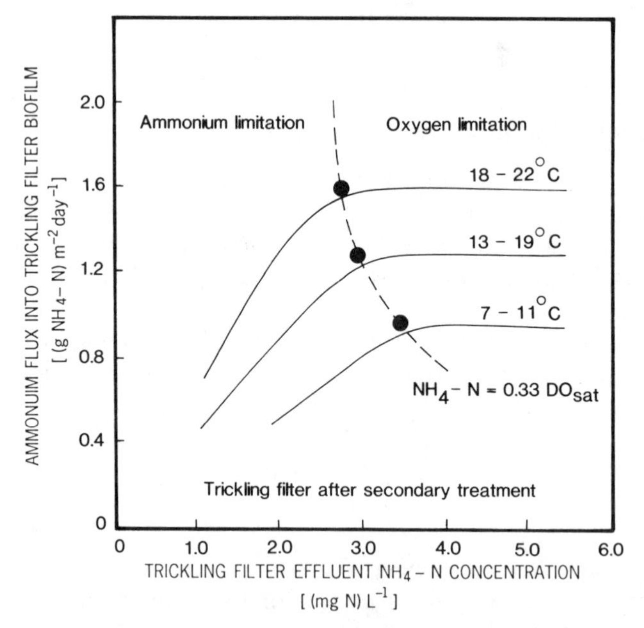

Figure 10.18 Nitrification rate in trickling filters receiving secondary effluent which contains negligible degradable organic carbon is influenced by temperature (U.S. Environmental Protection Agency, 1975). The dashed line reflects the boundary between nitrogen-limited and oxygen-limited biofilm activity. (Reprinted with permission from Characklis and Gujer, 1979.)

REFERENCES

Adamkiewicz, V.W. and D.R. Pilon, *Can. J. Microbiol.*, **29**, 464-467 (1983).

Alldredge, A.L. and Y. Cohen, *Science*, **235**, 689-691 (1987).

Amy, P.S., C. Pauling, and R.Y. Morita, *Appl. Environ. Microbiol.* **45**, 1041-1048 (1983).

Atkinson, B., *Biochemical Reactors*, Pion, London, 1974.

Azam, F. and J.W. Ammerman, "Cycling of organic matter by bacterioplankton in pelagic marine ecosystems: Microenvironmental considerations" in M.J.R. Fasham (ed.), *Flows of Energy and Materials in Marine Ecosystems*, Plenum Publ. Corp., New York, 1984, pp. 345-360.

Baier, R.E. "Substrata influences on adhesion of microorganisms and their resultant new surface properties," in G. Bitton and K.C. Marshall (eds.), *Adsorption of Microorganisms to Surfaces*, Wiley-Interscience, New York, 1980, pp. 59-104.

Bailey, J.E. and D.F. Ollis, *Biochemical Engineering Fundamentals*, 2nd ed., McGraw-Hill, New York, 1986.

Baker, J.H., *Can. J. Microbiol.*, **30**, 511-515 (1984).

Bakke, R., "Dynamics of biofilm processes: Substrate load variations," unpublished Master of Science thesis, Montana State University, Bozeman, MT, 1983.

Bakke, R. and P.Q. Olsson, *J. Microbiol. Methods*, **5**, 93-98 (1986).

Bakke, R., M.G. Trulear, J.A. Robinson, and W.G. Characklis, *Biotech. Bioeng.*, **26**, 1418-1424 (1984).

Beeftink, H.H. and P. Staugaard, *Appl. Env. Microbiol.*, **52**, 1139-1146 (1986).

Benedek, P. and P. Farkas, "The influence of temperature on the reactions of the activated sludge process," in *International Symposium on Water Pollution Control in Cold Climates*, Environmental Protection Agency Water Pollution Control Series, No. 16100 EXH 11/71, 1971, pp. 164-179.

Bird, R.B., W.E. Stewart, and E.N. Lightfoot, *Transport Phenomena*, Wiley, New York, 1960.

Borrow, A., *Can. J. Microbiol.*, **10**, 445-466 (1964).

Bott, T.L. and T.D. Brock, *Limnol. Oceanogr.*, **15**, 333-342 (1970).

Bright, J.J. and M. Fletcher, *Appl. Environ. Microbiol.*, **45**, 818-825 (1983).

Brock, T.D. "The study of microorganisms *in situ*: Progress and problems," in M. Fletcher, T.R.G. Gray and J.G. Jones (eds.), *Ecology of Microbial Communities*, Cambridge University Press, London, 1987, pp. 1-17.

Brown, A.D., *J. Gen. Microbiol.*, **17**, 640-648 (1957).

Brown, M.J. and J.N. Lester, *Wat. Res.*, **13**, 817-837 (1979).

Brown, M.J. and J.N. Lester, *Appl. Env. Microbiol.*, **40**, 179-185 (1980).

Busch, A.W., *Aerobic Biological Treatment of Waste Waters*, Gulf Publishing Co., Houston, TX, 1971.

Chang, H.T. and B.E. Rittmann, *J. WPCF*, **60**, 362-368 (1988).

Characklis, W.G., *J. Env. Eng. Div. ASCE*, **104**(EE3), 531-534 (1978).

Characklis, W.G., "Biofilm development: A process analysis," in K.C. Marshall (ed.), *Microbial Adhesion and Aggregation*, Springer-Verlag, Berlin, 1984, pp. 137-157.

Characklis, W.G. and W. Gujer, *Prog. Wat. Tech.*, Suppl. 1, 111-130 (1979).

Cole, J.J., *Ann. Rev. Ecol. Syst.*, **13**, 291-314 (1982).

Cooke, W.B., *Ecology*, **40**, 273-291 (1959).

Cooney, C.L. and Makiguchi, "Temperature as an engineering parameter in SCP production," in A.C.R. Dean, D.C. Ellwood, C.G.T. Evans, and J. Melling (eds.), *Continuous Culture 6: Applications and New Fields*, Ellis Horwood, Chichester, England, 1976, pp. 146-157.

Corpe, W.A., "Microfouling: The role of primary film forming bacteria," in R.F. Acker, B.F. Brown, J.R. De Palma, and W.P. Iverson, eds., *Proc. 3rd International Congress on Marine Corrosion Fouling*, Northwestern University Press, Evanston, IL, 1973, pp. 598-609.

Cowen, J.P. and M.W. Silver, *Science*, **224**, 1340-1342 (1984).

Curds, C.R., *Oikos*, **15**, 282-289 (1965).

Curds, C.R., *Wat. Res.*, **7**, 1269-1284 (1973).

Dempsey, M.J., *Mar. Biol.*, **61**, 305-315 (1981).

Downing, A.L., H.A. Painter, and G. Knowles, *J. Inst. Sew. Purif.*, **2**, 130-151 (1964).

Ellwood, D.C., C.W. Keevil, P.D. Marsh, C.M. Brown, and J.N. Wardell, *Phil. Trans. Roy. Soc. Lond.*, **B297**, 517-532 (1982).

Escher, A.R., "Bacterial colonization of a smooth surface: An analysis with image analyzer," unpublished doctoral thesis, Montana State University, Bozeman, MT, 1986.

Escher, A.R. and W.G. Characklis, *Biotech. Bioeng.*, **24**, 2283-2290 (1982).

Fenchel, T., *Adv. Microbial Ecol.*, **9**, 57-97 (1986).

Flegal, T.M. and E.D. Schroeder, *J. WPCF*, **48**, 2700-2707 (1976).

Fletcher, M., *J. Gen. Microbiol.*, **94**, 400-404 (1976).

Fletcher, M., "The question of passive versus active attachment mechanisms in nonspecific bacterial adhesion," in R.C.W. Berkeley, J.M. Lynch, J. Melling, P.R. Rutter, and B. Vincent (eds.), *Microbial Adhesion to Surfaces*, Ellis Horwood, Chichester, 1980, pp. 197-210.

Fletcher, M., "Comparative physiology of attached and free-living bacteria," in K.C. Marshall (ed.), *Microbial Adhesion and Aggregation*, Springer-Verlag, Berlin, 1984, pp. 223-232.

Fletcher, M. and K.C. Marshall, *Adv. Microbial Ecol.*, **6**, 199-236 (1982a).

Fletcher, M. and K.C. Marshall, *Appl. Environ. Microbiol.*, **44**, 184-192 (1982b).

Frank-Kamenetskii, D.A., *Diffusion and Heat Transfer in Chemical Kinetics*, Plenum, New York, 1969.

Friedman, A.A. and E.D. Schroeder, Temperature effects on growth and yield for activated sludge, in *Proc. 26th Industrial Waste Conference.*, Purdue University, Eng. Ext. Ser., No. 140, Part 2, 1971, pp. 1060-1073.

Geesey, G.G., and J.W. Costerton, *Can. J. Microbiol.*, **25**, 1058-1062 (1979).

Gerchakov, S.M., D.S. Marszalek, F.J. Roth, and L.R. Udey, "Succession of periphytic microorganisms on metal and glass surfaces," in V. Romanovsky (ed.), *Proc. 4th International Congress on Marine Corrosion Fouling*, Centre de Recherches et d'Etudes Oceanographiques, Boulogne, 1977, pp. 203-211.

Goldman, J.C. and J.H. Ryther, *Biotech. Bioeng.*, **18**, 1125-1144 (1976).

Grade, R. and B.M. Thomas, "The influence and control of algae in industrial cooling systems," in A.M. Pritchard (ed.), *Progress in the Prevention of Fouling in Industrial Plant,* Nottingham University, U.K., 1981, pp. 101-120.

Green, V.W. and J.J. Jezeski, *Appl. Microbiol.,* **2**, 110-117 (1954).

Güde, H., *Microbial Ecol.,* **5**, 225-237 (1979).

Gujer, W., *Prog. Wat. Tech.,* **9**, 323-336 (1977).

Gujer, W., *Wat. Sci. Tech.,* **19**, 495 (1987).

Haack, T.K. and G.A. McFeters, *Microb. Ecol.,* **8**, 115-126 (1982).

Haug, R.T. and P.L. McCarty, *Nitrification with the Submerged Filter,* Technical Report No. 149, Dept. of Civil Engineering, Stanford University, 1971.

Hermansson, M. and K.C. Marshall, *Microbial Ecol.,* **11**, 91-105 (1985).

Hirsch, P., "Life under conditions of low nutrient concentrations," in M. Shilo (ed.), *Strategies of Microbial Life in Extreme Environments,* Verlag Chemie, Weinheim, 1979, pp. 357-372.

Howell, J.A. and B. Atkinson, *Wat. Res.,* **10**, 307-315 (1976).

Hsieh, K.M., L.W. Lion, and M.L. Shuler, *Appl. Env. Microbiol.,* **50**, 1155-1161 (1985).

Humphrey, B.A. and K.C. Marshall, *Arch. Microbiol.,* **140**, 166-170 (1984).

Ingraham, J.L., *J. Bact.,* **76**, 75-80 (1958).

Jansen, J. and G.H. Kristensen, *Fixed Film Kinetics: Denitrification in Fixed Films,* Report 80-59, Department of Sanitary Engineering, Technical University of Denmark, 1980.

Jarman, T.R. and G.W. Pace, *Arch. Microbiol.,* **137**, 231-235 (1984).

Jensen, R.H. and C.A. Woolfolk, *Appl. Env. Microbiol.,* **50**, 364 (1985).

Johnson, F.H. and I. Lewis, *J. Cell. Comp. Physiol.,* **28**, 47-75 (1946).

Jones, R.C. and J.S. Hough, *J. Gen Microbiol.,* **60**, 107-116 (1970).

Jordan, T.L., and J.T. Staley, *Microbial Ecol.,* **2**, 241-251 (1976).

Jorgensen, B.B., "Mineralization and the bacterial cycling of carbon, nitrogen and sulfur in marine sediments," in D.C. Ellwood, J.N. Hedger, M.J. Latham, J.M. Lynch, and J.H. Slater (eds.), *Contemporary Microbial Ecology,* Academic Press, London, 1980, pp. 239-252.

Kefford, B., S. Kjelleberg, and K.C. Marshall, *Arch. Microbiol.,* **133**, 257-260 (1982).

Kjelleberg, S., B.A. Humphrey, and K.C. Marshall, *Appl. Environ. Microbiol.,* **43**, 1166-1172 (1982).

Kjelleberg, S., B.A. Humphrey, and K.C. Marshall, *Appl. Environ. Microbiol.,* **46**, 978-984 (1983).

Kjelleberg, S., K.C. Marshall, and M. Hermansson, *FEMS Microbiol. Ecol.,* **31**, 89-96 (1985).

Knowles, G., A.L. Downing, and M.J. Barrett, *J. Gen. Microbiol.,* **38**, 263-278 (1965).

Kornegay, B.H. and J.F. Andrews, *J. WPCF,* **40**, R460-R468 (1968).

LaMotta, E.J., *Biotech. Bioeng.,* **18**, 1359-1370 (1976a).

LaMotta, E.J., *Env. Sci. Tech.,* **10**, 765-769 (1976b).

Lawrence, A.W. and P.L. McCarty, *J. San. Eng. Div. ASCE,* **96**, 757-778 (1970).

Lawrence, J.R. and D.E. Caldwell, *Microbial Ecol.*, **14**, 15–27 (1987).

Lawrence, J.R., P.J. Delaquis, D.R. Korber, and D.E. Caldwell, *Microbial Ecol.*, **14**, 1–14 (1987).

Lee, S.S., A.P. Jackman, and E.D. Schroeder, *Water Res.*, **9**, 491–498 (1975).

Lewandowski, Z., *Wat. Res.*, **21**, 147–153 (1987).

Mack, W.N., J.P. Mack and A.O. Ackerson, *Microbial Ecol.*, **2**, 215–226 (1975).

Marshall, K.C., *Interfaces in Microbial Ecology*, Harvard University Press, Cambridge, MA, 1976.

Marshall, K.C. (ed.), *Microbial Adhesion and Aggregation*, Springer-Verlag, Berlin, 1984.

Marshall, K.C., R. Stout, and R. Mitchell, *J. Gen., Microbiol.*, **68**, 337–348 (1971a).

Marshall, K.C., R. Stout, and R. Mitchell, *Canad. J. Microbiol.*, **17**, 1413–1416 (1971b).

Marszalek, D.S., S.M. Gerchakov, and L.R. Udey, *Appl. Environ. Microbiol.*, **38**, 987–995 (1979).

Martin, P. and R.A. MacLeod, *Appl. Environ. Microbiol.*, **47**, 1017–1022 (1984).

Matin, A., "Microbial regulatory mechanisms at low nutrient concentrations as studied in chemostat," in M. Shilo (ed.), *Strategies of Microbial Life in Extreme Environments*, Verlag-Chemie, Weinheim, 1979, pp. 323–340.

McCoy, W.F. and J.W. Costerton, *Dev. Ind. Microbiol.*, **23**, 441 (1982).

Mennet, R.W. and T.O.M. Nakayama, *Appl. Microbiol.*, **22**, 772–776 (1971).

Michaelis, L. and M.L. Menten, *Z. Biochemie*, **49**, 333–369 (1913).

Mills, A.L. and R. Maubrey, *Microbial Ecol.*, **7**, 315–322 (1981).

Monod, J., *Recherches sur la Croissance des Cultures Bacteriennes*, Hermann, Paris, 1942.

Monod, J., *Ann. Rev. Microbiol.*, **3**, 371–394, (1949).

Morita, R.Y., *Adv. Microbial Ecol.*, **6**, 171–198 (1982).

Muck, R.E. and C.P.L. Grady, *J. Env. Eng. Div. ASCE*, **100**, 1147–1163 (1974).

Namkung, E. and B.E. Rittmann, *Wat. Res.*, **20**, 795–806 (1986).

Nasri, M., S. Sayadi, J.-N. Barbotin, P. Dhulster, and D. Thomas, *Appl. Env. Microbiol.*, **53**, 740–744 (1987).

Neihof, R. and G. Loeb, *J. Mar. Res.*, **32**, 5–12 (1974).

Nelson, C.H., J.A. Robinson, and W.G. Characklis, *Biotech. Bioeng.*, **27**, 1662–1667 (1985).

Ng, H., *J. Bact.*, **98**, 232–237 (1969).

Norrberg, A.B. and H. Persson, *Biotech. Bioeng.*, **26**, 239–246 (1984).

Novitsky, J.A. and R.Y. Morita, *Appl. Environ. Microbiol.*, **33**, 635–641 (1977).

Odum, E.P., *Fundamentals of Ecology*, 3rd ed., W.B. Saunders, Philadelphia, 1971.

Ou, L.-T. and M. Alexander, *Arch. Microbiol.*, **101**, 35–44 (1974).

Palumbo, S.A. and L.D. Witter, *Appl. Microbiol.*, **18**, 137–141 (1969).

Picologlou, B.F., N. Zelver, and W.G. Characklis, *J. Hyd. Div. ASCE*, **106**, 733–746 (1980).

Poindexter, J.S., *Adv. Microbial Ecol.*, **5**, 63–89 (1981).

Power, K. and K.C. Marshall, *Biofouling*, **1**, 163-174 (1988).

Quinlan, A.V., "Thermochemical optimization of microbial biomass production and metabolite excretion rates," in H.W. Blanch, E.T. Papoutsakis, and G. Stephanopoulos (eds.), *Kinetics and Thermodynamics in Biological Systems*, ACS Symposium Series, **207**, American Chemical Society, Washington, 1983, pp. 463-488.

Robinson, J.A., M.G. Trulear, and W.G. Characklis, *Biotech. Bioeng.*, **26**, 1409-1417 (1984).

Roszak, D.B. and R.R. Colwell, *Microbiol. Rev.*, **51**, 365-379 (1987).

Rubio, M. and P.A. Wilderer, *Env. Letters*, **8**, 87-94 (1986).

Rudd, T., R.M. Sterritt, and J.N. Lester, *Wat. Res.*, **18**, 379-384 (1984).

Ryu, D.Y. and R.I. Mateles, *Biotech. Bioeng.*, **10**, 383-397 (1968).

Savage, D.C. and M. Fletcher (eds.), *Bacterial Adhesion. Mechanisms and Physiological Significance*, Plenum, New York, 1985.

Schaechter, M., O. Maaløe, and N.O. Kjeldgaard, *J. Gen. Microbiol.*, **19**, 592-606 (1958).

Schaftel, S.O., "Processes of aerobic/anaerobic biofilm development," unpublished Master of Science thesis, Montana State University, Bozeman, MT, 1982.

Senez, J.C., *Bacteriol. Rev.*, **26**, 95-107 (1962).

Siebel, M.A., "Binary population biofilms," unpublished doctoral thesis, Montana State University, Bozeman, MT, 1987.

Siegrist, H., "Stofftransportprozesse in festsitzender Biomasse," unpublished doctoral dissertation, ETH, Zürich, 1985.

Sizer, I.W., *J. Biol. Chem.*, **154**, 461-473 (1944).

Soracco, R.J., H.K. Gill, C.B. Fliermans, and D.H. Pope, *Appl. Env. Microbiol.*, **45**, 1254-1260 (1983).

Stevenson, L.H., *Microbial Ecol.*, **4**, 127-133 (1978).

Tempest, D.W. and J.R. Hunter, *J. Gen. Microbiol.*, **41**, 267-273 (1965).

Topiwala, H. and C.G. Sinclair, *Biotech. Bioeng.*, **13**, 795-813 (1971).

Trulear, M.G., "Dynamics of biofilm processes in an annular reactor," unpublished Master of Science thesis, Rice University, Houston, TX, 1980.

Trulear, M.G., "Cellular reproduction and extracellular polymer formation in the development of biofilms," unpublished doctoral dissertation, Montana State University, Bozeman, MT, 1983.

Trulear, M.G. and W.G. Characklis, *J. Wat. Poll. Contr. Fed.*, **54**, 1288-1310 (1982).

Turakhia, M.H., "The influence of calcium on biofilm processes," unpublished doctoral dissertation, Montana State University, Bozeman, MT, 1986.

Turakhia, M.H. and W.G. Characklis, *Biotech. Bioeng.*, **33**, 406-414 (1989).

Uhlinger, D.J. and D.C. White, *Appl. Environ. Microbiol.*, **45**, 64-70 (1983).

U.S. Environmental Protection Agency, *Process Design Manual for Nitrogen Control*, U.S. Government Printing Office, Washington, 1975.

Valeur, A., A. Tunlid, and G. Odham, *Arch. Microbiol.*, **149**, 521-526 (1988).

Wanner, O. and W. Gujer, *Wat. Sci. Tech.*, **17**, 27-44 (1985).

Ward, D.M., *Appl. Env. Microbiol.*, **35**, 1019-1026 (1978).

Weise, W. and G. Rheinheimer, *Microbial Ecol.*, **4**, 175–188 (1978).

Wezernak, C.T. and J.J. Gannon, *Appl. Microbiol.*, **15**, 1211–1251 (1967).

Yoon, H., G. Klinzing, and H.W. Blanch, *Biotech. Bioeng.*, **19**, 1193–1210 (1977).

Zaidi, B.R., R.F. Bard, and T.R. Tosteson, *Appl. Env. Microbiol.*, **48**, 519–524 (1984).

ZoBell, C.E., *J. Bacteriol.*, **46**, 39–56 (1943).

PART IV

MODELING BIOFILM PROCESSES

MODELING MIXED POPULATION BIOFILMS

WILLI GUJER

Swiss Federal Institute for Water Resources and Water Pollution Control, Dubendorf, Switzerland

OSKAR WANNER

Swiss Federal Institute for Water Resources and Water Pollution Control, Dubendorf, Switzerland

SUMMARY

Biofilm systems are very complex. Their properties may be characterized by a large number of particulate and dissolved components. Their behavior is a result of numerous chemical, physical and biological processes, as well as environmental conditions. Some of these determinants have been elucidated in the past years, but many questions remain unanswered. Experimental investigations of biofilm systems are formidable. For the observation of the interactions between components and processes in the biofilm, its intact structure is a prerequisite and requires the analysis of changes that take place over a range of micrometers. Chapter 3 reports on progress with regard to experimental techniques. Still, more information is needed for a complete understanding of biofilm behavior.

Thus, complexity of the systems and experimental difficulties make mathematical models a very important tool in biofilm research and applications. Models help the researcher to state and test hypotheses, as well as represent and interpret data. In practical applications, models provide a means for prediction of biofilm behavior and for failure analyses.

Complexity and missing knowledge also are reflected by the biofilm models available today. Thus, this chapter attempts to derive fundamental equations for the mathematical modeling of a mixed population biofilm. The equations are based on as few *a priori* assumptions as feasible. Simplifying assumptions have been clearly identified, and general information on the significance and mathematical description of transport and transformation processes is presented.

The resulting set of equations primarily serves to identify where simplifying assumptions are introduced in biofilm models. Usually, these equations cannot be solved for a specific biofilm system, and additional simplifications are required. Thus, the concept of the *base biofilm model* is described. This model describes biofilm behavior as accurately and on as high a level of complexity as presently feasible.

11.1 INTRODUCTION

The behavior of a biofilm is determined by a variety of biological, chemical, and physical processes internal to the biofilm as well as by interactions between the biofilm and its environment. The biofilm and its environment form a very complex system, which is sometimes difficult to analyze experimentally. Thus, phenomenological mathematical models represent an important tool in biofilm research, because hypotheses are stated in a precise and unambiguous language, stated in a quantitative form, and stated in manner that permits their testing so they can be confirmed or rejected.

11.1.1 Deductive Modeling

Biofilm modeling has a long history of *inductive modeling*. Early model equations were developed for the design of trickling filters and were empirical or "black box" models. Later, models considered the heterogenous nature of biofilm systems by identifying the exposed surface area as an important parameter. Further developments included molecular diffusion of a substrate into the depth of the biofilm, which resulted in a relationship between intrinsic microbial kinetics and the observed kinetic behavior of biofilms. Including several nutrients in the analysis permitted discussion of multiple substrate limitations. Presently, distribution of different microbial species throughout the film are modeled (Table 11.1). If the mathematical model fits the experimental data, this is frequently considered "proof" of the underlying assumptions. Rarely are these models used to extend information from one type of biofilm system to another one.

Deductive modeling requires that a phenomenological model be developed first. The model is general in scope but too complex for application. However, the underlying mathematical derivations are consistent and clearly identify assumptions used. Further assumptions, specific for the problem to be described, can be made to simplify the model so that it still describes the phenomena of interest and at the same time is mathematically consistent. Deductive modeling has the advantage that each assumption must be clearly identified and its significance discussed. Clearly, some assumptions generally made in inductive biofilm modeling may, for some situations, prove invalid.

11.1.2 Preliminaries

A clear definition of the goal for the modeling effort and a verbal description of the model are important first steps and require significant attention. The goal of the modeling effort must be clearly defined to identify the anticipated results from the model before the model is developed and successfully applied.

The following information must usually be gathered prior to model development:

1. Components of the model system must be listed. Usually, the components constituting the solid matrix of the biofilm (e.g., microorganisms, inert solids) and components in the bulk water compartment that are affected by transformation processes are the necessary ingredients.
2. Process kinetic expressions and parameters as well as stoichiometric coefficients for all transformation, transport, and interfacial transfer processes that affect the model components must be known.
3. Rate expressions for all interfacial transfer processes (e.g., flocculation, sloughing, adsorption) must be obtained.
4. *A priori* assumptions to be included in the model must be determined. For example, is the distribution and density of active biomass in the film uniform?
5. A method for coupling the biofilm model with the biofilm reactor system model must be chosen.
6. *A priori* information on the state and the behavior of the biofilm should be determined. This information is required to define boundary conditions for the solution of model equations.

Mathematical modeling of biofilm systems is frequently quite complex. Systematic compilation of the above information makes the modeling effort more direct and more likely to succeed.

11.2 MATHEMATICAL DESCRIPTION OF BIOFILMS

The following mathematical description of biofilms is as fundamental and complex as presently feasible. The derived equations rely on mass balance principles and on the *a priori* assumption that all relevant components (e.g., chemical species, particulate material such as microorganisms) may be treated as a continuum (e.g., may be characterized by average quantities rather than the shape and behavior of individual particles). The resulting complex mathematical relationships will be simplified by introducing more specific assumptions and used to predict the behavior of some specific biofilm systems in Section 11.4. For simplicity the model equations are derived for only one dimension in space, but experienced modelers can easily expand the model to two or three dimensions (see Appendix to this chapter).

TABLE 11.1 Mixed Population Biofilm Models in the Literature[a]

Ref. No.[b]	Microbial Spatial Distribution	Basis for Spatial Distribution	Microbial System Studied	Reactor	Kinetics	Type of Model[c]		Biomass Detachment	Solution Technique	Experiments
						Substrates	Biomass			
1	Uniform	*A priori* assumption	3 groups of anaerobic organisms	Submerged anaerobic filter	Monod	Stst	Stst	—	Numerical	Yes
2	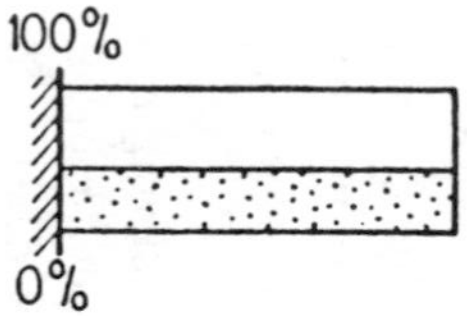 100%	*A priori* assumption	BOD degradation, nitrification	RBC	Multiple Monod	Dyn	Stst	Yes, production calculated	Numerical	Yes
3	0%	*A priori* assumption	BOD degradation, nitrification	RBC	Zero order	Stst	Stst	—	Analytical	Yes
4	Layered	Experimental results	Glucose degradation, nitrification	RBC	Zero order	Dyn	Stst	—	Numerical	Yes
5	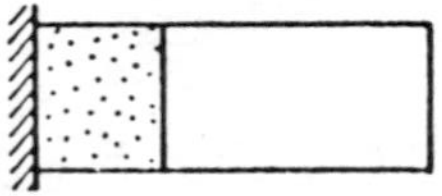	Microscopic observation	Allegedly Desulvovibrio and Beggiatoa	RBC	—	—	—	—	—	Yes
6	Linear 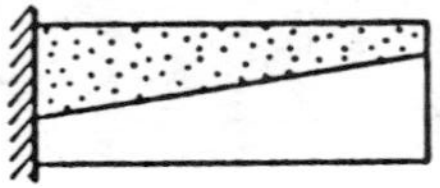	Experimental observations of microbial activity	Nitrosomonas, Nitrobacter	Fluidized bed	Double Monod	Dyn	Stst	—	Numerical	Yes

7	Exponential 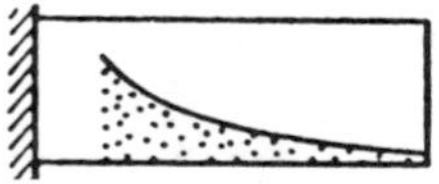	Analytical mathematical model	Organics degradation, nitrification	—	Zero order	Stst	Pseudo Stst	Not possible	Analytical	No
8	Hyperbolic 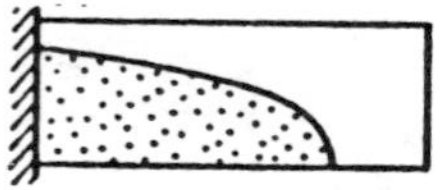	Data fitting	Organics degradation (aerobic, denitrification), nitrification	RBC	Zero order	Dyn	Stst	—	Numerical	Yes
9	Monolayer series 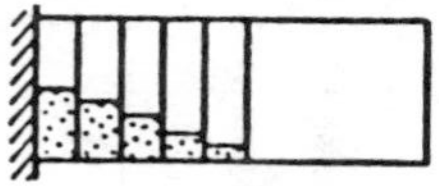	Conceptual model for replication of individual bacteria	Any two competing species	—	—	Dyn	Dyn	—	—	No
10	Variable 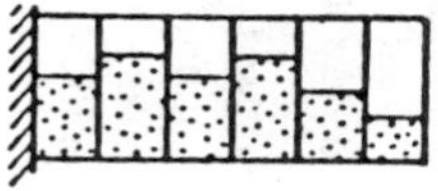	Microbial growth in a defined series of film segments	Organics degradation (aerobic, denitrification), nitrification	CSTR	Monod	Pseudo Stst	Dyn	May be included	Numerical	No

[a] Wanner and Gujer, 1986.

[b] Reference numbers refer to: (1) Young and McCarty (1968); (2) Mueller et al. (1978); (3) Goenenc (1982); (4) Watanabe et al. (1982); (5) Alleman et al. (1982); (6) Tanaka and Dunn (1982); (7) Harremoes (1982); (8) Watanabe et al. (1984); (9) McCarty et al. (1981); (10) Kissel et al. (1984).

[c] Stst = steady state; Dyn = dynamic; Pseudo stst = pseudo steady state.

11.2.1 Compartments and Phases

A biofilm system consists of different compartments such as a solid substratum, the biofilm, bulk water, and possibly a gas compartment (Figure 1.2, Chapter 1). Within the biofilm compartment, different phases can be distinguished (Figure 11.1):

1. A continuous liquid (water) phase, 1, which fills a connected fraction of the biofilm volume and contains different dissolved and suspended particulate materials. The suspended material consists of particles that can move in space independently of other suspended particles.
2. A series of solid phases, s, each composed of specific particulate material, such as a species of microorganisms, extracellular material, etc. The solids cannot move freely in space, because they are attached to each other. Movement of attached particles within one solid phase causes displacement of neighboring particles. Each type of attached solid constitutes a *solid phase,* which in addition may contain other components (e.g., sorbed components, components stored within microbial cells).

The transfer of a suspended particle from the liquid phase to a solid phase (e.g., attachment) must be modeled as a transformation process, since educt and product do not belong to the same phase. This transformation process must

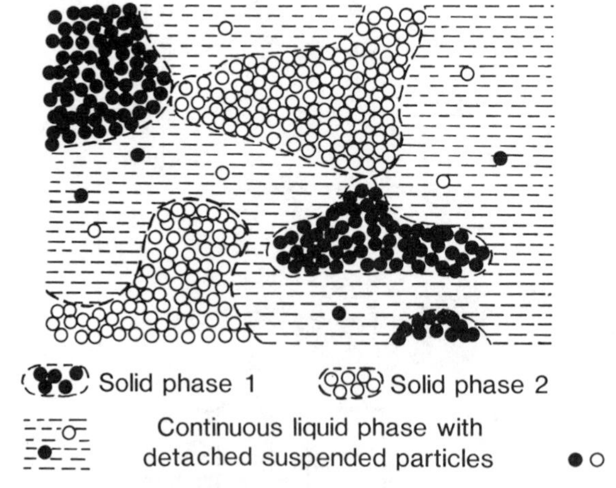

Figure 11.1 Schematic representation of the different phases within the biofilm compartment. The continuous liquid (water) phase may contain dissolved as well as suspended particles. The solid phases are bound to each other and individual attached particles cannot move independently. Only one liquid phase exists while several solid phases may be present.

be distinguished from the transport processes within the phase. Since the suspended particle and the attached particle are subject to different transport and transformation processes, different model equations (mass balance equations) must be written for these two components.

11.2.2 Component Concentrations

Transformation and some transport processes are controlled by local activities of the components involved. Activities have a thermodynamic definition, while concentrations do not. For a specific modeling effort, it may be possible to assume ideal behavior and work with concentrations.

Concentration is the mass of a component contained within a unit volume. In homogenous systems, where only one phase is present, the definition of unit volume is relatively easy. In a heterogeneous system where several phases may occur within any small finite control volume, the definition of unit volume is more complex.

In the biofilm matrix, k different phases exist, and each occupies a fraction ϵ_k of the total biofilm volume. Using the index l for the biofilm liquid (water) phase and the index s for the biofilm solid phases,

$$\sum_k \epsilon_k = \epsilon_l + \sum_s \epsilon_s = 1 \qquad [11.1]$$

where ϵ_k = local volume fraction of the total biofilm volume occupied by phase k (—)

In the subsequent derivation of model equations, the following definitions for concentrations are used:

1. *Concentration* C_{ki}*:* The rates of transport and transformation processes are governed by the concentration of a component i within one specific phase k. The concentration C_{ki} is the mass of component i contained within a unit volume of phase k. For the liquid phase, C_{li} is identical to the concentration of the component within the water squeezed from the biofilm matrix. This concentration should not be confused with the frequently used concentration based on a unit volume of total biofilm. This latter concentration is smaller than C_{ki} (since the total volume is increased by other phases) and may be predicted from the product $\epsilon_k C_{ki}$.

2. *Density* ρ_k*:* Each phase k consists of one major or bulk component such as water in the liquid phase or biomass in a specific solid phase. The concentration of this bulk component is its density (e.g., $\rho_l = \rho_{\text{water}} = 1000 \text{ kg m}^{-3}$ or $\rho_s = \rho_{\text{biomass}} = 200$ (kg dry solids) m^{-3} drained biomass volume). Since this bulk component plays an outstanding role for each phase, its concentration $C_{k,\text{bulk}}$ merits a special symbol, ρ_k. The amount of bulk material present per unit volume of

total biofilm is $\epsilon_k \rho_k$; this product is frequently used in the literature to express the biomass concentration within a biofilm.

For many aspects of biofilm modeling, it may appear to be irrelevant to distinguish between C_{ki} and $\epsilon_k C_{ki}$. This assumption should be tested in specific systems. For dynamic behavior of multispecies biofilms, this distinction must be considered.

11.2.3 The Mass Balance Equation

The mass balance equation for a one dimensional biofilm relates the change of the state of a system (C) to physical transport (J) and chemical transformation processes (r):

$$\frac{\partial \epsilon_k C_{ki}}{\partial t} = \epsilon_k \frac{\partial C_{ki}}{\partial t} + C_{ki} \frac{\partial \epsilon_k}{\partial t} = -\frac{\partial J_{ki}}{\partial z} + r_{ki} \qquad [11.2]$$

where J_{ki} = flux of component i within phase k per unit total cross-sectional area of biofilm ($M_i L^{-2} t^{-1}$)
r_{ki} = rate of production of component i within phase k per unit total volume of biofilm ($M_i L^{-3} t^{-1}$)

J and r are based on the total biofilm cross-sectional area or volume and not only the area or volume occupied by phase k, because the mass balance (Eq. 11.2) is for the total biofilm volume and not only the partial volume occupied by one phase.

For many solid phase bulk components, the density ρ_s may be assumed to be constant. After substitution of ρ_s for C_{si}, Eq.11.2 becomes

$$\rho_s \frac{\partial \epsilon_s}{\partial t} = -\frac{\partial J_s}{\partial z} + r_s \qquad [11.2a]$$

For the liquid phase, the volume fraction ϵ_l is frequently assumed to be constant. The mass balance for a dissolved component i would then read

$$\epsilon_l \frac{\partial C_{li}}{\partial t} = -\frac{\partial J_{li}}{\partial z} + r_{li} \qquad [11.2b]$$

Equation 11.2b is a form of Eq. 11.2, which has frequently been used in biofilm modeling, usually with the assumption $\epsilon_l = 1$ and $J_{li} = -D_i \partial C_{li} / \partial z$ (Fick's first law of diffusion).

The mass balance (Eq. 11.2) is a partial differential equation and may be written for all components i in each phase k. It relates the independent variables t and z to the two dependent variables ϵ_k and C_{ki} (two unknowns) and the two parameters J_{ki} and r_{ki}. In order to find a particular solution for the dependent variables, it is necessary to obtain:

1. Additional information to eliminate one dependent variable, either ϵ or C. For the liquid phase, it may be possible to specify the volume fraction $\epsilon_l(t, z)$ in explicit form (Eq. 11.2b) and substitute in Eq. 11.2. Eq. 11.2 can then be solved for C_{li}. For a specific solid phase s', it may be possible to specify the density $\rho_{s'}(t, z)$ of the bulk component to solve Eq. 11.2 for $\epsilon_{s'}$ as indicated in Eq. 11.2a. Once $\epsilon_{s'}$ is known, Eq. 11.2 is solved for all other concentrations $C_{s'i}$, in this phase.
2. Initial conditions for the remaining dependent variable, which is either $\epsilon_k(z, t_0)$, $C_{ki}(z, t_0)$, or $\rho_k(z, t_0)$. Initial conditions for concentrations must satisfy the electroneutrality condition see Eq. 11.13.
3. Information to determine the value of the transport parameter J_{ki} (Section 11.3.1).
4. Information to determine the value of the source and sink terms r_{ki} (Section 11.3.2).
5. Information (boundary conditions) to couple different forms of Eq. 11.2 at compartment boundaries (Section 11.2.4).

11.2.4 General Boundary Condition

Biofilm systems consist of different compartments (substratum, biofilm matrix, bulk liquid, gas) separated by interfaces where the physical characteristics have discontinuities (e.g., solid biofilm exists only on one side of the biofilm–bulk water interface). Efficient solution of the mass balances (Eq. 11.2) frequently requires different forms of this equation for different system compartments. In order to connect these various equations mathematically, a condition must be defined that relates the variables ϵ and C on both sides of the interface. This condition is called a boundary condition, but the term "continuity condition" is equally descriptive, since it is a special form of the mass balance equation written for the interface, where the discontinuity is located.

The derivation of this continuity condition is analogous to the derivation of Eq. 11.2, but for a differential volume element that includes an interface that may move with the velocity v_I relative to the fixed space coordinate z. Further allowance will be made for an interfacial transfer process r'', such as attachment or detachment (erosion) of particulate material (e.g., biomass) or adsorption of soluble components on the fixed matrix. This transfer process is a source or sink term for an interface rather than a volume. In all interfacial transfer processes, the transferred component does not change its chemical structure, but it is transferred from one phase to another and therefore always appears in two mass balances: once as a sink and once as a source term. The yield for an interfacial transfer processes is always ± 1.0.

The continuity condition is expressed as follows (Figure 11.2):

$$\frac{d\epsilon_{ki1} C_{ki1}\, \Delta z_1}{dt} + \frac{d\epsilon_{k2} C_{ki2}\, \Delta z_2}{dt}$$
$$= J_{ki1} - J_{ki2} + r_{ki1}\, \Delta z_1 + r_{ki2}\, \Delta z_2 + r''_{ki} \quad [11.3]$$

where r''_{ki} = interfacial transfer process rate
= amount of component i produced per unit total cross-sectional area of the interface ($M_i L^{-2} t^{-1}$)

The index 1 refers to the side of the interface with lower z coordinates, and 2 to the side with higher coordinates.

Allowing the interface to move relative to the z-coordinate at a velocity v_I and considering that $\Delta z = \Delta z_1 + \Delta z_2$ results in

$$\frac{d\Delta z_1}{dt} = -\frac{d\Delta z_2}{dt} = v_I \quad [11.4]$$

Applying the product rule to Eq. 11.3, substituting Eq. 11.4, and allowing $\Delta z \to 0$ results in the final form of the continuity condition:

$$v_I(\epsilon_{k1} C_{ki1} - \epsilon_{k2} C_{ki2}) = J_{ki1} - J_{ki2} + r''_{ki} \quad [11.5]$$

This continuity equation is a special form of Eq. 11.2 and must be satisfied by all components i in all phases k at any location z. It may also be used wherever

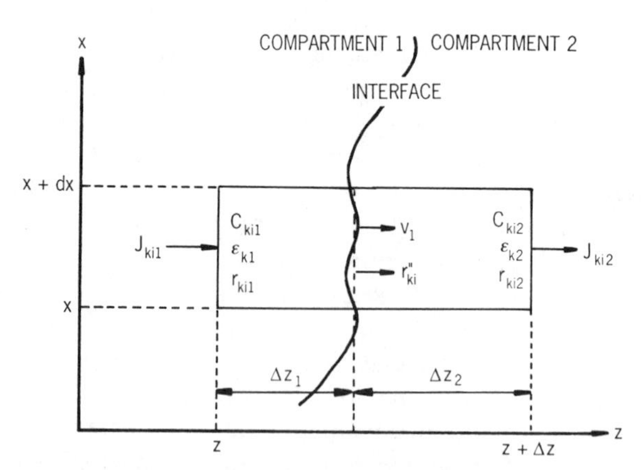

Figure 11.2 Differential volume element over the interface between the compartments 1 and 2. v_I = velocity of the interface relative to the fixed coordinate and r''_{ki} = interfacial transfer rate of component i in phase k. The C_{ki}'s are concentrations, the ϵ_k's volume fractions, and the r_{ki}'s production rates.

discontinuities in the state (ϵ, C) or transport properties (J) are expected. Specifically, it may be applied to interfaces between system compartments. Its application requires additional information to determine the interfacial transfer rate, r''_{ki} (Section 11.3.3).

For many applications to well-defined interfaces in biofilm systems, Eq. 11.5 can be simplified:

1. At the interface between a solid, impermeable substratum (index $1 = sb$), fixed in space at $z = 0$ and not subject to interfacial transfer processes (e.g., adsorption), and a biofilm (index $2 = f$), we have $v_I = 0$, $J_{ki,sb} = 0$ (there is no flux within the solid support), and $r''_{ki} = 0$. Thus,

$$J_{ki,f}(0, t) = 0 \qquad [11.5a]$$

which applies to the liquid as well as the solid phase. Eq. 11.5a is the *no flux* boundary condition, frequently applied in biofilm modeling.

2. At the interface between the biofilm (index $1 = f$) and the bulk water (index $2 = b$), three typical forms of Eq. 11.5 may be distinguished. Eq.11.5b applies for components in the biofilm liquid phase, which exists on both sides of the biofilm–bulk water interface but is not subject to an interfacial transfer process ($\epsilon_{l,b} = 1$, $r''_{li} = 0$). Equation 11.5c applies to components in a solid phase, which may occur in the biofilm only ($\epsilon_{s,b} = 0$, $J_{si,b} = 0$). Equation 11.5d applies to suspended solids, which are frequently assumed to occur in the bulk water compartment only ($\epsilon_{l,b} = 1$, $C_{li,f} = 0$):

$$v_I(\epsilon_{l,f}C_{li,f} - C_{li,b}) = J_{li,f} - J_{li,b} \qquad [11.5b]$$

$$v_I\epsilon_{s,f}C_{si,f} = J_{si,f} \qquad + r''_{si} \qquad [11.5c]$$

$$-v_IC_{li,b} = \qquad - J_{li,b} + r''_{li} \qquad [11.5d]$$

If applied to solid bulk material, Eq. 11.5c may be used to predict the velocity v_I at which a biofilm expands or shrinks. v_I is the result of two processes: (1) transport of solid bulk material towards the interface (J_s) and (2) production (e.g., attachment) or loss (e.g., detachment) of bulk material at the interface (r''_s).

3. At an interface between gas (index $1 = g$) − bulk water (index $2 = b$), fixed in space ($v_I = 0$), Eq. 11.5 for a dissolved component is as follows:

$$0 = -J_{li,b} + r''_{li} \qquad [11.5e]$$

where the interfacial transfer process rate r''_{li} represents the dissolution of the gaseous component i (a phase change or transformation process).

11.3 PROCESSES

The mass balance (Eq. 11.2), together with the continuity conditions (Eq. 11.5), can be solved only if additional information regarding transport (J_{ki}), transformation (r_{ki}), and interfacial transfer processes (r''_{ki}) is provided. Many aspects of the following discussion are specific for the biofilm compartment where solid phases are present. However, other compartments may be seen as a special case of a biofilm compartment without solid material present ($\epsilon_s = 0$).

11.3.1 Transport Processes in the Biofilm

Molecular diffusion (index M), eddy or turbulent diffusion (index T), and advection (index A) are relevant transport processes in biofilm systems. The total transport, expressed as a specific flux per unit total cross-sectional area (J_{ki}), is then the sum of the individual contributions:

$$\underset{\text{total flux}}{J_{ki}} = \underset{\substack{\text{molecular}\\\text{diffusion}}}{J_{Mki}} + \underset{\substack{\text{turbulent}\\\text{diffusion}}}{J_{Tki}} + \underset{\text{advection}}{J_{Aki}} \qquad [11.6]$$

The flux J_{ki} relates to the total cross-sectional area of a biofilm and not only to the area available for transport (e.g., the area filled with the liquid phase). This definition of flux is used in most biofilm models and requires rigorous definition of all terms (Chapter 9).

A number of other transport processes may exist: transport induced by the turbulence generated by motile organisms (bioturbation), active movement of microorganisms (motility), undulation of surface filaments, diffusion of suspended particulate material. Since we do not have generally accepted mathematical models to describe these processes, they are not discussed further. However, bioturbation and diffusion of particulate material might be modeled in analogy to molecular or turbulent diffusion as suggested by Bouwer (1987).

11.3.1.1 Molecular Diffusion in the Biofilm Molecular diffusion is driven by a potential energy gradient related to Brownian motion. It will be discussed only in the liquid phase. Within the biofilm matrix, two forms of potential energy are of prime importance: (1) the chemical potential and (2) an electrical potential for electrically charged particles. The sum of the two potentials is the electrochemical potential and may be predicted from

$$\bar{\mu}_{li} = \mu^{\circ}_{li} + RT \ln(\gamma_{li} C_{li}) + z_{li} F_A \Phi_l \qquad [11.7]$$

where $\bar{\mu}$ = electrochemical potential ($ML^2t^{-1}\ mol^{-1}$)
μ° = standard chemical potential ($ML^2t^{-2}\ mol^{-1}$)
γ = activity coefficient (—)

z = valence of an ion (charges): >0 for positive, <0 for negative charges (—)
Φ = electrical potential ($ML^2t^{-2}Q^{-1}$)
R = universal gas constant = 8.314×10^3 kg m^3 s^{-2} kg-mol^{-1} K^{-1}
T = temperature (T)
F_A = Faraday constant = 9.652×10^4 coulombs g-equivalent^{-1}

The transport due to molecular diffusion is given by the Nernst–Planck flux equation, which is usually applied to the liquid biofilm phase only:

$$J_{Mli} = -C_{li}U_i' \frac{\partial \bar{\mu}_{li}}{\partial z} \qquad [11.8]$$

where U' = apparent mobility.

Since a fraction of the cross-sectional area is not available for transport within the biofilm matrix, an apparent mobility U' is introduced and related to the apparent molecular diffusion coefficient D' as follows:

$$D_i' = U_i' RT\left[1 + \frac{d \ln \gamma_{li}}{d \ln C_{li}}\right] \qquad [11.9]$$

where D_i' = apparent diffusion coefficient of component i in the liquid phase (L^2t^{-1})

Frequently the apparent mobility U' and the diffusion coefficient D' are given as a constant fraction f of their values in pure water:

$$f = \frac{D_i'}{D_i} = \frac{U_i'}{U} \qquad [11.10]$$

f has been reported in the range of 0.5 to 0.8 (Siegrist and Gujer, 1985) for naturally grown biofilms. Substituting Eq. 11.7 into Eq. 11.8 and neglecting temperature gradients,

$$J_{Mli} = -U'RT\left[\frac{\partial C_{li}}{\partial z} + \frac{C_{li}}{\gamma_{li}}\frac{\partial \gamma_{li}}{\partial z}\right] + C_{li}U_i'z_{li}F_A\frac{\partial \Phi_l}{\partial z} \qquad [11.11]$$

Thus, molecular diffusion modeled by Eq. 11.11 is governed by three driving forces:

1. A concentration gradient expressed in the form of Fick's first law of diffusion.
2. A gradient of the activity coefficient γ_{li}, which is generally considered negligible. In biofilm systems, in the presence of solid material, such a gradi-

ent may be caused by physicochemical interactions between chemical species dissolved in the liquid phase and extracellular hydrophobic organic material attached in high density to the solid phases. Its importance is unclear.

3. A gradient in the electrostatic potential Φ_l, which may be due to electrical interactions between charges in solution (diffusion potential) or between charges in solution and charges fixed to the solid matrix (Donnan potential).

In Eq. 11.11, two new parameters have been introduced:

1. γ_{li} depends on the local chemical environment and may be estimated with the aid of the Debeye–Hückel equation for charged particles or based on solvent–solution interactions for neutral particles as outlined in the UNIFAC method (Fredenslund et al., 1977).
2. Φ_l depends on the distribution of charges in the liquid and solid phases. It may be obtained from Eq. 11.11 by considering that the net transport of charges must be balanced by an electrical current externally applied:

$$\sum_i J_{Mli} z_{li} = \text{current density, usually zero} \qquad [11.12]$$

Solution of Eq. 11.11 requires a boundary condition $C_{ki}(z_0, t)$, and if charged particles are considered, $\Phi_l(z_0, t)$ must be specified. Further, when fixing the initial conditions for Eq. 11.2, electroneutrality must be considered:

$$\sum_i [C_{li} z_{li} + C_i^* z_i^*] = 0 \qquad [11.13]$$

where C_i^* = concentration of component i attached to the solid phases per unit volume of liquid phase ($M_i L^{-3}$)

z_i^* = number of charges per unit mass of attached particles ($z > 0$ for positive, $z < 0$ for negative charges)

Equation 11.8 indicates that the gradient of the electrochemical potential ($\partial \bar{\mu}/\partial z$) must always have a finite value (or else the flux J_M would be infinite too), which may only be fulfilled if $\bar{\mu}$ is a continuous function in space. For an interface between two compartments, this may only be true if the electrochemical potential is equal on both sides of the interface:

$$\bar{\mu}_{li1} = \bar{\mu}_{li2} \qquad [11.14]$$

Equation 11.14 must be satisfied at any location z, especially at an interface between two phases. This allows the coupling of the mass balances (Eq. 11.2) for different compartments, which incorporate Eqs. 11.8 or 11.11 to describe transport. Equation 11.14 is analogous to the continuity condition (Eq. 11.5). However, Eq. 11.14 is used to satisfy the requirement for an additional boundary condition for application of Eq. 11.8. It is evident (Eqs. 11.12 and 11.13) that transport of electrically charged particles can only be described if all charged chemical species of significant concentration are considered in the modeling effort.

Care must be taken if a development of a Donnan potential is predicted ($C_i^* > 0$). Fixed charges need not be distributed evenly throughout the matrix. Since the effect of fixed charges may only reach over the approximate distance of the Stern double layer, the full Donnan potential may not accumulate as predicted with the above equations if the distribution of charges is not uniform (Siegrist and Gujer, 1985).

Molecular transport and interactions between liquid and solid biofilm phases allow for step changes of concentrations C_{li} at locations where a Donnan potential develops, or where hydrophobic interactions (γ_{li}) suddenly change. Step changes in concentrations in the liquid biofilm phase do however induce a step change in the osmotic pressure within the biofilm matrix. Do extreme gradients in osmotic pressure lead to the rupture of the solid matrix? This hypothesis may well be an important phenomenon in biofilm sloughing (Chapters 7 and 10).

With the assumption that no electrically charged particles are considered ($\partial\Phi/\partial z = 0$) and that no physicochemical interactions between different components are observed (ideal behavior, $\gamma = 1$, $\partial\gamma/\partial z = 0$), Eq. 11.11 after substitution of Eq. 11.9 results in Fick's first law of diffusion:

$$J_{Mli} = -D_i' \frac{\partial C_{li}}{\partial z} \qquad [11.15]$$

Equation 11.15 is the reduced form of Eq. 11.11 most frequently used in biofilm modeling. With the same assumptions, Eq. 11.14 may be rewritten as

$$C_{li1} = C_{li2} \qquad [11.16]$$

which may be used to satisfy the continuity condition for the electrochemical potential in the frequent cases where transport may be described with Fick's first law.

11.3.1.2 *Turbulent or Eddy Diffusion in the Biofilm* Turbulent diffusion is driven by eddies of different scale and results in a net transport of matter in the direction of the negative concentration gradient. If the scale of the eddies is considerably smaller than the scale of the biofilm system over which these gradients are of interest, turbulent diffusion is described by analogy to Fick's first law:

$$J_{Tli} = -D_T' \frac{\partial C_{li}}{\partial z} \qquad [11.17]$$

where $D_T'(z, t)$ = apparent eddy diffusion coefficient (L^2t^{-1}).

D_T' is an apparent coefficient because J_T relates to the entire biofilm cross-sectional area. However, turbulence applies only to the liquid phase within the biofilm. The eddy diffusion coefficient depends on the local hydrodynamic condition and is independent of the component transported. D_T is a function of the space coordinate, important when Eq. 11.17 is substituted into Eq. 11.b and further into Eq. 11.2. Use of Eq. 11.17 requires either C_{li} to be a continuous function in space (no step changes of C_{li} are possible) or D_T to be zero. This excludes the possibility of a Donnan potential which could develop in an area where turbulent diffusion occurs, since a Donnan potential is invariably related to a step change in a concentration.

11.3.1.3 Advection in the Biofilm The advective flux in the biofilm is given by

$$J_{Aki} = v_k \epsilon_k C_{ki} \quad [11.18]$$

where v_k = velocity at which a molecule or particle within phase k moves relative to the coordinate system (Lt^{-1}). The application of Eq. 11.18 requires the determination of $v_k(z, t)$. In some models, a fictitious approach velocity (apparent velocity without solids present) is used, which may be predicted from $\epsilon_k v_k$.

Advection in the Liquid Phase. For the liquid phase, v_l may be considered as a filtration rate governed by the permeability of the biofilm and the available pressure difference. Advection of water within a biofilm is not a frequently observed phenomenon, but it might play a role if the biofilm grows on a permeable substratum such as a lake sediment through which groundwater may infiltrate. Filtration rates may be estimated with the aid of Darcy's law, which predicts the flow velocity v_l to be proportional to the gradient of the potential energy available and the permeability of the solid material (Chapter 18).

Advection in the Solid Phases. All solid biofilm phases are subject to a common advection process. If any one attached particle moves, this causes displacement of its neighboring particles. Therefore, the advection velocity v_s is identical for all solid phases and is the result of the volume change of the bulk solid materials due to their production (e.g., biomass growth) or of a volume change of the liquid phase (Figure 11.3). The expansion process may be quantified with the aid of a mass balance (Eq. 11.2) over all bulk solid phases in the biofilm considered together.

Substituting Eq. 11.18 into Eq. 11.2 for bulk solid biofilm material and applying the product rule yields

$$\epsilon_s \frac{\partial \rho_s}{\partial t} + \rho_s \frac{\partial \epsilon_s}{\partial t} = -\epsilon_s \rho_s \frac{\partial v_s}{\partial z} - v_s \left(\epsilon_s \frac{\partial \rho_s}{\partial z} + \rho_s \frac{\partial \epsilon_s}{\partial z} \right) + r_s \quad [11.19]$$

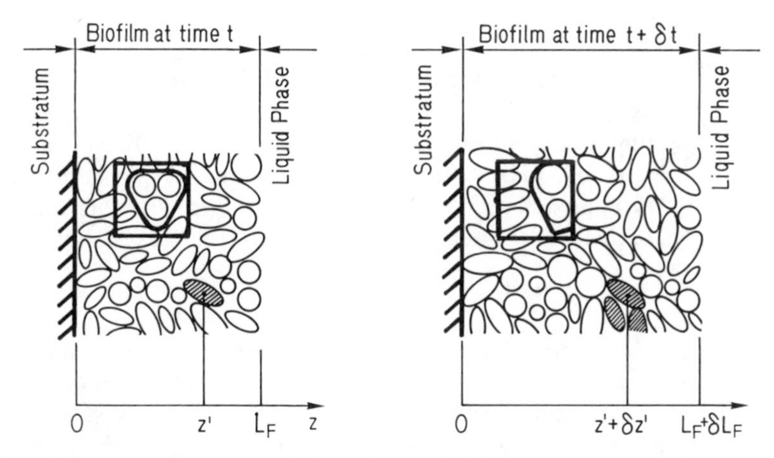

Figure 11.3 Advection of attached solid material is due to the expansion of biomass (growth). The increase in volume displaces neighboring organisms and causes the biofilm thickness to increase with time ($v_l = \delta L_f/\delta t$). The quadratic window relates to fixed coordinates and identifies advection as one mechanism that may cause local change in biomass composition; the coordinate z' is the location of a specific organism and permits calculation of the advection velocity $v_s = \delta z'/\delta t$.

Dividing Eq. 11.19 by ρ_s, summing over all solid biofilm phases, and substituting Eq. 11.1 ($\Sigma\, \epsilon_s = 1 - \epsilon_l$) yields

$$\sum_s \frac{\epsilon_s}{\rho_s}\frac{\partial \rho_s}{\partial t} - \frac{\partial \epsilon_l}{\partial t} = -(1-\epsilon_l)\frac{\partial v_s}{\partial z} - v_s\left[\sum_s\left(\frac{\epsilon_s}{\rho_s}\frac{\partial \rho_s}{\partial z}\right) - \frac{\partial \epsilon_l}{\partial z}\right] + \sum_s \frac{r_s}{\rho_s} \qquad [11.20]$$

With the two assumptions that led to Eqs. 11.2a and 11.2b (ϵ_l = constant, ρ_s = constant), Eq. 11.20 simplifies to

$$\frac{dv_s}{dz} = \frac{1}{1-\epsilon_l}\sum_s \frac{r_s}{\rho_s} \qquad [11.20a]$$

The solution of Eq. 11.20 or 11.20a requires one single boundary condition $v_s(z_0, t)$ for Eq. 11.20 or $v_s(z_0)$ for Eq. 11.20a, which may be $v_s = 0$ at the solid substratum. Equation 11.20 must be solved simultaneously with a set of mass balances (Eq. 11.2). An example of such a solution is given by Wanner and Gujer (1986).

The significance of v_s as predicted from Eq. 11.20 for solid attached material is similar to that of the diffusion coefficient for soluble material. Eq. 11.18 for biofilm solids is the equivalent of the law of diffusion (Fick's law) for soluble material.

11.3.2 Transformation Processes: Source and Sink Terms

Transformation processes in biofilms cannot be described in any detail without choosing a specific system. Microbial, biochemical, and chemical reactions occurring in a biofilm are dependent on the local chemical conditions (e.g., substrates, nutrients, redox, pH). A method for presenting kinetic information, consistent with the approach used in this chapter, was introduced in Chapter 2 and is expanded in this section. In addition, interfacial transfer processes will be discussed further.

11.3.2.1 Transformation Processes A *transformation process* is a chemical, biochemical, or microbial process in which a molecular rearrangement occurs. The process is characterized without reference to a specific system. A process description, therefore, contains only intensive properties. Interfacial transfer processes such as attachment (transformation of suspended to attached solid material) can be considered transformation processes, since the educts (suspended material) and products (attached material) do not belong to the same phase.

The *process rate equation* describes the influence of the immediate environment on the transformation process rate. A rate equation relies only on intensive properties and therefore can be stated without reference to a specific system. Process rate expressions always result in positive values.

An *observed transformation rate* indicates the rate at which a specific component is either produced (positive rate, source) or consumed (negative rate, sink) by all transformation processes occurring in the system. An observed transformation rate is an intensive property.

Stoichiometry relates transformation rates of different components within one process to the process rate. Stoichiometric coefficients are negative for educts and positive for products. They are multiplied by the process rate to obtain the transformation rate of one specific component in one specific process. Stoichiometric coefficients are intensive properties of a process.

11.3.2.2 Presentation of Process Kinetics A systematic and lucid method for presenting kinetic information on transformation processes was introduced by Petersen (1965) and has been refined specifically for the description of microbial transformation processes by Roels (1983) and Grady et al. (1986). In principle, the method requires definition of an array of process rate equations P_j over all transformation processes j and a stoichiometric matrix ν_{ji} with all stoichiometric coefficients for all components i over all processes j (Chapter 2). The array of observed transformation rates r_i is then obtained by matrix operations:

$$r_i = \sum_j \nu_{ji} P_j \qquad [11.21]$$

The information is conveniently presented in a manner illustrated in Table 11.2. The table includes the following: (1) as column headings, the components i with their mathematical symbols, with a short verbal description at the bottom of the table; (2) as row labels, the processes j with their names, and their process rate equations P_j in the rightmost column; and (3) in the body of the table, definitions of stoichiometric coefficients, with missing values representing zero. This tabular presentation for process kinetic information summarizes complex information in a concise manner.

11.3.2.3 General Remarks on Microbial Stoichiometry and Kinetics Stoichiometric (Chapter 6) and kinetic (Chapter 8) information on microbial processes has been presented elsewhere in this book. However, a few points relevant to mathematical modeling should be considered:

1. Stoichiometry and kinetics provide intensive information on a process (i.e., it does not depend on the definition of the system) and therefore can be used in widely differing systems. The information is not specific to biofilms, so information can be obtained on these topics in a variety of systems (e.g., chemostats).
2. Stoichiometric information arises from conservation principles. Conservation of mass, chemical elements, electrical charges, and equivalences (e.g., Chemical Oxygen Demand) provide additional *a priori* information, which may be used to further reduce the number of required empirical stoichiometric parameters. Furthermore, consistent use of conservation principles provides a check on the consistency of computer code, a problem that invariably arises when complex model equations are solved numerically.
3. Kinetic rate expressions for biological processes are abundant in the literature. They are all empirical and fit some data reasonably well (Spriet, 1982). In mathematical modeling, a fit to the data cannot be the only criterion for choosing an expression. Other criteria may frequently be more important when choosing the specific form of an equation, including the following: (1) the number of parameters in relation to available data or other system information, (2) the prospect for finding an analytical solution to the differential equations, (3) the stability of numerical solutions, or (4) the required computation time.

There has been little systematic treatment of these truly multidisciplinary problems in the literature. Some aspects are treated by Roels and Kossen (1978) and Roels (1980, 1983).

11.3.2.4 Physical Processes within the Biofilm Adsorption, absorption, and desorption are physical processes that occur within the biofilm matrix. Attachment processes may also occur there, but little is known about their importance. Attachment will be discussed as an interfacial transfer process in Section 11.3.3. Adsorption, absorption, and desorption are defined in Chapters 1,

TABLE 11.2 Stoichiometry and Rate Laws for the Competition of Heterotrophic and Autotrophic (Nitrifying) Microorganisms within a Biofilm[a]

j	Process	Component i = / Variable	1 ϵ_1	2 ϵ_2	3 ϵ_3	4 C_4	5 C_5	6 C_6	Process Rate P_j $[ML^{-3}T^{-1}]$
					Heterotrophic				
1	Growth		$+1$			$-\frac{1}{Y_1}$		$-\frac{1-Y_1}{Y_1}$	$\mu_{m1}\rho_1\epsilon_1 \frac{C_6}{K_{1,6}+C_6} \cdot \frac{C_4}{K_{1,4}+C_4}$
2	Endogenous Respiration		-1					-1	$b_1\rho_1\epsilon_1 \frac{C_6}{K_{2,6}+C_6}$
3	Inactivation		-1		$+1$				$k_1\rho_1\epsilon_1$
					Autotrophic				
4	Growth			$+1$			$-\frac{1}{Y_2}$	$-\frac{4.57-Y_2}{Y_2}$	$\mu_{m2}\rho_2\epsilon_2 \frac{C_6}{K_{4,6}+C_6} \cdot \frac{C_5}{K_{4,5}+C_5}$
5	Endogenous Respiration			-1				-1	$b_2\rho_2\epsilon_2 \frac{C_6}{K_{5,6}+C_6}$
6	Inactivation			-1	$+1$				$k_2\rho_2\epsilon_2$

Attached heterotrophic organisms [COD]	Attached autotrophic organisms [COD]	Attached inert solid material [COD]	Dissolved organic substrate [COD]	Dissolved ammonium [N]	Dissolved oxygen $[O_2]$

Observed Transformation Rate $[M_iL^{-3}T^{-1}]$

$$r_i = \sum_j \nu_{ji} P_j$$

Stoichiometric Parameters

Y_1 = yield coefficient for heterotrophic organisms $[M_1M_4^{-1}]$
Y_2 = yield coefficient for autotrophic organisms $[M_2M_5^{-1}]$
ν_{ij} = stoichiometric coefficient of process j and component i $[M_iM_{j*}^{-}]$

Kinetic Parameters

μ_{mi} = maximum growth rate of organism i $[T^{-1}]$
$K_{j,i}$ = saturation concentration for component i in process j $[M_iL^{-3}]$
b_i = endogenous respiration rate constant of organism i $[T^{-1}]$
k_i = inactivation rate constant of organism i $[T^{-1}]$

[a] Wanner and Gujer (1986).

7, and 9 for cells, and Chapter 12 describes a mathematical model for adsorption of cells on the substratum.

Sorption of a dissolved component refers to transfer of the component from the liquid phase into the solid phase without changing its molecular structure. Adsorption is a surface phenomenon, while absorption refers to a process by which the sorbed component enters the bulk solid phase volume. Desorption is the reverse of adsorption, where a sorbed component is released into the liquid phase. Educts and products of a sorption process do not belong to the same phase, and therefore appear in separate mass balance equations.

Sorption is frequently described as a reversible process for which two reaction rate laws apply: one for adsorption, one for desorption. Frequently, it is not possible to distinguish which fraction of the sorbed material is associated with which solid phase. Therefore, rate laws will be presented for an exchange of a component between the liquid phase l and the sum of all solid phases s. Table 11.3 presents sorption process kinetics in the format of Table 11.2.

Since sorption processes are frequently very fast (typical diffusion distances are on the order of the size of individual microorganisms and result in diffusion times $\ll 1$ s), it is frequently possible to assume an equilibrium between dissolved and sorbed component. The distribution coefficient, K_{eq}, is frequently known or may be estimated from physicochemical data, and is defined as follows:

$$K_{eq} = \frac{C_{si}}{C_{li}\rho^*_{\text{tot}}} = \frac{k_A}{k_D \rho^*_{\text{tot}}} \qquad [11.22]$$

where k_A, k_D = adsorption and desorption rate coefficients, respectively ($L^3M^{-1}t^{-1}$)

$\rho^*_{\text{tot}} = \Sigma_s \rho_s \epsilon_s$ = mass per unit volume of total biofilm (ML^{-3})

If the distribution coefficient is known, but no information is available on the sorption rate coefficients, it is frequently acceptable to assume a high rate constant for one process and use Eq. 11.22 to estimate the other one.

TABLE 11.3 Proposed Kinetic Expressions and Stoichiometry for Adsorption and Desorption Processes

Component	$i = 1$	2	Process Rate[a]
j Process	C_{s1}	C_{l2}	P_j ($ML^{-3}t^{-1}$)
1 Adsorption	+1	−1	$k_A \rho^*_{\text{tot}} C_{si}$
2 Desorption	−1	+1	$k_D \rho^*_{\text{tot}} C_{li}$

[a] $\rho^*_{\text{tot}} = \Sigma_s \rho_s \epsilon_s$ = total concentration of solid attached material within the biofilm matrix (mass per unit total volume of biofilm); k_A, k_D = adsorption and desorption rate coefficients, respectively ($L^3M^{-1}t^{-1}$).

11.3.3 Interfacial Transfer Processes

Interfacial transfer processes are those specific for the interface between the two bulk compartments in a biofilm system: (1) bulk water and (2) biofilm matrix. These interfacial transfer processes include attachment (deposition of particles on the biofilm) and detachment (e.g., erosion or sloughing). These are physical processes and consist in the exchange of attached solid material to material dissolved or suspended in the bulk water compartment or vice versa (see Chapter 7 for more discussion). These processes are confined to the interface and do not occur throughout the entire system. Characterization of interfacial transfer processes in biofilms has been severely limited, and no generally accepted rate expressions are available.

Attachment rates have been discussed generally in Section 9.8.3. Bouwer (1987) suggests that larger particles may settle out over horizontal parts of biofilms and uses Stokes's law to predict the flux of particulate material towards the surface of the biofilm:

$$r''_{si} = -r''_{li} = r_{\text{settling}} C_{li} \qquad [11.23]$$

In a theoretical analysis of attachment, Bouwer (1987) suggests, in analogy to flocculation theory, the following rate law for attachment:

$$r''_{si} = -r''_{li} = k\alpha_e C_{li} \qquad [11.24]$$

where k is a mass transfer coefficient that depends on system geometry, hydrodynamics, and particle size. α_e, the sticking efficiency, is the fraction of collisions between particles and biofilm interface that result in particle attachment. α_e is known to depend strongly on the chemical interactions between particle and surface. Realistically, Bouwer indicates a range of 0.001 to 0.1 for the value of α_e. Escher (1986) has observed a sticking efficiency ranging between 0.01 and 0.001 for adsorption of *Pseudomonas aeruginosa* on glass (Chapter 12). No predicitve theory is available for reliable estimates of α_e.

Erosion detachment has been studied by Trulear and Characklis (1982), who indicate that nutrient supply, biofilm thickness, and hydraulic shear are important factors (Chapters 7 and 9). Furthermore, the shear stress history of the biofilm apparently influences the rate and extent of erosion. Wanner and Gujer (1986), in a theoretical study, used the following rate expression for erosion:

$$r''_{si} = -r''_{li} = -\lambda L_f^2 C_{si} \qquad [11.25]$$

where λ is a rate coefficient that may depend on local hydraulic shear and biofilm strength, and L_f is the local biofilm thickness.

The cause of biofilm *sloughing* is poorly understood (Chapters 7 and 9). Howell and Atkinson (1976) modeled sloughing as a consequence of low nutrient concentration. Harremoes et al. (1980) report that sloughing is due to the development of nitrogen gas bubbles in the lower layers of a denitrifying biofilm. Many researchers speculate that sloughing occurs as a consequence of nutrient (substrate or oxygen) deficiency in the depth of the biofilm. Extreme osmotic pressure gradients may also be a factor. No mechanism for this process has been accepted, and kinetic information is lacking. Wanner and Gujer (1986) theoretically demonstrate, however, that biofilm sloughing has an effect on the performance of a multispecies biofilm. Their proposed kinetic expression for sloughing is as follows:

$$r''_{si} = -r''_{li} = \sigma_s C_{si} \qquad [11.26]$$

where σ_s = a velocity that is set equal to zero between sloughing events and assumes a high value during sloughing (Lt^{-1}). The expression permits the integration of the mass balance equations continuously in time, even though sloughing appears to introduce discontinuities. Since sloughing may affect small patches of the surface of a biofilm in a more or less random distribution, the prediction of the consequences of sloughing may require two and three dimensional modeling as well as the introduction of stochastic elements into the model.

11.4 MODELING MIXED POPULATION BIOFILMS

The previous sections have established a general mathematical framework for deterministic modeling of biofilm systems. The equations permit description of transport and transformation processes for dissolved and particulate components in any compartment or phase of a biofilm system. The equations have been derived from fundamental physical laws using few *a priori* assumptions. Thus, they are suited for fundamental investigations on biofilm systems. However, for direct application to practice, they yield a biofilm system model that is too complex. In most cases, it requires data and information that are not available. Thus, additional simplifications and abstractions from the complexity of natural biofilms have to be made to obtain practically useful biofilm models.

11.4.1 Assumptions

A biofilm consists of two different structures (Chapter 1): (1) a base film, in which the particulate components form a continuous solid matrix, and (2) a surface film, which has a rather discontinuous, filamentous structure (Figure 11.4a). In the latter structure, suspended particles (e.g., particulate substrates or microbial cells) are transported in the liquid phase, permitting direct exchange of particulate material between the bulk water and the biofilm depth. In the base film, the particulate components are packed very closely, so that there is not enough space in the liquid phase for free movement of suspended particles. Thus, only dissolved components are able to penetrate the base biofilm, while

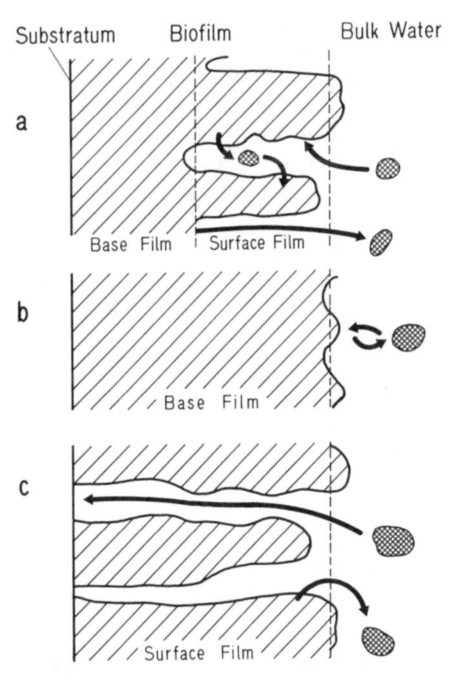

Figure 11.4 Schematic representations of biofilm structures and exchange of particulate material between the biofilm compartment and the bulk water compartment. In general, a solid base biofilm and a fluffy surface film are distinguished (a). However, in some cases, one of the two structures may be insignificant (b and c).

only at the surface film are particles exchanged between the biofilm and the bulk water. Electron micrographs indicate the existence of both types of structures. However, their mathematical modeling is not feasible, since no data are available on mechanisms by which particles in the biofilm are exchanged between the liquid and the solid phases. Thus, biofilm modeling today is restricted to base biofilms, i.e., biofilms where the surface film is insignificant (Figure 11.4b). Thus, the term "base biofilm" is used synonymously with the model in this section.

The model is a compromise that reduces the complexity to make it practically useful while retaining essential features of the biofilm structure and behavior. To do so requires the assumptions that are listed in Table 11.4. Various types of assumptions are distinguished. Assumption A1 states that the continuum concept applies and is a prerequisite for the chosen approach of deterministic biofilm modeling. Assumptions A3 to A6 and A16 are definitions rather than necessary simplifications. Assumptions A9 and A14 specify the mechanisms of transport in the liquid and solid phases and can readily be modified to include additional transport processes as the need arises (Section 11.3.1). Assumption A12 is made for the sake of simplicity; the complexity of the model would hardly be increased if assumption A12 were not made (Section 11.2.1). The remaining

TABLE 11.4 Simplifying Assumptions Underlying the Base Biofilm Model

A1	The biofilm is treated as a continuum: Components are not described by the structure and behavior of individual particles and molecules, but by averaging quantities such as concentrations and volume fractions.
A2	Gradients of system properties are orders of magnitude greater perpendicular to the substratum than in other directions: A one dimensional model in space applies.
A3	The biofilm consists of one liquid (water) phase and one or more solid phases.
A4	Each solid phase consists of the mass of one particulate component.
A5	The constituent particulate components are bound to each other and form a continuous solid matrix.
A6	The void spaces between the constituent particulate components are interconnected and occupied by the liquid phase.
A7	The volume fraction of the liquid phase, ϵ_l, is constant.
A8	The liquid phase contains only dissolved components.
A9	Transport of the dissolved components in the liquid phase is by molecular diffusion.
A10	The dissolved components in the liquid phase are electrically neutral and behave ideally: Fick's first law of diffusion describes molecular diffusion.
A11	The ratio of diffusion coefficients in the biofilm and in pure water, $f = D_i'/D_i$, is constant.
A12	A solid phase contains no dissolved or suspended particulate components.
A13	The density of the constituent particulate components, ρ_s, is constant.
A14	Advective transport (displacement) of the solid phases is the result of volume changes of the constituent particulate components.
A15	The advective velocities for all solid phases, v_s, are equal.
A16	Stoichiometry and kinetics of transformation processes do not depend on systems and compartments.

assumptions are restrictions prompted by the lack of relevant experimental observations. For example, the volume fraction of the liquid biofilm phase is very likely to change with time and space. However, the water content of the biofilm is usually assumed constant because few data on its variation are available. Bakke (1986) has provided some data on biofilm density changes with experimental run time (Chapter 13), which suggest that the liquid phase volume fraction decreases with time. Table 11.4 permits the modeler to keep track of assumptions as the model is developing. The validity of some assumptions listed in Table 11.4 may depend on the specific biofilm system being considered.

11.4.2 Mass Balance Equations for the Base Biofilm Model

All types of components that contribute to the mass of the solid biofilm matrix are included in the model as particulate components. The different types of components are attached to each other, and each constitutes a solid phase in the biofilm. To distinguish the particulate biofilm components from suspended particulate components, they also are referred to as "constituent particulate components." A constituent particulate component may be a single microbial species, a group of microbial species (e.g., nitrifiers), or just "biomass," depending on the level of structure desired or permitted by the experimental data. Dead cells, extracellular polymers, or inorganic particles embedded in the film could each be a particulate component. Thus, a solid phase may consist of microbial cell mass, or other organic or inorganic particulate material.

Those dissolved components that have an effect on the considered chemical, biological, or physical processes and that are themselves affected by these processes must be included in the model. Examples are electron donors and acceptors, nutrients, and educts and products of chemical reactions. If the concentration of a dissolved component in the biofilm does not significantly change with time or space, it should be treated as a model parameter and not as a variable, in order to minimize the complexity of the model.

Applying the simplifying assumptions of Table 11.4 to the general mass balance (Eq. 11.2) yields two sets of partial differential equations, one for the constituent particulate and one for the dissolved components considered in the model:

$$\rho_s \frac{\partial \epsilon_s}{\partial t} = -\frac{\partial J_s}{\partial z} + r_s \quad [11.27]$$

$$\epsilon_l \frac{\partial C_{li}}{\partial t} = -\frac{\partial J_{li}}{\partial z} + r_{li} \quad [11.28]$$

Using the transport rate expressions discussed in Section 11.3.1 and with ρ_s substituted for C_{si}, Eqs. 11.10, 11.15, and 11.18 yield

$$J_s = v_s \rho_s \epsilon_s \quad [11.29]$$

$$J_{li} = -D_i f \frac{\partial C_{li}}{\partial z} \quad [11.30]$$

The advective velocity v_s is the result of mass production and related volume change of the particulate components, which causes their subsequent displacement in space (Figure 11.3). v_s can be calculated from Eq. 11.20a using assumptions A2, A7, A13, and A15 (Table 11.4):

$$\frac{dv_s}{dz} = \frac{1}{1 - \epsilon_l} \sum_s \frac{r_s}{\rho_s} \quad [11.31]$$

Eq. 11.27 to 11.31 model the dynamics and the spatial profiles of the concentrations of the dissolved components and the volume fractions of the constituent particulate components as a function of transport and transformation processes in the biofilm. To use the model, relevant components must be chosen and their kinetics specified (Section 11.3.2).

11.4.2.1 Boundary Conditions To obtain a particular solution of Eqs. 11.28 and 11.30, two values of any of the dependent variables at any location must be known. Usually, the values are given as boundary conditions at the interfaces that link the biofilm to its environment, i.e., to the substratum and the hydrodynamic boundary layer or bulk water (Section 11.2.4). The number of boundary conditions needed for the solution of Eqs. 11.27 and 11.29 depends on the relationship between attachment of particulate components to the biofilm surface and detachment from the biofilm to the bulk water. As long as detachment is dominant, only one boundary condition is needed, usually Eq. 11.5a for the biofilm–substratum interface at $z = 0$. If attachment is the dominant process, the volume fractions of the particulate components at the biofilm-bulk water interface, $\epsilon_s(z = L_f)$, depend on the relative attachment strengths of the various components to the biofilm surface. For this case, a second boundary condition is needed to calculate $\epsilon_s(z = L_f)$. This boundary condition is provided by Eq. 11.5c, which after substituting Eq. 11.29 and setting $\rho_s = C_{si}$ yields

$$\epsilon_s|_{z=L_f} = \frac{r_s''}{\rho_s(v_I - v_s|_{z=L_f})} \qquad [11.32]$$

where r_s'' = interfacial transfer rate in the solid biofilm phase ($ML^{-2}t^{-1}$)
L_f = biofilm thickness-distance from the biofilm–bulk water interface to the substratum (L)
v_I = velocity of the biofilm–bulk water interface relative to the substratum (Lt^{-1})

Then, L_f and v_I are related as follows:

$$\frac{dL_f}{dt} = v_I \qquad [11.33]$$

This equation yields the progression of biofilm thickness. The velocity v_I is given by

$$v_I = v_s|_{z=L_f} + \frac{1}{1 - \epsilon_l} \sum_s \frac{r_s''}{\rho_s} \qquad [11.34]$$

which is obtained by substitution of Eq. 11.29 into 11.5c, summation over all constituent particulate components present in the biofilm, and substitution of $1 - \epsilon_l$ for $\Sigma\, \epsilon_s$ (Eq. 11.1). Once the transformation and interfacial transfer rates are known, $v_s(t,z)$ can be calculated by Eq. 11.31, $v_I(t)$ by Eq. 11.34, and $L_f(t)$ by Eq. 11.33.

The movement of the biofilm–bulk water interface is the result of volume expansion or contraction of the biofilm. It is an essential phenomenon of biofilm behavior, since a monolayer of microbial cells immobilized on a substratum may develop to a biofilm several hundred micrometers thick within a few days. In mathematical terms, the movement of the interface constitutes a moving boundary problem. This problem is already represented by the velocity v_I in the boundary condition Eq. 11.5. However, the numerical treatment of the moving boundary problem requires further consideration (Section 11.4.3).

11.4.2.2 Initial Conditions Modeling a nonsteady state situation also requires initial conditions. These include initial values for the biofilm thickness $L_f(t_0)$; the spatial profiles of the dissolved component concentrations, $C_{li}(t_0, z)$; the volume fractions of the particulate components, $\epsilon_s(t_0, z)$, and any boundary conditions that apply.

Since the concentrations of the dissolved components stabilize in a very short time, their steady state profiles can usually be used as initial conditions. Initial spatial profiles of the volume fractions of the particulate components may be obtained in several ways: (1) from experimental data such as electron micrographs of the biofilm composition, (2) as *a priori* assumptions about the spatial distribution of the constituent particulate components, or (3) by calculation of the steady state profiles of the volume fractions of these components with a base biofilm model.

11.4.2.3 Model Parameters Numerical values of parameters have to be provided for practical applications of the base biofilm model. Transformation rates of the dissolved and particulate components considered in the model have to be specified by rate expressions, stoichiometric coefficients, and kinetic parameters. Chapter 4 presents experimental determinations for diffusion coefficients in the biofilm. The densities of the particulate components (ρ_s), the volume fraction of the liquid phase (ϵ_l), and the interfacial transfer process rates r'' are system-specific parameters that have to be determined experimentally.

11.4.2.4 Microbial Competition and Coexistence in Mixed Population Biofilms If microorganisms are suspended as individual cells in water, they compete for the nutrients that are available in their environment. If they are attached to a substratum and form a biofilm, they are subject to an additional competition for the available space. In a mixed population biofilm, the ratios of the transformation rates of the various species determine whether they will coexist in the film. Since the values of the transformation rates may change drastically over the depth of the film, a mathematical model is usually needed to analyze whether one species outgrows the others, is displaced towards the biofilm surface, or is overgrown by other species and buried in the depth of the biofilm.

Wanner and Gujer (1986) present an illustration of the use of the base biofilm model for competition and coexistence of heterotrophic and autotrophic microbial species in biofilms. The authors included a heterotrophic and an autotrophic microbial species and inert material as particulate components. As relevant dissolved components, COD, NH_4, and O_2 were considered. The relevant processes were growth and endogenous respiration of the two microbial species and transformation of microbial cells to inert particulate material. Table 11.2 lists the stoichiometry and rate expressions describing the relations between components and processes. An example of biofilm growth as predicted by the model is presented in Figure 11.5. In this example, COD, NH_4, and O_2 concentrations in the bulk water were 3 (g COD) m^{-3}, 13 (g NH_4) m^{-3}, and 8 (g O_2) m^{-3}, respectively. The computation was started with an initial biofilm thickness of 5 μm, constant concentrations of the dissolved components over the biofilm depth equal to those in the bulk liquid, and constant volume fractions of the particulate components over the biofilm depth of 0.0 for inert material, 0.65 for the heterotrophic species, and 0.35 for the autotrophic species. The substratum was assumed to be impermeable, fixed in space at $z = 0$ and not subject to reaction (Eq. 11.5a):

$$J_s|_{z=0} = 0 \quad [11.35]$$

and

$$J_{li}|_{z=0} = 0 \quad [11.36]$$

The biofilm was assumed to grow in a large, vigorously mixed reactor, in which the concentrations of the dissolved components, $C_{li,b}$, could be kept constant. With this assumption, the boundary condition provided by Eq. 11.16 applies:

$$C_{li,f}|_{z=L_f} = C_{li,b} \quad [11.37]$$

Attachment of particulate material to the biofilm surface was neglected, so another boundary condition was not needed to obtain a particular solution. Detachment was also assumed to be zero. Figure 11.5a illustrates the predicted progression of biofilm thickness and volume fractions of the particulate components under these conditions, while Figure 11.5b is a snapshot of the dissolved component concentration profiles after six days of unrestricted biofilm growth.

Is there coexistence of the heterotrophic and the autotrophic species for the environmental conditions, or will one of the two species eventually disappear from the biofilm? This question can be discussed by use of the base biofilm model. Substitution of Eqs. 11.29, 11.31, and 11.35 into Eq. 11.27 yields for the biofilm–substratum interface

$$\frac{d\epsilon_s(t, 0)}{dt} = -\frac{\epsilon_s(t, 0)}{1 - \epsilon_l} \sum_s \frac{r_s(t, 0)}{\rho_s} + \frac{r_s(t, 0)}{\rho_s} \quad [11.38]$$

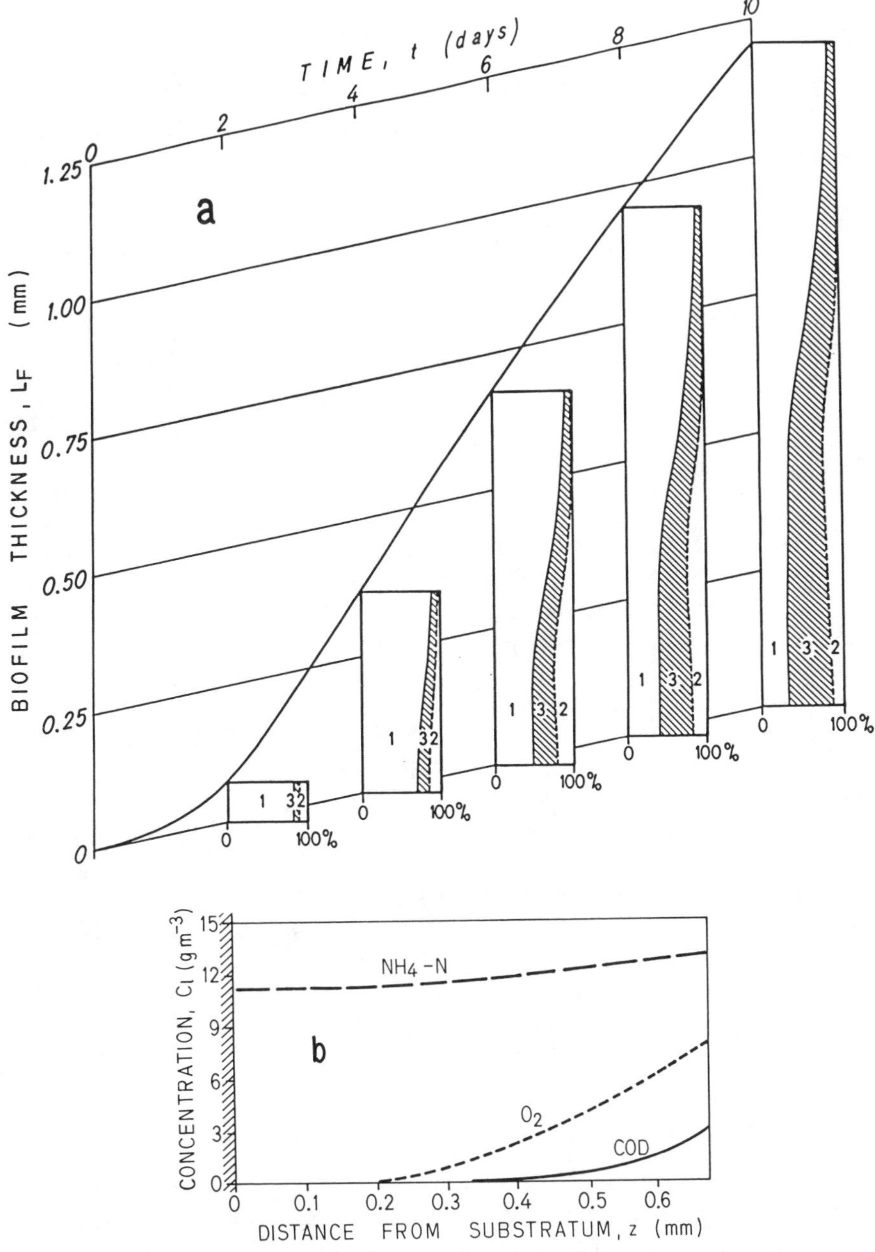

Figure 11.5 (a) Progression of biofilm thickness and particulate components in the biofilm. Bars show the fractions of heterotrophs (1), autotrophs (2), and inert material (3) as percent volume of the solid matrix from the substratum (bottom) to the biofilm–bulk water interface (top). During the first three or four days the heterotrophic species outgrows the autotrophic species throughout the biofilm. Then, due to limitation of COD supply (see b), growth conditions in the depth of the biofilm become more favorable for the autotrophic species. (b) Concentration profiles of the dissolved components in the biofilm after six days of growth.

The volume fraction $\epsilon_s(t, 0)$ of the constituent particulate species described in Eq. 11.38 is stable in time if the terms on the right hand side of the equation add up to zero:

$$\frac{1}{1-\epsilon_l} \sum_s \frac{r_s(t, 0)}{\rho_s} = \frac{r_s(t, 0)}{\rho_s \epsilon_s(t, 0)} \quad [11.39]$$

Since the term on the left hand side of the equation is the same for all constituent particulate species present in the biofilm, Eq. 11.39 holds for any of these species:

$$\frac{r_1(t, 0)}{\rho_1 \epsilon_1(t, 0)} = \frac{r_2(t, 0)}{\rho_2 \epsilon_2(t, 0)} \quad [11.40]$$

Consequently the model predicts that the volume fractions of the heterotrophic (1) and autotrophic (2) species at the substratum-biofilm interface will be stable in time if their observed specific transformation rates $r_s/\rho_s\epsilon_s$ are equal. The result can be interpreted to mean that steady state coexistence of particulate components in the biofilm depth is possible only if their specific observed transformation rates are equal. Otherwise, the component with the highest rate displaces the other components. It can be shown that this result, obtained for the biofilm–substratum interface, holds for the entire steady state biofilm if the kinetics of its constituent particulate components are such that the observed transformation rates are nonnegative. In such a case, the biofilm–substratum interface is the only location in the biofilm where $v_s = 0$. If the observed transformation rates of the particulate components, r_s, can have negative values (e.g., endogenous decay dominates in the biofilm depth), the velocity v_s may become zero at several locations which divide the biofilm into a number of layers. Then, competition and coexistence has to be analyzed separately in each of the layers. A layered biofilm structure may lead to numerical difficulties (Section 11.4.3.4).

11.4.3 Numerical Treatment of the Base Biofilm Model

11.4.3.1 Characteristic Times The base biofilm model consists of a set of nonlinear partial differential equations that describe the concentrations and volume fractions of components as function of time and space resulting from transport and transformation processes. The equations can only be solved numerically. In most cases, even their numerical solution is difficult. Very often, the complexity of the model can be significantly reduced by additional simplifications from analysis of characteristic times t_C (Table 11.5). Characteristic times are based on typical quantities such as characteristic length L (e.g., diameter), compartment volume V, volumetric flow rate Q, or mass transfer coefficient k.

The characteristic times for transformation and transport processes are compared with the time t_0, which characterizes the duration of an observation pe-

TABLE 11.5 Characteristic Times of Typical Processes in Biofilm Systems

Process	Characteristic Time	Estimate
Advection	t_A	L/v
Molecular diffusion	t_M	L^2/D
Turbulent diffusion	t_T	L^2/D_T
Reaction	t_R	C/r
Interfacial transfer	t_{SR}	LC/r''
Mass transfer	t_{MT}	L/k
Detention	t_D	V/Q

riod, an experiment, or a computer simulation. This comparison indicates possible simplifications to the partial differential Eqs. 11.27 and 11.28. Results of the characteristic time analysis, with significance to the base biofilm model, are as follows:

1. Mixing processes (i.e., processes that decrease spatial concentration gradients) with characteristic times much larger than those of other mixing processes in a mass balance equation may be omitted. Characteristic times may be used to identify the dominant transport process among the potential transport processes (Section 11.3.1).
2. If the characteristic times of transformation and mixing processes both are on the order of magnitude of t_0, the behavior of a component has to be described by partial differential equations such as Eq. 11.27 and Eq. 11.28.
3. Mass balance equations with transport and transformation processes that have characteristic times much smaller than t_0 may be treated as steady state equations. A typical example is the mass balance equation usually used for the hydrodynamic boundary layer between biofilm and bulk water.
4. If the characteristic time of the mixing processes is much smaller than that of the reaction and mass transfer processes, no spatial concentration gradients exist and the CFSTR description applies.
5. If the characteristic times of transport and transformation processes are much larger than t_0, then the changes of the variables with time are so small during the observation period that they may be neglected. For example, the microbial composition of the biofilm does not change significantly during an experiment that only lasts a few minutes.

Summarizing, an analysis of characteristic times indicates terms that may be neglected to simplify partial differential equations to ordinary differential equations in time or space, or even to algebraic equations. Since characteristic times yield order of magnitude estimates, simplifications are acceptable when estimates of comparable characteristic times are at least two orders of magnitude apart.

11.4.3.2 Stiffness of the Equation System The system of partial differential equations (Eqs. 11.27 and 11.28) for the base biofilm model has to be integrated numerically, which is frequently difficult because of the inherent stiffness of the equation system. An equation system is *stiff* if it includes certain variables that respond orders of magnitude faster to changes in the environment than other variables of the system. An instructive and lucid treatise on the problem of stiffness is given by May and Noye (1984). The numerical treatment of stiff equation systems usually requires a tremendous amount of computer time, and stiffness of the system necessitates the application of specific integration algorithms, such as backward differentiation formulas (Gear, 1971). However, for the base biofilm model the equations can be separated into two groups, each of which is only moderately stiff. One group includes the particulate, and the other the dissolved components. The separation is justified because the particulate components always respond orders of magnitude slower than the dissolved components. This can be illustrated by the following order of magnitude estimate based on the characteristic time for transformation processes, t_R (Table 11.5). For a particulate component with typical values of $\mu_m = 10^0$ day^{-1}, the estimate yields

$$t_{Rs} = \frac{\rho_s \epsilon_s}{r_s} = \frac{\rho_s \epsilon_s}{\rho_s \epsilon_s \mu_m} = \frac{1}{\mu_m} = 10^0 \text{ day} \qquad [11.41]$$

For a dissolved component, typical parameter values are $C_l = 10$ g m^{-3}, $Y = 10^{-1}$ for the yield coefficient, and $\epsilon_l = 1$ for the liquid phase volume fraction. For the solid phase, $\rho_s \epsilon_s = 10^4$ g m^{-3} for the concentration and $\mu_m = 10^0$ day^{-1} for the maximum specific transformation rate of the particulate component. Then, the order of magnitude estimate for $r = -r_l$ yields

$$t_{Rl} = \frac{\epsilon_l C_l}{-r_l} = \frac{\epsilon_l C_l}{\rho_s \epsilon_s \mu_m / Y} = 10^{-4} \text{ day} \qquad [11.42]$$

Comparison of t_{Rs} and t_{Rl} reveals that the characteristic times of the two groups of components are indeed four orders of magnitude apart. Thus, the analysis permits a simplified numerical treatment of the base biofilm model and the use of efficient integration methods. If the objective is to model the behavior of the particulate components, the concentrations of the dissolved components in the biofilm are described by steady state profiles. If the dynamics of the dissolved components is to be investigated, the mass of the particulate components produced or dissolved during a brief experiment is neglected.

11.4.3.3 Treatment of the Moving Boundary Problem There are several approaches to the numerical solution of the moving boundary problem describing changes in the biofilm thickness:

1. The biofilm is divided into a series of space elements. The element at the interface between biofilm and bulk water has a variable size, which is calculated by Eq. 11.33. If the size exceeds that of the other elements, an additional element is introduced. If the element size becomes negative, the number of space elements is reduced by one.

2. Kissel et al. (1984) divided the biofilm into a series of space elements with equal, but variable lengths. After each integration time step, the element lengths are recalculated according to the volume expansion or contraction obtained for the individual elements. The volume fractions of the particulate components are fitted by a cubic spline, which is then used to divide the film into a series of elements of equal length again.

3. An alternative to the numerical approaches is a coordinate transformation that eliminates the moving boundary. The model remains in analytical form, facilitates the numerical treatment, and enhances the analysis and discussion of the model. Wanner and Gujer (1986) introduced a space coordinate

$$\zeta(t) = \frac{z}{L_f(t)} \qquad [11.43]$$

which describes the distance from the substratum, normalized by the biofilm thickness L_f. Any function $f_1(t, z)$ may be expressed in ζ-coordinates as $f_2(t, \zeta)$ according to the following transformation rules:

$$f_1(t, z) = f_2(t, \zeta) \qquad [11.44]$$

$$\frac{\partial f_1(t, z)}{\partial z} = \frac{1}{L_f} \frac{\partial f_2(t, \zeta)}{\partial \zeta} \qquad [11.45]$$

$$\frac{\partial f_1(t, z)}{\partial t} = \frac{\partial f_2(t, \zeta)}{\partial t} + \frac{\partial \zeta}{\partial t} \frac{\partial f_2(t, \zeta)}{\partial \zeta}$$

$$= \frac{\partial f_2(t, \zeta)}{\partial t} - \frac{\zeta v_I}{L_f} \frac{\partial f_2(t, \zeta)}{\partial \zeta} \qquad [11.46]$$

For simplicity, functions in ζ-coordinates are expressed by the same symbols as have been used in z-coordinates. Then, applying these rules to Eqs. 11.27 to 11.30 yields

$$\frac{\partial \epsilon_s}{\partial t} = -\frac{1}{L_f}\frac{\partial v_s}{\partial \zeta}\epsilon_s + \frac{\zeta v_I - v_s}{L_f}\frac{\partial \epsilon_s}{\partial \zeta} + \frac{r_s}{\rho_s} \qquad [11.47]$$

$$\frac{\partial C_{li}}{\partial t} = \frac{1}{\epsilon_l L_f}\frac{\partial}{\partial \zeta}\left(\frac{D_i f}{L_f}\frac{\partial C_{li}}{\partial \zeta}\right) + \frac{\zeta v_I}{L_f}\frac{\partial C_{li}}{\partial \zeta} + \frac{r_{li}}{\epsilon_l} \qquad [11.48]$$

If ζ-coordinates are used, all equations for the biofilm phases, as well as the boundary conditions, have to be transformed according to Eqs. 11.44 to 11.46. The movement of the interface between biofilm and bulk water will also change the volume of the bulk water in a confined system. If the bulk water volume is large compared to the biofilm, the change may be neglected.

11.4.3.4 *Discontinuities within the Biofilm* For some problems, observed transformation rates of particulate components, r_s, may have negative values, leading to zones of contraction in the biofilm while the biofilm as a whole is expanding. Thus, the velocity v_s may be zero at locations in the biofilm other than the substratum-biofilm interface. At these locations, the biofilm is divided into layers that may be composed of different particulate components, and the volume fractions of the particulate components are discontinuous. Numerical solutions provide values of ϵ_s that change rather abruptly from one spatial grid point to the next if $v_s = 0$ between these points. With finer spatial discretization, the discontinuity becomes more localized and $\partial\epsilon_s/\partial z$ becomes steeper. In actual biofilms, layers are probably separated by transition zones rather than distinct interfaces at defined locations, thus justifying the approximation.

An analytical approach to the problem is to treat the two layers by separate sets of mass balances (Eqs. 11.27 and 11.28) and to link the latter with appropriate boundary conditions (Eqs. 11.16 and 11.5) with $r''_{ki} = 0$ and $v_I = 0$. Equation 11.27 for the particulate components can be treated separately in the various layers, as discussed previously (Section 11.4.2.4).

11.4.4 Simplified Forms of the Base Biofilm Model

Biofilm systems for which simplified forms of the base biofilm model apply include thin biofilms, biofilms at steady state, and monopopulation biofilms.

11.4.4.1 *Thin Biofilms* A simplified thin base biofilm model may be applied if a biofilm is only a few layers of microorganisms thick. This model, a *surface reaction model,* disregards spatial gradients and structure within the biofilm. The model may also be used for thick biofilms that are consistent with the following two assumptions:

A17. The dissolved component concentrations and the particulate component volume fractions at the biofilm surface, $C_{li}(t, L_f)$ and $\epsilon_s(t, L_f)$, are equal to their spatial averages over the biofilm depth, $\overline{C}_{li}(t)$ and $\overline{\epsilon}_s(t)$, respectively.

A18. The transformation rates $r(\overline{C}_{li}, \overline{\epsilon}_s)$, calculated from the spatial averages of concentrations and volume fractions, are equal to their spatial averages $\overline{r}$ over the biofilm depth.

For dissolved components, the validity of these assumptions may be tested by an order of magnitude estimation based on the following dimensionless equation:

$$\frac{C_l(z/L_f)}{C_l(z/L_f = 1)} = \frac{r_l L_f^2}{2DfC_l(z/L_f = 1)}\left[1 - \left(\frac{z}{L_f}\right)^2\right] + 1 \qquad [11.49]$$

This is the steady state solution for Eqs. 11.28 and 11.30 for a zero order transformation process, where $C_l(z/L_f)$ and $C_l(1)$ are the concentrations of the dissolved component in the biofilm and at its surface. Concentration profiles obtained by Eq. 11.49 with typical values of $C_l = 10$ g m^{-3}, $r_l = -10^5$ g m^{-3} day^{-1}, $D = 10^{-4}$ m^2 day^{-1}, and $f = 0.5$ indicate that, for $L_f < 10$ μm, only small errors are made if the thin base biofilm model is used and if spatial gradients are not considered (Figure 11.6). The same result may be obtained by an order of magnitude estimate based on characteristic times (Table 11.5). For L_f

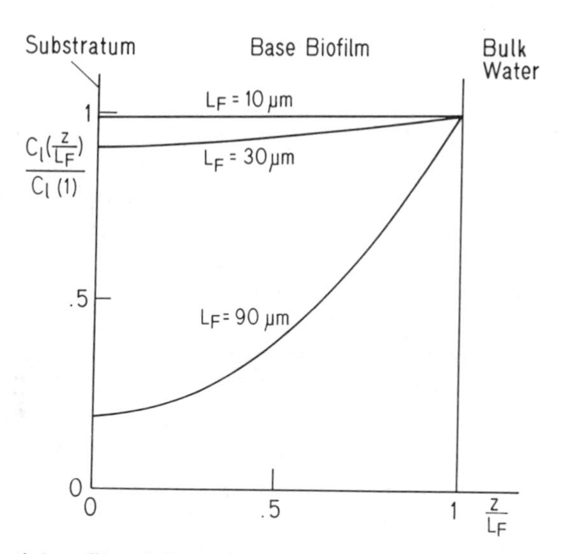

Figure 11.6 Spatial profiles of dimensionless concentration for a dissolved component, $C_1(z/L_f)/C_1(1)$, versus dimensionless distance from the substratum, z/L_f. Profiles for various values of the biofilm thickness L_f are calculated by Eq. 11.49 with assumptions and parameters as given in the text.

= 10 μm, the illustrative parameters, and $r = -r_l$, the characteristic times for transformation processes (t_R) and for transport by molecular diffusion (t_M) yield

$$\frac{t_R}{t_M} = \frac{-C_l/r_l}{L_f^2/D} = \frac{10^{-4}\text{ day}}{10^{-6}\text{ day}} = 10^2 \qquad [11.50]$$

The estimate indicates that mixing is so fast that spatial concentration gradients are negligible for $L_f = 10\ \mu$m. For particulate components, it usually is sufficient to show that the observed transformation rates r_s fulfill assumption A18.

The thin base biofilm model is derived from Eqs. 11.27 and 11.28 by integration of these equations over biofilm depth and replacement of the integrals of $\epsilon_s(t, z)$, $C_{li}(t, z)$, $r_{li}(t, z)$, and $r_s(t, z)$ with their spatial averages defined as

$$\overline{F}(t) \equiv \frac{1}{L_f(t)} \int_0^{L_f(t)} F(t, z)\, dz \qquad [11.51]$$

Averaging yields two sets of ordinary differential equations for the particulate and the dissolved components considered in the model, respectively:

$$\rho_s \frac{d\overline{\epsilon}_s}{dt} = -\frac{1}{L_f} [J_s(L_f) - J_s(0)] + \overline{r}_s \qquad [11.52]$$

$$\epsilon_l \frac{d\overline{C}_{li}}{dt} = -\frac{1}{L_f} [J_{li}(L_f) - J_{li}(0)] + \overline{r}_{li} \qquad [11.53]$$

During the first steps of the derivation of Eqs. 11.52 and 11.53, two additional terms are obtained. These terms originate from the time derivative of L_f and from the variable integral boundary, L_f. If assumption A17 is applied, the terms cancel each other. Assumption A18 is needed because the information required to calculate the transformation rates $\overline{r}_s$ and $\overline{r}_{li}$ by Eq. 11.51 is not available in the thin base biofilm model, and their values have to be approximated by $r_s(\overline{C}_{li}, \overline{\epsilon}_s)$ and $r_{li}(\overline{C}_{li}, \overline{\epsilon}_s)$, respectively.

The boundary conditions that link the biofilm to the bulk water are discussed in Section 11.4.2.1. For the interface between the biofilm and the substratum, boundary conditions are given by Eq. 11.5a. The boundary conditions yield $J_s(z = 0) = 0$ and $J_{li}(z = 0) = 0$. Eq. 11.5a is based on the assumptions that there are no interfacial transfer processes at the interface and that the substratum is impermeable and fixed in space at $z = 0$. If these conditions do not apply, Eq. 11.5a has to be replaced by the appropriate form of Eq. 11.5. At the interface between the biofilm and the bulk water, the boundary conditions for the dissolved components are given by Eqs. 11.5b and 11.16, and for the particulate components by Eq. 11.5c. Substitution of these boundary conditions into Eqs. 11.52 and 11.53 and application of assumptions A17 and A18 yields two sets of ordinary differential equations for the particulate and the dissolved components considered in the model, respectively:

$$\rho_s \frac{d\bar{\epsilon}_s}{dt} = -\frac{v_I}{L_f}\rho_s\bar{\epsilon}_s + \frac{r_s''}{L_f} + r_s(\bar{C}_{li}, \bar{\epsilon}_s) \quad [11.54]$$

$$\epsilon_l \frac{d\bar{C}_{li}}{dt} = -\frac{J_{li,b}(L_f)}{L_f} - \frac{v_I}{L_f}(\epsilon_l - 1)\bar{C}_{li} + r_{li}(\bar{C}_{li}, \bar{\epsilon}_s) \quad [11.55]$$

where $J_{li,b}(L_f)$ is the dissolved component flux in the bulk water at the biofilm surface. There are two unknowns in Eq. 11.55. Thus, additional information regarding $\bar{C}_{li}$ or $J_{li,b}(L_f)$ is needed to solve the equation. This information may be obtained in three alternative ways:

1. Assume $\bar{C}_{li}$ is equal to the concentration in the bulk water, $C_{li,b}$ (Figure 11.7a). Then, Eq. 11.55 may be replaced by a mass balance equation that holds for both the biofilm and the bulk water compartment, and $\bar{C}_{li}$ can be determined as follows:

$$\frac{d\bar{C}_{li}}{dt} = \frac{Q_{\mathrm{IN}}}{V_l}(C_{li,\mathrm{IN}} - \bar{C}_{li}) + \frac{A_f L_f}{V_l} r_{li}(\bar{C}_{li}, \bar{\epsilon}_s) + \frac{V_b}{V_l} r_{li}(\bar{C}_{li}, \bar{C}_{li,b}) - \frac{A_f v_I}{V_l}(\epsilon_l - 1)\bar{C}_{li} \quad [11.56]$$

where A_f = area of biofilm surface (L^2)
$\bar{C}_{li}$ = average concentration of the dissolved component i in the biofilm plus bulk water compartment (M_iL^{-3})
$\bar{C}_{li,b}$ = average concentration of dissolved or suspended particulate component i in the bulk water compartment (M_iL^{-3})
Q_{IN} = volumetric inflow rate (L^3t^{-1})
V_b = volume of bulk water (L^3)
V_l = volume of biofilm liquid phase plus bulk water (L^3)

Equation 11.56 is obtained if a mass balance equation is set up for the completely mixed bulk water compartment and is combined with Eq. 11.55 by use of the continuity conditions given by Eqs. 11.5b and 11.16. Analogously, a mass balance equation is set up for the particulate components suspended in the completely mixed bulk water:

$$\frac{d\bar{C}_{li,b}}{dt} = \frac{Q_{\mathrm{IN}}}{V_f}(C_{li,\mathrm{IN}} - \bar{C}_{li,b}) + \frac{A_f}{V_f} J_{li,b}(L_f) + r_{li}(C_{li}, \bar{C}_{li,b}) \quad [11.57]$$

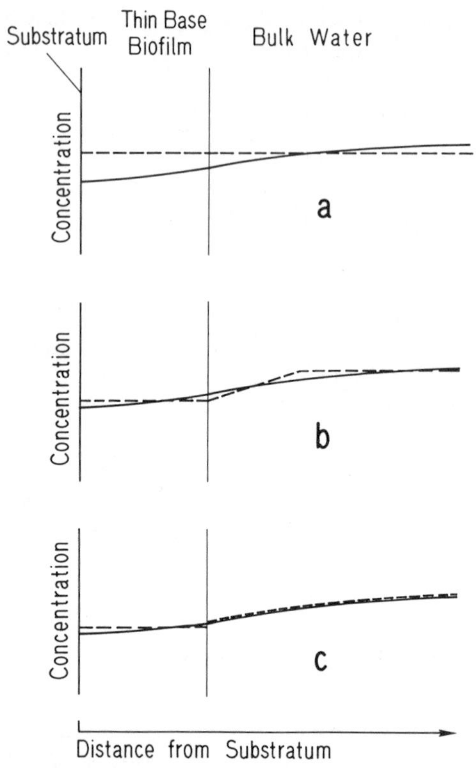

Figure 11.7 Spatial profile of dissolved component concentration as typically encountered in a biofilm and the bulk water compartment (solid lines) and approximations of the profile by the thin base biofilm model under various simplifying assumptions as described in the text (dashed lines): (a) assumption 1, (b) assumption 2, (c) assumption 3.

Substitution of the flux $J_{li,b}(L_f)$ by use of Eq. 11.5d, together with assumption A8 (no suspended particulate components in the liquid phase of the base biofilm), yields

$$\frac{d\bar{C}_{li,b}}{dt} = \frac{Q_{\text{IN}}}{V_f}(C_{li,\text{IN}} - \bar{C}_{li,b}) + \frac{A_f v_I}{V_f}\bar{C}_{li,b} + \frac{A_f}{V_f} r''_{li} + r_{li}(\bar{C}_{li}, \bar{C}_{li,b}) \quad [11.58]$$

where the interfacial transfer process rate, r''_{li} in Eq. 11.26, is equal to $-r''_s$ in Eq. 11.54. The velocity of displacement of the solid phases at the biofilm surface, $v_s(L_f)$, is obtained by integration of Eq. 11.31, by use of Eq. 11.5a and with assumption A18:

$$v_s(L_f) = \frac{L_f}{1 - \epsilon_l} \sum_s \frac{r_s(\bar{C}_{li}, \bar{\epsilon}_s)}{\rho_s} \qquad [11.59]$$

If the average concentration of the dissolved components, $\bar{C}_{li}$, in the biofilm is assumed to be equal to that in the bulk water, $C_{li,b}$, then Eqs. 11.54, 11.56, 11.58, and 11.59 form the model for a thin base biofilm. Equations 11.33 and 11.34 are used to calculate v_I and L_f.

2. Provide the flux of dissolved components in the bulk water compartment at the biofilm surface, $J_{li,b}(L_f)$, by means of a mass transfer coefficient k:

$$J_{li,b}(L_f) = k(\bar{C}_{li} - \bar{C}_{li,b}) \qquad [11.60]$$

The second form of the thin base biofilm model is based on Eqs. 11.33, 11.34, 11.54, 11.55, 11.58, 11.59, 11.60 and a mass balance for the dissolved components in the bulk water compartment (Figure 11.7b).

3. Obtain $J_{li,b}(L_f)$ by a complete mathematical description of transport and transformation of dissolved and suspended particulate components in the bulk water by mass balances (Eq. 11.2). This alternative, with the exception of Eq. 11.60 and the mass balance for the bulk water compartment, is based on the same equations as the second alternative (Figure 11.7c).

Another alternative to obtain the missing information would use effectiveness factors. However, in mixed population biofilm models these cannot be readily applied. For the implementation of the thin base biofilm model, the components and the kinetics have to be specified (Sections 11.1.2, 11.2.2, and 11.3.2). The interfacial transfer rates, r_s'' and r_{li}'', which are experimental parameters, are discussed in Section 11.3.3. For nonsteady state situations, initial conditions are required for Eqs. 11.33, 11.54, and 11.55 or 11.56.

In most cases the system of ordinary differential equations for the thin base biofilm model has to be treated numerically. The moving boundary does not provide a special numerical problem in this case.

11.4.4.2 Steady State Biofilms A biofilm is at steady state if none of its components or phases is changing with time. In practice, this condition will never be fulfilled exactly. However, as long as the deviations are small and occur only locally and infrequently, the assumption may be made that the time derivatives of all biofilm properties and state variables are zero.

A19. At steady state, no changes in time of biofilm properties and variables occur: $\partial/\partial t = 0$.

Mathematical modeling of a steady state base biofilm is based on Eqs. 11.27 to 11.31. If assumption A19 is applied to these equations, they may be rewritten as

$$\rho_s \frac{d\epsilon_s}{dz} = \frac{1}{v_s}\left(r_s - \rho_s \epsilon_s \frac{dv_s}{dz}\right) \qquad [11.61]$$

$$\frac{dC_{li}}{dz} = -\frac{J_{li}}{D_i f} \qquad [11.62]$$

$$\frac{dJ_{li}}{dz} = r_{li} \qquad [11.63]$$

Further, Eq. 11.33 leads to $v_l = 0$, which together with Eqs. 11.5c, 11.29, and 11.31 yields:

$$r_s'' = -\frac{\rho_s \epsilon_s(L_f)}{1 - \epsilon_l} \sum_s \int_0^{L_f} \frac{r_s}{\rho_s} dz \qquad [11.64]$$

Equation 11.64 relates the interfacial transfer process rate r_s'' to the biofilm thickness L_f. Both properties can be experimentally observed in order to derive additional information from Eq. 11.64.

Together with Eq. 11.31, Eqs. 11.61–11.63 form a system of first order, ordinary differential equations that relate the dependent variables ϵ_s, C_{li}, J_{li}, and v_s to the dependent variable z. This system of equations, together with Eq. 11.64, represents the steady state base biofilm model. To implement this model, the following must be specified: (1) the relevant components, (2) the relevant transformation processes, (3) the kinetics and stoichiometry of the transformation processes, and (4) *either* the thickness of the resulting steady state biofilm *or* the interfacial transfer rate. The boundary conditions that apply to the steady state base biofilm model are derived from Eqs. 11.5a–11.5d with $v_l = 0$ and Eq. 11.16, similarly to previous sections.

There are two alternatives for the numerical treatment of a steady state problem:

1. *Solution by Relaxation:* The problem is described by the general biofilm model as given by Eqs. 11.27 to 11.31. These equations are then integrated over time with constant boundary conditions until the steady state solution is reached.
2. *Solution by the Shooting Method:* Because in most cases not all boundary conditions are given at the same boundary of the biofilm, the steady state Eqs. 11.31 and 11.61 to 11.64 have to be solved by iteration. The equations are not stiff, and numerical problems are encountered only if $v_s = 0$ at locations other than $z = 0$ or $z = L_f$.

The shooting method is more efficient than relaxation, but the latter is usually simpler to handle. The assumption of steady state has the advantage that it considerably facilitates the analysis and discussion of a biofilm system. For given boundary conditions, it provides clearly defined solutions, whereas dynamic models result in a variety of solutions depending on the time history of the system.

The following example (Wanner and Gujer, 1984) illustrates the specific application of the steady state biofilm model to the microbial system described in Section 11.4.2.4. Steady state solutions for a biofilm thickness of 500 μm are calculated for numerous combinations of COD and NH_4 concentrations in the bulk water. For this thickness, substrate or oxygen limitation is predicted in the depth of the biofilm. As a consequence, the resulting substrate consumption does not depend on the biofilm thickness. The results are then stored in a data base, permitting generation of diagrams by interpolation. The regions of coexistence for heterotrophic (H) and autotrophic (A) organisms is predicted by the model for various combinations of COD and NH_4 (Figure 11.8). Furthermore, the predicted substrate removal rates (substrate flux into the biofilm) stored in

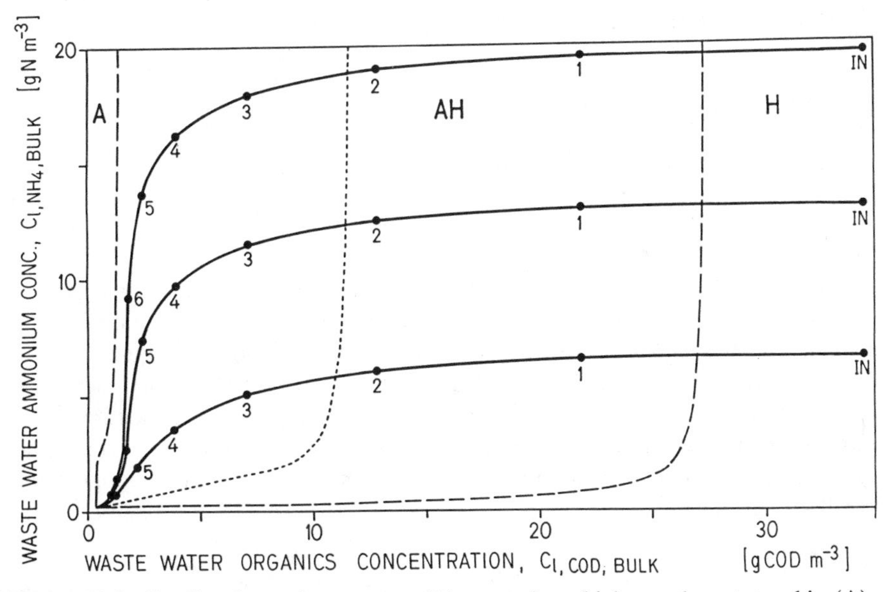

Figure 11.8 Predicted steady state conditions under which purely autotrophic (A), mixed (AH), or purely heterotrophic (H) base biofilms would accumulate as a function of bulk water ammonium and COD concentration (dashed lines separate these regions). (Reprinted with permission from Wanner and Gujer, 1984.) Trajectories of the composition of the bulk water along the depth coordinate of a trickling filter are presented for three different inlet (IN) concentrations. Full circles indicate locations equidistant over the depth. The dotted line indicates biofilms with equal oxygen consumption for nitrification and degradation of organics.

the data base permit calculation of a trajectory of the composition of the bulk water (COD versus NH_4) as a function of depth in a trickling filter biofilm system with the following equation (Wanner and Gujer, 1984):

$$\frac{dC_{l,\mathrm{NH_4}}(z = L_f)}{dC_{l,\mathrm{COD}}(z = L_f)} = \frac{J_{l,\mathrm{NH_4}}(z = L_f)}{J_{l,\mathrm{COD}}(z = L_f)} \tag{11.65}$$

The predictions of this model agree qualitatively with experimental observations. Such analyses may serve design purposes in the future.

11.4.4.3 Monopopulation Biofilms Monopopulation base biofilm models are a subset of the mixed population models. Chapter 13 extensively describes such a system. The following assumption holds:

A20. The solid matrix of the biofilm is constituted by only one solid phase (attached component, microorganisms).

Together with assumption A7 (ϵ_l = constant) and Eq. 11.1, assumption A20 leads to ϵ_s = constant. Equation 11.27 is therefore not required. Equations 11.31 and 11.34 may be written, with Eq. 11.5a, as follows:

$$v_s(z) = \frac{1}{\rho_s \epsilon_s} \int_0^z r_s \, dz' \tag{11.66}$$

$$v_I = v_s \big|_{z=L_f} + \frac{1}{\rho_s \epsilon_s} r_s'' \tag{11.67}$$

Equations 11.66 and 11.67, together with Eqs. 11.28 and 11.30 for dissolved components and Eq. 11.33 for L_f, represent the monopopulation base biofilm model. The model focuses on the dynamics of the dissolved components and biofilm thickness. This model is further simplified with the steady state assumption A19, which then leaves only Eqs. 11.62, 11.63, and 11.67 with $v_I = 0$ (Rittmann and McCarty, 1978, Harremoes, 1978, Grady and Lim, 1980). Equations 11.62 and 11.63 may be combined to form the steady state monopopulation biofilm equation:

$$\frac{d^2 C_{li}}{dz^2} = -\frac{r_{li}}{D_i f} \tag{11.68}$$

The appropriate boundary conditions (Eq. 11.5a) are $dC_{li}/dz = 0$ for $z = 0$ and (Eq. 11.16) $C_{li}(z = L_f) = C_{li,b}$.

In the simplest case, the complex set of equations introduced in Section 11.2 is now reduced to Eq. 11.68, which applies to one or more dissolved components. The list of assumptions (A1 to A20) necessary to obtain this simple model indicates that perhaps few biofilm systems may be satisfactorily described with it. Yet, this model has been found suitable for use in the wastewater treatment field for many years.

APPENDIX: THREE DIMENSIONAL BIOFILMS

The derivation of all model equations relied on the assumption that biofilms may be described by one dimensional models (e.g., gradients of all state variables parallel to a substratum would be negligibly small). This assumption is, however, not usually acceptable in multispecies biofilms. Patchy development of biofilms, segregation of different species of organisms (Gujer, 1987), filamentous growth at the biofilm surface, sloughing of macroscopic pieces of the biofilm, etc., all suggest that biofilms may not necessarily be described by one dimensional models.

11A.1 Three Dimensional Model Equations

Vector and scalar properties must be distinguished if the model equations derived in Sections 11.2 and 11.3 are to be written in three dimensional form. All transport rates (J_{ki}, v_l, v_s) and interfacial transfer rates (r''_{ki}) are vector properties; all state variables (C_{ki}, ρ_k, ϵ_k, Φ_l, μ, $\bar{\mu}$, γ) and reaction rates (r_{ki}) are scalar properties.

If the partial derivative operator $\partial/\partial z$, used in the one dimensional equations is replaced by the nabla operator ($\partial/\partial x$, $\partial/\partial y$, $\partial/\partial z$), then the divergence is obtained from vector properties and the gradient is obtained from scalar properties. With this change, all one dimensional equations may be written for three dimensions in space. All boundary, initial, and continuity conditions may be derived from the previous text.

Since only Eq. 11.20 is available for the determination of the vector $\mathbf{v}_s$ (which now counts for three unknowns), additional information must be available to determine this vector property. Most probably, a new state variable must be introduced, such as the potential energy available for the deformation of the solid film matrix (π_s), which may depend on the viscoelastic nature of the biofilm. Furthermore, a function $v_s(\text{grad } \pi_s)$ must be defined but presently no information is available for such a definition.

11A.2 Alternatives to Three Dimensional Modeling

Instead of solving the full three dimensional model equations, it might be possible to model biofilms as a two dimensional distribution of patches of one dimensional biofilms. This would exclude consideration of transport parallel to the substratum, but might nevertheless permit consideration of differences in interfacial transfer processes such as sloughing in two directions.

Simulations with three dimensional biofilms (Gujer, 1987) indicate that such an approach is acceptable if clearly identified biofilm patches exist with diameters at least 5 to 10 times larger than the penetration depth of different substrates.

REFERENCES

Alleman, J.E., J.A. Veil, and J.T. Canaday, *Wat. Res.*, **16**, 543 (1982).

Bakke, R., Biofilm detachment, unpublished doctoral dissertation, Montana State University, Bozeman, MT, 1986.

Bouwer, E.J., *Wat. Res.*, **34**, 1489-1498 (1987).

Escher, A., "Bacterial colonization of a smooth surface: An analysis with image analyzer," unpublished doctoral dissertation, Montana State University, Bozeman, MT (1986).

Fredenslund, A., J. Gmehling, and P. Rasmussen, *Vapor–Liquid Equilibria using UNIFAC*, Elsevier, New York, 1977.

Gear, C.W., *Numer. Math.*, **14**, 176 (1971).

Goenenc, I.E., "Nitrification on rotating biological contactors," LtH 82-42, Department of Environmental Engineering, The Technical University of Denmark, 1982.

Grady, C.P.L. and H. Lim, *Biological Wastewater Treatment: Theory and Applications*, Marcel Dekker, New York, 1980.

Grady, C.P.L., W. Gujer, M. Henze, G.V.R. Marais, and T. Matsuo, *Wat. Sci. Tech.*, **18**, 47 (1986).

Gujer, W., *Wat. Sci. Tech.*, **19**, 495 (1987).

Harremoes, P., "Biofilm kinetics," in R. Mitchell (ed.), *Water Pollution Microbiology*, John Wiley, New York, 1978, pp. 71–110.

Harremoes, P., J. La Cour Jansen, and G.H. Kristensen, *Prog. Wat. Tech.*, **12**, 253 (1980).

Harremoes, P., *Wat. Sci. Tech.*, **14**, 167 (1982).

Howell, J.A. and B. Atkinson, *Wat. Res.*, **10**, 307 (1976).

Kissel, J.C., P.L. McCarty, and R.L. Street, *J. Env. Eng.*, **110**, 393 (1984).

May, R. and J. Noye, "Computational techniques for differential equations," in J. Noye (ed.), *North-Holland Mathematics Studies*, **83**, Elsevier, Amsterdam, 1984.

McCarty, P.L., M. Reinhard, and B.E. Rittman, *Env. Sci. Technol.*, **15**, 40 (1981).

Mueller, J.A., P. Paquin, and J. Famularo, "Nitrification in rotating biological contactors," presented at the 51st Annual Conference Water Pollution Control Federation, Anaheim, CA, 1978.

Petersen, E.E., *Chemical Reaction Analysis*, Prentice-Hall, Englewood Cliffs, NJ, 1965.

Rittmann, B.E. and P.L. McCarty, *J. Env. Eng. Div. ASCE*, **104**, 889 (1978).

Roels, J.A., *Biotech. Bioeng.*, **22**, 2457 (1980).

Roels, J.A., *Energetics and Kinetics in Biotechnology*, Elsevier Biomedical Press, Amsterdam, 1983.

Roels, J.A. and N.W.F. Kossen, *Prog. Ind. Microbiol.*, **14**, 95–203 (1978).

Siegrist, H. and W. Gujer, *Wat. Res.*, **19**, 1369 (1985).

Spriet, J.A., "Modeling the growth of microorganisms," in S. Rinaldi (ed.), *Environmental Systems Analysis and Management,* North Holland, 1982.

Tanaka, H. and I. J. Dunn, *Biotech. Bioeng.*, **24**, 669 (1982).

Trulear, M.G. and W.G. Characklis, *J. Wat. Poll. Control Fed.*, **54**, 1288 (1982).

Wanner, O. and W. Gujer, *Wat. Sci. Tech.*, **17**, 27 (1984).

Wanner, O. and W. Gujer, *Biotech. Bioeng.*, **28**, 314–328 (1986).

Watanabe, Y., K. Nishidome, Ch. Thanantaseth, and M. Ishiguro, "Kinetics and simulation of nitrification in a rotating biological contactor," presented at 1st International Conference on Fixed-Film Biological Processes, Kings Island, OH, 1982.

Watanabe, Y., S. Masuda, K. Nishidome, and C. Wantawin, *Wat. Sci. Tech.*, **17**, 385 (1984).

Young, J.C. and P.L. McCarty, "The anaerobic filter for waste treatment," Technical Report No. 87, Department of Civil Engineering, Stanford University, Stanford, CA, 1968.

12

MODELING THE INITIAL EVENTS IN BIOFILM ACCUMULATION

ANDREAS ESCHER

Basel, Switzerland

WILLIAM G. CHARACKLIS

Montana State University, Bozeman, Montana

SUMMARY

The initial events of biofilm accumulation (colonization) at a substratum are the net result of transport, interfacial transfer (e.g., adsorption, desorption) and transformation (growth) processes. Diffusive or advective transport processes carry the cell to a point adjacent to the substratum. In laminar flow, only diffusive transport need be considered. In turbulent flow, advective transport generally dominates. Adsorption is the linking of the cell with the substratum. The cell is adsorbed to the substratum if it has a linkage to it and hence becomes immobilized for a discrete, minimum time. There is no apparent difference between the adsorption processes in laminar and in turbulent flow. Desorption and erosion, on the other hand, are strongly affected both by turbulence and by surface roughness. Multiplication of cells on the substratum is also unaffected by fluid flow regime, except as it influences substrate flux to the adsorbed cells.

Colonization consists of a number of processes, some occurring in series and some in parallel. The analysis in this chapter focuses on the various steps contributing to microbial colonization of a substratum and attempts to provide direction in methods for determining the rate-determining step. Three major concerns are reported regarding initial events in biofilm accumulation:

1. Experimental programs considering colonization of a surface must define and/or control the fluid dynamics in their experimental system. The fluid dynamics should simulate as well as the possible the fluid flow regime of relevance to the investigator.
2. The substratum roughness must be measured or controlled if at all possible. Roughness is an important variable influencing cell accumulation at the substratum.
3. The image analysis system permits the necessary measurements for conducting a process analysis of microbial colonization of a substratum.

Many questions remain unanswered regarding these initial events, and image analysis combined with mathematical modeling will be useful in the search for controlling mechanisms of adsorption in various systems.

12.1 INTRODUCTION

Mathematical models describing biofilm accumulation generally assume an existing and, frequently, uniform distribution of irreversibly adsorbed cells on the substratum as an initial condition for solving the system of equations (Chapter 11). However, any model describing biofilm accumulation at a "clean" substratum must include the transport of cells to the substratum as a first step, followed by cell adsorption and subsequent growth. This chapter focuses on a model for describing these initial events in biofilm accumulation in laminar flow, but some experimental data will be presented from turbulent flow systems.

12.1.1 Terminology

Terminology describing events occurring between a cell and the substratum has been pooled from various disciplines in this book to arrive at a consistent set of definitions. The terms reflect the cellular, as opposed to molecular, focus of this book. For this chapter, the terms *attachment, adhesion,* and *adsorption* require further emphasis. These terms are sometimes used casually and, in some cases, synonymously in reference to the intimate interaction of a cell with the substratum. More precise definitions are required for presentations in this chapter. *Sorption* is a process in which a molecule or cell moves from one phase to be accumulated in another, particularly when the second phase is a solid. Sorption is a relatively nonspecific term, which includes both adsorption and absorption. *Absorption* is the penetration of molecules or cells nearly uniformly among those of another phase to form a "solution" with the second phase. Absorption is a three dimensional process and is more suited to describing the interaction of the bulk water with biofilm, which usually has a more accessible three dimensional structure than the substratum. Therefore, absorption is more related to attachment as defined below. *Adsorption* is the concentration of molecules or cells on a substratum. Adsorption involves the interphase accumulation or concentration of substances at a substratum or interface. Adsorption is a two dimensional process. *Desorption,* the reverse of adsorption, is the movement of molecules or cells from the substratum back into the bulk liquid. *Attachment* is defined as the capture and/or entrapment of cells in a biofilm. Attachment refers to the interaction of bulk water components with the biofilm components, in contrast to adsorption, which occurs at the water-substratum interface. *Detachment,* the reverse of attachment, is the movement of cells or other components from a biofilm into the bulk liquid. Detachment is loss of components from the biofilm, in contrast to desorption, which is loss of components from the substratum. All of these processes have been discussed in detail qualitatively in Chapter 7 and quantitatively in Chapters 8 and 9.

12.1.2 Processes

The initial events of biofilm accumulation (colonization) at a substratum is the net result of transport, interfacial transfer (e.g., adsorption, desorption) and transformation (growth) processes. Colonization processes can be expressed in terms of two variables: *colony forming units* (CFU), and *cells*. This distinction is essential, since cells can adsorb in groups or as single cells. Thus, a single cell is a CFU, but an aggregate of five cells is also a CFU. Moreover, not all the cells accumulate at the substratum through transport or interfacial transfer processes alone. Cells form at the substratum by growth and reproduction, changing the number of cells per colony with time, or they may even glide away from their colony of origin to form a new colony. The following four processes *in terms of CFU* have been distinguished (Escher, 1986; Figure 12.1):

1. *Diffusive or advective transport* processes carry the CFU to a point adjacent to the substratum. In laminar flow, only diffusive transport is considered. In turbulent flow, advective transport generally dominates.
2. *Adsorption* is the linking of the CFU with the substratum. The cell is adsorbed to the substratum if it has a linkage to it and hence becomes immobilized for a discrete, minimum time.
3. *Desorption* is the breaking of the CFU-substratum linkage and the complete removal of the CFU from the substratum. Desorption is the reverse of adsorption.
4. *CFU separation,* although not related to adsorption or desorption, contributes to the accumulation of CFU at the substratum by changing the number of CFU adsorbed. A CFU with more than one cell can separate into two independent CFU as a result of fluid shear or even cell motility. CFU separation does not influence cell numbers on the substratum.

The four processes can also be described in terms of cells by determining the number of cells in each CFU. In addition to advective transport, adsorption, and

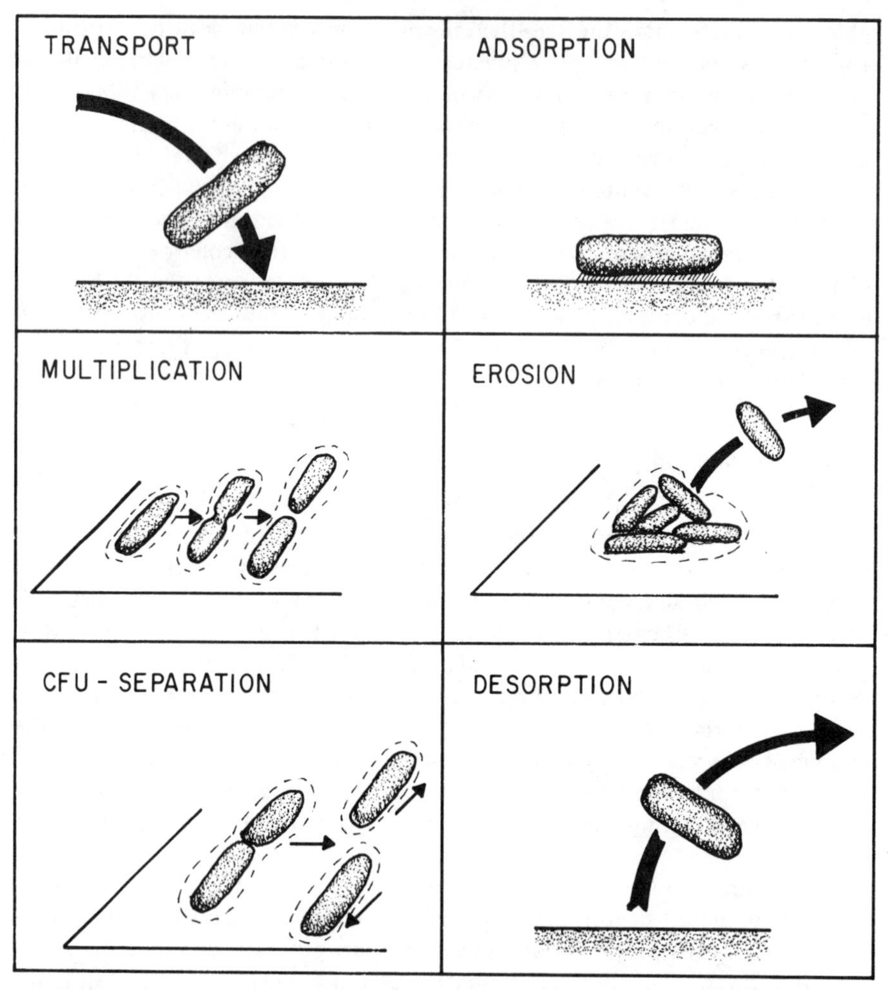

Figure 12.1 Schematic diagram of fundamental processes considered by the conceptual model for microbial colonization of a substratum.

desorption, two additional processes are necessary when *cell surface concentration* is the variable:

1. *Multiplication* is related to cellular growth. In contrast to growth, which includes the entire growth cycle of a cell, multiplication represents the recording of a singular event of a cell division. Cells within a CFU multiply, and the number of cells within this CFU increases. This does not change the accumulated number of CFU, but does change the accumulated number of cells.
2. *CFU erosion* arises from the fact that cells within a CFU can *detach* and hence reduce the cell number of the CFU. This process is the reverse of

multiplication in the sense that it is a nonselective "death" rate in the CFU. Erosion (i.e., detachment of cells from a CFU) must be contrasted to desorption, which is detachment of an entire CFU from the substratum.

When a process consists solely of steps in series, the slowest step(s) in the process network exerts the greatest influence and *controls* the rate; it is called the *rate-determining step* or *rate-controlling step*. If the process network consists only of parallel steps, the fastest step(s) controls the overall rate of the process. Colonization is described by a process network consisting of a number of processes, some occurring in series and some in parallel. Thus, the process analysis in this chapter focuses on the various steps contributing to microbial colonization of a substratum and attempts to provide direction in methods for determining the rate-determining step.

12.1.3 Image Analysis Methods

The individual processes contributing to substratum colonization have been observed with image analysis methods (Chapter 3). In the laminar flow experiments described in this chapter, microbial cells were grown in a chemostat and continuously flowed through a rectangular glass capillary. The chemostat operation minimized variations in cell physiological state and cell concentration throughout an experiment. The inner wall of the rectangular capillary is the substratum, and the processes occurring at the substratum were monitored continuously by a high resolution video camera mounted on a microscope. The video signal was transmitted to an image analyzer, which converted the grey image into a binary image. A single binary image of the specimen was stored on disk at constant time intervals (every five minutes) for subsequent analysis.

12.1.4 Experimental System

The experimental organism for these experiments was *Pseudomonas aeruginosa*, a Gram-negative, motile bacterium. The cells were grown in a chemostat, at a dilution rate approximately equal to half the maximum specific growth rate of the cells, in a mineral salts medium in which glucose was the sole carbon source and was the limiting nutrient. The substratum was a rectangular glass capillary which provided a smooth substratum for adsorption. In this case, "smooth" refers to the size of the roughness elements as compared to the cell size. The glucose concentration in the capillary was approximately 0.5 g m^{-3}. The water temperature was maintained at 25°C.

12.2 SUBSTRATUM COLONIZATION IN LAMINAR FLOW

The analysis of adsorption, desorption, and growth-related processes during the early colonization of a substratum have not been considered in traditional terms of continuum mass balances. Rather, population balances were used in which an observed cell is a member of the total population. Thus, the probability of an event is the key variable.

12.2.1 Population Balance in Terms of CFU

12.2.1.1 Transport Generally, particles or cells are transported by momentum and diffusion to the substratum. In laminar flow, however, there is no momentum transport perpendicular to the substratum, so the cells are transported to the substratum by diffusion processes. Assuming no chemotaxis, the diffusive transport results from either Brownian motion or the random motility of the cell (Section 9.9, Chapter 9). Hence, Brownian and non-Brownian diffusion are both responsible for transport to the substratum. The non-Brownian diffusion is calculated from the average free path length and velocity of the random movement, and can exceed the Brownian motion by orders of magnitude. Stanley (1983) observed that motile *Pseudomonas aeruginosa* cells adsorbed to stainless steel more rapidly than the same cells when their flagella were removed by blending. Stanley's experiments were conducted under quiescent conditions, so it is likely that the flagella influenced transport and not adsorption. Stanley's conclusions are the same. Turbulent conditions decreased the adsorption rate of the motile cells significantly while not influencing that of nonmotile cells. Thus, under quiescent conditions, transport to the substratum controlled the colonization rate. Under turbulent conditions, interfacial processes controlled it. Belas and Colwell (1982) suggest that temperature influences lateral flagellum production and, thus the adsorption rate. Diffusive transport depends on the cell diffusivity D_X, which is also a function of temperature. Thus, Belas may also have been using a transport-controlled experimental system.

The cells contact the substratum as a result of transport and come to a brief stop. As long as cells do not interact with the substratum, they are not considered adsorbed.

12.2.1.2 Offset Time Observations with the image analysis system indicate that adsorption of cells begins immediately upon exposure of the substratum. However, the first desorption event is only observed after a brief but finite period of time passes (approximately 300 s with *P. aeruginosa* on glass). The time between the onset of adsorption and the onset of desorption is termed the *offset time* t_r. It is the time interval between the start of adsorption ($t = 0$) and the onset of desorption events ($t = t_r$) in an experimental run. The offset time is probably dependent on the physiology of the adsorbing CFU, shear stress, and change in the organic conditioning film on the substratum. For example, Stanley (1983) observed very low adsorption rates of *Pseudomonas aeruginosa* in distilled water. In such conditions, a conditioning film may require a long time to accumulate significantly. Under Stanley's batch conditions, there may not be enough organic material present to adsorb to any extent. In continuous flow systems, the organic material is continuously flowing past the substratum and will ultimately establish a film. Escher (1986) has observed that the offset time is not dependent on the cell number in the bulk water.

12.2.1.3 *Adsorption/Desorption* Adsorption can be expressed as the probability that a CFU that has been transported to the substratum becomes immobilized. The probability of adsorption for a CFU, once transported to the substratum, can be defined as a *sticking efficiency* that depends on substratum properties, CFU characteristics, physiological state of organisms, and hydrodynamic shear stress at the substratum.

Two types of adsorption can be differentiated: (1) physical adsorption and (2) chemisorption. Physical adsorption is a *reversible* or equilibrium adsorption involving primarily physical forces (e.g., van der Waals, hydrogen bonds, protonation, coordination bonds, and water bridging), is characterized by a low heat of adsorption per chemical bond (20–50 kJ mol^{-1}), and exhibits low specificity between the adsorbent and the adsorbate. The net rate of physical adsorption is frequently described by saturation kinetics. Chemisorption is generally *irreversible* adsorption and is characterized by a high heat of adsorption per chemical bond (40–400 kJ mol^{-1}) and a more definitive chemical interaction (e.g., ionic or covalent bond). Chemisorption generally exhibits a high specificity of the adsorbate for the adsorbent and usually results in only single layer adsorption. In cell–surface (substratum) interactions, chemisorption is also referred to as *adhesion,* because it is more or less equivalent to irreversible adsorption, frequently mediated by an adhesin (Marshall, 1985) such as an extracellular polymer. Definition of chemisorption in a batch reactor system has generally required reference to an assay technique to reduce the concept to an operational level (e.g., Fletcher and Marshall, 1982).

The image analysis methods applied to a flowing system do not require a special, destructive assay for defining reversible adsorption. When CFU adsorb to a substratum, their bonds can be fairly weak, but strong enough to resist shear stress of the water for some time. The image analysis system recognizes the immobilized cell as being adsorbed. Due to the weak bonding, the cells have a probability to desorb within a short time. But with increasing time, the cells strengthen the adsorption bonds with the substratum and hence decrease their probability to desorb. During the period where their probability to desorb is greater than zero, they are reversibly adsorbed. As soon as their probability to desorb reaches zero, the cells are irreversibly adsorbed.

Desorption of a CFU refers to the detachment of an entire CFU from the substratum and reentrainment in the bulk water, i.e., all of the cells in the CFU detach. The probability of desorption of a CFU depends on the strength of adsorption and its size, which are probably changing with time. If a CFU at the substratum increases its bonding strength with increasing cell residence time at the substratum, then the probability of CFU desorption decreases. Thus, the probability of desorption is not equal for all adsorbed CFU, and the probability of desorption is related to the probability of adsorption rather than to the total number of CFU at the substratum (assuming constant shear force at the substratum). The rate of desorption can be defined as the product of the probability of desorption (β) and the rate of adsorption. β is probably a function of many variables, including the physiological state of the cells and characteristics of the substratum. Shear stress is clearly an important, if not the dominant, factor influencing desorption. Correlations between desorption rate and shear force have been reported (Beal, 1978) for inert particles and with microbial cells.

Stanley (1983) has observed the influence of pH and viability on the irreversible adsorption of *Pseudomonas aeruginosa* on stainless steel. The optimum pH for adsorption of viable cells was between 7 and 8. However, nonviable cells adsorbed more rapidly at pH 2.4. The more rapid adsorption at pH 7–8 may be due to active transport processes or simply greater metabolic activity at that pH. Stanley (1983) and MacRae and Evans (1984) have observed that increased Ca and Mg increase irreversible adsorption rates and conclude that the cations reduce desorption rates.

The influence of substratum roughness on accumulation rate of cells may have profound consequences. The effect of substratum roughness on accumulation of inert particles at the substratum has been addressed by Browne (1974). His calculations suggest that small changes in substratum roughness result in a large increase in accumulation rate. His calculations show reasonable agreement with experimental results of others.

12.2.1.4 CFU Separation CFU separation is the separation of one CFU into two (or more) independent CFU. It increases the number of CFU at the substratum, but not the number of cells, so its inclusion is necessary to the CFU population balance. All irreversibly adsorbed CFU have the same probability of separating, which presumably depends on the shear stress, the substratum characteristics, and the physiology of the CFU. Results of Escher (1986) suggest that the probability of separation is independent of the CFU concentration at the substratum.

12.2.2 Population Balance in Terms of Cells

The number of cells per CFU can be determined to obtain a population balance in terms of cells. Thus, all processes described in terms of CFU can be translated into terms of cells. However, CFU separation does not contribute to accumulation in terms of cells. On the other hand, growth-related processes contribute to the cell population balance.

Multiplication is a growth-related process at the substratum. Whenever a cell within a discrete CFU divides and the newly formed daughter cell remains within the same CFU, the process of multiplication is observed. If the newly formed cell is lost immediately (i.e., due to shear forces), no multiplication will be observed, since this process describes only the formation of new cells which remain within the same CFU. When an established cell within a CFU (with more than one cell) detaches and reenters the bulk flow, the CFU is eroding. If a CFU erodes, the cell number within the CFU decreases but the CFU is not removed from the substratum.

The probabilities of multiplication and erosion are defined in the same way as that of CFU separation. All irreversibly adsorbed cells are presumed to have the same probability for multiplication and subsequent erosion. The probability terms for both are defined as a probability per time (analogous to a first order rate coefficient). The rates for multiplication and erosion are expressed as number of cells per unit area per unit time.

12.2.3 Kinetic Expressions in Terms of CFU

The rate expressions for the model can be used more explicitly to derive the kinetic equations for the population balance. Accordingly, the stoichiometry of the CFU population balance can be written in the following way:

$$\underset{\substack{\text{suspended}\\ \text{or planktonic}\\ \text{CFU}}}{[\text{CFU}]_{\text{bulk}}} \rightleftarrows \underset{\substack{\text{reversibly}\\ \text{adsorbed CFU}}}{[\text{CFU}]_{\text{rev}}} \rightarrow \underset{\substack{\text{irreversibly}\\ \text{adsorbed CFU}}}{[\text{CFU}]_{\text{irrev}}} \qquad [12.1]$$

$$\underset{\substack{\text{Total adsorbed}\\ \text{CFU}}}{[\text{CFU}]_{\text{tot}}} = \underset{\substack{\text{reversibly}\\ \text{adsorbed CFU}}}{[\text{CFU}]_{\text{rev}}} + \underset{\substack{\text{irreversibly}\\ \text{adsorbed CFU}}}{[\text{CFU}]_{\text{irrev}}} \qquad [12.2]$$

Using this general stoichiometry, a model describing the rate of substratum colonization will be derived.

12.2.3.1 Transport from the Bulk Fluid to the Substratum Bowen et al. (1978) proposed an analysis based on a material balance for particles with a first order reaction approximation for the substratum–particle capture rate, which leads to an expanded Graetz solution. They base their development on the conservation equation for particle diffusion in a potential energy field $\phi = \phi(y)$ in a parallel plate channel (Figure 12.2):

$$\frac{3}{2}\left[1 - \frac{z^2}{h^2}\right] v_b \frac{\partial X}{\partial x} = \frac{\partial}{\partial z}\left[D \frac{\partial X}{\partial z} + \frac{D_X X}{k_B T}\frac{\partial \phi}{\partial z}\right] \qquad [12.3]$$

where h = half thickness of the channel (L)
v_b = average velocity of the fluid (Lt^{-1})
D_X = diffusivity (L^2t^{-1})
X = particle concentration (particles L^{-3})
k_B = Boltzmann constant, 1.3805×10^{-23} J K^{-1}
T = absolute temperature (T)
ϕ = potential energy between each particle and channel wall
= sum of electrical double layer and van der Waals interactions (Section 7.3.1.2, Chapter 7)

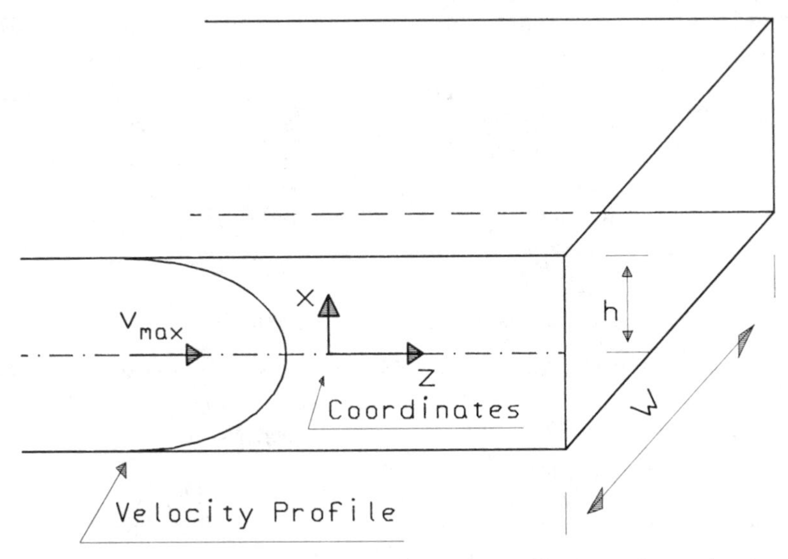

Figure 12.2 Schematic illustration of the rectangular capillary experimental system, coordinate system, and laminar velocity distribution (Eq. 12.3).

The flow is laminar, as reflected by the velocity distribution, which is a function of z and is comparable to the laminar flow distribution in a circular tube (Example 9B). Equation 12.3 is subject to the following boundary conditions:

$$\text{at } x = 0 \quad \text{for all } z, \qquad X(0, z) = X_0 \qquad [12.4]$$

$$\text{at } z = 0 \quad \text{for all } x, \qquad \left[\frac{\partial X(x, z)}{\partial z}\right]_{z=0} = \left[\frac{\partial X(x, 0)}{\partial z}\right] = 0 \qquad [12.5]$$

$$\text{at } z = b, \qquad X(x, b) = 0 \qquad [12.6]$$

The initial condition (Eq. 12.4) indicates that the suspension is homogeneous at the concentration X_0 at $x = 0$, the entrance to the observation region. Equation 12.5 is a symmetry condition about the median plane. Equation 12.6 implies that there exists an infinitely deep potential energy well (presumably due to molecular attraction) at infinitesimal cell–substratum separations. This assumption (i.e., the substratum behaves as a perfect sink for cells) is common to all deposition problems, regardless of the specific interaction being considered.

The solution of Eq. 12.3 with the boundary conditions converges well for large enough Peclet numbers (see Eq. 12.9) and proved to be accurate for inert particles adsorbing to charged substrata. The solution has the following form for microbial cells adsorbing to a smooth substratum:

$$r''_{a,\mathrm{CFU}} = \frac{X_{bC} D_X}{h} \left[\frac{\left(\frac{2}{9K_1}\right)^{1/3}}{\Gamma(\frac{4}{3}) + \frac{1}{\epsilon_{sp}}\left(\frac{2}{9K_1}\right)^{1/3}} \right] \qquad [12.7]$$

where $r''_{a,\mathrm{CFU}}$ = adsorption rate of CFU to the substratum (CFU $\mathrm{L^{-2}t^{-1}}$)
X_{bC} = CFU concentration in bulk liquid (CFU $\mathrm{L^{-3}}$)
D_X = diffusivity of cell or CFU ($\mathrm{L^2t^{-1}}$)
h = half thickness of channel (L)
K_1 = dimensionless distance from channel inlet (—)
ϵ_{sp} = substratum–particle capture factor (—)
Γ = gamma function [$\Gamma(\frac{4}{3}) = 0.89338$]

K_1, the dimensionless length, is calculated as follows:

$$K_1 = \frac{1}{Pe}\left(\frac{8L}{3h}\right) \qquad [12.8]$$

where h = half-thickness of channel (L)
L = length from inlet (L)
Pe = Peclet number (—)

The Peclet number is the ratio of advective mass transport to diffusive mass transport:

$$Pe = \frac{4 v_b h}{D_X} \qquad [12.9]$$

D_X has been described in more detail in Section 9.7.1.1 (Chapter 9).

12.2.3.2 *CFU Adsorption and Desorption Rates* Equation 12.7 indicates that the adsorption rate of CFU increases linearly with increasing CFU concentration in the bulk water. Escher (1986) has observed the same first order kinetics in experiments with *Pseudomonas aeruginosa* using image analysis techniques (Figure 12.3a). Equation 12.7 implicitly includes the CFU transport rate to the substratum as well as adsorption (i.e., the cell–substratum interaction), so the rate of adsorption may well be controlled by the transport rate. Experimental results also indicate that the desorption rate of CFU is also linear in CFU concentration in the bulk water after the offset time. However, more importantly, the desorption rate is directly proportional to the adsorption rate (Figure 12.4). The proportionality coefficient is termed β_C (see Eq. 12.14).

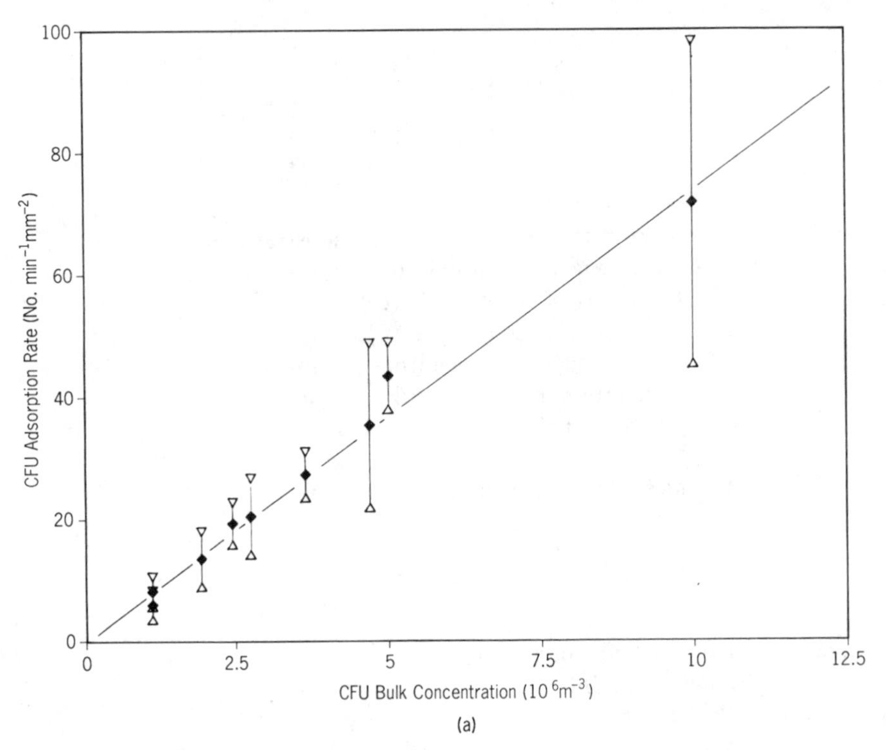

Figure 12.3 Adsorption rate for *Pseudomonas aeruginosa* is proportional (first order) to cell concentration in the bulk water for CFU (a) and for cells (b). Shear stress was 0.5 N m^{-2}. (Escher, 1986.)

12.2.3.3 *Significance of the Substratum–Particle Factor* ϵ_{sp} The dimensionless substratum–particle capture factor, ϵ_{sp}, may be particularly significant in the analysis of experimental data. ϵ_{sp} is independent of the bacterial transport rate to the substratum and describes the interaction of the cells with the substratum during the process of adsorption. Any factor affecting the probability of successful adsorption of a cell to the substratum will influence ϵ_{sp}. Thus, ϵ_{sp} serves as a quantitative parameter for comparing the affinity for adsorptive interactions between specific cells and substrata. Escher (1986) observed that ϵ_{sp} decreased with increasing hydrodynamic shear stress in the rectangular capillary experiments (Figure 12.5). An exponential regression in Figure 12.5 was chosen to represent the data because ϵ_{sp} varies from ∞ to 0, not because it provided the best fit to the data. ϵ_{sp} was observed to be independent of bulk water cell concentration.

12.2.3.4 *CFU Population Balance at Substratum* CFU are transported from the bulk liquid to the substratum, where they reversibly adsorb. With time,

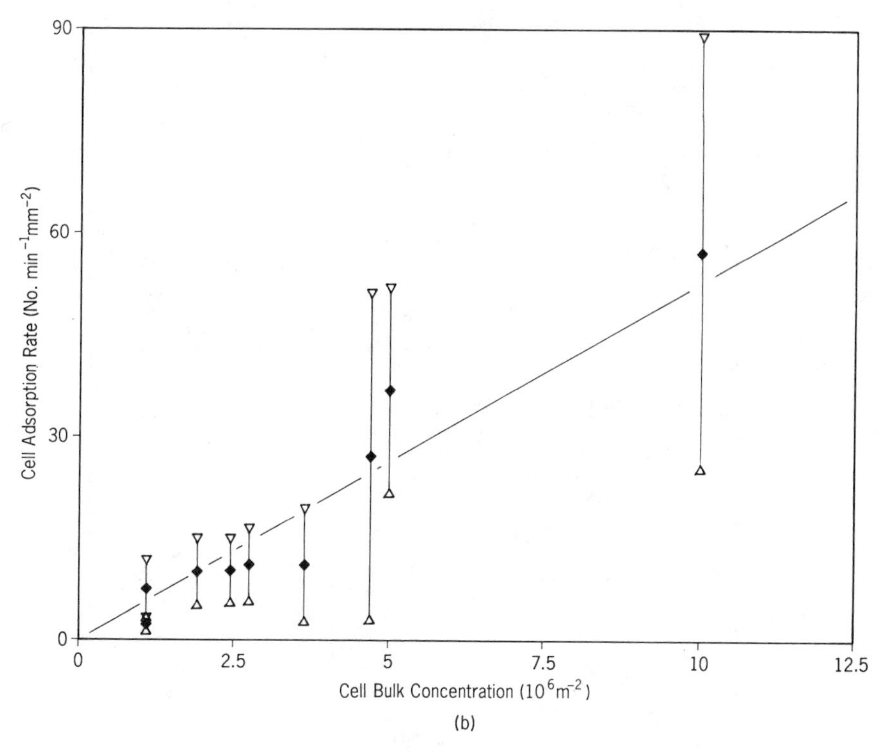

Figure 12.3 Continued

some of the reversibly adsorbed cells will irreversibly adsorb (Eq. 12.1). The population balances for the two forms are as follows:

Reversible Adsorption (CFU):

$$\underset{\substack{\text{rate of}\\\text{accumulation}\\\text{of reversibly}\\\text{adsorbed cells}}}{\frac{dX''_{fC,\mathrm{rev}}}{dt}} = \underset{\substack{\text{rate of CFU}\\\text{adsorption}}}{r''_{a,\mathrm{CFU}}} - \underset{\substack{\text{rate of CFU}\\\text{desorption}}}{r''_{d,\mathrm{CFU}}} - \underset{\substack{\text{rate of}\\\text{transformation}}}{r''_{t,\mathrm{CFU}}} \qquad [12.10]$$

where $X''_{fC,\mathrm{rev}}$ = concentration of reversibly adsorbed cells at substratum (CFU L^{-2})

$r''_{d,\mathrm{CFU}}$ = desorption rate of CFU from substratum (CFU $L^{-2}t^{-1}$)

$r''_{t,\mathrm{CFU}}$ = transformation rate of reversibly adsorbed CFU to irreversibly adsorbed CFU at the substratum (CFU $L^{-2}t^{-1}$)

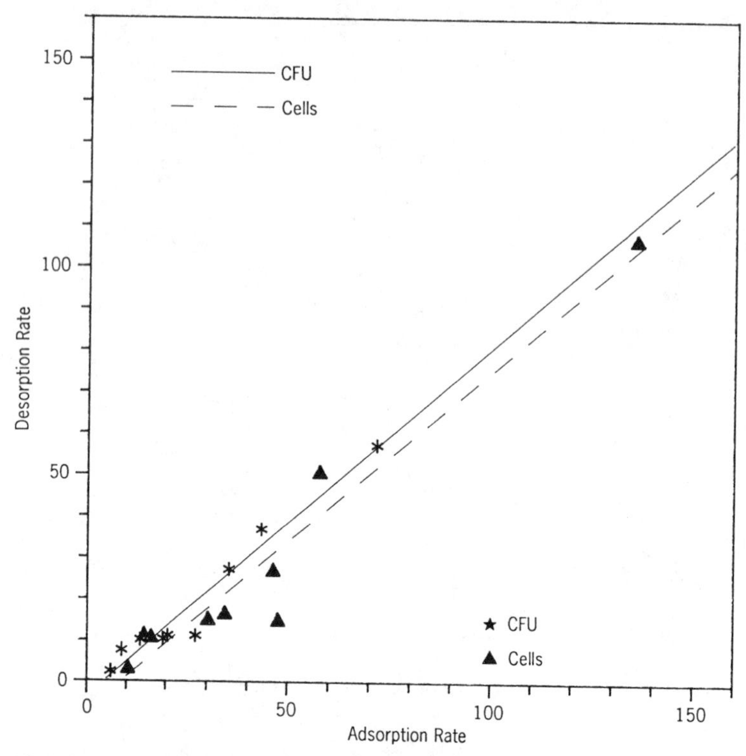

Figure 12.4 Rate of cellular desorption for *Pseudomonas aeruginosa* is linearly related to cellular adsorption rate at a shear stress of 0.5 N/m^2. The slope of the regression line represents the probability of desorption for an adsorbed CFU or cell. Rates are in units of number of cells per min per mm^2. (Escher, 1986).

Irreversible Adsorption (CFU):

$$\underset{\substack{\text{rate of}\\\text{accumulation}\\\text{of irreversibly}\\\text{adsorbed cells}}}{\frac{dX''_{fC,\text{irrev}}}{dt}} = \underset{\substack{\text{rate of}\\\text{transformation}}}{r''_{t,\text{CFU}}} + \underset{\substack{\text{rate of}\\\text{CFU separation}}}{k''_{s,\text{CFU}} X''_{fC,\text{irrev}}} \qquad [12.11]$$

where $k''_{s,\text{CFU}}$ = probability of CFU separation (t^{-1})

CFU separation rates were found by Escher (1986) to be zero order with respect to CFU at the substratum (Figure 12.6). The CFU separation rate did not significantly change with increased shear stress.

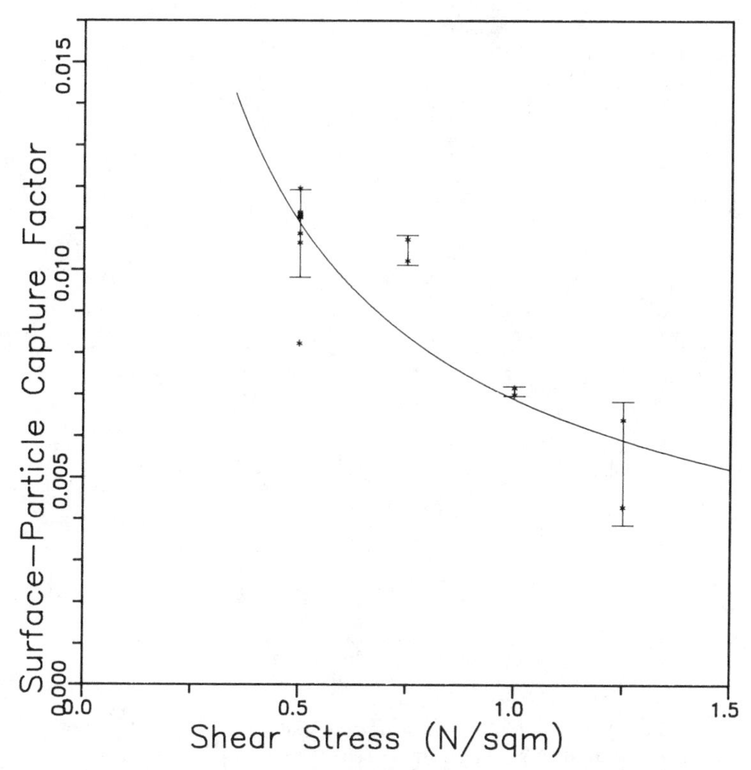

Figure 12.5 Surface-particle capture factor ϵ_{sp} for *Pseudomonas aeruginosa* CFU decreases with increasing shear stress (Escher, 1986).

Total CFU:

$$\frac{dX''_{fC,\text{tot}}}{dt} = \frac{dX''_{fC,\text{rev}}}{dt} + \frac{dX''_{fC,\text{irrev}}}{dt} \qquad [12.12]$$

Equation 12.10 can be integrated with the following boundary conditions:

$$\begin{aligned} &\text{at} \quad t = 0, \qquad X''_{fC,\text{rev}} = 0 \\ &\text{at} \quad t \le t_r, \qquad r''_{d,\text{CFU}} = 0, \quad r''_{t,\text{CFU}} = 0 \end{aligned}$$

where $t = 0$ is the time of the first adsorption event. This boundary condition leads to a discontinuous function with respect to t:

$$X''_{fC,\text{rev}} = \begin{cases} r''_{a,\text{CFU}}t & t \le t_r \\ r''_{a,\text{CFU}}t - (r''_{d,\text{CFU}} + r''_{t,\text{CFU}})(t - t_r), & t \ge t_r \end{cases} \qquad [12.13]$$

where t_r = offset time of reversible adsorption (t^{-1})

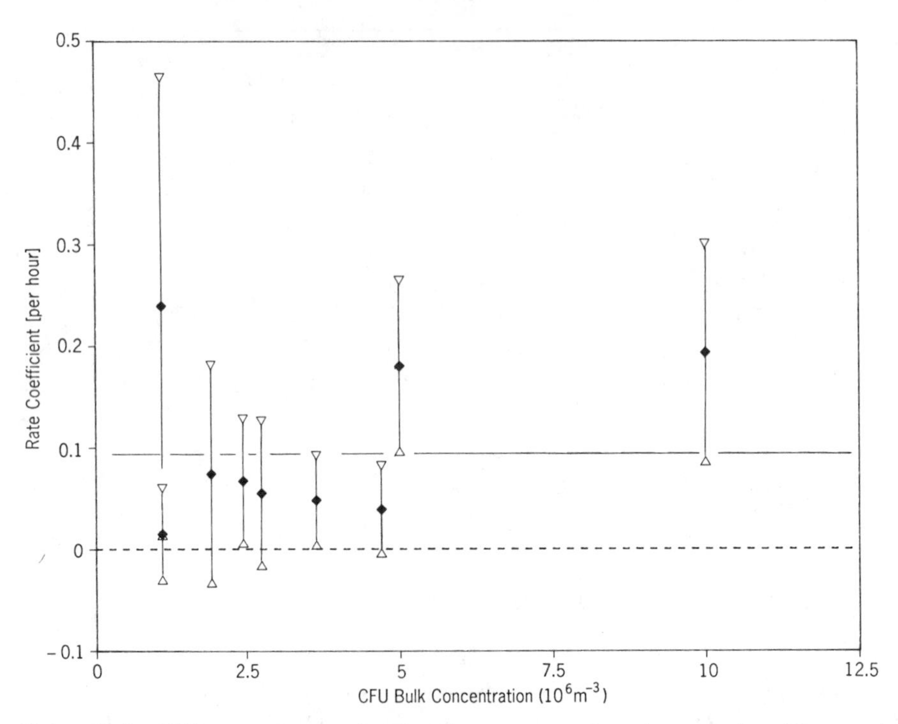

Figure 12.6 CFU separation rate for *Pseudomonas aeruginosa* is zero order with respect to CFU concentration at the substratum (Escher, 1986).

The rates $r''_{d,\text{CFU}}$ and $r''_{t,\text{CFU}}$ (for $t \geq t_r$) can be expressed by the probability term β_C (see Section 12.2.3.2) for desorption:

$$X''_{fC,\text{rev}} = r''_{a,\text{CFU}}\{t - [\beta_c + (1 - \beta_c)][t - t_r]\} \qquad [12.14]$$

Escher (1986) observed that β_C does not change with run time, so the population balance of reversibly adsorbed CFU in Eq. 12.14 (for $t \geq t_r$) can be reduced to

$$X''_{fC,\text{rev}} = r''_{a,\text{CFU}} t_r \qquad [12.15]$$

Thus in Escher's experiments, Eq. 12.15 states that the number of reversibly adsorbed CFU per area is independent of time.

The population balance for the irreversibly adsorbed CFU (Eq. 12.11) can be integrated with the following condition:

$$\text{at} \quad t = t_r, \qquad X''_{fC,\text{irrev}} = 0$$

This boundary condition indicates that, for any time smaller than t_r, no transformation from $X''_{fC,\text{rev}}$ to $X''_{fC,\text{irrev}}$ will occur; it leads to a discontinuous function for $X''_{fC,\text{irrev}}$ with the discontinuity at $t = t_r$:

$$X''_{fC,\text{irrev}} = \begin{cases} 0, & t \leq t_r \\ \dfrac{r''_{t,\text{CFU}}}{k_{s,\text{CFU}}} \left(e^{k''_{s,\text{CFU}}(t-t_r)} - 1\right), & t \geq t_r \end{cases} \quad [12.16]$$

In Eq. 12.16, the term for $r''_{t,\text{CFU}}$ can be replaced with the probability term in the following way:

$$r''_{t,\text{CFU}} = r''_{a,\text{CFU}}(1 - \beta_c) \quad [12.17]$$

Thus, the accumulation of irreversibly adsorbed cells is directly proportional to the adsorption rather than the transformation rate.

By combining Eqs. 12.2, 12.15, 12.16, and 12.17, an expression for the total CFU adsorbed on the substratum can be obtained:

$$X''_{fC,\text{tot}} = \begin{cases} r''_{a,\text{CFU}} t & t \leq t_r \\ r''_{a,\text{CFU}} \left[t_r + \left(\dfrac{1 - \beta_C}{k_{s,\text{CFU}}} \left(e^{k''_{s,\text{CFU}}(t-t_r)} - 1\right)\right)\right], & t \geq t_r \end{cases} \quad [12.18]$$

Equation 12.18 describes the progression of CFU colonization at the substratum (Figure 12.7). The cumulative number of CFU collecting at the substratum due to adsorption increases linearly with time. The cumulative number of CFU leaving the substratum due to desorption also increases linearly after the offset time is reached. The net accumulation of cells at the substratum follows a saturation trajectory similar to that described in Eq. 9.76 (Chapter 9) for molecular adsorption (Figure 12.8). The net accumulation of cells is obviously influenced by the flow regime. The substratum area covered by cells is approximately 1% after 300 min (Figure 12.7b).

12.2.4 Sticking Efficiency

The upper limit of transport (without adsorption) can be calculated by assuming the substratum–particle capture rate approaches infinity (i.e., $\epsilon_{sp} = \infty$, Eq. 12.7). This upper limit assumes that the substratum is an infinite sink for CFU and reduces Eq. 12.7 to the following:

$$B''_{\text{CFU}} = \frac{X_{bC} D_X}{h} \left[\frac{(2/9K_1)^{1/3}}{\Gamma(\frac{4}{3})} \right] \quad [12.19]$$

where B''_{CFU} = flux of CFU to substratum [CFU $L^{-2}t^{-1}$].

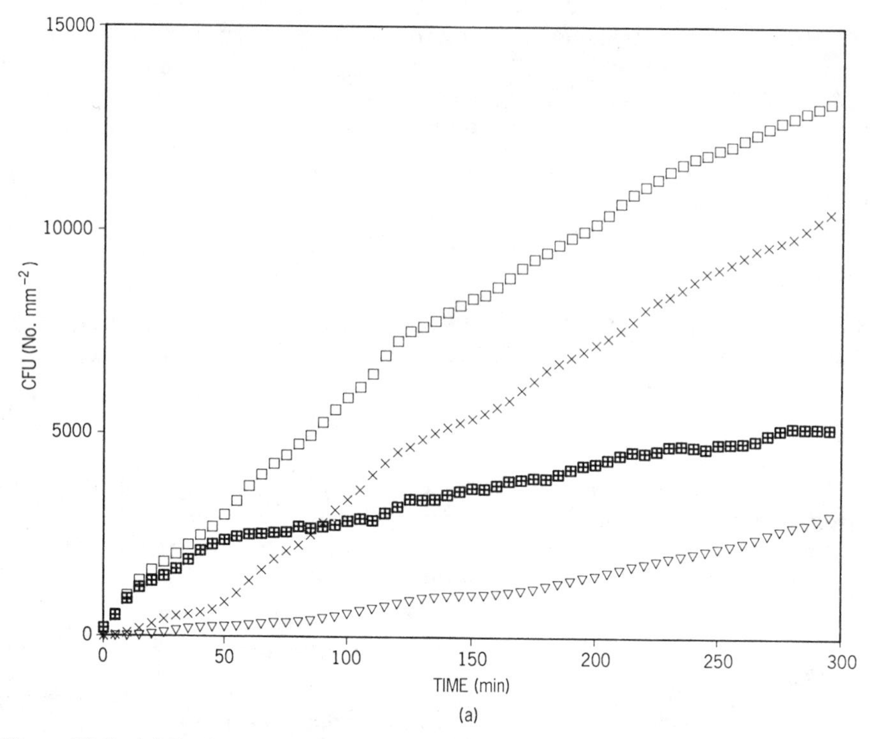

Figure 12.7 (a) Typical colonization progression for *Pseudomonas aeruginosa* in terms of CFU. Adsorption (□), desorption (×), CFU separation (▽) and accumulation (⊞) are displayed. (b) Area coverage of adsorbed cells on glass substratum. Shear stress was 0.5 N m^{-2}, and bulk water CFU concentration was 5 × 10^{12} cells m^{-3}. (Escher, 1986.)

Sticking efficiency has been used to describe the ratio of cell adsorption rates to the cell flux to the substratum. Thus, it describes the probability that a cell striking the substratum will adsorb. The sticking efficiency is not the same as the substratum-particle capture factor ϵ_{sp}, but rather a function of it. In terms of CFU, the sticking efficiency can be described by the ratio of Eq. 12.7 to 12.19, and for cells by Eqs. 12.21 and 12.22:

$$\alpha_e = \frac{\Gamma(\frac{4}{3})}{\Gamma(\frac{4}{3}) + (1/\epsilon_{sp})(2/9K_1)^{1/3}} \qquad [12.20]$$

where α_e = sticking efficiency [—].

Thus, the sticking efficiency α_e has a range of zero to one. It is the probability that a CFU transported to the substratum will adsorb, i.e., the dimensionless ratio of adsorption rate to flux to the substratum.

The sticking efficiency has been calculated by Characklis (1985) for experimental systems reported in the literature. Three reports were chosen for comparison. Fletcher (1976) used *P. fluorescens* as the test organism and polystyrene as

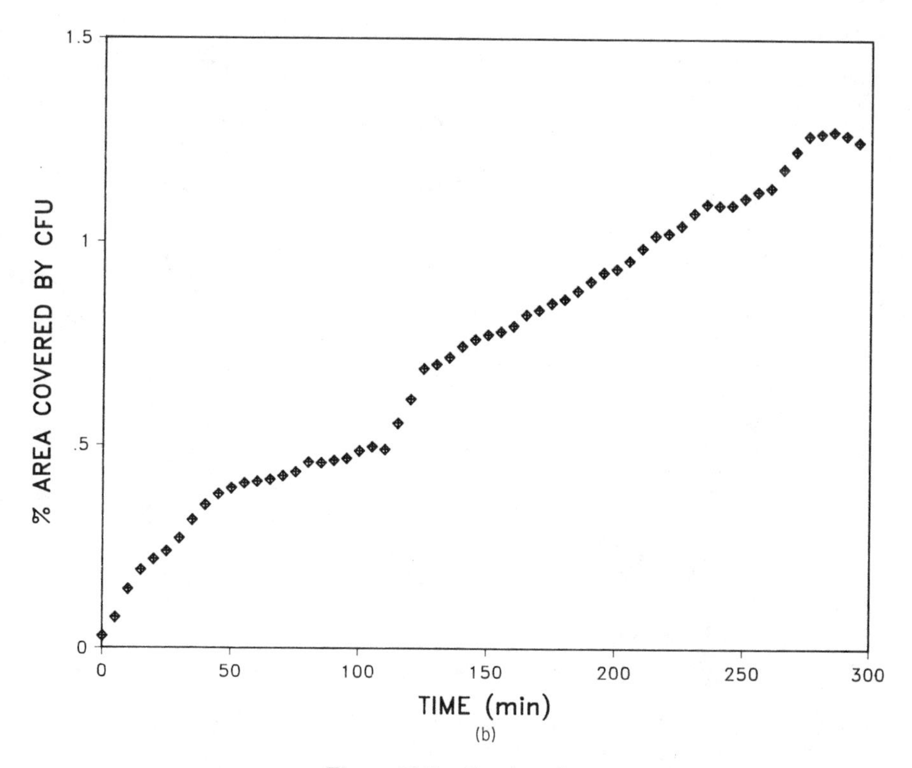

Figure 12.7 Continued

the substratum in a stagnant flow system where sedimentation was assumed to be the dominant transport mechanism. Powell and Slater (1983) used *Bacillus cereus* as their test organism and a glass rectangular capillary as the substratum in a laminar flow system comparable to the one used by Escher (1986) with *Pseudomonas aeruginosa*. Comparing the data from Fletcher, Powell and Slater, and Escher for different shear stresses (Figure 12.9), it is evident that shear stress has a significant effect on microbial colonization at a smooth substratum. The data of Hermanowicz et al. (1989) for *Pseudomonas* spp. in a turbulent flow are also included in the figure. The trends observed for sticking efficiency are similar to those for the substratum–cell capture factor ϵ_{sp} (Figure 12.5).

The calculation of sticking efficiency by Eq. 12.20 assumes that transport of cells to the substratum occurs only by diffusion. Other mechanisms are possible and, since migration of particles in a tube occurs from the wall to the tube center, an annulus of increased concentration may occur (Bungay and Wiggert, 1965). This *tubular pinch effect* has no clear explanation and demonstrates that further measurements are needed to verify the transport characteristics of cells in flow fields. The image analysis technique can provide a lower bound on the number of cells striking the substratum, but is not sensitive enough to provide quantitative data.

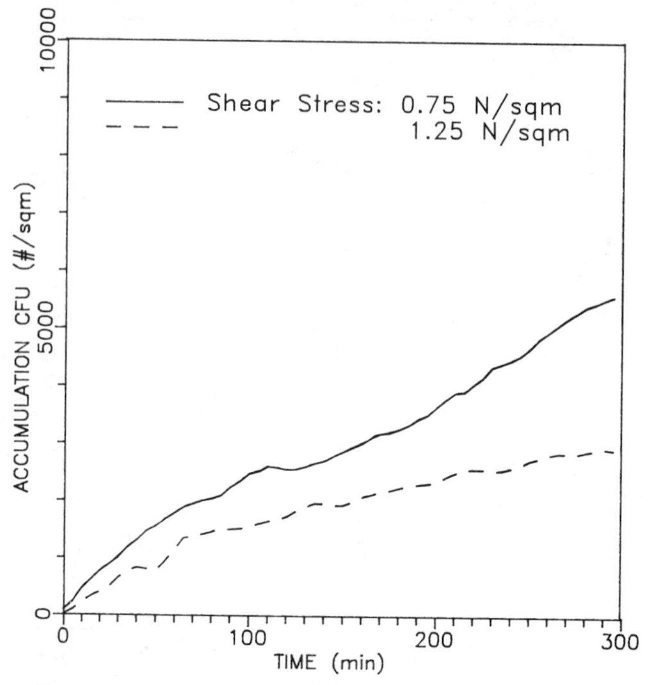

Figure 12.8 The net accumulation of *Pseudomonas aeruginosa* cells at the substratum follows a saturation rate trajectory (Eq. 9.76, Chapter 9). The net accumulation rate decreases with increasing shear stress. The bulk water CFU concentration was 6.7–6.8 × 10^{12} CFU m^{-3} (Escher, 1986).

12.2.5 Kinetic Expressions in Terms of Cells

The number of cells in each CFU can be determined in the image analysis system. Thus, all rates expressed in terms of CFU can be converted into units of cell numbers. Additionally, changes in the number of cells per individual CFU can be observed (see Figure 12.13 in Section 12.2.6.3 below). Hence, the processes of cell multiplication and erosion can also be quantified. CFU separation does not contribute to the change of total cells adsorbed, so it is omitted from kinetic expressions in terms of cells.

12.2.5.1 *Transport from Bulk to Substratum* The rate of transport can be defined in the same way as for CFU (Eq. 12.7):

$$r''_{aX} = \frac{X_b D_X}{h}\left[\frac{\left(\dfrac{2}{9K_1}\right)^{1/3}}{\Gamma(\frac{4}{3}) + \dfrac{1}{\epsilon_{sp}}\left(\dfrac{2}{9K_1}\right)^{1/3}}\right] \qquad [12.21]$$

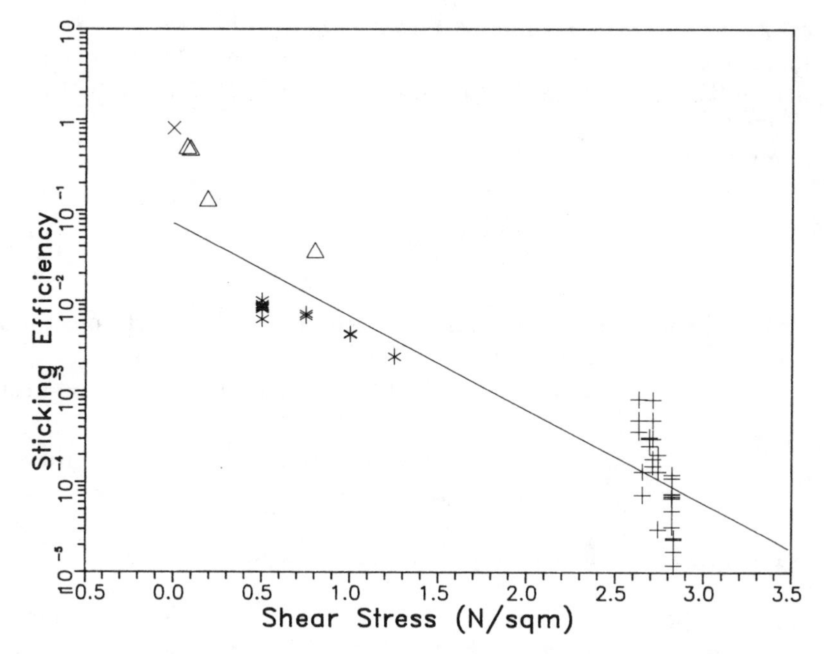

Figure 12.9 Sticking efficiency α_e decreases with increasing Reynolds number. Data of Fletcher (1976) ($\times$) are for a marine pseudomonad in a quiescent Petri dish in which sedimentation is probably the major mechanism for transport. Powell and Slater (1983) ($\triangle$) and Escher (1986) ($*$) conducted experiments in laminar flow in a rectangular capillary with *Bacillus cereus* and *Pseudomonas aeruginosa*, respectively. The primary mechanism of transport is probably diffusion. Hermanowicz et al. (1989) ($+$) used an enriched population dominated by *Pseudomonas* spp. on an inclined plate flow reactor with turbulent flow.

where r''_{aX} = adsorption rate of cells to the substratum (cells $L^{-2}t^{-1}$)
X_b = cell concentration in bulk water (cells L^{-3})
D_X = diffusivity of cells in bulk water (L^2t^{-1})

The main difference between Eq. 12.7 and Eq. 12.21 is in the bulk water concentration variable. Equations 12.4 and 12.5 require no other changes to describe the processes in terms of cells. The adsorption rate of cells increases linearly with increasing cell concentration in the bulk water as predicted by Eq. 12.21 (Figure 12.3b). As observed with CFU, the desorption rate of cells is directly proportional to their adsorption rate (Figure 12.4) with the proportionality coefficient β_X. The rate of transport (not adsorption) can be determined by setting $\epsilon_{sp} = \infty$ in Eq. 12.21:

$$B''_X = \frac{X_b D_X}{h}\left[\frac{(2/9K_1)^{1/3}}{\Gamma(\frac{4}{3})}\right] \qquad [12.22]$$

where B''_X = flux of cells to substratum (cells $L^{-2}t^{-1}$).

12.2.5.2 *Population Balance at the Substratum* Cells at the substratum undergo a transformation from reversibly to irreversibly adsorbed. The population balances for cells are similar to those for CFU (Eqs. 12.10–12.12).

Reversible Adsorption (Cells):

$$\frac{dX''_{f,\text{rev}}}{dt} = r''_{aX} - r''_{dX} - r''_{tX} \qquad [12.23]$$

where $X''_{f,\text{rev}}$ = reversibly adsorbed cell concentration at substratum (cells $L^{-2}t^{-1}$)

r''_{dX} = desorption rate of cells from the substratum (cells $L^{-2}t^{-1}$)

r''_{tX} = transformation from reversibly adsorbed cells to irreversibly adsorbed cells (cells $L^{-2}t^{-1}$)

The initial conditions for Eq. 12.23 are as follows:

$$\text{at} \quad t = 0, \qquad X''_{f,\text{rev}} = 0 \qquad [12.24]$$

$$\text{at} \quad t \leq t_r, \qquad r''_{dX} = 0, \quad r''_{tX} = 0 \qquad [12.25]$$

Solution of Eq. 12.23 leads to a discontinuous function of t:

$$X''_{f,\text{rev}} = \begin{cases} r''_{aX}t, & t \leq t_r \\ r''_{aX}t - (r''_{dX} + r''_{tX})(t - t_r), & t \geq t_r \end{cases} \qquad [12.26]$$

If the probability β_X of desorption is independent of time, then Eq. 12.26 can be simplified (analogously to Eqs. 12.14 and 12.15) for the case of $t \geq t_r$:

$$X''_{f,\text{rev}} = r''_{aX}t_r \qquad [12.27]$$

Thus, $X_{f,\text{rev}}$ is independent of time.

Irreversible Adsorption (Cells):

$$\frac{dX''_{f,\text{irrev}}}{dt} = r''_{tX} + (r''_{mX} - r''_{eX})X''_{f,\text{irrev}} \qquad [12.28]$$

where r''_{mX} = specific multiplication rate of cells on the substratum (t^{-1})

r''_{eX} = specific erosion rate of cells from the substratum (t^{-1})

For the irreversible adsorbed cells, Eq. 12.16 can be adapted after combining r''_{mX} and r''_{eX} in Eq. 12.28:

$$r''_{aX} = r''_{mX} - r''_{eX} \tag{12.29}$$

Using the same initial conditions as for Eq. 12.23 (Eqs. 12.24 and 12.25), Eq. 12.28 has the following solution:

$$X''_{f,\mathrm{irrev}} = \begin{cases} 0, & t \le t_r \\ (r''_{tX}/r''_{aX})(e^{r''_{aX}(t-t_r)} - 1), & t \ge t_r \end{cases} \tag{12.30}$$

Total Cells:

$$\frac{dX''_{f,\mathrm{tot}}}{dt} = \frac{dX''_{f,\mathrm{rev}}}{dt} + \frac{dX''_{f,\mathrm{irrev}}}{dt} \tag{12.31}$$

Using an analogous step to Eq. 12.17 and combining Eqs. 12.27, 12.29, and 12.30, the equation for total cell concentration at the substratum can be written analogously to the total CFU concentration at the substratum (Eq. 12.18):

$$X''_{f,\mathrm{tot}} = \begin{cases} r''_{aX}t, & t \le t_r \\ r''_{aX}\left[t_r + \left(\dfrac{1-\beta_X}{r''_{aX}}(e^{r''_{aX}(t-t_r)} - 1)\right)\right], & t \ge t_r \end{cases} \tag{12.32}$$

Thus, Eqs. 12.18 and 12.32 estimate the progression of both CFU and cell accumulation at the substratum. The cumulative numbers of cells accumulating due to adsorption and leaving the substratum due to desorption and erosion increase linearly with time (Figure 12.10a and b).

However, cells at the substratum increase exponentially due to multiplication (Figure 12.10b). Multiplication rates, as measured by the image analysis method, only includes those daughter cells that remain on the substratum. Thus, the sum of multiplication rate and erosion rate is a minimum estimate of the total replication rate at the substratum. Then for *P. aeruginosa* (Escher, 1986), a minimum estimate of the total multiplication rate at the substratum was approximately 0.8–1.0 h^{-1}. The specific growth rate (i.e., dilution rate) of *P. aeruginosa* in the chemostat was 0.18 h^{-1} and was the source of cells for the rectangular capillary where colonization was being observed. How are the multiplication rate and the specific growth rate related? The specific growth rate describes the frequency of population doubling of an entire cell population, whereas the multiplication rate is the frequency of observed individual cell divi-

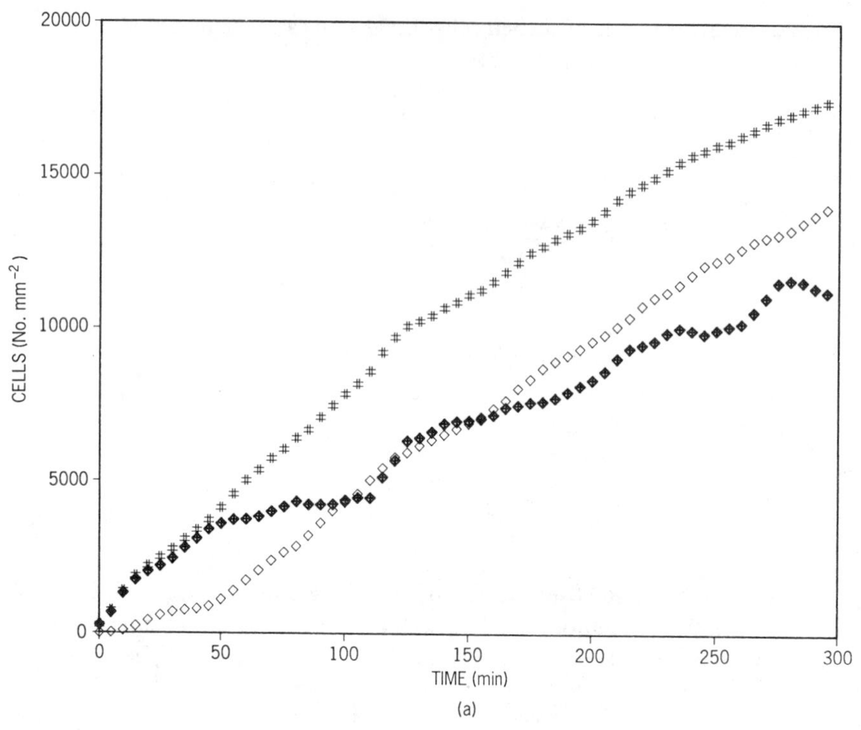

Figure 12.10 Typical results from a *Pseudomonas aeruginosa* colonization progression in terms of cells on a smooth glass substratum at a shear stress of 0.5 N m^{-2} as observed by image analysis. (a) Sorption-related processes include adsorption (#), desorption (◇), and accumulation (◈). (b) Growth-related processes include multiplication (◇), erosion (△), and accumulation (◈). (Escher, 1986.)

sions, ignoring the period a cell requires to pass through an entire division cycle. Thus, the multiplication rate, in this instance, is a characteristic of an individual cell, and the specific growth rate is a population parameter. The two rates are related as follows (Eq. 2.14, Chapter 2):

$$\mu = (\ln 2)/g \qquad [12.33]$$

where μ = specific growth rate (t^{-1})
g = cellular generation time (t^{-1})

The mean maximum specific growth rate, μ_{max}, for *P. aeruginosa* in a chemostat in a glucose–mineral salts medium is 0.37 h^{-1} ± 0.03 (s.d.) (Table 8.2, Chapter 8). If the cells at the substratum are multiplying at their maximum rate based on the specific growth rate measured in a chemostat, then their multiplication rate is approximately 1.9 h^{-1}. Then, since total multiplication rates (in-

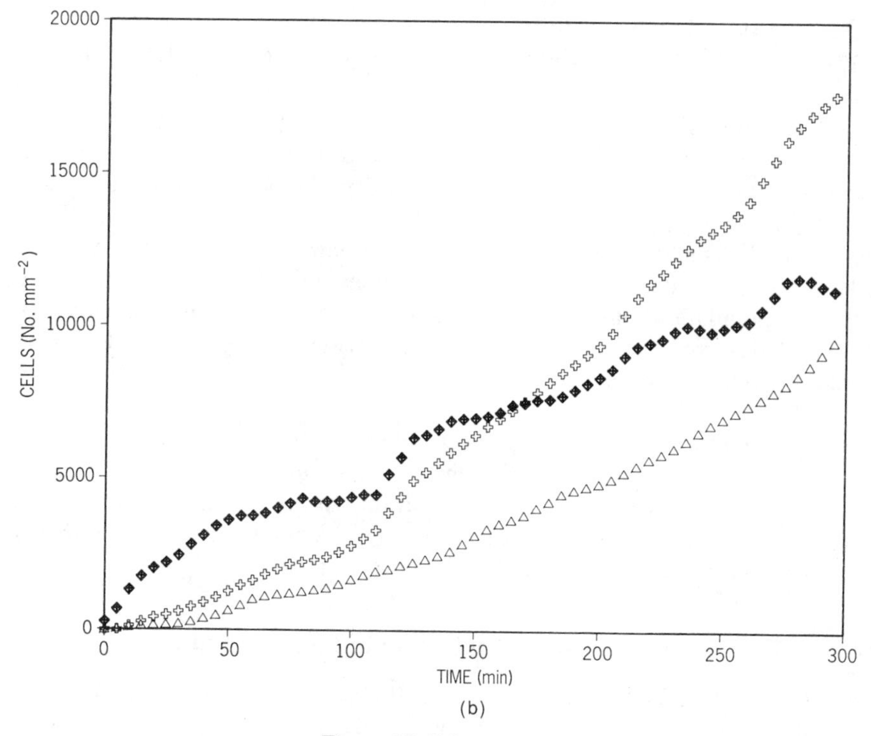

Figure 12.10 Continued

cluding erosion rates) of approximately 0.8–1.0 h^{-1} were observed, the specific growth rate for the adsorbed *P. aeruginosa* population is estimated to be about one-half the maximum specific growth rate, or 0.18 h^{-1}. The chemostat in the experiments was operated at 0.18 h^{-1}.

There is reason to consider the possibility that cells adsorbed at the substratum grow at a higher rate than those in the chemostat. For example, Marshall and coworkers have observed relatively high growth rates for adsorbed cells (Kjelleberg et al., 1982). Any observed increased growth rate at the substratum in a continuous flow system could be attributed to several possible phenomena:

1. CFU adsorbing to the substratum are in the process of cell division when they adsorb. Thus, the cell division cycle began prior to colonization.

2. Due to immobilization of the cells and continuous flux of nutrients to the substratum, the organisms "see" more nutrients than they did in the chemostat (no diffusive boundary layer). The substrate concentration in the chemostat was somewhat higher than in the rectangular capillary. However, the substrate flux experienced by the cell may have been higher than that experienced by the cells in the chemostat.

3. The substrate adsorbs to the substratum and is consumed by the organisms adsorbed to the substratum faster than it is consumed by the planktonic cells.

4. Adsorption is a selective process (sticking efficiency $\alpha_e \leq 0.008$). It is conceivable, even in a monopopulation, that the distribution of cell properties through the population leads to differential adsorption. For example, in the case of motile *P. aeruginosa*, perhaps only highly active cells are adsorbing, since more active cells may be more motile, leading to an advantage over less active cells in terms of adsorption. Alternatively, there may be genetic differences within the population that lead to enhanced adsorption. For example, Valeur et al. (1988) observed different surface characteristics in adsorbed cells versus planktonic cells in their experiments.

Clearly, more research is necessary in controlled, continuous flow reactors using techniques like the image analysis system to observe *in situ* the various processes contributing to colonization of the substratum. In addition, rigorous process analysis is necessary to evaluate experimental results in relation to the numerous processes that may be contributing to the observations.

12.2.6 Cell Behavior at the Substratum in Laminar Flow

Cell behavior reflects activities of CFU at the substratum such as movement, orientation of adsorbed CFU, cell number per CFU, and net growth at the substratum. Selected observations with the image analyzer using *P. aeruginosa* at a hydrodynamic shear stress of 0.5 N m^{-2} will be presented in this section as illustrations of the capability of the image analysis system (Escher, 1986). The observations of cell behavior at the substratum also may lead to better insight into the colonization process.

12.2.6.1 *Residence Time* CFU are transported from the bulk liquid to the substratum, where they adsorb reversibly. With time, some of the reversibly adsorbed cells will absorb irreversibly (Eq. 12.1). Thus, a finite residence time appears to be a prerequisite for irreversible adsorption or adhesion. The image analysis system permits the determination of the cell residence time at the substratum. The determination of two residence time distributions is possible. The *finite residence time distribution* includes only those CFU for which adsorption *and* desorption was observed during a 5 hour experiment (see Figure 7.4, Chapter 7). Thus, the distribution represents only the residence times of the reversibly adsorbed CFU. The *overall residence time distribution* includes the residence times of all cells that adsorbed, including those that never desorbed throughout the experiment. Only the finite residence time distribution will be considered below.

The longer the cell remains at the substratum, the greater becomes the probability that it will adsorb irreversibly. For example, the probability of desorbing between 0 and 5 min at 0.5 N m^{-2} is 48%; between 5 and 10 min it reduces to 12%, and so on. The mean cell residence time at the substratum also increases with decreasing hydrodynamic shear stress at the substratum (Figure 12.11), which clearly indicates the influence of shear on desorption processes.

12.2.6.2 Orientation of CFU During Adsorption The orientation of the CFU during adsorption in a flow field, in some cases, may reveal the mode of adsorption. For example, a cell adsorbing by means of a polar flagellum may be restricted in its adsorbed orientation. The distribution of *Pseudomonas aeruginosa* orientation at its first observation after adsorption is bimodal (Figure 12.12). One peak is at 0° and one at 90° (the flow direction is 90°). In some cases in which CFU orientation was uncertain, the orientation was reported as 0°. Essentially, 30 to 35% of the CFU had a true 0° orientation. Therefore, *Pseudomonas aeruginosa* adsorb either perpendicular or parallel to the flow direction with a very small probability for any values in between.

12.2.6.3 Cell Number per CFU The cell numbers in a CFU will vary during an experiment, primarily due to multiplication and erosion processes. The distribution of the number of *Pseudomonas aeruginosa* cells in a CFU as a func-

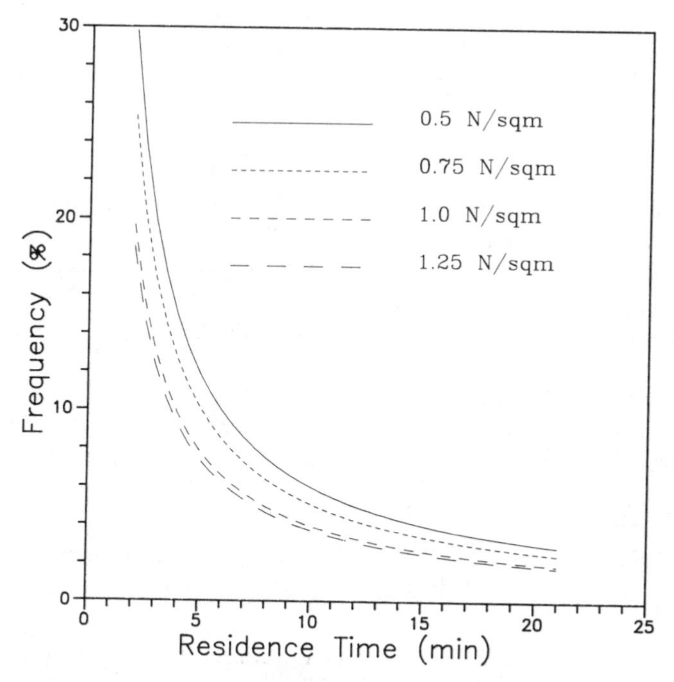

Figure 12.11 Residence time at the substratum for reversibly adsorbed *Pseudomonas aeruginosa* CFU as a function of shear stress (Escher, 1986).

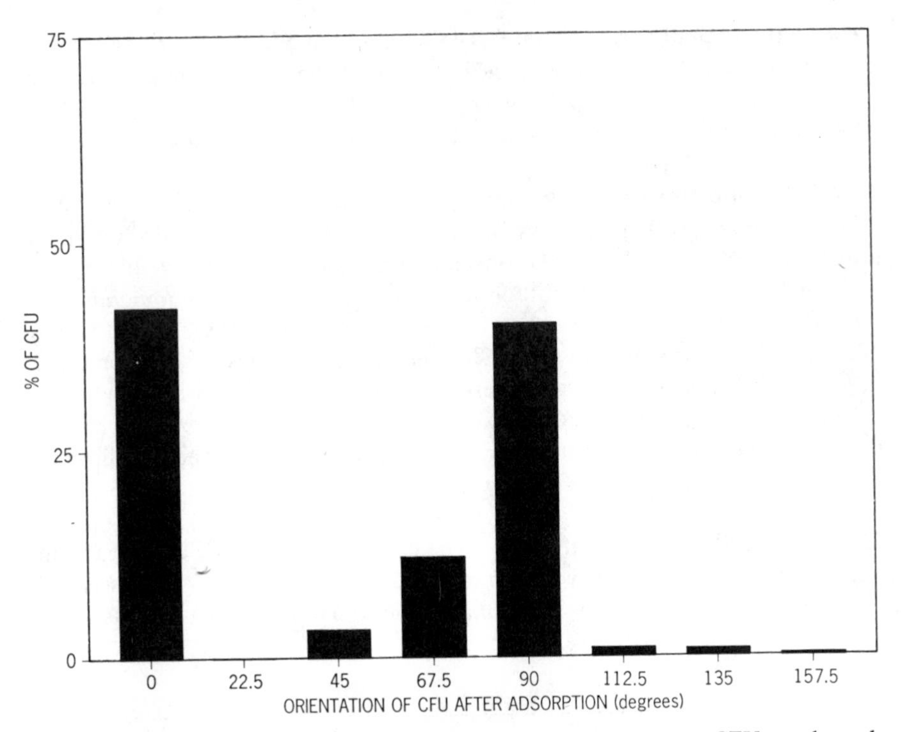

Figure 12.12 Orientation of adsorbed *Pseudomonas aeruginosa* CFU at the substratum. Direction of flow was 90°. (Escher, 1986.)

tion of time in an experiment changes, as does the number of CFU (Figure 12.13). These values have been calculated by dividing the measured area of each CFU by the average cell size in it. A comparison of the first and last observations indicates net growth of the CFU during their residence time at the substratum, most probably a result of multiplication within the CFU.

12.2.6.4 Spatial Distribution of Adsorbed Cells The spatial distribution of the adsorbed bacteria is easy to observe with the image analyzer (Escher, 1986). Three different interactions between the established and newly adsorbed CFU are possible:

1. *No Influence*. The CFU is transported to the substratum and may adsorb at any location regardless of the other adsorbed CFU. The resulting distribution of adsorbed cells will be random.
2. *Positive Influence*. The adsorbed CFU collect the cells being transported to the substratum. The resulting distribution of adsorbed cells will be aggregated.

3. *Negative Influence*. The adsorbing CFU are inhibited by the CFU already at the substratum. The resulting distribution of adsorbed cells will be uniform.

Spatial distribution analyses are generally conducted using a nearest neighbor analysis. The analysis results in an index that classifies the distribution as random, aggregated, or uniform. Escher (1986) has developed a technique for calculating a dimensionless nearest neighbor number that is effective in determining the spatial distribution characteristics. The results of his analysis with respect to *Pseudomonas aeruginosa* on glass indicate that adsorbed cell distribution is random during the early stages of colonization when cell concentration in the bulk water is low ($1 \times 10^{12}\ m^{-3}$). With increasing cell concentration in the bulk water and increasing cell concentration at the substratum, the adsorbed cell distribution becomes more uniform. Thus, adsorbed *Pseudomonas aeru-*

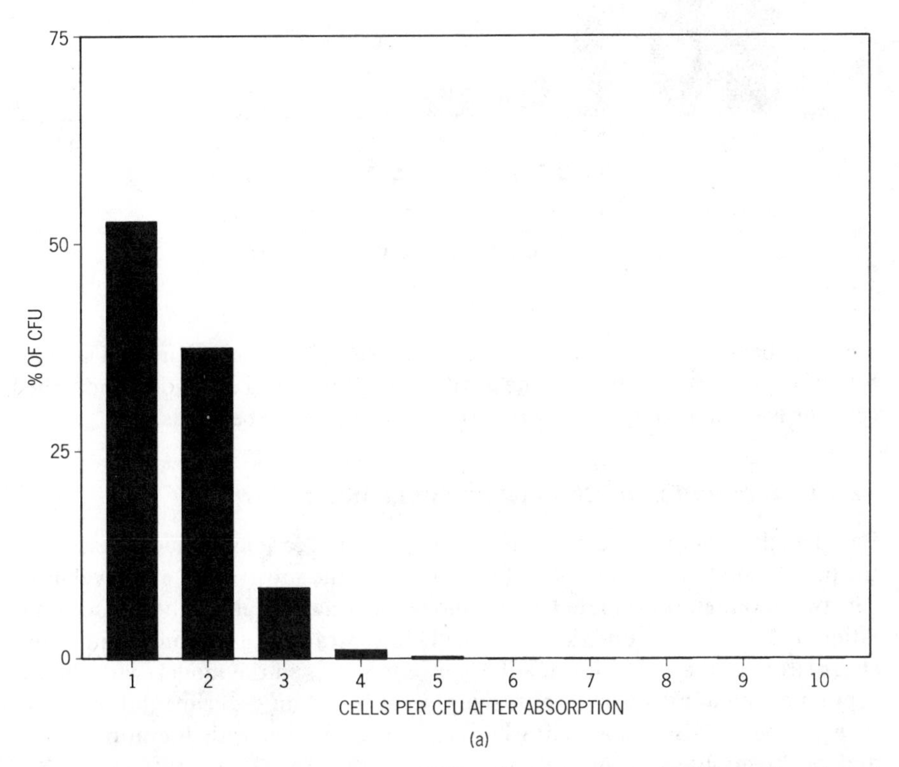

Figure 12.13 *Pseudomonas aeruginosa* CFU and cell number per CFU as a function of time during an experiment. The cell number per CFU increases with increasing residence time of the CFU at the substratum at a shear stress of $0.5\ N\ m^{-2}$. (a) Distribution immediately after the CFU adsorbs; (b) distribution after 5 h exposure. (Escher, 1986.)

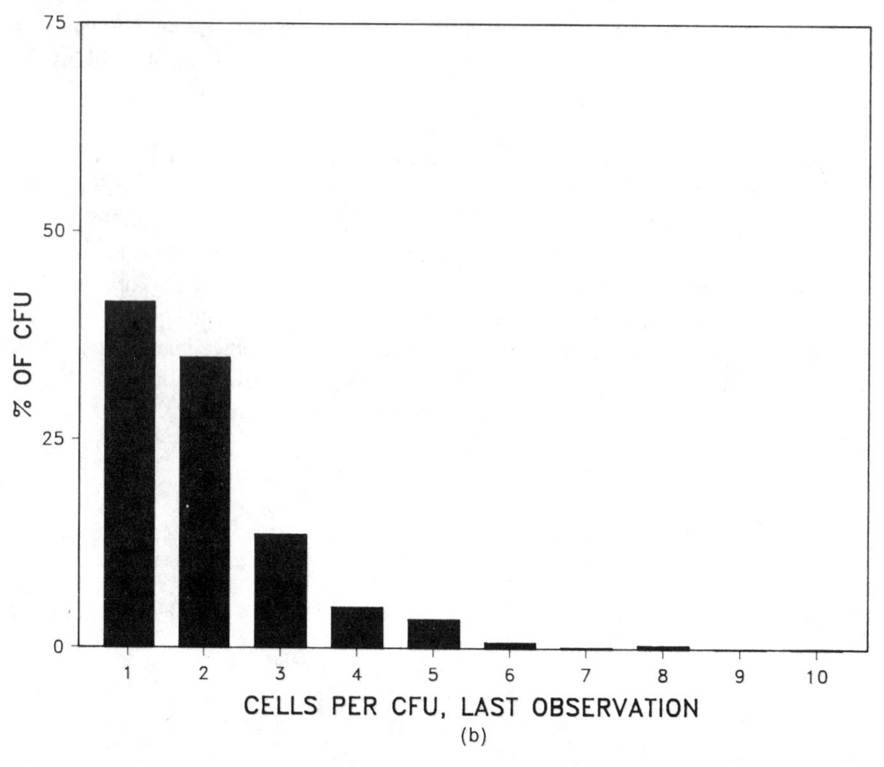

Figure 12.13 Continued

ginosa appear to inhibit adsorption of other like cells in their vicinity. The inhibition may be due to effects of the conditioning film, EPS around the adsorbed cells, or hydrodynamics at the microscale near the adsorbed cells.

12.2.7 Simulation of Colonization in Laminar Flow

Based on the kinetic expressions derived in previous sections, a model describing the population balance for both CFU and cells at the substratum was developed. The two parameters evaluated by this model that may influence the net accumulation of cells at the substratum are cell concentration in the bulk liquid and shear stress at the substratum. All other parameters in the model, with the exception of the non-Brownian diffusivity, were based on experimental measurements. The non-Brownian diffusivity can be estimated with literature values and/or direct observations.

Accumulation of cells at the substratum was simulated over a period of 10 hours. The simulation clearly indicates the increase in net accumulation rate of cells at the substratum with increasing bulk water cell concentration (Figure 12.14) at constant shear stress (0.5 N m^{-2}). The predicted progression compares

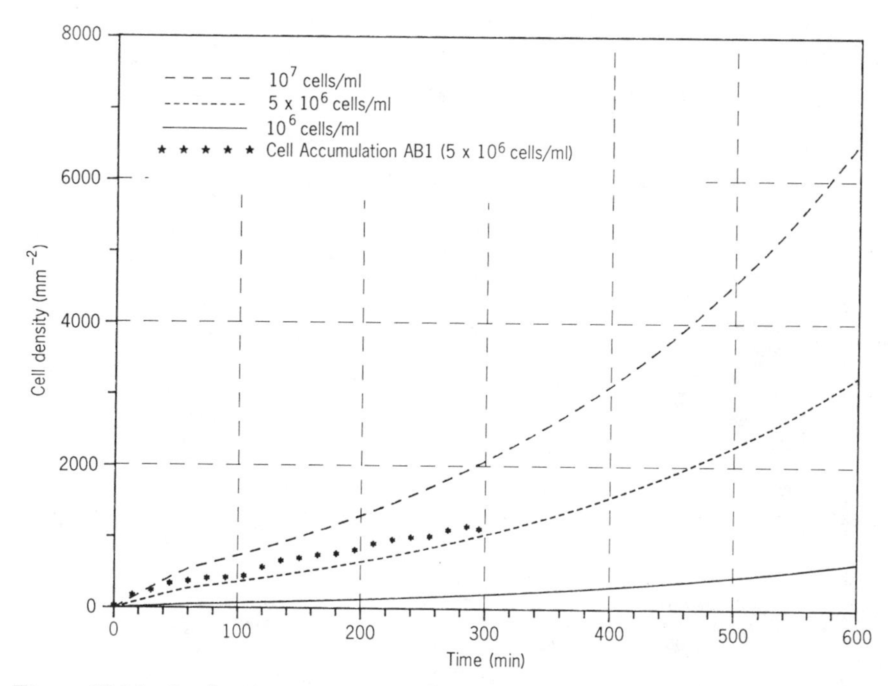

Figure 12.14 Predicted accumulation of *Pseudomonas aeruginosa* cells at constant shear stress (0.5 N/m^2) for a range of bulk water CFU concentrations. The simulated results (curves) are compared with cell accumulation data (asterisks) obtained by Escher (1986).

well with experimental results (marked as data points). The simulation with a CFU concentration of 5×10^{12} CFU m^{-3} has a cell concentration at the substratum in the same range as the experiment. Sorption-related processes (transport, adsorption, desorption) dominate the accumulation processes for approximately 100 min, after which growth-related processes (multiplication and erosion) begin to dominate. The rate of growth, however, is a function of the CFU concentration at the substratum which initially is due to adsorption. Variation of shear stress at a constant CFU concentration in the bulk flow also influences the net accumulation rate of cells at the substratum. At low shear, the accumulation rate is greater (Figure 12.15).

The experimental results indicate that sorption-related processes are dependent on shear stress and bulk CFU concentration, whereas growth-related processes are a direct function of the cell (or CFU) concentration at the substratum. Sorption-related rate processes are zero order, while growth-related rate processes are first order, with respect to the cell (or CFU) concentration at the substratum. The image analysis methods permit direct measurement of essential independent processes contributing to colonization of substrata. Sorption-related processes depend on the shear stress and bulk CFU concentration, whereas

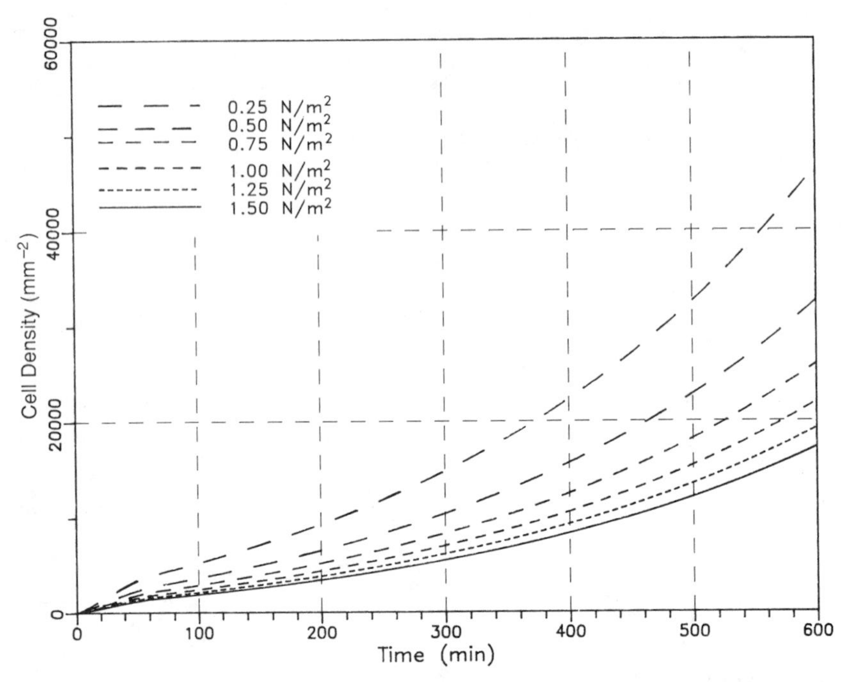

Figure 12.15 Predicted accumulation of *Pseudomonas aeruginosa* cells at constant bulk water CFU concentration (5×10^{12} cells m^{-3}) over a range of shear stresses (Escher, 1986).

growth-related processes depend on the cell concentration at the substratum alone.

12.3 SUBSTRATUM COLONIZATION IN TURBULENT FLOW

Previous chapters have described biofilm accumulation as the net result of several physical, chemical, and biological processes, including (1) transport and adsorption of organic molecules to the substratum, (2) transport and adsorption of microbial cells to the substratum, (3) microbial growth and product formation within the biofilm, and (4) detachment of portions of the biofilm. This chapter has focused on the initial events in biofilm accumulation, namely transport and adsorption of cells to the substratum. However, up to now the emphasis has been solely on laminar flow regimes. Turbulent flow is the rule rather than the exception in nature as well as in technological systems. Therefore, a discussion of microbial colonization in turbulent flow is of great relevance. What influence, if any, will turbulence have on the biofilm accumulation?

Turbulence primarily influences the rate of transport and rate of some interfacial transfer processes. Since colonization consists of a number of processes, some occurring in series, the slowest step in the process network exerts the greatest influence and controls the rate of colonization. This step is called the "rate-determining step" or "rate-controlling step." When turbulence increases the rate of a process that is rate-controlling in laminar flow systems, the mechanism dominating colonization changes. The experiments of Stanley (1983) present a good illustration of the effect of turbulence on microbial colonization. Stanley observed that motile (by means of flagella) *Pseudomonas aeruginosa* cells adsorbed to stainless steel more rapidly under quiescent conditions than they did when flagella were removed by blending. When the experiments were repeated under turbulent conditions, the adsorption rate of the motile cells decreased significantly while that of the nonmotile cells was unaffected. Stanley concluded that the loss of flagella influenced transport but not the adsorption rate. Thus, under quiescent conditions, transport to the substratum controlled the colonization rate. Under turbulent conditions, interfacial processes controlled the colonization rate. Thus, the difference between turbulent and laminar flow regimes will be found primarily in the rate of transport and interfacial transfer processes. Beal (1970) and Lister (1981) present good analyses for transport and adsorption of particles to pipe walls in turbulent flow.

12.3.1 Transport

When a clean surface is immersed in a microbial suspension, transport controls the rate of initial deposition because no colonization can occur without microbes at the substratum. In very dilute suspensions of microbial cells and nutrients, transport of microbial cells to the surface may be the rate-controlling step for long periods of times, since the driving force for mass transport is a concentration gradient. Biofilm development in open ocean waters or distilled water storage tanks may be illustrative of these cases. Thus, concentration is a critical factor influencing mass transport.

In laminar flow, diffusion is a dominant mechanism for transport of cells. In turbulent flow, many mechanisms can influence and/or control the rate of transport of cells to the substratum (Section 7.2.2, Chapter 7). Consider a model microbial cell suspension consisting of spherical cells 1 μm in diameter suspended in a turbulent flow (Table 12.1). The cells have a specific gravity of 1.1. A number of mechanisms can be considered for transporting the cells from the turbulent bulk liquid to the substratum and have been described in Section 7.2.2 (Chapter 7): thermophoresis, diffusion, gravity, motility, and fluid mechanical forces such as inertia, lift, drag, drainage, and downsweeps. Interfacial forces keeping the microbes at the wall include electrostatic, London–van der Waals, interfacial tension, and covalent bonding. Desorption, which results in reentrainment of the microbes in the bulk fluid, is caused by various forces including fluid mechanical forces, shear forces, lift (upsweeps), and motility.

Cells in turbulent flow are transported to within short distances (approximately the thickness of the viscous sublayer) of the substratum by eddy diffusion. The cells are then propelled into the viscous (or laminar) sublayer under

TABLE 12.1 Characteristics of a Model Microbial Cell Suspension Consisting of Spherical Cells with Two Different Diameters

Particle diameter, d_p (μm)	1		5
Specific gravity	1.1		1.1
Particle mass, m_p (g)	6×10^{-13}		7×10^{-11}
Particle number concentration, n (m^{-3})		$\leq 10^{10}$	
Particle mass concentration, X (g m^{-3})		≤ 1	
Particle volume, V_p (m^{-3})	5×10^{-19}		6.5×10^{-17}
Particle terminal velocity, v_t (mm s^{-1})	7×10^{-5}		2×10^{-5}
Brownian diffusion coefficient, D_B (m^2 s)	6×10^{-13}		1×10^{-13}
Diffusion velocity, v_D (mm s^{-1})	1.1		0.1

their own momentum. Turbulent eddies supply the initial impetus, while frictional drag slows down the cell as it penetrates the viscous sublayer. For microbial cells, *inertial* forces are very small because of their small volume and density (similar to that of water). Thus, their momentum is small, and the cell will stop before it penetrates the viscous sublayer to any great distance.

If the cell is traveling faster (v_p) than the fluid (v) in the region of the wall, the *lift* force directs the particle toward the wall. This will normally be the case if particle density is greater than the fluid density (specific gravity $\approx$ 1.0) and the particle is moving toward the wall:

$$F_{\text{lift}} \sim (v_p - v)_{\text{axial}} \text{ pN} \qquad [12.34]$$

Frictional drag forces can be significant, especially in the viscous sublayer region. Drag slows down the particle as it approaches the surface. For a microbial cell,

$$F_{\text{drag}} \sim (v_p - v) \qquad [12.35]$$

The structure of the viscous sublayer in turbulent flow is visualized as containing *downsweeps,* or turbulent bursts of fluid from the turbulent core, which penetrate all the way to the substratum (Cleaver and Yates, 1975). Particles in the bulk fluid are transported all the way to the wall by these convective downsweeps. Aside from lift, the turbulent bursts are the only fluid mechanical force directing the particle to the wall. Downsweeps are apparently quite important in particle transport to the wall in turbulent flow. Assuming a 1.27 cm ID pipe at 30°C, the bursts resulting from the downsweeps have the following characteristics (Cleaver and Yates, 1975):

Burst diameter	1.1 mm
Average axial distance between bursts	5.0 mm
Mean time between bursts	0.006 s

The minimum transport rate of cells will be observed when the cell diameter is approximately 0.1 μm. At this diameter, Brownian diffusion starts exerting a significant effect. The calculated cell flux to the pipe wall for a bulk fluid cell concentration of 10^{10} cells m^{-3} is approximately 1000 cells m^{-2} s^{-1}.

If the mass density of the cell, ρ_X, differs substantially from the fluid density ρ_b, the resulting *gravity* force is

$$F_{\text{gravity}} = (\rho_x - \rho_b)V_x g \qquad [12.36]$$

where V_x = volume of particle (L^3).

For our cells, $F_{\text{gravity}} \approx -1 \times 10^{-4}$ pN and will be negligible in turbulent flow.

Thermophoresis is only relevant when cells are being transported in a temperature gradient, as in a heat exchanger. If the substratum is hot and the bulk fluid is cold, the thermophoretic force will repel the particle from the substratum:

$$F_{\text{thermo}} \approx -1 \times 10^{-1} \text{ pN} \quad \text{for} \quad \Delta T = 10°\text{C} \qquad [12.37]$$

Eddy diffusion may be instrumental in dispersing particles in the turbulent core region, thus maintaining a relatively uniform concentration in that region. Gudmundsson and Bott (1977) present an analysis of particle diffusivity in turbulent pipe flow. However, eddy diffusion will not be significant in transporting cells to the wall.

Brownian diffusion is not a significant transport process in turbulent flow for microbial cells. However, certain microbes are capable of motility through their own internal energy and can generate velocities as high as 40 μm s^{-1}. The force generated by this movement in the viscous sublayer for our cells is as follows:

$$F_{\text{motile}} \approx 1.5 \times 10^{-1} \text{ pN} \qquad [12.38]$$

The forces of these various relevant forces influencing microbial cell transport in laminar flow and in quiescent systems are presented in Table 12.2 for comparison. Diffusion dominates the transport mechanisms in each case for our hypothetical cell (taxis is a contribution to diffusion, Eq. 9.65). Diffusion has been assumed to be the dominant mechanism in the analyses presented in this chapter regarding laminar flow regimes.

12.3.2 Desorption

Turbulence may result in different mechanisms or modes by which cells desorb from the substratum and are transported back into the bulk fluid. Thus, the sticking efficiency of cells in turbulent flow, measured by Hermanowicz et al. (1989), is an important parameter for any predictive modeling (Beal, 1978).

TABLE 12.2 The Estimated Force Generated by Various Mechanisms of Transport for Two Hypothetical Microbial Cells 1 and 5 μm in Diameter[a]

	Laminar Flow ($Re = 1250$)		Quiescent	
	$d_p = 1\ \mu m$	$5\ \mu m$	$d_p = 1\ \mu m$	$5\ \mu m$
$F_{inertia}$ (pN)	0	0	0	0
F_{drag} (pN)	−0.5	−0.05	−0.001	−0.1
F_{lift} (pN)	0.001	0.0001	0	0
$F_{gravity}$ (pN)	0.0005	0.06	0.0005	0.06
$F_{drainage}$ (maximum) (pN)	−0.002	−0.2	−0.2	−0.2
$F_{diffusion}$ (pN)	0.8	0.03	0.03	0.03
F_{taxis} (pN)	0.03	0.15	0.15	0.15

[a] See Table 12.1.

For steady flow in the viscous sublayer where viscous forces predominate, the *drag* force on our cells when adsorbed is approximately

$$F_{drag} \approx 1 \times 10^3 \text{ pN} \qquad [12.39]$$

The *lift* force is zero in this calculation because steady flow has been assumed in the viscous sublayer. *Upsweeps* are analogous to the downsweeps as discussed in the context of transport to the substratum. Upsweeps result in turbulent bursts that move away from the substratum into the bulk fluid (Cleaver and Yates, 1976). Upsweeps generate a lift force, which can influence detachment:

$$F_{lift} \approx 0.1\text{–}12 \text{ pN} \qquad [12.40]$$

It is obvious that drag forces are much more important in detaching our cells than are lift forces. Thus, viscous shear forces parallel to the substratum may dislodge a cell, but, unless a lift force is present, the cell will presumably roll over until another surface adsorption site is contacted.

Motility is insignificant with respect to desorption, in view of the force calculated for motile organisms in the section on transport (Section 12.3.1).

12.3.3 Substratum Roughness

Substratum roughness may significantly influence transport rate and colonization for several reasons, including the following: (1) increased convective mass transport near the substratum, (2) more shelter from shear forces for cells, resulting in lower desorption rates, and (3) increased surface area for adsorption. If the surface roughness elements are larger than the viscous sublayer, the roughness can be measure quantitatively by hydraulic means, i.e., frictional resistance (Section 9.5.2, Chapter 9). If the surface roughness elements are smaller than the viscous sublayer (i.e., microroughness), roughness measurements are difficult and the results are difficult to interpret (Figure 12.16). Nevertheless, adsorption rates of particles are strongly influenced by microroughness (Browne, 1974).

12.3.4 Experimental Observations

Nelson et al. (1985) observed the accumulation of *Pseudomonas aeruginosa* on glass in a turbulent flow system. There was no exogenous substrate fed into the continuous flow adsorption reactor and no observable growth occurred on the substratum. The accumulation rate increased with increasing cell numbers in the bulk water. A chemostat provided the cells for the experiments. The cell accumulation rate at the substratum increased with decreasing specific growth rate of the cells in the chemostat. Bryers and Characklis (1982) observed that the accumulation rate of biomass at the substratum increased with increasing specific growth rate of the mixed population in the chemostat. Nelson et al. also observed that the spatial distribution at saturation deviated from random in the direction of uniformity.

Figure 12.17 shows Bryers's (1980) data on relative rates of accumulation in a glass tubular reactor (12.7 mm ID) as a function of biomass concentration in the

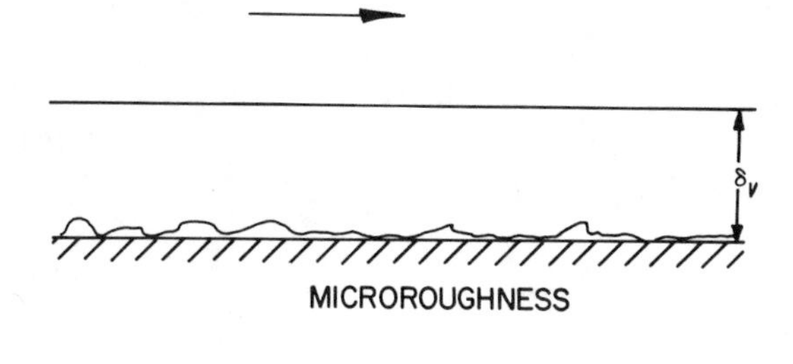

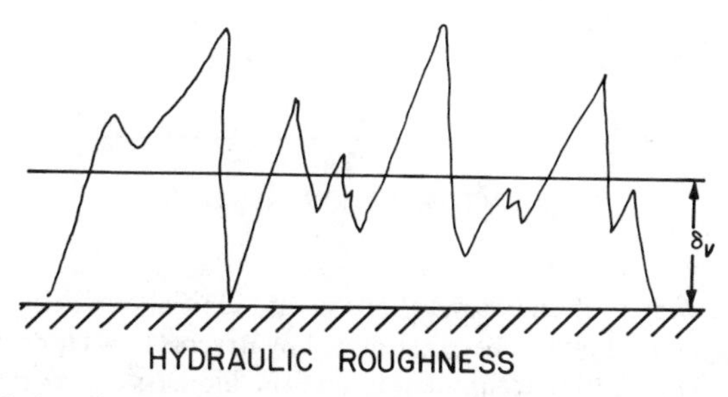

Figure 12.16 Hydraulic roughness, as reflected by increased friction factor measurements, is only evident if the roughness peaks exceed the viscous sublayer thickness δ_v. Microroughness does not influence the friction factor, but may be important in influencing deposition of cells at the substratum.

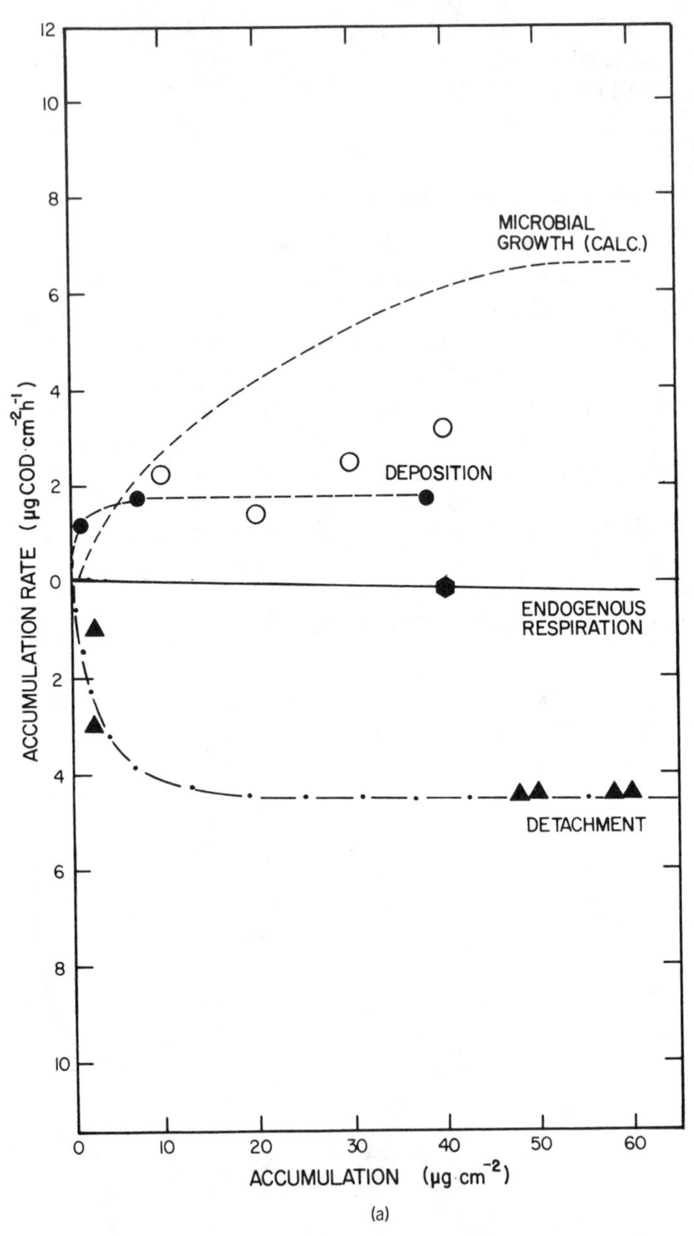

Figure 12.17 The contribution of several processes to initial events in the accumulation of a mixed population biofilm has been estimated by Bryers (1980) for turbulent flow in a glass tube at 25C. The biofilm accumulation and the biomass concentration in the bulk water were determined by a Chemical Oxygen Demand (COD) method. (a) Re = 13,000, and biomass concentration in the bulk water = 20 g m^{-3}. (b) Re = 26,000, and biomass concentration in the bulk water = 20 g m^{-3}. (c) Re = 13,000, and biomass concentration in the bulk water = 3.5 g m^{-3}. The open circles (○) are measured accumulation.

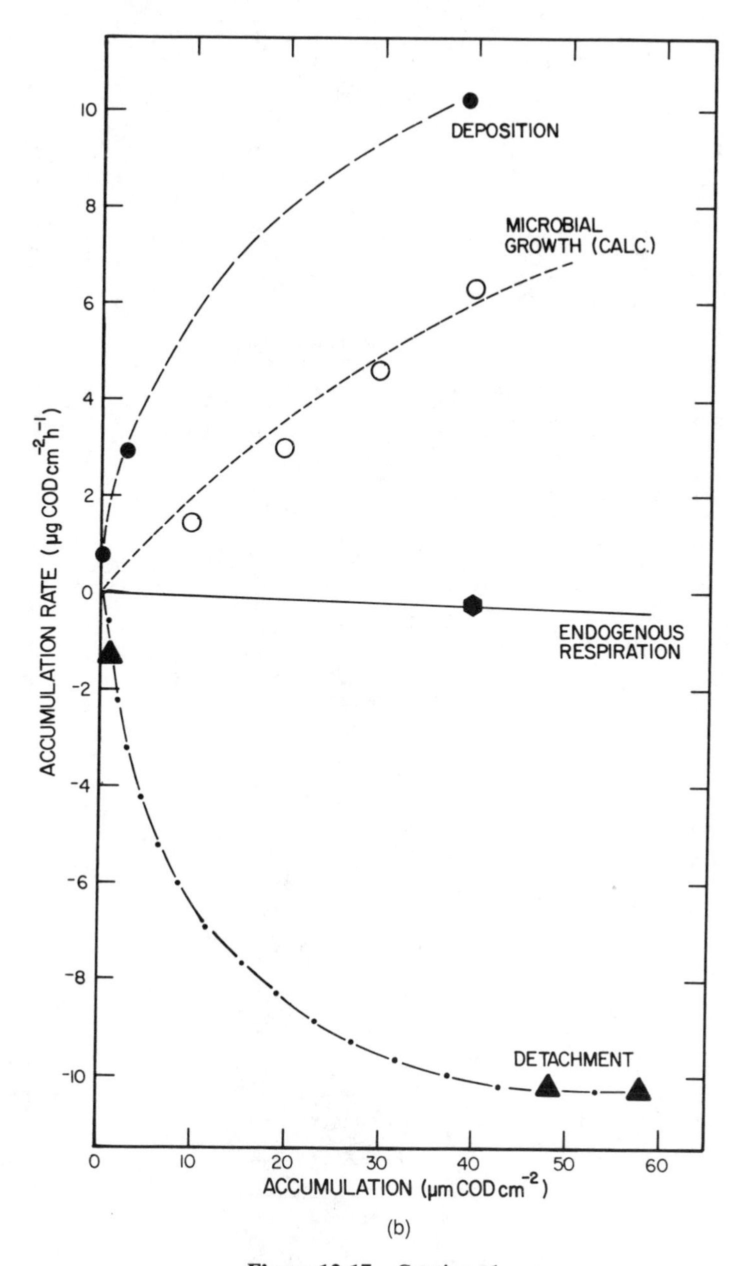

(b)

Figure 12.17 Continued

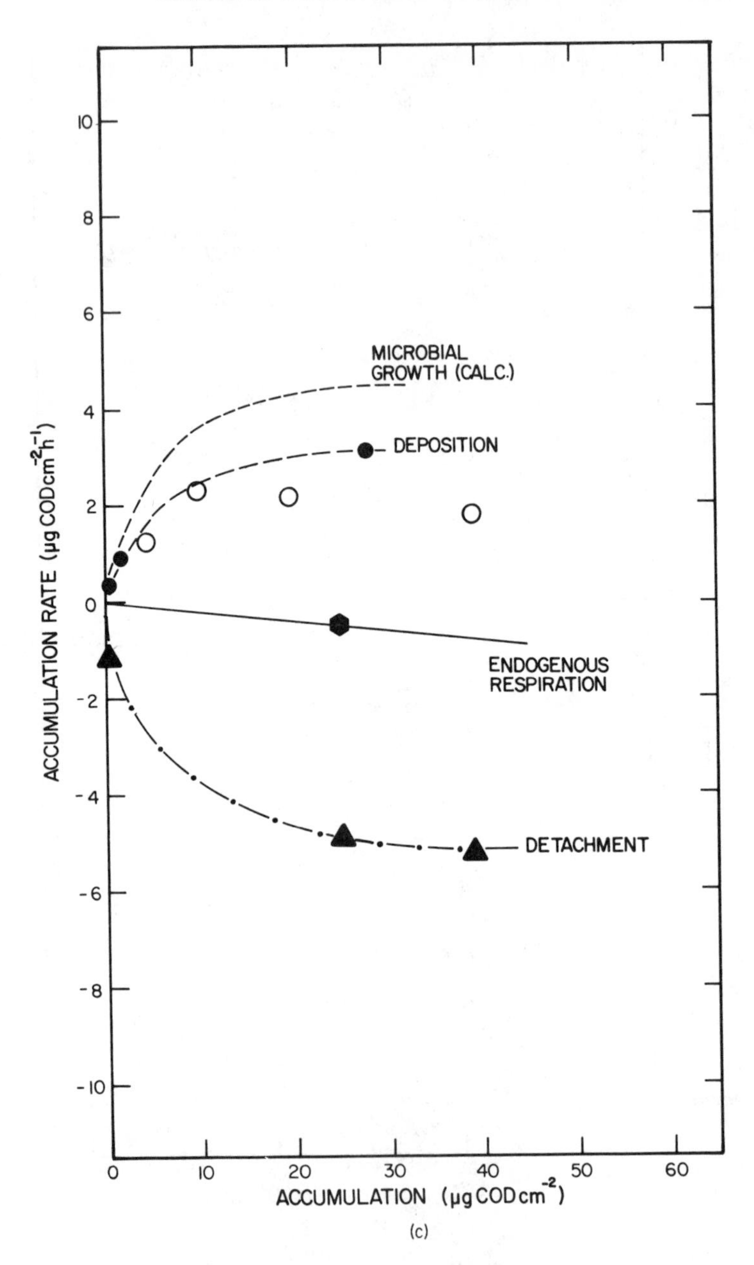

Figure 12.17 Continued

bulk fluid and of Reynolds number (i.e., fluid velocity), as an example of the way in which various processes influence early events in biofilm formation. The inlet substrate (1:1 weight ratio of trypticase soy broth and glucose) concentration was 20 g m^{-3}. The temperature was 31°C. The overall accumulation was approximately 0.6 g m^{-2} at the end of the experiments, which is equivalent to a biofilm thickness of approximately 20 μm, significantly smaller than the viscous sublayer thickness for this flow.

Certain observations are pertinent:

1. Deposition refers to the combined processes of *transport, adsorption,* and *attachment*. When *Re* increased from 13,000 to 26,000, the deposition increased by a factor of about 5. The detachment, on the other hand, increased by a factor of about 2.5 (Figure 12.17a and b). The microbial growth did not change much.
2. The deposition almost doubled when the biomass concentration in the bulk water increased from 3.5 to 20 mg/L (Figure 12.17a and c). The microbial growth, endogenous respiration, and detachment did not change to any great extent.

The results suggest that deposition, net detachment, and microbial growth are significant, even in the early stages of biofilm accumulation. Clearly, rates of deposition and detachment are strongly influenced by the Reynolds number (i.e., fluid velocity).

The fluid dynamics in experimental systems for observing biofilm processes are important for any flow regime (zero, laminar, or turbulent). Substratum surface characteristics (e.g., roughness) may also be important in determining the rate-controlling step in the overall process of colonization.

REFERENCES

Beal, S.W., *Nucl. Sci. Eng.*, **40**, 1–11 (1970).

Beal, S.W., *J. Aerosol Sci.*, **9**, 455–461 (1978).

Belas, M.R. and R.R. Colwell, *J. Bact.*, **151**, 1568–1580 (1982).

Bowen, B.D., S. Levin, and N. Epstein, *J. Coll. Interface Sci.*, **54**, 375–390 (1978).

Browne, L.W.B., *Atm. Environ.*, **8**, 801–816 (1974).

Bryers, J.D., "Dynamics of early biofilm formation in a turbulent flow system," unpublished doctoral dissertation, Rice University, Houston, TX, 1980.

Bryers, J.D. and W.G. Characklis, *Biotech. Bioeng.*, **24**, 2451–2476 (1982).

Bungay, H.R. and J.M. Wiggert, *Biotech. Bioeng.*, **7**, 323–325 (1965).

Characklis, W.G. "Biofilm development: A process analysis," in K.C. Marshall (ed.), *Microbial Adhesion and Aggregation,* Springer-Verlag, 1984, pp. 137–157.

Cleaver, J.W. and B. Yates, *Chem. Eng. Sci.*, **30**, 983–992 (1975).

Cleaver, J.W. and B. Yates, *Chem. Eng. Sci.*, **31**, 147–151 (1976).

Escher, A.R., "Bacterial colonization of a smooth surface: An analysis with image analyzer," unpublished doctoral thesis, Montana State University, Bozeman, MT, 1986.

Fletcher, M., *J. Gen. Microbiol.*, **94**, 400–404 (1976).

Fletcher, M. and K.C. Marshall, *Appl. Environ. Microbiol.*, **44**, 184–192 (1982).

Gudmundsson, J.S. and T.R. Bott, *J. Aerosol Sci.*, **8**, 317–319 (1977).

Hermanowicz, S.H., R.E. Danielson, and R.C. Cooper, *Biotech. Bioeng.*, **33**, 157–163 (1989).

Jang, L.K. and T.F. Yen, "A theoretical model of convective diffusion of motile and nonmotile bacteria toward solid surfaces," in J.E. Zajic and E.C. Donaldson (eds.), *Microbes and Oil Recovery*, Vol. 1, International Bioresources Journal, 1985, pp. 226–246.

Kjelleberg, S., B.A. Humphrey, and K.C. Marshall, *Appl. Environ. Microbiol.*, **43**, 1166–1172 (1982).

Lister, D.H., "Corrosion products in power generating systems," in E.F.C. Somerscales and J.G. Knudsen (eds.), *Fouling of Heat Transfer Equipment*, Hemisphere, Washington, 1981, pp. 135–200.

MacRae, I.C. and S.K. Evans, *Water Res.*, **18**, 1377–1380 (1984).

Marshall, K.C., "Introduction to the Dahlem Conference," in K.C. Marshall (ed.), *Microbial Adhesion and Aggregation*, Springer-Verlag, 1985, pp. 1–3.

Nelson, C.H., J.A. Robinson, and W.G. Characklis, *Biotech. Bioeng.*, **27**, 1662–1667 (1985).

Powell, M.S. and N.K.H. Slater, *Biotech. Bioeng.*, **25**, 891–900 (1983).

Stanley, P.M., *Can. J. Microbiol.*, **29**, 1493–1499 (1983).

Valeur, A., A. Tunlid, and G. Odham, *Arch. Mikrobiol.*, **149**, 521–526 (1988).

13

MODELING A MONOPOPULATION BIOFILM SYSTEM: *PSEUDOMONAS AERUGINOSA*

RUNE BAKKE
Hogskolesenteret i Rogaland, Stavanger, Norway

WILLIAM G. CHARACKLIS
Montana State University, Bozeman, Montana

MUKESH H. TURAKHIA
Houston, Texas

AN-I YEH
National Taiwan University, Taipei, Taiwan, R.O.C.

SUMMARY

Efforts to simulate biofilm accumulation and activity in a CFSTR and a plug flow reactor are presented for monopopulation biofilms. For many years environmental engineers have been modeling biofilms as if they consisted of single populations, so technological applications of "effective" monopopulation biofilms exist. The major impetus for modeling monopopulation biofilms, however, is the need to elucidate the fundamental processes contributing to biofilm accumulation and persistence without the confounding factors introduced by population dynamics and interactions.

Predicting the rate and extent of biofilm processes is necessary and useful for ecosystem analysis, design and operation of heat exchangers and pipelines subject to fouling, wastewater treatment plant design and operation, and determining the feasibility of biofilm reactors for biotechnological applications. However, many hypotheses remain to be verified before accurate predictions of long term biofilm behavior can be made with mathematical models.

13.1 INTRODUCTION

Biofilms are biologically active matrices of cells and noncellular material accumulated on solid surfaces. Predicting the rate and extent of biofilm processes would be useful in ecosystem analysis, design and operation of heat exchangers and pipelines subject to fouling, wastewater treatment plant design and operation, and determining the feasibility of biofilm reactors for biotechnological applications. For example, (1) a predictive model may be used to determine the feasibility of biofilm reactors for extracellular polymer production; (2) the required wastewater treatment reactor volume and surface area to attain effluent water quality standards can be estimated (Chapter 17). The model may include adsorption processes leading to accumulation of biofilm, so optimization of substratum colonization and process startup can be predicted. The model may also be used to predict the onset and extent of biofouling problems, such as excessive heat transfer resistance and fluid frictional losses (Chapter 14). Thus, feedback control systems may be designed to activate fouling biofilm removal procedures (Chapter 15) prior to observing equipment performance deterioration.

This chapter describes a model for simulating monopopulation biofilms. Although such biofilms rarely exist in nature, models of them do have applications. For example, environmental engineers have been modeling biofilms as if they consisted of single populations for many years. However, the major impetus for research and modeling of monopopulation biofilms has been the need to elucidate the fundamental processes contributing to biofilm accumulation and persistence without the confounding factors introduced by population dynamics and interactions.

13.1.1 System and Process Description

This chapter presents a model for a biofilm consisting of two components: *Pseudomonas aeruginosa* cells and their extracellular polymeric substances (EPS). The biofilm was observed to consist only of a base film (Figures 1.2 and 1.4, Chapter 1). The soluble substrate, glucose, is the rate-limiting nutrient and sole carbon and energy source. Substrate is supplied continuously to the biofilm in the overlying turbulent bulk water compartment of the reactor. Results from chemostat (Table 8.1) and RotoTorque experiments with *P. aeruginosa* have produced estimates of stoichiometric and rate coefficients for the fundamental processes described in Chapters 7 through 9. Processes considered in the model include (1) advective transport of substrate, cells, and products in the bulk water compartment, (2) diffusive transport of substrate from the bulk water into the biofilm, (3) attachment and detachment (erosion) of cells and polymers between the bulk water and the substratum, and (4) cellular reproduction and product formation in the bulk water and in the biofilm. The model predicts progression of substrate removal, cellular accumulation, and extracellular polymer accumulation in the bulk water and in the biofilm.

The assumptions leading to the mathematical model for a monopopulation biofilm have been specified in Chapter 11, eventually leading to the description of the model for monopopulation biofilms in Section 11.4.4.3.

13.1.2 Mathematical Preliminaries

Models concerned with the prediction of concentration variations within a single phase or compartment are obtained with the help of the following sets of equations (Chapter 9): (1) the balance equations (or conservation laws), (2) the flux laws, and (3) the rate equations describing production and depletion (transformation terms). A general review of overall balance equations was presented in Chapter 2. The general expression for the balance of a conservative quantity within a single compartment is expressed as follows (see Eq. 9.1):

$$\frac{d}{dt}\int_V S\,dV = \int_V r_S\,dV + \int_A \mathbf{n}\cdot\mathbf{J}_S\,dA \qquad [13.1]$$

accumulation rate of S in control volume V	volumetric transformation rate of S in control volume V	transport rate into control volume V through the surface boundary A

where r_S = volumetric reaction rate of S in volume V ($ML^{-3}t^{-1}$)
$\mathbf{J}_S$ = surface flux vector for S across the control volume boundary ($ML^{-2}t^{-1}$)
$\mathbf{n}$ = inward normal unit vector on the control volume surface

For the bulk liquid of a CFSTR in which S and r_S are uniform in the bulk fluid (volume V) of the reactor, the conservation equation for reactive species S (Eq. 13.1) reduces to the following:

$$V\frac{dS}{dt} = Q(S_i - S) + r_S V \qquad [13.2]$$

where S = concentration (ML^{-3})
Q = volumetric flow rate into the reactor (L^3t^{-1})

The conservation equation may have to be written for every compartment or phase where S appears. Overall or macroscopic balances, as illustrated in Chapter 2 (Eq. 2.1), are useful especially when the velocity, temperature, and concentration are uniform in the system, e.g., a CFSTR or chemostat. However, the nature of biofilm processes usually results in *gradients*, which significantly influence the microenvironment of the biofilm organisms. In such cases, *differential* models or *lumped parameter* models must be employed rather than macroscopic balances. Thus, the model is based on fundamental conservation principles and classic process analysis techniques presented in Chapter 2.

The stoichiometric and kinetic components of the model are presented in matrix and vector notation in this chapter for at least two reasons: (1) to systematically represent the large number of processes and parameters required to describe biofilm accumulation and activity, and (2) to further illustrate the matrix notation introduced in Chapter 2 and used in Chapter 11.

13.2 A CONTINUOUS FLOW STIRRED TANK BIOFILM REACTOR (CFSTR)

CFSTR biofilm systems occur in technology, as illustrated by a recirculating cooling tower system, which acts as such a system by virtue of its high recycle rate and low dilution rate. A fixed film rotating biological contactor (Chapter 17) also is suitably modeled as a CFSTR.

The CFSTR model also describes phenomena occurring in the laboratory RotoTorque reactor (Chapter 3). The RotoTorque reactor is operated as a CFSTR in which bulk water concentration gradients do not exist. Thus, Eq. 13.2 suitably describes the bulk liquid compartment of the RotoTorque reactor.

13.2.1 Compartments

The system must be partitioned into a number of distinct compartments depending on physical characteristics, requirements of model resolution, and available computing capacity. The simplest case of one compartment, $m = 1$, is appropriate for ideal CFSTR, or chemostats, in which all activity resides in a homogeneous liquid phase. Plug flow reactors can be modeled as several CFSTRs in series, and then $m > 1$ (Chapter 2). When a biofilm exists in a CFSTR, then $m \geq 2$, since a bulk liquid compartment and one or more biofilm compartments may exist (Chapter 1). Layers in a biofilm, with distinct biological properties (physiological and genetic), may result in mixed population biofilms requiring several biofilm compartments each with several phases (e.g., different microbial species).

13.2.2 Components

Accumulation of components in the bulk water is described by material balances similar to Eq. 13.2. Accumulation of biofilm components can be described by constitutive equations of the same general form as Eq. 13.1 with advective transport generally considered negligible. Use of the component vector **a** permits the left hand side of Eq. (13.2) to be described as $d\mathbf{a}/dt$:

$$\mathbf{a} = [S \quad X_f' \quad P_f' \quad X \quad P] \qquad [13.3]$$

where S, X, P = bulk water substrate, cell mass, and EPS concentrations, respectively (ML^{-3})
X_f', P_f' = biofilm cell mass and EPS concentration per unit reactor volume, respectively (ML^{-3})

In Chapter 11 (Section 11.2.2), a distinction is made between the concentration and the density of a component in the biofilm. Thus,

$$X_f = \frac{X_f''}{L_f} = \epsilon_X \rho_X \qquad [13.4]$$

where X_f'' = biofilm cell mass per unit substratum area (ML^{-2})
X_f = biofilm cell mass per unit biofilm volume (ML^{-3})
L_f = biofilm thickness (L)
ρ_X = specific gravity or density of the cells (ML^{-3})
ϵ_X = fraction of biofilm volume occupied by cells ($L^3_{cells} L^{-3}_{biofilm}$)

The specific gravity of cells, ρ_X, has been determined by numerous investigators (Section 5.5.3, Chapter 5) to be on the order of 1050–1070 kg m^{-3} (i.e., specific gravity $\approx$1.05–1.07). Estimates of ϵ_X, based on biofilm water content (Section 4.5, Chapter 4), indicate that a biofilm is approximately 87–99% water. Then, $\epsilon_X \rho_X \approx$ 10.5 kg m^{-3}, which compares well with measured values reported in Table 4.2 (Chapter 4). The variable X_f is related to X_f' as follows:

$$X_f' = \frac{X_f'' A}{V} = \epsilon_X \rho_X \frac{L_f A}{V} \qquad [13.5]$$

or

$$X_f' = \epsilon_X \rho_X \frac{\text{biofilm volume}}{\text{reactor volume}}$$

so

$$X_f' = \frac{X_f A L_f}{V}$$

where A = substratum area (L^2)
V = reactor volume (L^3)

X_f can be determined if the biofilm thickness is measured. The variable X_f'' is frequently needed when a measurement of the biofilm thickness cannot be accomplished. The variable P_f' is defined in analogous manner to X_f'.

13.2.3 Net Transport into the Reactor

The net transport into the reactor is a result of bulk water flow through it. The net transport terms are extensive, not intensive, variables and thus depend on the size of the system. There is negligible input of suspended biomass after the inoculation period (i.e., when sterile feed is injected), and the input product carbon concentration is zero. The transport vector, $\mathbf{a}'$, then becomes

$$\mathbf{a}' = [D(S_i - S) \quad 0 \quad 0 \quad -DX \quad -DP] \tag{13.6}$$

where D = dilution rate (t^{-1})
S_i = inlet substrate concentration (ML^{-3})

The influent concentration S_i is a parameter needed for integrating Eq. 13.2. The effluent concentration is a dependent variable determined by the integration. For constant bulk flow rate, the dilution rate D varies with the biofilm volume:

$$D = D_0\left(1 - L_f \frac{A}{V}\right)^{-1} \tag{13.7}$$

where D_0 = dilution rate at $L_f = 0$

Frequently, this correction is quite small and can be ignored.

13.2.4 Stoichiometry

Stoichiometric relations describing the processes are expressed in terms of intensive variables in the system and are as follows:

Process 1. Net adsorption:

$$X \rightarrow X_f' \tag{13.8}$$

Process 2. Growth:

$$a_{21}S \rightarrow a_{22}X_f' \tag{13.9}$$

Process 3. Product (EPS) formation:

$$a_{31}S \rightarrow a_{33}P_f' \tag{13.10}$$

Process 4. Cellular erosion:

$$X_f' \rightarrow X \qquad [13.11]$$

Process 5. EPS erosion:

$$P_f' \rightarrow P \qquad [13.12]$$

where a_{ij} = a stoichiometric coeficient for the amount of component j reacted in process i (M)

Variables with a subscript f refer to the biofilm compartment. Then, the stoichiometric matrix (5 × 5) is as follows:

$$\mathbf{a}'' = \overset{\text{Component}}{\begin{array}{c} \begin{array}{ccccc} S & X_f' & P_f' & X & P \end{array} \\ \begin{bmatrix} 0 & 1 & 0 & -1 & 0 \\ -a_{21} & a_{22} & 0 & 0 & 0 \\ -a_{31} & 0 & a_{33} & 0 & 0 \\ 0 & -1 & 0 & 1 & 0 \\ 0 & 0 & -1 & 0 & 1 \end{bmatrix} \end{array}} \begin{array}{l} \text{Process 1} \\ \text{Process 2} \\ \text{Process 3} \\ \text{Process 4} \\ \text{Process 5} \end{array}$$

$$= \begin{bmatrix} 0 & 1 & 0 & -1 & 0 \\ -1/Y_{X/S} & 1 & 0 & 0 & 0 \\ -1/Y_{P/S} & 0 & 1 & 0 & 0 \\ 0 & -1 & 0 & 1 & 0 \\ 0 & 0 & -1 & 0 & 1 \end{bmatrix} \qquad [13.13]$$

where $Y_{X/S} = a_{22}/a_{21}$, the growth yield
$Y_{P/S} = a_{33}/a_{31}$, the EPS yield

The yields represent stoichiometric ratios between the product (X or P) formed and the amount of reactant (S) consumed for that particular process, and are fundamentally characteristic of (i.e., intrinsic to) aerobic *P. aeruginosa* glucose metabolism. Observed yields do not have the fundamental significance of intrinsic yields and should not be confused with them. The difficulty in discriminating between observed and intrinsic yields for EPS and cells in a *P. aeruginosa* biofilm has been described in Sections 6.3.3 and 6.3.4 (Chapter 6). The interfacial transfer processes, such as detachment, always have stoichiometric coefficients equal to ± 1.

13.2.4.1 *Biofilm Elemental Composition* The carbon, hydrogen, and nitrogen content of the *P. aeruginosa* biofilm are consistent with those of microbial cells reported in the literature (Turakhia, 1986), i.e., approximately 50% carbon, 20% oxygen, 10–15% nitrogen, and 8–10% hydrogen on a dry, ash-free weight basis (Table 4.8).

13.2.4.2 *Cellular and EPS Yields* Bakke et al. (1984) and Turahkia (1986) observed the linear relationship between specific substrate removal rate and specific growth rate in a *P. aeruginosa* biofilm (Figure 13.1a). Robinson et al. (1984) observed a similar relationship for *P. aeruginosa* in a chemostat (Figure 8.7). Robinson et al. proposed a method for calculating the intrinsic yields for cell and EPS mass. Based on this calculation, the cellular mass yield ($Y_{X/S}$) was 0.30–0.34 (g cell carbon) (g glucose carbon)$^{-1}$. The EPS mass yield ($Y_{P/S}$) was approximately 0.56 (g polymer carbon) (g glucose carbon)$^{-1}$. The observed biomass (cells plus EPS) yield was approximately 0.61 (g biomass carbon) (g glucose carbon)$^{-1}$. Thus, there is a significant error in the calculation of the intrinsic yield. Assuming the cellular mass yield is 0.34 g g^{-1} and the observed biomass yield (0.61 g g^{-1}) is correct, the EPS yield is approximately 0.27 g g^{-1}.

The results presented in Figure 13.1a clearly indicate a similar energy metabolism for suspended and biofilm *Pseudomonas aeruginosa* cells, i.e., immobilization does not induce any dramatic change in metabolism. These results correspond to the average metabolism for all cells in the biofilm. They relate only to cell carbon in the biofilm; the EPS carbon in the biofilm is ignored when determining the specific substrate removal rate and specific growth rate. If only the biofilm mass (cell plus EPS carbon) had been measured, biofilm metabolism would have borne no consistent relationship to chemostat metabolism. The inconsistent behavior is illustrated in Figure 13.1b, which compares the specific biomass production rate (as opposed to specific cellular growth rate) in the biofilm and the substrate removal rate per unit biofilm mass (as opposed to substrate removal rate per unit cell carbon) with their values in a chemostat. Thus, a more structured biofilm model is needed to relate chemostat rate and stoichiometric coefficients to those in a biofilm.

The RotoTorque reactors were operated at high dilution rates ($D = 6\ \text{h}^{-1} \gg \mu_{max}$), so cellular reproduction and product formation by suspended organisms were negligible.

13.2.4.3 *Oxygen Uptake* The stoichiometry of the aerobic degradation of glucose by *P. aeruginosa* is described by the following (Turahkia, 1986):

Energy Reaction:

$$C_6H_{12}O_6 + 6O_2 \rightarrow 6CO_2 + 6H_2O - 120.11\ \text{kJ}\ (\text{mol e}^-)^{-1} \qquad [13.14]$$

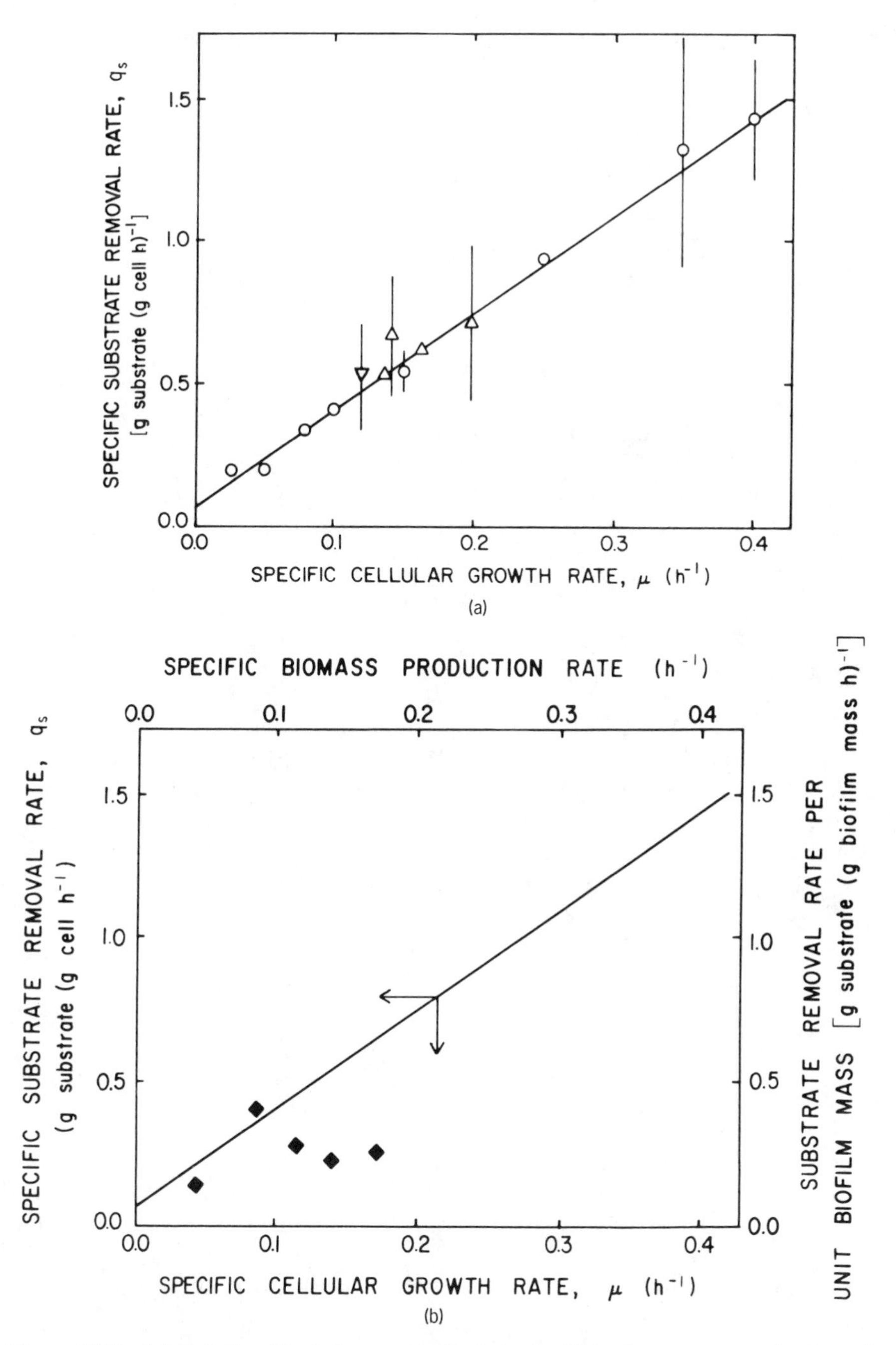

Figure 13.1 (a) Relationship between steady state specific substrate removal rate and specific growth rate of *P. aeruginosa* in a chemostat (○) and in a biofilm (△, $D = 6\ h^{-1}$; ▽, $D = 3\ h^{-1}$) (reprinted with permission from Bakke et al., 1984). (b) The relationship between substrate removal rate per unit biofilm mass and specific biomass production rate in the biofilm is compared with the regression line representing the chemostat data in Figure 13.1a. No consistent trend is observed.

Synthesis Reaction:

$$C_6H_{12}O_6 + 1.2NH_3 \rightarrow 6CH_{1.4}O_{0.4}N_{0.2} + 3.6H_2O \quad [13.15]$$

Overall Reaction:

$$C_6H_{12}O_6 + 2.5O_2 + 0.7NH_3 \rightarrow 3.5CH_{1.4}O_{0.4}N_{0.2} + 2.5CO_2 + 4.6H_2O \quad [13.16]$$

Busch (1971) described the same stoichiometry for mixed microbial populations (Eq. 2.24, Chapter 2). According to the stoichiometry, one gram of oxygen is required for biochemical conversion of 0.91 grams of glucose carbon. Turakhia established the validity of this stoichiometry for a *P. aeruginosa* biofilm by plotting specific substrate removal rate versus specific oxygen removal rate in a RotoTorque (Figure 6.8).

13.2.5 Kinetics

13.2.5.1 *Net Adsorption* Net adsorption of bacteria to the substratum may be modeled by a saturation function (Langmuir, 1918). In turbulent CFSTR studied by Nelson et al. (1985), the specific net adsorption rate r'_{aX} was proportional to the suspended cell concentration X and was a saturation function of the attached cells, X'_f (Chapter 9):

$$r''_{aX} = k''_{aX} X \left(1 - \frac{X'_f}{k'_{aX}}\right) \quad [13.17a]$$

or

$$r_{aX} = r''_{aX} \frac{A}{V} = k_{aX} X \left(1 - \frac{X'_f}{k'_{aX}}\right) \quad [13.17b]$$

where r''_{aX} = net cellular specific adsorption rate per unit substratum area ($ML^{-2}t^{-1}$)

r_{aX} = net cellular specific adsorption rate per unit reactor volume ($ML^{-3}t^{-1}$)

X'_f = biofilm (or adsorbed) cell concentration per unit reactor volume (ML^{-3})

k''_{aX} = net adsorption rate coefficient (Lt^{-1})

k_{aX} = net adsorption rate coefficient (t^{-1})

k'_{aX} = saturation coefficient (ML^{-3})

This net adsorption process refers to the interaction of the cells with the substratum. When the substratum becomes saturated with cells, the net adsorption goes to zero. Contrast the net adsorption process with the net detachment process, which describes interactions of suspended cells with the biofilm. Chapter 12 describes the adsorption process in more detail.

13.2.5.2 Specific Growth Rate The growth rate of microorganisms depends on the cell concentration in the following way:

$$r_X = \mu X_f' \tag{13.18}$$

where r_X = population growth rate ($ML^{-3}t^{-1}$)
μ = specific growth rate of the population (t^{-1})

The specific growth rate μ is presumed to be a saturation function of reactor substrate concentration (Chapter 8):

$$\mu = \frac{\mu_{\max} S}{K_S + S} \tag{13.19}$$

The coefficients $\mu_{\max} = 0.37\ h^{-1}$ and $K_S = 2\ g\ m^{-3}$ are reported in Table 8.2 (Chapter 8). The expression is useful as long as the diffusional resistance is negligible. If significant diffusional resistance exists, the expression must be modified. Substrate diffusion between the bulk liquid and the biofilm is driven by the substrate concentration gradient from the bulk liquid to the biofilm–bulk water interface. Bulk flow into the base film of the biofilm can be neglected, since $\mathbf{v} = 0$ in the biofilm. For biofilm exhibiting a rough morphology, bulk flow into the film may be significant, but monopopulation *Pseudomonas aeruginosa* biofilms are assumed to be base films, since no significant hydraulic roughness has ever been detected due to their accumulation. Microroughness, however, may be important (see Section 13.2.8).

The substrate concentration decreases with increasing biofilm depth, due to transformation processes that consume substrate as it diffuses through the biofilm. The average substrate concentration in the biofilm is therefore less than S, which results in a specific substrate consumption rate lower in the biofilm than in the bulk liquid. Atkinson (1974) derived an effectiveness factor (using a Thiele modulus), which is the ratio of the actual specific substrate consumption rate in the biofilm to the value it would have without diffusional resistance. Trulear (1983) used the relationships derived by Atkinson and determined that no significant diffusional resistance occurs in *P. aeruginosa* biofilms in the RotoTorque system. The effectiveness factor was never smaller than 0.9 in Trulear's experiments (Figure 13.2). Wanner (unpublished results) has modeled the *Pseudomonas aeruginosa* biofilm and also reports a negligible diffusion resistance within the biofilm.

13.2.5.3 Specific EPS Formation Rate The EPS formation rate obeys the Luedeking–Piret equation for product formation (Section 8.4.1, Chapter 8):

$$r_P = k\mu X_f' + k' X_f' \tag{13.20}$$

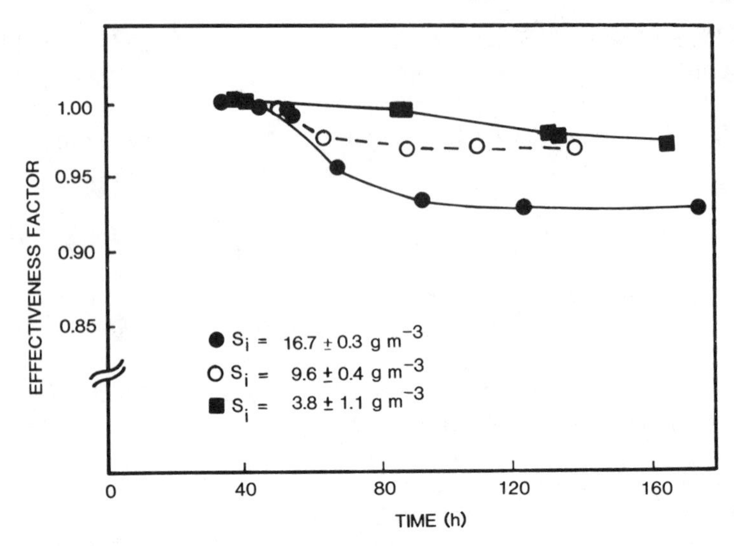

Figure 13.2 Progression of effectiveness factor for a *P. aeruginosa* biofilm at three different substrate loading rates in a turbulent RotoTorque reactor (Trulear, 1983). The effectiveness factor is always greater than 0.9, indicating minimal diffusion resistance. The mean biofilm thickness did not exceed 40 μm.

where r_p = EPS formation rate ($ML^{-3}t^{-1}$). Then the specific EPS formation rate q_P is

$$q_P = k\mu + k' \qquad [13.21]$$

13.2.5.4 *Specific Detachment (Erosion) Rate* Erosion from the biofilm is presumed to depend on biofilm mass and is independent of substrate loading rate, growth rate, and dilution rate. More specifically, the cellular erosion rate from the biofilm is first order in the biofilm cell density (Section 9.8.3, Chapter 9):

$$r''_{eX} = k''_{eX} X'_f \qquad [13.22a]$$

and

$$r_{eX} = r''_{eX} \frac{A}{V} = k_{eX} X'_f \qquad [13.22b]$$

where r''_{eX} = cellular erosion rate per unit biofilm area ($ML^{-2}t^{-1}$)
r_{eX} = cellular erosion rate per unit reactor volume ($ML^{-3}t^{-1}$)
k''_{eX} = cellular erosion rate coefficient (Lt^{-1})
k_{eX} = cellular erosion rate coefficient (t^{-1})

The EPS erosion rate is

$$r''_{eP} = k''_{eP} P'_f \qquad [13.23a]$$

and

$$r_{eP} = r''_{eP} \frac{A}{V} = k_{eP} P'_f \qquad [13.23b]$$

where r''_{eP} = EPS erosion rate per unit biofilm area ($ML^{-2}t^{-1}$)
r_{eP} = EPS erosion rate per unit reactor volume ($ML^{-3}t^{-1}$)
k''_{eP} = EPS erosion rate coefficient (Lt^{-1})
k_{eP} = EPS erosion rate coefficient (t^{-1})

The erosion rates of the various biofilm components may also be related to the hydrodynamic shear force at the biofilm-water interface, τ, by a first order expression:

$$k_{ei} = k^*_{ei} \tau \qquad [13.24]$$

where k_{ei} = erosion rate coefficient for component i in the biofilm (Lt^{-1})
k^*_{ei} = shear-related erosion rate coefficient (L^2tM^{-1})
τ = shear stress at the biofilm-bulk water interface ($ML^{-1}t^{-2}$)

According to Blasius's law of friction for a tube, valid for $4000 < Re < 10^5$, the friction factor f can be related to the Reynolds number Re:

$$f = 0.0791\, Re^{-0.25} \qquad [13.25]$$

$$Re = vd/\nu \qquad [13.26]$$

where ν = kinematic viscosity (L^2t^{-1})

Then, by substituting for Eq. 13.26 in Eq. 13.25,

$$f = 0.0791(vd/\nu)^{-0.25} \qquad [13.27]$$

But the friction factor is, by definition,

$$f \equiv 2\tau/\rho v^2 \qquad [13.28]$$

Substituting Eq. 13.27 into Eq. 13.28 and rearranging yields

$$\tau = 0.04(d/\nu)^{-0.25} \rho^{0.75} v^{1.75} \qquad [13.29]$$

Then from Eq. 13.24 for turbulent systems,

$$k_{ei} = k_T v^{1.75} \qquad [13.30]$$

where k_T = erosion rate coefficient characteristic of turbulent flow ($L^{-0.75}t^{0.75}$)

Trulear and Characklis (1982) report that the total (cells and EPS) erosion rate for a mixed population biofilm increases with increasing water velocity, v. Equation 13.30 fits the data well (Figure 9.34), supporting the contention that the erosion rate is proportional to the hydrodynamic shear force (Eq. 13.24). Rittmann (1982), using the data of Trulear and Characklis, presented equations by which the erosion rate can be calculated for various experimental systems and conditions.

Intuitively, the cellular erosion rate coefficient, k_{eX}, would be expected to depend on the biofilm cell concentration. However, limited observations on a *P. aeruginosa* biofilm (Bakke et al., 1984) indicate that k_{eX} is independent of X_f (Figure 13.3). It is likely that k_{eX} indirectly depends on substrate loading, since

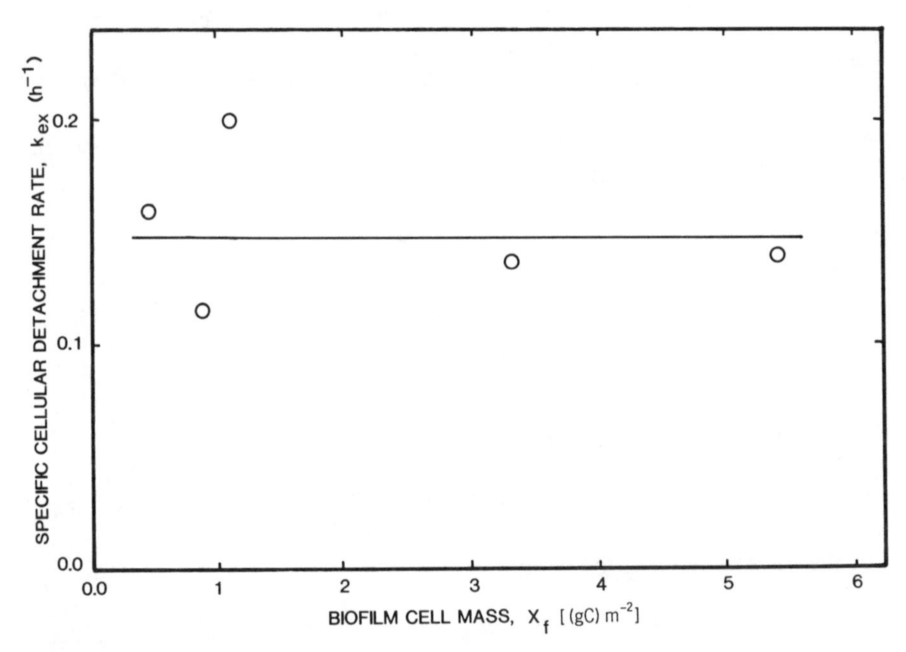

Figure 13.3 There is no significant influence of biofilm areal concentration on specific cellular erosion rate for a *P. aeruginosa* biofilm in a turbulent RotoTorque reactor (Bakke et al., 1984).

erosion rate equals growth rate at steady state. Then in those systems where substrate loading rate is proportional to velocity, the observed dependence of erosion on velocity may be related to growth processes.

13.2.5.5 *Process Rate Vector* The resulting process rate vector (1 × 5) is

$$\mathbf{r} = \left[k_{aX}X\left(1 - \frac{X_f'}{K_{aX}}\right) \quad \mu X_f' \quad q_P X_f' \quad k_{eX}X_f' \quad k_{eP}P_f' \right] \qquad [13.31]$$

13.2.6 Material Balances

The stoichiometric matrix ($\mathbf{a}''$) and the process rate vector ($\mathbf{r}$) can be combined in a transformation matrix (1 × 5):

$$\mathbf{r}\cdot\mathbf{a}'' = [-r_X - r_P \quad r_{aX} + r_X - r_{eX} \quad r_P - r_{eP} \quad -r_{aX} + r_{eX} \quad r_{eP}] \qquad [13.32]$$

Equation 13.2 can now be expressed in terms of Eqs. 13.6, 13.13, and 13.32 as

$$\frac{d\mathbf{a}}{dt} = \mathbf{a}' + \mathbf{r}\cdot\mathbf{a}'' \qquad [13.33]$$

Complete material balances and constitutive equations derived from Eq. 13.33 can now be listed.

Substrate Carbon

$$\frac{dS}{dt} = D(S_i - S) - \left[\frac{\mu}{Y_{X/S}} + \frac{q_P}{Y_{P/S}}\right]X_f' \qquad [13.34]$$

X_f' is the biofilm cell carbon per unit volume in the reactor, i.e., has units of mass per unit volume.

Suspended Cellular Carbon

$$\frac{dX}{dt} = -DX + k_{eX}X_f' - k_{aX}X(1 - X_f'/k_{aX}') \qquad [13.35]$$

Suspended Polymer Carbon

$$\frac{dP}{dt} = -DP + k_{eP}P_f' \qquad [13.36]$$

Biofilm Cell Carbon

$$\frac{dX_f'}{dt} = \mu X_f' - k_{eX} X_f' + k_{aX} X\left(1 - \frac{X_f'}{k_{aX}'}\right) \qquad [13.37]$$

Biofilm Polymer Carbon

$$\frac{dP_f'}{dt} = q_P X_f' - k_{eP} P_f' \qquad [13.38]$$

Since RotoTorque experiments were conducted with a dilution rate significantly greater than μ_{max} ($D = 6\ h^{-1}$), growth and product formation in the bulk water were negligible.

13.2.7 Steady State Material Balances

Biofilm accumulation reaches a steady state when $d\mathbf{a}/dt = 0$. Eq. 13.33 then simplifies to:

$$0 = \mathbf{a}' + r \cdot \mathbf{a}'' \qquad [13.39]$$

Transformation rates can then be determined from Eq. 13.39, since all quantities in the transport vector $\mathbf{a}'$ are measurable. Assume net adsorption is negligible at steady state.

Substrate Carbon

$$\frac{D(S_i - S)}{X_f'} = \mu\left(\frac{1}{Y_{X/S}} + \frac{k}{Y_{P/S}}\right) + \frac{k'}{Y_{P/S}} \qquad [13.40]$$

but the specific substrate removal rate in the biofilm, q_S, is

$$q_S = \frac{D(S_i - S)}{X_f'} \qquad [13.41]$$

The yield coefficients represent the intrinsic (as opposed to observed) stoichiometry for *P. aeruginosa* and are presumed constant. Then, Eq. 13.40 predicts a linear relation between q_S and μ:

$$q_S = \mu\left[\frac{1}{Y_{X/S}} + \frac{k}{Y_{P/S}}\right] + \frac{k'}{Y_{P/S}} \qquad [13.42]$$

Experimental results indicate that q_S is a linear function of μ (Figure 13.1a) in a chemostat with *P. aeruginosa*. The results from a *P. aeruginosa* biofilm in a RotoTorque (also Figure 13.1a) are consistent with the chemostat data, at least for the relatively narrow range of specific growth rates observed in the biofilm.

Suspended Cell Carbon

$$X = \frac{X_f' k_{eX}}{D} \quad [13.43]$$

Suspended Product Carbon

$$P = \frac{P_f' k_{eP}}{D} \quad [13.44]$$

Biofilm Cell Carbon

$$(\mu - k_{eX})X_f' = 0 \quad [13.45]$$

or

$$\mu = k_{eX} \quad [13.46]$$

Biofilm Polymer Carbon

$$P_f' = \frac{q_P X_f'}{k_{eP}} \quad [13.47]$$

Thus, production rate (cells and EPS) equals erosion rate at steady state.

13.2.8 Other Model Considerations

13.2.8.1 Conditioning Film Formation of an organic conditioning film at the substratum may be a prerequisite for cellular adsorption. Quantitative data regarding organic adsorption and its influence on cellular adsorption are presented in Chapter 9. However, the mass of the conditioning film is negligible compared to the steady state biofilm mass.

13.2.8.2 Endogenous Respiration and Maintenance Energy Endogenous respiration and maintenance energy requirements have been neglected.

However, chemostat results (Figure 8.7, Chapter 8) for *Pseudomonas aeruginosa* indicate that one of the two processes may be significant. However, graphical procedures for determining maintenance energy (Figure 8.7) suggest that the maintenance energy requirement for combined biomass (cells plus EPS) is negligible. The line for cell carbon (Figure 8.7) has a nonzero intercept, but that substrate can be accounted for by EPS production.

13.2.8.3 *Spatial Distribution of Biofilm Components* A homogeneous distribution of cells and EPS in both the bulk liquid and the biofilm has been observed and has been predicted by mathematical models (Chapter 11). Transmission electron micrographs of a *P. aeruginosa* biofilm at steady state also suggest a cell distribution somewhere between uniform and random (Figure 1.4). The biofilm thickness is also presumed to be uniform. Bakke (1986), however, have observed significant changes in *P. aeruginosa* biofilm morphology with time in a laminar flow system. In the early stages of biofilm accumulation (50 h), the variation in biofilm thickness is small (Figure 10.6, Chapter 10). As the biofilm matures (272 h), "channels" form in the biofilm and the variation in film thickness becomes significant (Figure 10.7, Chapter 10). A continuous increase in the variation in biofilm thickness was observed in these experiments and reflects changing surface film morphology (Figure 10.8, Chapter 10). The increasing roughness of the biofilm undoubtedly influences mass transport at the biofilm–water interface. The morphology changes in biofilms accumulated in turbulent flow are also significant, as shown by observations of changing equivalent sand roughness (Figure 9.15, Chapter 9).

Changes in biofilm roughness also change the surface area of the biofilm–bulk water interface. The increased surface area may be reflected in increased attachment and detachment rates as well as increased substrate flux into the biofilm. Increased interfacial area also raises the probability that advective mechanisms will influence the transport of materials *within* the biofilm.

13.2.8.4 *Density of Biofilm Components* Constant cellular and EPS concentration in the biofilm are generally assumed for the biofilm. Consider the following relationship between cell concentration in the biofilm and biofilm thickness:

$$X_f = \epsilon_X \rho_X \qquad [13.48]$$

Now

$$\frac{dX_f'}{dt} = \frac{d(X_f L_f A/V)}{dt} \qquad [13.49]$$

Since the specific gravity of cells does not change ($d\rho_X/dt = 0$) and A/V is constant, the product rule for differentiation yields

$$\frac{dX_f'}{dt} = \frac{A}{V}\left(\rho_X L_f \frac{d\epsilon_X}{dt} + \epsilon_X \rho_X \frac{dL_f}{dt}\right) \qquad [13.50]$$

ϵ_X is the fraction of biofilm volume occupied by cells or, in other words, the ratio of cell volume in the biofilm, V_{fc}, to biofilm volume, AL_f:

$$\epsilon_X = V_{fc}/AL_f \qquad [13.51]$$

where V_{fc} = cell volume in the biofilm (L^3)

Bakke (1986) has shown (in laminar flow) that the thickness of a *P. aeruginosa* biofilm reaches a constant value very rapidly (24–40 h). However, the cell concentration in the biofilm, X_f, continues to increase (Figure 13.4). Then, since L_f is constant, X_f' continues to increase solely because the cell volume in the biofilm, V_{fc}, continues to increase:

$$\frac{dX_f'}{dt} = \frac{\rho_X}{V}\frac{dV_{fc}}{dt} \qquad [13.52]$$

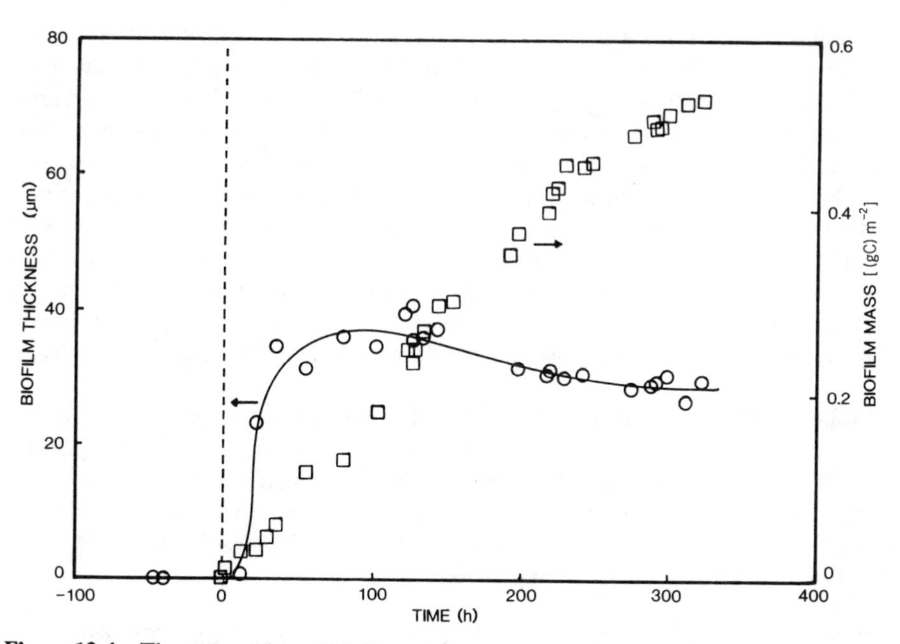

Figure 13.4 The progression of biofilm thickness (○) and biofilm cellular density (□) in a rectangular duct reactor in laminar flow. The biofilm thickness reaches steady state within 24 h. However, the biofilm cell density continues to increase. (Bakke, 1986.)

or, since the cell mass in the biofilm is

$$M_{fc} = \rho_X V_{fc} \qquad [13.53]$$

and ρ_X is constant,

$$\frac{dX_f}{dt} = \frac{1}{AL_f} \frac{dM_{fc}}{dt} \qquad [13.54]$$

where $M_{fc} = \rho_X V_{fc}$, the biofilm cell mass (M).

In undefined mixed populations, the biofilm density has also been observed to increase with increasing biofilm depth (Table 4.3, Chapter 4), increasing hydrodynamic shear stress (Figure 4.7, Chapter 4), and increasing substrate loading rate (Figure 4.8 and Table 4.4, Chapter 4).

13.2.8.5 Environmental Factors Effects of environmental factors, such as the temperature and the ionic composition of the medium, are not explicit in the model. Temperature, pH, and other parameters may influence both the kinetic and the stoichiometric coefficients. The influence of temperature and pH on biofilm accumulation is addressed in Chapter 14 for mixed population biofilms, but not for *P. aeruginosa* biofilms.

13.2.8.6 Calcium Concentration Turakhia and Characklis (1988) have observed decreasing erosion rate in a *P. aeruginosa* biofilm with increasing free calcium concentration in the growth medium (Figure 13.5). The decreased erosion rate resulted in higher biofilm accumulation with increased calcium concentration (Figure 13.6). Increased biofilm cellular concentration was responsible for the increased accumulation with increasing calcium concentration, since biofilm EPS changes were insignificant (Figure 6.5, Chapter 6). The calcium content of the biofilm also increased with increased calcium in the growth medium (Figure 6.4, Chapter 6). The observations indicate that calcium concentrates in the biofilm and probably improves the biofilm cohesive strength, possibly by strengthening intermolecular bonding between EPS molecules.

13.2.8.7 Hydrodynamic Shear Stress Bakke (1986) varied the hydrodynamic shear stress in a rectangular duct reactor containing *P. aeruginosa* biofilm. The shear stress, within a laminar flow regime ($Re \approx 5$–30), caused no observable lasting effect on biofilm thickness or biofilm density. In a mixed population biofilms, shear stress does affect biofilm thickness (Chapter 14) and biofilm density (Figure 4.7, Chapter 4).

13.2.8.8 Mass Transfer at the Biofilm–Water Interface The substrate concentration at the biofilm–water interface is assumed to equal the bulk water concentration (no external mass transfer limitation). This is a valid assumption

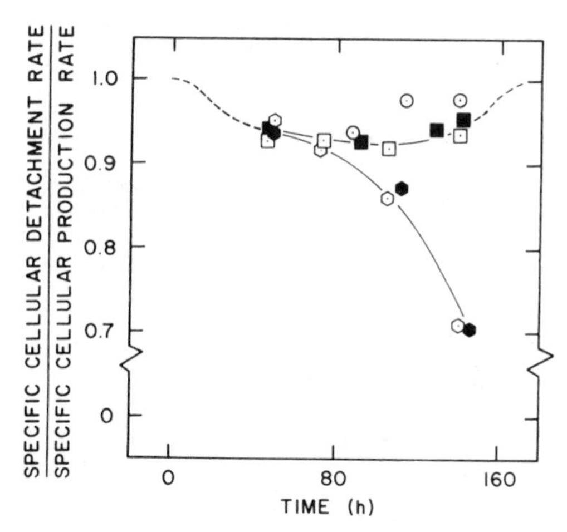

Figure 13.5 Ratio of specific cellular erosion rate to specific cellular growth rate in a *P. aeruginosa* biofilm accumulating in a turbulent RotoTorque reactor. Specific cellular erosion rate is always within 90% of specific growth rate for 0.4 (○) and 25 (□, ■) g m^{-3} calcium. However, for 50 g m^{-3} calcium (⬡, ⬢) the specific growth rate is approximately 30% higher, and consequently the biofilm accumulation rate is higher than at lower calcium concentrations (Turakhia and Characklis, 1988).

for the RotoTorque system at rotational speeds above 150 rpm, due to the vigorous bulk water compartment mixing (Figure 9.32, Chapter 9).

13.2.8.9 Numerical Integration Simultaneous numerical integration of the material balances is troublesome because the equations are inherently stiff (Chapter 11). Therefore, steady state is assumed for the biofilm substrate material balance: $dS/dt = 0$ throughout the progression. This is reasonable because of the short characteristic time for substrate diffusion in biofilms compared to the other processes. The time scale for diffusive transport is on the order of L_f^2/D, which, for a 100 μm thick biofilm and a diffusion coefficient of 2×10^9 m^2/s, gives 5 s, while the bulk liquid residence time is typically orders of magnitudes greater (Kissel, et al., 1984; this book, Chapter 11). In the RotoTorque experiments, the mean hydraulic residence time was 600 s.

13.2.9 Comparison of Model and Experiment The kinetic and stoichiometric coefficients used in the computer simulation were derived from experimental apparatus different from that in the experiments conducted in the RotoTorque reactors (Bakke et al., 1984). Adsorption coefficients were determined by Nelson et al. (1985) in a turbulent CFSTR (Section 12.3.4, Chapter 12). Trulear (1983) and Robinson et al. (1984) determined growth and product formation coefficients in chemostat experiments (Table 8.2, Chapter 8). Erosion coefficients were derived from steady state biofilm data (Trulear, 1983; Bakke et

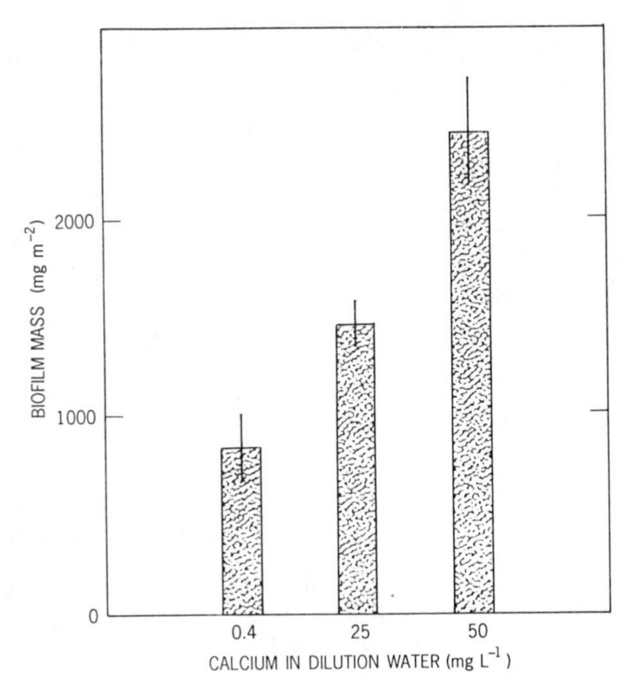

Figure 13.6 Biofilm areal concentration (after 140 h accumulation) increases with increasing calcium concentration in the growth medium in a turbulent RotoTorque reactor (Turakhia and Characklis, 1988).

al., 1984). Thus, coefficients were obtained from a distinct set of experimental systems (except for the erosion data) and were used to predict biofilm behavior in the RotoTorque system.

Experimentally observed bulk water substrate, cell, and EPS concentrations in the RotoTorque experiments are compared with the model predictions in Figure 13.7a, and biofilm cell and EPS concentrations are compared with the model predictions in Figure 13.7b. Based on the model, the influence of system variables on progression and steady state behavior in bioreactors can be predicted.

13.2.10 Results of the Modeling: Unsteady State

The operation of bioreactors can be controlled by regulating cell, substrate, and hydraulic loading rates through regulation of S_i and/or D. The influence of these parameters on biofilm cell progression is illustrated in Figures 13.8 and 13.9. Biofilm cell numbers are presented on a logarithmic scale to increase the range of S_i and D considered and to emphasize the subtle effect of the initial events in biofilm accumulation. The EPS and suspended cell concentration progressions are similar to those for X_f.

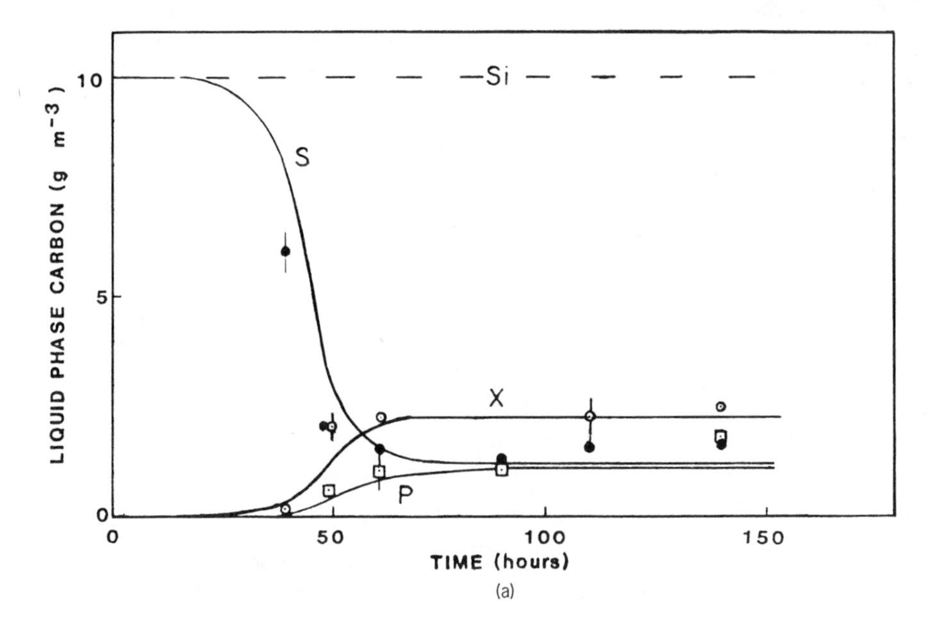

(a)

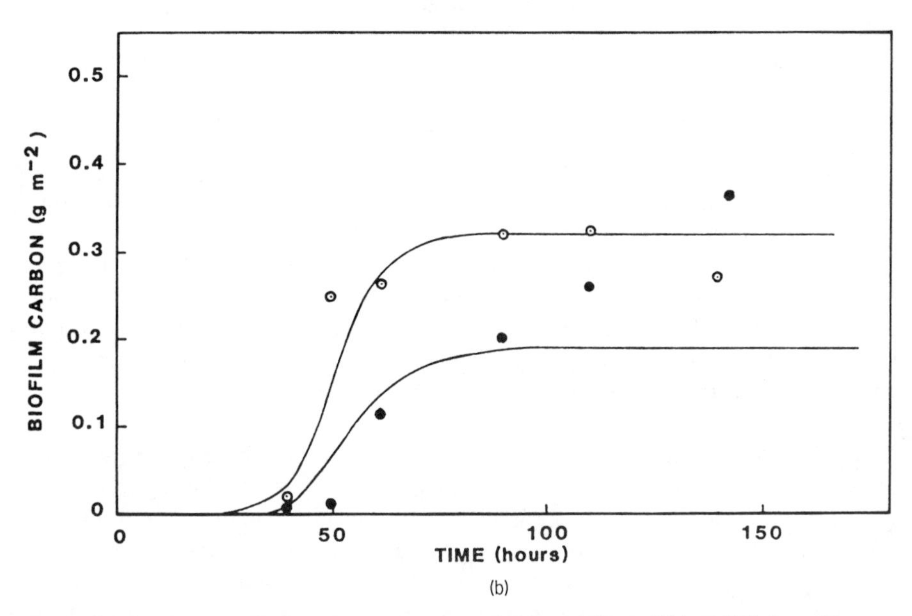

(b)

Figure 13.7 (a) Progression of water phase variables S (●), X (○), P (□) for a *P. aeruginosa* biofilm in a turbulent RotoTorque reactor as measured by Trulear (1983) and predicted by the model. (b) Progression of biofilm phase variables (X_f'' (○) and P_f'' (●)) for a *P. aeruginosa* biofilm in a turbulent RotoTorque reactor as measured by Trulear (1983) and predicted by the model. The dilution rate was 6 h^{-1}, and the temperature was 25°C.

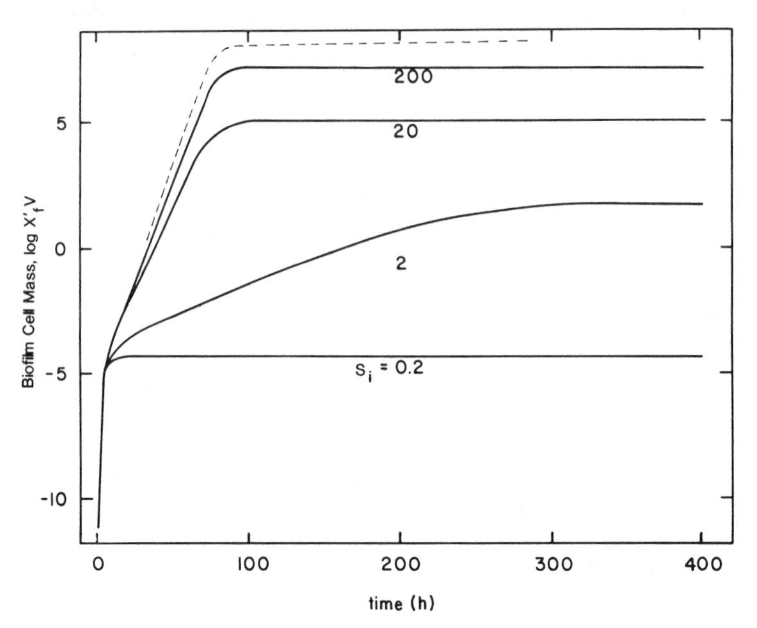

Figure 13.8 Predicted progression for biofilm cell concentration as a function of influent substrate concentration (S_i) in g m^{-3}. The dilution rate D was 10 h^{-1}, and the inlet cell concentration was 10^{12} cells m^{-3}. The dashed line reflects the influence of the potential diffusion resistance in the biofilm for $S_i = 200$ g m^{-3}.

13.2.10.1 Influent Substrate Concentration The influent substrate concentration S_i influences both the rate and the extent of biofilm accumulation (Figure 13.8). The rate of biofilm cell accumulation increases with increasing S_i at constant X_i and D. The substrate loading rate is proportional to S_i for constant D. Effects of diffusion limitations on biofilm accumulation are minimal (assuming oxygen is in excess), as indicated by the dashed lines in Figure 13.8, which represent the progression with no diffusion limitations.

13.2.10.2 Dilution Rate The dilution rate also affects both rate and extent of biofilm accumulation (Figure 13.9). The range of substrate loading rates simulated is the same in Figures 13.8 and 13.9. The resulting plateau biofilm accumulation is the same for a given loading rate in both illustrations. The progression, on the other hand, depends significantly on the manner in which loading rate is varied. For example, the biofilm cell progression for $S_i = 2$ in Figure 13.8 requires four times longer to reach "plateau" than the progression for $D = 1$ in Figure 13.9, even though the substrate loading for the two cases is the same. The initial phase of the biofilm accumulation, when cell adsorption is the dominating process, is not influenced much by D or S_i for the inlet cell concentrations used in the modeling.

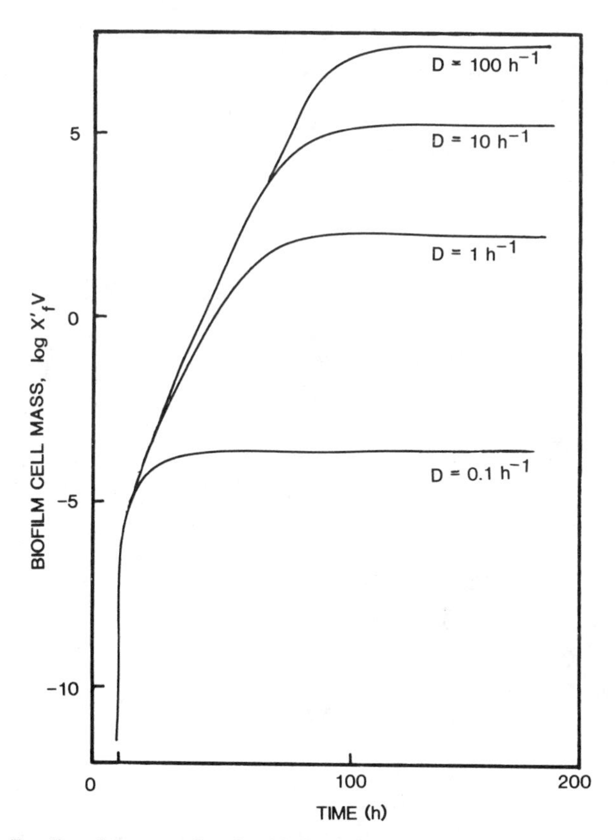

Figure 13.9 Predicted progression for biofilm areal cell concentration as a function of dilution rate D in h^{-1}. The influent substrate concentration S_i was 20 (g carbon) m^{-3}, and the inlet cell concentration was 10^{12} cells m^{-3}.

13.2.11 Results of the Modeling: Steady State

Many continuous flow biofilm reactors are maintained at steady state or pseudosteady state (accumulation rate is small compared to process rates). The definition of steady state, however, may not be readily apparent. Bakke (1986) observed steady state with respect to biofilm thickness within 24 h in experiments with *P. aeruginosa* in a laminar flow tube. Biofilm cell numbers, on the other hand, continued to increase for many days (Figure 13.4). Thus, the concentration of cells in the biofilm continues to increase while the biofilm thickness remains constant (Section 13.2.8).

D and S_i have a significant effect on biofilm reactor performance, and both variables can be controlled easily by the investigator. Cellular production rate in the reactor is a function of inflow substrate concentration for different dilution rates (Figure 13.10). The broken lines in the figure represent constant substrate loading rates into the reactor. The logarithm of the cell production rate increases

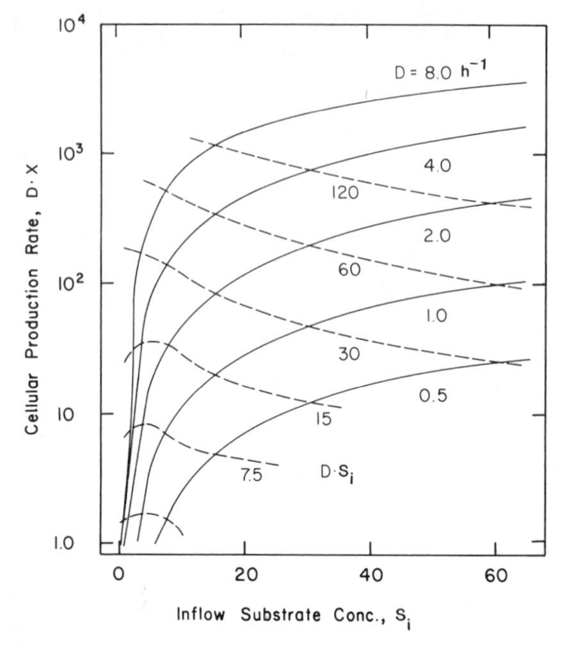

Figure 13.10 Predicted influence of substrate loading rate (dashed lines) and dilution rate (solid lines) on cellular production rate in a *P. aeruginosa* biofilm reactor at steady state.

with increasing S_i and dilution rate D. Thus, there are advantages to operating bioreactors at high dilution rate D, where biofilm, as opposed to dispersed biomass, is favored (Figure 13.10). For $D > 1$, washout of any dispersed population in a chemostat will occur. Both rate of production and effluent concentration of a product can be greatly increased for a given S_i by increasing D. Consider the cellular production rate for $S_i = 20$ in a chemostat with $D = 2$ in Figure 13.10. If S_i is doubled, the cellular production rate is approximately doubled as well. However if dilution rate is doubled, the cellular production rate increases by as much as five times. The results demonstrate that immobilized cell systems, such as biofilm reactors, could increase the productivity of some bioprocesses.

13.2.12 Summary of CFSTR Model

A mathematical model for accumulation and activity of a *P. aeruginosa* biofilm in a CFSTR (RotoTorque) has been presented, which is based on conservation principles, transport phenomena, and fundamental kinetic and stoichiometric relationships. The validity of the model was tested by comparing its predictions with experimental data. Experimental data from chemostat experiments were used to predict the progression of several variables in a biofilm reactor with reasonable success. However, there is a need to consider spatial and temporal variations within the biofilm if further progress is to be made. For example, biofilm density changes with time and undoubtedly influences biofilm activity and biofilm effects on other transport processes. Density may also change with biofilm depth. The *P. aeruginosa* biofilm thickness is not uniform especially after significant accumulation. With other microbial species such as *Klebsiella pneumoniae* (Section 10.5, Chapter 10), the biofilm thickness is extremely variable. The influence of variable biofilm thickness on fluid frictional resistance and mass transfer has been determined (Chapter 9), but other effects may be important.

Diffusional limitations within the *P. aeruginosa* biofilm never became significant. Is this characteristic of monopopulation biofilms? Indications are that other monopopulation biofilms can exhibit significant diffusional resistance. For example, Siebel (1987) experimented with *Klebsiella pneumoniae* and suggests that resistance to oxygen diffusion influences the metabolic activity of these biofilms. Perhaps the thickness of *Pseudomonas aeruginosa* biofilms is limited by oxygen diffusion, since *Pseudomonas aeruginosa* is an obligate aerobe.

Considerable more effort is required to provide a model that simulates all of the important processes and characteristics of biofilms and permits the accurate prediction of long term biofilm behavior.

13.3 A PLUG FLOW BIOFILM REACTOR (PFR)

There is an increasing need for modeling of biofilm processes in turbulent plug flow reactors. Pipelines, pipe distribution networks, heat exchangers, contaminated groundwater formations, and certain fixed film wastewater treatment reactors (e.g., trickling filters or packed towers) all exhibit plug flow characteristics. The building blocks for a plug flow biofilm reactor model will not differ much from the CFSTR model. The compartments, components, stoichiometry, and kinetics all remain the same. The only major difference is in the advective transport term. In the plug flow model, concentrations vary with space in the reactor in the direction of flow.

13.3.1 The Material Balances

Less attention has been given to the fundamental biofilm processes as they occur in plug flow reactors. Kirkpatrick et al. (1980) modeled simultaneous mass and heat transfer in a circular tube containing a biofilm. They considered both laminar and turbulent flow. However, they only considered two processes in their model: (1) substrate transport in the bulk water (diffusive and advective) and biofilm (diffusive) compartments and (2) microbial growth within the biofilm at the expense of substrate leading to biofilm accumulation. No detachment processes were considered. The differential material balance for a plug flow reactor, in contrast to the CFSTR, contains the advective terms. In cylindrical coordinates with axial symmetry, Eq. 13.2 becomes

$$\frac{1}{r}\frac{\partial}{\partial r}\left(Dr\frac{\partial S}{\partial r}\right)+\frac{\partial}{\partial z}\left(D\frac{\partial S}{\partial z}\right)+r_s=v_z\frac{\partial S}{\partial z}+v_r\frac{\partial S}{\partial v}+\frac{\partial S}{\partial t} \qquad [13.55]$$

The flow is presumed to be fully developed upon entering the biofilm-coated tube section. The biofilm is very thin compared to the inner radius of the tube, and net bulk flow of water into the biofilm is negligible. Axial diffusion is assumed to be negligible, since radial concentration gradients are expected to be much higher than axial gradients. It is also reasonable to presume that the concentration gradients establish themselves much faster than the biofilm thickness changes (Section 11.3.4.1, Chapter 11). Thus, the time derivative in Eq. 13.55 is zero. The following equation results:

$$\frac{1}{r}\frac{\partial}{\partial r}\left(Dr\frac{\partial S}{\partial r}\right)+r_s=v_z\frac{\partial S}{\partial z} \qquad [13.56]$$

The following dimensionless variables can be defined (Figure 13.11):

$$S^*=\frac{S}{S_i},\qquad r^*=\frac{r}{r_t},\qquad z^*=\frac{zD_{Sf}}{r_t^2 v_m} \qquad [13.57]$$

where r_t = inner radius of the tube (L)
v_m = mean velocity of the water (Lt^{-1})
S_i = inlet concentration (ML^{-3})
D_{Sf} = molecular substrate diffusivity in the biofilm (L^2t^{-1})

Then, inserting the dimensionless variables in Eq. 13.56,

$$\frac{1}{Dr^*}\frac{\partial}{\partial r^*}\left(Dr^*\frac{\partial S^*}{\partial r^*}\right)+\frac{r_s r_t^2}{S_i D}=\frac{v_z}{v_m}\frac{D_{Sf}}{D}\frac{\partial S^*}{\partial z^*} \qquad [13.58]$$

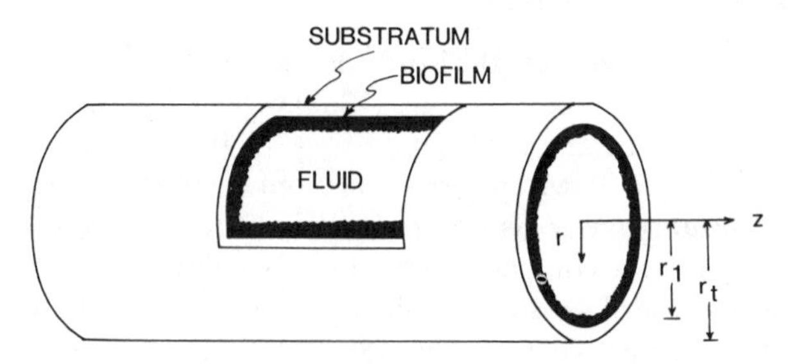

Figure 13.11 A schematic diagram of the plug flow reactor system.

Presuming the reaction only occurs in the biofilm, Eq. 13.58 for the water phase becomes

$$\frac{1}{r^*}\frac{\partial}{\partial r^*}\left[D_{Tb}r^*\frac{\partial S^*}{\partial r^*}\right] = \frac{v_z}{v_m}\frac{D_{Sf}}{D_{Tb}}\frac{\partial S^*}{\partial z^*} \tag{13.59}$$

where D_{Tb} = turbulent and molecular substrate diffusivity in water (L^2t^{-1}).

In the biofilm, mass transfer is only by molecular diffusion, so, presuming the biofilm is homogeneous, the diffusivity is constant and Eq. 13.46 becomes

$$\frac{1}{r^*}\frac{\partial}{\partial r^*}\left(r^*\frac{\partial S^*}{\partial r^*}\right) = -\frac{r_s r_t^2}{S_i D_{Sf}} \tag{13.60}$$

Using saturation kinetics,

$$r_S = \frac{k_1 S}{1 + k_2 S} \tag{13.61}$$

where $k_1 = \mu_{max} X_f / L_f Y_{X/S} K_S$ (t^{-1})
$k_2 = 1/K_S$ (L^3M^{-1})

In dimensionless form, Eq. 13.61 becomes

$$\frac{r_S r_t^2}{D_{Sf} S_i} = -\frac{k_1^* S^*}{1 + k_2^* S^*} \tag{13.62}$$

where $k_1^* = k_1 r_t^2 / D_{Sf}$
$k_2^* = k_2 S_i$

The water enters the tube with concentration S_i, which determines the initial condition:

$$S = S_i \quad \text{at} \quad z = 0 \tag{13.63}$$

The assumption of axial symmetry indicates a no flux boundary condition at the centerline:

$$\frac{\partial S}{\partial r} = 0 \quad \text{at} \quad r = 0 \tag{13.64}$$

Given the pseudosteady state approximation, the fluxes into and out of the biofilm–water interface ($r = r_1$) must be equal:

$$\left[D_{Sb}\frac{\partial S}{\partial r}\right]_{-r_1} = \left[D_{Sf}\frac{\partial S}{\partial r}\right]_{+r_1} \quad \text{at} \quad r = r_1 \tag{13.65}$$

The concentrations on each side of the interface must also be in equilibrium, so

$$S_{-r_1} = \alpha_P S_{+r_1} \quad \text{at} \quad r = r_1 \tag{13.66}$$

where α_P = partition coefficient (—). The tube wall is impermeable, so

$$\frac{\partial S}{\partial r} = 0 \quad \text{at} \quad r = r_t \tag{13.67}$$

13.3.2 Parameter Values

Glucose is the limiting nutrient and sole carbon source. Thus, the general stoichiometry described in Section 13.2.4 applies. The following parameters were used in the simulations:

$$k_1 = 0.06 \text{ s}^{-1}$$

$$k_2 = 0.008 \text{ m}^3 \text{ g}^{-1}$$

$$S_i = 100 \text{ g m}^{-3}$$

$$\alpha_P = 1$$

$$D_{Sb} = 6 \times 10^{-10} \text{ m}^2 \text{ s}^{-1} \qquad \text{(Table 9.6, Chapter 9)}$$

$$D_{Sf} = 1.5 \times 10^{-10} \text{ m}^2 \text{ s}^{-1} \qquad \text{(Table 4.7, Chapter 4)}$$

A thin biofilm (25 μm) was presumed initially present. Biofilm detachment was not considered, so the biofilm accumulation did not reach a steady state. Thus, computation ended when biofilm thickness reached 250 μm. The tube radius was 11 mm. For turbulent calculations, the mean velocity was 1.5 m s^{-1}.

13.3.3 Results of the Modeling

13.3.3.1 Laminar Flow In laminar flow, only diffusive transport is considered for the bulk water. The axial substrate concentration gradient remains relatively constant with time and increasing biofilm thickness (Figure 13.12). The calculated profile (Figure 13.12) does not change significantly and is valid for biofilm thicknesses ranging from 25 μm (zero time) to 250 μm. The low diffusivity in the bulk water results in biofilm accumulation being controlled by bulk water transport of the glucose to the biofilm as reflected by the radial concentration profiles (Figure 13.13). Crawford (1987) has largely verified this observation in a capillary biofilm experimental system.

13.3.3.2 Turbulent Flow Bulk water transport in a turbulent flow system is dominated by eddy diffusivity. In contrast to the laminar flow system, there is a change in the axial substrate concentration profile as the biofilm thickness in-

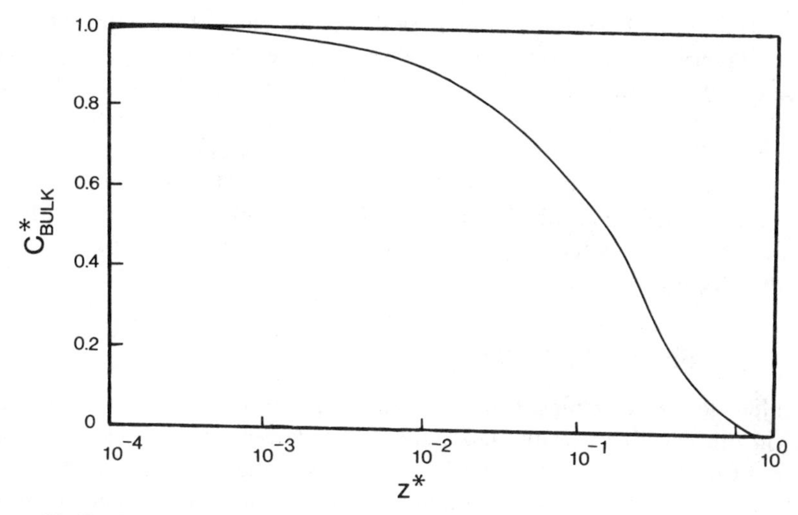

Figure 13.12 The change in dimensionless bulk water substrate concentration C^*_{BULK} with dimensionless axial distance z^* for laminar flow (reprinted with permission from Kirkpatrick et al., 1980).

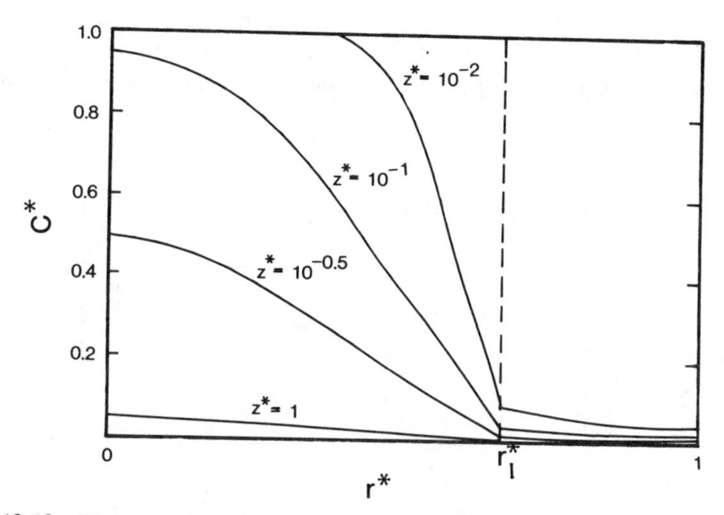

Figure 13.13 Dimensionless radial concentration [$C^*(r^*)$] profiles as a function of axial distance z^* for laminar flow (reprinted with permission from Kirkpatrick et al., 1980).

creases (Figure 13.14) in the turbulent flow system. The results indicate that the principal resistance to substrate utilization resides in the biofilm, as illustrated by the radial concentration profiles (Figure 13.15), which indicate that the water is well mixed while the steep concentration gradients are in the biofilm. The results also indicate that the bulk water substrate concentration decreases less than 1% over the first 30 m ($z^* = 10^{-4}$) of tube. Thus, the biofilm thickness is essentially constant over that distance. The biofilm thickness increases with time and decreases with increasing axial distance, since substrate is depleted at long distances from the entrance (Figure 13.16).

13.3.4 Summary of PFR Model

The plug flow reactor model indicates the primary resistances to transport of mass in a tube with an accumulating biofilm. In laminar flow, substrate utilization is diffusion-limited in the water phase, since radial mass transport is solely by molecular diffusion. In turbulent flow, radial mass transport is much higher, due to eddy diffusion, and the principal resistance lies in the biofilm. Initially, biofilm substrate utilization is limited by reaction kinetics. As the biofilm thickness increases, the bulk substrate concentration profile changes, since the thicker film utilizes more substrate. If growth continues without biofilm removal (e.g., by detachment), substrate only penetrates a portion of the biofilm.

Many of the concerns raised for the CFSTR model are also relevant for the PFR. In addition, the PFR model is restricted to growth processes and the resistances to substrate mass transfer. Other processes, especially detachment processes, must be incorporated to simulate biofilm behavior in a plug flow reactor

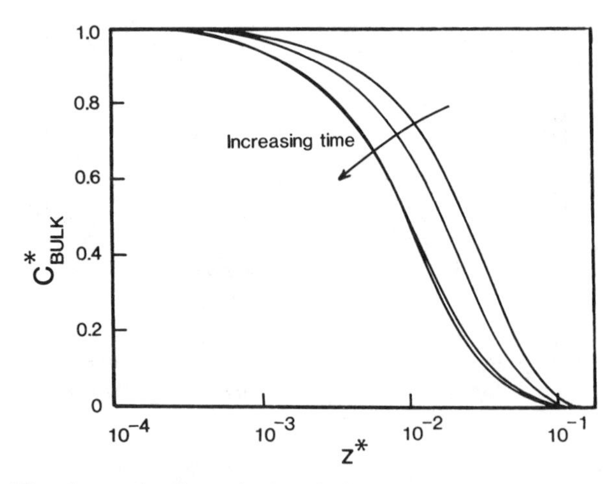

Figure 13.14 The change in dimensionless bulk water substrate concentration C^*_{BULK} with dimensionless axial distance Z^* for turbulent flow (reprinted with permission from Kirkpatrick et al., 1980).

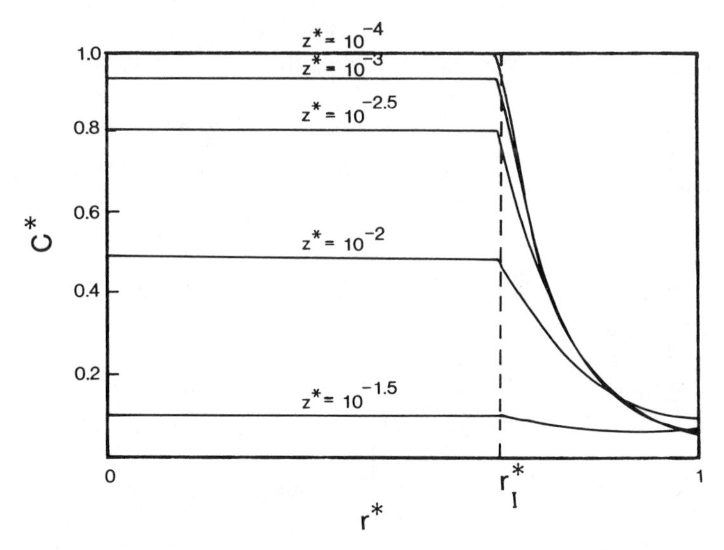

Figure 13.15 Dimensionless radial concentration profiles [$C^*(r^*)$] as a function of axial distance z^* for turbulent flow (reprinted with permission from Kirkpatrick et al., 1980).

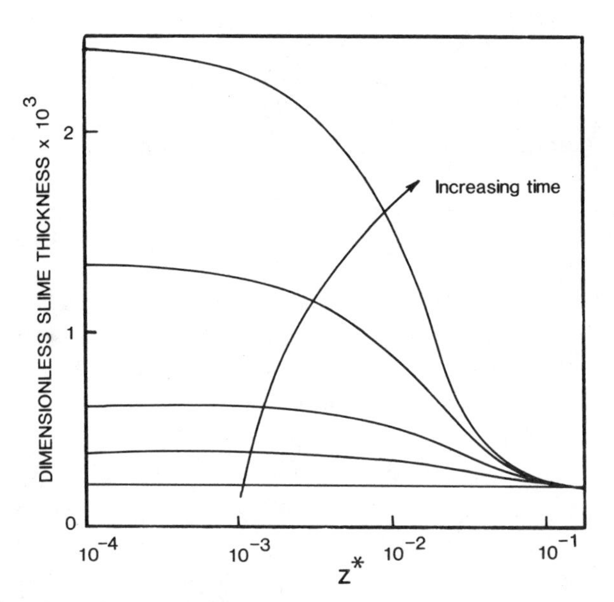

Figure 13.16 Dimensionless biofilm thickness as a function of axial distance z^* and time in turbulent flow (reprinted with permission from Kirkpatrick et al., 1980).

more effectively. Certainly a second reactant—oxygen—should be considered, since oxygen transport will frequently limit substrate utilization and growth before substrate transport.

REFERENCES

Atkinson, B., *Biochemical Reactors,* Pion Ltd., London, 1974.

Bakke, R., "Biofilm detachment," unpublished doctoral thesis, Montana State University, Bozeman, MT, 1986.

Bakke, R., M.G. Trulear, J.A. Robinson, and W.G. Characklis, *Biotech. Bioeng.,* **26**, 1418-1424 (1984).

Busch, A.W., *Aerobic Biological Treatment of Waste Waters,* Gulf Publishing Co., Houston, TX, 1971.

Crawford, D., "Hydraulic effects of biofilm accumulation in simulated porous media flow system," unpublished master's thesis, Montana State University, Bozeman, MT, 1987.

Kirkpatrick, J.P., L.V. McIntire, and W.G. Characklis, *Wat. Res.,* **14**, 117-127 (1980).

Kissel, J.C., P.L. McCarty, and R.L. Street, *J. Env. Eng.,* **110**, 393 (1984).

Langmuir, I., *J. Am. Chem. Soc.,* **40**, 1361 (1918).

Nelson, C.H., J.A. Robinson, and W.G. Characklis, *Biotech. Bioeng.,* **27**, 1662 (1985).

Rittmann, B.E., *Biotech. Bioeng.,* **24**, 501-506 (1982).

Robinson, J.A., M.G. Trulear, and W.G. Characklis, *Biotech. Bioeng.,* **26**, 1409-1417 (1984).

Siebel, M.A., "Binary population biofilms," unpublished doctoral thesis, Montana State University, Bozeman, MT, 1987.

Trulear, M.G., "Cellular reproduction and extracellular polymer formation in the development of biofilms," unpublished doctoral thesis, Montana State University, Bozeman, MT, 1983.

Trulear, M.G. and W.G. Characklis, *J. Wat. Poll. Contr. Fed.,* **54**, 1288 (1982).

Turakhia, M.H. and Characklis, W.G., *Biotech. Bioeng.,* **33**, 406-414 (1989).

Turakhia, M.H., "The influence of calcium on biofilm processes," unpublished doctoral thesis, Montana State University, Bozeman, MT, 1986.

PART V

BIOFILM TECHNOLOGY

14

MICROBIAL FOULING

WILLIAM G. CHARACKLIS
Montana State University, Bozeman, Montana

SUMMARY

Biofouling is the undesirable accumulation of a microbiological deposit on a surface. The fouling biofilm consists of an organic film composed of microorganisms embedded in a polymer matrix of their own making. Complex biofouling deposits, like those found in industrial environments, often consist of biofilms in intimate association with inorganic particles, crystalline precipitates or scale, and/or corrosion products. These complex *deposits* often form more rapidly and are more tightly bound than biofilm alone. Since the biofouling accumulation is not often as well defined as those described in previous chapters, the accumulation is referred to as a *deposit* to reflect its varied composition, which includes a significant amount of inorganic as well as organic and biotic substances.

The biofouling process is characterized by a *log rate* and an *extent,* each influenced by a variety of environmental and operating variables. This chapter focuses on the influence of several important variables on deposit accumulation measured as deposit thickness, deposit mass, hydrodynamic frictional resistance, and heat transfer resistance.

A modeling approach, such as that described in Chapters 11–13, is necessary to examine the effects of operating variables such as temperature, flow rate, and differences in nutrient supply on the dynamics of fouling deposit accumulation. Experimental and operational fouling monitors, preferably with more sensitivity than those available today, will serve to calibrate the model for the given application (season and site). Successful application of modeling depends on adequate representations of the processes, whose elucidation will necessarily be subjects for further laboratory research and field testing.

14.1 INTRODUCTION

Biofouling refers to the undesirable accumulation of a biotic deposit on a surface. The deposit may contain micro- and macroorganisms. The focus of this chapter is microbial fouling biofilms, which consist of an organic film composed of microorganisms embedded in a polymer matrix of their own making. The composite of microbial cells and EPS is termed a *biofilm*. Previous chapters have described the behavior of biofilms in well-defined experimental systems. This chapter includes many more data from field systems in which the experimental conditions are not as well defined. As a result, the individual processes contributing to biofilm accumulation frequently cannot be delineated. In addition, the surface accumulation often contains significant quantities of inorganic materials. Complex fouling deposits, like those found in industrial environments, often consist of biofilms in intimate association with inorganic particles (Zelver et al., 1982), crystalline precipitates or scale (Turakhia and Characklis, 1988), or corrosion products (Characklis et al., 1984). These complex *deposits* often form more rapidly and are more tightly bound than biofilm alone.

Thus, while Chapters 7 through 10 described biofilm processes, their kinetics, and their stoichiometry in terms of intensive variables, this chapter must generally describe observations in terms of performance parameters such as heat transfer resistance or fluid frictional resistance.

14.2 THE OPERATING PLANT ENVIRONMENT

An industrial operation contains numerous environments where corrosion and fouling can occur, including cooling water systems (recirculating and once through), storage tanks, water and wastewater treatment facilities, filters, and piping (Figure 14.1). Microbial fouling and corrosion also occur on ship hulls, reverse osmosis membranes (Ridgway et al., 1984), porous media (e.g., groundwater or oil-bearing formations), ion exchangers (Flemming, 1987), and drinking water distribution systems (Characklis, 1988), to name a few. Biofouling has been reported in turbulent flows and stagnant waters, on smooth surfaces and crevices, and on metals, concrete, and numerous other substrata.

Biofouling frequently occurs in conjunction with other types of fouling including crystalline or precipitation fouling (e.g., scaling) and particulate fouling (frequently due to sedimentation). Figure 14.2 schematically illustrates processes occurring in a hypothetical drinking water conduit—in the bulk water phase as well as at the liquid-solid interface. The diagram also depicts "under-deposit" corrosion that is frequently reported in relation to many types of deposit (Chapter 16). Microbial activity has been found in calcareous deposits, in tubercles, and in deposits of particulate material resulting from sedimentation or adsorption. Increased corrosion rates are frequently reported in these circumstances.

A recirculating cooling tower system (Figure 14.3) provides an illustration of

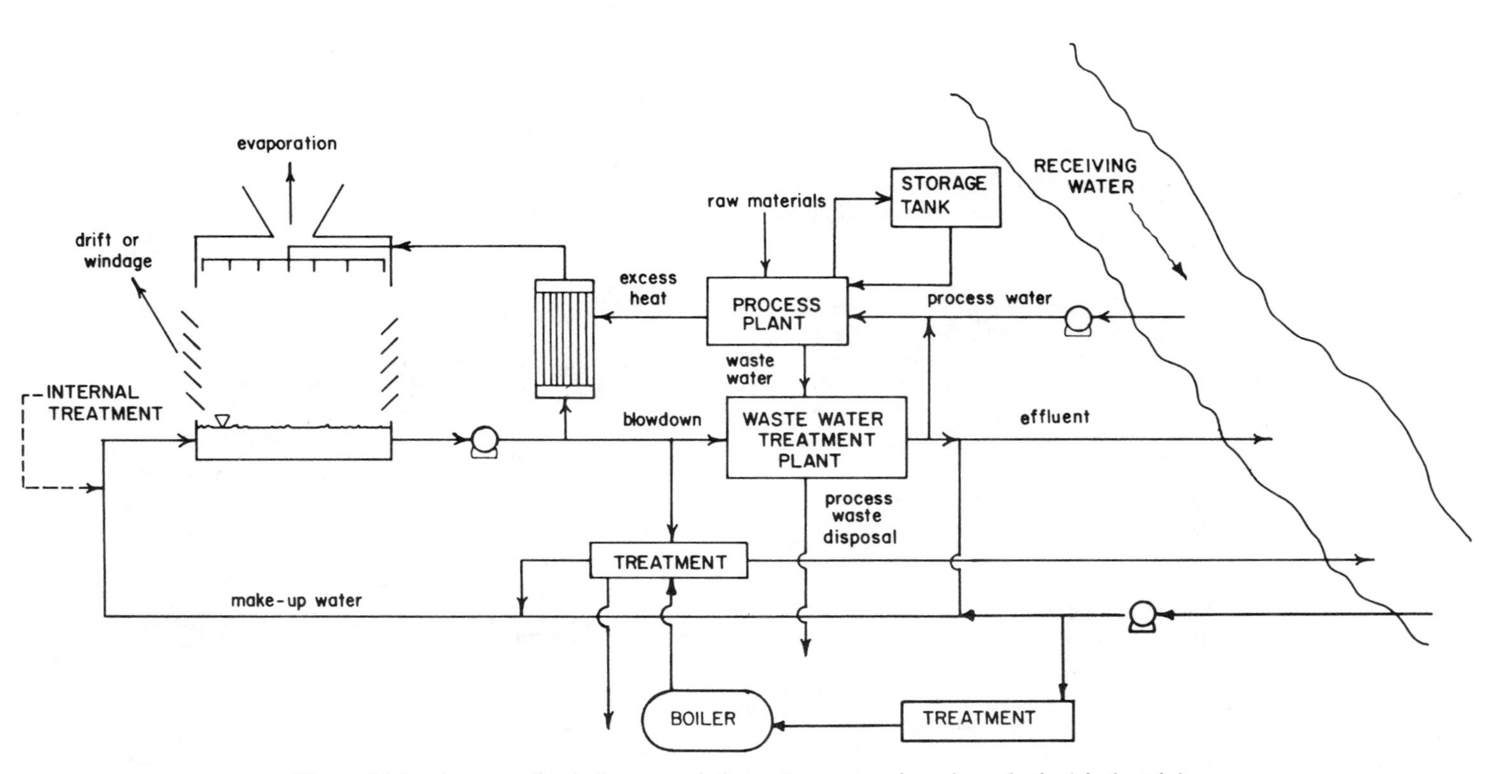

Figure 14.1 A generalized diagram of the water system in a hypothetical industrial facility.

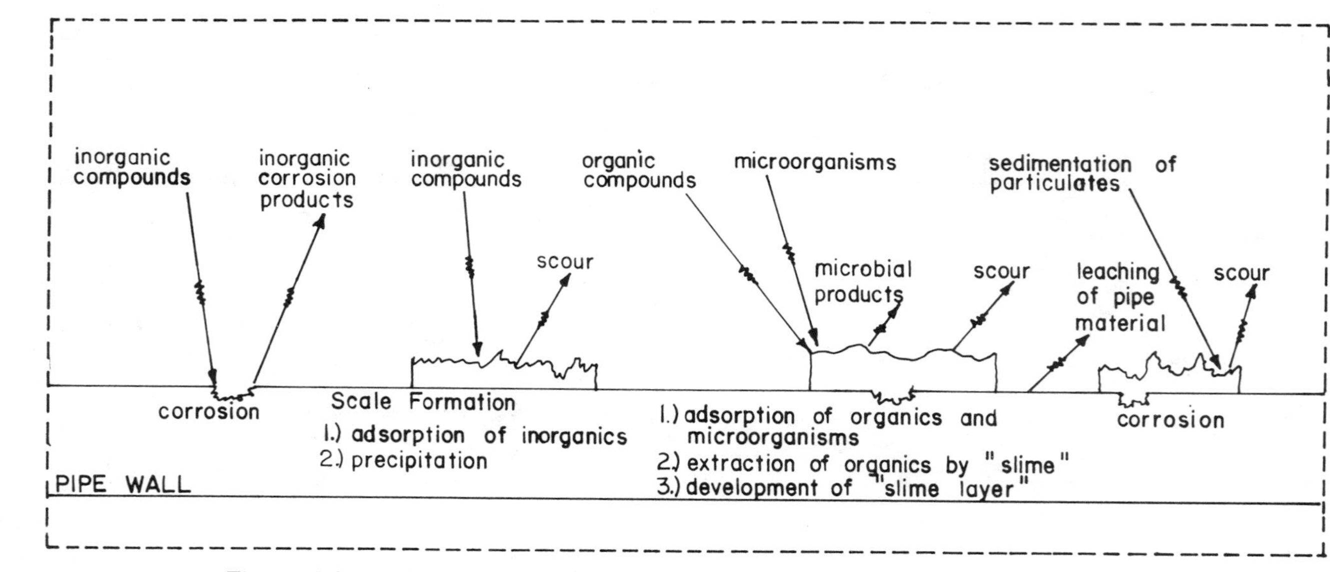

Figure 14.2 A schematic representation of fouling processes occurring in a water conduit. Transport of inorganic materials, organic compounds, and microorganisms to the pipe wall leads to their adsorption and accumulation. Transformations (e.g., corrosion, "underdeposit" corrosion, and microbial growth) result in products that further contribute to the accumulation of the fouling deposit. Detachment of portions of the fouling deposit leads to deterioration of water quality.

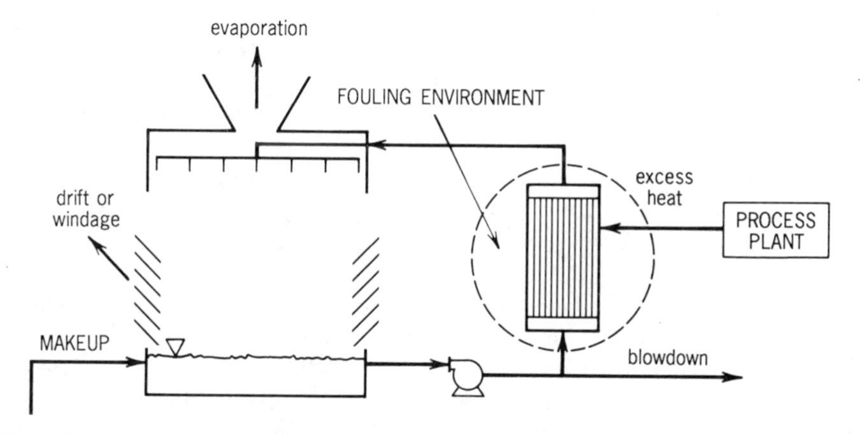

Figure 14.3 A schematic diagram of a forced draft, recirculating cooling tower system indicating the various water flows into and out of the system.

a plant environment in which fouling and corrosion can occur. The evaporative losses of water results in a concentration effect that increases nutrient concentration in the bulk water. The hydraulic residence time, water temperature, and ratio of surface area to volume are all high. As a consequence, microbial growth rates and cell numbers can be very high. If the fill material is wood, its degradation rate can be significant due to microbial degradation (fungi) or measures to control microbial activity (chlorine). In heat exchangers, scaling (under certain conditions) and microbial film development on the tubes and areas around flow obstructions can increase heat transfer resistance and (metal) corrosion rates.

14.3 FOULING: DEFINITIONS AND DESCRIPTION

Fouling is the formation of deposits on equipment surfaces that significantly degrades equipment performance and/or the useful life of the equipment. Several types of fouling, and their combinations, may occur (Epstein, 1981):

1. *Biological Fouling*: The accumulation and metabolism of macroorganisms (macrobial fouling) and/or microorganisms (microbial fouling).
2. *Chemical Reaction Fouling*: Deposits formed by chemical reaction in which the substratum (e.g., condenser tube) is not a reactant. Polymerization of petroleum refinery feedstocks is an important example of this type fouling.
3. *Corrosion Fouling*: The substratum itself reacts with compounds in the liquid phase to produce a deposit.
4. *Freezing Fouling*: Solidification of a liquid or some of its higher melting point constituents on a cooled surface.

5. *Particulate Fouling*: Accumulation on the equipment surface of finely divided solids suspended in the process fluid. *Sedimentation fouling* is an appropriate term if gravity is the primary mechanism for deposition.
6. *Precipitation Fouling*: Precipitation of dissolved substances on the equipment surface. This process is termed *scaling* if the dissolved substances have inverse temperature-solubility characteristics (e.g., $CaCO_3$) and the precipitation occurs on a superheated surface.

In most operating plant environments, more than one type of fouling will be occurring simultaneously. For example, microbial fouling is not limited to processes related to biological activity. It also includes the *combined* result of microbial activity and physicochemical processes in the associated slime layer with the chemical changes at the equipment surface and chemical reactions within the bulk fluid. The interaction can enhance some of the more commonly observed phenomena such as particulate, sedimentation, and corrosion fouling. Because of its complicated composition, the accumulated material will be termed a *deposit* unless data are available to classify it further.

The interactions between the various types of fouling are poorly understood and consequently provide a challenge in diagnosis and treatment.

14.4 PROBLEMS CAUSED BY FOULING

Fouling biofilms impair the performance of process equipment. They can form on any surface in contact with a process fluid. The economic consequences of fouling are the essential reason for industrial interest in the fouling of operating equipment. To assess the importance of a fouling situation, the economic and energy penalties arising from the operation of equipment subject to fouling must be considered. The deleterious effects of fouling include the following:

1. Energy losses due to increased fluid frictional resistance (e.g., in pipelines, on ship hulls and propellers, and in porous media such as oil and water wells or filters) and increased heat transfer resistance (e.g., power plant condensers and process heat exchangers). Zelver et al. (1982) have documented a case of biofouling at a nuclear power plant in which heat transfer rate in a fan cooler decreased by 30% due to biofouling in a 30 day period.

2. Increased capital costs for excess equipment capacity (e.g., excess surface area in heat exchangers) to allow for fouling. At a Canadian power plant site, biofouling decreased the heat transfer rate in a condenser by 30% over a two month period (Characklis and Robinson, 1984). The plant design had allowed for a 15% decrease due to fouling.

3. Increased capital costs for premature replacement of equipment experiencing severe underdeposit corrosion. Recently in a nuclear power plant, a con-

denser had to be replaced after approximately 6 years of operation because of severe corrosion attributed partially to microbial activity. The condenser had a design lifetime of approximately 20 years.

4. Unscheduled turnarounds or downtime, resulting in loss of production, to clean equipment that fouled at an unanticipated rate. Downtime can cost a power plant as much as $1 million per day because the utility must purchase the power from elsewhere to serve its customers. In oil production, downtime relates directly to the amount of product being shipped, which could amount to $10 million per day.

5. Quality control problems resulting from fouling of heat exchange equipment or fouling of product stream (e.g., sliming of paper mills or rolled steel).

6. Safety problems. Fouling of service water systems in nuclear power plants is a major concern because it reduces the heat transfer capacity available during an emergency or during an accident. Fouling in drinking water distribution systems may lead to high concentrations of microbes in the drinking water, which may affect public health. Deaths due to Legionnaire's disease have been attributed in several instances to fouling in cooling towers.

The anticipated presence of significant fouling can alter the size and other design features of operating equipment. The operation of equipment subject to fouling is constrained by the need to formulate economically justifiable cleaning schedules and internal treatment programs.

An estimate of the economic consequences of fouling was presented by Pritchard (1981) for fouling in Britain and suggests the cost was $600–1000-million per year (about 0.5% of the British 1976 GNP). van Nostrand et al. (1981) have estimated that for petroleum refining in the non-Communist countries, the total cost of fouling is $4.4 billion per year. The costs are bound to increase with increasing fuel and material costs. Fouling in heat exchangers alone may cost the United States billions of dollars annually (Lund and Sandu, 1981).

In industrial equipment, fouling of surfaces can be the main cause of progressive reduction in performance and efficiency. Accumulation of slime, dirt, and debris in the industrial environment rarely causes concern, but has accounted for whatever attention has been paid to fouling in the past. However, the costs of problems related to fouling are now more than significant and must be emphasized to designers and operators of equipment.

14.5 A RATIONAL APPROACH TO SOLVING FOULING PROBLEMS

Fouling is a complex phenomenon resulting from several processes occurring in parallel and in series. The *rate* and *extent* of these processes, in turn, are influenced by numerous physical, chemical, and biological factors in the immediate environment of the substratum. Many laboratory experiments and field observations have resulted in volumes of data without leading to relationships of wide, general use. A conceptual framework for describing fouling processes would be beneficial in interpreting available historical data and be invaluable in designing future experimental fouling tests. If the conceptual framework could be stated in mathematical terms, more benefits would accrue, including the ability to mathematically simulate fouling processes on the computer—computer "experiments" frequently being less expensive than laboratory experiments. The simulation can be supported by a large database and a logic program to form an *expert system*.

A rational approach, as contrasted with an empirical approach, develops the conceptual framework by resorting to fundamental processes that are reasonably well understood.

A rational approach to fouling entails a process analysis that identifies the contributing fundamental processes and determines the influence of process variables on process rate and extent. The approach requires experimental measurement techniques that permit the elucidation of the specific processes. Many of these fundamental processes have been described mathematically in Chapter 9. The mathematical description of the individual processes can be combined to develop models to extrapolate and generalize experimental results.

The difficulty in generalizing or extrapolating experimental fouling data is related to the complexity of the process, which frequently involves heat transfer, mass transfer, momentum transfer, and physical, chemical, and biological processes at the surfaces. The goal of a rational approach is to elucidate the fundamental processes that contribute to the fouling. Once these fundamental processes are properly understood, they are incorporated into a mathematical model of the overall fouling process. The model validation requires experiments designed specifically to investigate particular fundamental processes, rather than experiments that investigate only the overall deposit accumulation process. Such fundamental experiments frequently require more effort than ones that simply observe the overall fouling, but they ultimately lead to results that can be applied with greater confidence to a wider range of fouling situations.

The tools of process analysis have been described in Chapter 2. Burrill (1981) presents an illustration of this technique in explaining a practical problem of fouling and the decisions necessary to minimize its effects. These procedures can also lead to computer simulation "experiments," which can be used to test design concepts such as the influence of metallurgy, shear stress, heat flux, geometry, etc., on fouling processes and their influence on equipment performance. Simulation can also be used to test operating and maintenance procedures such as internal treatments and cleaning schedules. The process analysis technique may also lead to a more systematic method for developing and evaluating fouling control techniques, regardless of whether they employ physical, chemical, or even biological methods.

14.5.1 Defining Fouling Rate and Extent

Fouling biofilm accumulation can be considered the net result of the following physical, chemical, and biological processes (Chapter 7): (1) transport of soluble and particulate components to the wetted substratum, (2) their net adsorption to the wetted substratum, (3) chemical or microbial reaction at the substratum or within the deposit, and (4) net detachment, sloughing, or spalling of portions of the deposit from the wetted substratum.

The overall result is a sequence of events generally characterized by a sigmoidal-shaped progression including three identifiable periods (Figure 14.4): (1) an induction period in which very small changes in accumulation are detectable, (2) an exponential increase period, which is characterized by the *logarithmic rate* (or log rate) of accumulation and is used to represent the kinetics of net accumulation, and (3) an asymptotic or plateau period, which is a criterion for the *extent* of net accumulation.

Chapters 7–10 have described the progression of biofilm accumulation as the result of several distinct processes. The experimental methods and techniques necessary to distinguish the rate of individual contributing processes as described in Chapter 7 are sometimes difficult. Yet, characterization of the *rate*

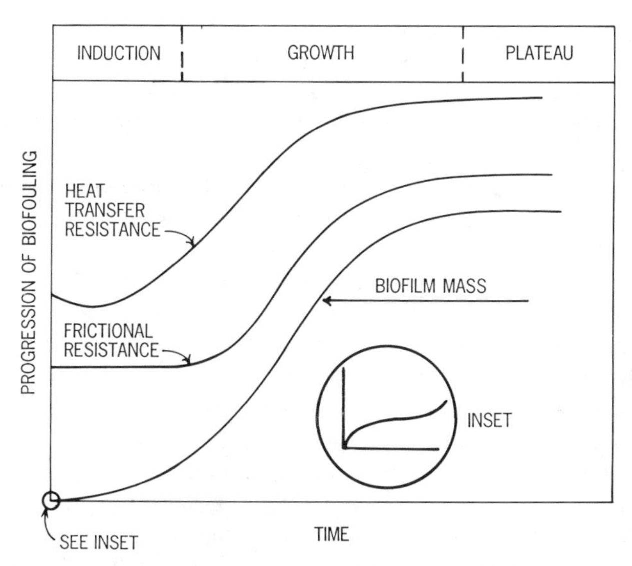

Figure 14.4 The progression of biofouling is generally described by a sigmoidal function (Eq. 14.1) when expressed in terms of deposit accumulation, hydrodynamic frictional resistance, or heat transfer resistance. Three periods or phases can be arbitrarily defined: (1) the *induction* period, during which little change in accumulation occurs, (2) the *log rate* accumulation period, and (3) the *plateau* period, when accumulation is constant and represents the extent of accumulation.

and *extent* of net accumulation is necessary for evaluation of many field tests in which the more fundamental processes cannot be determined.

The sigmoidal progression can be described by the logistic equation (Eq. 8.10, Chapter 8):

$$\frac{dX_f}{dt} = kX_f(1 - k'X_f) \quad [14.1]$$

where X_f = deposit concentration (arbitrary units)
k = rate coefficient (t^{-1})
k' = saturation coefficient (arbitrary units)

Using the logistic equation, the *log rate* of biofouling deposit accumulation can be defined in terms of the rate coefficient k. The *log rate* dominates early in the progression when $k'X_f \ll 1$. The *extent* of biofouling deposit accumulation can be represented by $1/X_{f,\max}$ or the coefficient k' and is determined at steady state when $k' = 1/X_f$. Thus, k and k' (or $1/X_{f,\max}$) are used in this chapter as coarse indicators of system kinetics and stoichiometry to describe factors that influence net biofouling deposit accumulation. The terms *log rate* and *extent,* when used in the specific context of net biofouling deposit accumulation, are italicized to emphasize the distinction between these two characteristics of biofouling deposit accumulation.

Several variables are used to represent net accumulation and can serve as the variable X_f in Eq. 14.1, including the biofilm thickness, biofilm mass, fluid frictional resistance, and heat transfer resistance resulting from biofilm accumulation (Section 14.5.2).

The rate and extent of fouling can be controlled by many process variables and can vary considerably with location and process environment. Hence, the length of the three periods (induction, exponential increase, plateau) may vary also. For example, the induction period may not be detectable in water containing very high concentrations of microorganisms, since the adsorption rate will be very high (Figure 14.5b). On the other hand, a very hard precipitate may form in parallel with the biofilm, so that the detachment rate is negligible and no plateau period is discernible (Figure 14.5a). The sensitivity of the measurement technique will also influence the length of the various periods. For example, an insensitive technique will yield a long induction period.

The deleterious effects of fouling can be experienced at drastically different levels of deposit accumulation. For example, deposition of very few cells in a computer chip manufacturing process can cause a significant decrease in yield. Approximately 50% of manufactured computer chips are wasted, many because of microbial contamination. In drinking water, on the order of 1000 to 10,000 coliform cells m^{-3} are sufficient to disturb the operation of the distribution system. On the other hand, a biofilm thickness of 40–60 μm ($\approx 10^{16}$ cells m^{-3}) may be necessary before an increase in fluid frictional resistance is detected.

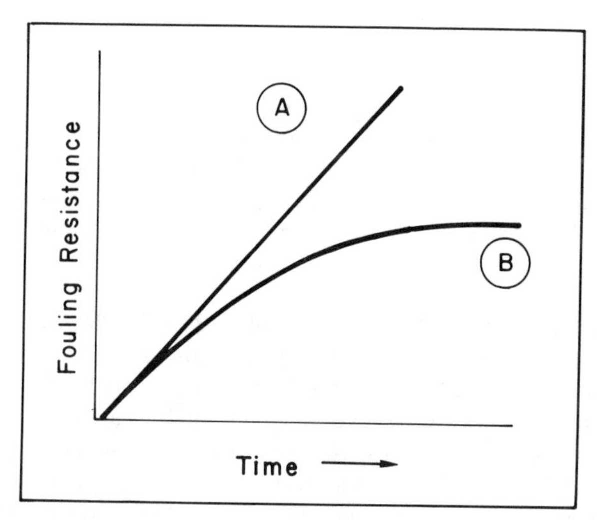

Figure 14.5 The progression of fouling resistance may not always obey a sigmoidal function. Curve B will be observed if the induction period is very short or negligible. Curve A will be observed if the induction period is negligible and the detachment rate is essentially zero.

14.5.1.1 *Transport and Interfacial Transfer Processes*

Transport. Transport of dissolved and particulate materials from the liquid to the wetted substratum is the first step in the overall fouling process and has been described in detail in Chapter 9. Transport rates for soluble materials can be predicted reasonably well with accepted empirical relationships. Predicting transport rates of particulates from the bulk fluid to the wetted substratum is more difficult (Chapter 9).

Deposition. Advective transport has frequently been combined with adsorption/desorption processes at the substratum into a process termed *deposition,* primarily because techniques to distinguish between these processes are difficult. Deposition is easier to measure, but is the net result of several processes. Frequently, investigators focus on advective transport rates and incorporate a sticking efficiency to rationalize measurements of deposition rates (Chapters 9, 12). Experimental observations, even with inert particulates (Beal, 1970), indicate that the sticking efficiency changes as deposition alters the characteristics of the wetted substratum.

Advective transport may be the rate-limiting process in some cases of particulate fouling. However, it appears doubtful that advective transport of cells to the substratum will limit the rate of deposition for biofouling except in the very early stage of accumulation (i.e., the induction period in Figure 14.4) or in very dilute environments.

Detachment. Detachment from the deposit is a significant process, especially as the deposit becomes thicker and fluid shear stress increases (Characklis, 1981). The detached material often deposits in other parts of the system, as has been observed with corrosion products (Lister, 1981). Detachment processes are mostly responsible for the asymptotic phase of a typical fouling progression (see Figure 14.4). During this phase, the detachment rate is equal to the combined effects of transport, adsorption/adhesion, and reaction on deposit accumulation.

14.5.1.2 Reactions at the Substratum or within the Deposit Chemical and biochemical reactions at the substratum or within the deposit can contribute significantly to the accumulation of the fouling deposit (Chapter 8). In some cases, there will be significant interactive effects such as are observed with microbially mediated corrosion, which is a combination of microbial growth and/or product formation as well as deterioration of the substratum (e.g., metal dissolution). In this case, microbial (metabolic) products may diffuse to the metal substratum and cause deterioration (corrosion) of the substratum.

Relevant reactions include microbial growth, extracellular polymer formation, endogenous respiration, and death/lysis. Microbial cells (chemosynthetic organisms), once attached to the substratum, grow, multiply, and form products using chemical energy derived from the bulk water. In most cases, organic compounds (heterotrophs) provide the energy, but reduced inorganic compounds (e.g., Fe^{+2}, NH_3, S^{-2}) can also be used (chemolithotrophs). Photosynthetic organisms can sometimes provide the organic compounds for heterotrophic growth, as has been observed in recirculating cooling tower systems. The relative extent to which cells or product (primarily extracellular polymer substances, EPS) dominate the deposit may significantly influence the effectiveness of any chemical treatment in preventing or removing a biological fouling deposit. The biological deposit, an adsorptive material, will frequently enhance the accumulation of other fouling materials such as particles (Zelver et al., 1984), precipitation deposits (Characklis et al., 1981c), and corrosion products (Characklis et al., 1984).

14.5.2 Measurement and Simulation

14.5.2.1 Need and Purpose Experimental investigation of the fouling process is a necessary and important route to minimizing its effects on performance. If reliable fouling data can be obtained with full scale equipment using the actual process fluids, other measurements and simulation are unnecessary. However, many of the parameters of interest vary considerably throughout the equipment and, in addition, vary with time. As a result, interpretation of operating data is difficult and measurements in the laboratory and on site are necessary so that some measure of control is possible. Laboratory observations under carefully controlled conditions provide the framework for evaluating and interpreting field results where control of all parameters is not possible. On the other hand, field results indicate inconsistencies in the models and provide the impetus and experimental hypotheses for further laboratory work.

Laboratory Tests. Laboratory tests are generally more cost-effective than field tests when goals and objectives are concisely defined. The laboratory provides the proper environment for conducting defined tests that lead to useful models in terms of simulating real systems. Physical, chemical, and biological factors can be controlled at desired levels. Valuable information can be obtained related to design concepts such as influence of alloy (Zelver et al., 1984), fluid velocity, tube geometry (e.g., enhanced heat transfer surfaces), and temperature profiles (Characklis, 1980). The effect of water quality (chemical and biological) on these factors can also be evaluated, as can operation and maintenance procedures. For example, laboratory tests can evaluate the effectiveness of a remedial treatment (chemical or physical) procedure applied at various frequencies (Norrman et al., 1977; Characklis et al., 1980) and can frequently identify the operating conditions that are best for a given treatment procedure. Laboratory experimentation has also resulted in the development of novel measurement techniques useful for field tests and monitoring.

Field Tests. Ultimately, measurements at the process site are essential. Measurements with fouling monitors simulate the process environment and can be used to evaluate the potential for fouling as well as the effectiveness of treatment programs. Fouling monitors have been used extensively at power plant sites. For example, sidestream fouling monitors presently are capable of qualitatively simulating fouling processes in a condenser. In addition, sidestream monitors are useful for testing the effects of various design, operating, and environmental variables on fouling processes. However, *in situ* fouling monitors are needed for regulating fouling control programs in the condenser. More sensitive fouling monitors (sidestream and *in situ*) are needed to detect very thin biofilms (and/or their effects) and to permit rapid feedback for fouling control treatments. Frequently, however, on site fouling tests are only conducted for several months, starting with a clean tube, even though there may be slow processes that only manifest themselves in the fouling dynamics after longer periods of time. Monitors installed *in situ* will supply long term data that can serve adequately for calibrating models.

How do results from fouling monitor tests relate to operation of technical scale equipment? Fouling monitors, at present, are capable of *qualitatively* simulating fouling processes as they occur in operating equipment. For example, an experimental program using monitors in an on site sidestream at a nuclear power plant have resolved an important qualitative question: Is the effectiveness of a chlorine dose in a condenser proportional to concentration times the duration? The answer is No. Under realistic conditions of temperature and flow, chlorine concentrations, and daily exposure, the accumulation of a biofouling deposit on titanium tubes is much more dependent on the chlorine concentration than on the duration of the treatment (Section 15.4.4, Chapter 15). Accordingly, by using higher concentrations and shorter durations, it is possible to achieve a greater degree of fouling control for the same total chlorine dosage (i.e., product of concentration and duration), within regulatory limits. The essential qualitative result, that there is an optimal combination of concentration and duration (which is almost certainly different from the simple regulatory constraint of 2 hours per day at a dose that just meets the discharge limit), must hold true for operating condensers as well as for the test systems. The test system is similar enough to the operating condenser that there is no reason this should not be true. The experimental program also identified an optimal daily dose schedule under the conditions of the test.

Quantitatively, however, the specific values defining the optimal chlorine dosage (i.e., the most efficient combination of concentration and duration) in an operating condenser is bound to be somewhat different from that identified in a test system at that site, and will doubtless vary with the seasons and the site. Some of the problems related to seasonal variations in fouling are dependent on the design of the equipment (e.g., the heat transfer capacity of a condenser) and the changes in operation throughout the year (e.g., change in heat rate, Section 14.9.4.1). The least tractable causes of difference from season to season and from site to site are differences in substratum material, water chemistry, nutrient supply, and source microbial species. The variability owing to season and to site conditions can, of course, be taken into account by conducting measurements with a test system at that site and that time. What is not so simply accomplished is the lesser correction between the test system and the operating equipment.

The key to the differences between the test system and the operating equipment is scale. For example, an operating power plant condenser tube bundle is fed by a water box, with the result that different tubes experience different temperatures and flow regimes. Each condenser tube is generally an order of magnitude longer than the test section of the fouling monitor. Because of the gradient in heat exchange rate down the condenser tube and because the long tube is a plug flow reactor (Chapter 13), there will be systematic differences along the length of the condenser tube with respect to wall temperature, water temperature, residual chlorine, nutrient availability, and possibly other chemical changes wrought by the upstream deposit. Power plant operations personnel sometimes observe more fouling at the exit of the condenser than in the entrance tube region. Thus, a monitored test section can be assembled to mimic any particular short section of the tube, but the same test section cannot mimic the entire span simultaneously. Analysis of fouling deposits from a dismantled nuclear power plant condenser (admiralty brass) indicates, however, that the differences between tubes overwhelmed whatever pattern might have existed in the fouling gradient along the tubes (McCaughey et al., 1987b).

What limits our ability to transfer conclusions from fouling monitors to operating equipment? The extrapolation is constrained both by the amount of effort and attention that can be devoted to monitoring in an operational mode (*in situ*) and by the available scientific information for modifying fouling predictions to accommodate varying environmental or operating conditions. Studies with the fouling monitors at operating plants (e.g., Garey et al., 1980) have been conducted in an *experimental* mode (sidestream) and may require considerable resources for maintaining controlled conditions. An *in situ* or *operational* monitoring program places far fewer demands on personnel and other resources than an experimental monitoring program. However, operational monitoring will probably not be as precise as experimental monitoring, nor can the operational monitoring be as extensive. Experimental monitoring may identify an optimal chlorination schedule empirically, while an operational monitoring program will only observe responses of fouling rates to treatment, because operations will rarely be modified to observe their effect on fouling. Then a modeling effort is required to translate the results from an operational monitoring program into a recommended satisfactory treatment.

Modeling. The objective of developing a model is to determine the coefficients that describe the effects of various operating variables such as temperature, flow rate, and differences in nutrient supply on the dynamics of fouling accumulation. Then, operational monitoring will provide data to calibrate the model for the given application (season and site). This modeling objective is realistic and, with adequate representations of the component processes, will lead to control systems for optimizing biofouling control.

A mathematical model for biofilm accumulation is necessary for relating fouling monitor data to the operating equipment. For example, in a power plant condenser, the plug flow nature of the condenser tube, the seasonal variation of fouling, and variations of fouling between different tubes in the same condenser are concerns that could be resolved by a validated mathematical model. Field tests have been used to evaluate the influence of physical factors such as substratum composition (e.g., metal alloy) and water velocity on the *log rate* and *extent* of fouling. For example, the performance of different alloys has been tested at a power plant site using an instrument that simulates the heat exchange equipment (Zelver et al., 1982; this book, Section 14.9.1). Field tests are also useful in evaluating the performance of contemplated changes in treatment programs and/or process modifications (Characklis, et al., 1981c; Matson and Characklis, 1983). Of all factors influencing fouling, biological variables and chemical water quality appear to be most difficult to simulate and control in the laboratory. Thus, field tests are also necessary.

14.5.2.2 Measurements Related to Fouling Methods for monitoring the progress of the fouling process and methods for characterizing the fouling deposit can be conveniently classified as follows:

1. Direct measurement of deposit quantity and/or composition (Chapter 3)
2. Indirect measurement of deposit quantity by monitoring the effects of the deposit on transport processes (e.g., heat transfer and fluid frictional resistance) (Chapter 9)

Direct measurements, including deposit mass and deposit thickness, are essential for several reasons. Calibration of any indirect method involves comparison with an actual quantity of accumulated deposit. Direct measurements are a necessity when using mass conservation equations to establish process relationships and are useful in relating deposit accumulation to fluid frictional resistance and heat transfer resistance.

Indirect methods may provide significant benefits, including increased sensitivity (Section 3.5.2, Chapter 3). For example, organic carbon analysis of biofilm is as much as 25 times more sensitive than biofilm mass measurement (Chapter 3). In this case, the specific constituent of the deposit (i.e., organic carbon) provides an excellent measure of accumulation. However, if the deposit contains a large amount of silt and sediment, organic carbon will not be representative of the total fouling deposit accumulation. Light absorbance techniques are even more sensitive than carbon analyses but have other limitations (Section 3.5.2.3, Chapter 3).

Indirect methods include monitoring the influence of deposits on heat transfer and fluid frictional resistance. The extent of the influence of deposits on heat and momentum transfer depends strongly on deposit characteristics (e.g., composition, thermal conductivity, roughness).

14.6 FLUID FRICTIONAL RESISTANCE

Fouling deposits cause increased fluid frictional resistance by decreasing the effective diameter of tubes and by increasing the effective roughness of the substratum (Chapter 9). This section will focus on flow in conduits and flow past submerged surfaces (e.g., ship hulls). Frictional resistance measurement has several advantages as an indicator of fouling:

1. It is relatively simple and inexpensive.
2. In conduit flow, porous medium flow, and ship propulsion, frictional resistance is the quantity of most concern, i.e.,head loss, loss in carrying capacity, clogging, or propulsion efficiency.

However, in some situations, frictional resistance measurements alone are of limited value and sometimes misleading:

1. Frictional resistance measurements in turbulent flow are relatively insensitive until the fouling deposit thickness exceeds a certain value, approximately the thickness of the viscous sublayer (Chapter 9). The thickness of the viscous sublayer is inversely proportional to the flow velocity in a given geometry.

2. Some deposits, such as $CaCO_3$ scale, exhibit a low relative roughness and have a low thermal conductivity (Table 9.5). Therefore, their frictional resistance will be low even though their heat transfer resistance is significant.

3. In some instances, frictional resistance is not the major parameter of interest (e.g., in heat exchangers or condensers the major concern is heat transfer resistance).

Frictional resistance can increase significantly as a result of even a small increase in roughness. On ship hulls, an equivalent sand roughness height (densely packed) as small as 25 μm can increase drag by 8%, and a roughness element of 50 μm will increase drag by as much as 22%. The drag effects become more severe as the water velocity increases.

Fouling also causes increased friction losses in porous media by decreasing the effective porosity of the formation (Chapter 18).

14.7 HEAT TRANSFER RESISTANCE

14.7.1 Combined Heat Transfer and Fluid Friction Measurements

The overall heat transfer resistance is the sum of the conductive and the advective heat transfer resistance. Advective heat transfer resistance results from fluid motion and generally decreases as the fouling deposit accumulates, since the roughness of the deposit increases turbulence in the interfacial region. Conductive heat transfer resistance results from insulating layers formed by the deposits and generally increases as the fouling deposit accumulates. The relative changes in advective and conductive heat transfer resistance will depend on the following:

- Deposit thickness, deposit roughness, and deposit thermal conductivity
- Fluid flow rate
- Radial temperature gradient in the tube

Characklis et al. (1981a) have reported the influence of fouling deposits on conductive and advective heat transfer resistance in tubes in the laboratory and also in plant scale equipment (Chapter 9). If the deposit thickness is measured along with the overall heat transfer resistance and fluid frictional resistance, the effective deposit thermal conductivity and effective deposit roughness can be determined. Effective deposit thermal conductivity and roughness have been determined (Table 9.5, Chapter 9).

Frictional resistance and heat transfer resistance measurements indicate the effect of fouling on system performance. They can indicate the *extent* of fouling, but do not yield information on the type of the deposit. At a plant site, the deposits are rarely homogenous, but are typically a combination of biofilm, scale, and corrosion products. Results obtained by Characklis et al. (1981c) indicate significant differences in specific properties (*in situ* thermal conductivity and relative roughness) between deposits that vary in composition (Table 9.5). The different deposit transport properties have been used as a basis for a proposed *in situ*, nondestructive method for identifying the type of deposit accumulated on the heat transfer surface at a plant site (Characklis et al., 1981b).

14.7.2 Heat Transfer Fouling Factor

Heat transfer fouling results are customarily reported in terms of *fouling resistance* or *fouling factor* (R_f), which is defined as follows:

$$R_f = U_f^{-1} - U_0^{-1} \qquad [14.2]$$

where R_f = fouling resistance (t^3TM^{-1})
U_f = overall heat transfer coefficient for the fouled surface ($Mt^{-3}T^{-1}$)
U_0 = overall heat transfer coefficient for the clean surface ($Mt^{-3}T^{-1}$)

The use of R_f is sometimes misleading and frequently conceals valuable diagnostic information:

1. The influence of fouling on the heat exchange rate in engineering design is generally expressed as the thermal (conductive) resistance of the deposit (Kreith, 1973). In fact, measured R_f represents the *net* increase in the heat transfer resistance (conductive plus advective resistance) and not in the thermal (conductive) resistance of the deposit.

2. The conductive resistance of the deposit, in most cases, will be higher than R_f. The extent to which it is so will depend on the roughness of the deposit accumulated on the heat transfer surface. For example, the progression of heat transfer resistance due to accumulation of biofilm in a laboratory experiment (Characklis et al., 1981a) was observed to be sigmoidal (Figure 9.25, Chapter 9). In terms of R_f, an increase in heat transfer resistance of 0.00009 m^2 K/W was observed, but the increase in conductive resistance was 0.00023 m^2 K/W, 2.5 times higher than R_f.

3. The heat exchanger design values for R_f are generally selected from tables of questionable accuracy with vague information as to the operating condition (e.g., shear stress) and the type of deposit (scale, biofilm, etc.) for which the R_f-values were determined.

4. The calculation of R_f is directly related to the overall heat transfer resistance (i.e., advective resistance) at clean condition (U_0^{-1}). But the water velocity influences the advective resistance at clean condition, and hence the calculation of R_f. Thus, R_f-values at two different velocities are difficult to compare.

5. Most mathematical models (Kern and Seaton, 1959; Taborek et al., 1972; Watkinson and Epstein, 1969) describing the influence of fouling processes on heat transfer are based on the following relationship:

$$\frac{dR_f}{dt} = r_{acr} - r_{det} \qquad [14.3]$$

where r_{acr} = rate of accretion (t^2TM^{-1})
r_{det} = rate of detachment (t^2TM^{-1})

This relationship expresses the accretion rate and detachment rate of the fouling deposit in terms of energy units rather than in term of mass units. The model (Eq. 14.3) defines R_f as

$$R_f = L_f/k_{Td} \qquad [14.4]$$

where L_f = thickness of deposit (L)
k_{Td} = thermal conductivity of deposit ($MLt^{-3}T^{-1}$)

This relationship is valid *only* when the advective resistance remains constant and R_f equals the conductive resistance of the deposit. Fouling, in many cases, is accompanied by an increase in pressure drop or decrease in advective resistance. At constant R_f (i.e., $dR_f/dt = 0$) in the model, the deposition rate equals the detachment rate (i.e., the thickness remains constant as indicated in Eq. 14.4). However, constant R_f can also result when the increase in conductive resistance (or thickness) of the deposit equals the decrease in advective resistance (due to deposit roughness). The model (Eq. 14.3) cannot predict a negative R_f. Yet, negative R_f has been observed in numerous studies during initial biofilm accumulation (Figure 9.25, Chapter 9), when the increase in conductive resistance (or thickness) of the deposit is *less* than the decrease in advective resistance. Negative R_f in the early stages of biofilm accumulation has also been observed in the field when new surfaces are initially exposed to a fouling environment. At constant deposit (biofilm) thickness, a change in density or composition of the deposit can influence the conductive resistance R_f of the deposit (Section 13.2.8, Chapter 13). Asymptotic (or steady state) R_f can only result when the conductive and the advective resistance of the deposit remain constant. However, biofouling deposit properties such as thickness, density, and/or roughness may be varying.

Therefore, use of R_f may result in ambiguities, especially as related to simplistic models. Nevertheless, R_f has been used to evaluate fouling processes in many field tests when more detailed information was not available. R_f will be encountered in further discussion in this chapter and in Chapter 15 because of the numerous studies that have reported results in those terms. However, R_f will be used more generally to also include the fluid friction resistance resulting from fouling.

14.8 DEPOSIT PROPERTIES

Fouling monitors frequently measure the increase in frictional resistance and/or heat transfer resistance. These measurements indicate the effect of fouling and do not yield information on the type of deposit. Information regarding deposit properties and composition, however, is often useful in selecting an appropriate treatment procedure.

14.8.1 Deposit Mass and Composition

In addition to fouling monitors, analytical methods have been developed to assess the contribution of biological and chemical processes to overall fouling deposit accumulation. As an illustration, results from physical, chemical, and biological analysis of deposits removed from fouling monitors *and* operating condensers at several power plant sites will be given in succeeding sections. Results of deposit analyses from three different sites are presented in Table 14.1. The differences in deposit composition at different exposure times and between sites are evident.

14.8.2 Deposit Thickness

Deposit thickness is difficult to determine experimentally and has rarely been measured in field tests. However, the thickness is extremely important, because it is required to determine the deposit density and distribution of other constituents in the deposit. In addition, the usefulness of deposit roughness depends on the thickness measurement. In some cases, the standard deviation in the thick-

TABLE 14.1 Comparison of Deposit Characteristics from Fouling Monitors at Three Power Plant Sites[a]

Exposure Time (days)	Tube Material	Deposit Mass ($g\ m^{-2}$)	Composition of Deposit Mass (%)			
			Volatile	C	N	H
		Hudson River				
40	Titanium	1.15	6.28	4.07	0.41	1.27
61	Titanium	6.30	13.45	5.55	0.75	1.35
		Atlantic Estuary				
35	Titanium	14.89	26.65	9.50	0.03	1.43
62	Titanium	37.79	26.70	10.03	1.02	1.85
		Canadian Estuary				
60	Admiralty brass	7.77		14.07	0.55	2.59

[a]Water velocity in the tubes containing Hudson River water (McCaughey et al., 1987b) ranged from 1.34 to 2.23 $m\ s^{-1}$ on a seasonal basis to simulate the plant operating schedule. Site B, on an Atlantic estuary, used a water velocity of 2.16 $m\ s^{-1}$. Water velocity at the estuarine Canadian plant (Characklis and Robinson, 1984) was 1.97 $m\ s^{-1}$.

ness measurement represents an estimate of biofilm roughness (Section 13.2.8, Chapter 13).

14.8.3 Deposit Transport Properties

Industrial fouling deposits are rarely homogenous and can exhibit a wide range of chemical and microbial composition. Not surprisingly, then, two deposits of equal thickness can influence heat and momentum transfer in drastically different ways, since the deposit thermal conductivity and relative roughness can vary widely (Table 9.5, Chapter 9). If the *in situ* deposit thickness were determined, an effective deposit thermal conductivity and roughness could be determined by measurement of conductive and advective heat transfer resistance and fluid frictional resistance and by reference to engineering correlations. This *in situ* diagnostic method could be incorporated into an on line fouling monitor. Such a diagnostic tool would find advantage as a feedback control instrument in an operating plant. As more data accumulate, the deposit composition could be estimated from thermal conductivity and roughness determinations in much the same way as chemical composition is determined from "libraries" of spectral data associated with gas chromotography and mass spectrometry.

14.9 FACTORS INFLUENCING FOULING DEPOSIT ACCUMULATION

Overall fouling biofilm accumulation has been the subject of numerous laboratory and field investigations focusing on various physical, chemical, and biological factors that influence the rate of accumulation as reflected by the *log rate* and *extent*. The variables influencing fouling biofilm accumulation that are addressed in this chapter include (1) substratum composition and roughness, (2) selected water quality parameters including organic carbon concentration, (3) water velocity, (4) temperature, and (5) selected biological variables.

14.9.1 Substratum Properties

14.9.1.1 Composition

Laboratory Tests. The composition of the substratum may influence the *log rate* and the *extent* of deposition. There have been numerous tests that have evaluated the fouling and/or corrosion potential of various alloys. Table 14.2 reports the *extent* of biofouling deposition on copper–nickel and titanium in a laboratory seawater system (Characklis et al., 1984). If only the total deposit mass is considered, the copper–nickel alloy fouled to a much greater *extent* than titanium. However, consideration of volatile deposit mass (a measure of organic carbon) suggests that biofouling was essentially the same in the two alloys at low

TABLE 14.2 The Extent of Biofouling Deposition on Copper–Nickel and Titanium in a Laboratory Seawater System[a]

Sample (alloy)	Exposure Time (days)	Condition	Deposit Mass ($g\ m^{-2}$)	Volatile Deposit Mass (%)	Corrosion Loss ($\mu m\ y^{-1}$)	Deposit Cells (10^6 cells cm^{-2})		Bulk Water Cells (10^6 cells cm^{-3})	
						STP[b]	EPI[c]	STP[b]	EPI[c]
Ti	10	Control	3.3 ± 1.4(2)	—	—	6.1 ± 5.4(2)	118 ± 59(2)	6.3 ± 2.3(2)	30 ± 17(2)
Ti	10	5 mg L^{-1} Cl_2 (periodic for 1 h each 24 hr)	5.0 ± 0.5(2)	—	—	62 ± 75(2)	168 ± 185(2)	8.4 ± 0.8(2)	39 ± 6.3(2)
CuNi	45	5 mg L^{-1} Cl_2 (continuous)	10.0 ± 6.4(3)	10(1)	13 ± 10(4)	85(1)	17(1)	0.96 ± 0.61(5)	3.2 ± 1.3(10)
CuNi	45	Control	12.4 ± 6.1(3)	13(1)	16 ± 9(4)	80(1)	86(1)	2.7 ± 2.0(5)	6.3 ± 4.3(10)
Ti	37	5 mg L^{-1} Cl_2 (continuous)-	4.5 ± 1.3(8)	—	—	0.10(1)	6.2(1)	0.37 ± 0.39(4)	4.0 ± 3.2(7)
Ti	37	Control	22.1 ± 1.2(8)	—	—	—	150(1)	0.90 ± 0.50(4)	3.6 ± 1.5(7)

[a] Characklis et al. (1984). Values given are the mean ± one standard deviation, with the number of samples given in parentheses.
[b] Standard plate count (viable cells).
[c] AODC by epifluorescence (total cells).

substrate loading rate (trypticase soy broth at 26.3 mg m^{-2} h^{-1}) and that much of the fixed (inorganic) deposit mass on copper-nickel could be attributed to corrosion products (Characklis et al., 1984).

Field Tests. Field tests on the Hudson River (Zelver et al., 1984) comparing fouling in copper-nickel 90–10 and AL-6X stainless steel tubes indicate that the heat transfer resistance increases slightly faster in the stainless steel tube (Figure 14.6) at low water velocities (0.3–0.5 m s^{-1}). The tube wall temperature was 35°C, and the water temperature increased during the 100 day test from 17 to 24°C (Figure 14.7). The relative roughness of the tubes was not determined. The influent water contained approximately 3.9×10^{11} cells m^{-3}. The deposit obtained from the stainless steel tube at water velocity 0.3 m s^{-1} contained 1.6×10^{11} cells m^{-2} after 120 days exposure. The volatile content of the deposit was approximately 13% for both the AL-6X and the Cu–Ni tube.

Results from tests at higher water velocities with titanium and admiralty brass tubes and Hudson River water indicated similar trends (McCaughey et al., 1987b). Four tubes were operated in parallel (Figure 14.8). Tubes 1, 2, and 4 were operated at a water velocity that varied according to the season (1.34–2.23 m s^{-1}), thus simulating actual plant operation. Tubes 1 and 2 were titanium, while Tube 4 was admiralty brass. Tube 3 was titanium and was operated continuously at 1.34 m s^{-1}. The initial heat transfer resistance of the admiralty brass tube was lower because of the higher thermal conductivity of the admiralty

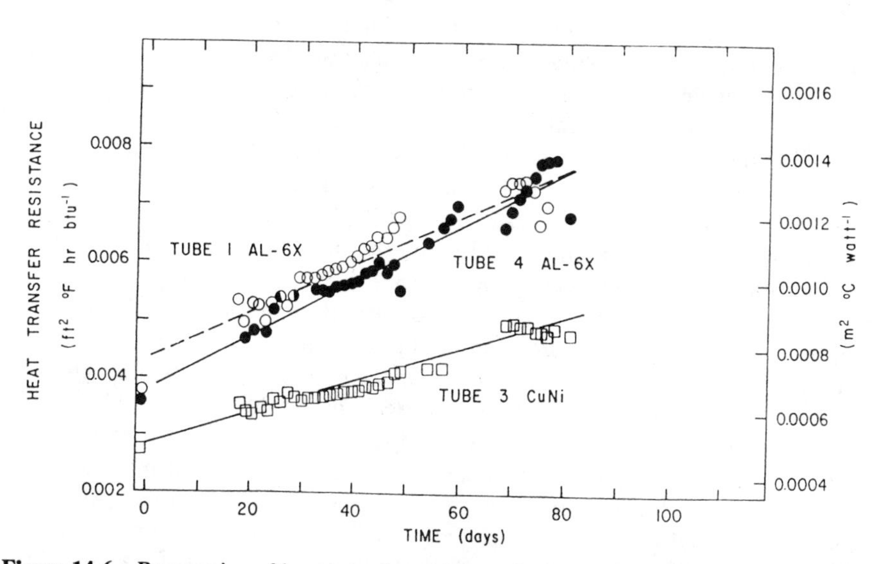

Figure 14.6 Progression of heat transfer resistance in AL-6X stainless steel (○, ●) and copper–nickel 90–10 (□) at water velocity 0.5 m s^{-1}. The results indicate a significant but small difference between the fouling rate in the AL-6X stainless steel tubes and in the Cu–Ni. Tests were conducted at a Hudson River site. (Reprinted with permission from Zelver et al., 1984.)

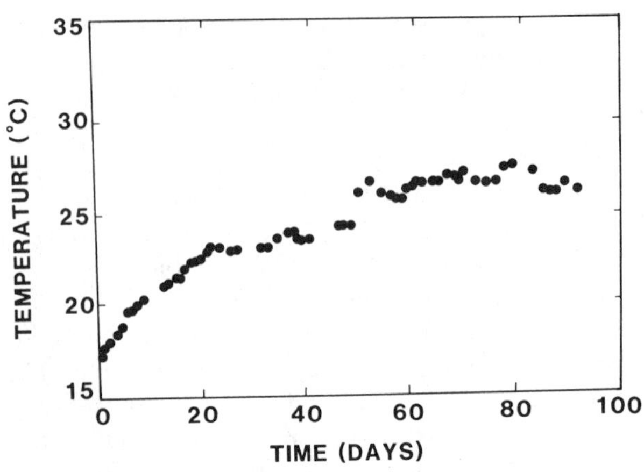

Figure 14.7 Progression of water temperature in the Hudson River during the period of tests reported by Zelver et al. (1984).

brass. Tubes 1, 2, and 4 were operated under identical conditions for nearly the entire period of the study, but the progression of heat transfer resistance was not the same in tube 4 as in tubes 1 and 2. The *log rate* of increase of heat transfer resistance attributed to the fouling deposit in the admiralty brass tube was significantly lower than in the titanium tubes, although the difference in rate was rather small. Thus, in view of the relatively high conductivity of the admiralty brass tube, anticipated fouling resistance in an admiralty brass condenser would be significantly lower than in a titanium condenser of the same design.

Garey (1979) reports virtually no difference in biofilm accumulation over a 16 day period on stainless steel and titanium as determined by ATP content in the deposit. The water velocity was 1.5 m s^{-1}. Accumulation on 90–10 copper–nickel was essentially zero for 8 days, but by the 16th day was the same as on the other alloys. Thus, copper–nickel extended the induction period but otherwise fouled to the same extent as the other alloys.

The unique biofilm progressions on copper alloys are not surprising, since biofilms have been shown to accumulate more slowly on copper alloys than on other metal surfaces such as titanium. Copper is presumed to be inhibitory to bacteria, although Busch (1971) reports excellent growth of mixed planktonic bacterial populations in the presence of copper after acclimation. Eaton et al. (1980) suggest that higher polysaccharide production by organisms colonizing copper substrata shields them from the inhibitory effects of copper. Although reduced microbial fouling on copper alloys is believed partially due to copper toxicity, another reason is possible. Copper alloys corrode faster than titanium and tend to spall or slough their corrosion products. The spalled corrosion products may be carrying much of the biofouling deposit into the bulk fluid also, thus reducing accumulation of the biofouling deposit (Characklis et al., 1984).

The deposit formed on copper alloys has a characteristically lower fraction of

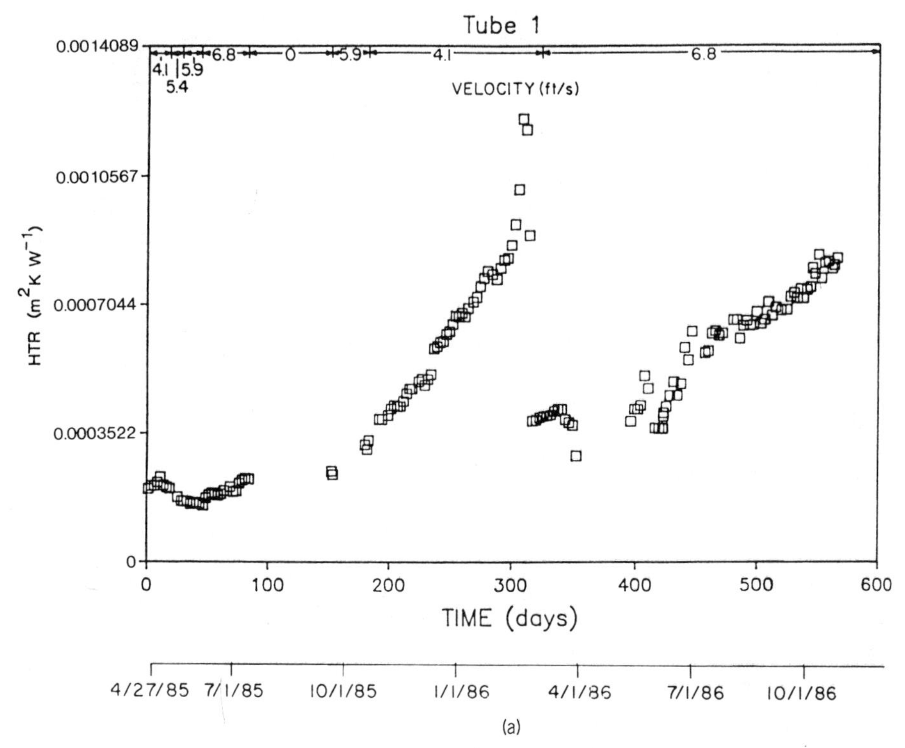

Figure 14.8 The progression of heat transfer resistance (HTR) at the Hudson River site for approximately 16 months. The heat transfer resistance increased continuously through the winter months. Tubes 1, 2, and 4 were operated at varying water velocity (1.34-2.2 m s^{-1}) through the seasons to simulate actual plant operation. Tube 4 was operated at 1.34 m s^{-1} continuously. Tubes 1, 2, and 3 were titanium, and tube 4 was admiralty brass. The fouling monitors were not operative for about thirty days in July 1985. (Reprinted with permission from McCaughey et al., 1987b.)

volatile solids. Characklis et al. (1984) observed fouling on copper-nickel 70-30 in seawater. The deposit generally contained a volatile solids content of 15-25%. The volatile content from titanium tubes was significantly greater. The volatile content of deposits from an admiralty brass power plant condenser was approximately 15-20% (McCaughey et al., 1987b), while the carbon content of the deposits was approximately 6%. At an estuarine power plant site in Canada, the carbon content (an indicator of volatile solid content) of deposits on admiralty brass tubes was generally 10-15% (Characklis and Robinson, 1984).

Little et al. (1979), using heat transfer resistance as a measure of accumulation, observed a more rapid increase in heat transfer resistance on 90-10 Cu-Ni than on AL-6X stainless steel, titanium, or aluminum 5062. Most of the heat transfer resistance in the Cu-Ni system was attributed to rapidly forming corrosion products. After 72 days exposure in the Gulf of Mexico, which included

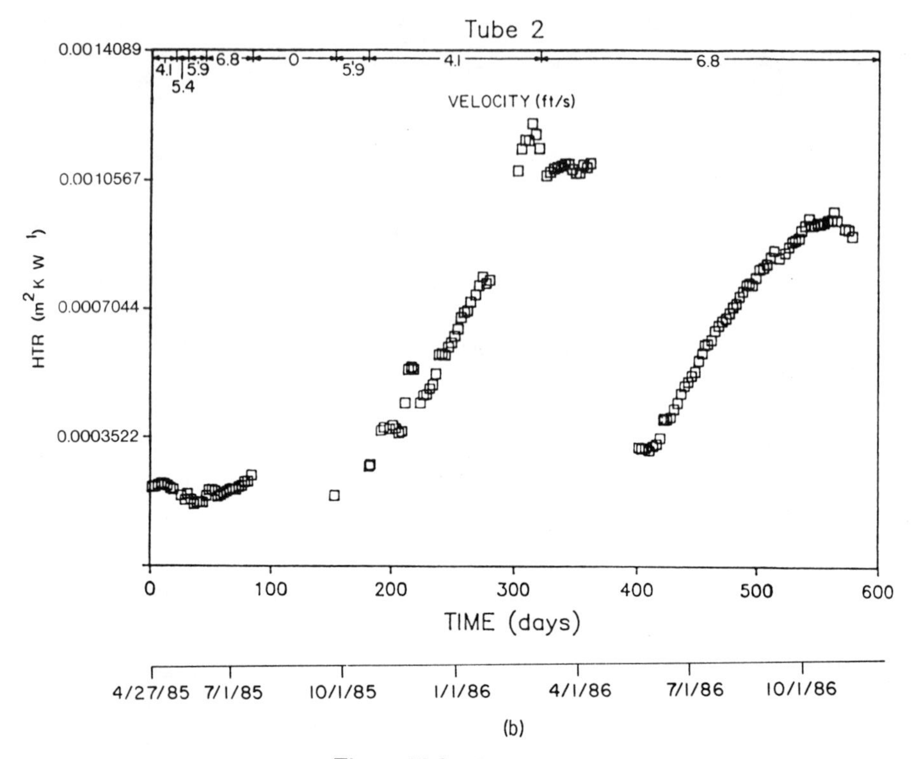

Figure 14.8 Continued

several flow excursions, the Cu–Ni alloy had the lowest heat transfer resistance, while the aluminum and stainless steel had the highest. Thus, while the initial rate of fouling on Cu–Ni was high, the relative extent of accumulation (measured at 72 days) was low.

Substratum composition apparently has the most influence on biofilm accumulation during the induction period.

14.9.1.2 Substratum Roughness The influence of substratum roughness, as contrasted with substratum chemical composition, is not always clear. Two types of substratum roughness can be distinguished: (1) optical roughness (e.g., glass lenses), which is generally small and two dimensional as a result of the machine tools used on the substratum, and (2) engineering roughness, which is three dimensional and is more difficult to characterize. Thus, glass is an optical surface and rolled stainless steel is an engineering surface (Figure 7.9, Chapter 7).

Loeb (1981), using torque on a rotating disk as a criterion for fouling, reports that the roughness of the substratum increased the *extent* of fouling on titanium

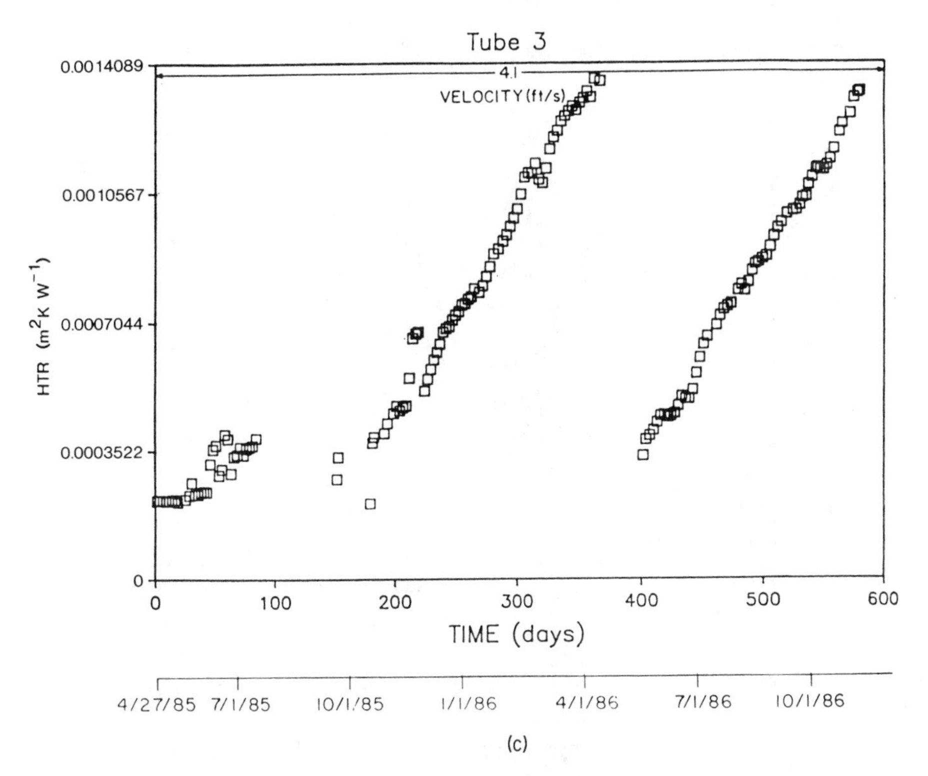

Figure 14.8 Continued

disks. The roughness element height of the smooth titanium disk was 0.05 ± 0.01 μm, and on the rough titanium disks approximately 0.4 μm and 11 μm. The drag was more than 15% greater on the rough disks than on the smooth disk in a Reynolds number range of 1×10^6–2×10^6. There was no evidence that biofilms can decrease drag by filling in the gaps between roughness elements. Picologlou et al. (1980), however, have observed a decrease in drag on a rough surface due to biofilm accumulation during the induction period (Figure 9.20, Chapter 9). The roughness element height, however, was 220 μm, and *Re* was considerably lower ($\approx$23,000).

The ratio of roughness element height to viscous sublayer thickness, e/δ_v, is an important variable affecting the flow behavior near the substratum. It was approximately 36 in Picologlou's experiments, while in Loeb's experiments it varied with radius and *Re*. e/δ_v will be small at high rotational speeds and large radial distance from the disk center. In all probability, the effective e/δ_v in Loeb's experiments was considerably smaller than 36. Harty and Bott (1981) report that the roughness of a brass substratum has only a small effect on biofilm thickness after 200 h in a laboratory reactor (Figure 14.9). No quantitative measure of the roughness was reported.

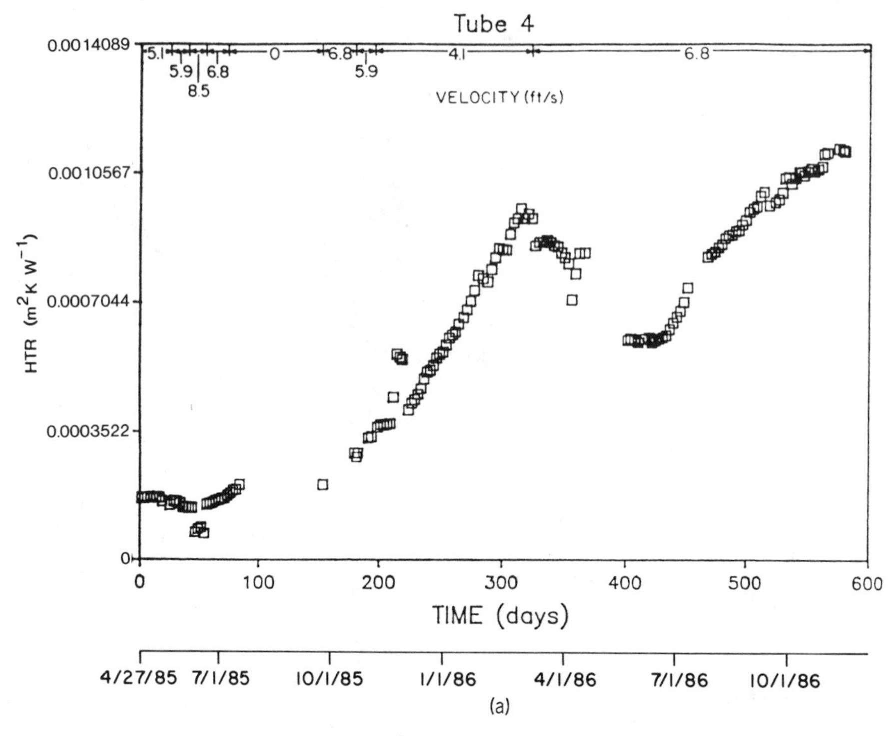

Figure 14.8 Continued

Substratum roughness has the greatest effect on biofouling accumulation rate during the induction period.

Turakhia and Characklis (1983) have also demonstrated the accelerated accumulation of biofilm on the downstream side of roughness elements (940 μm) simulating those in *enhanced* heat transfer tubes. These tubes are designed to increase advective heat transfer rates by means of baffles. However, in so doing, the rough tubes also enhance mass transfer (Chapter 9). No definitive experiments have been reported on the relative biofouling rates in these tubes. Watkinson et al. (1974) have observed differences in calcium carbonate fouling accumulation in various enhanced tube geometries, with some geometries performing better and some worse than a smooth tube.

Kornbau et al. (1983) tested a spirally enhanced AL-6X stainless steel heat exchange tube in seawater and determined that there was no difference in the rate of biofouling between the enhanced tube and a smooth tube over the 30 day test period. Longer tests may indicate a difference between enhanced and smooth tubes. Substratum roughness may also strongly influence the susceptibility of the deposit to detachment, but no conclusive experimental evidence has been reported.

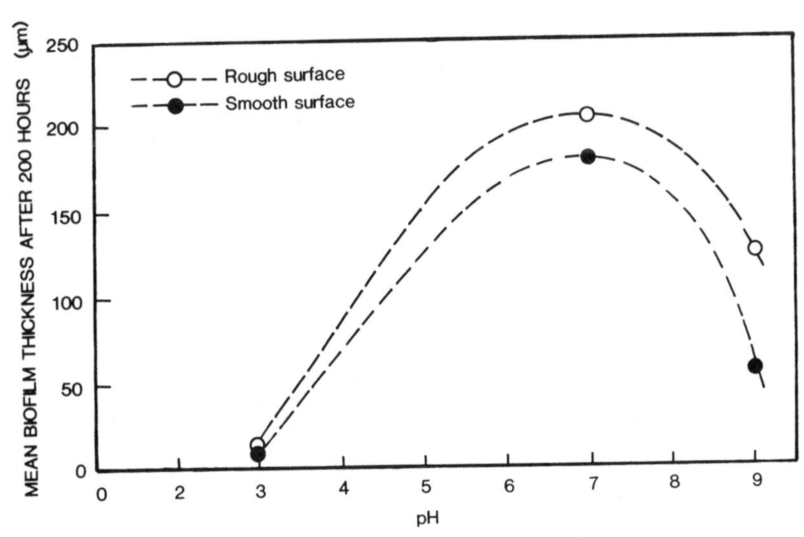

Figure 14.9 The influence of substratum roughness and pH on biofilm thickness after 200 h exposure at 31°C and Re = 15,000. A maximum mean biofilm thickness occurs at a pH ≈ 7. Roughness appears to have a small effect on biofilm thickness. (Reprinted with permission from Bott, 1979.)

14.9.2 Water Quality

Fouling biofilms accumulate to a troublesome *extent,* partially as a result of microbiological metabolic activities, including cellular reproduction and product formation (e.g., EPS formation) within the biofilm. The adsorbed cells reproduce and produce EPS, some of which remains in the biofilms while the remainder detaches. The driving force for the microbial metabolic processes is chemical or radiant energy. In most industrial systems, only chemical (not photosynthetic) energy is available; most often, it is in the form of organic compounds, from which the microorganisms (heterotrophs) obtain both energy and carbon.

14.9.2.1 Organic Carbon (OC) Energy from OC drives virtually all heterotrophic microbial growth observed in fouling processes. OC is present in virtually all waters. It is frequently summarized as total organic carbon (TOC), which equals dissolved organic carbon (DOC) plus particulate organic carbon (POC). The amount of TOC found in natural and modulated waters ranges widely: seawater off the coast of Hawaii had 1 g m^{-3}; cooling water from an industrial, "zero discharge," recirculating cooling tower system had 500 g m^{-3} (Matson and Characklis, 1983); distilled water has 1–2 g m^{-3}; oil field injection water from the Beaufort Sea had 3–4 g m^{-3} (Characklis, unpublished results); oil field produced water had 150 g m^{-3} (Characklis, unpublished results); and untreated municipal sewage had 200 g m^{-3}. Table 14.3 lists organic carbon concentrations reported for several power plant cooling waters. The seasonal variation of the

TABLE 14.3 Organic Carbon Concentration of Some Power Plant Cooling Waters[a]

Power Plant	Water Source	TOC ($g\ m^{-3}$)		Reference
		Range	Average	
Wisconsin Electric Valley Power Station Milwaukee, WI	Menomonee River	3–20	10	Battaglia, 1976
Wisconsin Electric North Oak Creek Milwaukee County	Lake Michigan	1–9	3	Battaglia, 1976
Duquesne Light Co. Phillips Power Station Pittsburgh, PA	Ohio River	3–4	4	Battaglia, 1976
Connecticut Light & Power Co. Devon Power Station Bridgeport, CT	Long Island Sound	3–6	4.5	Battaglia, 1976
Houston Lighting & Power Co. Deepwater Power Station Houston, TX	Houston Ship Channel	21–40[b]	33[b]	Characklis, 1980

Houston Lighting & Power Co. P.H. Robinson Plant Thompson's Corners, TX	Cooling Lake	14–25[b]	21[b]	Characklis, 1980
GPU Nuclear Corporation Oyster Creek Nuclear Generating Station Forked River, NJ	Estuary	3–15 2.7–11[b]		Marine Biocontrol Corp. and CCE, Inc., 1988
New York Power Authority Indian Point 3 Nuclear Generating Station Indian Point, NY	Hudson River	3–15 2.5–11[b]		McCaughey et al., 1987a
Bergen Generating Station Bergen, NJ		22–32		Sugam, 1980

[a] Seasonal variation in organic carbon concentration for the Hudson River is presented in Figure 14.10.
[b] After filtration through a 0.45 μm filter.

various forms of organic carbon can also be considerable (Figure 14.10) and may contribute to seasonal variation in biofouling.

The concentration of organic carbon, however, is not the only important consideration. The OC loading rate to the biofouling deposit, B_{AS}, is a measure of the energy flux through the system. It is defined as follows:

$$B''_{AS} = \frac{QS_i}{A} = v_m S_i \qquad [14.5]$$

where S_i = inlet organic carbon concentration (ML^{-3})
Q = volumetric flow rate (L^3t^{-1})
A = substratum surface area (L^2)
v_m = Q/A = mean velocity (Lt^{-1})
B''_{AS} = organic carbon loading rate ($ML^{-2}t^{-1}$)

Thus, even water with very low concentration of nutrients may provide a nutrient-saturated environment for growth and reproduction in the biofilm if its volumetric flow rate is high.

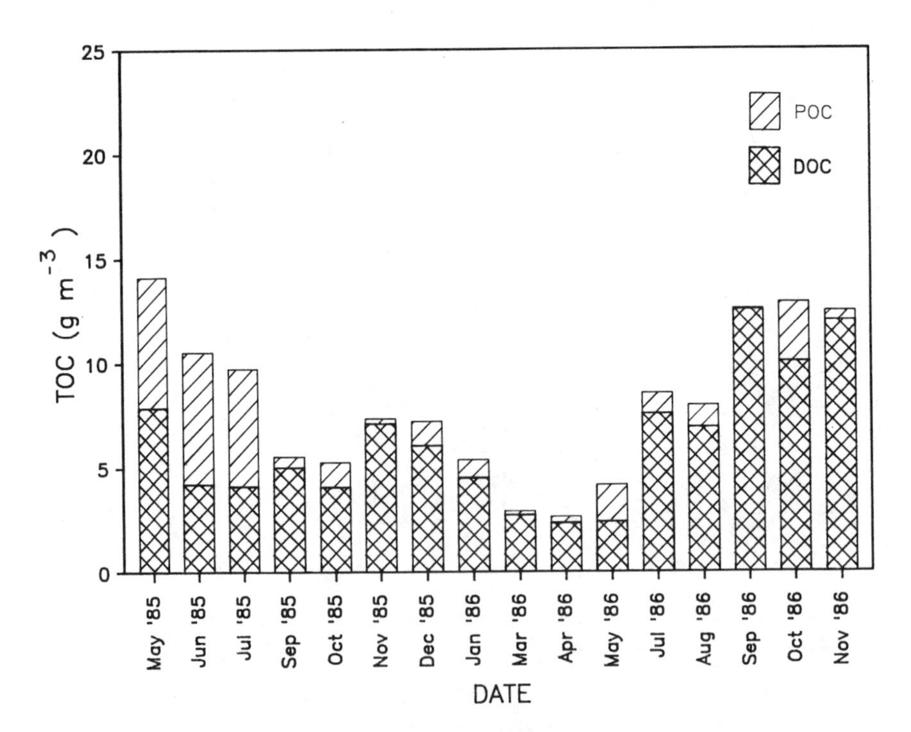

Figure 14.10 Total, dissolved, and particulate organic carbon vary considerably with season in the Hudson River (McCaughey et al., 1987b).

Laboratory Data. The effect of varying inlet OC concentrations has been modeled (Figure 13.8, Chapter 13). Increasing the OC loading rate may increase biofilm accumulation, as indicated by results of the modeling in Figure 13.10 (Chapter 13). At high enough OC loading rates, oxygen may become limiting and portions of the biofilm may become anaerobic.

Characklis et al. (1981b) report that increasing the glucose loading rate increases the *log rate* and *extent* of biofilm accumulation in a heated aluminum tube with water velocity 0.8 m s^{-1} (Figure 14.11). The wall temperature was maintained at 35°C while the water temperature was 27–32°C. Increasing the substrate loading rate may also increase biofilm density (Figure 4.8, Chapter 4). The increased density is a result of increased cell density in the biofilm (Table 4.4, Chapter 4), possibly due to the higher growth rates in the biofilm at higher substrate loading.

Organic carbon, partially responsible for the increased rate of biofouling, also represents a significant oxidant (e.g., chlorine) demand in the process water. Thus, an increase in OC content in cooling water not only contributes to

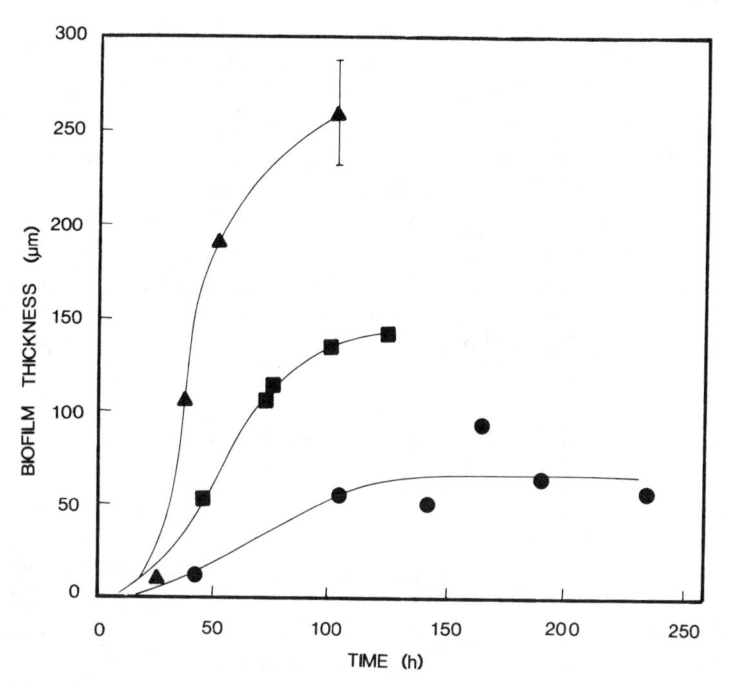

Figure 14.11 An increase in substrate (glucose) loading rate (1.0 (▲), 0.4 (■), 0.2 (●) g m^{-2} h^{-1}) increases both rate and extent of biofilm accumulation. Experiments were conducted in an aluminum tube with a water velocity of 0.8 m s^{-1}. The substratum temperature was 35°C while the bulk water temperature varied between 27 and 32°C. (Reprinted with permission from Characklis et al., 1981a.)

increased biofouling, but also may reduce the effectiveness of a given oxidant concentration in controlling biofouling.

Field Data. There are some data that suggest that biofouling correlates with TOC in cooling waters. Battaglia et al. (1981) measured TOC concentrations at several power plants and related them qualitatively to the biofouling problem experienced in the condensers at the site. No significant quantitative conclusions resulted, possibly due to confounding factors. For example, the OC loading rate may be more indicative of microbial fouling potential than the OC concentration. Therefore, it may be necessary to include the water velocity in the calculation. There is also the possibility that the residence time in a power plant condenser (<10 s) is so short that organic carbon or nutrients are not limiting and that the water velocity controls the *extent* of accumulation. However, even if it does, the substrate loading rate probably influences the *log rate* of biofilm accumulation.

The quality of the OC will also influence the *log rate* of biofouling and increase variance in any attempted correlation. For example, the rate and extent of OC biodegradability will influence biofouling processes. Glucose (dissolved) will be consumed much more rapidly than cellulose (insoluble or polymerized glucose) at the same OC concentration. Particulate organic carbon (POC) will generally be degraded slower than the dissolved organic carbon (DOC). Adsorption and/or attachment of colloidal or particulate organics contributes to the accumulation of the organic content of the fouling deposit, but is not a result of metabolic activity in the biofilm, further complicating the interpretation of deposit TOC results. An assay analogous to the biochemical oxygen demand (BOD) test may be necessary to more accurately assess the influence of OC quality on biofouling. Table 14.4 reports the BOD from various cooling waters. van der Kooij et al. (1981) has developed an *assimilable organic carbon* (AOC) test as a means for establishing the degradable organic carbon in drinking water systems, which may be as high as 2–3 (g AOC) m^{-3}.

Organic carbon concentration, flux, and biodegradability are important considerations in any attempt to predict the rate and extent of biofouling.

However, a simple correlation of biofouling with OC concentration in cooling water may not exist. The OC loading rate probably influences the *log rate* of biofouling and may be a good indicator of the *potential* for biofouling. In systems with high residence times, such as a recirculating cooling tower system, the OC loading rate may control the *extent* of biofouling.

14.9.2.2 Nutrients Carbon may not always be the limiting nutrient. For example, Bott and Gunatillaka (1983) have indicated that the biofilm thickness increases with increasing carbon : nitrogen ratio (C/N) in the water. A medium with C/N greater than 7–10 is considered nitrogen-limited for microbial growth (Section 6.4, Chapter 6), and the cells in it generally produce copious quantities

TABLE 14.4 Water Quality at Various Test Sites Reported in Figure 14.20

	Bulk Water Temp (°C)	Viable Cell Numbers (m^{-3})	pH	DOC ($g\ m^{-3}$)	BOD ($g\ m^{-3}$)	Susp Solids ($g\ m^{-3}$)
Deepwater plant on the Houston Ship Channel	16	1×10^9	6.4	33	19	100
Cooling lake at Thompson's Corner	12	1×10^9	9.0	21	4	—
Laboratory tapwater, Rice University			7.0	8.6	0	0
Laboratory tapwater with trypticase soy broth (TSB),[b]						
TSB concentration						
5.0 $g\ m^{-3}$	30–40	2×10^{10}	7.0	2.2	2.5[c]	0
12.5	30–40	6×10^{12}	7.0	5.6	6.2[c]	0
20.0	30–40	6×10^{12}	7.0	9.0	10[c]	0
100.0	30–40	16×10^{12}	7.0	45.0	50[c]	0

[a] Characklis (1980).
[b] Chemical oxygen demand of TSB: 1.70 (g COD) $(g\ TSB)^{-1}$.
[c] Estimated value.

of extracellular polysaccharides (Wilkinson, 1958). Characklis and Dydek (1976) also observed a higher biofilm accumulation at high C/N. The higher accumulation, however, probably contains fewer cells than a carbon-limited biofilm. Brown et al. (1977) have observed reduced adhesion when cells are grown in a nitrogen-limited medium and produce excess polymer.

Phosphorus may also be limiting in some aquatic systems. For example, Lake Michigan is phosphorus-limited in the summer (Scavia and Laird, 1987). Phosphorus limitation is also suspected in produced water (source was seawater) at an oil production unit where $C:P \approx 500$ (Characklis, unpublished results). Phosphate limitation takes on added significance if phosphorus-based corrosion or scale inhibitors are being added to the water system.

Nutrient limitation, aside from organic carbon, may limit microbial fouling accumulation. In such systems, chemical additives (e.g., corrosion and scale inhibitors) must be screened carefully lest they increase the biofouling.

14.9.2.3 *Inorganic Particles* The concentration of suspended solids provides another confounding factor influencing the *log rate* and *extent* of biofouling. Battaglia et al. (1981) indicate that high suspended solids content frequently accompanies low biofouling *log rates* even in the presence of significant OC in power plant condensers. The authors suggest that suspended solids can scour the tube surfaces and thus inhibit biofilm accumulation.

Suspended solids are observed in many fouling deposits from untreated surface waters. The effect of suspended solids on the rate and extent of fouling deposit accumulation will probably depend on the particle characteristics (size, specific gravity) and the water velocity.

Lowe et al. (1984) observed the effect of suspended kaolin (2–5 μm) on the accumulation of a *Pseudomonas fluorescens* biofilm (Figure 14.12). At a water velocity of 1.2 m s^{-1}, 50 g m^{-3} kaolin significantly extended the induction period for biofilm accumulation. After 48 h exposure, biofilm cell numbers were only 8% of those in the control with no kaolin. The initial colonization may have been inhibited by the abrasive action of the kaolin on the adsorbed cells. However, the ultimate *extent* (after 350 h) of biofilm accumulation, as measured by dry weight, was greater with the kaolin present. Presumably, kaolin attaches to the biofilm, which is more adsorptive than the stainless steel substratum used in the experiments. Kaolin did not affect the growth of the cells in shake flask culture.

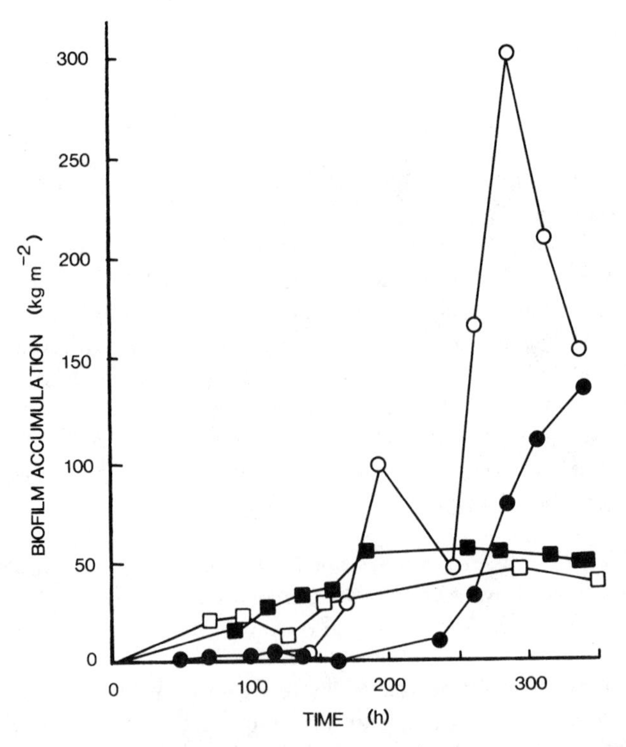

Figure 14.12 Kaolin (size range 2-5 μm) at 50 g m^{-3} increases the accumulation of a *Pseudomonas fluorescens* biofilm in a stainless steel tube. ○ and ● contained kaolin, and □ and ■ contained no kaolin. (Lowe et al., 1984.)

14.9.2.4 Summary Water quality is a critical factor in assessing or controlling biofouling in virtually any system. Organic carbon and other nutrients provide energy and building blocks for cell synthesis. Dissolved organic carbon is generally more degradable. However, particulate organic and inorganic material may add to the accumulation by collecting in the adsorptive biofilm. In some cases, inorganic particulates may reduce the biofouling deposit by scouring.

14.9.3 Water Velocity

Hydrodynamic shear stress influences advective transport, interfacial transfer phenomena, and reactions. Shear stress is critical in fouling environments, since it strongly influences transport and detachment rates. Shear stress is generally related to the flow rate through the system (except in recycle flow systems). As a result, the substrate loading rate is also dependent on shear stress. Since the substrate loading rate also depends on the surface area to volume ratio in the reactor, there may be a strong interactive influence on fouling biofilm development between OC concentration, water velocity, and available surface area.

14.9.3.1 Laboratory Observations Characklis (1980) has demonstrated the combined influence of OC and water velocity on the *log rate* of biofilm accumulation in a laboratory tubular reactor using an undefined microbial population (Figure 14.13). The reactor was operated at a high recycle rate, so that the substrate loading rate B''_{AS} was independent of flow velocity through the tube. Thus, B''_{AS} was dependent only on the inlet substrate concentration S_i. The *log rate* of accumulation, based on friction factor measurements, was a saturation function of the glucose loading rate B''_{AS}. The maximum *log rate,* as reflected by the asymptotic values of the *log rate,* is also dependent on the hydrodynamic shear stress. The influence of shear stress on biofilm accumulation *log rate* is minimal at low B''_{AS}, whereas increasing B''_{AS} strongly influences the *log rate* of accumulation. However, at higher B''_{AS}, the *log rate* of accumulation is independent of B''_{AS}. Harty and Bott (1981) and Characklis (1980) have observed the influence of hydrodynamic shear stress on biomass accumulation rate. The accumulation rate was lower (Figure 14.14) at the higher flow rate suggesting that detachment, not nutrient supply, probably controlled the *extent* of biofilm accumulation in their systems.

Characklis (1980) reports an optimum in *log rate* of biofouling accumulation as a function of water velocity (Figure 14.15). The accumulation *log rate* was determined from friction factor determinations in laboratory experiments as well as field tests with natural waters. Table 14.4 compares the water quality at the various field sites and laboratory systems for these tests. The fluid velocity (shear stress) strongly influences two rate processes in a biofouling context: (1) mass transfer increases with increasing fluid velocity and (2) detachment rate of biofilm increases with increasing velocity (Figure 9.34, Chapter 9). Characklis's results suggest that at lower velocities, mass transfer of nutrients to the deposit is

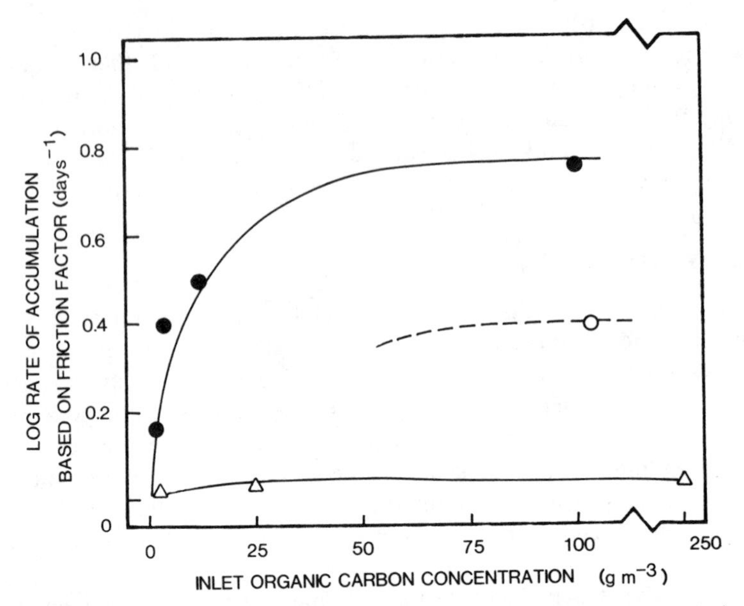

Figure 14.13 Increasing inlet organic carbon concentration and water velocity [0.92 (●), 1.07 (○), and 1.92 (△) m s^{-1}] increases the *log rate* of fouling biofilm accumulation in a circular tube (12.7 mm ID). A tubular recycle reactor was used, and the mean water velocity did not influence the organic carbon loading rate. Thus, the organic carbon loading rate was proportional to the inlet organic carbon concentration (Characklis, 1980).

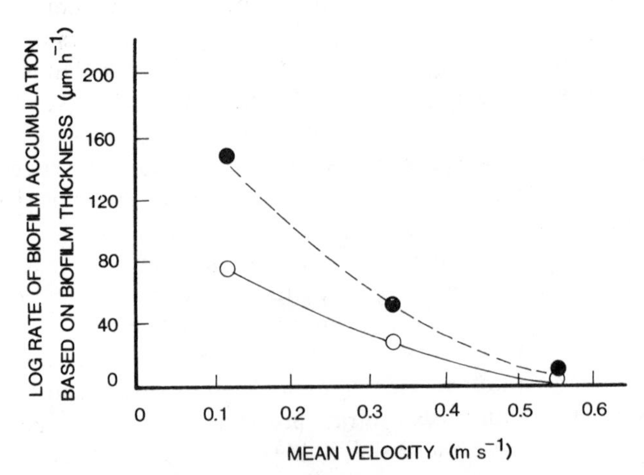

Figure 14.14 Maximum biofilm accumulation rates in a brass duct (0.01 × 0.02 × 2.0 m) decrease with increasing water velocity and decreasing water temperature [42°C (●) and 37°C (○)]. (Reprinted with permission from Harty and Bott, 1981.)

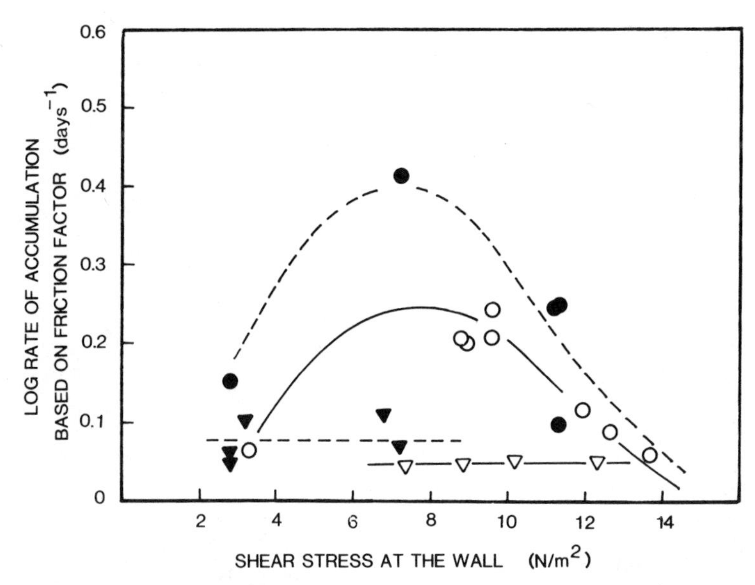

Figure 14.15 Increasing water velocity in a 12.7 mm ID tube, combined with higher carbon loading rates, increases log rate of fouling biofilm accumulation in the range of low velocities and decreases fouling biofilm accumulation in the higher velocity range. Increasing water velocity at low organic carbon loading rates does not influence rate of accumulation. Laboratory data are from a tubular reactor at high (0.25 g m^{-2} h^{-1}, ●) and low 0.06 g m^{-2} h^{-1}, ▼) organic carbon loading rate. Field data are from a cooling lake at Thompson's Corners, an oligotrophic (▽) site, and the Houston Ship Channel, a eutrophic (○) power plant site. The laboratory system and field sites are described in Table 14.4.

rate-limiting for biofilm accumulation, since increasing velocity increases accumulation rate. In the high velocity region, detachment controls biofilm accumulation, so increasing velocity decreases the *log rate* of biofouling. The combined result is an optimum velocity for biofouling.

The water velocity influences the mass transfer rates from the bulk water to the biofilm and the detachment rate of material from the biofouling deposit.

The combined effect of hydrodynamic shear and substrate loading rate on *extent* of biofouling (as measured by the maximum biofilm thickness attained) has been demonstrated by Characklis (1980). As the biofilm accumulates, the fluid forces shear off more cells until biofilm production equals biofilm detachment and a pseudosteady state is established. Shear stress has a great effect on the *extent* of biofilm accumulation at high nutrient loading rates (Figure 14.16). However, as the nutrient loading decreases to very low levels, shear stress has little effect on the maximum biofilm thickness since the system is nutrient-limited (Figure 14.16). The effect of shear stress was also observed by Harty and

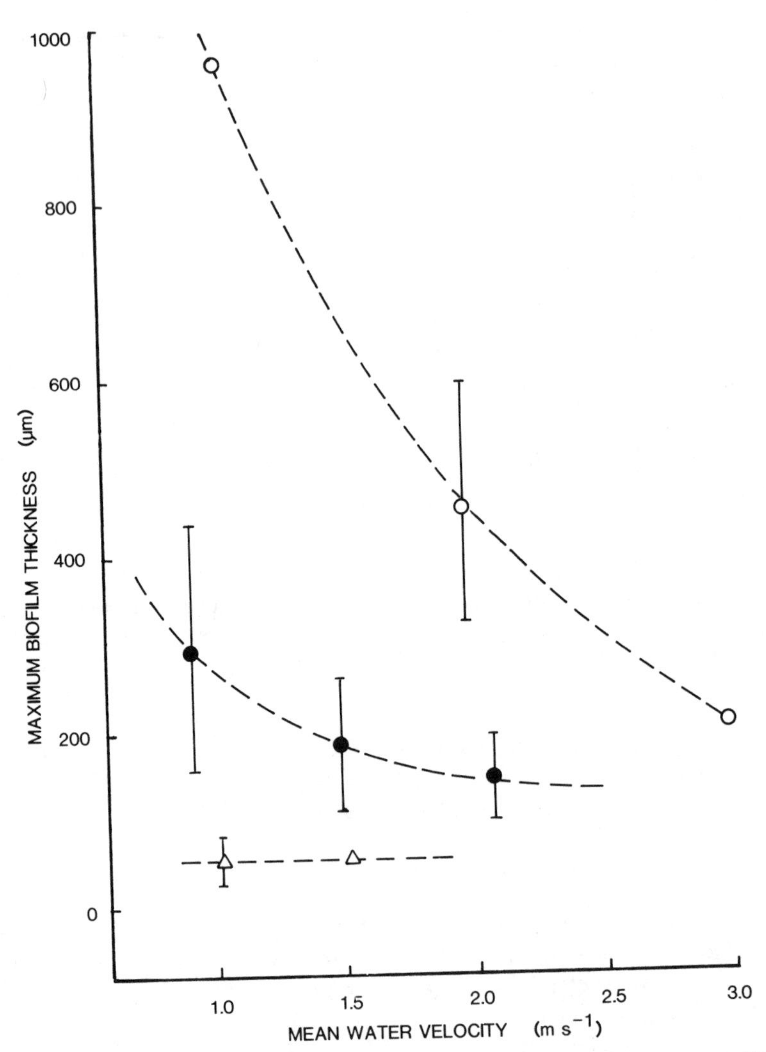

Figure 14.16 The extent of biofilm accumulation in a circular tube (12.7 mm ID) is a function of organic carbon loading rate [2.4 (○), 0.5 (●), and 0.1 (△) g m^{-2} h^{-1}] and water velocity (shear stress). At high organic carbon loading rates, the water velocity controls the extent of fouling biofilm accumulation, as determined by the maximum thickness. At low organic carbon loading rates, the influence of water velocity is considerably less and organic carbon controls the extent of accumulation. (Characklis, 1980.)

Bott (1981) in a rectangular duct reactor (Figure 14.17) and by Characklis (1980) in the RotoTorque system (Figure 14.18).

The extent *of accumulation is dependent on both organic carbon loading rate and water velocity. The chemical energy (organic carbon) leads to production while the mechanical energy (water velocity) leads to removal of fouling deposit.*

Pedersen (1982) observed a significant increase in rate of biofouling in seawater when the water velocity was increased from 0.005 to 0.15 m s^{-1}. At these low velocities, biofouling is probably mass-transfer-limited (substrate), and an increase in velocity would be expected to increase the biofouling rate.

14.9.3.2 Field Observations The influence of water velocity on fouling deposit accumulation has been observed by Zelver et al. (1982) in brackish water. AL-6X stainless steel heat exchanger tubing was maintained at an inside tube wall temperature of 35°C. A rapid accumulation of a microbial matrix containing considerable silt debris occurs. The progression of heat transfer resistance and fouling factor for a flow velocity of 0.50 m s^{-1} and a constant inside wall temperature of 35°C indicated no flow excursion or other perturbations occurred during the 80 day test period (Figure 14.19). The progression followed a typical sigmoidal pattern which is frequently observed. However, during the 0.3 m s^{-1} test, momentary flow excursions to a flow of 0.90 m s^{-1} occurred on day 24 and resulted in sloughing of the fouling deposit and resultant decrease in heat

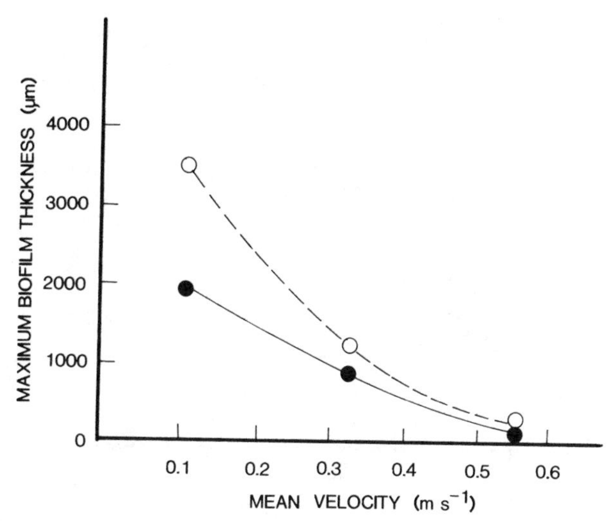

Figure 14.17 Increasing hydrodynamic shear stress (water velocity) decreases the extent of biofilm accumulation as determined by maximum biofilm thickness on a brass plate. Increasing temperature [42°C (○) and 37°C (●)] also increases extent of accumulation. (Reprinted with permission from Harty and Bott, 1981.)

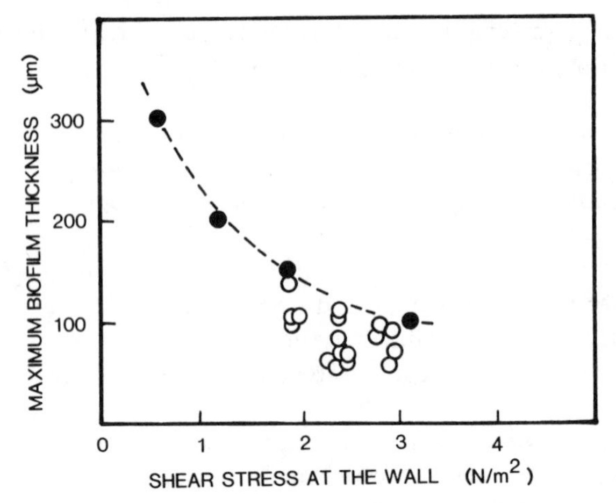

Figure 14.18 The influence of hydrodynamic shear stress and substrate (glucose) loading rate on maximum (plateau) biofilm thickness in a RotoTorque reactor [reprinted with permission from Kornegay and Andrews, 1967 (●); Characklis, 1980 (○)]. The experiments reported by Characklis were conducted at significantly lower substrate loading rates.

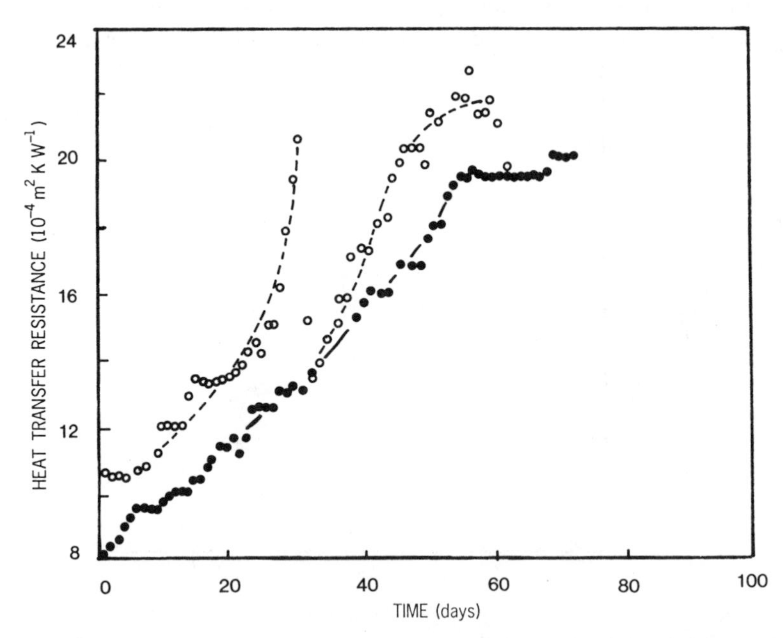

Figure 14.19 Progression of heat transfer resistance in an AL-6X stainless steel tube (14.5 mm ID) operated with Hudson River water for a water velocity of 0.3 (○) and 0.5 (●) m s^{-1} and a constant wall temperature 35°C (reprinted with permission from Zelver et al., 1982).

transfer resistance. The *log rate* of fouling at 0.50 m s^{-1} is very similar to that observed at 0.30 m s^{-1}.

The influence of shear stress on the *extent* of deposit accumulation was observed by Characklis (1980) in laboratory experiments (with undefined microbial populations) and at various field locations using frictional resistance as a measure of accumulation (Figure 14.20). The laboratory data were from experiments with low glucose loading rates, $B''_{AS} < 0.25$ g m^{-2} h^{-1}, consistent with those at the field sites (Table 14.4). The results suggest that the *extent* of accumulation is controlled by shear stress at low B''_{AS}, since one curve describes the range of experiments. This result is consistent with laboratory results described in the previous section.

Water velocity influences the biofouling accumulation in partly filled flow conduits as well. Bland et al. (1978) observed the influence of water velocity in unglazed clay sewer pipe (100 mm ID) carrying domestic wastewater under pressure. The relation between velocity, deposit accumulation, and hydraulic roughness is presented in Table 14.5. A small decrease in deposit accumulation and hydraulic resistance results from increasing the water velocity from 0.76 to 1.1 m

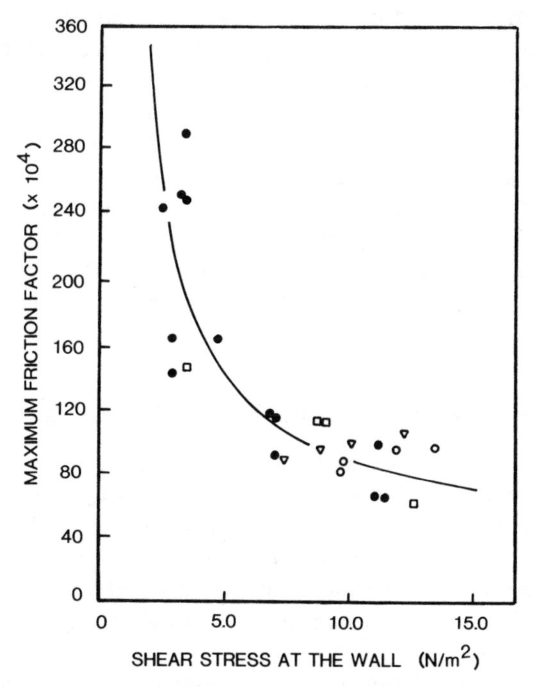

Figure 14.20 Increasing hydrodynamic shear stress decreases the extent of biofilm accumulation as determined by frictional resistance measurements in a laboratory tubular reactor (●) and two power plant sites: Houston Ship Channel before chlorination (○), after chlorination (□), and Thompson's Corners (▽). The organic carbon loading rate $B_{AS} < 0.25$ (g C) m^{-2} h^{-1} for all systems. (Characklis, 1980.)

TABLE 14.5 The Influence of Water Velocity on Deposit Accumulation and Hydraulic Roughness in Unglazed Clay Sewer Pipe (100 mm ID) Carrying Domestic Wastewater under Pressure[a]

Water Velocity ($m\ s^{-1}$)	Deposit Mass ($g\ m^{-2}$)	Hydraulic Roughness (mm)
0.76	72	4.62
1.1	70	2.45
1.1	48	1.68
1.1	80	3.68
1.5	28	0.71
2.1	5	0.05

[a] Bland et al. (1978).

s^{-1}. The decrease in accumulation and hydraulic resistance is considerably more as the velocity increases to 1.5 and 2.1 m s^{-1}.

14.9.3.3 *Summary* Hydrodynamic shear stress influences numerous processes contributing to biofouling, including advective transport, adsorption phenomena, detachment, and reactions *at the interface*. The water velocity influences the mass transfer rates from the bulk water to the biofilm and the detachment rate of material from the biofouling deposit. As a result, an optimum in biofouling rate may be observed with respect to water velocity. In addition, the *extent* of accumulation is dependent on both organic carbon loading rate and water velocity. At high substrate loading rates, the shear stress will control the *extent* of biofouling accumulation (e.g., maximum biofilm thickness). At low substrate loading rates, the shear stress will control the *extent* of accumulation.

14.9.4 Temperature and Seasonal Effects

Biofouling is generally considered to be more of a problem in summer months because higher temperatures generally increase the rate of biological processes. However, other factors may also be indirectly causing the perceived increase in biofouling intensity during summer. For example, the water velocity in heat exchangers (e.g., a power plant condenser) may be increased to provide a higher cooling capacity in the summer months.

14.9.4.1 Power Plant Condensers Another reason fouling biofilms reasonably receive more attention from power utilities in the summer is that inlet and outlet cooling water temperatures are considerably higher than in the winter. This leads to condenser operation near the design capacity (i.e., at a much higher steam back pressure) in the summer than in the winter, even when the condenser is clean. Results of biofouling tests at two power plants, however, indicate that biofouling is significant in the winter months (McCaughey et al., 1987a).

Utility surface condensers are designed as high performance heat transfer equipment. The overall heat transfer coefficient for a new unit with brand-new, clean tubes will vary from about 2200 in extremely cold water to over 4000 W m^{-2} K^{-1} in the peak heat of summer when inlet temperatures may approach 38°C. The amount of fuel required to produce a kilowatt-hour of electrical power varies throughout the year for even a clean condenser; it is termed the *heat rate,* and has units of energy consumed per kilowatt hour of power produced. Operation at the minimum possible unit heat rate has become increasingly critical as fuel costs have risen. Condenser performance—especially as affected by microbial biofilm, which inhibits heat transfer—directly affects the unit heat rate and operating costs for this reason. The expected steam back pressure and net heat rate for a new condenser vary seasonally (Figure 14.21) with the inlet cooling water temperature (Ferguson, 1981). The data in Figure 14.21 are for a condenser in the northeast, where the inlet cooling water temperature is generally low. In the south, seasonal variation in heat rate would be considerably higher. The expected back pressure and heat rates for the unit with an effective surface area reduced by 34% are also presented in Figure 14.21. The steam back pressure increases due to decreased condenser performance are comparable between the summer and winter conditions. The *heat rate penalty* for the same back pressure increase, however, is about three times as high in the summer as in the winter. The effect of condenser fouling on back pressure and condenser performance can vary even more for many units. In some cases, the winter penalty will be negligible, there being an extremely high penalty for the same degree of fouling in the summer. The winter condenser performance decrease may not be noticed for this reason unless careful measurements are made.

Effects of fouling deposition in equipment will not be detected as readily when the equipment is operating at a fraction of its maximum capacity.

Some power plants do not use any fouling control measures during the winter. As a result, the accumulated deposit and increases in water temperature may result in very rapid increases in steam back pressure in early spring.

14.9.4.2 *Seasonal Effects: Field Tests at Power Plant Sites* Fouling factor increases were observed at various times of the year at two power plant sites with an electrically heated fouling monitor that simulated condenser tube conditions. The fouling deposit mass and composition were determined at both sites (Table 14.1) as well as the influence of the deposits on heat transfer and hydrodynamic frictional resistance. The progression in heat transfer resistance (Figure 14.8) and deposit mass (Figure 14.22) continuously increased throughout the winter months at site A. Variable water velocity mimicked the operation of the plant condenser, which reduces its water velocity in the winter months

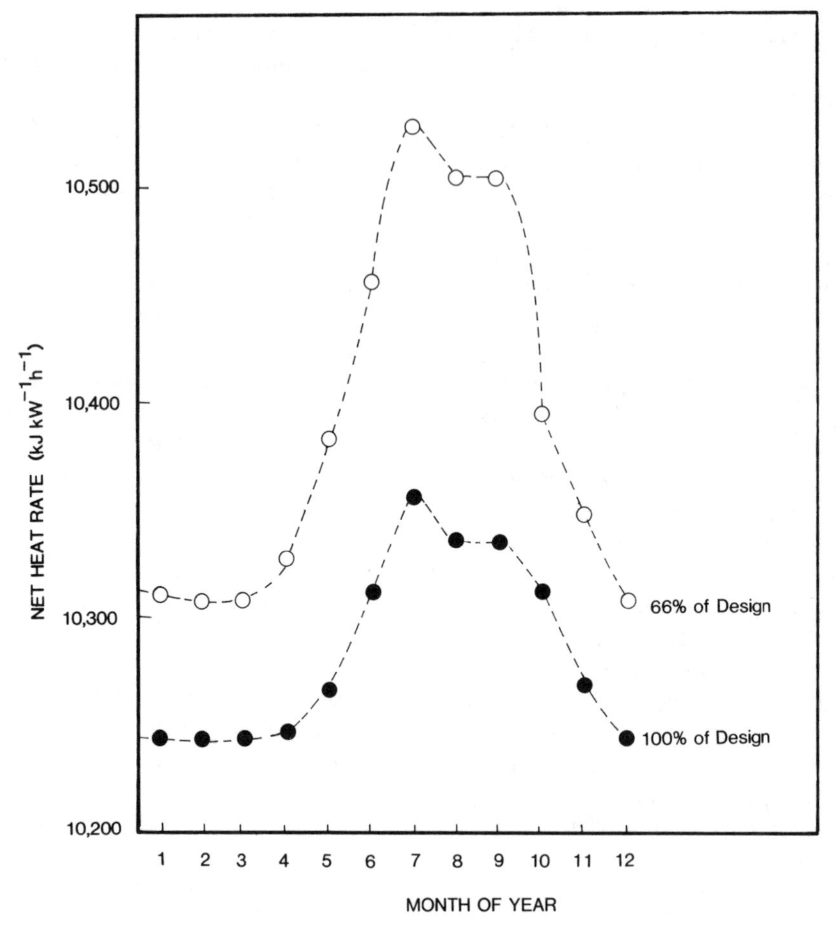

Figure 14.21 The heat rate profile for a 350 MW power plant condenser unit at full load, indicating the seasonal effect on heat rate (reprinted with permission from Ferguson, 1981).

because of reduced water temperatures. The cooling water temperature (Figure 14.23) during the winter period of this test dropped as low as 1°C (34°F).

The progression of heat transfer resistance at the Atlantic estuary site was observed during late autumn and winter (Figure 14.24). The continual increase in heat transfer resistance clearly indicates that fouling accumulation continues unabated throughout the winter months. The deposit mass also increases continuously through the winter months.

Biofouling deposit accumulation can be significant during the winter.

During the tests at the Atlantic estuary site, a set of tests was conducted with two test heat exchangers, which were cleaned monthly to bare metal. Thus, a new progression in heat transfer resistance was observed each month with these two tubes. Thus, the length of the induction period, the *log rate* of increase, and the *extent* of heat transfer resistance were determined for each month (Figure 14.25). Clearly, *log rate* of fouling decreased each month throughout the winter, reaching a minimum of zero in February. Thus in the long term tests (Figure 14.24), the fouling accumulation continued to increase and the fouling rate never decreased to zero. Yet in the monthly tests, no fouling was observed in February. The apparent discrepancy can be explained by referring to the fundamental biofouling processes. In the long term test, a fouling deposit containing viable microorganisms was present at the beginning of February. The microbial growth rate is proportional to the number of viable cells, and since the cells at the substratum were experiencing the wall temperature of 35°C (not the bulk temperature), the fouling accumulation continued by microbial growth at the expense of nutrients in the cooling water. In the monthly test, no microorganisms were present at the beginning of February, the water temperature was considerably lower than the wall temperature, and microbial growth was inhibited. Therefore, fouling accumulation continued in the long term test but not in the monthly test.

The field test results demonstrate the significant *log rate* of biofouling occurring on condenser tubes in the winter as well as the summer, even in colder waters representative of the northeast. Thus, biological activity and/or fouling processes may not be as dependent on cooling water temperatures as previously believed. The substratum temperature may influence fouling processes more strongly than the water temperature. Furthermore, these results indicate that there may be a reward for fouling biofilm control in the colder months so that a relatively clean surface exists when higher temperatures arrive.

14.9.4.3 Bulk Water Temperature The bulk water temperature influences the rate of most chemical and biochemical reaction processes as well as transport processes. The influence of temperature on biofilm growth processes is much greater when the cells are nutrient-saturated, i.e., when cells are growing near their maximum specific growth rate (Section 10.6.1.6, Chapter 10).

Laboratory Tests. Stathopoulos (1981) conducted tests in a stainless steel tubular recycle reactor (Chapter 3) using a defined mixed microbial population with temperature as the only variable. He observed that the *log rate* and *extent* of accumulation, based on biofilm thickness, increased with increasing temperature (15–40°C) (Figure 14.26a). The effect was also observed using biofilm mass as a criterion for accumulation (Figure 14.26b). The progression of biofilm accumulation demonstrates the characteristic sigmoidal function for both thickness and mass. The two variables also permit the calculation of the biofilm density, which was not very sensitive to temperature between 15 and 40°C but decreased dramatically at 45°C, where *log rate* and *extent* of accumulation also decreased. Thus, the density of the biofilm at 45°C was probably reduced as a result of decreased microbial growth rate within the biofilm. In a natural aquatic system,

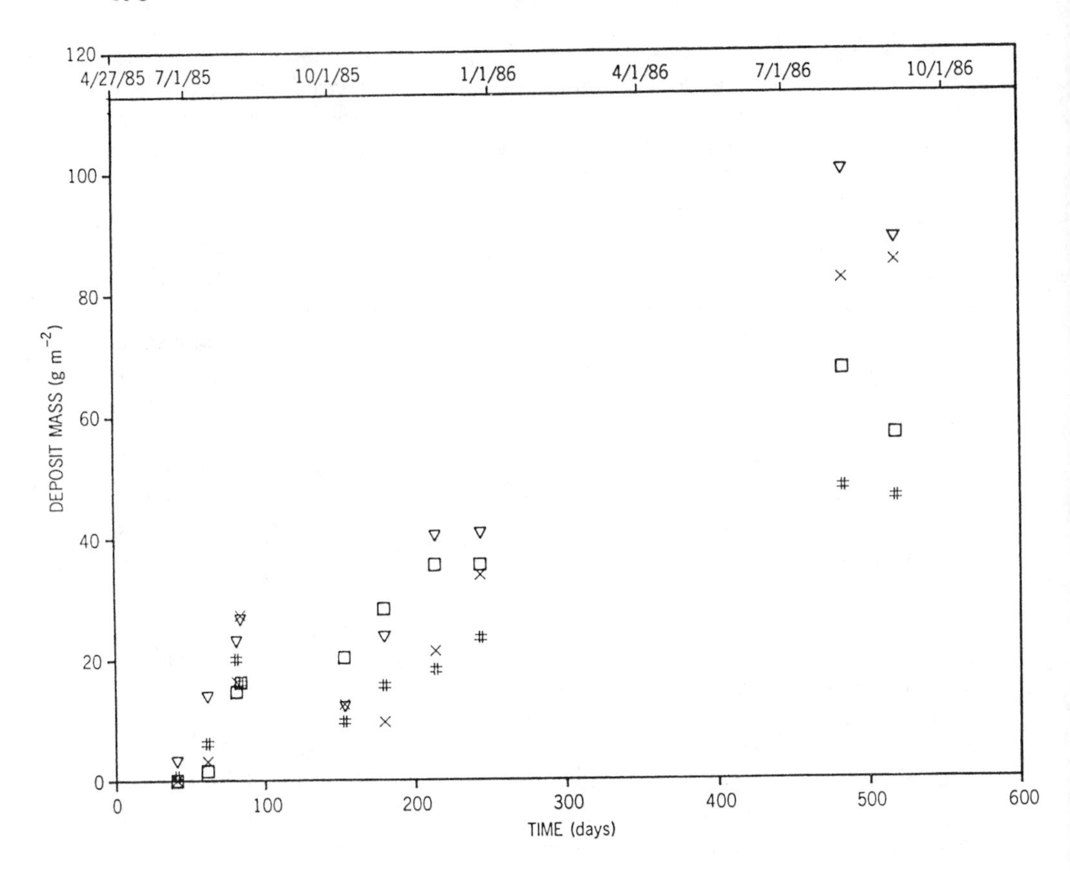

Figure 14.22 The progression of deposit mass at the Hudson River site for approximately 16 months. The heat transfer resistance increased continuously through the winter months from 1 Ocober 1985 until approximately 1 April 1986. The deposit mass increased from 1 October to 1 January continuously. Tubes 1, 2, and 4 were operated at varying water velocity (1.34–2.2 m s^{-1}) through the seasons to simulate actual plant operation. Tube 4 was operated at 1.34 m s^{-1} continuously. Tubes 1 (□), 2 (×), and 3 (▽) were titanium, and tube 4 (#) was admiralty brass. The fouling monitors were not operative for about 30 days in July 1985. (McCaughey et al., 1987b.)

continuous exposure to 45°C would probably lead to selection of a population very tolerant and able to grow rapidly at that temperature. However, the population was restricted to the inoculum in these experiments.

The bulk water temperature influences the rate and extent of biofouling deposit accumulation.

In tests with an undefined population in a RotoTorque, the highest *log rate* of biofilm accumulation occurred at 35°C (Figure 14.27). The biofilms in these

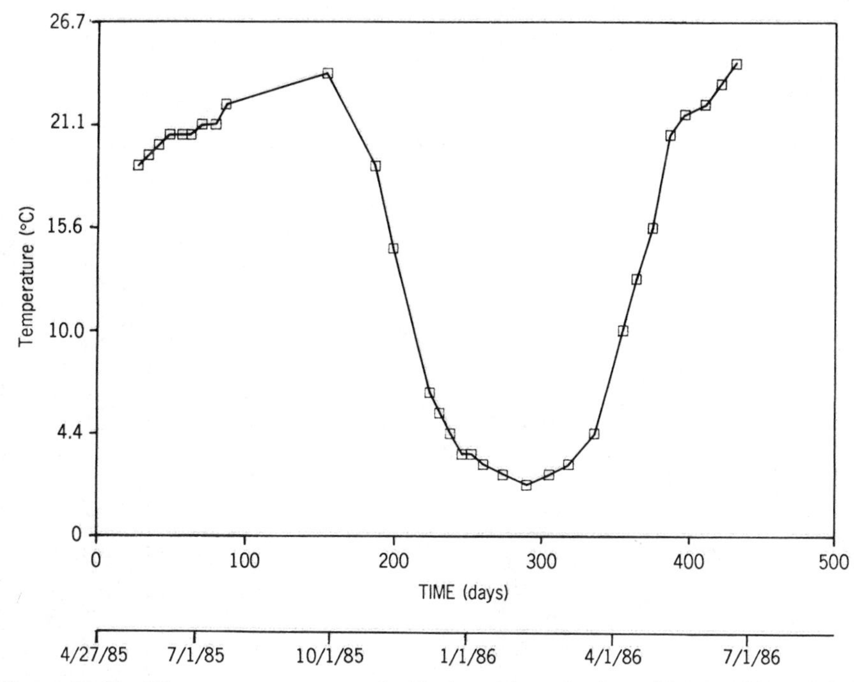

Figure 14.23 Water temperature at the Hudson River site from 27 July 1985 to 1 July 1986, the testing period described by the data in Figure 14.22. (McCaughey et al., 1987b.)

experiments were probably nutrient-saturated and thus were very sensitive to temperature effects (Section 10.6.1.6, Chapter 10).

Field tests. In field tests, the microbial populations that grow fastest are generally selected. Thus as temperature changes during the seasons, the population distribution probably changes in response to the environmental variations. Two tests were conducted at a the Hudson River site, each in a different season. The first test was conducted from 1 October 1985 to 1 April 1986, when the water temperature decreased from 24 to 4.5°C. The second test was conducted from 1 April 1986 to 15 October 1986, when the temperature increased from 4.5 to 24°C (see Figure 14.23). The progression of heat transfer resistance was almost identical for both tests (see Figure 14.22). Thus, the water temperature did not significantly influence the *log rate* of increase in heat transfer resistance due to biofouling. The wall temperature was maintained constant at 35°C and may be a greater influence on the metabolism of the fouling microorganisms.

Pedersen (1982) conducted biofouling experiments in seawater at very low water velocities and observed significant decreases in the induction period and large increases in rate of accumulation with the seasons. The temperature ranged from 2 to 25°C in his experiments. The wall temperature was the same as the water temperature.

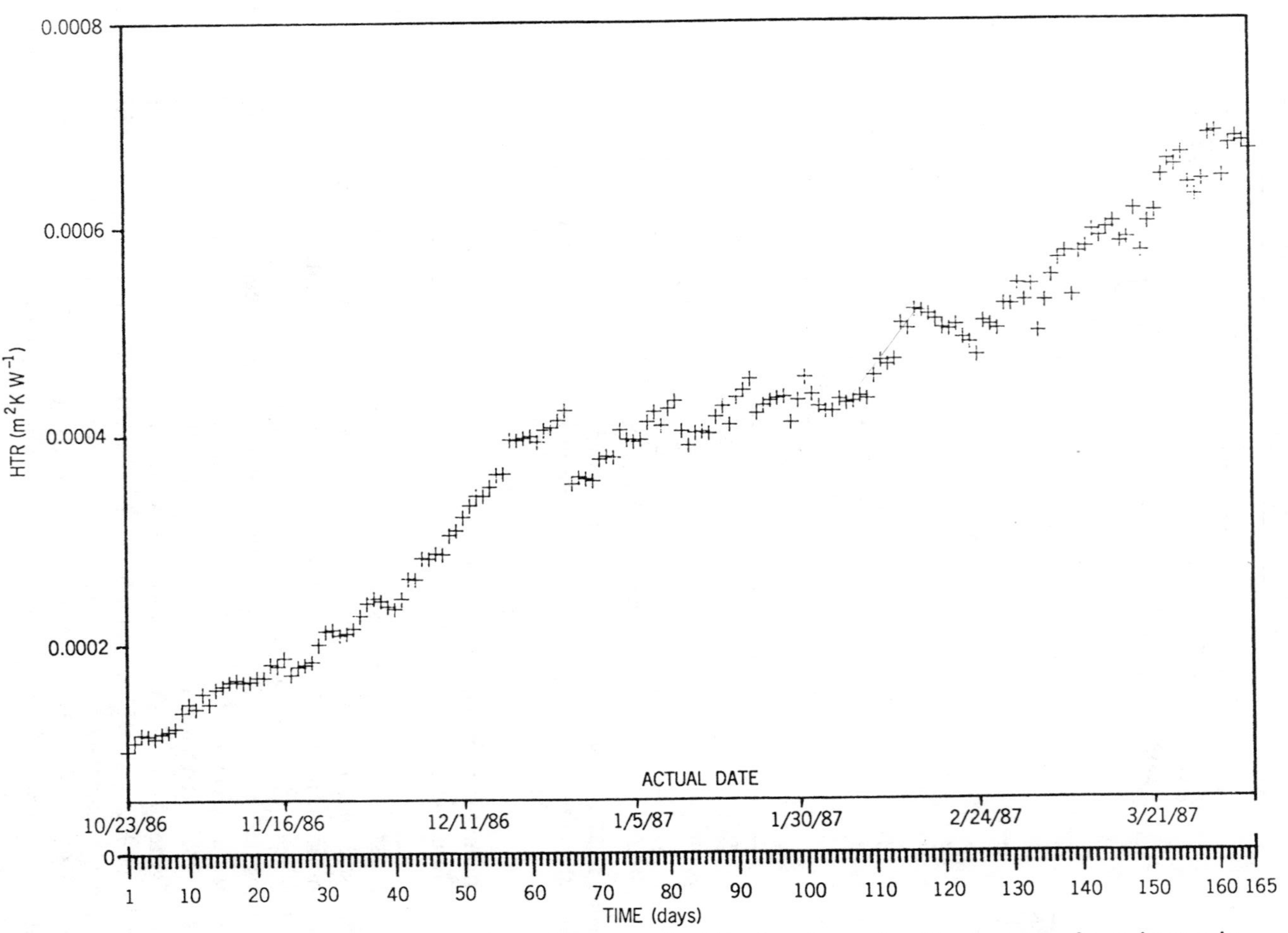

Figure 14.24 The progression of heat transfer resistance at the Atlantic estuary site. The heat transfer resistance increased continuously through the winter months. (McCaughey et al., 1987b.)

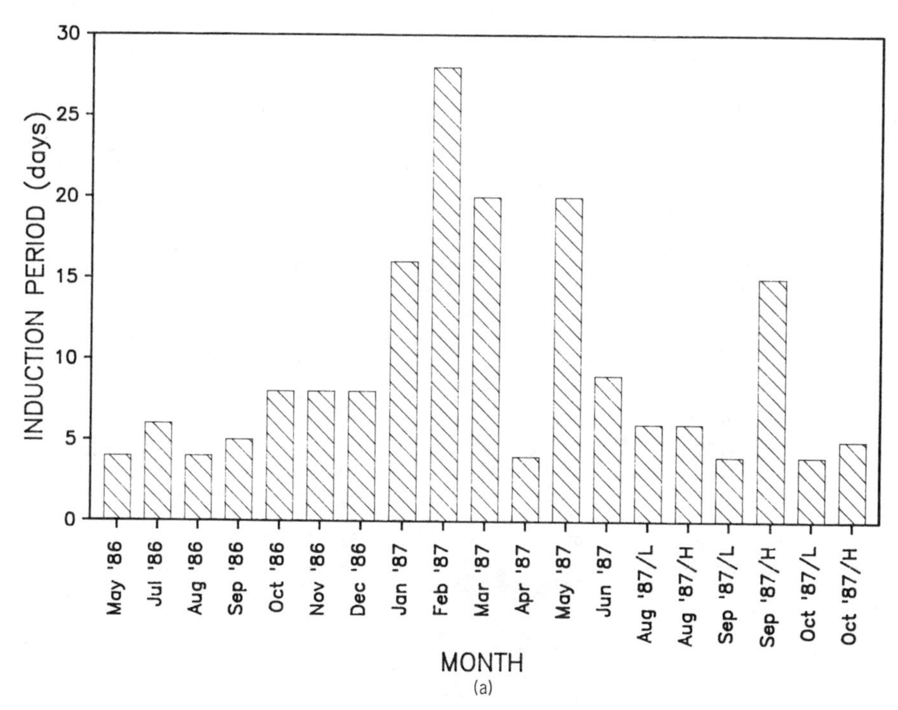

Figure 14.25 Two fouling monitors were operated in parallel at the Atlantic estuary site for one month and then cleaned to bare metal. The monitors were then started again for another month. The heat transfer resistance results are presented in terms of the (a) length of the induction period, (b) *log rate,* and (c) *extent* of deposit accumulation for each month. The three periods are analogous to the three periods defined in Figure 14.4. (Marine Biocontrol and CCE, Inc., 1988.)

14.9.4.4 Substratum Temperature

Laboratory Tests. The substratum temperature also influences biofilm processes. The *extent* of biofilm accumulation, as measured by the maximum biofilm attained, is a function of substratum temperature and nutrient loading rate in an aluminum (6061-T6) tube (13.1 mm ID) operating at a velocity of 0.81 $m\ s^{-1}$ (Figure 14.28). There is a significant interaction between substratum temperature and substrate loading rate. For example, the surface temperature has a minimal effect on the maximum biofilm thickness at a substrate loading rate of 0.06 $g\ m^{-2}\ h^{-1}$. However, at a high substrate loading rate, the substratum temperature has a strong influence on the maximum biofilm thickness. The bulk water temperature was maintained at 35°C during these experiments.

Field Tests. A field test conducted with AL-6X and copper-nickel 90/10 tubes in the Hudson River indicates that the tube wall temperature significantly influences the accumulation of fouling deposits (Characklis and Zelver, 1983).

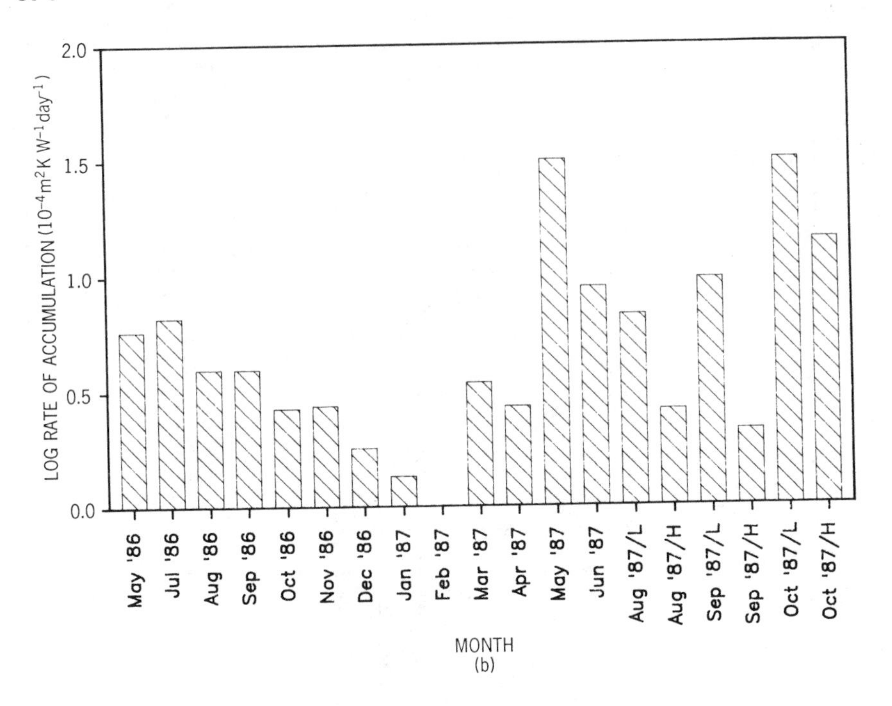

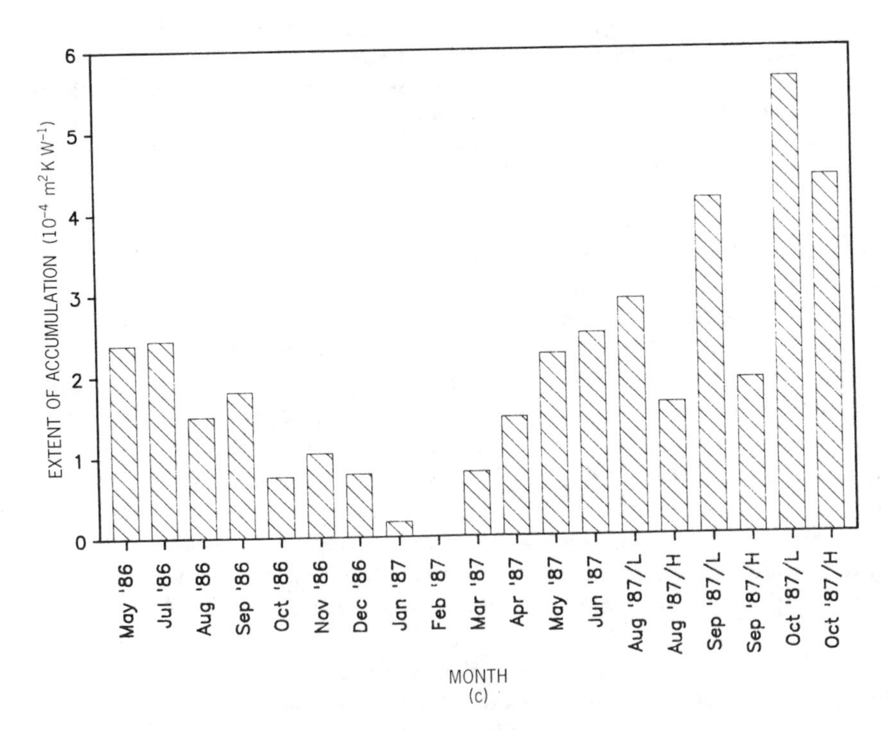

Figure 14.25 Continued

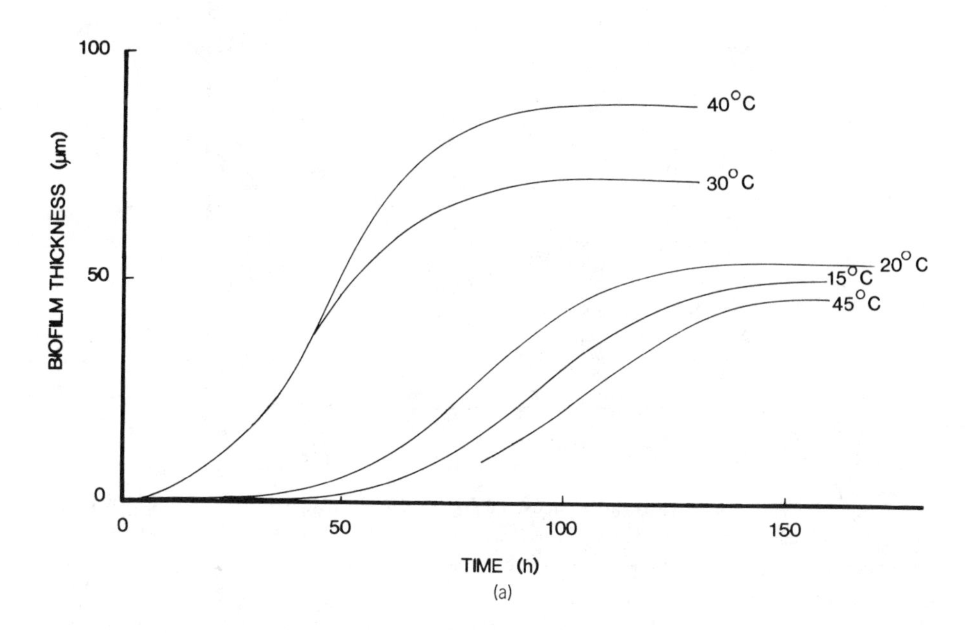

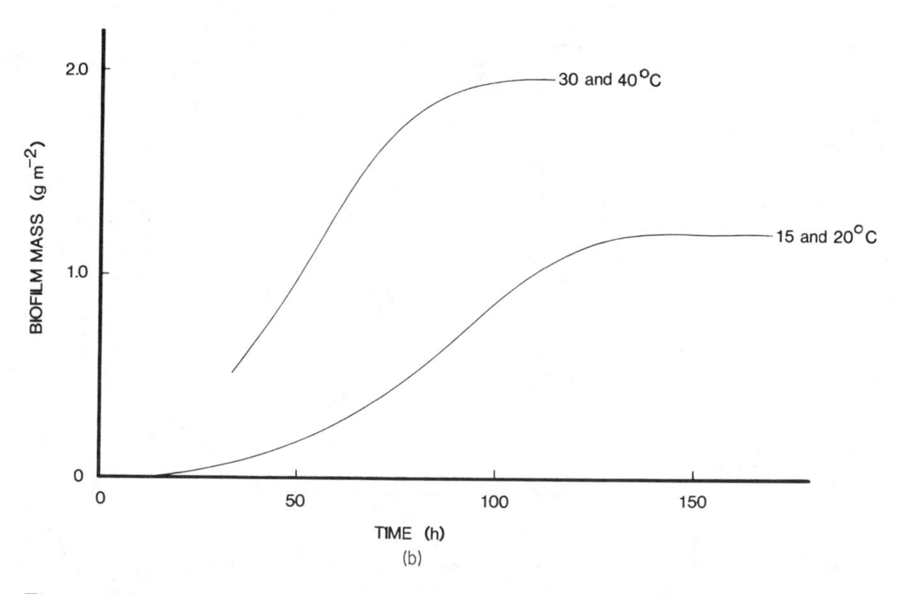

Figure 14.26 Temperature has a significant influence on biofilm accumulation as measured by (a) biofilm thickness and (b) biofilm mass. Experiments were conducted in a tubular (steel tube, 12.7 mm ID) recycle reactor with water velocity 0.9 m s^{-1}. Nutrient (trypticase soy broth) loading rate was low at 0.1 mg m^{-2} h^{-1}. (Stathopoulos, 1981.)

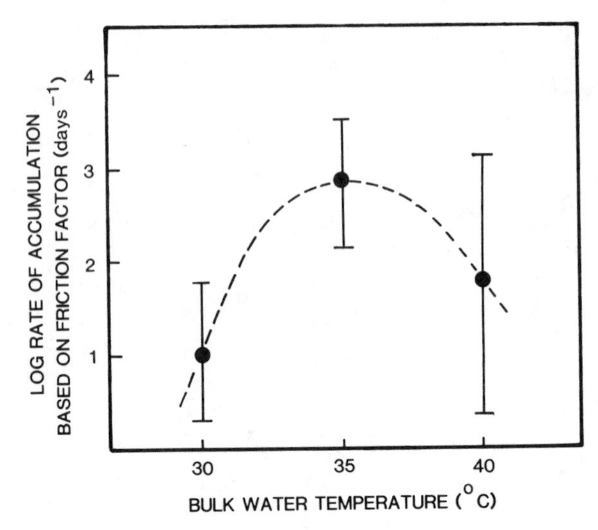

Figure 14.27 Temperature influences the log rate of biofilm accumulation in a RotoTorque. Shear stress was 1.6 N m^{-2} (150 rpm), and substrate loading rate (glucose) ranged from 0.1 to 0.3 mg m^{-2} h^{-1}. (Characklis, 1980.)

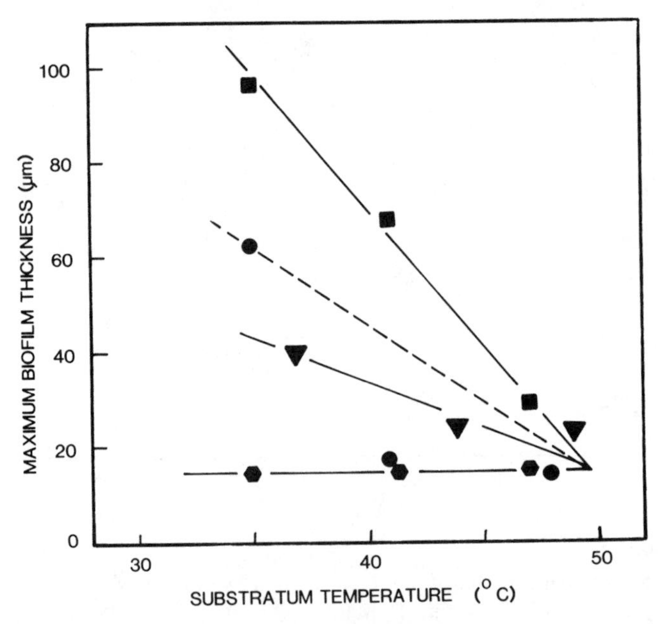

Figure 14.28 Substratum temperature and substrate (glucose) loading rate [(■) 2.36, (●) 0.5, (▼) 0.3, (⬢) 0.06 g m^{-2} h^{-1}] influence the extent of biofilm accumulation in an aluminum tube with a water velocity of 0.8 m s^{-1} (shear stress was 4.1 N m^{-2}). Bulk water temperature was 35°C. (Characklis, 1980.)

TABLE 14.6 Deposit Accumulation in the Heated (T_w = 35°C) and the Unheated Section of an AL-6X Stainless Steel Tube and a Copper-Nickel 90–10 Tube Receiving Hudson River Water[a]

		Deposit Mass (g m^{-2})	
Tube Material	Water Velocity	Heated Section	Unheated Section
AL-6X	1.7	48.5	21.0
AL-6X	1.7	45.0	26.0
AL-6X	1.0	34.5	41.5
Cu-Ni	1.7	22.5	15.5

[a] Water temperature varied from 17 to 24°C, and water velocity was 0.5 m s^{-1}. (Zelver et al., 1984.)

Significantly more deposit accumulated in the heated section of a tube (T_w = 35°C) than in the unheated section; the water temperature varied from 17 to 24°C when the water velocity was 0.5 m s^{-1} (Table 14.6). For water velocity 0.3 m s^{-1}, deposit accumulation was greater in the unheated tube. Similar results (Table 14.7) are reported by McCaughey et al. (1987b) for Hudson River water flowing in titanium condenser tubes (2.2 m s^{-1}). Novak (1982) has also observed the influence of biofouling deposit temperature on *log rate* of biofouling deposit accumulation in the Rhine River (Figure 14.29). The maximum rate of fouling was observed around 30°C.

TABLE 14.7 Deposit Mass, Volatile Deposit Mass, and Deposit Cell Number Accumulation in the Heated and the Unheated Section of a Titanium and an Admiralty Brass Tube (T_w = 35°C) Receiving Hudson River Water[a]

	Heated		Unheated	
	Ti	AB[b]	Ti	AB[b]
Mass (g m^{-2})	140.85	167.04	76.86	47.09
Volatile mass (%)	13.30	15.52	19.70	19.30
Standard plate count (number m^{-2})	1×10^8	6×10^9	9×10^{10}	3×10^{11}
Pseudomonas (number m^{-2})	$>3 \times 10^7$	8×10^5	$>4 \times 10^7$	$>4 \times 10^7$
Sulfate-reducing bacteria (number m^{-2})	3×10^7	3×10^7	$>4 \times 10^7$	$>4 \times 10^7$

[a] Water temperature varied from 1 to 24°C. Water velocity in the admiralty brass tube varied from 1.34 to 2.2 m s^{-1} as necessary to simulate plant operating conditions. Water velocity in the titanium tube was 1.34 m s^{-1} throughout the test. (McCaughey et al., 1987b.) 582 days exposure. > or < indicates above or below detection limits, respectively.

[b] Admiralty brass.

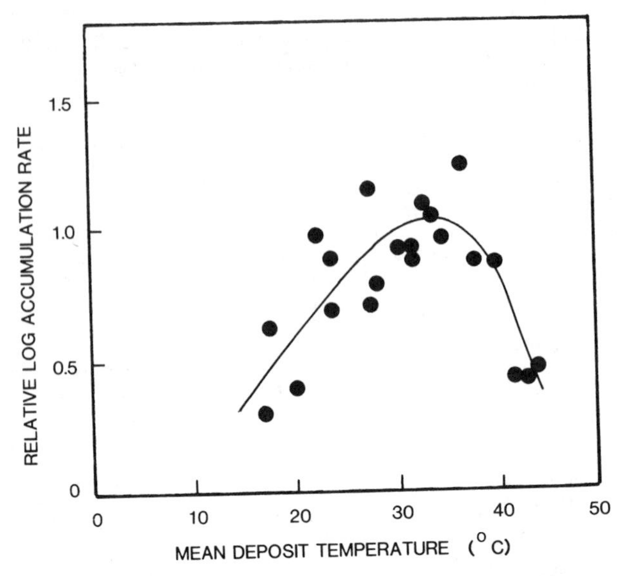

Figure 14.29 Biofilm deposit temperature influences the log rate of fouling accumulation in the Rhine River. The relative log rate of accumulation is the ratio of the log rate of accumulation at temperature T to the rate at 30°C. The maximum rate occurs at approximately 35°C. (Reproduced with permission from Novak, 1982.)

The substratum temperature influences the rate and extent of biofouling accumulation.

14.9.5 Biological Factors

Biological factors have their greatest influence on biochemical reactions within the deposit, where bacterial cell numbers, the physiological state of the cells, and the microbial community structure determine the influence of the deposit on equipment performance. However, biological factors may also influence interfacial adsorption/attachment as well as detachment. For example, microbial cell accumulation may increase the adsorptive capacity of the wetted substratum and serve as an effective trap for particulates, as indicated by the progression of fouling deposit *composition* at a power plant site. The mass of cells relative to inorganic matter in a deposit is a function of deposit accumulation and may be different at different sites (Figure 14.30). At the Atlantic estuary site, cell numbers increase with deposit accumulation and occupy a constant fraction of the deposit. At the Hudson River site, cells appear to initially colonize the substratum followed by accumulation of inorganic material which dominates the deposit thereafter. Microbial activity within the deposit may also increase the cohesiveness of the deposit (Duddridge et al., 1981), forming an organic matrix that provides resistance to hydrodynamic shear forces. Biological factors probably influence the effectiveness of chemical treatment programs in minimizing

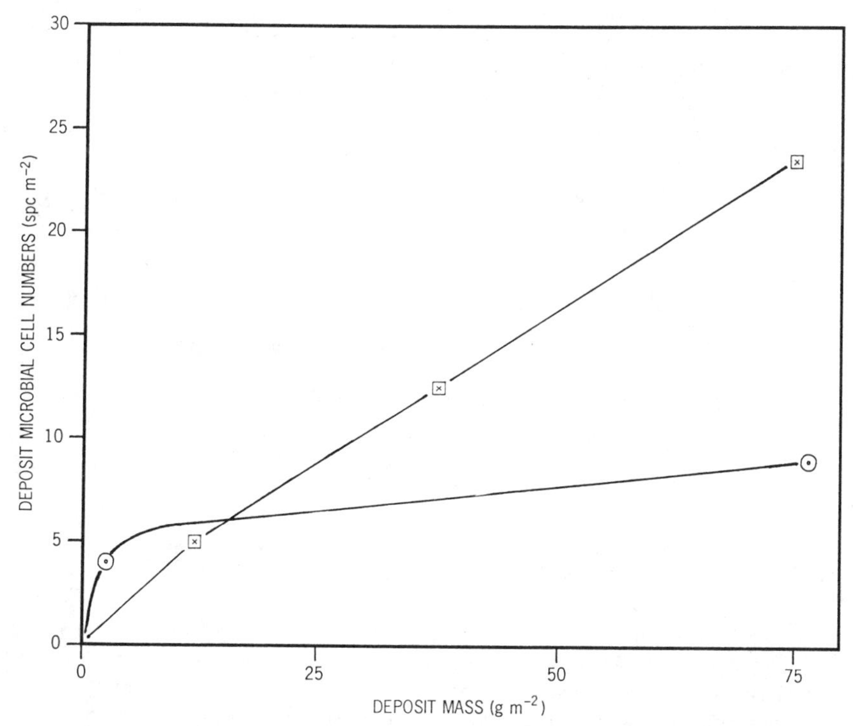

Figure 14.30 The relative amount of viable cells (standard plate count, spc) in a deposit as a function of deposit accumulation is different at different sites. At the Atlantic estuary site (⊠), cell numbers increase with deposit accumulation and occupy a constant fraction of the deposit (McCaughey et al., 1987b). At the Hudson River site (⊙), cells appear to initially colonize the substratum, followed by accumulation of inorganic material, which dominates the deposit thereafter (McCaughey et al., 1987a).

other types of fouling (e.g., precipitation and corrosion fouling). For example, microbial activity increases the rate of hydrolysis of complex phosphates used to inhibit corrosion and scale.

Fouling is a heterogeneous process in that it requires both a liquid and solid phase. Consequently, conditions at the water–substratum interface control the process rates that contribute to fouling deposition. However, interfacial concentrations are extremely difficult to measure, and so attempts are made to relate interfacial processes to bulk water conditions. As an illustration, consider biofouling in an open recirculating cooling tower (RCT) system. The extent of biofouling, an interfacial process, is frequently correlated with the colony-forming units (CFU) in the bulk water as measured by a plate count procedure. Characklis et al. (1981c) report, however, that no consistent relationship exists between CFU in the bulk water and biofouling accumulation in a recirculating cooling tower system (Figure 14.31). The difficulty in relating CFU in the bulk

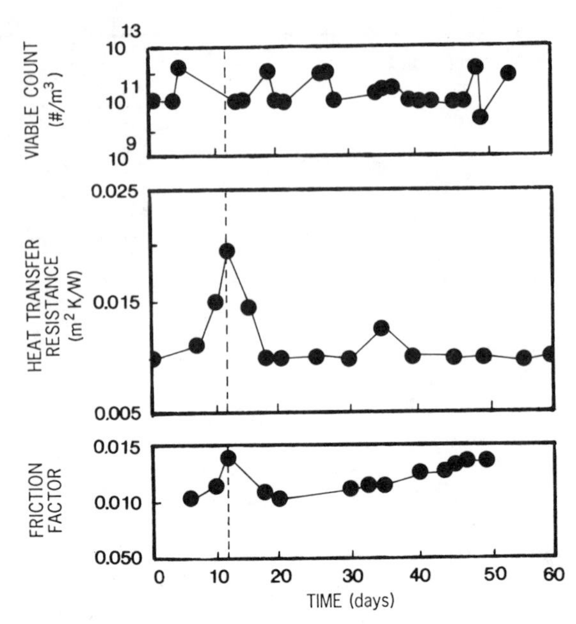

Figure 14.31 The relationship between suspended (planktonic) cell numbers, heat transfer resistance, and friction factor in a tube within a recirculating cooling tower system. Chlorine was added continuously for the first 12 days of the test. Thereafter, a bromochlorodimethylhydantoin was added. The total cell count in the bulk water did not correlate with the biofilm-related variables, heat transfer resistance, and friction factor. (Characklis et al., 1981c.)

water to biofouling accumulation stems from the numerous processes by which microorganisms may enter and leave the bulk water: they may (1) enter with the make-up water, (2) enter by growth and reproduction within the bulk water, (3) enter by being washed from the air, (4) enter by detachment from biofouled surfaces, (5) leave in the blowdown and drift, (6) leave by adsorption to surfaces and (7) leave by death, at least partially due to biocide application. Obviously, the number of CFU in the cooling water is not strictly related to the *extent* of biofouling accumulation in the RCT system. The friction factor also increased in these tests, with no measurable increase in heat transfer resistance, suggesting that accumulation of a hydraulically rough biofilm was continuing throughout the test period.

The *log rate* of fouling accumulation may also be influenced by microbial growth rate and microbial cell number concentration, as reported by Bryers and Characklis (1981) for a glass laboratory tubular reactor continuously inoculated with an undefined population growing in a chemostat. Increasing the specific growth rate (dilution rate) in the chemostat resulted in an increased *log rate* of biofilm accumulation (see Figure 7.7, Chapter 7). Increasing the cell number concentration also increased the biofilm accumulation *log rate* (Figure 14.32b).

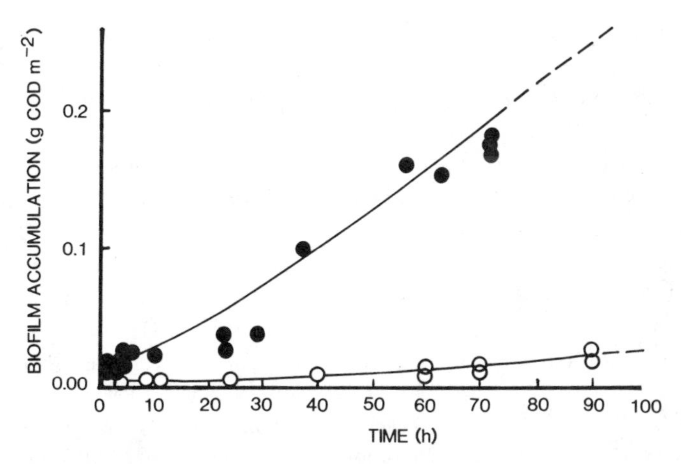

Figure 14.32 The influence of bulk water biomass concentration [48 (●) and 4.4 (○) g m^{-3}] on the accumulation of a mixed population biofilm in a glass tube. The specific growth rate in the chemostat was 0.28 h^{-1}, and the Reynolds number 17,200. (Reprinted with permission from Bryers and Characklis, 1981.)

The difference in physiological states between cells in a chemostat (actively growing) and cells in an oligotrophic or nutrient-deficient natural environment is not clear, but it may be significant in processes leading to the accumulation of biofilms and must be addressed in more detail.

REFERENCES

Battaglia, P.J., D.P. Bour, and R.M. Burd, *Biofouling Control Practice and Assessment,* Final Report, EPRI CS-1976, Project 1132-1, Electric Power Research Institute, Palo Alto, CA, 1981, pp. 221.

Beal, S.K., *Nucl. Sci. Eng.,* **40**, 1-11 (1970).

Bland, C.E.G., R.W. Bayley, and E.V. Thomas, *J. Wat. Poll. Contr. Fed.,* **50**, 134-143 (1978).

Bott, T.R., "Biological fouling of heat transfer surfaces," presented at Fouling—Science or Art, University of Surrey, England, 1979.

Bott, T.R. and M. Gunatillaka, "Nutrient composition and biofilm thickness," in R.W. Bryers and S.S. Cole (eds.), *Fouling of Heat Exchanger Surfaces,* United Engineering Trustees, New York, 1983, pp. 727-734.

Brown, C.M., D.C. Ellwood, and J.R. Hunter, *FEMS Microbiol. Lett.,* **1**, 163-166 (1977).

Bryers, J.D. and W.G. Characklis, "Kinetics of biofilm formation within a turbulent flow system," in E.F.C. Somerscales and J.G. Knudsen (eds.), *Fouling of Heat Transfer Equipment,* Hemisphere, Washington, 1981, pp. 313-334.

Burrill, K.A., "Laboratory studies to resolve fouling of heat exchangers and sieve trays in Canadian girdler-sulfide heavy water plants", in E.F.C. Somerscales and J.G. Knudsen (eds.), *Fouling of Heat Transfer Equipment,* Hemisphere, Washington, 1981, pp. 227-247.

Busch, A.W., *Aerobic Biological Treatment of Waste Waters,* Gulf Publishing Co., Houston, TX, 1971.

Characklis, W.G., *Biofilm Development and Destruction,* Final Report, Project RP902-1, Electric Power Research Inst., Palo Alto, CA, 1980.

Characklis, W.G., *Biotech Bioeng.,* **23**, 1923-1960 (1981).

Characklis, W.G., *Bacterial Regrowth in Distribution Systems,* Final Report, American Water Works Association Research Foundation, Denver, CO, 1988.

Characklis, W.G. and S.T. Dydek, *Wat. Res.,* **10**, 515-522 (1976).

Characklis, W.G. and J.A. Robinson. *Development of a Fouling Control Program.* Final Report, Canadian Electrical Association, No. 219 G 388, 1984.

Characklis, W.G. and N. Zelver, *Reactor Containment Fan Cooler Biofouling Test,* Final Report, Westinghouse Corporation, PO 544-MXX-438527-SP, 1983.

Characklis, W.G., M.G. Trulear, N.A. Stathopoulos, and L.C. Chang, "Oxidation and destruction of microbial films," in R.L. Jolley et al. (eds.), *Water Chlorination,* Ann Arbor Press, Ann Arbor, MI, 1980, pp. 349-368.

Characklis, W.G., M.J. Nimmons, and B.F. Picologlou, *J. Heat Transfer Eng.,* **3**, 23-37, 1981a.

Characklis, W.G., N. Zelver, M.H. Turakhia, and M.J. Nimmons, "Fouling and heat transfer," in J.M Chenoweth and M. Impagliazzo (eds.), *Fouling of Heat Exchange Equipment,* Amer. Soc. Mech. Eng., New York, 1981b, pp. 1-15.

Characklis, W.G., N. Zelver, M.H. Turakhia, and F.L. Roe, "Energy losses in water conduits: Monitoring and diagnosis," in *Proc. 42nd International Water Conference,* Engineers' Society of Western Pennsylvania, 1981c, pp. 229-235.

Characklis, W.G., N. Zelver, C.H. Nelson, R.O. Lewis, D.E. Dobb, and G.K. Pagenkopf, "Influence of biofouling and biofouling control techniques on corrosion of copper-nickel tubes," Nat. Assoc. Corrosion Eng., Paper No. 250, CORROSION/83, Anaheim, CA, 1984.

Duddridge, J.E., C.A. Kent, and J.F. Laws. "Bacterial adhesion to metallic surfaces," in *Progress in the Prevention of Fouling in Industrial Plant,* Corrosion Science Division, Institution of Corrosion Sciences and Technology, Nottingham University, Nottingham, England, 1981, pp. 127-153.

Eaton, A., C. Chamberlain, and M. Cooney, "Colonization of Cu-Ni surfaces by microfouling organisms," in J.F. Garey, R.M. Jorden, A.H. Aitken, D.T. Burton, and R.H. Gray, (eds.), *Condenser Biofouling Control,* Ann Arbor Science, Ann Arbor, MI, 1980, pp. 105-120.

Epstein, N., "Fouling: Technical aspects," in E.F.C. Somerscales and J.G. Knudsen (eds.), *Fouling of Heat Transfer Equipment,* Hemisphere, Washington, 1981, pp. 31-53.

Ferguson, R.J., "Determination of seasonal variations in microbiological fouling factors and average slime thickness," CTI paper no. TP-237A, Cooling Tower Inst., Houston, TX, 1981.

Flemming, H., *Wat. Res.,* **21**, 745-756 (1987).

Garey, J.F., "Biofouling control investigations," Interim Report, EA 1082, Research Project 1372-1, Electric Power Research Institute, 1979.

Garey, J.F., R.M. Jorden, A.H. Aitken, D.T. Burton, and R.H. Gray (eds.), *Condenser Biofouling Control,* Ann Arbor Science, Ann Arbor, MI, 1980.

Harty, D.W.S. and T.R. Bott. "Deposition and growth of microorganisms on simulated heat exchanger surfaces," in E.F.C. Somerscales and J.G. Knudsen (eds.), *Fouling of Heat Transfer Equipment,* Hemisphere, Washington, 1981, pp. 334–344.

Kern, D.Q. and R.E. Seaton, *Brit. Chem. Eng.,* **4**, 258–262 (1959).

Kornbau, R.W., C.C. Richard, and R.O. Lewis, "Seawater biofouling countermeasures for spirally enhanced condenser tubes," Event No. 279, in *Proc. European Fed. Chem. Eng.,* Pergamon Press, London, 1983.

Kornegay, B.H. and J.F. Andrews, *Characteristics and Kinetics of Biological Film Reactors,* Final Report, Grant WP-01181, Fed. Wat. Poll. Contr. Admin., U.S. GPO, Washington, 1967.

Kreith, F., *Principles of Heat Transfer*, 3rd ed., Harper and Row, New York, 1973, pp. 571–573.

Lister, D.H., "Corrosion products in power generating systems," in E.F.C. Somerscales and J.G. Knudsen (eds.), *Fouling of Heat Transfer Equipment,* Hemisphere, Washington, 1981, pp. 135–200.

Little, B.J., J. Morse, G.I. Loeb, and F. Spiehler, "A biofouling and corrosion study of ocean thermal energy conversion (OTEC) heat exchanger candidate metals," in *Proc. 6th OTEC Conference, Ocean Thermal Energy for the 80's,* Vol. 2, 1979, 12.13.1–12.13.9.

Loeb, G.I., *Drag Enhancement of Microbial Slime Films on Rotating Discs,* Naval Research Memorandum Report 4412, Naval Research Lab., Washington, 1981.

Lowe, M.J., J.E. Duddridge, A.M. Pritchard, and T.R. Bott, Biological-particulate fouling interactions: Effects of suspended particles on biofilm development, *in Proc. First National UK Heat Transfer Conference,* Leeds, 1984, pp. 391–400.

Lund, D. and C. Sandu, "Chemical reaction fouling due to foodstuffs," in E.F.C. Somerscales and J.G. Knudsen (eds.), *Fouling of Heat Transfer Equipment,* Hemisphere, Washington, 1981, pp. 437–476.

Marine Biocontrol Corporation and CCE, Inc., *Oyster Creek Nuclear Generating Station Chlorine Optimization Study,* Final Report, General Public Utilities Nuclear Corporation, Forked River, NJ, 1978.

Matson, J.V. and W.G. Characklis, *J. Cooling Tower Inst.,* **4**, 24 (1983).

McCaughey, M.S., W.L. Jones, W.G. Characklis, and D. Goodman, *Condenser Fouling Test Program,* Final Report, Contract No. 025240-84, New York Power Authority, New York, 1987a.

McCaughey, M.S., A. Thau, W.G. Characklis, and W.L. Jones, "An evaluation of condenser tube fouling at an estuarine nuclear power plant," presented at Joint Power Generation Conference, ASME, Miami, FL, 1987b.

Norrman, G., W.G. Characklis, and J.D. Bryers, *Dev. Ind. Microbiol.,* **18**, 581–590 (1977).

Novak, L., *J. Heat Transfer,* **104**, 663–669 (1982).

Pedersen, K., *Appl. Environ. Microbiol.,* **44**, 1196–1204 (1982).

Picologlou, B.F., N. Zelver, and W.G. Characklis, *J. Hyd. Div. ASCE,* **106**, 733–746 (1980).

Pritchard, A.M. "Fouling—science or art? An investigation of fouling and antifouling measures in the British Isles," in E.F.C. Somerscales and J.G. Knudsen (eds.), *Fouling of Heat Transfer Equipment,* Hemisphere, Washington, 1981, pp. 513–523.

Ridgway, H.F., C.A. Justice, C. Whittaker, D.G. Argo, and B.H. Olson, *J. AWWA,* **76**, 94 (1984).

Scavia, D. and G.A. Laird, *Limnol. Oceanogr.,* **32**, 1017–1033 (1987).

Stathopoulos, N., "Influence of temperature on biofilm processes," unpublished Master of Science dissertation, Rice University, Houston, TX, 1981.

Sugam, R., *Biofouling Control with Ozone at the Bergen Generating Station,* Final Report CS 1629, Project No. 733-1, Electric Power Research Institute, Palo Alto, CA, 1980.

Taborek, J., J.G. Knudsen, T. Aoki, R.B. Ritter, and W.J. Palen, *Chem. Eng. Prog.,* **68**, 59–67 (1972).

Turakhia, M.H. and W.G. Characklis, *Can. J. Chem. Eng.,* **61**, 873–875 (1983).

Turakhia, M.H. and W.G. Characklis, *Biotech. Bioeng.,* **33**, 406–414 (1989).

van der Kooij, D., A. Visser, W.A.M. Hijnen, *J. Am. Wat. Wks. Assoc.,* **74**, 540–545 (1981).

van Nostrand, W.L., S.H. Leach, and J.L. Haluska, "Economic penalties associated with the fouling of refining heat transfer equipment," in E.F.C. Somerscales and J.G. Knudsen (eds.), *Fouling of Heat Transfer Equipment,* Hemisphere, Washington, 1981, pp. 619–643.

Watkinson, A.P. and N. Epstein, *Chem. Eng. Prog. Symp. Series,* **65**(92), 84 (1969).

Watkinson, A.P., L. Louis, and R. Brent, *Can. J. Chem. Eng.,* **52**, 558–562 (1974).

Wilkinson, J.F., *Bact. Rev.,* **22**, 46–73 (1958).

Zelver, N., J.R. Flandreau, W.H. Spataro, K.R. Chapple, W.G. Characklis, and F.L. Roe, "Analysis and monitoring of heat transfer fouling," 82-JPGC/Pwr-7, ASME, 1982.

Zelver, N. W.G. Characklis, J.A. Robinson, F.L. Roe, Z. Dicic, K.R. Chapple, and A. Ribaudo, "Tube material, fluid velocity, surface temperature and fouling: A field study," CTI paper no. TP-84-16, Cooling Tower Institute, Houston, TX, 1984.

15

MICROBIAL BIOFOULING CONTROL

WILLIAM G. CHARACKLIS

Montana State University, Bozeman, Montana

SUMMARY

The choice of a method for controlling fouling biofilm accumulation and/or activity in a specific operating environment is based on the overall cost, which includes costs related to environmental control, corrosion losses, necessary plant modification, and safety. The extent to which these factors influence fouling control varies with the season, process operation, and other variables. Thus, process considerations are important and must influence the appropriate choice of chemical concentration, duration of treatment, and frequency of treatment. In addition, a means to monitor the continuing effectiveness of the control program is necessary. Part of the motivation for developing a rational approach to biofilm accumulation in preceding chapters is the expectation that such an approach will lead to more satisfactory methods for preventing and/or controlling biofouling.

Chlorine is the most commonly used chemical for controlling biofouling and hence is used in this chapter to illustrate the mode of action of biocides on fouling biofilms. The general principles, however, are just as relevant to other oxidizing biocides and nonoxidizing chemical treatments. The effectiveness of biocides depends on their ability to inactivate biofilm organisms and/or detach significant portions of the biofilm matter. Acceptable kinetic expressions have yet to be established for relating rate of inactivation or rate of detachment to biocide concentration. As a result, modeling of biocide effects on biofilm accumulation is still a rather empirical activity.

Frequently, the effectiveness of biofouling control procedures varies with environmental or operating variables. Consequently, sensors are needed to provide feedback so that process adjustments can be made. More sensitive monitors are needed in some instances, as well as instruments to assess the deposit composition. However, the major requirement at present is a mathematical model for simulating the action of a biocide (or mechanical treatment) on a biofilm. The model will serve to distill the convoluted methods presently being used to assess kill efficiency into rational process parameters that can be easily interpreted in the context of the operating equipment. The model will also enable fouling monitors to provide feedback response to the treatment process. Such a model is needed urgently.

15.1 INTRODUCTION

The deleterious effect of biofilms on various technological systems was described in Chapter 14. As a consequence of these economic losses, numerous methods have been used for minimizing the accumulation of biofilms on surfaces, including the following: (1) external water treatment to reduce the number of bacteria entering the system, (2) coatings or other treatments that alter the surface energy of the substratum or release toxic components into the aqueous phase, (3) addition of biocides (oxidizing and nonoxidizing) to the bulk water, which inactivate organisms entering the system or reduce the growth rate of microorganisms within the biofilm, and (4) internal chemical treatments (e.g., biocides, dispersants, surfactants) and mechanical treatments (e.g., sponge balls, brushes) that result in removal (detachment) of biofilm from the substratum (Table 15.1). Each method is directed at influencing one or more of the fundamental biofilm processes (Section 15.2).

A major decision regarding the choice of treatment technique is related to whether *prevention* or *control* of biofilm accumulation is desirable. Prevention requires sterilization of the incoming water, continuous flow of a biocide at high concentration, and/or treatment of the substratum that completely inhibits microbial adsorption. Some measure of control is generally possible with all the methods described. The extent to which any treatment can be applied depends on environmental, process, and economic considerations.

Significant success has been achieved in controlling certain specific types of fouling deposition when they are homogeneous enough. For example, calcium carbonate crystallization fouling can be controlled quite effectively by phosphonate compounds (or other inhibitors). Laboratory and field observations, however, indicate that these same compounds are not as effective in preventing surface precipitation when biofouling is not controlled.

Little is known about the interactive effects *between different types of fouling deposits and effective means for their control.*

For example, do polyphosphate corrosion inhibitors in drinking water systems remain effective when significant biofouling is occurring on the metal surface? Does the added phosphate enhance microbial biofilm activity?

TABLE 15.1 Common Fouling Control Methods Can Be Categorized According to Their Effect on Fundamental Processes Contributing to Fouling Accumulation[a]

	Control Methods		
Fundamental Process	Biofouling	Scaling	Corrosion
Water phase transport	Filtration	Water softening	Deaeration
Water-substratum interactions	Biodispersants	Dispersants	Coatings Surface treatments
Transformations in the deposit	Biocides Biocidal coatings	Crystal growth inhibitor	Coatings Alloy (Cu) Electrolytic protection
Detachment	Biocides Mechanical methods	Chelants Inhibited acid	—

[a] Roe and Characklis, 1983.

Chlorination is a common means of controlling biofilm formation in utilities and in the process industry. The use of chlorine as a treatment for biofouling will be described in detail in this chapter as an illustration of the use of biocides and concerns related to their use. Numerous proprietary and nonproprietary biocides as well as other chemicals have been useful for inhibiting biofilm accumulation, and their effectiveness will be described by illustrating the effects of chlorine on biofilms.

The effectiveness of various treatments in actual field applications must be assessed in a meaningful, accurate, and sensitive manner, because expensive treatment programs can be ineffective and in some cases may even enhance fouling. "Cleanliness" factors are sometimes reported for power plants using heat transfer rate measurements from full scale heat exchangers, but these data are useless for assessing the extent of biofouling, because of the noise in the measurement. Some simplistic sidestream devices depend on visual observations of the deposit, while others measure the frictional resistance and/or heat transfer resistance resulting from the deposition. In the oil field, viable cell counts are being used to determine the effectiveness of biocide treatments, even though such counts may seriously underestimate overall microbial activity in the system.

Quantitative measures of equipment performance or lifetime are necessary to provide the operations engineer with a measure of the effects of fouling, influence of operating variables on fouling rates, and effectiveness of treatment programs.

Without quantitative measurements, very little of long term value will be learned from any sampling and monitoring program.

15.2 PROCESS CONSIDERATIONS IN FOULING CONTROL

Fouling biofilm accumulation can be considered the net result of several physical, chemical, and biological processes (Chapter 7). In Table 15.1, common fouling control methods are categorized according to their effect on fundamental processes (Roe and Characklis, 1983).

Fouling control involves interrupting any of the first three fundamental fouling processes in Table 15.1 and/or enhancing the fourth process.

15.2.1 Transport

The rate of initial adsorption of cells to the substratum is directly proportional to the concentration of cells in the bulk water. Thus, reducing the cell concentration in the bulk water will decrease the transport rate (and hence the adsorption rate) of cells to the substratum. The reduced rate of transport will reduce the rate of biofouling. For example, the water can be filtered to remove microorganisms, thus reducing the concentration of the deposit-promoting material in the feed water. However, filtration to remove bacteria can be a very expensive if large volumes of high quality water must be produced. Disinfection of the incoming water, as in drinking water distribution systems, can be relatively effective in minimizing biofilm accumulation.

Nevertheless, the accumulation of an established biofilm (e.g., after a chlorine treatment) is due primarily to growth processes, and the contribution of transport and attachment of cells is negligible.

15.2.2 Adsorption

Interrupting the adsorption process will also reduce biofouling and can be accomplished by reducing the attraction between the cells and the substratum. Dexter (1979) and Baier (1972) suggest that the forces of attraction can be minimized by modifying the substratum surface energy, usually by applying a specific coating to the substratum. However, if the solution were so simple, biofouling would not continue to be a problem. In addition, cell surface charge can be modified by adding polyelectrolytes or surfactants to the water (Drew Chemical Corp., 1979). By modifying bonding sites on the cells or substratum or both, the fouling process may be inhibited, but prevention is doubtful.

15.2.3 Transformations

Chemical additives (biocides) are widely used in industry to control biofouling by inhibiting microbial growth and/or metabolism. Oxidizing (e.g., chlorine) and nonoxidizing biocides (e.g., quaternary ammonium compounds) are added periodically to inactivate (i.e., "kill") the biofilm cells. However, a significant number of the biofilm organisms will usually survive treatment. Since the microbial growth rate is proportional to the number of viable cells, the biocide reduces the rate of growth processes leading to biofilm accumulation. The biocide is sometimes trapped in a paint or coating and released slowly from the substratum, inhibiting growth at the substratum–water interface.

15.2.4 Detachment

Both mechanical and chemical methods are used to disrupt the bonds between biofilm cells and the substratum or to destroy the cohesiveness of cells within the deposit, resulting in detachment of deposit material. Mechanical methods include periodic increases in flow rate as well as entrainment of proprietary sponge balls or brushes that scrape the substratum clean. In larger pipelines, scrapers or "pigs" are used to mechanically remove biofilm and other deposits from the pipe surfaces. Certain chemicals, added periodically, can also cause detachment of accumulated fouling deposits. For example, chlorine is applied in "shock" doses to strip microbial slimes from fouled surfaces. However, dispersants, acids, and chelants may also be effective.

15.3 SYSTEM CONSIDERATIONS

Part of the impetus for developing a rational approach to biofilm accumulation in the preceding chapters is the expectation that such an approach will lead to more satisfactory methods for preventing and/or controlling biofouling. There are both physical and chemical methods available for this purpose (Table 15.1). However, none of these methods is universally cost-effective, or biofouling would no longer be a problem.

The decision regarding a method for controlling fouling biofilm in a specific operating environment must consider the following system factors:

1. Environmental control
2. Water quality
3. Influence of treatment on corrosion losses (cost)
4. Cost of treatment program or process
5. Cost for plant modification
6. Cost, cost, cost.

The extent to which these factors influence fouling control varies with season, process operation, and other variables. Hence, the cost effectiveness of a fouling control program will generally be site-specific.

15.3.1 Environmental Quality

Some chemical treatments are quite effective but are undesirable due to their impact on environmental quality. Chlorine has been used for years because it is a reasonably effective method for controlling biofouling. But concern over the toxicity of chlorine and its reaction products has spurred the search for alternatives. Biocides, in general, are toxic materials and must be evaluated thoroughly prior to their large scale use. However, thoughtful engineering approaches may permit the use of compounds such as chlorine while restricting the environmental concentrations within satisfactory limits.

Intermittent chlorination has been the most widely used control process for heat exchanger biofouling in power station condensers and process plant heat exchangers in the United States. A small number of utilities and many process plants employ continuous low level chlorination to control macrofouling and microfouling when other control options are not feasible. However, the use of chlorine has been restricted by the effluent limitations imposed because of potential adverse environmental impact on aquatic organisms.

Total chlorine residual in the discharge of power plants in the United States is restricted to 0.2 g m^{-3} for two hours per day because of its environmental/health effects in the receiving waters. The effluent limitations on chlorine can be met either by control of the chlorination process or by the use of other methods, including dechlorination. Control of the chlorination process to ensure that effluent limitations are not exceeded can be difficult and may not allow flexibility to increase the chlorine dosage as required to control biofouling at every plant site. Operating experience at a number of power plants has proven that the EPA effluent limitations on total residual chlorine (TRC) for once-through cooling water systems cannot be met in all cases while maintaining satisfactory plant performance (Rice, 1980; Gaulke et al., 1980; Schumacher and Lingle, 1980). Frequently, the criterion for dosage is the maintenance of a residual at the outlet. Thus, the chlorine demand in the water is presumably nullified. Typical chlorine residuals in the outlet range from 0.1 to 0.5 mg/L for effective performance. However, chlorine concentrations as high as 20 mg/L applied several times a day may be needed to effectively control biofouling at a power plant (Battaglia et al., 1981). "Targeted chlorination," treating only a fraction of the system at high chlorine concentration and low duration, provides an alternative for maintaining performance while remaining within the chlorine effluent regulations (Mussalli et al., 1985; Moss et al., 1985) because dilution with untreated effluent occurs prior to discharge into the receiving water.

15.3.2 Water Quality

Frequently, other reacting components interfere with the intended control procedure. Chlorine is a useful biofouling control compound but, in heavily contaminated waters, is consumed in side reactions (chlorine demand reactions) and is rendered ineffective. Even copper-nickel alloys possess a significant chlorine demand (Characklis et al., 1983). Therefore, water quality and substratum composition are two of the factors that must be considered in choosing a treatment program to minimize fouling. In oil field water floods, the injection water is maintained anaerobic to minimize corrosion of the mild steel distribution system. In this case, oxidizing biocides with high rates of demand may be at a decided disadvantage. As a consequence, nonoxidizing biocides are predominant in oil field practice.

15.3.3 Influence of Treatment on Corrosion Losses

Some equipment materials are vulnerable to corrosion induced by high chlorine concentrations. For example, chlorine and quaternary ammonium compounds increase the corrosion rate of copper alloys (Characklis et al., 1983). When chlorine is reduced, chloride forms. In recirculating cooling tower systems operating at high cycles of concentration, organic compounds concentrate to high levels and microbial cells accumulate, leading to severe biofouling problems. If chlorine is added in the large quantities necessary to overcome high chlorine demand, increased chloride levels result. Chloride can increase the corrosion potential. High chlorine residuals may also lead to rapid deterioration of wood fill in cooling towers.

15.3.4 Cost of Treatment Program or Process

The ultimate goal of any biofouling control program is to maintain equipment performance and lifetime at the lowest cost. Some biocides are expensive, and high dosages of them increase operating costs significantly. However, biocides applied in suboptimum dosages often result in rapid biofilm *recovery,* i.e., the rate of biofilm accumulation after dosing with the biocide is greater than before treatment. Further increases in biocide dosage may increase the chances of discharging the toxic material into the environment, resulting in ecosystem damage and substantial costs.

15.3.5 Cost for Plant Modification

Chlorine usually requires little in the way of plant modifications, is inexpensive, and is frequently effective, all of which contributes to its popularity. Other biocides may require special feeding equipment or require safety precautions because of their toxic or otherwise hazardous properties. The added cost for plant modification may be substantial. However, chlorine is a dangerous chemical, and its transport and handling may be expensive.

15.4 BIOFOULING CONTROL WITH CHLORINE: A PROCESS ANALYSIS

In this section, I will present a process analysis for biofouling control with chlorine. I have chosen chlorine because it is a nonproprietary chemical and because it serves as a useful illustration of the effect of biocides in biofilms. The process analysis is equally applicable to other biocides and in other biofilm environments.

Chlorine has been widely used as a disinfectant to kill microorganisms in water treatment. Originally, chlorine was thought to control biofouling by killing microbial cells (i.e., disinfecting). Recently, however, experimental results have indicated that chlorine also oxidizes and depolymerizes the EPS in the biofilm, causing destruction of the biofilm matrix and its detachment from the surface (Characklis and Dydek, 1976).

Transport and reaction of chlorine in a tube with turbulent flow are perceived to occur as a result of the following processes (Figure 15.1):

1. Chlorine species entering the system react with chlorine-demanding components (viable cells, equipment surfaces, and chemical compounds) in the bulk water.
2. Chlorine species are transported through the bulk fluid to the water-biofilm interface.
3. Chlorine species diffuse and react within the biofilm.
4. Chlorine reacts with the substratum.

Since both "solids" (i.e., the biofilm and the substratum) and a liquid phase are present, the chlorine-biofilm reaction occurs in a heterogenous system. In heterogenous reaction systems, it is always possible for a transport process to limit the overall process rate. This includes transport of reactants and products within each compartment (biofilm and bulk water) and across compartment boundaries.

15.4.1 Transport of Chlorine in the Water Phase

The rate at which chlorine is transported through the water phase to the biofilm depends on the concentration of chlorine in the bulk water and the intensity of the turbulence. The chlorine concentration in the bulk water is the net result of the chlorine addition rate minus the chlorine demand rate of the water. The chlorine concentration at the biofilm-water interface drives the reactions of chlorine within the biofilm. If the chlorine reacts with the biofilm rapidly, the concentration at the interface will be low and transport of chlorine to the interface may limit the rate of the overall process within the biofilm.

By increasing the intensity of turbulence through increased flow rate, both the transport rates in the bulk water and the concentration at the biofilm-water interface will increase. In the bulk water, chlorine is transported to the substratum primarily by diffusion (molecular or eddy). Several investigators have observed that increasing turbulence results in increased chlorine reaction rates with the biofilm (Norrman et al., 1977; Novak, 1982; Miller and Bott, 1981). Novak (1982) observed a first order rate (with respect to chlorine concentration) for the removal of biofilm, which indicates either a first order reaction or a mass

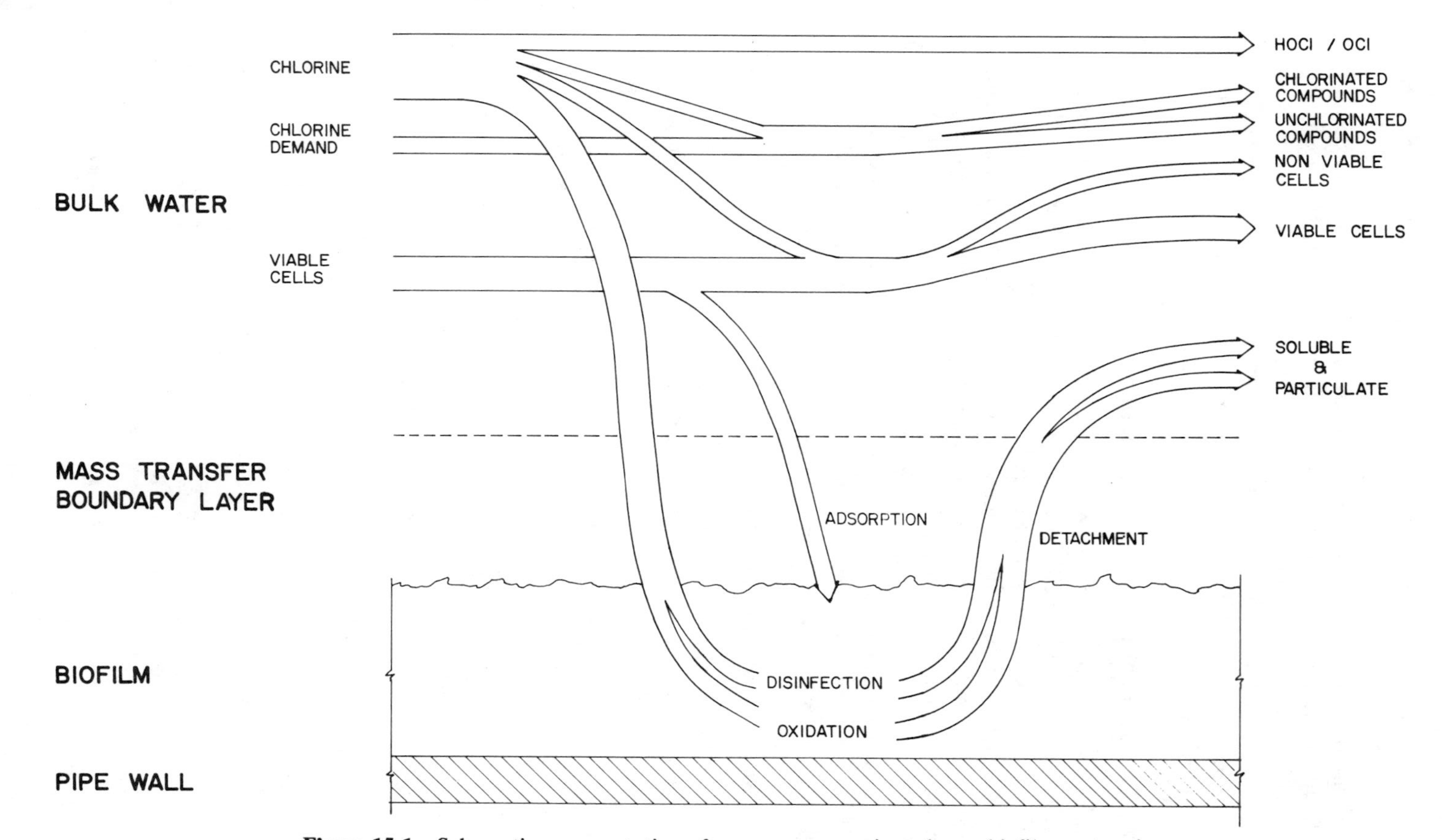

Figure 15.1 Schematic representation of processes occurring when a biofilm system is chlorinated.

transfer limitation (Figure 15.2). Under what circumstances is chlorine delivery to the fouling biofilm rate-limited by mass transfer in the water, diffusion in the biofilm, or reaction in the biofilm or at the substratum? The answer to this question significantly affects the manner in which any biocide should be applied (e.g., continuous application versus shock or periodic application) and the choice of biocide.

15.4.2 Transport of Chlorine in the Biofilm

The transport of chlorine within biofilms occurs primarily by molecular diffusion. Since the composition of biofilms is from 96 to 99% water, the diffusivity of chlorine in biofilm is probably some large fraction of its diffusivity in water (Table 4.7, Chapter 4). In biofilms of higher density, or in those containing microbial matter associated with inorganic scales, tubercles, or sediment deposits, diffusion of chlorine may be relatively slow. Diffusion and reaction of chlorine in a biofilm determine its penetration and hence its overall effectiveness. The same penetration phenomenon has been observed with nonoxidizing biocides (Section 15.8.2). Chapters 9 and 11 have described reaction–diffusion models, which are equally applicable to the biocide penetration problem.

15.4.3 Reaction of Chlorine in the Water Phase

The reaction (depletion) of chlorine in the bulk water is generally referred to as the *chlorine demand* of the water. The chlorine demand is due to soluble oxidiz-

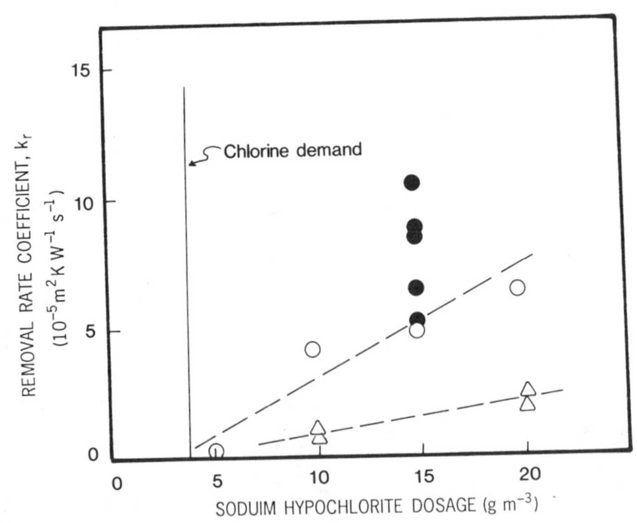

Figure 15.2 The removal rate coefficient k_r resulting from chlorination in Rhine River water at 27°C (biofilm temperature) appears first order with respect to chlorine. Symbols represent different runs. The intercept at $k_r = 0$ represents the chlorine demand of the bulk water. (Reprinted with permission from Novak, 1982.)

able inorganic compounds, soluble organic compounds, microbial cells, substratum, and other particulates in the bulk water. These oxidizable species compete with the biofilm for available chlorine and often reduce the effectiveness of chlorine for biofouling control. Chlorine, in the bulk water, can also inactivate microbial cells (disinfect) and oxidize microbial nutrients, reducing their nutritive value.

Both the extent and the rate of chlorine demand determine the amount of chlorine available for biofouling control.

Typical methods for measuring chlorine demand measure only *how much* chlorine is consumed, i.e., a stoichiometric quantity. The *rate of chlorine demand* (i.e., *how fast* chlorine is consumed) is often a more useful quantity in assessing chlorine effectiveness in biofouling control.

Bongers and Burton (1977) measured the rate of chlorine demand in estuarine waters and found it to occur in three phases. During phase 1, free chlorine decreased to about half its initial concentration in 30 s. Rates of decay were significantly slower in two distinct subsequent phases. The data imply that reactions in the water phase cannot necessarily be modeled by a simple first order rate expression. On the other hand, if the total hydraulic residence time in the system (e.g., condenser tube) is less than 30 s, only the first phase chlorine demand may be relevant.

Temperature has been shown to significantly affect both the rate and the extent of chlorine consumption in sea water (Wong, 1980). Consequently, the bulk water temperature may affect fouling control where there is significant chlorine demand. To what extent does temperature or a temperature gradient affect biofouling control with chlorine? Does temperature, which varies significantly in heat exchanger tubes (radially and axially), significantly influence the rate of chlorine demand?

Demand in the water phase or due to reactions in the biofilm also have been observed with nonoxidizing biocides. Eagar et al. (1988) report a depletion of glutaraldehyde (approximately 20%) in a biofouled laboratory reactor (high surface area to volume ratio and a residence time of 10 min) as well as in an oil field water injection system.

15.4.4 Reaction of Chlorine with the Biofilm

Chlorine reacts with various organic and reduced inorganic components within the biofilm. It can disrupt cellular material (detachment) and inactivate cells (disinfection). In a mature, thick biofilm, significant amounts of chlorine may react with EPS (primarily polysaccharides), which are responsible for the physical integrity of the biofilm (Characklis and Dydek, 1976).

15.4.4.1 Rate and Stoichiometry A series of studies (Characklis, 1971; Characklis and Dydek, 1976; Characklis et al., 1980) have estimated the process rate and stoichiometry of the reaction of chlorine with biofilm. In one study (Characklis et al., 1980), samples of biofilm were scraped from a RotoTorque, suspended in chlorine-demand-free water, and homogenized. The samples were then exposed to varying concentrations of sodium hypochlorite for 24 h. Filtration of the control and treated samples indicated that chlorine destroys the structure of the cells and associated biofilm material, since in several samples no filterable material remained after exposure to chlorine. The results, based on a gravimetric analysis of the residual biomass, indicate a stoichiometric ratio of approximately 1 g biomass destroyed per 1 g chlorine reacted (Figure 15.3). It would be surprising if this reaction were to go to completion in a diffusion-limited regime such as observed in a biofilm. However, the stoichiometric data are important to modeling of the process.

Bacterial flocs and biofilms that are rich in EPS exhibit a more rapid, ultimately greater, chlorine demand than cells with little associated EPS (Characklis and Dydek, 1976). One possible explanation is derived from the observation that hypochlorite ion attacks glucose polysaccharides with extensive oxidation at the C_2 and C_3 positions of D-glucose units, which results in cleavage of the C_2–C_3 bond (Hullinger, 1963; Whistler et al., 1956). Depolymerization can result from the inductive effects of this oxidation, from direct oxidative cleavage of the glucosidic bond, or from degradation of the intermediate carbonyl compound.

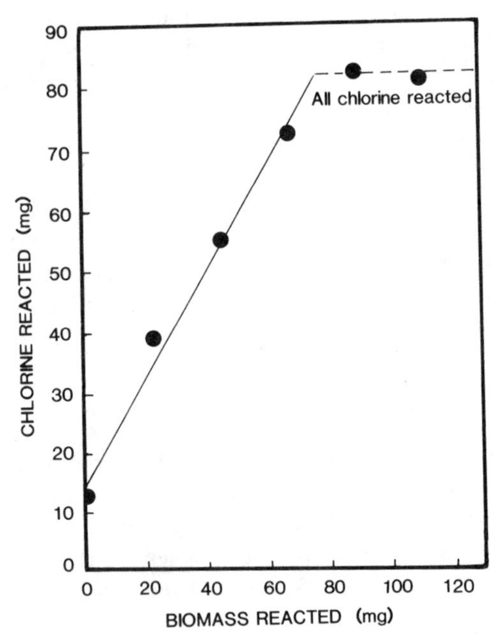

Figure 15.3 Results of the reaction of chlorine with dispersed biofilm, based on a gravimetric analysis of the residual biomass, indicate a stoichiometric ratio of approximately 1 g biomass destroyed per 1 g chlorine reacted. (Characklis et al., 1981.)

15.4.4.2 Concentration versus Duration Based on experience with disinfection, the product of chlorine concentration and treatment duration (CD) has been presumed to yield equivalent performance. However, laboratory measurements indicate that high chlorine concentrations at low durations are more effective than low concentration, high duration treatment for constant CD (Norrman et al., 1977; Characklis, 1980). Recently (Marine Biocontrol Corp. and CCE, Inc., 1988), field measurements have corroborated the laboratory measurements (Figure 15.4). The field results were obtained from a system of fouling monitors fed with power plant condenser cooling water and indicate that high chlorine concentration is more effective at equal CD. The results suggest that the major reaction of chlorine with biofilm is extremely rapid, so that typical chlorine residence times are much longer than the required duration. Thus, extended durations may result in diminishing returns in performance. The relationship between concentration, duration, and period (reciprocal of frequency) are discussed further in Section 15.7.

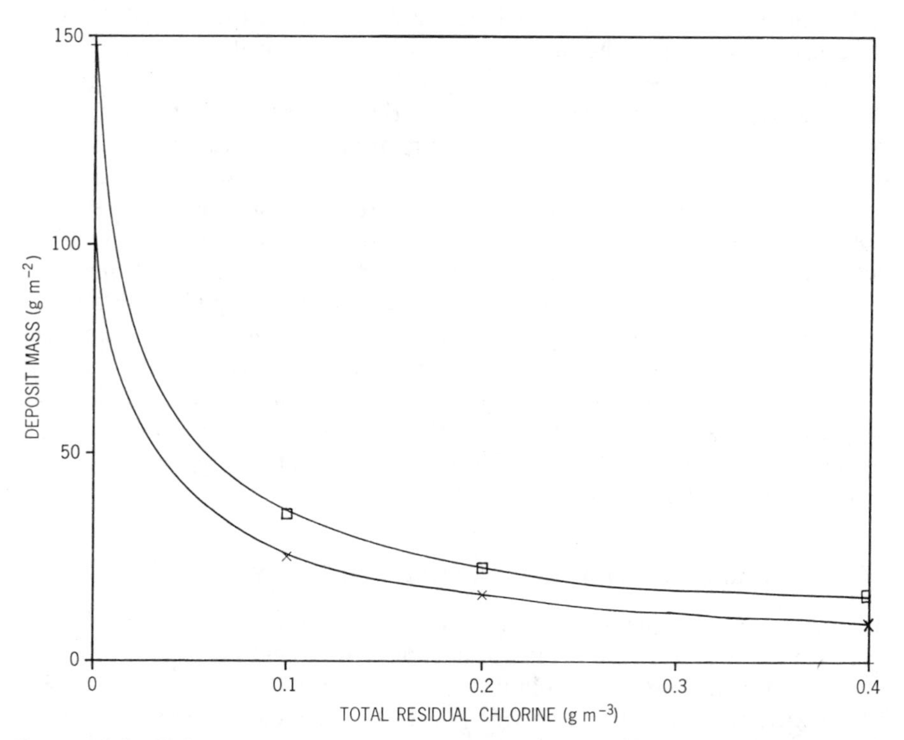

Figure 15.4 Field measurements in a heated (35°C wall temperature), once-through, tubular reactor indicate that high chlorine concentrations at low durations are more effective than low concentrations at high durations if the product of concentration and duration (in hours) is constant (i.e., CD = constant). The data were obtained at an estuarine power plant site with CD = 0.033 (□) and 0.066 (×) with concentration in units of g m^{-3}. (Marine Biocontrol Corp. and CCE, Inc., 1988).

15.4.4.3 Influence of pH Most of what is known about chlorine disinfection is derived from drinking water treatment studies. It is generally agreed that chlorine disinfects by oxidizing key components within the bacterial cell (e.g., proteins, nucleic acids). Chlorine has been found to be most effective as a disinfectant at pH 6.0 to pH 6.5, a range in which hypochlorous acid predominates. It is postulated that the hypochlorous acid molecule, which is electrically neutral and has a smaller solvation shell than the hypochlorite ion, can readily pass through cellular membranes, while the hypochlorite cannot. Characklis et al. (1980) found that high pH improved biofilm destruction (as opposed to disinfection) by chlorine, suggesting that hypochlorite ion is more effective at depolymerizing the biofilm. This explanation is consistent with hypohalite depolymerization of industrial polysaccharides (Section 15.4.6).

15.4.5 Reaction of Chlorine with the Substratum

The substratum may also exert a chlorine demand. Characklis et al. (1983) have documented the high chlorine demand of a copper-nickel alloy as contrasted with titanium, which exerted no such demand. Lewis (1982) also found that Cu-Ni reacts with chlorine. The chlorine demand may be directly related to the significant corrosion observed when copper-nickel alloys are chlorinated regularly in seawater. This may be due to stoichiometric reaction of the alloy with chlorine, or removal of a protective oxide coating from the alloy, rendering the underlying metal vulnerable to other corrosion reactions. However, the chlorine demand is not necessarily related directly to corrosion, since metal ions such as Cu(II) have been shown to catalyze chlorine decomposition (Lister, 1956) to chloride. How much does metallurgy affect the chlorine demand and chlorine's effectiveness in destroying biofilm? How do chlorination and biofouling affect corrosion rates in heat exchangers? Drinking water distribution pipes are often coated with a bituminous compound to prevent corrosion. The bituminous coating is in a very reduced state and consumes chlorine at a very high rate (Characklis, 1988). Does the hypothesized chlorine deficit at the pipe wall enhance biofilm accumulation? Does the chlorine reaction contribute significantly to chlorinated organics in the bulk water?

15.4.6 Biofilm Removal and/or Destruction

The reaction of chlorine with biofilm results in a significant removal of biofilm, as evidenced by a decrease in biofilm thickness (Characklis et al., 1980; Miller and Bott, 1981), decrease in fluid frictional resistance (Characklis et al., 1980; Characklis and Dydek, 1976; Norrman et al., 1977), and decrease in heat transfer resistance (Lewis, 1982; Zelver et al., 1981; Marine Biocontrol Corp. and CCE, Inc., 1988). The chlorine-biofilm reaction also produces an immediate response in the bulk water, as evidenced by increased turbidity and dissolved organic constituents (Characklis and Dydek, 1976). Laboratory results also suggest that hypochlorite oxidizes EPS within the biofilm resulting in EPS depolymerization, dissolution, and detachment. Characklis et al. (1980) reported that biofilm detachment resulting from chlorination is much higher between pH 7.5 and 8.5 than between pH 6 and 7, where disinfection is optimum. These data are consistent with data on depolymerization of polysaccharides by chlorine (e.g., starch), which is optimum at pH > 7 (Whistler and Schweiger, 1957).

Chlorine "enhancers," generally dispersants, are sometimes added to improve chlorination performance. Presumably, these compounds increase the penetration of chlorine into the biofilm and improve the effectiveness of the chlorine, possibly by loosening deposits and allowing chlorine contact with lower layers of the deposit. Are chlorine "enhancers" really effective? What circumstances improve their effectiveness?

Nonoxidizing biocides do not necessarily result in biofilm detachment. Inactivation of biofilm with mercuric chloride (a nonoxidizing biocide) did not result in any biofilm removal (Characklis and Dydek, 1976). On the other hand, glutaraldehyde (another nonoxidizing biocide) addition sometimes resulted in significant detachment of a *Pseudomonas fluorescens* biofilm (Eagar et al., 1988).

15.4.6.1 Chlorine Residual The criterion for chlorine dosage is frequently the residual concentration in the water. By maintaining a residual in the water, chlorine demand in the water is presumably overwhelmed. However, mass transfer considerations suggest that, despite a chlorine residual in the bulk water, the region near the bulk water–biofilm interface may be devoid of chlorine. If so, biofouling control effectiveness may be poor even when a measurable chlorine residual is measured in the outlet. Under what conditions is maintenance of a chlorine residual in the outlet sufficient to eliminate the effect of chlorine demand in the condenser? Is the chlorine residual in the outlet sufficient as a criterion for biofouling control effectiveness?

Characklis (1988a) reported that a *continuous* chlorine residual as high as 0.8 g m^{-3} applied to finished drinking water was not sufficient to prevent biofilm accumulation in the distribution system. Results indicate that a significant biofilm accumulated after approximately 35 days in the simulated pipeline system (four RotoTorques in series) carrying treated drinking water. The biofilm cell population was dominated by pseudomonads. Typical chlorine residuals in a drinking water distribution system are less than 0.1–0.3 g m^{-3}.

15.4.6.2 Water Velocity Hydrodynamic shear forces may play a major role in the enhanced removal of biofilm when chlorine is being added. Several investigators (Norrman et al., 1977; Miller and Bott, 1981; Novak, 1982; Characklis et al., 1980) measured a higher rate of chlorine uptake by biofilms and a greater amount of biofilm detachment when the fluid shear stress was increased (Figures 15.5 and 15.16 [*sic*]). The increased shear force has the following effects during chlorination:

1. Increased mass transfer of chlorine from the bulk water to the biofilm.
2. Disruption of the biofilm during chlorination, reentraining reaction prod-

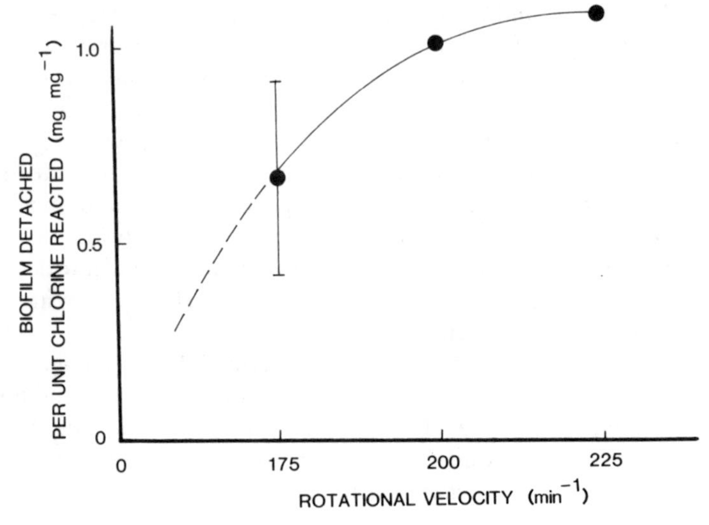

Figure 15.5 A higher rate of chlorine uptake and a greater amount of biofilm detachment by biofilms was observed when the fluid shear stress was increased. These data were obtained in a laboratory RotoTorque system operated at 30°C with an inlet chlorine concentration of 5 g m^{-3} and a mean hydraulic residence time of 150 s. (Reprinted with permission from Characklis et al., 1980.)

ucts in the bulk water and exposing new biofilm surfaces for chlorine attack.

3. Generally, decreased thickness of the viscous sublayer (Figure 9.5, Chapter 9), which is the region of low shear force near the biofilm–water interface. Biofilm lying within the viscous sublayer will not be disrupted to a great extent unless the shear force is very large. Timperley (1981) has commented on the effect of mean flow velocity on the cleaning of biofouled tubes.

Observations indicate that a residual biofouling deposit remains even after repeated chlorine treatment. The residual deposit probably has a significant effect on recovery rate of the biofilm.

15.4.7 Biofouling Control through Disinfection

Low level chlorination can significantly reduce viable microbial cells in water and destroy some of the nutrients necessary for cell growth. Although chlorine can be effective in controlling biofilm accumulation in heat exchanger systems, low level chlorination is not sufficient to prevent attachment of surviving viable cells and subsequent biofilm accumulation (Bongers and Burton, 1977). Chlorine is used to maintain acceptable bacterial quality at residual concentrations as low as 0.03 mg/L in many drinking water systems (Biros and Lamblin, 1980).

However, Characklis (1988a) observed the accumulation of biofilm in treated drinking water with continuous chlorination at a relatively high concentration (0.8 g m^{-3}). Concentrations of chlorine high enough to make systems aseptic may be prohibitively expensive and create unacceptable environmental damage. Thus, biofouling control by continuous disinfection may be effective for only limited periods or in special circumstances.

15.5 BIOFILM RECOVERY

Numerous investigators and plant operators have observed a rapid resumption of biofouling immediately following chlorine treatment and have termed this phenomenon "regrowth." *Recovery* is a more appropriate term, since growth may only be one of the processes contributing to reestablishment of the biofilm. Thus, recovery may be due to one or all of the following:

1. The remaining biofilm contains enough viable organisms to preclude any lag phase in biofilm accumulation as observed on clean tubes. Thus, fouling biofilm accumulation after shock chlorination is more rapid than on a clean surface.
2. The residual biofilm imparts an increased relative roughness to the tube surface and thus enhances transport and sorption of microbial cells and other compounds to the surface. The roughness of the deposit may provide a stickier surface.
3. The chlorine preferentially removes EPS and not biofilm cells, thus leaving biofilm cells more exposed to the nutrients when chlorination ceases.
4. EPS is rapidly created by surviving organisms [perhaps injured organisms, as reported by McFeters and Camper (1985)] as a protective response to irritation by chlorine.
5. There is selection for organisms (e.g., one that normally produces excess amounts of EPS) less susceptible to chlorine, which proliferate between successive chlorine applications.

Norrman et al. (1977) observed rapid recovery after chlorination in a laboratory system (Figure 15.6), Novak (1982) observed it in a model heat exchanger operated with Rhine River water (Figure 15.7), and Lewis (1982) observed it in a test heat exchanger operated on an estuary (Figure 15.8). Rapid regrowth has also been observed on walls of drinking water distribution systems (Ridgway et al., 1984; Berg et al., 1982). These authors demonstrated that bacteria previously exposed to chlorine were more resistant to it than those never exposed, an observation consistent with hypotheses 4 and 5.

After extended periods of repeated chlorination, a refractory material has sometimes been observed accumulating at the substratum (Characklis, 1980; Mangum and Shepherd, 1973; Lewis, 1982). The deposit increases in ash con-

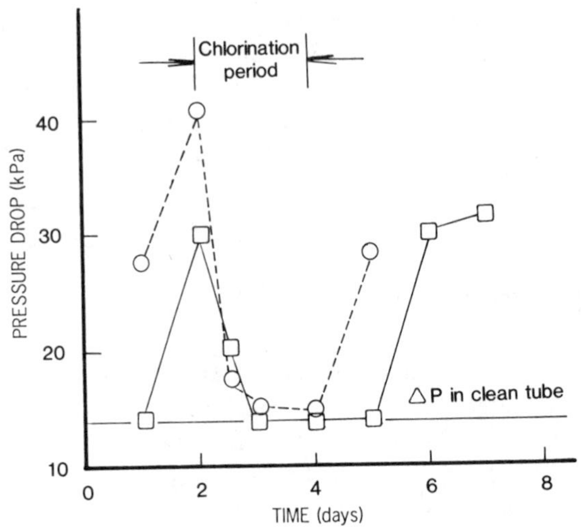

Figure 15.6 Rapid biofilm recovery was observed after chlorination in a laboratory tubular (acrylic) reactor system. Biofilm accumulated (as reflected by increased pressure drop) without chlorine application for two days. Then chlorine [loading rate = 0.17 (□) and 0.08 (○) g h^{-1}] was added continuously for 48 h. Biofilm accumulation was reduced to almost zero in that period, but recovered rapidly when chlorination ceased. (Reprinted with permission from Norrman et al., 1977.)

tent with repeated chlorination, and in several cases an inordinately high content of manganese has been observed. In one case, the refractory deposit had an apparent inhibitory effect on *macrobial* fouling (Mangum and Shepherd, 1973). However, the refractory material may accumulate to such an extent that it reduces performance (e.g., impedes heat transfer) and continued chlorine application has no effect. Manganese and iron deposition has also been observed in a glass laboratory tubular reactor after repeated chlorination (Characklis, 1980) when groundwater was the source water. What is the rate and extent of accumulation of refractory material, and how is it influenced by substratum composition (e.g., tube alloy) and water quality? How does the refractory material affect subsequent biofilm development? In what form do the Mn and Fe enter the system? In dissolved state (reduced ions or in association with high molecular weight organics) or particulate state (oxidized or adsorbed on inorganic or organic particles)? What is the most effective way to inhibit or remove the refractory material? If the refractory material inhibits further biofouling, how can a thin film be maintained that does not significantly influence heat transfer or frictional resistance?

Observations at an estuarine power plant site (Marine Biocontrol Corp. and CCE, Inc., 1988) indicate that chlorine dosage influences the extent of mineral deposition and the relative amounts of different minerals at a titanium substratum (Figure 15.9). The tube wall temperature was 35°C, while the water

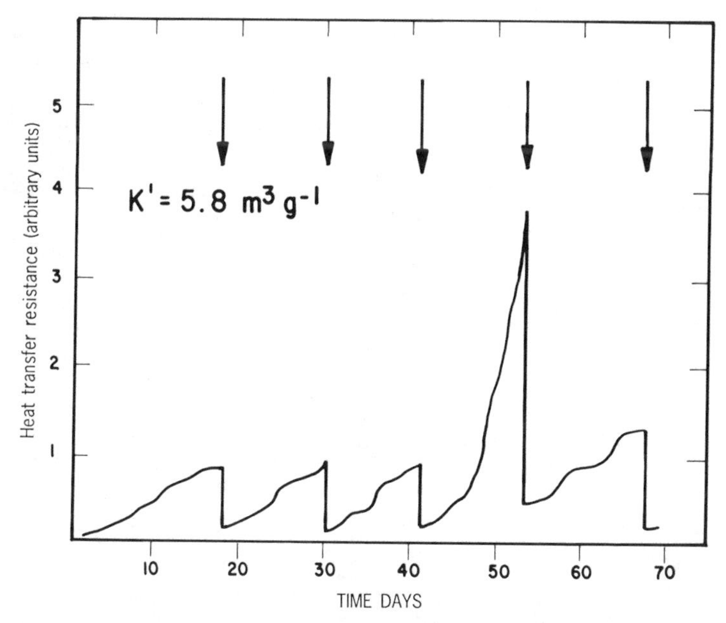

Figure 15.7 Rapid recovery of biofilm accumulation (as reflected by heat transfer resistance) was observed after each chlorine treatment (indicated by arrows) in a test (titanium) heat exchanger (27°C wall temperature) operated with Rhine River water. Chlorine application concentration was 15 g m^{-3} in all cases, and duration was 3.5, 4.5, 4.0, 4.5, and 4.4 h, respectively. (Reprinted with permission from Novak, 1982.)

temperature varied over approximately 10–25°C. The highest amounts of iron and manganese occurred in association with the highest biofilm accumulation (other than the control). Thus, the net accumulation of manganese and iron may be result from the absorption of the metals (or metal-organic complexes in which the metal ion is in a reduced state) in the biofilm. Periodic metal oxidation by chlorine results in immobilization of the metals at the tube wall.

15.6 CONTROL OF FOULING BIOFILMS BY CHLORINATION

Chlorine is used in a variety of ways to control biofouling. It is applied as chlorine gas, hypochlorite ion, and chlorine dioxide, and is electrochemically generated in sea water. Chlorine is used in once-through systems or in recirculating cooling towers; it is applied continuously or in shock dosages. It is used in freshwater systems, and in seawater (generally producing oxidized bromine species); in clean water, and in contaminated water containing high levels of oxidizable materials.

Although biofouling can be controlled by addition of chlorine, its use has dis-

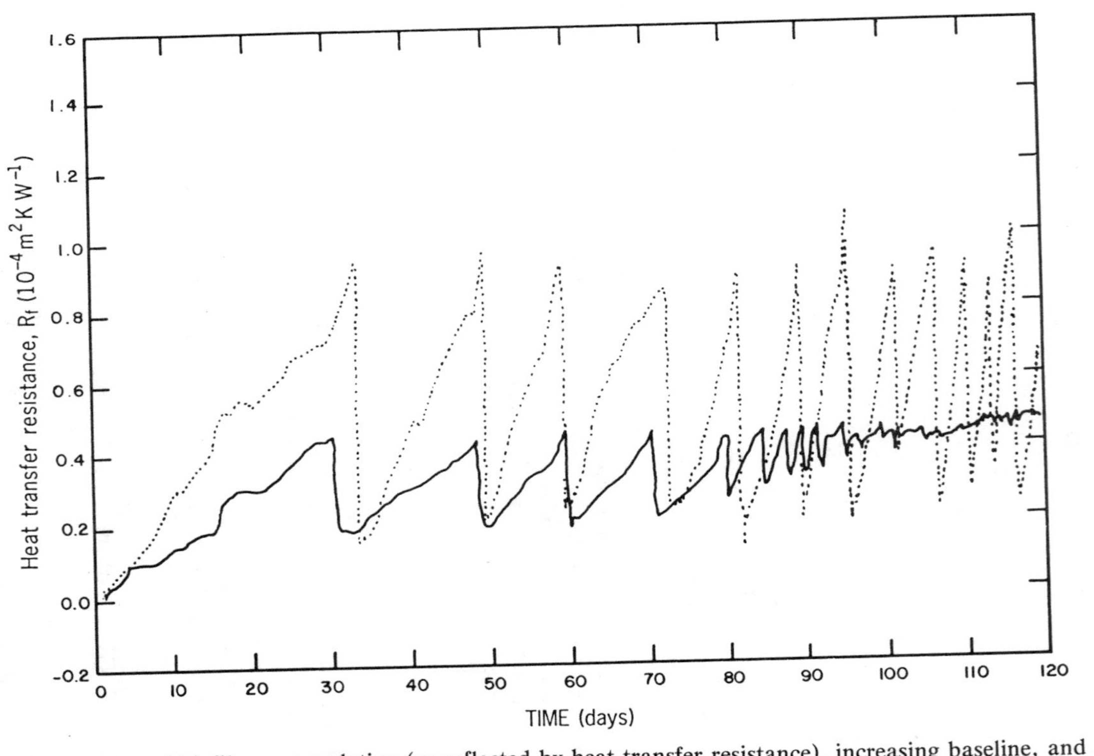

Figure 15.8 Rapid recovery of biofilm accumulation (as reflected by heat transfer resistance), increasing baseline, and a significant decrease in susceptibility to treatment was observed in a copper–nickel test heat exchanger operated on an estuary and treated with sponge balls and 0.25 g m^{-3} hypochlorite for 0.4 h each day (solid line). Treatment was initiated when $R_f = 0.4 \times 10^{-4}$ m^2 K W^{-1}. Recovery after treatment with brushes (dotted line) also resulted in rapid recovery, but baseline increase was minimal after the first 30 days. Brush treatment was initiated when $R_f = 0.9 \times 10^{-4}$ m^2 K W^{-1}. (Reprinted with permission from Lewis, 1982.)

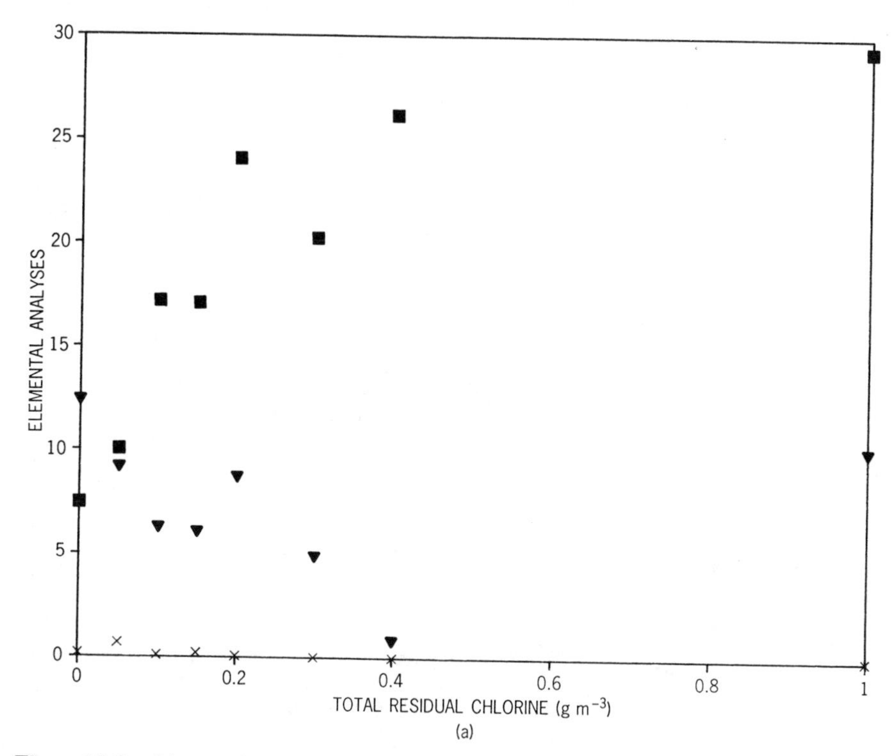

Figure 15.9 Observations at an estuarine power plant site indicate that (a) mineral content [Mn (×), Si (▼), and Fe (■) in % of dry deposit mass] in the fouling deposit and (b) total deposition (units of g m^{-2}) on a heated (35°C wall temperature) titanium tube vary with chlorine dosage (Marine Biocontrol Corp. and CCE, Inc., 1988).

advantages. Constraints placed on the use of chlorine have often limited its success in biofouling control. These constraints include:

1. *Cost Effectiveness*. This may be a factor in once-through or recirculating cooling systems where extremely high chlorine demand exists.

2. *Corrosion*. Some condenser materials are vulnerable to corrosion induced by high chlorine concentrations or chloride resulting from its reduction.

3. *Environmental Regulations*. Environmental concerns limit the concentration of chlorine that can be released to receiving waters. Recently, concern over the toxicity of chlorine (Hall et al., 1981) and some of its reaction products (Trabalka and Burch, 1978) has stimulated a search for alternative chemicals (Battaglia et al., 1981; Waite and Fagan 1980; Burton, 1980). Nevertheless, chlorination is still widely practiced in power plants and in process plants for minimizing biofouling.

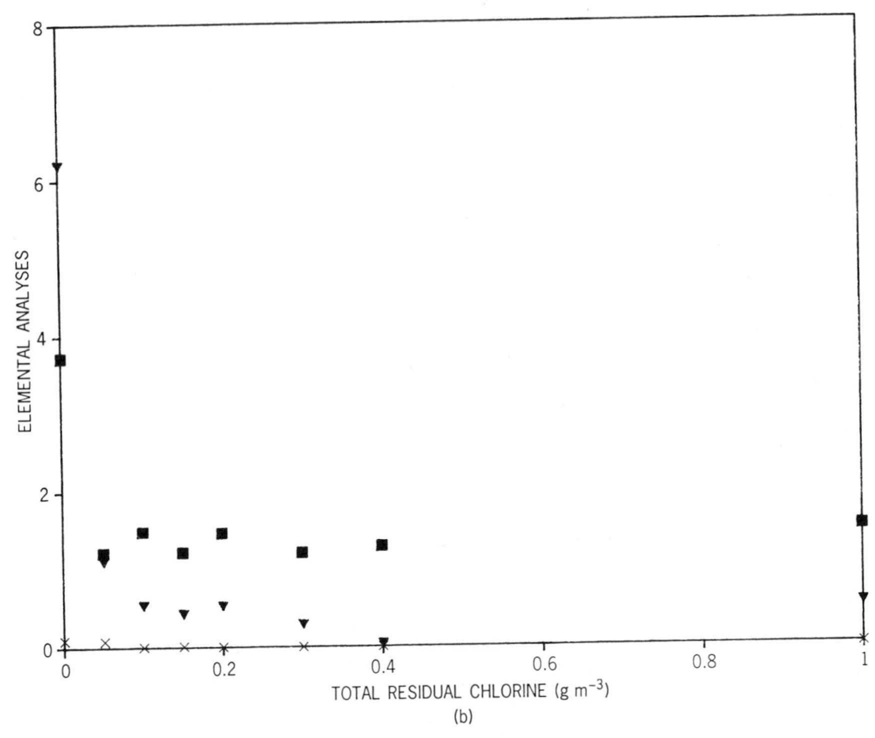

Figure 15.9 Continued

Chlorine and other strong oxidants (chlorine dioxide, ozone, bromine compounds) are effective microbial fouling control agents because they disrupt and loosen biofouling deposits. When chlorine contacts biofilm, the following processes are observed to occur (Characklis, 1971; Characklis et al., 1980):

1. Detachment of biofilm
2. Dissolution of biofilm components
3. Disinfection

Chlorine is reduced in all of these reactions, producing chloride and organochlorine compounds. Limited observations suggest that the following factors influence the rate and extent of the chlorine–biofilm reaction:

1. *Turbulent Intensity*. Transport of bulk water chlorine to the water–biofilm interface is the first step in the chlorine–biofilm interaction. The transport rate increases with increasing bulk water chlorine concentration and turbu-

lence. Practically, turbulent intensity may be increased by an increased flow rate, as proposed in some targeted chlorination systems.

2. *Chlorine Concentration at the Water–Biofilm Interface.* The transport of chlorine within the biofilm or deposit is a direct function of the chlorine concentration at the interface. The transport rate (or diffusion) in the biofilm can be increased by increasing the chlorine concentration at the bulk water–biofilm interface. High chlorine concentration for short durations is more effective than low concentration for long durations, assuming the same long term chlorine application rate for both cases; i.e., the product of treatment concentration and duration, CD, does not characterize the performance (Norrman et al., 1977; Characklis, 1980; Marine Biocontrol Corp. and CCE, Inc., 1988). However, environmental regulations frequently place a limit on the effluent chlorine concentration. *Targeted* chlorination systems may permit the use of a short duration, high concentration treatment, while minimizing chlorine the concentration reaching the receiving waters.

3. *Composition of the Fouling Biofilm.* Reaction of chlorine within the biofilm is dependent on the organic and inorganic composition of the biofilm as well as its thickness or mass. Little published information is available regarding the chlorine demand of biofilms other than in laboratory experiments (Characklis et al., 1980). This is surprising, since the main objective of biofouling control is to react the chlorine with the biofilm. Disinfection (i.e., killing of microbial cells) in potable water systems is effective at low chlorine concentrations. However, in well-developed biofilms much of the material is extracellular and may compete effectively for available chlorine within the biofilm thereby, reducing the chlorine available for killing cells. The substratum may also consume chlorine, and so may also compete for it.

4. *Fluid Shear Stress at the Water–Biofilm Interface.* Detachment and reentrainment of biofilm, primarily due to fluid shear stress, accompanies the reaction of biofilm with chlorine. Detachment of biofilm due to chlorine treatment has been observed, and the rate and extent of removal depend on the chlorine application and on the fluid shear stress at the bulk water–biofilm interface.

5. *pH.* The hypochlorous acid–hypochlorite ion equilibrium may be critical to performance effectiveness. OCl^- apparently favors detachment, while HOCl enhances disinfection. The equilibrium can be manipulated in recirculating cooling water systems where the pH is controlled.

6. *Salinity.* In saline waters with high bromide ion concentration (e.g., seawater), chlorine oxidizes the bromide to bromine. Thus, the oxidants attacking the biofilm are bromine species. The relative effectiveness of bromine versus chlorine will probably be site-specific.

Summarizing the above statements, a chlorination treatment program may be improved by the following measures:

1. *Increase the Chlorine Concentration at the Water–Biofilm Interface.* Most practically, this is accomplished by increasing the chlorine con-

centration in the bulk water. However, another possibility is to target chlorination at the water-biofilm interface. For example, reduction of the bulk water chlorine demand is generally a difficult, costly task. But if chlorine can be injected in the viscous sublayer region, the chlorine demand of the water will have significantly less effect on the process.

2. *Increase the Fluid Shear Stress at the Water–Biofilm Interface.* Increasing the water velocity, especially during chlorination periods, appears to increase the effectiveness of chlorination in removing biofilm.

3. *Use pH Control.* High pH favors hypochlorite-ion-promoted detachment of mature biofilms, and low pH enhances hypochlorous acid disinfection of thin films. An interesting alternative would be to alternate between continuous chlorination at pH 6.5 and "shock" chlorination at pH 8.

Although chlorination has been effective in many applications, the processes contributing to the short and to the long lived effectiveness of a chlorination program for biofouling control are not well understood.

The work that has appeared in the literature on predicting the fate of the chlorine molecule in water is primarily based on chemical studies and equilibrium or kinetic models (Zielke and Macy, 1980; Sugam and Helz, 1981; Haag and Lietzke, 1980; Helz et al., 1978). These studies have concentrated on equilibrium or kinetic measurements of the chlorine demand of cooling water and have generally ignored the reaction of chlorine with the fouling biofilm at the substratum. These studies have also ignored the influence of possible mass transfer limitations at the biofilm–water interface. Consequently, application of the models has not always been successful (Helz et al., 1978), and expensive chlorine minimization programs have been used in most cases (Moss et al., 1979) for determining chlorination application rates. However, even chlorine minimization studies cannot take account of seasonal changes in water quality and variations in plant operation, which influence the effectiveness of chlorine for biofouling control.

Despite the importance of biofouling control, as well as the apparent effectiveness of chlorine, the stoichiometry and kinetics of biofouling control with chlorine (and other oxidants) have never been adequately addressed.

For example, a recently proposed chlorination field test at a power plant site was stymied by lack of dependable rate and stoichiometric information on the chlorine–biofilm reaction. If the nature of the stoichiometry and kinetics of the chlorine–biofilm process were known, procedures for establishing a continuing, efficient, cost-effective, environmentally compatible chlorination program could be developed. This same data must also be obtained for other biocides and will be invaluable in planning future test programs. Such information would also be useful in comparing novel physical, chemical, and even biological methods for biofouling control.

15.7 A PROCEDURE FOR DETERMINING OPTIMAL CHLORINE DOSAGE: PERIODIC ADDITIONS

Due to environmental constraints, optimum chlorination for biofouling control is the minimum *dosage* that will suffice to maintain the fouling resistance within acceptable levels. The dosage consists of the product of treatment chemical concentration (C_b), treatment duration (D), and period (P) between treatment (reciprocal of treatment frequency). A procedure for determining the required dosage of chlorine is presented here, which is a modification of that presented by Novak (1982). The procedure is not restricted to use with chlorine, but may be used with any biocide or mechanical treatment.

The treatment cycle can be simply represented as the sum of two phases: (1) accumulation of fouling resistance R_f over the period P between treatments (reciprocal of treatment frequency), and (2) removal of R_f during and briefly after treatment at concentration C_b and duration D (Figure 15.10). The fouling resistance R_f can be measured in terms of one or more of the following variables (Figure 14.4, Chapter 4): (1) biofilm mass, biofilm thickness, biofilm cell number, biofilm TOC, etc.; (2) hydrodynamic frictional resistance; (3) heat transfer resistance. In the following development, linear or other simplistic functions are considered. Obviously, more complicated functions for the increase or decrease of fouling deposit are possible within operating equipment.

15.7.1 Accretion

The accretion (positive accumulation) rate of the fouling deposit—the rate of R_f increase—is reflected by a specific rate coefficient describing the increment of R_f increase per unit R_f present per unit time:

$$\frac{dR_f}{dt} = k_a R_f^n \qquad [15.1]$$

where k_a = specific rate coefficient
n = exponent

Hypothetical progressions for $n = 0$ and $n = 1$ are presented in Figure 15.11. Equation 15.1 results in a continuous increasing R_f, and no plateau is reached, although the typical biofilm accumulation progression includes a plateau phase (Figure 14.4, Chapter 4). This presumption is nevertheless acceptable in many applications. For example, the treatment frequency in power plants is one per day, too short a period for a plateau to be attained. In several field studies (Novak, 1982; Lewis, 1982; Zelver et al., 1984; Marine Biocontrol Corp. and CCE, Inc. 1988) and in the laboratory (Norrman et al., 1977), n ranged from

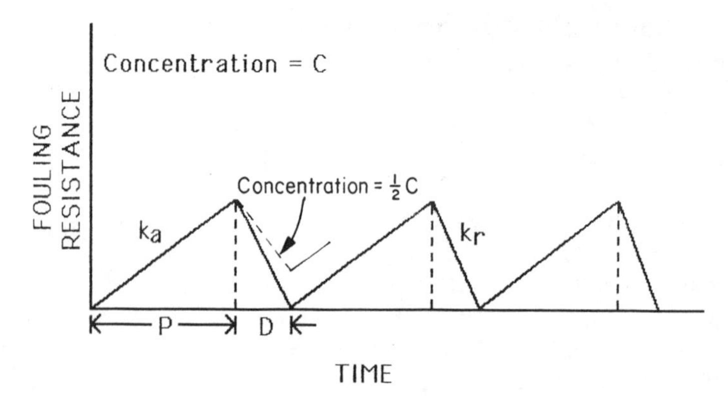

Figure 15.10 A biocide (e.g., chlorine) treatment cycle can be approximated by the sum of two components: (1) accretion (positive accumulation) of the fouling resistance R_f over the period P between treatments (reciprocal of treatment frequency), and (2) removal of R_f during and briefly after chlorination at concentration C and duration D. If the concentration is reduced by half, the treatment effectiveness decreases and subsequent biofilm accumulation increases.

zero to 0.8 (Figures 15.4–15.6). For simplicity, n will be presumed to be zero for the remainder of this discussion. So, at time P after the previous treatment,

$$R_f = k_a P \qquad [15.2]$$

where P = period between chlorination treatments (t).

Environmental parameters (e.g., TOC in the cooling water) and operating parameters (e.g., water velocity) influence k_a, as described in detail in Chapter 14. Immediately after treatment is terminated, R_f accretion resumes, resulting in recovery. The rate of recovery may be influenced by a residual fouling resistance, R_f', after treatment.

15.7.2 Detachment

Detachment of biofilm (fouling resistance) resulting from chlorination is also described by a specific rate coefficient:

$$\frac{dR_f}{dt} = -k_r C_a (R_f - R_f')^m \qquad [15.3]$$

where k_r = specific detachment rate coefficient
m = exponent
C_a = chlorine concentration at the bulk water–biofouling deposit interface (ML^{-3})

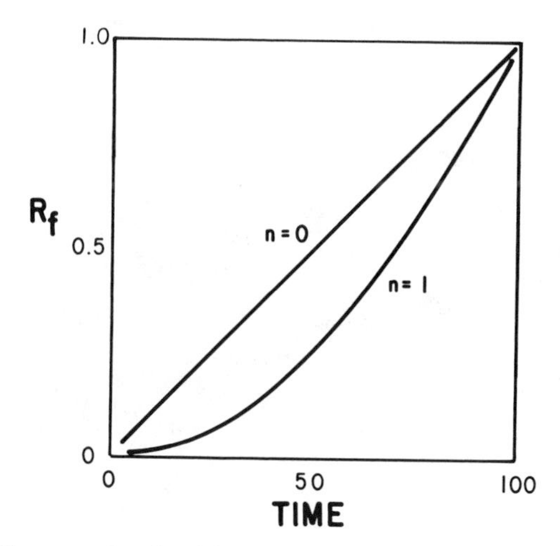

Figure 15.11 Biofilm accretion (positive accumulation) rate is reflected by specific rate coefficients, k_a and n (Eq. 15.1). Hypothetical progressions for $n = 0$ and $n = 1$ are presented.

Thus, the rate of R_f removal due to chlorine application is proportional to the R_f accumulated (Figure 15.12), although a certain refractory accumulation, R_f', may remain after treatment. Reports have indicated that m varies between 0 and 1 (Novak, 1982). The extent of biofilm removal due to chlorination appears directly proportional to the amount of biofilm accumulated prior to chlorination. Data from Characklis (1980) and Norrman et al. (1977) under similar experimental conditions indicate that the biofilm thickness removed as a result of chlorine application is linearly related to the initial biofilm thickness (Figure 15.13). The data also indicate an intercept at approximately 40 μm, the viscous sublayer thickness for these experiments. Conceivably, the biofilm material as close as 40 μm to the substratum does not experience the shear necessary for removal, even though chlorine may have penetrated the biofilm to that depth.

Equation 15.3 requires that C_a be determined. Novak (1982) has plotted k_r versus C_d (Figure 15.2) for the Rhine River and obtained an intercept of $C_d \approx 4$ g m^{-3}, which Novak suggests is the chlorine demand of the water. At a macroscopic level, the chlorine *dosed* to the system has the following fate:

$$C_d = C_a + C_w + C_e \qquad [15.4]$$

where C_d = chlorine concentration dosed to the water (ML^{-3})
C_w = chlorine demand of the water (ML^{-3})
C_e = effluent chlorine concentration (ML^{-3})

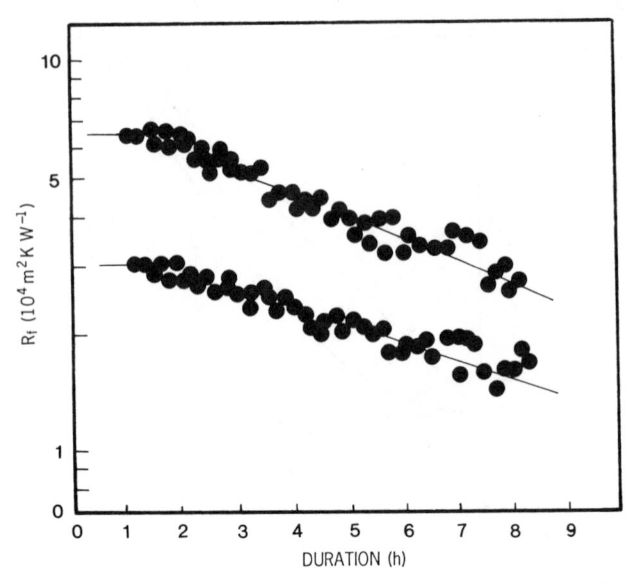

Figure 15.12 R_f removal rate resulting from chlorination is apparently first order in R_f. Two tests were done with Rhine River water at a chlorine concentration $= 20$ g m^{-3}. (Reprinted with permission from Novak, 1982.)

C_a thus determined, Eq. 15.4 can be integrated (for $m = 0$ and $R_f' = 0$) to determine the fouling resistance removed (detached), $R_{f,\mathrm{mem}}$:

$$R_{f,\mathrm{rem}} = -k_r C_a D \qquad [15.5]$$

where D = duration of chlorine treatment (t).

According to Eq. 15.5, C_a remains constant for the duration of the treatment. Characklis (1980) presents experimental evidence (R_f measured in terms of friction factor) for using this assumption for short durations and low chlorine concentrations (Figure 15.14). The first chlorine treatment was applied to a heavily fouled tube (initial $f \approx 0.4$) in which $C_d = 27.1$ g m^{-3}. For the first 12 min of treatment, $C_e = 0$, but C_e increased to approximately 5 g m^{-3} by 25 min. R_f decreased to almost clean conditions after 40 min, with biofilm thickness decreasing from 240 to 40 μm. The second chlorine treatment was conducted in a tube with $R_f \approx 0.015$ and biofilm thickness ≈ 80 μm, a relatively low level of deposit accumulation. $C_d = 10.6$ g m^{-3}, and C_e remained at approximately 5 g m^{-3} throughout the chlorination, which reduced the R_f to almost clean conditions and reduced the biofilm thickness to approximately 30 μm. Under these conditions, the reaction is probably restricted to the upper surface layers of the biofilm, and chlorine reduction is limited by surface area of the biofilm. Thus,

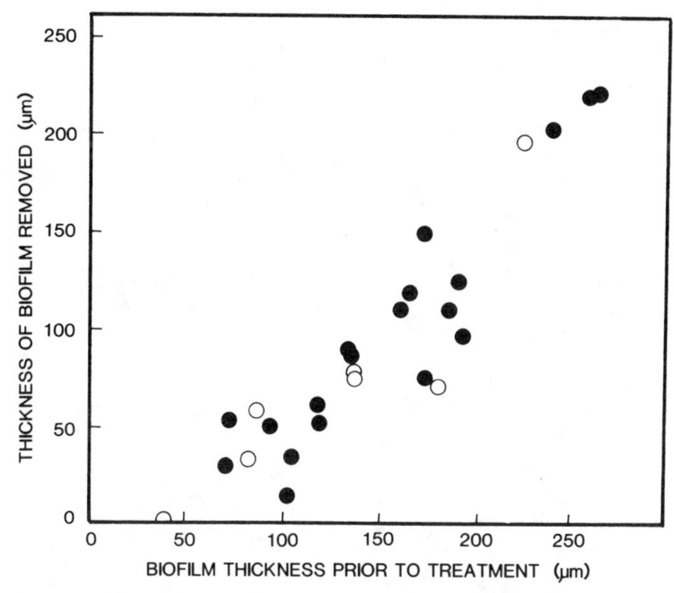

Figure 15.13 The extent of biofilm removal due to chlorination is proportional to the amount of biofilm present prior to chlorination. The intercept when removal is zero represents a "refractory" thickness, which may be related to the viscous sublayer thickness [Characklis, 1980 (○); Norrman et al., 1977 (●)].

C_a remains nearly constant throughout the test. Again, removal of the fouling deposit beyond the viscous sublayer thickness (approximately 40 μm in these tests) appears difficult.

Novak (1982) has indicated that increasing temperature will increase the removal rate coefficient $k_r C_a$ (Figure 15.15). Miller and Bott (1981) have indicated that removal rate (and extent) is also influenced by the water velocity (Figure 15.16).

15.7.3 Calculating the Optimal Dosage

Presuming all of the fouling resistance that accumulates in the period P is removed by the chlorine treatment, then

$$R_f = R_{f,\text{rem}} \qquad [15.6]$$

So Eqs. 15.2 and 15.5 can be combined to give

$$k_a P = k_r C_a D \qquad [15.7]$$

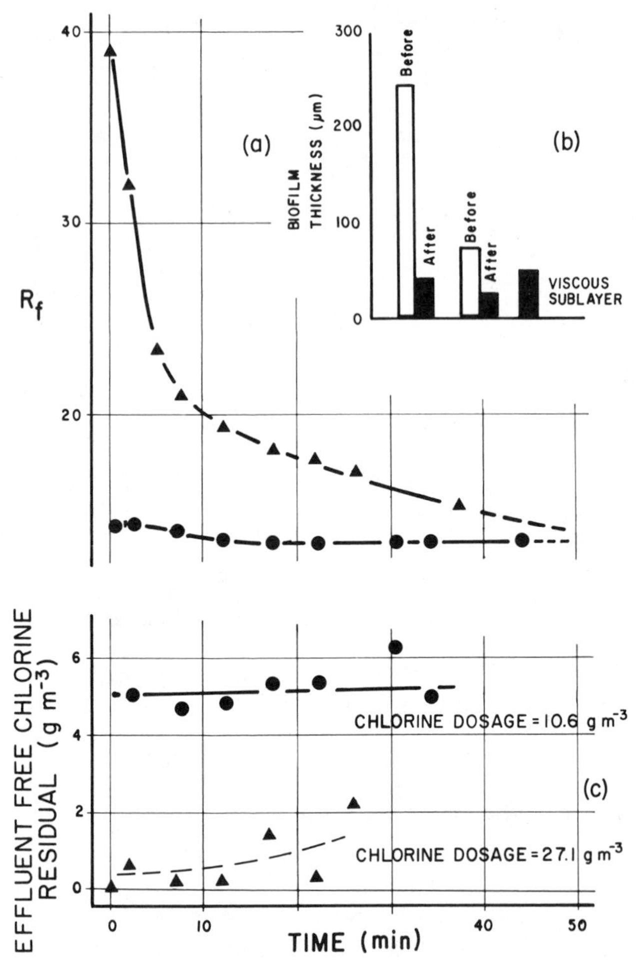

Figure 15.14 Two laboratory experiments (▲, ●) were conducted in a tubular reactor in which biofilm accumulated and then exposed to hypochlorite. (a) Reduction in R_f (as reflected by frictional resistance) resulting from chlorination continues throughout the chlorination period although treatment does not completely remove the deposit. (b) The residual R_f may reflect the viscous sublayer thickness in these experiments. (c) Effluent chlorine residual, however, remains relatively constant throughout the treatment. (Characklis, 1980.)

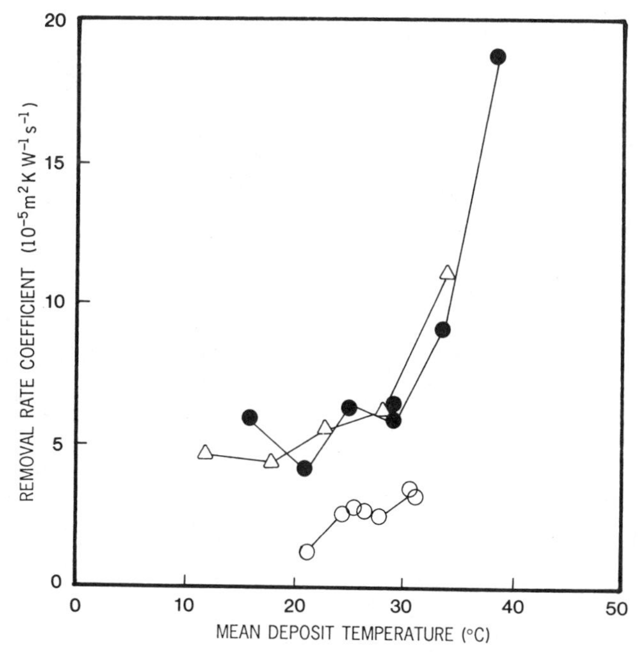

Figure 15.15 Increasing temperature increases the R_f removal rate k_r as a result of chlorine treatment in the Rhine River. (Reprinted with permission from Novak, 1982.) Symbols refer to different runs.

or

$$K_{a/r} = \frac{k_a}{k_r} = \frac{C_a D}{P} \tag{15.8}$$

where $K_{a/r}$ = a rate coefficient (—).

$K_{a/r}$ may be a useful operating parameter. For example if $K_{a/r}$ is known, $C_a D$ can be calculated for a given P. Then, $K_{a/r}$ can be used to determine P in an operating system if a sufficiently sensitive fouling monitor with a set upper limit on R_f (see Figure 15.8) is used. The upper limit represents the maximum acceptable fouling resistance for equipment performance. The measurement of $K_{a/r}$ and its variation with time and site-specific operating and environmental variables is the critical focus of this analysis, since $K_{a/r}$ represents a confounding of all the operating variables influencing dosage. Novak (1982) determined $K_{a/r}$ = 5.8 m^3 g^{-1} for the Rhine River (see Figure 15.7). Characklis et al. (1980) determined $K_{a/r}$ = 3.2 m^3 g^{-1} in a laboratory reactor (Figure 3.8). Lewis (1982) observed a significant change in $K_{a/r}$ with time in his tests, which suggest that k_r and R_f' were changing.

The coefficients k_a and k_r can be determined by monitoring R_f with a fouling

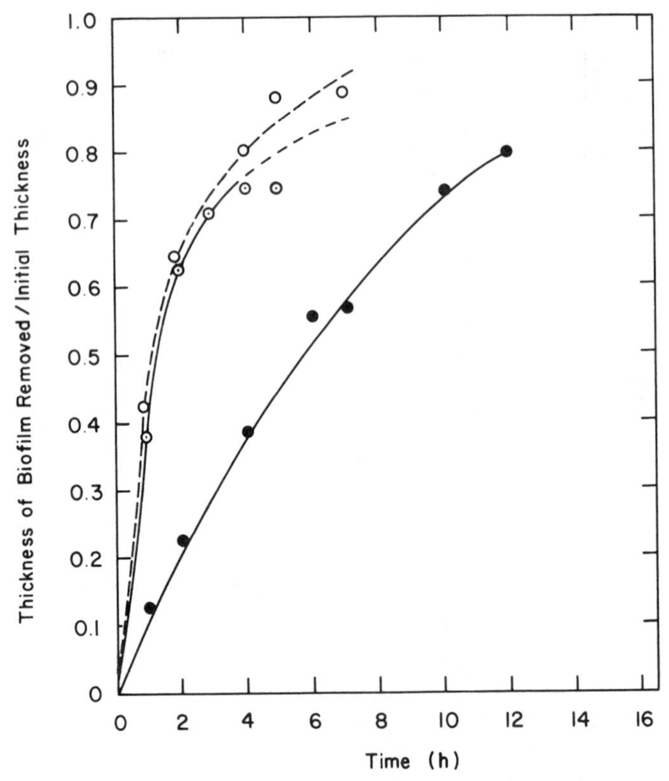

Figure 15.16 Biofilm removal rate by chlorination increases with increasing water velocity [0.66 (○), 0.50 (⊙), and 0.12 (●) m s^{-1}]. All tests were conducted at 25°C with chlorine concentration 2.5 g m^{-3}. (Miller and Bott, 1981.)

monitor. The model can easily accept the increased complexity of nonlinear kinetic expressions if desirable. The purpose of this simplistic analysis was to demonstrate a rational approach for analyzing biocide treatment of a biofouled system and to permit determination of an effective biocide dosage. The simplistic analysis results in a parameter that includes the effects of all the relevant rate coefficients in the process.

15.8 OTHER CHEMICALS

15.8.1 Oxidizing Biocides

Other oxidizing biocides besides chlorine have proved effective in controlling biofouling. In marine or estuarine waters, bromide ion may represent a significant fraction of the chlorine demand. Bromine (hypobromite ion) is rapidly formed from reaction of chlorine with bromide ion:

$$HOCl + Br^- = HOBr + Cl^- \quad [15.9]$$

Hypobromite is a strong disinfectant and is probably effective at destroying biofilm EPS as well. Bromine is also generated for biofouling control in low salinity waters by adding hypochlorite to sodium bromide.

Ozone can be effective in controlling biofouling and may also result in reduced corrosion rates (Edwards, 1987). Sugam (1980) compared ozone and chlorine and found chlorine to be more effective than ozone at the same applied concentration, but ozone was still effective at concentrations <0.5 g m^{-3}. The source water for Sugam's tests carried a significant organic load, as evidenced by TOC and BOD concentrations as high as 32 and 12 g m^{-3}, respectively. Ozone is consumed rapidly in side reactions (ozone demand), so its usefulness in heavily contaminated waters is dramatically reduced. Garey (1979) reported that ozone was more effective than chlorine, at equal concentrations, in controlling biofouling in low salinity waters. In seawater, ozone oxidizes bromide ion and its effectiveness is approximately the same as chlorine. Ozone requires special generators, which are costly and consume much power. Ozone may also react with organics in water to form epoxides, which are an environmental problem.

Halogenated organic compounds have been used effectively to control biofouling, sometimes in heavily contaminated waters. The performance of bromochlorohydantoin (BCR) clearly surpassed that of chlorine in a field study conducted at a recirculating cooling tower system (Matson and Characklis, 1983). The cooling tower operated at approximately 40 cycles of concentration with a feed or makeup water consisting partially of treated wastewater effluent. The result was a cooling water organic carbon concentration of approximately 500 g m^{-3}, which exerted a substantial oxidant demand. The field measurements of biofouling were conducted with a heat transfer fouling monitor and began with an assessment of chlorine effectiveness (Figure 14.31, Chapter 14). Within twelve days, despite the continuous addition of chlorine, fouling had increased to an unsatisfactory level. The addition of BCR resulted in an immediate decrease in heat transfer resistance and an increase in planktonic cells, presumably a result of massive detachment of biofilm cells. BCR maintained the system at a satisfactory operating condition for the remainder of the test.

The amines are a class of compounds, ubiquitous in nature, that are often responsible for chlorine demand in cooling water sources. Ammonia, the parent compound for the amines, reacts with chlorine to form mono-, di-, and trichloramine:

$$HOCl + NH_3 \rightarrow NH_2Cl + H_2O \quad [15.10]$$

$$HOCl + NH_2Cl \rightarrow NHCL_2 + H_2O \quad [15.11]$$

$$HOCl + NHCl_2 \rightarrow NCl_3 + H_2O \quad [15.12]$$

Where other amines are present, or in seawater where bromide is present, the number of possible reactions and cross reactions increases very quickly (Haag and Lietzke, 1980). It is not known to what extent halogenated amines, often referred to as "combined chlorine," can disrupt microbial films. LeChevallier et al. (1987), however, suggests that chloramines can be more effective against biofilms because they react slower in side reactions and thus are able to penetrate further into the film.

Chlorine dioxide has been successfully used to control biofouling in several industrial environments, including an oil field water flood operation (Knickrehm et al., 1987). Chlorine dioxide has been used for many years in the paper industry, and is receiving consideration for drinking water and industrial cooling water (Pacheco et al., 1987).

Hydrogen peroxide has the advantage that only oxygen and water are formed when it is reduced. Peroxide has been used effectively in special applications, including control of microbial activity in nuclear reactor cooling water. Characklis (1980) reports that hydrogen peroxide is rather ineffective in detaching biofilm in a laboratory reactor.

15.8.2 Nonoxidizing Biocides

15.8.2.1 Transport Considerations The performance of nonoxidizing biocides, like that of oxidizing biocides, is frequently limited by transport processes, i.e., penetration of the biocide into the biofilm. Eagar et al. (1988) report that the minimal concentration of glutaraldehyde necessary for complete inactivation of planktonic *Pseudomonas fluorescens* is approximately 25 g m^{-3}. When a *Pseudomonas fluorescens* biofilm was scraped from the substratum and the cells thoroughly dispersed, approximately 25–30 g m^{-3} glutaraldehyde was necessary for complete kill. When glutaraldehyde was applied to an intact *Pseudomonas fluorescens* biofilm, approximately 200 g m^{-3} was necessary. Thus, the diffusional resistance and other unknown factors resulting from the biofilm structure resulted in a concentration factor of approximately 8 for effective performance against dispersed versus intact biofilm cells. In oil field applications, partitioning of the biocide in phases, such as hydrocarbons, may also influence its effectiveness.

Generally, biocide demand is only considered important when evaluating oxidizing biocides. Recent reports, however, indicate that monitoring of nonoxidizing biocide residual may be necessary, especially in expansive systems such as oil field water flood operations. Eagar et al. (1988) report a 20% decrease in glutaraldehyde concentration in a laboratory RotoTorque with a 10 min residence time and 500 g m^{-3} inlet concentration. Similar percentage decreases have been reported in the field (Eagar et al., 1988). Gaylarde and Johnston (1982) report that a sterile steel coupon can reduce the effectiveness of a biocide in a planktonic test, suggesting that biocides may also react with the substratum.

15.8.2.2 Varying Susceptibility Microorganisms frequently display decreasing susceptibility to nonoxidizing biocides over time. The decreased susceptibility is sometimes believed to be the result of an accumulated resistance as observed with more specific antibiotics. However, because the microbial populations are composed of numerous microbial species, decreased susceptibility may be due to the emergence of a dominant species with less sensitivity to the broad spectrum biocide. Sanders and Stott (1987) describe an oil field water flood system in which the initially effective biocide had to be abandoned after one year because of its decreased effectiveness. As a consequence, alternation of biocides is clearly a strategy to be considered.

15.8.3 Other Chemicals

Turakhia et al. (1983) indicate that biofilm treatment with a calcium-specific chelant, ethylene glycol-bis(β-aminoethyl ether)-N,N-tetraacetic acid (EGTA), results in significant detachment. EGTA apparently attacks the calcium–EPS bonds, resulting in weakening of the polymer matrix and detachment. The chelant had little effect on cell viability.

Biocidal paints and coatings are used primarily for control of macrobial fouling. Generally, the paints leach an inhibitory heavy metal such as copper or tin into the fouling deposit. These paints have a significant environmental impact, and their use may be severely restricted in the future. Systematic tests to determine the rate-limiting step in their biocidal effectiveness against microbial films have not been reported.

15.9 MECHANICAL METHODS

Fouling control strategies can include a combination of mechanical, chemical, and thermal treatments. In power plants, backwashing of condensers and heat exchangers helps clear impinged macrofouling organisms and other aquatic debris from tube sheets. Traveling water screens and, in some cases, debris filters are used to prevent transport of material to power plant cooling water systems. Some power generating stations use recirculating sponge balls to control condenser tube biofouling. Off line mechanical cleaning techniques include passing air and water, brushes, or scrapers through condenser tubes to remove microfouling organisms from condenser surfaces. The configuration of some power plants allow backwashing of heated discharge water through sections of intake piping, providing thermal control of macrofouling on precondenser surfaces.

In oil field water flood operations and in drinking water distribution systems, periodic (e.g., quarterly) *pigging* treatments are used to reduce the biofouling accumulation on distribution system piping (Dewar, 1986; Galbraith and Lofgren, 1987). The pigs typically have steel brushes mounted on them to scrape the deposit from the pipe surface. Biocide treatment immediately after a pigging is often recommended as an effective method for reducing biofilm recovery (Galbraith and Lofgren, 1987).

Mechanical treatment may also have a subtle, yet critical, effect on biofilm accumulation and associated microbial corrosion (Chapter 16). The damaging corrosive effects of microorganisms are believed to occur only after significant colony formation within the deposit. In other words, a critical number of microorganisms is necessary to cause damage. Costerton (personal communication) believes the critical cell number in a sulfate-reducing bacterial (SRB) colony to initiate corrosion is approximately 500 cells. Mechanical treatment certainly does not remove all the biofilm organisms from the substratum. However, the biofilm organisms remaining after mechanical treatment may be in a different spatial distribution than before treatment, i.e., they have been stirred up. This more heterogeneous population distribution undoubtedly reduces the symbiotic relationships between species which accelerate corrosion and may also contribute to accelerated biofilm accumulation. There is evidence that colony formation within the biofilm also reduces biocide effectiveness (Costerton, 1984).

Experiments at a Canadian power plant site tested fouling reduction on admiralty brass tubes by passing soft sponge balls through the tubes (Characklis and Robinson, 1983). The power plant cooling water was from an Atlantic estuary in New Brunswick, and the monitoring began in April 1983. The overall heat transfer resistance at this site was measured in a fouling monitor over approximately 200 days. The maximum fouling factor R_f (at approximately day 60) was 7.04×10^{-5} m^2 K W^{-1}. The total deposit mass was around 8 g m^{-2}. The cleanliness factor determined in the fouling monitor was less than 70%, as compared to a condenser design value of 85%. The fouling monitor only measured the influence of the biofilm (i.e., water side fouling) on the cleanliness factor, while the condenser design value incorporates other contributions to performance. Three passes with a soft sponge ball was initially effective in reducing the effects of fouling (Figure 15.17). Even a second treatment with 12 ball passes was effective. However, subsequent treatments with the soft balls resulted in the formation of a refractory deposit, partly the result of continuous application of ferrous ion for corrosion inhibition. Treatment with abrasive balls (carborundum) was then initiated but was ineffective. More frequent treatment with abrasive balls may have been successful, but there was concern that erosion of the soft heat exchanger tubing could be significant.

Berger and Berger (1986) also indicate that the passage of sponge rubber balls through titanium, stainless steel (AL-6X), and aluminum (3004) tubing removes loose surface biofouling material but was inadequate for long term control.

Even increases in water velocity can result in significant biofilm detachment. Zelver et al. (1981, 1984) describe field experiments in which water velocity was accidentally or purposely increased. In each case, biofilm detachment occurred (Figure 15.18). Characklis (1980) demonstrated the positive influence of reversing flow on biofilm detachment in a laboratory tubular reactor (Figure 15.19). Timperley's experiments on this technique have been presented in Section 7.3.2.8 (Chapter 7).

In summary, mechanical methods for the control of soft biofouling deposits will generally be very effective in maintaining a clean system. However, when the deposits also consist of inorganic scale and/or tubercles, mechanical treatments may be inadequate.

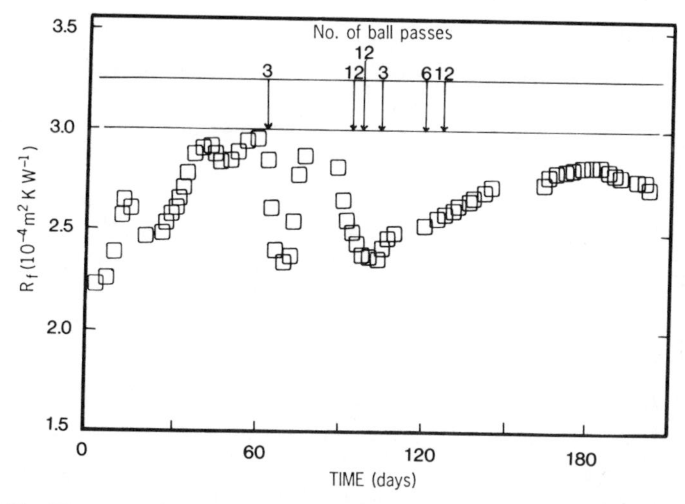

Figure 15.17 Treatment by passing soft sponge balls through a heated admiralty brass tube was initially effective in reducing the effects of fouling. However, subsequent treatments resulted in the formation of a refractory deposit. Further treatment with abrasive balls (carborundum) was ineffective. The refractory deposit was probably iron oxide. The numbers refer to the number of ball passes through the tube. (Characklis and Robinson, 1983.)

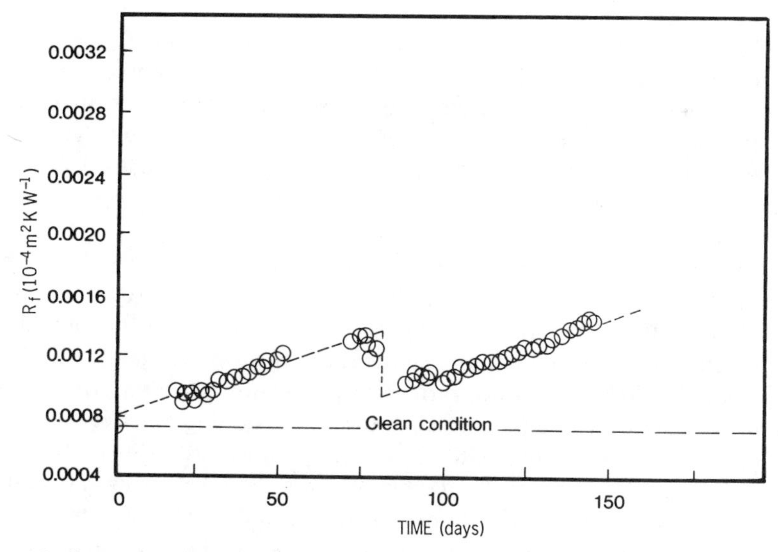

Figure 15.18 An increase in water velocity (i.e., flow surge) from 0.6 to 1.8 m s^{-1} in a heated (35°C wall temperature) AL-6X stainless steel tube operated for approximately 80 days on Hudson River water resulted in significant biofilm removal as indicated by heat transfer resistance. Rapid biofilm recovery is also evident. (Reprinted with permission from Zelver et al., 1984.)

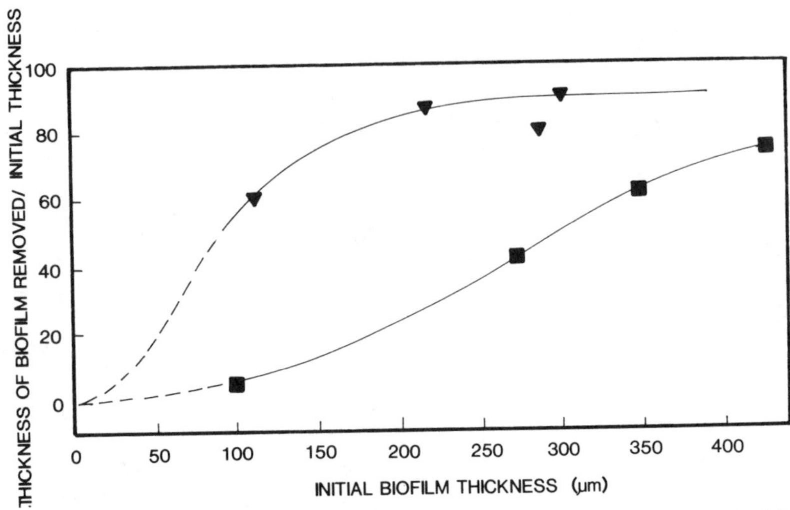

Figure 15.19 Flow surges (from approximately 1.6 to 2.7 m s^{-1}), combined with flow reversal (■), resulted in significant biofilm detachment in a laboratory tubular (aluminum) reactor. Increases in bulk water temperature (▼) from 30 to approximately 60°C also resulted in significant biofilm removal (Characklis, 1980).

15.10 USE OF FOULING MONITORS

Presume you are asked to control pH in a recirculating cooling tower system in which pH varies considerably with the seasons and even the weather. You are given a storage tank of acid and a storage tank of base with the ancillary pumps, flow meters, valves, etc. Before any effective pH control is possible, however, you will need a pH electrode. The electrode reading will provide feedback (manual or automatic) so that you can assess the necessary flow rate or amount of acid or base addition.

In a similar manner, presume you are asked to determine the necessary amount of treatment chemical (e.g., chlorine) needed to control biofouling and still operate within the environmental regulations. Numerous factors influence (1) the rate and extent of biofouling and (2) the chemical demand of the cooling water and fouling biofilm. Both variables significantly affect the efficient operation of the operating equipment. Therefore, determining an effective chemical dosage (e.g., chlorine minimization) can be a complex task. The problem is further complicated by seasonal (and other uncontrollable) changes in the effective chemical dosage. A fouling monitor (analogous to the pH electrode above) will permit the frequent evaluation of a dosing frequency, duration, and concentration. The costs for fouling monitors vary considerably depending on their level of sophistication and the accessories provided. However, the cost of the most expensive fouling monitor pales in relation to energy losses, downtime, and even chemical costs related to fouling processes.

15.10.1 Sidestream or *in Situ* Measurements?

Sidestream monitors are attractive for optimization studies, where flexibility in experimental design and accuracy are required, while risk to the operating unit is eliminated. In addition, using multiple sidestream test units provides a means for evaluating various control treatments simultaneously. Parallel testing of different treatments is important, since the fouling characteristics of the source water may be constantly changing. Using the operating unit to develop treatment strategies only provides information on a single treatment during one period of time. Although simulating operating equipment in every detail with a sidestream test apparatus is virtually impossible, important parameters (water flow velocity, tube material, heat flux) can be matched to assure that a realistic *effective* treatment regime can be developed.

In situ monitoring of fouling and corrosion and evaluating failures caused by these processes in industrial water systems is difficult. Several factors contribute to this situation. First, collecting representative samples of water, deposits, and system materials is expensive and very time-consuming. Typically, samples are rendered useless after lengthy delays or through decontamination actions. Second, portions of the systems suitable for visual inspection are quite limited. And third, postfailure evaluations are often limited by the lack of trended fouling and/or corrosion data. In other words, without an indication of fouling or corrosion rates, tendencies, or potentials, it is difficult to determine whether a failure was caused primarily by recent operating conditions or preoperational/outage conditions that existed years ago. Without the necessary information, monitoring results may thus be more theoretical than empirical, and remedial action recommendations more reactive than proactive. A means of monitoring the extent and progression of fouling within the pipes is necessary. The following monitoring methods could be used:

1. Install on-line fouling monitors in the water system.
2. Establish a small sidestream facility utilizing a fouling monitor.

The above methods should incorporate the following objectives in their design:

1. Monitor water quality in the water system. Establish the basis for flushing frequencies and wet layup treatments.
2. Design a facility for testing the impact of proposed water treatment programs on various system components and water system materials.
3. Determine biofouling/biocorrosion rates in water system under stagnant, low flow, and high flow conditions. Address both short and long term rates for general and localized pitting corrosion.
4. Monitor corrosion in various valves, and localized galvanic corrosion at welds.
5. Provide for visual inspection and removal of samples for destructive analysis.

6. Provide for expansion of sidestream facility to include additional fouling/corrosion testing equipment.

A sidestream facility dedicated to continual monitoring of fouling and corrosion tendencies is a necessity for mitigating the costly effects of fouling and corrosion. The investment in such a facility should be proportional to the value of the water system and will be remunerated many times over in production savings and extended system lifetime.

15.10.2 Monitor Sensitivity

One difficulty with present monitors is their lack of sensitivity. Consider that power plant condensers are generally chlorinated once a day. In one day, very little fouling deposit accumulates and no significant response is received from the monitor. However, some fouling deposition does occur. Several weeks may pass before a significant fouling resistance is observed as reflected by heat transfer or friction factor. In the meantime, the fouling deposit accumulation may have become refractory to the treatment. An example of a refractory deposit is the manganese accumulation that frequently results from repeated chlorination at some sites. The deposit is generally not detectable by the friction factor or heat transfer resistance measurements used in present fouling instruments.

Several special methods are under development that may provide better monitoring techniques in the near future. These techniques are highly sensitive, and in many cases the sensors are small and inexpensive, so that a number of them can be installed in the actual condenser. One such sensor is a printed circuit of approximately 1 cm^2 area and measures heat transfer resistance (Stenberg et al., 1988). Another one is even smaller in area and uses optical methods to detect deposit accumulation. It has detected a biofilm as thin as 5 μm (Siebel and Characklis, unpublished data).

15.10.3 Fouling Monitor Tests versus Equipment Operation

Fouling monitors at present are qualitatively simulating fouling processes as they occur in operating equipment. For example, an experimental program using monitors in an on site sidestream at a nuclear power plant has resolved an important qualitative issue: Is the effective chlorine dose a simple product of concentration times duration? The answer is no. Under realistic (but not necessarily representative) conditions of temperature and flow, over a realistic range of chlorine concentrations and daily exposure, accumulation of a biofouling deposit on titanium tubes is much more dependent on chlorine concentration than on duration of the treatment (Section 15.4.4). Accordingly, by implementing higher concentrations and shorter durations, a greater degree of fouling control may be achieved for the same total chlorine dosage (i.e., the product of concentration and duration, CD), within regulatory limits. The experimental program has identified the optimal dosage under the conditions of the test.

The essential qualitative result, that there is an optimal combination of concentration and duration (which is almost certainly different from the simple regulatory constraint of 2 hours per day at a dose that just meets the discharge limit), must hold true for operating condensers as well as for the test systems, since the test systems are similar enough to operating condensers in all important respects.

Quantitatively, however, the specific values defining the optimal combination (i.e., the most efficient CD) in an operating condenser is probably somewhat different from that identified in a test system at that site, and will doubtless be even more different during other seasons and at other sites. For example, some of the seasonal variations observed in performance and attributed to fouling are dependent on the design of the condenser and the change in heat rate throughout the year (Ferguson, 1981). The least tractable causes of difference from season to season and from site to site are differences in tube material, water chemistry, nutrient supply, and source microbial species composition. The variability owing to season and to site conditions can be determined by conducting measurements with a test system at the specific site and at a specific time. The lesser correction between the test system and the operating condenser is not so simply accomplished.

15.10.3.1 Geometry and Scale The key difference between a test system and operating equipment is scale. Consider a power plant condenser. The operating condenser bundle is fed from a water box, with the result that different tubes experience different temperatures and flow regimes. The condenser tube is at least an order of magnitude longer than a typical test section of a fouling monitor. Because of the heat exchange process in the condenser and the action of the long tube as a plug flow reactor, there are systematic differences along the length of the condenser tube with respect to wall temperature, water temperature, residual chlorine, and possible other chemical changes wrought by the upstream deposit. Although a monitored test section could be set up to mimic any particular short section of the tube (beginning or end), the same test section cannot mimic the entire span simultaneously.

15.10.3.2 Plug Flow Effects Observations on spatial variations in power plant condensers are conflicting. Analysis of deposits from a dismantled condenser at a nuclear power plant (admiralty brass) indicate that differences in deposit accumulation between tubes overwhelmed variation in deposit accumulation with length along the tubes (McCaughey et al., 1987). This condenser received no treatment or chemical addition for fouling control. At other facilities, operations personnel frequently report visual observations of more fouling at the exit of the condenser than in the entrance tube region, mostly in condensers receiving chlorination. Since visual observations can only extend a short distance into the condenser tube, higher turbulence in the entrance length may be responsible for the reduced amount of deposit observed there.

The variable deposition in a chlorinated condenser tube can be explained by calculations based on a plug flow model of the condenser tube. The model describes the chlorine–biofilm reaction in a circular tube under turbulent flow conditions and includes transport and reaction of chlorine in the water and in the biofilm. The principal relationships in the model are derived from the equation of continuity expressed in cylindrical coordinates (Figure 13.11, Chapter 13). The derivation assumes that radial concentration gradients of chlorine are much greater than axial concentration gradients and that mass transport in the bulk water is solely by radial eddy diffusion and axial advection. The assumptions result in Eq. 15.13 for processes occurring in the bulk water compartment and Eq 15.14 for processes occurring in the biofilm compartment:

$$\frac{1}{D_b r^*}\frac{\partial}{\partial r^*}\left(D_b r^* \frac{\partial C^*}{\partial r^*}\right) + \frac{r_b r_1^2}{C_0 D_m}\left(\frac{D_m}{D_b}\right) = \frac{v_z D_m}{v_m D_b}\frac{\partial C^*}{\partial z^*} \quad [15.13]$$

$$\frac{1}{r^*}\frac{\partial}{\partial r^*}\left(r^* \frac{\partial C^*}{\partial r^*}\right) = \frac{-r_f r_1^2}{C_0 D_f} \quad [15.14]$$

where C = chlorine concentration (ML^{-3})
C^* = C/C_0, dimensionless concentration (—)
r_1 = tube radius (L)
r_I = radial distance to the biofilm–bulk water interface (L)
r^* = r/r_1, dimensionless radial distance (—)
z^* = $zD_m/r_1^2 v_m$, dimensionless axial distance (—)
D_f, D_b = effective diffusivity of chlorine in the biofilm and bulk water, respectively (L^2t^{-1})
D_m = molecular diffusivity of chlorine in water (L^2t^{-1})
v_z = axial time-smoothed fluid velocity (Lt^{-1})
v_r = radial time-smoothed fluid velocity (Lt^{-1})
v_m = mean water velocity in the tube (Lt^{-1})
r_f, r_b = reaction rate of chlorine in the biofilm and bulk water, respectively ($ML^{-3}t^{-1}$)

Reactions in the water and in the biofilm were approximated by first order (with respect to chlorine concentration) kinetic expressions. Rate coefficients were estimated from laboratory results (Characklis et al., 1980). The boundary conditions are as follows:

1. $C = C_0$ at $z = 0$, the entrance to the tube.
2. $dC/dr = 0$ at $r = r_1$, the tube wall.
3. $dC/dr = 0$ at $r = 0$, due to symmetry of the tube.
4. $D_f\, dC/dr = D_b\, dC/dr$: the fluxes into and out of the biofilm–fluid interfaces are equal.
5. $C|_{-r_I} = \alpha_p C|_{+r_I}$ at $r = r_I$, i.e., the concentration of chlorine in the fluid must be in equilibrium with that in the film (interface partition coefficient $\alpha_p = 1$).

Biofilm removal was described by a simple mass balance, i.e., a constant proportion of chlorine entering the biofilm results in removal of biomass from the surface.

The results of the computer simulation clearly indicate the significant effect that chlorine demand in the water can have on the performance of a chlorination program intended to minimize biofouling (Figure 15.20). The model results indicate that chlorine residual decreases rapidly with distance through the tube. At short exposure times, the biofilm thickness has not been reduced very much, even at the beginning of the tube, where the chlorine concentration is high. However, the biofilm thickness in the tube with no chlorine demand in the bulk water remains low for a significantly longer distance down the tube, since more chlorine is available for reaction with the biofilm. As time increases, the biofilm thickness decreases, and larger downstream removals are observed in the tube with no chlorine demand in the bulk water. Thus, the depletion of chlorine with distance down the tube may well explain the elevated amounts of biofouling deposits observed at the exit of some power plant condenser tubes.

15.10.4 Experimental versus Operational Monitoring

The extrapolation from fouling monitors to operating equipment is constrained both by the amount of effort and attention that can be devoted to monitoring in

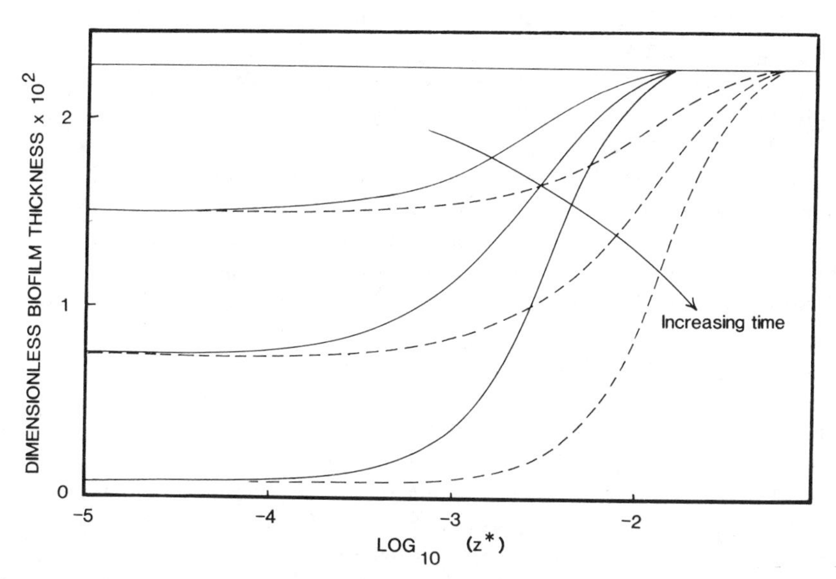

Figure 15.20 Computer simulations indicate increasing biofilm thickness with time and with increasing axial distance z^* through the tube as a result of chlorine treatment. When the bulk water has no chlorine demand (dashed curves), the biofilm is reduced for a significantly longer distance down the tube. When chlorine demand in the bulk water is significant (solid curves), the biofilm remains high closer to the inlet.

an operational mode and by the available scientific information for adjusting fouling monitor predictions to take account of environmental and/or operational changes. Experimental studies with fouling monitors require considerable manpower and resources. These studies maintain controlled conditions and explore a substantial number of treatment combinations in search of the optimum treatment program. An operational (as opposed to an experimental) monitoring program at a plant must place far fewer demands on personnel and budgets than an experimental research program can. Therefore, operational monitoring may not be as precise or extensive. Experimental monitoring can identify an optimal treatment schedule by varying appropriate environmental or operating variables during the testing. However, an operational monitoring program will only show measured fouling rates and the manner in which the equipment responded to the treatment. The translation of the operational information contained in the measured data into a recommended efficient treatment is a modeling problem.

The modeling problem consists in determining the various rate coefficients that describe the effects of operating variables (e.g., substratum temperature, flow rate) and environmental variables (e.g., nutrient supply, water temperature) on the dynamics of fouling deposit accumulation. The dynamics must also include variables such as biocide demand and susceptibility of the biofilm to the biocide as influenced by biocide concentration, frequency, and duration of treatment.

The operational monitoring will serve essentially to calibrate the model for the given application (season and site). This modeling objective is practical even though more adequate descriptions of some of the contributing processes are still needed. The elucidation of the processes contributing to biofouling deposit accumulation will necessarily be subjects for further testing with fouling monitors in the experimental mode.

Frequently, experiments with fouling monitors are only conducted for several months. However, there may be slow processes that only manifest themselves in the fouling dynamics after longer periods of time (e.g., increasing deposit density with time, as in tubercle formation). Monitors installed in operating mode will supply long term data that can serve adequately for calibrating the model.

15.10.5 Summary

Sidestream fouling monitors presently are capable of qualitatively simulating fouling processes in operating equipment. In addition, sidestream monitors will continue to be useful in testing the effects of various design, operating, and environmental variables on fouling processes. However, *in situ* fouling monitors are needed for regulating fouling control programs in the operating equipment. In addition, more sensitive fouling monitors (sidestream and *in situ*) are needed to detect very thin biofilms (and/or their effects) and thus permit rapid feedback for fouling control treatments.

Consider a hypothetical fouling control system incorporated in a cooling system (Figure 15.21). A sidestream of the cooling water supply enters a fouling

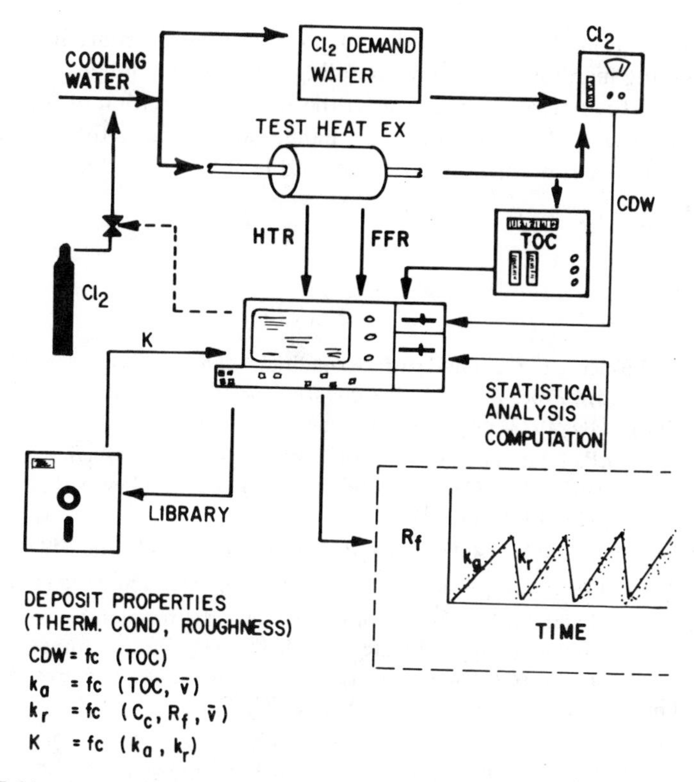

Figure 15.21 A hypothetical fouling control system incorporating a fouling monitor, which measures heat transfer resistance (HTR), fluid frictional resistance (FFR), and water quality [chlorine demand (CDW) and total organic carbon (TOC) of the bulk water]. A computer conducts statistical calculations to determine appropriate rate coefficients and operating parameters, which are stored for analyzing historical trends. Calculated rate coefficients determine the chlorine dosage.

monitor [measures heat transfer (HTR) and fluid frictional resistance (FFR)] and a flowthrough reactor to determine the chlorine demand of the bulk water (CDW). A small sample is also withdrawn periodically for determination of water quality [e.g., total organic carbon (TOC)]. These quantities all are entered in a computer, where statistical computations determine k_a and k_r from the accumulation data. The statistical estimates of k_a and k_r are entered (and stored) in the computer so that long term correlations between the rate coefficients and environmental variables (e.g., TOC, water temperature) and operating variables (e.g., chlorine concentration C, water velocity v) can be made. $K_{a/r}$ can also be calculated and used in the correlations. Deposit properties (e.g., thermal conductivity and hydraulic roughness) can also be calculated from the fluid frictional resistance and heat transfer resistance. Finally, changes in chlorine concentration, duration, and period can be determined rapidly for real time control of biofouling.

Mathematical models for biofouling deposit accumulation are necessary for relating fouling monitor data to the operating equipment. The plug flow behavior of some of the operating equipment, the seasonal variation of fouling, and variations of fouling due to operational changes are concerns that could be resolved by a validated mathematical model and lead to more effective control of biofouling.

REFERENCES

Baier, R.E., "Influence of the initial surface conditions of materials on bioadhesion," in R.F. Ackers (ed.), *Proc. International Congress on Marine Corrosion and Fouling,* Gaithersburg, MD, 1972, pp. 633–640.

Battaglia, P.J., D.P. Bour, and R.M. Burd, *Biofouling Control Practice and Assessment,* Final Report, EPRI CS-1976, Project 1132-1, Electric Power Research Institute, Palo Alto, CA, 1981, pp. 221.

Berg, J.D., A. Matin, and P. Roberts, *Appl. Env. Microbiol.,* **44**, 814–819 (1982).

Berger, L.R. and J.A. Berger, *Appl. Environ. Microbiol.,* **51**, 1186–1198 (1986).

Biros, J.L. and H. Lamblin, "Reduction of residual chlorine by intensive monitoring of bacterial populations in supplied water," in R.L. Jolley, W.A. Brungs, and R.B. Cumming (eds.), *Water Chlorination: Environmental Impact and Health,* Vol. 3, Ann Arbor Science, Ann Arbor, MI, 1980, pp. 743–755.

Bongers, L.H. and D.T. Burton, *Bromine Chloride—An Alternative to Chlorine for Fouling Control in Condenser Cooling Systems,* Final Report, U.S. Environmental Protection Agency, EPA-600/7-77-053, Washington, 1977.

Burton, D.T., "Biofouling control procedures for power plant cooling water systems," in J.F. Garey, R.M. Jorden, A.A. Aitken, D.T. Burton, and R.H. Gray, (eds.), *Condenser Biofouling Control,* Ann Arbor Science, Ann Arbor, MI, 1980, pp. 251–266.

Characklis, W.G., "Effect of hypochlorite on microbial slimes," in *Proc. 26th Industrial Wastes Conference,* Purdue University, 1971, pp. 171–181.

Characklis, W.G., *Wat. Res.,* **7**, 1249–1258 (1972).

Characklis, W.G., *Biofilm Development and Destruction,* Final Report, EPRI CS-1554, Project RP902-1, Electric Power Research Inst., Palo Alto, CA, 1980.

Characklis, W.G., *Bacterial Regrowth in Distribution Systems,* Final Report, American Water Works Association Research Foundation, Denver, CO, 1988.

Characklis, W.G. and S.T. Dydek, *Wat. Res.,* **10**, 515–522 (1976).

Characklis, W.G. and J.A. Robinson, *Development of a Fouling Control Program,* Final Report, Canadian Electrical Association, No. 219 G 388, 1983.

Characklis, W.G., M.G. Trulear, N.A. Stathopoulos, and L.C. Chang, "Oxidation and destruction of microbial films," in R.L. Jolley, W.A. Brungs, and R.B. Cumming (eds.), *Water Chlorination: Environmental Impact and Health,* Vol. 3, Ann Arbor Science, Ann Arbor, MI, 1980, pp. 349–368.

Characklis, W.G., N. Zelver, C.H. Nelson, R.O. Lewis, D.E. Dobb, and G.K. Pagenkopf, "Influence of biofouling and biofouling control techniques on corrosion of copper-nickel tubes," Paper No. 250, presented at CORROSION 83, Nat. Assoc. Corrosion Eng., Houston, TX, 1983.

Costerton, J.W., *Dev. Ind. Microbiol.*, **25**, 363-354 (1984).

Dewar, E.J., *Matls. Perf.*, **25**, 39-47 (1986).

Dexter, S.C., *J. Colloid. Interface Sci.*, **70**, 346-354 (1979).

Drew Chemical Corporation, *Principles of Industrial Water Treatment*, Drew Chemical Corporation, Boonton, NJ, 1979.

Eagar, R.J., J. Leder, J.P. Stanley, and A.B. Theis, *Matls. Perf.*, **27**, 40-45 (1988).

Eaton, A., C. Chamberlain, and M. Cooney, "Colonization of Cu-Ni surfaces by microfouling organisms," in J.F. Garey, R.M. Jorden, A.H. Aitken, D.T. Burton, and R.H. Gray (eds.), *Condenser Biofouling Control*, Ann Arbor Science, Ann Arbor, MI, 1980, pp. 105-120.

Edwards, H.B., *J. Cooling Tower Inst.*, **8**, 10-21 (1987).

Ferguson, R.J., "Determination of seasonal variations in microbiological fouling factors and average slime thickness," CTI Paper No. TP-237A, Cooling Tower Institute, Houston, TX, 1981.

Galbraith, J.M. and K.L. Lofgren, "An update on monitoring of microbial corrosion in Prudhoe Bay's produced water and sea water floods," Paper No. 379, presented at Corrosion/87, Nat. Assoc. Corrosion Eng., San Francisco, 1987.

Garey, J.F., *Biofouling Control Investigations: 18 Month Summary Report*, Electric Power Research Institute, 1979.

Gaulke, A.E., T.E. Webb, and F.L. Stokes, "A chlorine minimization program for an electric generating facility located on the Ohio River," in J.F. Garey, R.M. Jorden, A.H. Aitken, D.T. Burton, and R.H. Gray (eds.), *Condenser Biofouling Control*, Ann Arbor Science, Ann Arbor, MI, 1980, pp. 339-354.

Gaylarde, C.C. and J.M. Johnston, *Int. Biodet.*, **18**, 111 (1982).

Graham, J., *Biofouling Control Assessment—A Preliminary Data Base Analysis*, Final Report, EPRI CS-2469, Electric Power Research Institute, Palo Alto, CA, 1982, p. 235.

Haag, W.R. and M.H. Lietzke, "A kinetic model for predicting the concentrations of active halogen species in chlorinated saline cooling waters," in R.L. Jolley, W.A. Brungs, and R.B. Cumming (eds.), *Water Chlorination: Environmental Impact and Health*, Vol. 3, Ann Arbor Science, Ann Arbor, MI, 1980, pp. 415-426.

Hall, L.W., G.R. Helz, and D.T. Burton, *Power Plant Chlorination: A Biological and Chemical Assessment*, Ann Arbor Science, Ann Arbor, MI, 1981.

Helz, G.R., R. Sugam, and R.Y. Hsu, "Chlorine degradation and halocarbon production in estuarine waters," in R.L. Jolley, H. Gorchev, and D.H. Hamilton (eds.), *Water Chlorination: Environmental Impact and Health*, Vol. 2, Ann Arbor Science, Ann Arbor, MI, 1978, p. 209.

Hullinger, C.H., "Hypochlorite-oxidized starch," in R.L. Whistler and M.L. Wolfram (eds.), *Methods in Carbohydrate Chemistry*, Academic Press, New York, 1963, pp. 313-315.

Knickrehm, M., E. Caballero, P. Romualdo, and J. Sandidge, "Use of chlorine dioxide in a secondary recovery process to inhibit bacterial fouling and corrosion," Paper No. 383, presented at CORROSION 87, National Association of Corrosion Engineers, Houston, TX, 1987.

LeChevallier, M.W., T.M. Babrock, and R.G. Lee, *Appl. Env. Microbiol.*, **53**, 2714-2724 (1987).

Lewis, R.O., *Materials Performance,* **21**, 31–38 (1982).

Lister, M.W., *Can. J. Biochem.,* **34**, 479–488 (1956).

Mangum, D.C. and B.P. Shepherd, *Methods of Controlling Marine Fouling in Seawater Desalination Plants,* Final Report, Contract No. 14-30-2829, Office of Saline Water, U.S. Dept. of the Interior, 1973, p. 124.

Marine Biocontrol Corp. and CCE, Inc., "Oyster Creek nuclear generating station chlorine optimization study," Final Report, General Public Utilities Nuclear Corp., Forked River, NJ (1988).

Marine Research, Inc., *Possible Alternatives to Chlorination for Controlling Fouling in Power Station Cooling Water Systems,* Final Report, MRI, Sandwich, MA, 1976.

Matson, J.V. and W.G. Characklis, *J. Cooling Tower Inst.,* **4**, 27 (1983).

McCaughey, M.S., A. Thau, W.G. Characklis, and W.L. Jones, "An evaluation of condenser tube fouling at an estuarine nuclear power plant," presented at Joint Power Generation Conference, ASME, Miami, FL, 1987.

McFeters, G.A. and A.K. Camper, *Adv. Appl. Microbiol.,* **29**, 177–193 (1985).

Miller, P.C. and T.R. Bott, "The removal of biological films using sodium hypochlorite solutions," in *Progress in the Prevention of Fouling in Industrial Plant,* Nottingham University, Nottingham, England, 1981, pp. 121–136.

Moss, R.D., H.B. Flora, R.A. Hiltunen, C.V. Seaman, N.D. Moore, and S.H. Magliente, "A chlorine minimization/optimization study for condenser fouling control," in R.L. Jolley, H. Gorchev, and D.H. Hamilton (eds.), *Water Chlorination: Environmental Impact and Health,* Vol. 2, Ann Arbor Science, Ann Arbor, MI, 1979.

Moss, R.D., P.A. March, and S.P.Gautrey, "The flexibility of a manifold-type targeted chlorination system," in W. Chow and Y.G. Mussalli (eds.), *Proc. Condenser Biofouling Control—State-of-the-Art Symposium,* Electric Power Research Institute, Palo Alto, CA, 1985, pp. 4.39–4.50.

Mussalli, Y.G., J. Kaspar, G. Hecker, M. Padmanabhan, and W. Chow, "Targeted chlorination," in W. Chow and Y.G. Mussalli (eds.), *Proc. Condenser Biofouling Control—State-of-the-Art Symposium,* Electric Power Research Institute, Palo Alto, CA, 1985, pp. 4.12–4.38.

Norrman, G., W.G. Characklis, and J.D. Bryers, *Dev. Ind. Microbiol.,* **18**, 581–590 (1977).

Novak, L., *J. Heat Transfer,* **104**, 663–669 (1982).

Pacheco, A.M., T.D. Dishinger, and J.L. Tomlin, "Experiences in controlling microbially-induced corrosion," Paper No. 376, presented at CORROSION 87, Nat. Assoc. Corrosion Eng., Houston, TX, 1987.

Rice, J.K., "Chlorine minimization—An overview," in J.F. Garey, R.M. Jorden, A.H. Aitken, D.T. Burton, and R.H. Gray (eds.), *Condenser Biofouling Control,* Ann Arbor Science, Ann Arbor, MI, 1980, pp. 295–300.

Ridgway, H.F., C.A. Justice, C. Whittaker, D.G. Argo, and B.H. Olson, *J. AWWA,* **76**, 94 (1984).

Roe, F.L. and W.G. Characklis, "A systematic approach to fouling control," in R.W. Bryers (ed.), *Fouling of Heat Exchange Surfaces,* United Engineering Trustees, New York, 1983, pp. 51–68.

Sanders, P.F., and J.F.D. Stott, "Assessment, monitoring, and control of microbiological corrosion hazards in offshore oil production systems," National Association of Corrosion Engineers, Houston, TX, paper no. 367 (1987).

Schumacher, P.D. and J.W. Lingle, "Chlorine minimization studies at the Wisconsin Electric Power Co. Valley and Oak Creek power plants," in J.F. Garey, R.M. Jorden, A.H. Aitken, D.T. Burton, and R.H. Gray (eds.), *Condenser Biofouling Control*, Ann Arbor Science, Ann Arbor, MI, 1980, pp. 301–324.

Stenberg, M., G. Stemme, G. Kittilsland, and K. Pedersen, *Sensors and Actuators*, **13**, 203–221 (1988).

Sugam, R., *Biofouling Control with Ozone at the Bergen Generating Station*, Report No. CS-1629, EPRI Project No. 733-1, Electric Power Research Institute, Palo Alto, CA, 1980.

Sugam, R. and G.R. Helz, "Seawater chlorination: A description of chemical speciation," in R.L. Jolley, W.A. Brungs, and R.B. Cumming (eds.), *Water Chlorination: Environmental Impact and Health*, Vol. 3, Ann Arbor Science, Ann Arbor, MI, 1981, pp. 427–433.

Timperley, D., "Effect of Reynolds number and mean velocity on cleaning-in-place of pipelines," in B. Hallstrom, D.B. Lund, and C. Tragardh (eds.), *Fundamentals and Applications of Surface Phenomena Associated with Fouling and Cleaning in Food Processing*, Lund University, Tylosand, Sweden, 1981, pp. 402–412.

Trabalka, J.R. and M.B. Burch, "Investigations of the effects of halogenated organic compounds produced in cooling systems and process effluents on aquatic organisms," in R.L. Jolley, H. Gorchev, and D.H. Hamilton (eds.), *Water Chlorination: Environmental Impact and Health*, Vol. 2, Ann Arbor Science, Ann Arbor, MI, 1978, pp. 163–173.

Turakhia, M.H., K.E. Cooksey, and W.G. Characklis, *Appl. Env. Microbiol.*, **46**, 1236–1238 (1983).

Waite, T.D. and J.R. Fagan, "Summary of biofouling control alternatives," in J.F. Garey, R.M. Jorden, A.H. Aitken, D.T. Burton, and R.H. Gray (eds.), *Condenser Biofouling Control*, Ann Arbor Science, Ann Arbor, MI, 1980, pp. 441–462.

Whistler, R.L. and R. Schweiger, *J. Am. Chem. Soc.*, **79**, 6460–6464 (1957).

Whistler, R.L., E.G. Linke, and S. Kazeniac, *J. Am. Chem. Soc.*, **78**, 4704–4709 (1956).

Wong, G.T.F., "The effect of temperature on dissipation of chlorine in seawater," in R.L. Jolley, W.A. Brungs, and R.B. Cumming (eds.), *Water Chlorination: Environmental Impact and Health*, Vol. 3, Ann Arbor Science, Ann Arbor, MI, 1980, pp. 395–406.

Zelver, N., W.G. Characklis, and F.L. Roe, "Discriminating between biofouling and scaling in a deposition monitor," CTI Paper No. TP-239-A, Cooling Tower Institute, Houston, TX, 1981.

Zelver, N., W.G. Characklis, J.A. Robinson, F.L. Roe, Z. Dicic, K.R. Chapple, and A. Ribaudo, "Tube material, fluid velocity, surface temperature and fouling: A field study," CTI Paper No. TP-84-16, Cooling Tower Institute, Houston, TX, 1984.

Zielke, R.L. and S.K. Macey, "Validation of a kinetic model to predict residual chlorine in fresh water," in R.L. Jolley, W.A. Brungs, and R.B. Cumming (eds.), *Water Chlorination: Environmental Impact and Health*, Vol. 3, Ann Arbor Science, Ann Arbor, MI, 1980, pp. 445–451.

16

MICROBIAL CORROSION

BRENDA J. LITTLE
Naval Ocean Research and Development Activity, NSTL, Mississippi

P. A. WAGNER
Naval Ocean Research and Development Activity, NSTL, Mississippi

WILLIAM G. CHARACKLIS
Montana State University, Bozeman, Montana

WHONCHEE LEE
Montana State University, Bozeman, Montana

SUMMARY

Biofilm formation and metallic corrosion have traditionally been evaluated as separate, independent processes that occur simultaneously or sequentially on metal surfaces. The corrosion rate of metals was thought to be determined exclusively by metallurgical properties, the chemical composition of the electrolyte, the temperature, and the flow velocity. Water chemistry, water temperature, pressure, shear stress, substratum composition, and design features were recognized as determinants for the surface colonization of microorganisms. This chapter stresses the interdependence of the two processes and the techniques needed to quantify their interaction.

A clearer understanding of microbial corrosion will be attained when biofilm models can be developed that integrate electrochemical processes occurring at the biofilm-substratum interface with microbial processes in the biofilm and chemical quality of the bulk water.

16.1 INTRODUCTION

Microbial films of varied composition and thickness develop on surfaces in contact with aqueous environments. Metabolic reactions mediated by certain microorganisms residing in biofilms can promote (Miller and Tiller, 1970; Obuekwe et al., 1981) or impede the biodeterioration of materials, including metals, concrete, and plastics. This chapter is limited to discussing the effects of microorganisms on metal surfaces.

Most confirmed cases of microbial corrosion (MC) are localized. Discrete mounds or columns related to tuberculation can develop on metal surfaces as a result of microbial activities. The morphology and location of deposits are sometimes indicative of the microbial species that caused the deposit. For example, distinctive hemispherical or conical tubercles on the surface of steel and subsurface pitting are characteristics of iron-oxidizing bacteria. Sulfate-reducing bacteria (SRB) produce open pitting, or *gouging,* on stainless steel. When SRB are active along the edges of gasketed joints, shallow crevice corrosion is often found under adjacent gasket areas. SRB attack on cast iron typically produces graphitization: the corroded areas are filled with a soft skeleton of graphite. On nickel and cupronickel alloys, SRB are reported to produce conical pits containing concentric rings (Kobrin, 1976). These types of observations have been used to document MC problems. Despite the recognition and the documentation, the identification of specific mechanisms for MC has remained elusive because of the complexity of microbiological processes and the lack of analytical techniques to quantify localized corrosion. In this chapter, we shall review electrochemical terms and techniques that have been applied to the investigation of MC, the microbial activities known to affect biodeterioration, as well as the effect of corrosion on biofilm formation.

16.2 ELECTROCHEMISTRY

Since metallic corrosion is generally an electrochemical process, most MC nomenclature and techniques are electrochemically derived. The nomenclature will be introduced by considering a model corrosion system (Figure 16.1) in which a drop of an electrolyte is in contact with a clean metal surface (M) and a patch of impurity, such as a colony of microorganisms (M′). The model demonstrates that a corrosion system is really a short-circuited electrochemical cell, the electrodic areas M and M′ being in contact with each other and in contact with an electrolyte. Potential differences developing between M and M′ will cause an electric current, which will result in corrosion. The metal sites, at which oxidation of metal atoms is the predominant reaction, are defined as anodes. Simultaneously, electrons are driven back to sites, termed cathodes, where reduction is the predominant reaction. The tendency for a metal to corrode can be expressed in terms of electromotive force (emf). The greater the value of emf for any cell and the resulting current, the greater the tendency for the overall corrosion reaction to occur. The interface between the metal surface and the electrolyte bears

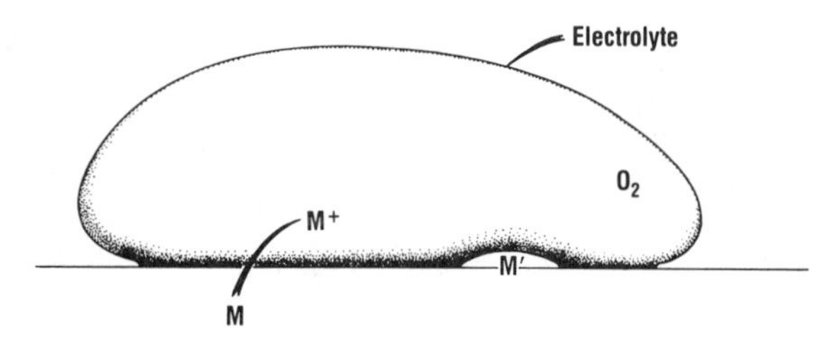

Figure 16.1 Model corrosion system. Clean metal surface (M) and a colony of microorganisms (M′) in contact with a drop of electrolyte.

an unequal distribution of electrical charge, which gives rise to an electrical double layer. Electrochemical methods for evaluating corrosion involve the determination of specific properties of the double layer.

In the absence of biofilms with complexing or precipitating ions, the following anodic reactions can be predicted for metals in aqueous electrolytes:

$$M \rightarrow M^{z+}(aq) + ze^- \qquad [16.1]$$

$$M + zH_2O \rightarrow M(OH)_z\,(solid) + ze^- + zH^+ \qquad [16.2]$$

$$M + zOH^- \rightarrow M(OH)_z\,(solid) + ze^- \qquad [16.3]$$

$$M + zH_2O \rightarrow MO_{\frac{z}{2}}^{z-}(aq) + 2zH^+ + ze^- \qquad [16.4]$$

$$M + zOH^- \rightarrow MO_{\frac{z}{2}}^{z-}(aq) + zH^+ + ze^- \qquad [16.5]$$

Electrons that arrive at cathodes can be removed by either hydrogen ion or oxygen reduction. Alternative reaction pathways for the cathodic production of hydrogen are delineated in Figure 16.2. Each reaction pathway has a specific activation energy that is a function of the metal and the electrolyte composition.

If the reduction product accumulates at the cathode, the corrosion cell is cathodically polarized and metal dissolution continues until the emf decreases and corrosion ceases. If, however, the product is continuously removed from the cathode or if other driving forces control the rate of removal of metal ions from the anode, corrosion will continue. A decrease in the production of metal cations slows down the corrosion rate.

In neutral aerobic electrolytes, the accumulation of electrons at the cathodes (*cathodic polarization*) is prevented by the intervention of oxygen as follows:

$$O_2 + 4e^- + 2H_2O \rightarrow 4OH^- \qquad [16.6]$$

The cathode is depolarized, and oxygen is the depolarizer. In many abiotic environments, diffusion of oxygen to the metal surface will be the rate-controlling step in the corrosion process.

The rate of corrosion is the rate of metal dissolution. The rate of charge trans-

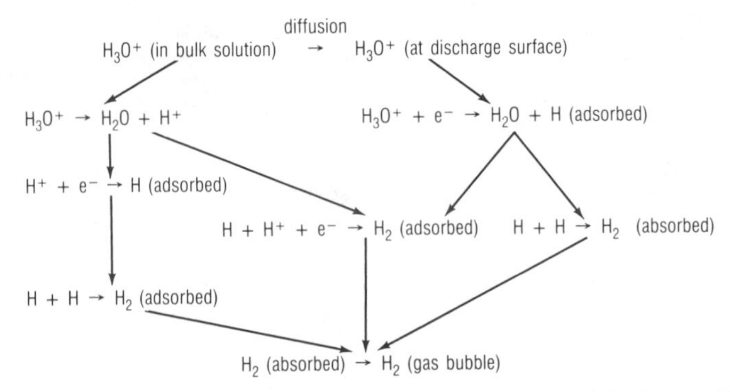

Figure 16.2 Reaction pathways for the cathodic production of hydrogen (Duquette, 1985).

fer, which may be regarded as either the rate of transfer of metal cations in one direction or the rate of transfer of electrons in the opposite direction, in an electrochemical reaction is the current I.

A dynamic equilibrium exists between the cathodic current and the anodic current for a single metal ion electrode reaction ($M = M^{+2} + 2e^-$) when there is no net current across the electrode (Figure 16.3):

$$I_{\text{net}} = I_a + I_c = 0 \quad [16.7]$$

$$I_a = -I_c = I_0 \quad [16.8]$$

where I_a, I_c, I_0 = anodic, cathodic, and exchange current, respectively.

The anodic and cathodic currents are equal in value but opposite in sign. The absolute value of each current is called the exchange current (I_0), and the corre-

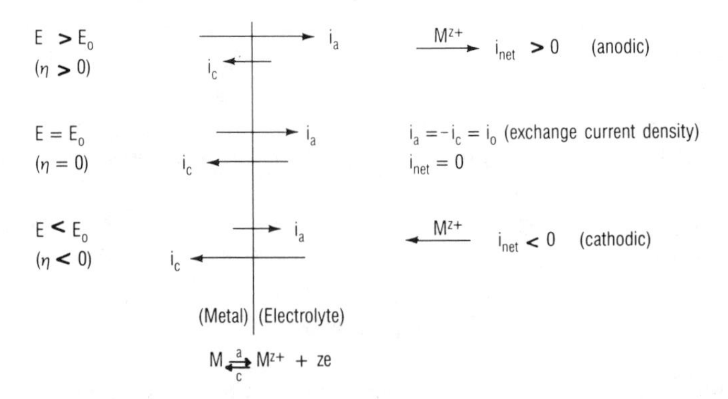

Figure 16.3 Charge transfer polarization for the reaction $M = M^{z+} + ze^-$ under the condition of anodic or cathodic overvoltage.

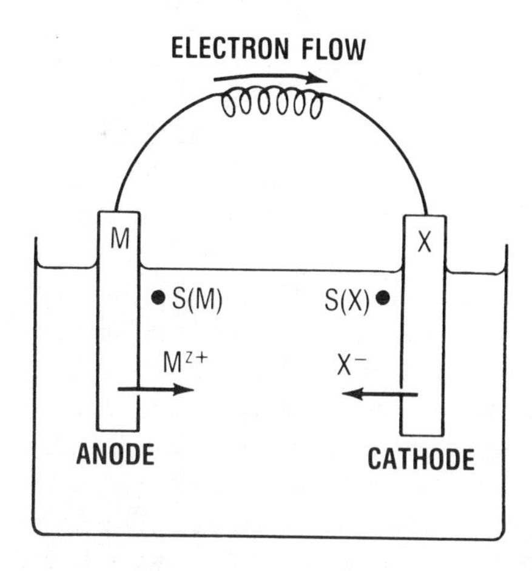

Figure 16.4 Corrosion cell of a more noble electrode (cathode) and a more base electrode (anode).

sponding potential is referred to as the open circuit potential (potential for zero current flow). If two different electrodes are connected (Figure 16.4), they will adopt a common potential that is a compromise between their open-circuit potentials. In this case the more noble electrode (cathode) can consume the electrons generated by the more base electrode (anode). As a consequence, a spontaneous corrosion reaction happens. The current (I_{corr}) generated by the metal dissolution process ($M \rightarrow M^{+2} + 2e^-$) is exactly balanced by the current ($-I_{corr}$) consumed by the cathodic process ($X + 2e^- \rightarrow X^{-2}$) (Figure 16.5). At

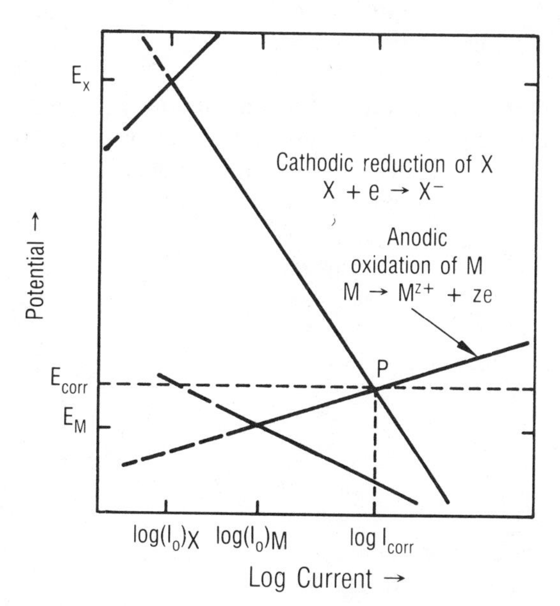

Figure 16.5 Spontaneous corrosion reaction resulting from the connection of electrodes M and X. E_{corr} and I_{corr} are the corrosion potential and corrosion current, respectively.

point P, I_{corr} is the corrosion current and is equal to I'_a or $-I'_c$, and the corresponding potential is the corrosion potential E_{corr}. The potential difference across the electric double layer cannot be defined in absolute terms, but only with respect to another charged interface, i.e., a reference electrode. Potential measurements can be made using either a potentiometer or an electrometer and a reference electrode.

The net current density is defined as the difference between the oxidation and reduction current densities. The partial anodic and cathodic current densities cannot be determined directly by means of an ammeter unless the anodic and cathodic areas can be separated physically. If the metal is polarized, a net current density, i_c for cathodic polarization and i_a for anodic polarization, can be measured with an ammeter. Galvanostatic polarizations can be made with constant direct current power units or banks of accumulators used in conjunction with a variable resistance; potentiostatic polarizations, with potentiostats of varying output currents.

An isolated metal immersed in a solution of its cations will develop the following equilibrium:

$$\underset{\text{in solution}}{\underset{\text{hydrated ions}}{M^+(aq) + e^-}} \underset{\overleftarrow{i}_1}{\overset{\overrightarrow{i}_1}{\rightleftharpoons}} \underset{\text{in lattice}}{\underset{\text{metal ions}}{M^+ + e^-}} \quad [16.9]$$

where $\overrightarrow{i}_1$ = rate of the cathodic process
$\overleftarrow{i}_1$ = rate of the anodic process

At equilibrium the rates of dissolution and discharge become equal. The potential at which the net exchange current density equals zero is the reversible, or equilibrium, potential. The displacement of the potential of an electrode from its reversible value is the overpotential η, and

$$\eta = E_p - E_r \quad [16.10]$$

where E_p = polarized potential
E_r = reversible or equilibrium potential

Further,

$$E_r = \eta_c + \eta_a + I R_{soln} \quad [16.11]$$

where η_c = cathode overpotential
η_a = anode overpotential
I = rate of charge transfer or current
R_{soln} = electrolytic resistance of the solution

Thus, the rate of a corrosion reaction depends on both the thermodynamic parameter E_r and the kinetic parameters η_a and η_c.

For any given electrode process, charge transfer at a finite rate will involve an activation overpotential, η_a, which provides the activation energy required for the reactant to surmount the energy barrier that exists between the energy states of the reactant and the product. In corrosion, the resistance of the metallic path to charge transfer is negligible; therefore, resistance overpotential η_r may be defined as

$$\eta_r = I(R_{\text{soln}} + R_f) \qquad [16.12]$$

where R_{soln} = electrical resistance of the solution
R_f = resistance due to coatings applied or formed on the surface

R_{soln} depends on the electrical resistivity of the solution and the geometry of the corroding system.

16.2.1 Localized Corrosion

Corrosion can range from highly uniform to highly localized. In uniform corrosion, the anodic and cathodic sites are physically inseparable. Uniform corrosion can be described by general theories and can be predicted with some degree of accuracy using direct current (dc, Section 16.2.2) and alternating current (ac, Section 16.2.3) techniques. Crevice corrosion, filiform corrosion, pitting, selective leaching, and erosion–corrosion are typical forms of localized corrosion. No general theories apply to all forms. Nevertheless, the following factors are frequently important in localized attack:

1. Cathode/anode surface area ratio
2. Differential aeration
3. pH changes at anodic and cathodic sites

A fundamental principle of all corrosion is expressed by the following equation:

$$\Sigma I_a = \Sigma I_c \qquad [16.13]$$

The sum of the rates of the cathodic reactions must equal the sum of the rates of the anodic reactions. If attack is uniform and there are predominant anodic and cathodic reactions, then

$$\frac{I_a}{A_a} = \frac{I_c}{A_c}, \quad \text{or} \quad i_a = i_c \qquad [16.14]$$

since the area of the cathode, A_c, equals the area of the anode, A_a. If attack is localized, $A_a < A_c$ and $i_a > i_c$. The larger the i_a:i_c ratio, the more intense the attack.

Such an effect can be produced if a crevice is present such that oxygen can diffuse readily to the metal surface outside it while there is little oxygen available to the area within. When the concentration of oxygen is higher at one part of the metal surface and lower at another, the metal exposed to the lower concentration will become the anode of a corrosion cell, with consequent localized attack. The large cathode/anode area ratio leads to localized attack of the metal within the crevice. For localized attack to proceed, there must be a continuous supply of electron acceptor species at the cathode surface and the anodic reaction must not be stifled.

Cathodic reduction of dissolved oxygen may result in an increase in pH of the solution in the vicinity of the metal:

$$\tfrac{1}{2}O_2 + H_2O + 2e^- \rightarrow 2OH^- \qquad [16.15]$$

At the anodic site, the metal will form cations:

$$M \rightarrow M^{+2} + 2e^- \qquad [16.16]$$

which can be hydrolyzed by water to form H^+ ions:

$$M^{+2} + 2H_2O \rightarrow M(OH)_2 + 2H^+ \qquad [16.17]$$

If the cathodic and anodic sites are separated from one another, and if the electrolyte is stagnant, the pH of the anolyte will decrease and the pH at the catholyte will increase. Cells of this type are referred to as occluded cells. The acidity that develops within pits was used by Hoar (1947) to explain the pitting of passive metals in solutions containing Cl^- ions. Brown (1970) reported that pitting, crevice corrosion, intergranular attack, filiform corrosion, and hydrogen cracking are accompanied by local acidification due to hydrolysis of metal ions.

As mentioned in the introduction, MC is most often characterized as localized corrosion. The effect of colonizing microorganisms on cathode/anode area ratios, differential aeration, and local pH changes will be discussed in Section 16.3.1.

16.2.2 Direct Current Methods

The most commonly used electrochemical techniques for evaluating uniform corrosion are (1) determination of the steady state corrosion potential (E_{corr}); (2) determination of the variation of E_{corr} with time; (3) determination of the E–i relationships during polarization at constant current density (galvanostatic potential is the variable), and (4) determination of the E–i relationships during polarization at constant potential (potentiostatic current is the variable).

An Evans diagram is a graphical method for demonstrating the relationship between corrosion rate and the extent of polarization of anodic and cathodic reactions in which both reactions are presented as linear E–I (or E–i) curves that converge and intersect. The intersection defines the corrosion potential E_{corr} and the corrosion current I_{corr}. A typical Evans diagram for the corrosion of a single metal is shown in Figure 16.6. Extrapolation of Tafel lines from Evans diagrams is a popular dc technique for predicting the corrosion rate. As previously discussed, E_{corr} can be determined by means of a reference electrode, but if the anodic and cathodic sites are inseparable, direct determination of I_{corr} by means of an ammeter is not possible. The Evans diagram has been used to examine various types of corrosion, including uniform corrosion of a single metal and bimetallic corrosion. A recent modification of the Evans diagram has been the replacement of the linear anodic E–I curve with the discontinuous potentiostatic curve in order to provide a more fundamental representation of electrode kinetics (Shrier, 1979).

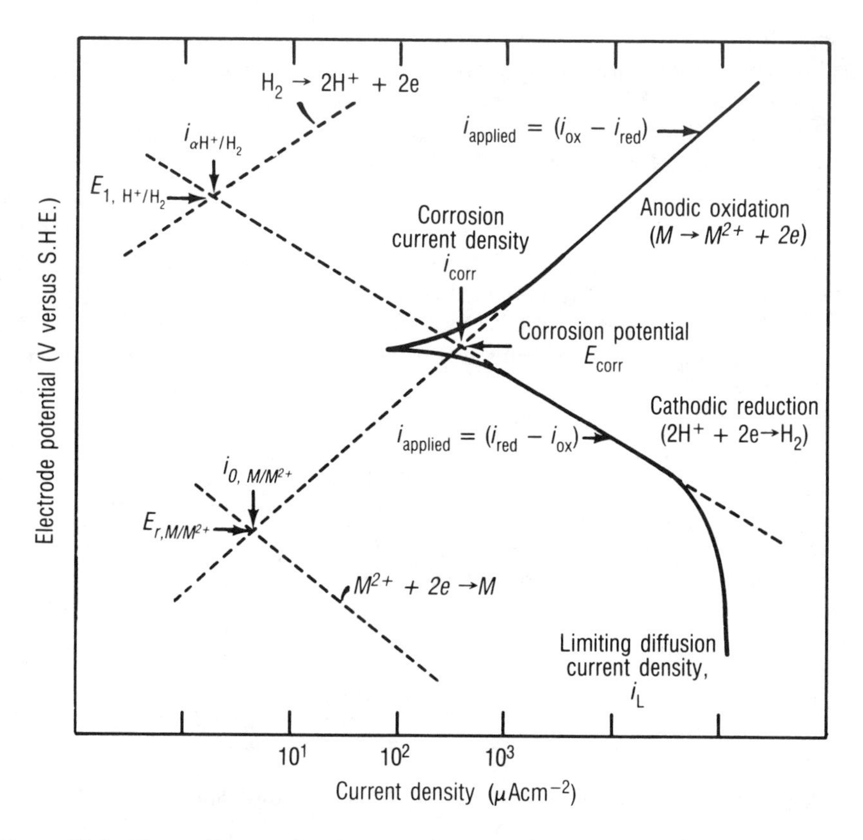

Figure 16.6 Evans diagram for the corrosion of a single metal in a reducing acid in which there are two exchange processes involving the oxidation of M and the reduction of H^+ (Shrier, 1979).

16.2.2.1 ***Tafel Equation*** The potential difference across an interface at which an electrochemical reaction is occurring changes linearly with the logarithm of the current density (Tafel, 1905). A Tafel plot is generated by polarizing a metal specimen anodically and cathodically from the corrosion potential E_{corr}. The corrosion current I_{corr} is obtained from a Tafel plot by extrapolating the linear portions of the curves to E_{corr}. The anodic and cathodic portions of the plot are described by the Tafel equation:

$$\eta = \beta \log\left(\frac{I}{I_{corr}}\right) \quad [16.18]$$

where η = overpotential, the difference between the potential of the specimen and the corrosion potential
β = Tafel coefficient
I_{corr} = corrosion current density
I = current density at overpotential (i.e., $\eta = \beta(\log i - \log i_{corr})$)

This equation has the form $y = mx + b$, so a plot of η versus $\log i$ is a straight line with slope β. The slope of the linear portions of the anodic and cathodic portions are termed Tafel coefficients and are as designated β_a and β_c. The units for Tafel coefficients are mV/decade or V/decade. A decade of current is one order of magnitude. Current density can be equated to the weight loss per unit time according to Faraday's law (Shrier, 1979):

$$\frac{m}{t} = \frac{Mit}{zF} \quad [16.19]$$

where m = mass of metal corroded (M)
M = molecular weight (M mol^{-1})
z = number of electrons involved in one act of the corrosion reaction (# e)
F = Faraday's constant, 96487 coulombs
i = current density (AL^{-2})
t = time (t)
ρ = density (ML^{-3})
A = surface area of metal involved (L^2)
L = thickness of metal removed (L)

and

$$m = \rho AL \quad [16.20]$$

By rearranging and substituting, the rate of penetration is given by

$$\frac{d}{t} = \frac{Mi}{zF} \quad [16.21]$$

Conversion factors (Uhlig and Revie, 1985) can be used to express the weight loss in millimeters of penetration per year (mm y^{-1}), milligrams per square decimeter per day (mdd), or other convenient units.

16.2.2.2 *Polarization Resistance* The measurement of polarization resistance is another commonly used technique for quantifying uniform corrosion. A linear relationship between potential and applied anodic and cathodic current densities is established when the applied potentials are low (+10 mV of E_{corr}). Skold and Larson (1957) first observed this behavior in studies of the corrosion of steel and cast iron in natural water. The slope of the linear curve, E–I or E–i, is the polarization resistance R_p (in resistance units, e.g., ohms). Stern and Geary (1957) derived the following expression showing that the corrosion rate is inversely proportional to R_p at potentials close to E_{corr}:

$$\frac{1}{R_p} = \left[\frac{\Delta i}{\Delta E}\right]_{E_{corr}} = 2.3 \frac{\beta_a + |\beta_c|}{\beta_a|\beta_c|} i_{corr} \qquad [16.22]$$

where R_p = polarizatio resistance. R_p is determined at potentials close to E_{corr}. Thus, the corrosion rate is inversely proportional to R_p, and i_{corr} can be evaluated if the Tafel coefficients are known.

The use of direct current electrochemical measurements for evaluating microbial corrosion is described in Section 16.2.2. These techniques have the advantage that many observations can be made on a single specimen in a matter of hours. However, there are several disadvantages. Standard dc techniques rely on the Stern–Geary assumptions, which state that for small polarization (<10 mV), the interface behaves as a resistance, with magnitude inversely proportional to the corrosion current. However, the interface really behaves as a parallel resistance–capacitance combination. Additional processes, such as chemical reactions, solvation, and adsorption of reaction intermediates, further complicate the system. The presence of organic coatings or biofilms on the metal introduces additional electrical and electrochemical complexities (e.g., see Section 11.3.1.1, Chapter 11, for diffusion of charged species in biofilms). Direct current techniques measure the overall or net result of all of the aforementioned processes, from the slowest to the most rapid, and provide an overall indication of corrosion.

When a reaction rate is so high that the electroactive species cannot reach or be removed from the electrode surface, concentration polarization occurs. The reaction rate becomes diffusion-controlled. Polarization measurements include an ohmic potential drop. This contribution to the polarization is equal to iR, where i is the current density, and R is equal to l/K (ohms), where l is the resistance path length and K is specific conductivity. The formation of a biofilm results in an uncompensated iR drop and nonlinear Tafel behavior.

A potentially more serious problem with dc methods is the assumption of uniform corrosion. Polarization techniques assume that the metal is uniformly corroding by a Wagner–Traud mechanism (Wagner and Traud, 1938) in which oxidation and reduction reactions occur randomly over a surface with regard to both space and time. Patchy biofilms, localized colonization, and localized biochemical reactions give rise to anodes and cathodes that are distinct in space and may be stable with time.

16.2.2.3 *Pitting Potential* Potentiostatic techniques can be used to evaluate the susceptibility of metals to localized corrosion. The breakdown potential E_b is a characteristic of the pitting of a number of metals that rely on passivity for corrosion resistance. The potentiostatic anodic curves (Leckie and Uhlig, 1966) for a stainless steel in NaCl solutions containing various concentrations of SO_4^{-2} ions are presented in Figure 16.7. In all cases, there is a sudden increase in current at some potential E_b, which may be regarded as the most negative potential to cause the initiation and propagation of pits. Alternatively, E_b may be regarded as the most positive potential that results in a decrease in current due to passivation of the entire surface. E_b for a particular metal varies with the electrolyte and with the technique used for its determination, and cannot be considered a property of the metal.

16.2.3 Alternating Current Methods

Impedance or resistance measurements at an electrochemical interface are ac methods that can be used to evaluate the corrosion rate. With ac methods, only those processes fast enough to occur during the alternation of the electric field are measured. Slower processes are negligible. The measurement of the imped-

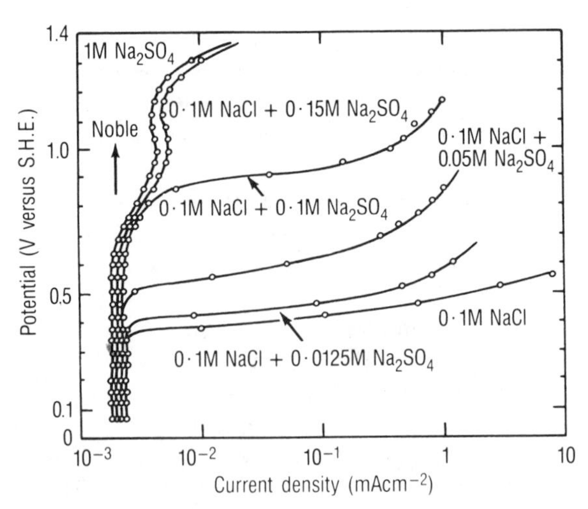

Figure 16.7 Breakdown potentials of stainless steel ($Fe_{18}Cr_8Ni$) in 0.1 M NaCl plus Na_2SO_4 during anodic polarization (Shrier, 1979, after Leckie and Uhlig, 1966).

ance of an electrode over a wide range of frequencies (from a few mHz to several kHz) provides analytical discrimination of the various processes occurring between the metal and the electrolyte (or coating). Specific electrochemical processes can be distinguished by their relaxation times using a variety of frequencies, provided that the relaxation times are not too similar.

Interfacial impedance is usually presented as an equivalent circuit, a model approximating the real system. The equivalent circuit for a bare metal electrode surface on which a reaction occurs can be represented as a reaction resistance in parallel with a double layer capacitance (Figure 16.8). Impedance is commonly plotted on the complex plane as illustrated in Figure 16.9. R_{ct} is the charge transfer resistance, which is equivalent to the linear polarization resistance under pure activation control, and C_{dl} is the double layer capacitance. The frequency at the top of the semicircle corresponds to the time constant RC, and the frequency limit ω corresponds to R_{ct}.

One disadvantage of the complex plane is that it is difficult to determine the best fit semicircle through the experimental points. A more accurate way of plotting the impedance over a wide range of frequencies is the Bode plot (Figure 16.10). For the equivalent circuit in Figure 16.8, the Bode plot consists of two frequency domains. At high frequencies, log Z is a straight line with slope -1, which corresponds to the impedance of the capacitance. At low frequencies, the

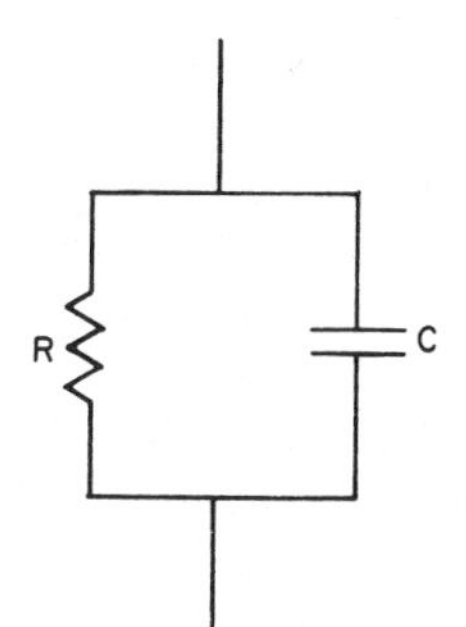

Figure 16.8 The equivalent circuit for a bare metal electrode surface on which a reaction occurs can be represented as a reaction resistance (R) in parallel with a double layer capacitance (C).

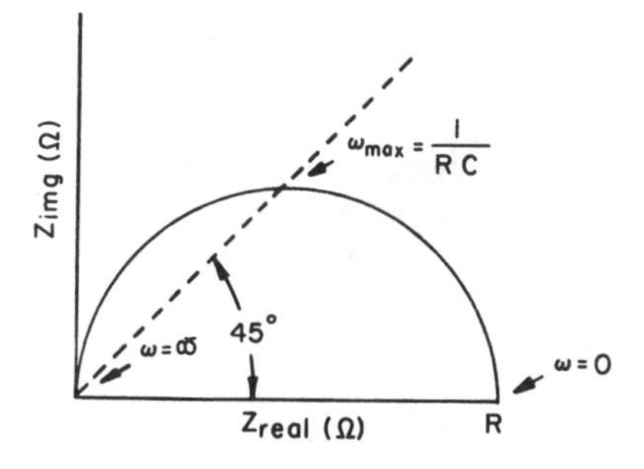

Figure 16.9 The complex plane for the circuit in Figure 16.8 is characterized by frequency $\omega_{\max}$ at the top of the semicircle which corresponds to the RC time constant. The low frequency limit ($\omega = 0$) corresponds to R_{ct}.

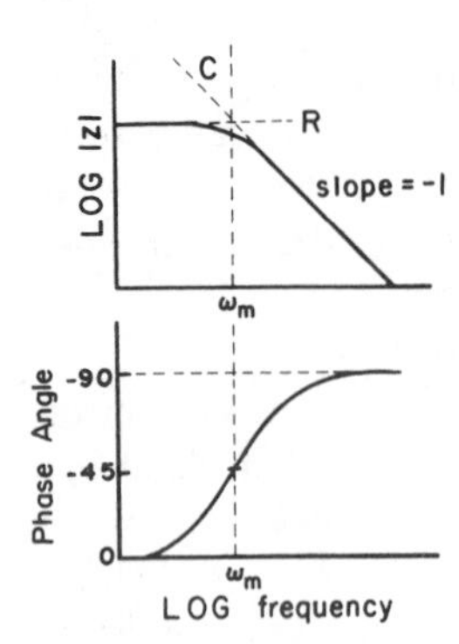

Figure 16.10 The Bode plot for the circuit in Figure 16.8 is characterized by ω_m which corresponds to R_{ct}.

line with slope zero corresponds to the resistance. The region between the two lines is curved and corresponds to the time constant of the circuit. An advantage of the Bode plot is the ease of separating effects by varying the values of individual components in the network. For example, the impedance plot of Figure 16.10 is shown as a solid line (curve 1) in Figure 16.11. An increase in C causes a shift of the sloping line to curve 2 (dotted line). An increase of R shifts the line to curve 3, and an increase of R and a decrease of C shift the line to curve 4 (Bode, 1945).

Ideal semicircle behavior in the complex plane or a single time constant in a Bode plot is rarely observed. The equivalent circuit can be more complex, as in the case of a reaction that involves an adsorption pseudocapacitance C_ϕ (Figure 16.12), where R_f and R_s represent the fast and slow step reaction resistances. The line segments corresponding to C_{dl}, R_f, C_ϕ, and R_s are clearly separated in the frequency domain (Van Meirhaeghe, et al., 1976). The situation can be even more complicated if reactions occur on an oxide or on an organic film or biofilm at the surface.

To illustrate the principles by which impedance data are interpreted, consider an electrochemical equivalent circuit complicated by solution resistance (R_{sol})

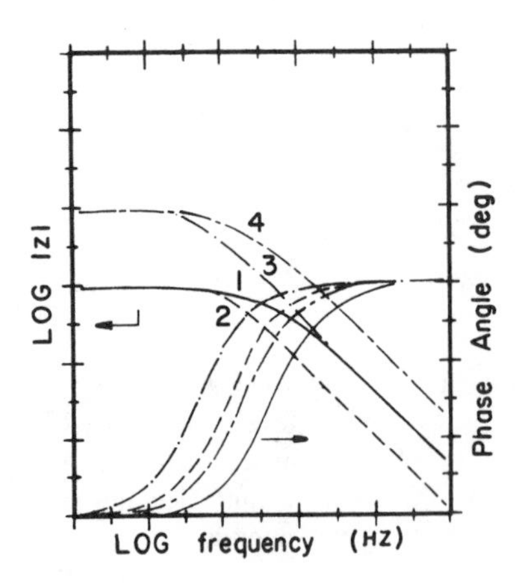

Figure 16.11 The impedance plot of Figure 16.10 is shown as a solid line (curve 1). An increase in C causes a shift of the sloping line to curve 2 (dashed line). An increase of R shifts the line to curve 3, and an increase of R and a decrease of C shift the line to curve 4. (Bode, 1945.)

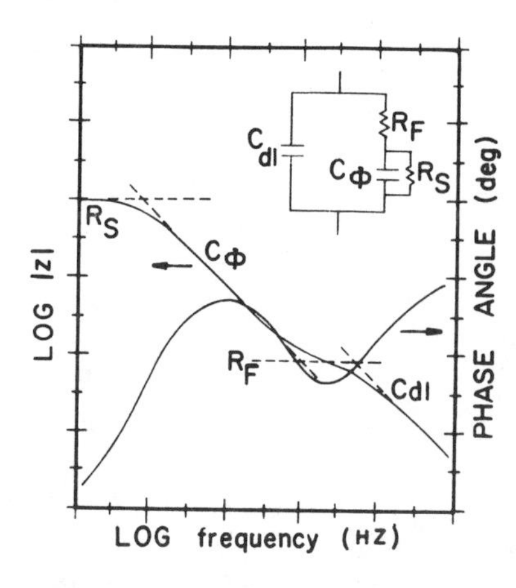

Figure 16.12 The equivalent circuit for the case of a reaction that involves an adsorption pseudocapacitance C_ϕ, where R_f and R_s represent the fast and slow step reaction resistances. The line segments corresponding to C_{dl}, R_f, C_ϕ, and R_s are clearly separated in the frequency domain. (Van Meirhaeghe, et al., 1976.)

and a film that will have, in general, a resistive (R_f) and capacitive (C_f) component (Figure 16.13). The most general Bode plot for this circuit (Cahan and Chen, 1982) is presented in Figure 16.14. By definition, $R_f \ll R_s$, and C_ϕ is usually $\gg C_{dl}$. R_f has been chosen such that $R_{\text{sol}} \ll R_f \ll R_F$. C_f is usually $<$ C_{dl} for a film thicker than a few angstroms having any reasonable dielectric constant. For $R_f \ll R_{\text{sol}}$, line segment 3 moves down until segment 2 disappears and R_f and/or C_f are no longer experimentally accessible (curve a). For $R_f \gg R_s$, line segment 3 moves up to curve b (with segment 2 extended correspondingly) and the kinetics are masked. Variation of other components in the network causes similar displacements of corresponding line segments. The interposition of any resistive or capacitive circuit element in series with this network causes a displacement upward and/or to the right, while a parallel addition causes displacement to the lower left. With a family of curves, it is possible to differentiate between a parallel and a series addition by the direction of shift produced in the curve. For example, when C_ϕ disappears outside its potential window, the curve shifts to position c as a limiting value. For the case when R_s is very close to infinite (Figure 16.12), the log Z curve extends upward to the left and the phase angle decreases to $-90°$, yielding a Bode plot similar to that in Figure 16.15. The value of R_f is found at the frequency where the phase shift is minimum, even

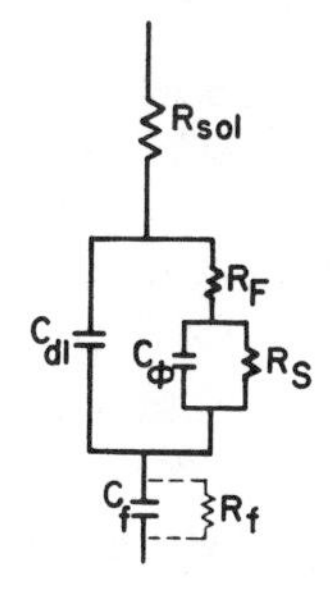

Figure 16.13 An electrochemical equivalent circuit is complicated by solution resistance (R_{sol}) and a film that will have, in general, a resistive (R_f) and capacitive (C_f) component.

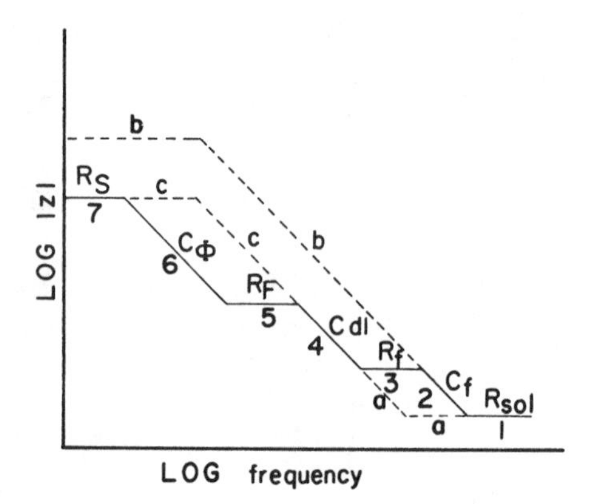

Figure 16.14 The most general Bode plot for the circuit (Cahan and Chen, 1982) presented in Figure 16.13. By definition, $R_f \ll R_s$ and C_ϕ is usually $\gg C_{dl}$. R_f has been chosen such that $R_{sol} \ll R_f \ll R_F$. C_f is usually $< C_{dl}$ for a film thicker than a few angstroms having any reasonable dielectric constant. For $R_f \ll R_{sol}$, line segment 3 moves down until segment 2 disappears and R_f and/or C_f are no longer experimentally accessible (curve a). For $R_f \gg R_s$, line segment 3 moves up to curve b (with segment 2 extended correspondingly) and the kinetics are masked.

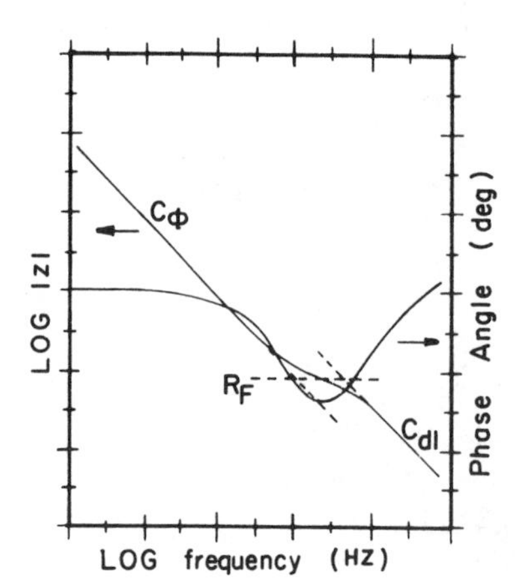

Figure 16.15 When R_s is very close to infinite (Figure 16.12), the log $|Z|$ curve extends upward to the left and the phase angle decreases to $-90°$, yielding a Bode plot.

though no clear horizontal line segment is evident. When R_s is finite, the low frequency is horizontal and the phase angle goes to zero at low frequencies. The phase angle starts to decrease at frequencies about one decade higher than the corresponding RC time constant.

In practice, there are problems in applying ac impedance techniques to the study of *MC*. The only measurement leading to the determination of the metal corrosion rate is the charge transfer resistance (R_{ct}). The other values obtained from ac impedance (solution resistance, film resistance, and dielectrical capacities) are related to the reaction rates between the electrode and the electrolyte.

Once the corrosion product becomes bulky, as sometimes observed with microbial corrosion, the impedance at low frequencies becomes very high, and the extrapolations needed for calculations of corrosion rate are more difficult and less accurate. Since the measurement of the interfacial impedance is continuous, short term fluctuations can be detected. These are usually caused by localized attack on the test electrode and are not easy to detect when the full frequency sweep measurement is conducted. Since microbial corrosion is generally localized, the noise may be a more useful indicator of corrosion in biologically active environments.

16.2.4 Other Corrosion Measurement Techniques

Several innovative electrochemical techniques have been developed to overcome problems in the discrimination of MC from other types of corrosion. A two compartment cell was developed by Little et al. (1986) to enable experimental separation of anodic and cathodic processes and to measure corrosion currents from separated electrodic areas (Figure 16.16). Briefly, metal electrodes of the same area and identical electrochemical properties are placed in each of the two compartments separated by a 0.1 μm cellulose nitrate–cellulose acetate membrane and connected via a zero resistance ammeter. When both compartments are

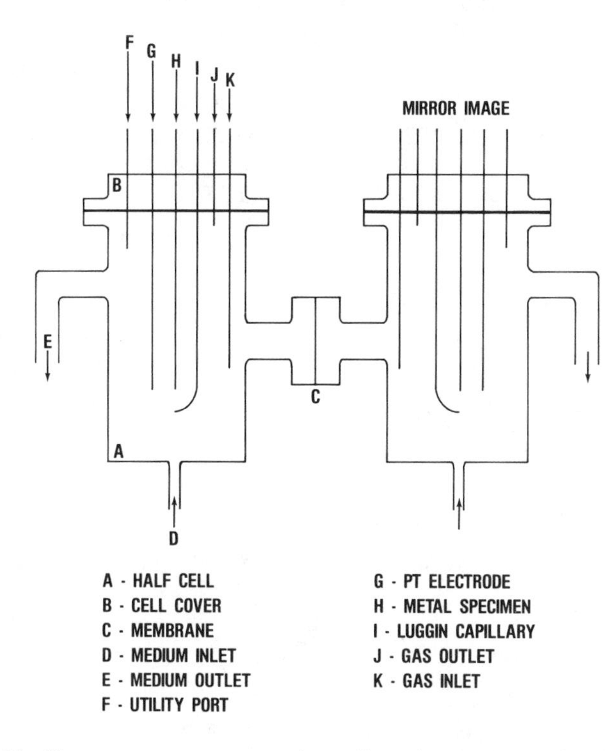

Figure 16.16 Two compartment corrosion cell used to evaluate microbial corrosion.

maintained under identical conditions, homogeneous corrosion of the metals takes place, but because the metals are at the same potential, no current flows between the compartments. However, when one of the compartments is chemically or biologically perturbed, nonuniform corrosion results, which creates finite anodic or cathodic sites. The direction of the current indicates whether anodic or cathodic reactions are occurring. The magnitude of the current indicates the intensity of that reaction. The corrosion current obtained with this method can be converted to rate of penetration using equations such as Eq. 16.21.

The effects of a marine pseudomonad, an obligate thermophilic filamentous bacterium, and an iron-oxidizing stalked bacterium, were independently examined using copper, nickel, and mild steel electrodes, respectively, in the previously described measurement cell with electrolytes similar to those from which they were isolated (Gerchakov et al., 1986). The bacterium B-3 (tentatively identified as *Pseudomonas waxmanii* sp. nov.) was isolated from a pitted metal surface painted with a coating containing cuprous oxide and tributyl tin oxide exposed to seawater. *Thermus aquaticus,* an organism that requires temperatures from 60 to 80°C for growth, was isolated from a failed nickel 201 heat exchanger that had been maintained with distilled water at 60°C. The third organism, an iron-oxidizing bacterium, was isolated from a 1018 cold-rolled mild steel water box filled with estuarine water.

In the three experimental studies, the metal specimen in the microbial compartment became the anode of the couple; currents ranged from 1.00 to approximately 10.00 μA cm^{-2} (Figure 16.17). The optimum growth temperature for *T. aquaticus* is 60–70°C, and the organism is completely inactive below 50°C. The dual compartment cell holding nickel 201 specimens was submerged in a water bath maintained at 60°C, and one compartment was inoculated with the organism. After about 20 h, the current at the inoculated electrode began to increase

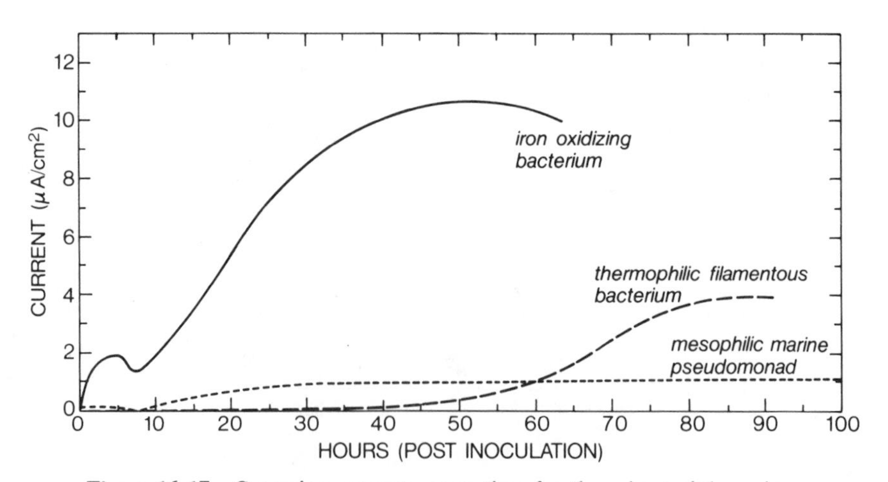

Figure 16.17 Corrosion current versus time for three bacterial species.

anodically and finally stabilized at 3.80 μA cm^{-2}. At ambient temperature, where *T. aquaticus* is inactive, no current was observed between the compartments. Similarly, no current was observed when the organisms colonizing the nickel electrode were heat-killed. These results suggest that these organisms must be metabolically active to effect corrosion of nickel 201.

Bacterium B-3 was inoculated into one side of the two compartment cell holding copper specimens. After a 10 h incubation time at ambient temperature, the current at the inoculated electrode began to increase anodically to about 1.00 μA cm^{-2}. Similar results were obtained when the stalked iron-oxidizing bacterium was inoculated into the system containing mild steel electrodes. An anodic current of 10.40 μA cm^{-2} was measured after approximately 40 h. The iron-oxidizing bacteria on the mild steel surface produced tubercles of ferric hydroxide. When these microorganisms were heat-killed, the anodic current remained stable at 8.60 μA cm^{-2}. The tubercles prevented diffusion to and from the surface and created differential aeration cells that were independent of the biochemical activity of the bacteria.

16.3 EFFECT OF MICROORGANISMS ON CORROSION

16.3.1 Influence of Gradients and Patchiness on Corrosion

As previously discussed, prior to the colonization of a surface by microorganisms, a film of macromolecules is adsorbed. This spontaneous adsorption of organic material from the aqueous phase alters the interfacial free energy of the solid, as well as the corrosion potential of metal surfaces. The physical presence of microbial cells on a metal surface, as well as their metabolic activities, modifies electrochemical processes. The adsorbed cells grow and reproduce, forming colonies that constitute physical anomalies on a metal surface, resulting in the formation of local cathodes or anodes (Figure 16.18). Nonuniform (patchy) colonization by bacteria results in the formation of differential aeration cells where areas under respiring colonies are depleted of oxygen relative to surrounding noncolonized areas (Figure 16.19). Colony formation gives rise to potential differences and consequently to corrosion currents. Under aerobic conditions, the areas under the respiring colonies become anodic and the surrounding areas become cathodic. If microroughness of the substratum is considered (Figure 16.20), corrosion currents can flow between the peaks and valleys of the roughness elements.

A mature biofilm composed of microorganisms and their extracellular secretions prevents the diffusion of oxygen to cathodic sites and the diffusion of aggressive anions, such as chloride, to anodic sites. Outward diffusion of metabolites and corrosion products is also impeded. If areas within the biofilm become anaerobic, i.e., if the aerobic respiration rate within the film is greater than the oxygen diffusion rate through the film, a change in the cathodic reaction mechanism occurs.

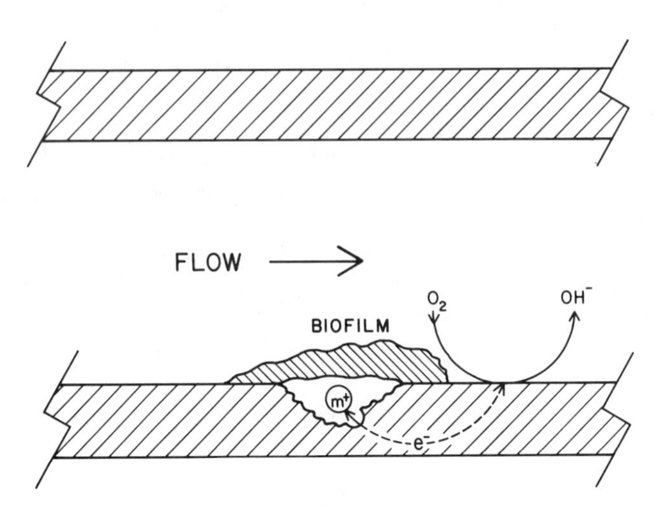

Figure 16.18 The physical presence of microbial cells on a metal surface, as well as their metabolic activities, modifies electrochemical processes. The adsorbed cells grow and reproduce, forming colonies that constitute physical anomalies on a metal surface, resulting in the formation of local cathodes or anodes.

Metabolic processes within the biofilm significantly affect corrosion. It is traditional to discuss specific mechanisms for MC in terms of aerobic and anaerobic conditions and to further discuss selected mechanisms for specific microorganisms. However, microorganisms form synergistic communities that conduct combined processes that individual species cannot. For example, anaerobic and aerobic microorganisms coexist in naturally occurring biofilms in oxygenated environments (Figure 16.21). Thus, aerobic bacteria and sulfate-reducing bacteria, obligate anaerobes, can proliferate in the same biofilm along with other anaerobic heterotrophs. Furthermore, a single type of microorganism can simultaneously affect electrochemical processes via several mechanisms. The relationship between anaerobic heterotrophs, sulfate reducers, and methanogens in a biofilm has been described by Parkes (1987) and is diagrammatically presented in Figure 16.22. Clearly, the interaction among the species within the biofilm community is an important consideration in a MC analysis. Sulfate-reducing and iron-oxidizing microorganisms are the most frequently cited causative agents for MC. However, all microorganisms colonizing metal surfaces have the potential for effecting electrochemical processes.

16.3.2 Low Molecular Weight Extracellular Products

Most heterotrophic bacteria secrete organic acids during the fermentation of organic substrates. The types and amounts of acids produced depend on the kinds

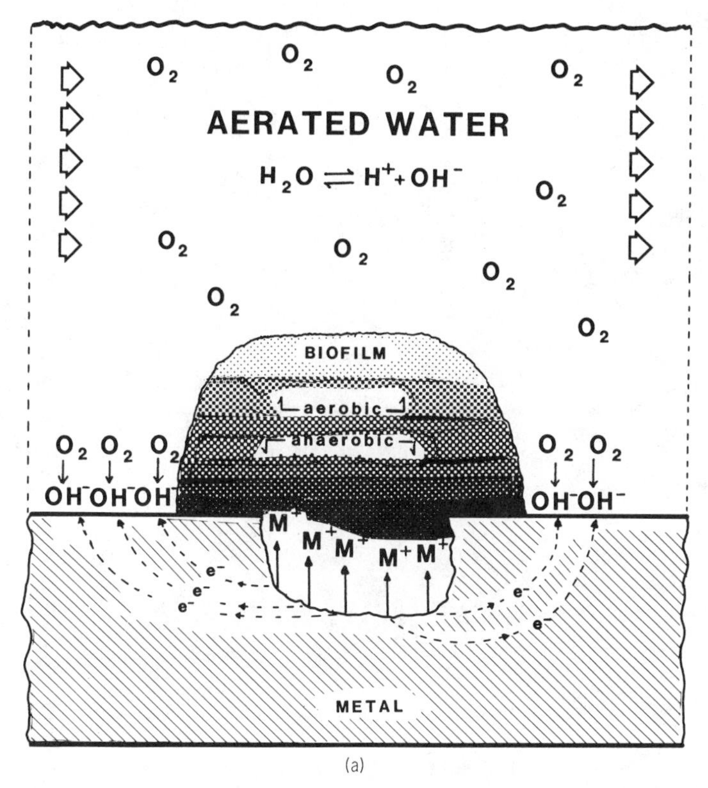

Figure 16.19 (a) Nonuniform (patchy) colonization by bacteria results in the formation of differential aeration cells where areas under respiring colonies are depleted of oxygen relative to surrounding noncolonized areas. (b) Scanning electron micrograph shows (patchy) colonization of mild steel by *Vibrio alginolyticus* in a simulated marine environment (Gaylarde and Videla, 1987). (c) When the metal is cleaned, intense pitting is noted where the bacterial colonies had accumulated (Gaylarde and Videla, 1987).

of organisms and the available substrate molecules. Organic acids may result in a physical shift in the tendency for corrosion to occur as measured by the potential shift between anodes and cathodes. The impact of metabolites secreted by microorganisms is intensified as they are trapped at the colony-metal interface. Corrosive metabolic products, such as hydrogen sulfide (H_2S) from *Desulfovibrio desulfuricans*, acetic acid (CH_3COOH) from *Clostridium aceticum*, and sulfuric acid (H_2SO_4) from *Thiobacillus thiooxidans*, are obvious contributors to corrosion processes. In addition, it has been demonstrated that the organic acids of the Krebs cycle can promote the electrochemical oxidation of a variety of metals by preventing or removing an oxide film. Burnes et al. (1967) showed that under aerobic conditions, solutions of citric, fumaric, ketoglutaric, glutaric, maleic, malic, itaconic, pyruvic, and succinic acids formed metallic salts when incubated with copper, tin, and zinc. Acetic, ketoglutaric, succinic, and lactic

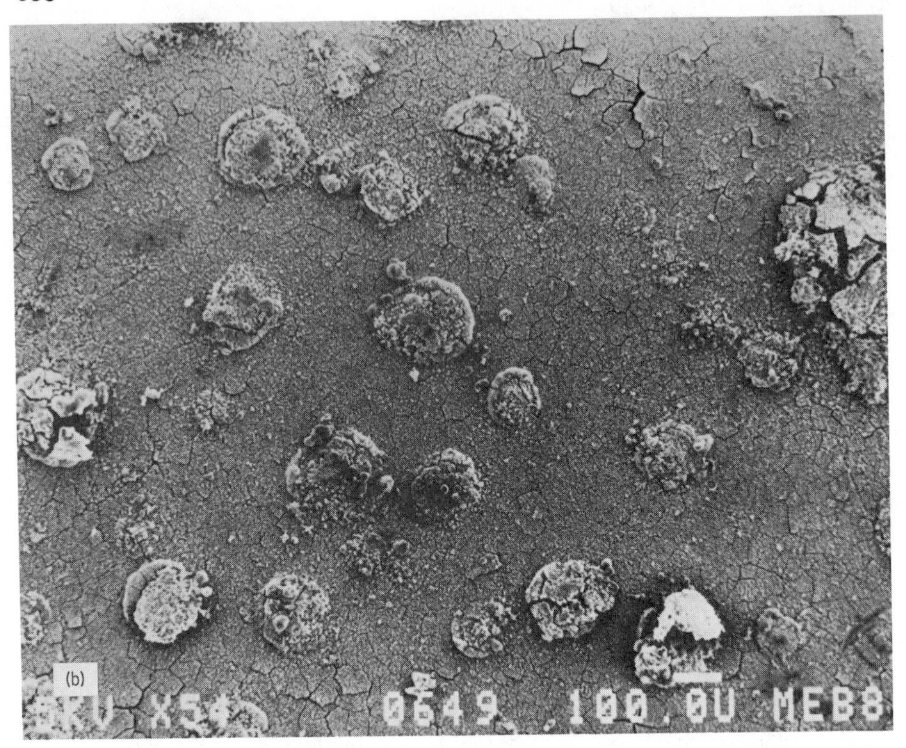

Figure 16.19 (b)

acids were isolated from an anaerobic culture of *E. coli* in which enhanced corrosion of carbon steel had been documented. Little et al. (1986) have demonstrated that isobutyric and isovaleric acids accelerate nickel corrosion in concentrations that are environmentally relevant. Gerchakov and Udey (1984) have suggested that such metabolites as amino and dicarboxylic acids may also be aggressive ions to some metal substrata, such as copper. Little et al. (1987) have demonstrated that under laboratory conditions an aerobic acetic-acid-producing bacterian can accelerate the corrosion of a cathodically protected stainless steel electrode in synthetic salt solution. The acetic acid destabilizes or dissolves the calcareous film that formed during cathodic polarization.

16.3.3 Hydrogen Embrittlement

Walch and Mitchell (1983) have proposed mechanisms for the role of microorganisms in hydrogen embrittlement of metals: (1) production of molecular hydrogen during fermentations, which may be dissociated into atomic hydrogen and adsorbed into the metals; (2) production of hydrogen ions via organic or mineral acids, which may be reduced to form hydrogen atoms at cathodic areas; (3) production of hydrogen sulfide, which may stimulate the adsorption of

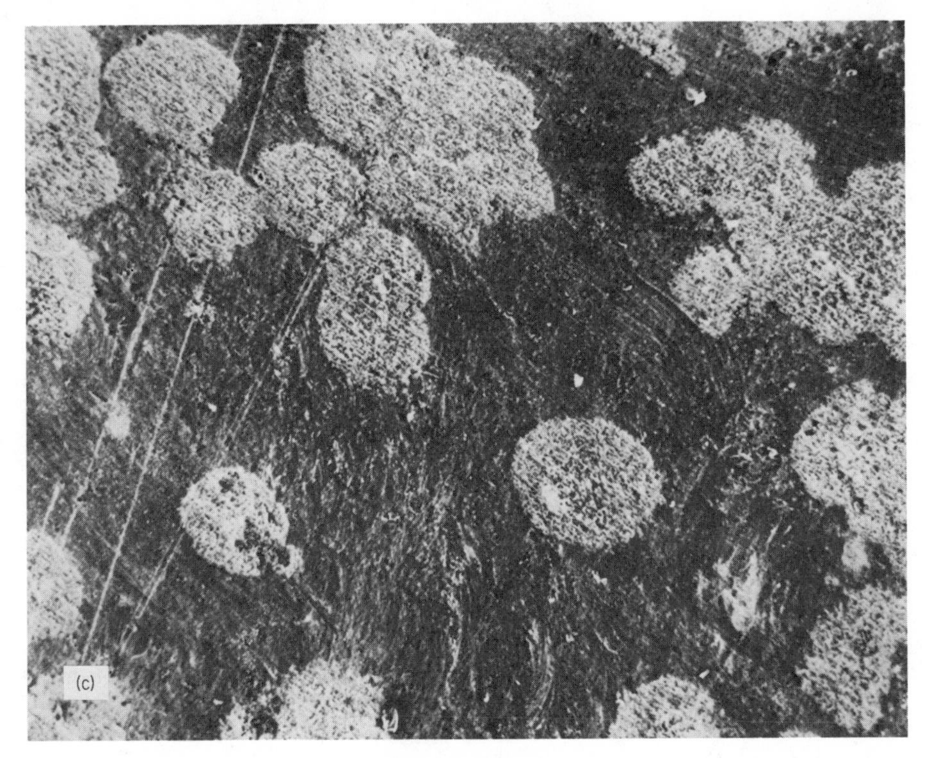

Figure 16.19 (c)

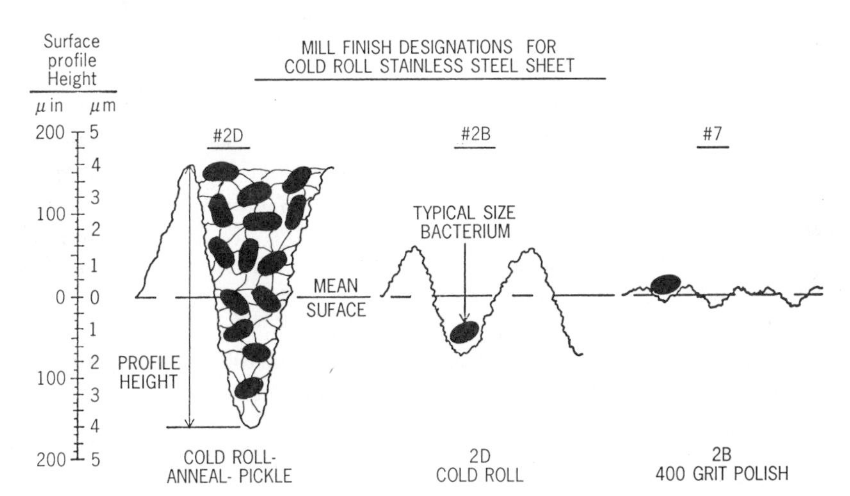

Figure 16.20 When microroughness of the substratum is considered, microbial activity in the valleys may cause currents to flow through the metal between the peaks and valleys of the roughness elements.

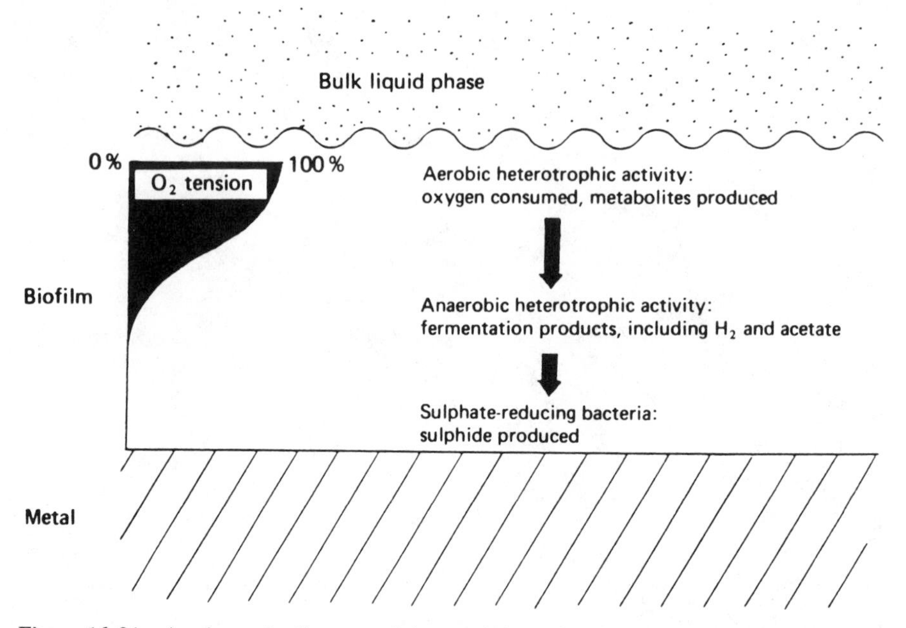

Figure 16.21 A schematic diagram of the spatial relationship between aerobes, heterotrophic anaerobes, and sulfate-reducing bacteria in a biofilm accumulated on a metal substratum (Hamilton, 1985).

atomic hydrogen into metals by preventing its recombination into hydrogen molecules; and (4) destablization of metal oxide films.

16.3.4 Inactivation of Corrosion Inhibitors

Organic compounds such as aliphatic amines are used as corrosion inhibitors. These compounds are degraded by microorganisms, decreasing the effectiveness of the compounds and increasing the microbial populations. Some bacteria reduce NO^{-3} or NO^{-2} to N_2 gas, which escapes the system. Characklis (unpublished results) analyzed a closed recirculating cooling system that was requiring unusually high amounts of nitrite (NO_2^-) corrosion inhibitor. The system was anaerobic, and the pH was continually increasing despite attempts to control it. The cause was microbial denitrification.

Denitrification may be described conceptually by the following stoichiometric equation using methanol as an illustrative electron donor:

$$NO_2^- + \tfrac{1}{2}CH_3OH \rightarrow \tfrac{1}{2}N_2\uparrow + OH^- + \tfrac{1}{2}CO_2 + \tfrac{1}{2}H_2O \qquad [16.23]$$

The methanol is the electron donor (energy source), while nitrate and nitrite are the electron acceptors.

Many organisms produce NH_3 from the metabolism of amino acids or the

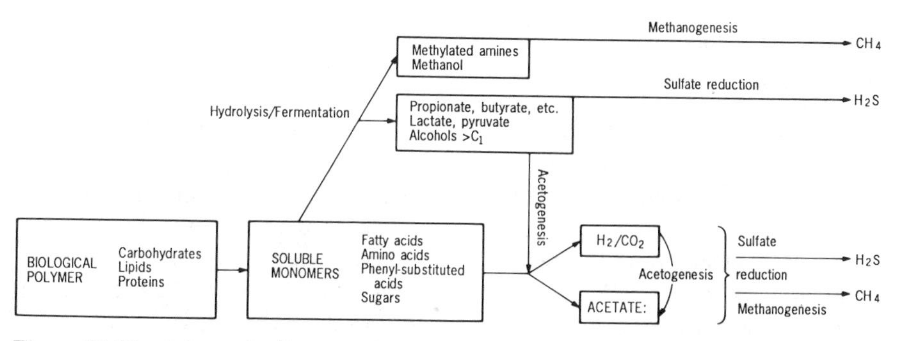

Figure 16.22 Schematic diagram of the hypothetical interrelationship between heterotrophic anaerobes, sulfate-reducing bacteria, and methanogens in a biofilm. (Reprinted with permission from Parkes, 1987.)

reduction of nitrate or nitrite, forming NH_4^+ in solution, which accelerates the corrosion of copper alloys (Pope et al., 1984).

16.3.5 Sulfate-Reducing Bacteria (SRB)

Anaerobic bacteria, particularly sulfate reducers, have been the focus of most microbiological corrosion investigations. The early work of Von Wolzogen Kuhr and Van der Vlugt (1934) suggested the following electrochemical reactions:

$$4Fe \rightarrow 4Fe^{+2} + 8e^- \quad \text{(anodic reaction)} \quad [16.24]$$

$$8H_2O \rightarrow 8H^+ + 8OH^- \quad \text{(water dissociation)} \quad [16.25]$$

$$8H^+ + 8e^- \rightarrow 8H\ (ads) \quad \text{(cathodic reaction)} \quad [16.26]$$

$$SO_4^{-2} + 8H \rightarrow S^{-2} + 4H_2O \quad \text{(bacterial consumption)} \quad [16.27]$$

$$Fe^{+2} + S^{-2} \rightarrow FeS \quad \text{(corrosion products)} \quad [16.28]$$

$$4Fe + SO_4^{-2} + 4H_2O \rightarrow 3Fe(OH)_2 + FeS + 2\,OH^- \quad \text{(overall reaction)} \quad [16.29]$$

The overall process was described as *depolarization*, based on the theory that these bacteria remove hydrogen that accumulates on the surface of iron. The electron removal as a result of hydrogen utilization results in cathodic depolarization and forces more iron to be dissolved at the anode. The direct removal of hydrogen from the surface is equivalent to lowering the activation energy for hydrogen removal by providing a *depolarization* reaction. The enzyme, hydrogenase, synthesized by many species of *Desulfovibrio*, may be involved in this specific depolarization process. Booth et al. (1968), using polarization techniques and weight loss measurements versus hydrogenase activity, have provided additional evidence to substantiate this theory. However, Iverson (1966) first presented direct evidence for cathodic depolarization, using benzyl viologen as an indicator of reduction. Bacteria, by removing adsorbed hydrogen to produce sulfide and water, increase the rate of dissolution of Fe_2S. Non-hydrogenase-producing strains of *Desulfovibrio* can also stimulate corrosion.

Miller and Tiller (1970) have proposed cathodic depolarization induced by microbially produced FeS. King et al. (1973) and Booth et al. (1968) demonstrated that weight losses of steel were proportional to the concentrations of ferrous sulfide present and depended on the stoichiometry of the particular ferrous sulfide minerals. They concluded that the accelerated corrosion of mild steel in the presence of sulfate-reducing bacteria was due principally to the formation of iron sulfide. Duquette (1985) has reviewed the possible electrochemical consequences of the formation of FeS and concluded that if FeS is the cathodic site for hydrogen reduction, the activation energy for hydrogen evolution may be reduced. In such an instance, a simple increase in the effective area of a sulfide film would also lead to an increase in the cathodic reaction rate.

Salvarezza and Videla (1980), using potentiostatic polarization techniques, evaluated the breakdown of passivity of mild steel in seawater in the presence of sulfate-reducing bacteria. The experiments were performed in a synthetic medium in the presence and absence of *Desulfovibrio.* Figure 16.23a is the anodic polarization curve for mild steel in sterile, aerated and deaerated artificial seawater. Both curves show a pitting potential, which is virtually unaffected by aeration or deaeration. Polarization curves obtained in the presence of sulfate-reducing bacteria of different ages and sulfide concentrations (Figure 16.23b) showed pitting potentials more active than those corresponding to sterile media. The progression of total sulfides and redox potential was as follows:

Time (h)	Total Sulfides (mol)	pH	Redox Potential (mV)
72	1.4×10^{-4}	7.5	−500
96	1.0×10^{-3}	7.8	−510
240	8.0×10^{-4}	7.2	−510

The addition of sulfate-reducing bacteria and sodium sulfide resulted in pitting potentials that were 100–200 mV more active than in seawater alone, and pits were physically observed on the surfaces. The results indicate that the effect of sulfate-reducing bacteria is to add sulfide to the system and the sulfide species behave similarly to chemically added sulfide. Furthermore, the authors demonstrated that deaerated solutions required lower levels of sulfide or metabolic products to induce pitting (Figure 16.24).

The impact of oxygen on obligate anaerobic, sulfate-reducing bacteria was examined by Hardy and Bown (1984) using a synthetic seawater medium and a *Desulfovibrio* strain. Corrosion rates were determined by weight loss measurements and by electrical resistance probe measurements. They were low under totally anaerobic conditions, but increased with the addition of oxygen. Successive aeration–deaeration shifts caused variation in the corrosion rate. High rates were observed during periods of aeration. The attack was confined to areas be-

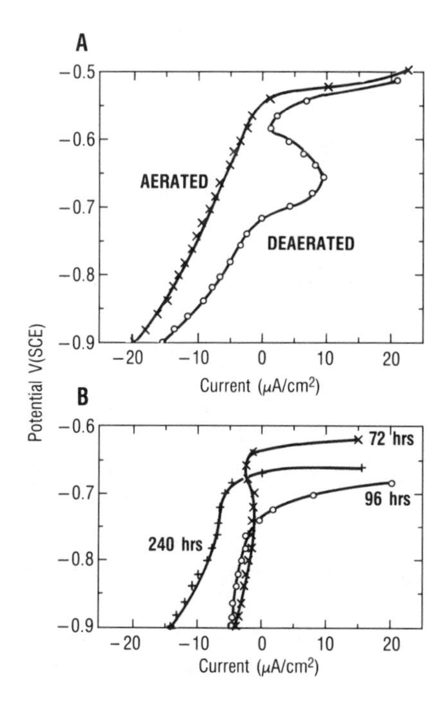

Figure 16.23 Potentiostatic polarization curves for 1020 SAE steel. (a) Sterile artificial seawater, pH 8.0. (b) Artificial seawater plus sulfate-reducing bacteria inoculated for 72, 96, and 240 hours (reprinted with permission from Salvarezza and Videla, 1980).

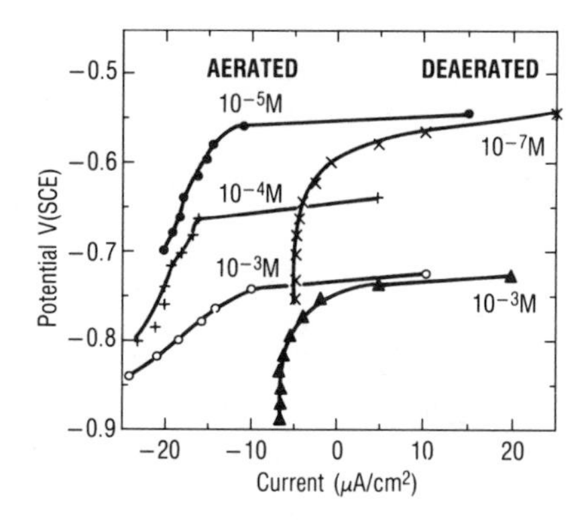

Figure 16.24 Potentiostatic polarization curves of mild steel in aerated and deaerated 0.5 M NaCl solutions at different sulfide concentrations (reprinted with permission from Salvarezza and Videla, 1980).

neath tubercles that consisted of loosely adherent material, as opposed to the hard, tightly adherent films on uncorroded metal. The authors concluded that the presence of tubercles fixed the anode and forced the cathodic reaction to occur on the adherent sulfide film. Since significant corrosion rates were only observed when oxygen was present, some of the reported laboratory tests with SRB may have been contaminated with oxygen.

16.3.6 Metal Oxidation by Bacteria

In recent years, the role of metal-oxidizing bacteria in MC has been emphasized. Ghiorse (1984) has pointed out that metal oxidation has not been demonstrated in some cases and that certain microorganisms can catalyze the oxidation of metals. Other microorganisms accumulate abiotically oxidized metal precipitates. The iron-oxidizing genera most often cited are the filamentous forms of *Sphaerotilius*, *Crenothrix*, and *Leptothrix* (which may be different forms of the same organism), and the stalked organism *Gallionella*. These organisms oxidize ferrous ions to ferric ions or manganous to manganic ions to obtain energy for growth. There are also reports of microbial oxidations and reductions of chromium. Metal-oxidizing organisms create environments for the accumulation of chloride ions (to maintain charge neutrality) and form acidic ferric chloride and manganic chloride, which are highly corrosive to stainless steel. Further pit development is enhanced as an oxygen concentration cell develops. Duquette (1985) has summarized these developments on a schematic anodic polarization diagram for a passive metal or alloy (Figure 16.25). In this diagram, curve 1 represents sufficient cathodic reduction of oxygen to passivate the alloy, while curve 2 shows the result of decreasing the oxygen concentration to a level that will not support passivity. A stably passive alloy is indicated by the intersection of the anodic and cathodic curves at point 3. When chlorides are present, pitting occurs. The pitting potential shifted in the active direction by chloride ion as

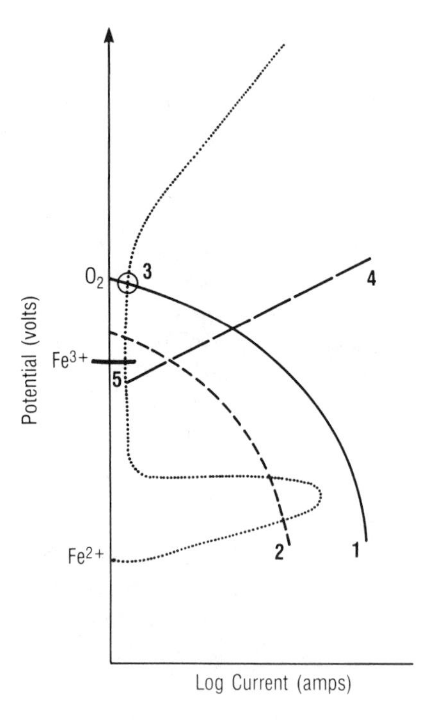

Figure 16.25 Schematic diagram of anodic polarization of a passive alloy under varying conditions (Duquette, 1985).

indicated by line 4. The potential may be fixed above the pitting potential by an Fe^{+2}-Fe^{+3} redox couple, illustrated by line 5. Curve 3 demonstrates the effect of increasing chloride concentration, which lowers the pitting potential.

Metal-oxidizing organisms efficiently scavenge oxygen and therefore provide conditions for the growth of obligate anaerobic bacteria. Numerous reports document the presence of sulfate-reducing bacteria in the tubercles formed by metal-oxidizing species (Tatnall, 1981; Postgate, 1979; Miller and Tiller, 1970).

16.3.7 Metal Reduction by Bacteria

Obuekwe et al. (1981) examined the propensity of a *Pseudomonas* sp., an iron-reducing bacterium, to affect the corrosion behavior of mild steel in synthetic media. A typical polarization curve for 1018 steel in sterile medium at room temperature (Figure 16.26) was used to construct Tafel plots, which displayed well-defined Tafel regions. Extrapolation of the anodic and cathodic branches to the intersection point yielded a corrosion potential of −0.655 V. The polarization characteristics of the mild steel coupon in the sterile growth medium and in the presence of the bacterium under microaerobic conditions are presented in Figures 16.27 and 16.28. The anodic polarization curves demonstrate that the metal becomes more active when incubated with populations of the iron-reducing organism. In the absence of the organism the coupon became more resistant to corrosion with exposure. Both the anodic and the cathodic reaction are time-dependent, since both reactions are polarized with increasing time from inoculation. The anodic polarization curve is almost vertical after 120 h, which indicates that the solution is less corrosive as time passes. In the presence of the bacteria, the cathodic behavior of the steel is virtually unchanged by the organism. The

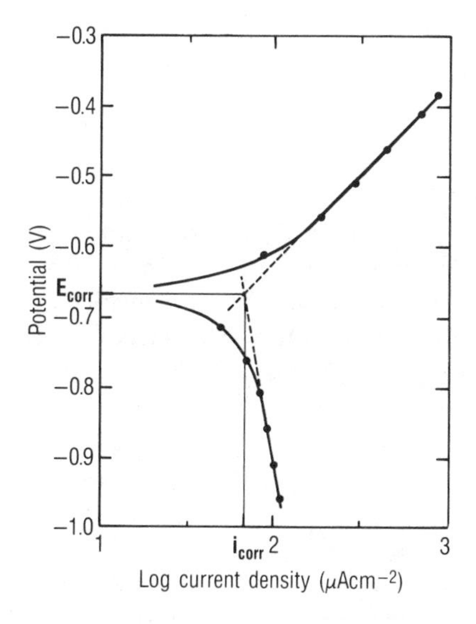

Figure 16.26 Tafel plot constructed from data obtained during polarization of mild steel coupons in sterile medium (reprinted with permission from Obuekwe et al., 1981).

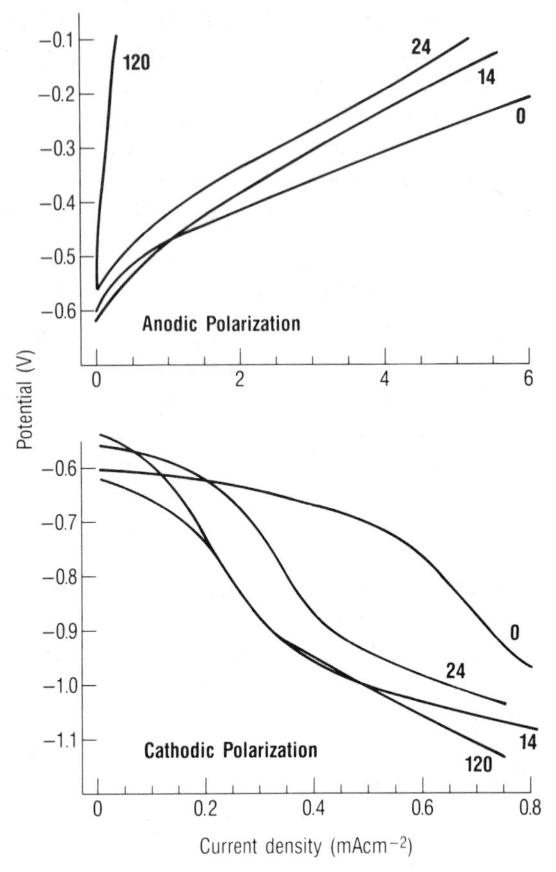

Figure 16.27 Anodic and cathodic polarization curves of mild steel in growth medium at (25 ± 2)°C. Numbers associated with each curve represent time in hours of incubation prior to polarization. (Reprinted with permission from Obuekwe et al., 1981.)

anodic polarization curves showed a tendency toward depolarization with increasing time from incubation in the synthetic medium with the organism. The authors concluded that bacteria decreased the activation energy for anodic dissolution or, conversely, that a different film was formed on the steel surface in the presence of bacteria leading to enhanced corrosion rates. The authors did not attempt to determine the relationship of the microorganisms to the mild steel electrode surface. It is therefore impossible to relate these results specifically to biofilm formation.

16.3.8 Extracellular Polymeric Substances (EPS)

All microorganisms that colonize surfaces secrete extracellular polymers forming a gel matrix on the metal substratum. This extracellular gel can have numer-

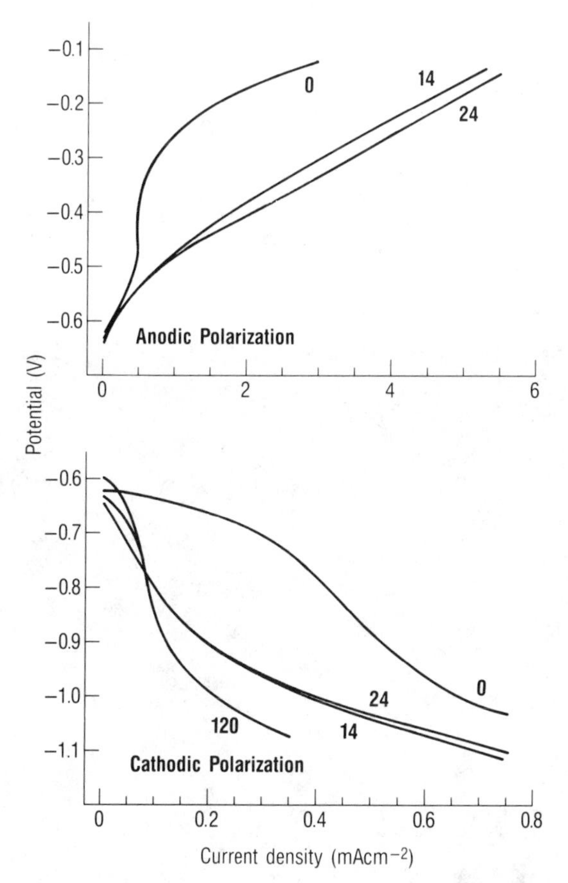

Figure 16.28 Anodic and cathodic polarization curves of mild steel in the presence of iron-reducing bacteria at (25 ± 2)°C. The numbers associated with each curve represent incubation time in hours prior to polarization. (Reprinted with permission from Obuekwe et al., 1981.)

ous effects on interfacial processes, including the following: (1) the gel immobilizes water at the biofilm–substratum interface, (2) it can entrap metal species (e.g., copper, manganese, chromium, and iron) and corrosion products at the substratum, (3) it can decrease diffusion rates toward and away from the substratum, (4) it can immobilize corrosion inhibitors and/or biocides.

Metal deposits have been reported in association with biogenic polymers. These deposits can establish galvanic currents between the metal species and the substratum metal. Dobb (1985) reported that corrosion products from a copper-nickel alloy in seawater exhibit different morphology depending on flow regime. In stagnant water, the deposit consists of spherical CuO particles at the substratum overlying an NiO layer (Figure 16.29a and b). In flowing water, no such particles exist (Figure 16.29c). Dobb simulated a biofilm by coating the Cu–Ni coupons with agar and biogenic polymer gel (approximately 10 μm thick layer).

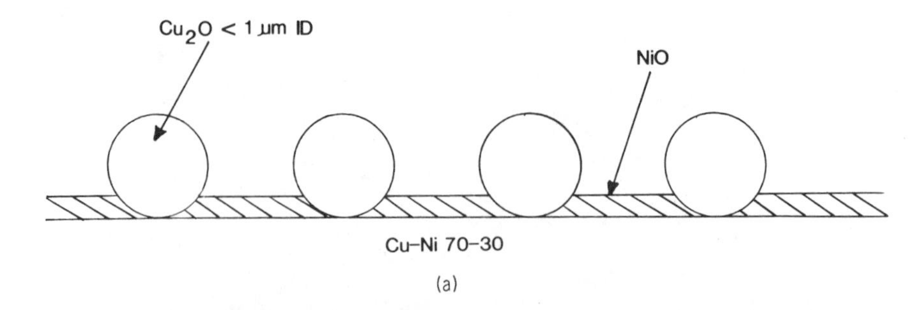

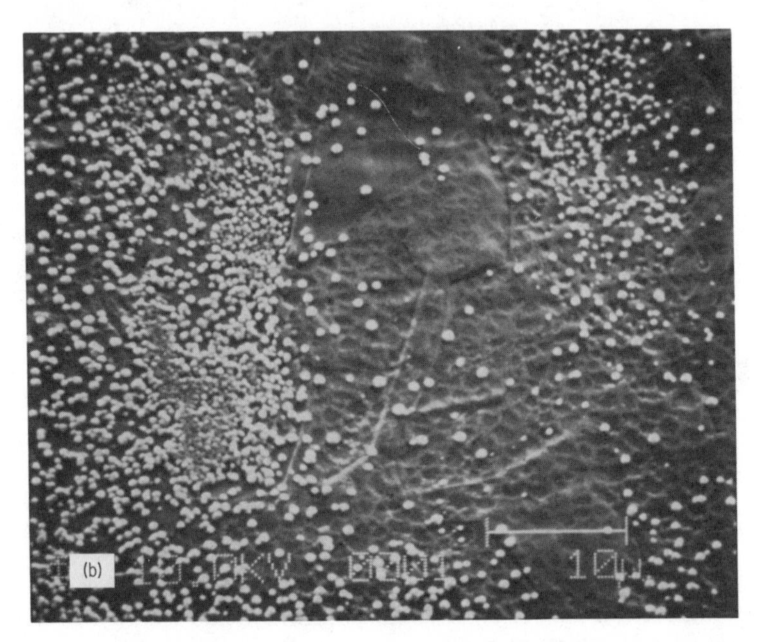

Figure 16.29 Corrosion products from a copper–nickel alloy in seawater exhibit different morphology depending on the flow regime. In stagnant water, the deposit consists of spherical CuO particles at the substratum overlying an NiO layer (a and b) while in flowing water, no such particles exist (c). (Dobb, 1985.)

When the gel-coated coupon was immersed in flowing seawater, CuO spherical particles formed as in the experiment with stagnant water. The polymer gel immobilized the water at the substratum. Interestingly, mechanical removal of the polymer gel (similar to sloughing) also resulted in removal of the CuO particles. Consequently, sloughing of biofilm may also influence the corrosion rate by periodically exposing the substratum to and shielding it from the bulk water conditions.

Nivens et al. (1986) demonstrated that *Vibrio natriegens* increased the corrosion current density of 304 stainless steel coupons in a marine growth medium. The corrosion current density i_{corr}, calculated from Tafel coefficients and polar-

Figure 16.29 Continued

ization resistance data, increased from 0.20 to 2.90 μA cm^{-2} during a six day incubation period. The corrosion current density began to increase when colonies of organisms were detectable on the surface of the coupon by microscopy with epifluorescent illumination and scanning electron microscopy. The most rapid increase in corrosion current density correlated with the formation of extracellular material.

16.3.9 Cell Death

Cell death, or lysis, within a well-developed biofilm does not necessarily mean a cessation of the influence on electrochemical processes. Miller and Tiller (1970) have confirmed that iron-oxidizing microorganisms such as *Gallionella* oxidize ferrous ions to ferric ion to obtain energy for growth reactions. This oxidation results in thick deposits of ferric hydroxide. Pitting corrosion can proceed under these deposits independently of the biochemical activity of the bacteria. Similarly, Booth and Tiller (1962) found that microbiologically generated FeS was corrosive in the absence of viable cells. Thus, the deposits containing extracellular products create differential aeration that may persist after cell death, a most important conclusion relevant to measures for MC prevention and/or control. Control of MC may require the complete removal of the biofilm, rather than killing or inactivating the cells within it.

16.4 EFFECT OF CORROSION ON BIOFILM FORMATION

Abiotic corrosion processes probably influence the rate, extent, and distribution of colonizing microbial species, as well as the chemical composition and physical properties (e.g., cohesive strength) of the resulting biofilm. It has been demonstrated that the composition of a metal substratum influences the rate and cell distribution of microfouling films in seawater (Marszalek et al., 1979; Zachary et al., 1980). Nonuniform corrosion (localized anodes and cathodes) promotes patchy adsorption of microorganisms. The pH and electrolyte concentration increase at the interface of corroding surfaces at cathodic sites and decrease at anodic sites, thus influencing bacterial adsorption (Daniels, 1980). Similarly, different inorganic ions produced at the two electrodic areas may affect adsorption (Corpe, 1970; Kaneko and Colwell, 1975). The depletion or reduction of oxygen at cathodic sites will also influence settlement. The presence of hydrated oxide or hydroxide passivating films on metal surfaces (Figure 16.30) provides bacteria with sites for film attachment (Kennedy et al., 1976). Titanium hydroxides, for example, are insoluble over the normal physiological pH range and have been used for cell immobilization matrices with no inhibitory effect on biologically active molecules. The microbial cell immobilization process for a number of metal hydroxides presumably involves replacement of hydroxyl groups on the metal hydroxide surface by suitable ligands from the cell, resulting in the formation of partial covalent bonds. Kennedy et al. (1976) have demonstrated that cells become firmly adsorbed to the metal hydroxide and are not just loosely trapped in the gelatinous oxide matrix.

Spalling or sloughing of corrosion products results in the detachment of biofilm patches associated with the corrosion products (Characklis et al., 1983). Copper-based alloys have long been considered as toxic or inhibitory surfaces, since biofilm accumulation on copper alloys is usually less than on titanium or stainless steel alloys. An alternate explanation for reduced biofilm accumulation on copper alloys is that copper-based alloys corrode faster and thus carry away biofilm with the spalled corrosion products. Characklis et al. (1983) observed significant Cu and Ni in biofilm deposits accumulated on Tygon surfaces in their experimental apparatus, indicating the mobility of the corrosion products from Cu-based alloys.

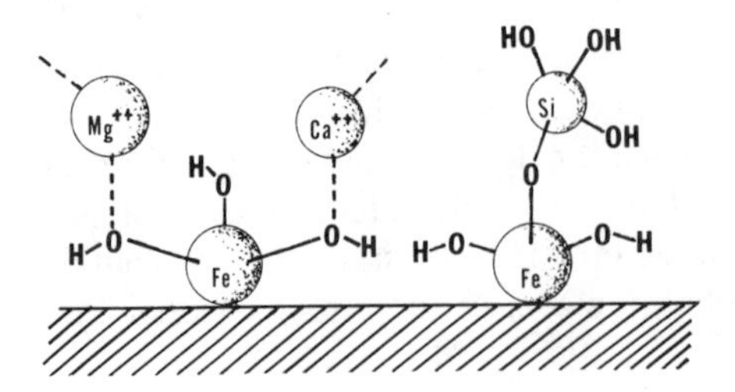

Figure 16.30 The presence of hydrated oxide or hydroxide passivating films on metal surfaces provides bacteria with sites for film attachment.

ACKNOWLEDGMENTS

This work was supported in part by NORDA program element 61153N (Herbert Eppert, program manager), NORDA Special Project 001:333:87. Special thanks to Dr. David Duquette, Rensselaer Polytechnic Institute, Troy, New York, for his review and constructive criticism of this chapter.

REFERENCES

Bode, H.W., *Network Analysis and Feedback Amplifier Design,* Van Nostrand, 1945.

Booth, G.H. and A.K. Tiller, *Trans. Faraday Soc.,* **58**, 2510–2516 (1962).

Booth, G.H., L. Elford, and D.S. Wakerly, *Brit. Corr. J.,* **3**, 242–245 (1968).

Brown, B.F., *Corrosion,* **26**, 249 (1970).

Burnes, J.M., E.E. Staffeld, and O.H. Calderon, *Develop. Ind. Microbiol.,* **8**, 327–334 (1967).

Cahan, B.D. and C.T. Chen, *J. Electrochem. Soc.,* 17–26 (1982).

Characklis, W.G., F.L. Roe, M.H. Turakhia, and N. Zelver, "Microbial fouling and its effect on power generation," Final Report, Office of Naval Research, N00014-80-C-0475, (1983).

Corpe, W.A., "Attachment of marine bacteria to solid surfaces," in R.S. Manley (ed.), *Adhesion in Biological Systems,* Academic Press, New York, 1970, pp. 73–87.

Daniels, S.L. "Mechanisms involved in adsorption of microorganisms to solid surfaces," in G. Bitton and K. Marshall (eds.), *Adsorption of Microorganisms to Surfaces,* Wiley, New York, 1980, pp. 7–58.

Dobb, D.E., "Influence of organic gels on the corrosion of 70/30 Cupronickel alloy in seawater," unpublished doctoral dissertation, Montana State University, Bozeman, MT, 1985.

Duquette, D.J., in H.A. Videla (ed.), *USA/Argentina Workshop on Biodeterioration CONICET-NSF,* LaPlata, Argentina, 1985, pp. 15–32.

Gaylarde, C.C. and H.A. Videla, *Int. Biodet.,* **23**, 91–104 (1987).

Gerchakov, S.M., B.J. Little and P.A. Wagner, *Corrosion* **42,** 632–695 (1986).

Gerchakov, S.M. and L.L. Udey, "Microfouling and corrosion," in J.D. Costlaw and R.C. Tipper (eds.), *Marine Biodeterioration: An Interdisciplinary Study,* Naval Institute Press, Annapolis, MD, 1984, pp. 83–87.

Ghiorse, W.C., *Ann. Rev. Microbiol.,* **38**, 515–550 (1984).

Hamilton, W.A., *Ann. Rev. Microbiol.,* **39**, 195–217 (1985).

Hardy, J.A. and J.L. Bown, *Corrosion,* **40**, 650–654 (1984).

Hoar, T.P., *Disc. Faraday Soc.,* **1**, 299 (1947).

Iverson, W.P., *Science,* **151**, 986–988 (1966).

Kaneko, T. and R.R. Colwell, *Appl. Microb.,* **29**, 269–274 (1975).

Kennedy, J.F., S.A. Barker, and J.D. Humphreys, *Nature,* **261**, 242–244 (1976).

King, R.A., J.D.A. Miller, and D.S. Wakerly, *Brit. Corros. J.,* **8**, 89–93 (1973).

Kobrin, G., *Material Performance,* **15**(7), 38–43 (1976).

Leckie, H. and H.H. Uhlig, *J. Electrochem. Soc.*, **113**, 1262 (1966).

Little, B.J., P.A. Wagner, S.M. Gerchakov, M. Walch, and R. Mitchell, *Corrosion,* **42**(9), 533-536 (1986).

Little, B.J., P.A. Wagner, and D. Duquette, "Microbiologically induced cathodic depolarization," Paper No. 370, in *Proc. National Association of Corrosion Engineers, Corrosion '87,* 1987.

Marszalek, D.S., S.M. Gerchakov, and L.R. Udey, *Appl. Environ. Microbiol.*, **38**, 987-995 (1979).

Miller, J.D.A., and A.K. Tiller, in J.D.A. Miller (ed.), *Microbial Aspects of Metallurgy,* Elsevier, New York, 1970, pp. 61-106.

Nivens, D.E., P.D. Nichols, J.M. Henson, G.G. Geesey, and D.C. White, *Corrosion,* **42**(4), 204-210 (1986).

Obuekwe, C.O., D.W.S. Westlake, J.A. Plambeck, and F.D. Cook, *Corrosion,* **37**(8), 461-467 (1981).

Parkes, R.J., *Soc. Gen. Microbiol. Symp.*, **41**, 147-177 (1987).

Pope, D.H., D.J. Duquette, A.H. Johnannes, and P.C. Wagner, *Materials Performance,* **23**(4), 14-18 (1984).

Postgate, J.R., *The Sulphate-Reducing Bacteria,* Cambridge University Press, Cambridge, England, 1979, p. 105.

Salvarezza, R.C. and H.A. Videla, *Corrosion,* **36**(10), 550-554 (1980).

Shrier, L.L., *Corrosion 1,* Newnes-Butterworths, Boston, 1979.

Skold, R.V. and T.E. Larson, *Corrosion,* **13**, 69-72 (1957).

Stern, M. and A.L. Geary, *J. Electrochem. Soc.*, **104**, 33-63 (1957).

Tafel, J.Z., *Z.f. physik. Chem. Bd.*, **50**, 641-712 (1905).

Tatnall, R.E., *Materials Performance,* **20**(9), 32-38 (1981).

Uhlig, H.H. and R.W. Revie, *Corrosion and Corrosion Control,* Wiley, New York, 1985, pp. 428-429.

Van Meirhaeghe, R.L., E.C. Dutoit, F. Cardon, and W.P. Gormes, *Electrochim. Acta,* **21**, 39-43 (1976).

Von Wolzogen Kuhr, C.A.H. and L.S. Van der Vlugt, *Water* (The Hague), **18**, 147-165 (1934).

Wagner, C. and W. Traud, *Elecktrochem.*, **44**, 391 (1938).

Walch, M. and R. Mitchell, "Role of microorganisms in hydrogen embrittlement of metals," Paper No. 249, in *Proc. of NACE Corrosion '83,* 1983.

Zachary, A., M.E. Taylor, F.E. Scott, and R.R. Colwell, "Marine microbial colonization of material surfaces," in Z.A. Oxley, D. Allsopp, and G. Becker (eds.), *Biodeterioration: Proc. 4th International Symposium,* Berlin, 1980, pp. 171-178.

17

BIOFILMS IN WATER AND WASTEWATER TREATMENT

JAMES D. BRYERS
Duke University, Durham, North Carolina

WILLIAM G. CHARACKLIS
Montana State University, Bozeman, Montana

SUMMARY

Biofilm and biomass support particle (BSP) systems for wastewater treatment depend on mixed populations of microorganisms, predominantly bacteria, which become immobilized on or within inert supports and form biofilms or biomass aggregates with the following advantages: (1) cheap, available support material (i.e., sand, quartz, plastic), (2) high reactor biomass concentrations, (3) proven capacity to handle full scale flow rates, and (4) the option to operate with mixed populations and without aseptic conditions.This last advantage allows biofilm and BSP systems to operate in a way that permits microbial growth while mediating a variety of biological reactions simultaneously (e.g., carbon oxidation, nitrification, methane production). A major disadvantage of fixed biofilm reactors is the lack of control over the biofilm thickness which contributes to mass transfer limitations and excessive biofilm sloughing. BSP, to a certain extent, permit control of biofilm thickness. However, in both biofilm and fixed biomass systems, slow-growing microorganisms (e.g., autotrophic nitrifiers, methane formers) create the problem of lengthy reactor startup periods (several months).

Artificially immobilized whole-cell systems have the unique advantage that a biological process can be stored indefinitely and then used, without a waiting period or startup problems. Disadvantages of immobilized whole cell catalysts are their short half-lives ($\approx$20–40 days), the requirement of alternating non-growth and growth periods of operation to maintain reactivity, and the physical breakage of encapsulating materials and resultant contamination of down-stream systems.

All immobilized whole-cell systems discussed in this chapter contain mono-populations. Immobilization of defined mixed populations that mediate multiple reactions may be one way for artificially immobilized cell systems to become more competitive with biofilm and BSP systems.

In general, biofilm and BSP systems are used more on all scales of water purification, from fundamental laboratory systems to technical scale municipal and industrial treatment facilities. In contrast, artificially immobilized whole cell systems find their most significant applications in specific water and wastewater analysis using microbial cell electrodes, and in water purification. To increase the potential uses of artificially immobilized whole-cell systems in water and wastewater treatment, research is necessary to answer questions regarding scaleup, extended useful lifetime of the active particles, and aseptic operation.

17.1 INTRODUCTION

Microorganisms will grow at a rate dictated by their physiological status and prevailing growth conditions. If the organisms are growing in suspension in a continuous flow reactor (e.g., a chemostat), the growth rate of the population is dictated by the hydraulic residence time. The cell concentration at a fixed residence time is a function of the influent concentration of the growth-limiting substrate. Should the reactor residence time decrease below the maximum generation time of the microbial culture—or should the microbial generation time suddenly increase—cells will be washed out of the reactor. The washout, or critical, residence time limits the maximum substrate removal rate in a chemostat.

The objective of biological wastewater treatment is to remove the maximum amount of a contaminant (substrate) with the minimum reactor residence time. Consequently, biological treatment processes are not conducted in chemostats, but rather in suspended growth systems where biomass is separated from the effluent and recycled to the reactor. Cell recycle increases the biomass concentration in the reactor and thus increases the overall substrate removal rate. External recycle forms the basis for the activated sludge process and its many variations. Biomass recycle systems with either external recycle (e.g., activated sludge, extended aeration) or internal recycle (e.g., anaerobic sludge blanket reactors) are discussed elsewhere (Grady and Lim, 1980).

This chapter reviews existing and proposed system alternatives for retaining biomass in a biological wastewater treatment process. All systems use a means of capturing active biomass and can be considered biofilm reactors. There are other advantages provided by biofilms in wastewater treatment besides retaining biomass and mechanically preventing washout, and some are listed in Table 17.1.

TABLE 17.1 Advantages of Immobilized or Captured Microbial Cell Reactors

- Increased reactor biomass concentration
- Increased overall substrate conversion rates due to higher biomass concentrations
- Independence of washout restrictions (i.e., reactor operation is independent of physiological restrictions on growth rate)
- Biomass recovery can occur at high solids concentrations
- Reduced reactor volumes
- Reduced susceptibility to shocks or transients (e.g., temperature, inhibitors)
- Possible elimination of the clarification/separation stages

17.2 TYPES OF BIOFILM WASTEWATER TREATMENT SYSTEMS

Three types of biofilm systems will be presented and are relevant to wastewater treatment:

1. Fixed biofilms
2. Biomass support particles (BSP)
3. Artificially immobilized intact cells

17.2.1 Fixed Biofilms

Biofilm, or *fixed film,* reactors depend on the natural tendency of mixed microbial populations to adsorb to surfaces and to accumulate in biofilms. Adsorbed microorganisms grow, reproduce, and produce extracellular polymeric substances that frequently extend from the cell, forming the gelatinous matrix called a biofilm (Chapter 7). The accumulation and persistence of a biofilm is the net result of several physical and biological processes that occur simultaneously, although their relative rates will change through the various stages of biofilm development.

Much biofilm wastewater treatment research has focused upon the metabolic capabilities of biofilm, i.e., prediction of the overall substrate removal kinetics of a biofilm. Only recently has research focused upon those processes contributing to the accumulation and persistence of a specific biofilm or its structural stability. The search for more structure in biofilm reactor models is being motivated by the increased desire for specificity in biological treatment of wastewater, including nitrification, denitrification, and removal of toxic substances.

The reactor geometries employed in fixed biofilm wastewater treatment systems are as follows (Figure 17.1):

1. Completely mixed reactor systems, where biofilm is exposed to a uniformly mixed bulk liquid. In biological treatment, the rotating biological contactor (RBC) is the most common example of completely mixed conditions.

2. Fixed bed or packed bed reactor systems. These are more commonly known as "fixed film" reactors or "trickling bed" reactors and have been used in wastewater treatment for more than a century. The mixing in these reactors is typical of plug flow unless significant recycle flow occurs.

3. Fluidized bed technology has only recently been used in biological treatment (Cooper and Atkinson, 1981). Here, influent passes through a bed of particles coated with biofilm at sufficient velocity to fluidize the bed. The mixing in these reactors is dependent on the degree of fluidization.

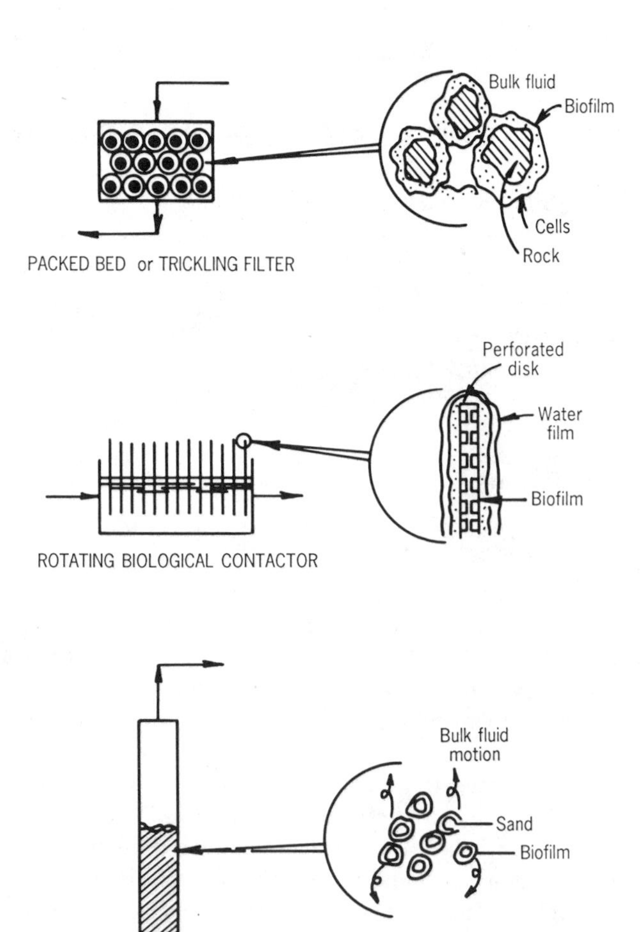

Figure 17.1 Schematic diagrams of biofilm entrapped cell systems.

4. Other biofilm systems include packed or fluidized bed systems operated with some degree of effluent biomass recycle. Operating either system with recycle provides more flexible control over influent concentrations, and in fluidized beds, recycle reduces influent volumetric flow rate requirements to achieve fluidization.

Fixed biofilm reactors are the systems currently used in most wastewater treatment applications. However, other novel reactor geometries are being promoted and may have advantages, at least in special applications.

17.2.2 Biomass Support Particle (BSP) Systems

In fixed film systems, biofilm thickness is generally not controlled and excessive biofilm accumulation leads to mass transfer limitations and system failures (e.g., excessive sloughing of biomass resulting in plugging of a packed bed reactor). Novel biomass support particles (BSP) were conceived in the late 1970s by Atkinson and coworkers to permit biofilm thickness control and thus cure this problem.

BSP, like the fixed biofilm reactors, depend on the natural tendency of microbes to adsorb or aggregate. BSP are composed of various materials in a random three dimensional lattice (Figure 17.2). BSP can be constructed from a variety of materials [e.g., poly(ester) foams, poly(propylene) lattice, stainless steel wire, fritted glass particles] in a range of sizes (both external and internal dimensions) and shapes. Microorganisms remain in the particle by adsorption to the surfaces and aggregation to each other, forming a biofilm matrix throughout the interstices of the support particle. When the particles are in motion relative to one another, as in a fluidized bed, any growth that exceeds the support lattice is removed by attrition. Thus, a biological particle of predetermined size results, containing a constant amount (holdup) of biomass that can be maintained indefinitely within a reactor operated at steady state. Table 17.2 provides a summary of the physical properties of clean and biomass-filled BSP of different construction. Details on BSP fabrication and the fundamentals of biological reactivity can be found in the articles by Atkinson and coworkers.

BSP can be effectively employed in either fixed or fluidized bed reactor systems, although their advantages are more commonly realized in fluidized systems. Preliminary studies by Atkinson et al. (1984) and Walker and Austin (1981) indicate the following advantages for BSP:

1. BSP can be used indefinitely at constant biomass holdup.
2. There are virtually no physical limitations on the throughput to the reactor, i.e., washout of the biomass is eliminated.
3. The biomass concentration in a BSP can be as high as 20 kg m^{-3} bed volume.
4. Gaseous products in a BSP do no apparent damage to the retained biomass.

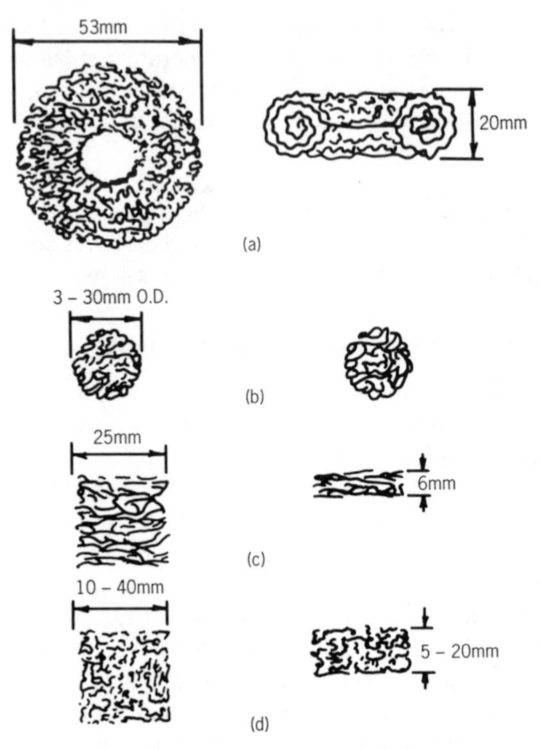

Figure 17.2 Schematic illustrations of biomass support particles. (a) toroid, (b) spheres of knitted stainless steel wire, (c) mats, and (d) porous foam pads.

5. BSP maintain biomass separate from soluble metabolic by-products and provide new methods for biomass recovery and subsequent concentration (Walker et al., 1984).
6. BSP exploit diffusional limitations to enhance biomass yields and control reaction rates in the presence of substrate inhibition (Atkinson, 1985) and increase the overall productivity from slow microbiological reactions (e.g., nitrification, methanogenesis).

Biomass support particles can be employed in either fixed, expanded, or fluidized bed geometries, or within agitated tanks. BSP can maintain a controlled, preselected particle size of reactive biomass. Consequently, reactor operation is focused on preventing excessive biomass accumulation outside the BSP to avoid blockage of the bed interstices, which is more likely to occur in fixed and expanded bed reactors. Periodic backwashing with high air or water velocities can remove excess biomass, which however requires that the excess sludge be collected and discarded in an appropriate manner. In fluidized bed operation, biomass is continually removed by abrasion (particle–particle contact). Both fixed

TABLE 17.2 Properties of Biomass Support Particles[a]

Type and Material	Characteristic Size	Material Density (kg m^{-3})	Porosity	BSP Particle Density (kg m^{-3})	
				Clean	With Biomass
Stainless steel spheres	6 mm OD	7700	0.8	2340	2420
Polypropylene toroids	Overall OD = 53 mm torus OD = 20 mm	900	0.9	990	1080
Reticulated polyester foam	Cubes $10 \times 10 \times 2$ mm	1200	0.9	1020	1110
Matted reticulated polypropylene sheets	Cubes $25 \times 25 \times 6$ mm	900	0.95	990	1090

[a] Cooper and Atkinson (1981).

and fluidized beds eliminate excess biomass, but periodic backwashing of a fixed bed results in a reactor with variable hydrodynamic and biological properties, while a fluidized bed operates at relatively uniform biomass conditions. BSP can also be used in a CFSTR arrangement with relatively small reactor volumes, in which the BSP are much like activated sludge flocs (except that they are considerably larger) and may not require conventional clarification by gravity settling. Use of poly(ester) sponge cubes allows biomass separation via roller-squeezing action, returning the sponges to the vessel and wasting the excess biomass. Thus, BSP systems may reduce bulking and other biomass separation problems that hamper the operation of conventional activated sludge systems. The sponges are produced by several manufacturers, which also promote certain process geometries.

17.2.3 Artificially Immobilized Whole Cell Systems

Reactor systems and operating strategies related to fixed biofilm and BSP reactors depend on microorganisms' natural tendency to adsorb and accumulate as biofilms. Artificially immobilized whole cell systems differ from fixed biofilms and BSP in at least two ways: (1) immobilization is an engineered process, with cells captured within or bound to a support by a variety of physical and/or chemical techniques, and (2) growth and replication of the immobilized cells is not necessary and is often undesirable. Immobilized whole cell biocatalysts have found use primarily in applications where immobilized enzyme biocatalysts have been unsatisfactory because of the following: (1) tedious and lengthy enzyme extraction and purification processes, (2) lack of *in vitro* enzyme structural stability, (3) requirements for cofactors in multiple enzyme reaction sequences, and (4) decaying and irreversible enzyme reactivity. Figure 17.3 demonstrates the various classes of immobilization techniques available and their interrelationship. The characteristics of the immobilized cells resulting from the various types of physical and chemical immobilization techniques currently available are compared in Table 17.3.

Immobilized whole cell catalysts can be fabricated in a variety of reaction geometries. In water and wastewater treatment, the two main applications are diagnosis and treatment. Consequently, whole cells may be immobilized in flat membranes (sheets) for use in biosensors, or as spherical or cylindrical pellets for use in packed bed treatment reactors. Fluidized beds of immobilized biocatalysis are not practical at present, due to their mechanical resiliency.

17.3 BIOFILM SYSTEMS IN WASTEWATER APPLICATIONS

17.3.1 Fixed Film Systems

17.3.1.1 Packed Bed Reactors In packed bed reactors, biofilms accumulate on solid substrata or media packed within a tall tower or large bed. In trick-

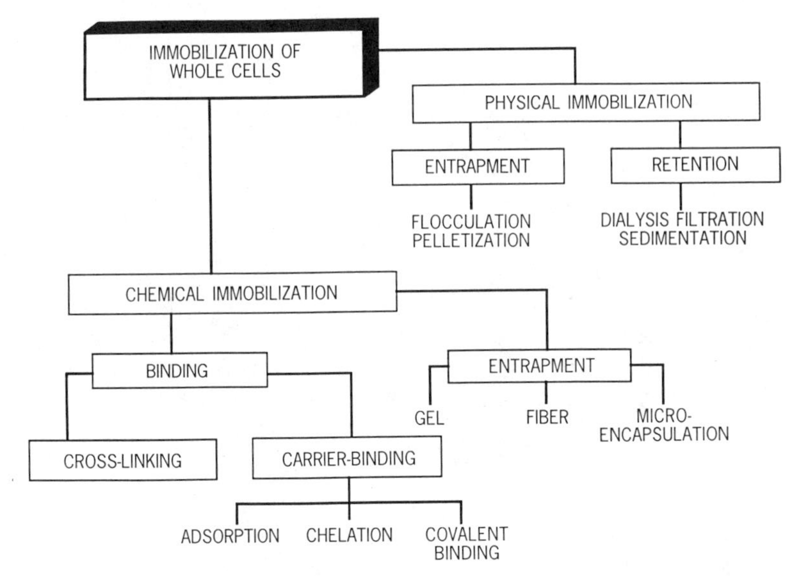

Figure 17.3 A classification of the many kinds of physical and chemical immobilization techniques available.

ling filters, influent liquid is distributed uniformly over the top of the media by a sprinkler system and trickles down through the bed in a thin water layer over the biofilm (Figure 17.1). Requisite oxygen for aerobic processes such as carbon or nitrogen removal is drawn up the bed by natural convection. Trickling filters (using rock media as support) have been used for biological wastewater treatment since the 1870s and (using wood chips) in industrial fermentation to produce vinegar since 1823 (Nickol, 1979). Pilot scale and laboratory scale packed bed reactors can be operated totally submerged if required O_2 is transferred to the influent liquid.

In full scale wastewater treatment trickling filters, crushed granite or limestone is used, but any rock or stone medium would work. Specifications for an ideal support media are (1) low cost, (2) large surface area per unit volume, and (3) sufficient void space to allow for air flow and removal of excess biofilm. For example, a 5.0 cm diameter rock packing will have 90 m^2 surface area per m^3 reactor volume with $\approx 50\%$ void space. Recently, plastic support media have been applied with marked success in packed bed applications. Typically, unsupported loose plastic media have a specific surface area of 98 $m^2\ m^{-3}$ and a void volume of 93–95%. Plastic media are made of either poly(vinyl chloride) (PVC), high density poly(propylene), or high density poly(ethylene), although any reasonably inert material would serve.

Trickling filters are mainly used in the removal of organic pollutants and dissolved ammonia and nitrogen by microbial oxidation. They are simple to operate, making them ideal for remote rural sites and small communities.They ap-

TABLE 17.3 A Comparison of Different Immobilization Techniques[a]

Characteristic	Cross-linking	Adsorption	Chelation	Covalent binding	Entrapping
Preparation	Intermediate	Simple	Simple	Difficult	Intermediate
Binding force	Strong	Weak	Intermediate	Strong	Intermediate
Retention of activity	Low	High	Intermediate	Low	Intermediate
Regeneration of carrier	Impossible	Possible	Possible	Rare	Impossible
Cost of immobilization	Intermediate	Low	Low	High	Low
General application	No	Yes	Yes	No	Yes
Protection from microbial attack	No	Yes	Yes	No	Yes
Viability	No	Yes	Yes	No	Yes

[a] Reprinted with permission from Kennedy and Cabral (1983).

pear less susceptible than suspended culture processes to failure due to shock increases in influent toxic material. However, once they are constructed, there is very little control over the process under changing operating conditions. Packed bed systems are not well suited to conditions of very low or intermittent flow, since the biofilms require a certain flow rate to remain uniformly wet. Ganczarczyk and Hutton (1984) present a thorough historical perspective on microbiological packed bed technology as well as a detailed process analysis of the trickling filter in the context of evaluating various support media.

17.3.1.2 Rotating Biological Contactors Rotating (disk) biological contactors (RBC) consist of parallel circular disks fixed perpendicularly to a horizontal shaft which passes through the center of the disk. The disks are generally positioned within a vessel so that the water level is just below the central shaft (Figure 17.1). Oxidation is performed by mixed populations of bacteria suspended in the water and entrapped within biofilms on the disc surface. The disk assembly is rotated (peripheral velocity ≈ 0.3 m s^{-1}) to promote mixing and, more importantly, oxygen transfer. Typically, RBC are used for aerobic processes, i.e., carbon oxidation and nitrification, although examples of their use in laboratory and pilot scale anoxic and anaerobic systems have been reported (Tait and Friedman, 1980; Rusten and Odegaard, 1982). The disk design and construction material vary. The most common materials are high density poly(propylene) or poly(ethylene) in a corrugated sheet or lattice design.

The advantages of RBC in full scale operations include the following: (1) low power requirements for mixing as compared to activated sludge systems, (2) less susceptibility to low flow conditions than trickling filters, (3) flexibility in oxygen transfer and biofilm control by rotational speed, and (4) low susceptibility to shock loading of toxic compounds. Since highly concentrated wastewaters require oxygen transfer rates not attainable by increased rotation of the RBC disks, supplemental aeration of the bulk water has been used. Several RBC in parallel also can increase the oxygen transfer capacity of the system. Grady (1982) has presented a review of modeling fixed film reactors in general with some emphasis on RBC. Watanabe et al. (1982, 1984) have presented detailed models of nitrification in a RBC.

17.3.1.3 Biological Fluidized Bed Reactors Biological fluidized bed reactors have been applied to aerobic (carbon removal, nitrification), anoxic (denitrification), and anaerobic (methanogenesis) biological wastewater treatment with success. However, only a few technical scale biological fluidized bed wastewater treatment plants exist. Nevertheless, development of the process is progressing rapidly from the laboratory scale to large pilot plants. The potential of fluidized bed reactors was reported as early as 1972 (Beer et al., 1972), and an extensive compilation of research and development has been prepared by Cooper and Atkinson (1981).

A fluidized bed is a bed of small particles (0.2 to 2.0 mm in diameter) freely suspended in the upward flow of water or a water and gas mixture. At the point of minimum fluidization, the pressure gradient across the bed just equals the total weight of the particles. Further increase in fluid flow rate expands the bed and increases the void volume. Biological fluidized bed reactors experience gradual bed expansion due to a change in overall particle density as biomass develops on the support particles.

Biological fluidized bed reactors are relatively simple in construction (Figure 17.1). However, the solids collection, biomass separation, and support media return systems can be quite elaborate. One common variation of the cylindrical bed design is a tapered fluidized bed column (Figure 17.4), which allows for higher loadings on the bed than possible in cylindrical volumes while still retaining the biomass support particles (Tanaka et al., 1981). The fluid velocity (and thus buoyant forces) decreases as the cross-sectional area increases with increasing distance up the bed. A variety of solid biofilm support media have been used in laboratory and pilot scale studies (e.g., quartz sand, anthracite coal, nylon, polypropylene). The selection of a specific solid support medium depends on its

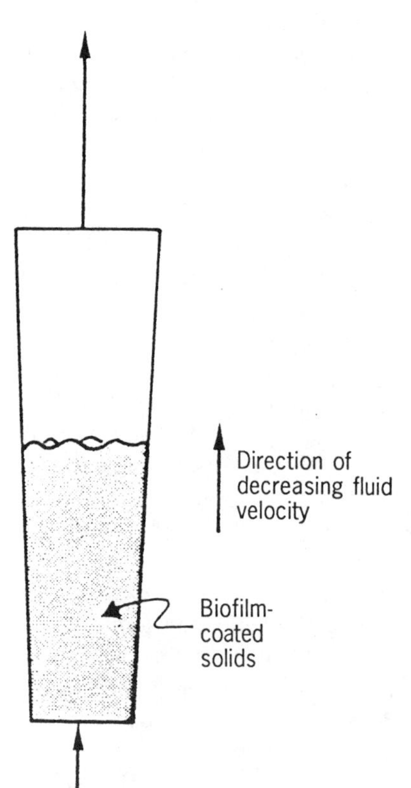

Figure 17.4 A frequent variation of the cylindrical fluidized bed design is a tapered column, which permits higher loadings on the bed than possible in cylindrical volumes while still retaining the biomass support particles.The fluid velocity and thus the buoyant forces decrease as the cross-sectional area increases with increasing distance up the bed.

specific availability and other economic considerations. Although no one support appears to have an operational advantage over the others, quartz sand appears to be most frequently employed, due to its low cost and availability.

Beeftink and Staugaard (1986) present a unique model for the formation of the biomass-sand aggregate, which is supported by electron micrograph evidence (Figure 17.5). Hermanowicz and Ganczarczyk (1983) present a process analysis for microbiological fluidized beds. Bousfield and Hermanowicz (1984) give a more detailed analysis for determining the biomass distribution in a microbiological fluidized bed.

17.3.1.4 Summary Carbon and nitrogen removal from wastewater are the most prevalent applications of fixed film systems. Packed bed or trickling filters have been the most common alternatives to aerobic activated sludge systems for

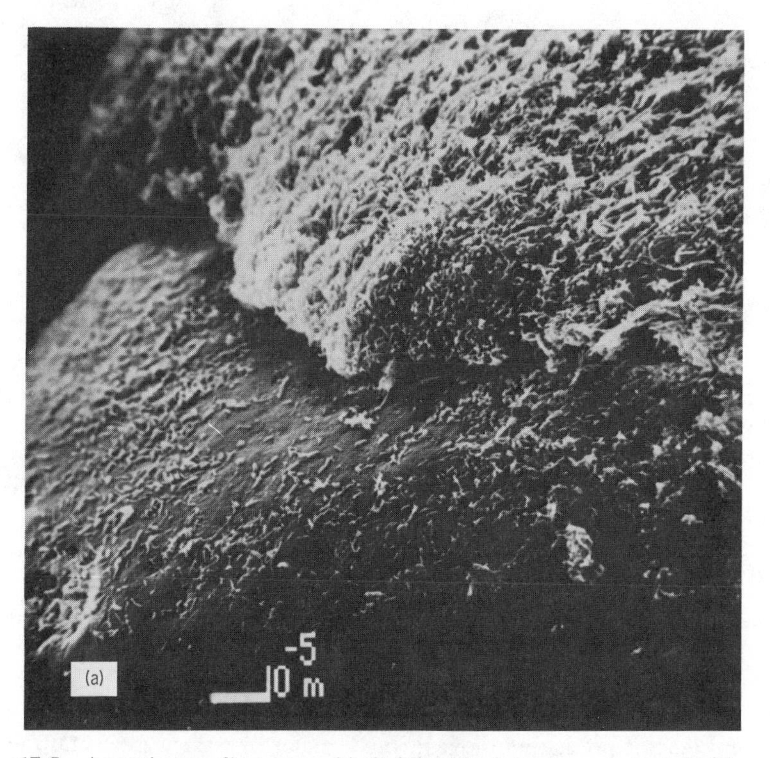

Figure 17.5 A continuous flow anaerobic fluidized bed reactor was started with sand as a support medium. The glucose feed concentration was approximately 10 kg m^{-3}. (a) Scanning electron micrograph of a sand grain partly covered by a biofilm. The exposed surface is a result of biofilm sloughing. Sand was continuously lost in the effluent and was not replaced. (b) Microbial aggregates formed from sloughed biofilm and continued to combine into larger aggregates. (Reprinted with permission from Beeftink and Staugaard, 1986.)

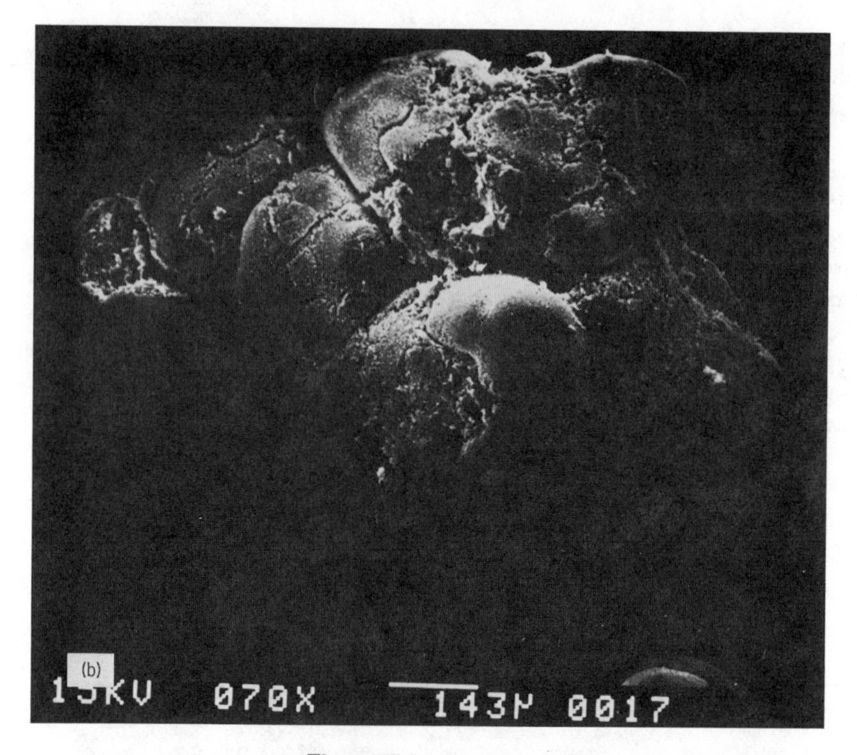

Figure 17.5 Continued

at least 50 years. More recently, RBC have proven to be efficient substitutes for classical suspended culture processes.

High rate trickling filters and RBC are also increasingly used for nitrification processes, but such systems, employing air only, are often oxygen-transfer-limited. Another problem with fixed film nitrification is the long startup times (3–6 months) required to establish a nitrifying biofilm. Further research on the enhancement of biofilm adsorption and accumulation is needed to overcome these problems.

Anaerobic treatment and methane production in packed beds (Kennedy and van den Berg, 1981; van den Berg and Lentz, 1981) and expanded beds (Jewell et al., 1981; Switzenbaum and Jewell, 1980) is a more stable, efficient process than suspended population reactors, and a technical scale fluidized bed biological treatment reactor producing biogas from a molasses wastewater is now operating (Heijnen, 1984). Fluidized bed reactors are not affected by problems of excessive gas bubble production (due to anaerobic methane production, denitrification, etc.) observed in packed beds containing thick biofilms and leading to biofilm detachment and sloughing.

The biofilms in wastewater treatment systems also appear to be resistant to inhibitory and toxic materials. Hollo et al. (1979) report on the removal of heavy metals during denitrification of an industrial wastewater (N as NO_3^-, 120 g m^{-3}) by a *Pseudomonas aeruginosa* biofilm accumulated on poly(vinyl chloride) plastic chips in a two column packed bed system. Methanol was the exogenous carbon source. At a concentration of 5.0 g m^{-3} heavy metals, the biofilm columns could effectively denitrify (85–90% removal) and collect heavy metals for up to 400 h before denitrification ceased. A biofilm consisting of an enriched culture of five specific bacteria (three *Pseudomonas* and two *Klebsiella* species) was observed to degrade an s-triazine herbicide in a highly saline waste stream within a fluidized bed system (Hogrefe et al., 1985). The biofilm community structure within the reactor column was fairly uniform despite the observed specific roles of each bacterial species in the degradation of the triazines to ammonium. Organic carbon was added periodically to stimulate bacterial growth and activity.

17.3.2 Biomass Support Particles

Most BSP reactors are pilot scale, but one technical scale BSP reactor has been reported (Walker et al., 1984). More than half of the pilot scale BSP reactors are being used in wastewater treatment applications. Suspended growth biological wastewater treatment processes (e.g., activated sludge) have traditionally been distinguished from fixed film biological processes (e.g., trickling filters). BSP has introduced an innovative set of biological processes that combine the two types of processes. One of the emerging technologies uses porous biomass supports to supplement the activity of suspended microbial activity. Another employs a fibrous plastic medium strung throughout the aeration basin.

The BSP technologies have the major advantage that a very high mean cell residence time (Θ_c) or low microbial growth rate can be accomplished with very short hydraulic residence times. Generally, the high Θ_c can be accomplished without major solids separation problems. In addition, very high reactor biomass concentrations can be achieved as compared to suspended biomass systems. Thus, the overall reaction rate (specific reaction rate times biomass concentration) can generally be increased over that in suspended growth systems.

Walker and Austin (1981) report a pilot scale assessment of three different types of plastic BSP that are more economical than stainless steel spheres for wastewater treatment. Polypropylene toroids, polypropylene mats, and polyester sponge-foam cubes (Figure 17.2) were each considered in a fluidized bed configuration.

A maximum of 8000 toroids m^{-3} reactor volume could be effectively fluidized. Higher toroid packing created uneven circulation and large clumps of toroids which resulted in an increase in the air/water flow rate. Required aeration within the column resulted in no decrease in BSP biomass accumulation. During startup, the toroids entrapped air and floated; consequently, a retaining screen was required. However once biomass filled the toroids, aeration created no such problems. The following observations were reported from the toroid test with a municipal wastewater:

1. The biomass concentration was as high as 11 g m^{-3}.
2. Nitrification was observed.
3. The biomass within the toroid became increasingly mineralized with time, possibly due to pH gradients in the BSP.

Polyester sponges, like the polypropylene toroids, entrapped air and floated prior to accumulating biomass. Once biomass filled the sponges, a low level of turbulence was required to maintain circulation. Over a long period of operation, biomass in the interior of the sponges became mineralized and required wastage from the system. Biomass can be removed from either the polypropylene pads or the polyester sponges by a mechanical roller device and gentle washing, while rather harsh washing procedures are required to remove biomass from the plastic toroids. Some nitrification tests with pads and sponges suggest that solids wasting is unnecessary.

Wetzel et al. (1986) modeled the sponge support system as an aid in evaluating the technology as compared to conventional suspended growth systems. Attachment and detachment of microorganisms from the biomass support particle is a distinguishing feature of the new technologies as compared to conventional suspended growth systems and certainly influences the *mean cell residence time* Θ_c, a most critical process parameter. Thus, Θ_c was the basis for one of the models. Operating data were provided by the U.S. Environmental Protection Agency and the manufacturers of the support particles. Previous technical scale tests had indicated complete nitrification with a hydraulic residence time of 1.67 h. Model analyses, however, indicated $\Theta_c = 14$ days. Thus although nitrifiers are slow growing organisms, they accumulate in the porous support particles to such an extent that nitrification occurs very rapidly. The measured biomass concentration was approximately 20 g m^{-3}. In a test for heterotrophic carbon removal, the biomass support particles did not perform any better than the conventional system. However, the substrate loading on the system was low, and under such circumstances the reaction time was always shorter than the residence time (Section 10.6, Chapter 10).

A laboratory comparison of a conventional activated sludge system with a BSP-polyester sponge system for treating a composite industrial wastewater (combined effluents from oil refineries, petrochemical plants, and municipal services) was reported by Nesaratnam and Ghobrial (1985). The activated sludge systems were two 0.005 m^{-3} (5 liter) internal recycle reactors with an adjustable baffle to effect internal settling of biomass. Two 0.005 m^{-3} square cross section vessels were used to retain 150 sponges (18 $\times$ 18 $\times$ 12 mm) which were freely suspended. Both systems were operated at residence times of 12.0 and 7.0 h. Nitrogen and chemical oxygen demand removal in the conventional activated sludge was 80–95% and 60–90%, respectively. The sponge systems removed 75–85% of the nitrogen and 50–70% of the COD. The authors attribute the lower performance of the sponges to their tendency to accumulate oil, which influenced the biological activity.

Messing (1982) captured methanogenic bacteria within the pores of a glass ceramic material for a two stage anaerobic system designed to convert sewage organics to methane gas. Reactors were operated at 20, 30, and 40°C at residence times of 2–5.5 h. COD reduction varied between 60 and 90%, with the resultant gas consisting of 90% CH_4 and 5% CO_2. The ceramic cell support was extruded Cordierite™ (Corning Glass Co., average pore diameter 3 μm, pore volume 4400 $mm^3\ g^{-1}$, porosity 57%) or extruded brick (average pore diameter 6 μm, pore volume 290 $mm^3\ g^{-1}$, porosity 34%). Both reactors were operated as once-through packed bed systems with no backwashing.

17.3.3 Immobilized Whole Cell Systems

Artificially immobilized cells have been used in wastewater treatment for both analytical monitoring and specific contaminant removal.

17.3.3.1 Analytical Applications Karube and coworkers (Karube et al., 1971, 1977a,b, 1980; Hikuma et al., 1979, 1980) detail the construction of captured microbial (bacteria and yeast) probes for analysis of wastewater biochemical oxygen demand (BOD) (Figure 17.6). A small volume of culture broth of the yeast *Trichosporon cutaneum* (AJ 4816) was vacuum-filtered onto porous acetylcellulose membrane (type HA, pore size 0.45 μm, diameter 47 mm, thickness 150 μm) and air-dried. The porous microbial membrane is placed on the Teflon membrane of an oxygen probe so the yeast is sandwiched between the two membranes. Upon injection of a solution containing equal amounts of glucose and glutamic acid into the sensor system, current decreases with time and reaches a steady state within 18 min. A linear relationship was observed between the current decrease and the 5-day BOD below a maximum concentration of 60 mg/L. The current output of the yeast probe was constant over a 17 day period and for 400 BOD tests.

A bacterial probe was also fabricated by attaching a strip of nylon net (20 mesh, 700 μm thickness, 3.4 mm diameter) to the platinum electrode with an adhesive. *Clostridium butyricum* is suspended in an acrylamide solution (0.36 g of 90% acrylamide and 10% N,N-methyleneactrylamide in 10 ml saline buffer) in an ice bath. Then 0.5 ml of the solution is cast upon the nylon net covered electrode, and the polymerization is allowed to proceed anaerobically. On placing the probe in a sample containing organic matter, the immobilized microbes convert the organic carbon into hydrogen and formate, which diffuses to the cathode and develops a high current density. The response time is dependent upon the concentration of dissolved organics. A linear relationship between current density and the BOD was observed for BOD values $\leq 300\ g\ m^{-3}$. The useful life of the electrode was $\approx$30–40 days. The BOD probes were calibrated against soluble BOD components of low molecular weight that are easily metabolized. The probes are not useful for samples containing either high molecular weight components or particulate organics.

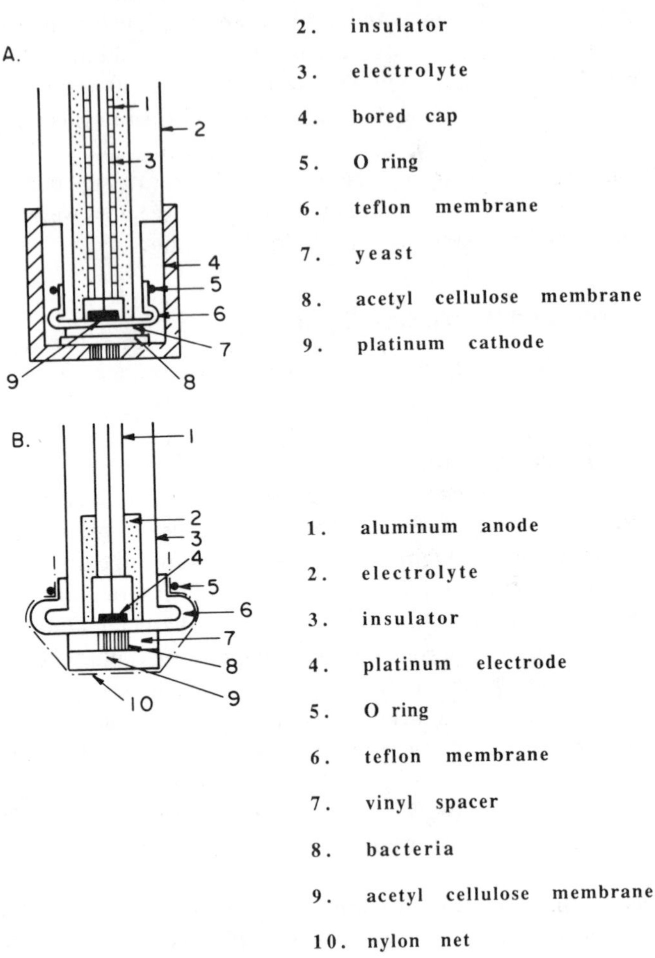

Figure 17.6 A diagrammatic illustration of the construction of captured microbial (bacteria and yeast) cell probes for analysis of wastewater biochemical oxygen demand.

Ammonia can be determined by a microbial sensor consisting of captured nitrifying bacteria (e.g., *Nitrosomonas europaea*) and a modified oxygen probe. Hikuma et al. (1980) report entrapping the nitrifying bacteria between a porous acetylcellulose membrane and the Teflon membrane of a standard oxygen electrode. Calibration is by the difference between the oxygen demand when the nitrifiers are immersed in a sample without ammonia and in one with ammonia. A linear response between current and ammonia concentration was reported by Hikuma et al. for concentrations of 0–1.5 g m^{-3} N as NH_3. The lifetime of the electrode was 14 days and 1400 assays.

17.3.3.2 Treatment Applications Karube et al. (1980) evaluated the production of methane gas from various wastewaters by methanogenic bacteria immobilized in three different artificial carriers: agar gel, polyacrylamide gel, and collagen membrane. Agar gel provided the highest methane production activity [450 (μmol CH_4) (g dry cell mass)$^{-1}$ h^{-1}; 50% of freely suspended cell activity] of the three carriers.

Nilsson et al. (1980) used *Pseudomonas denitrificans* cells immobilized in an alginate gel to denitrify water. In the presence of an exogenous carbon source, the immobilized bacteria reduced nitrate to nitrite and gaseous products. Captured cells retained 75% of their initial nitrate reduction capacity after 21 days of storage at 4°C. Operational stability of the catalysts was studied in both batch and continuous column reactors. The column reactor reportedly (Nilsson and Ohlson, 1982) produced 0.07 (m^3 denitrified water) (kg gel)$^{-1}$ h^{-1} for two months from a high nitrate [22 (g N as NO_3^-) m^{-3}] water. Nitrate reduction activity could be regenerated by periodically feeding organic nutrient to the column. Lysis of pellets and release of cells into the water did occur.

Nakajima et al. (1982) report uranium recovery from seawater and fresh water employing both *Streptomyces uiridochromogenes* and *Chlorella regularis* captured in polyacrylamide gels. Uranium adsorption was not pH-dependent. Adsorbed uranium could be desorbed and recovered with Na_2CO_3.

Klein et al. (1979) report phenol degradation by *Candida tropicalis* yeast captured within ionic polymer networks. The yeast was isolated and enriched for high phenol degradation rate in suspended culture. Polymer beads were formed using styrene–maleic acid copolymer. A fixed bed recycle reactor system employed in the kinetic study consisted of a catalyst bed of approximately 140 g of beads, an oxygen saturation vessel, and phenol wastewater injection. The half-life for the column was estimated at 19 days. Incubation of the beads within the enrichment medium facilitates regrowth of cells and prolongs the life of the catalyst column.

Sequential reduction of nitrate and/or nitrite in water by whole *Micrococcus denitrificans* encapsulated in a liquid–surfactant membrane (Figure 17.7) was demonstrated by Mohan and Li (1975). In batch reactors, denitrification was studied as a function of substrate and cell concentration and pH in both the internal and the external phase. The rate of denitrification was substrate-diffusion-limited. The captured cell system retained 80% of its original activity after 120 hours versus 0% for freely suspended cells after 16 hours.

Immobilization of *C. butyricum* cells has been applied to a biochemical fuel cell, which produced 1.2 mA continuously for 15 days with glucose as substrate. Suzuki et al. (1978) report the alternative use of an industrial alcoholic wastewater as carbon source for the anaerobic production of H_2 by *C. butyricum* entrapped within polyacrylamide gel. H_2 produced in the packed column passed to a fuel cell to generate a current of 8–9 mA for 20 days at an inlet BOD of 660 g m^{-3} (Figure 17.8). The fuel cell consisted of an anode chamber and a cathode chamber separated by an anion exchange membrane. The anode material was platinum black, and the cathode was a carbon electrode in 50×10^{-6} m^3 (50 ml)

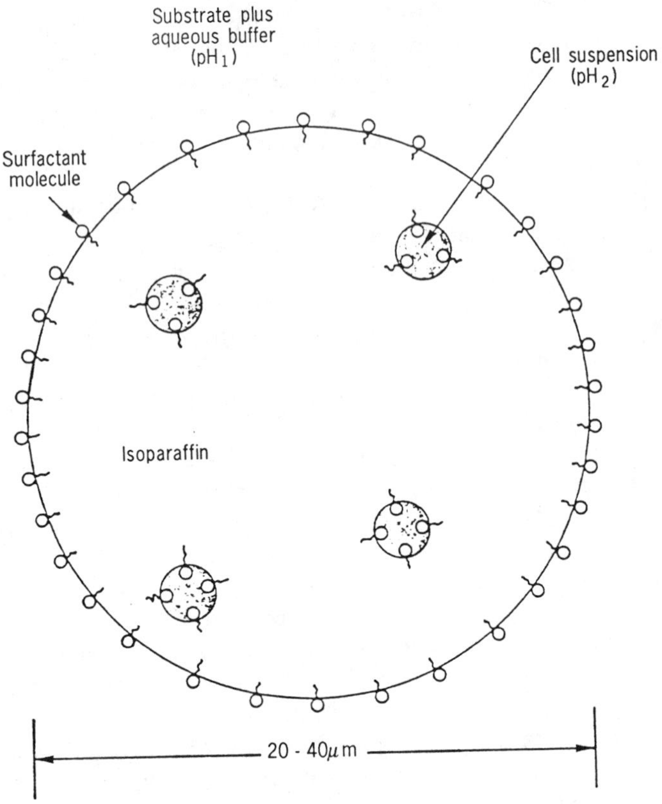

Figure 17.7 A diagrammatic illustration of whole *Micrococcus denitrificans* cells encapsulated in a liquid-surfactant membrane used for sequential reduction of nitrate and/or nitrite in water (reprinted with permission from Mohan and Li, 1975).

0.1 M phosphate buffer. Effluent (BOD = 100 g m^{-3}) leaving the fuel cell passed directly to an aerobic biological reactor, employing a gel-immobilized aerobic mixed bacterial culture, to further reduce its organic concentration.

17.4 DESIGN CONSIDERATIONS

Three types of reactor systems have been presented: (1) fixed or packed bed reactors, (2) fluidized bed reactors, and (3) continuous flow stirred tank reactors (CFSTR). Clearly, there are differences in the fluid–biomass contact patterns between a rotating biological contactor and a fluidized bed reactor. However, the approach to the process design of any heterogeneous reactor, whether it be a CFSTR or a packed bed reactor, is the same. Any design equation or model for a reactor employing heterogeneous reactions includes two components: (1) a reactor design equation describing flow patterns and residence time of the reactor

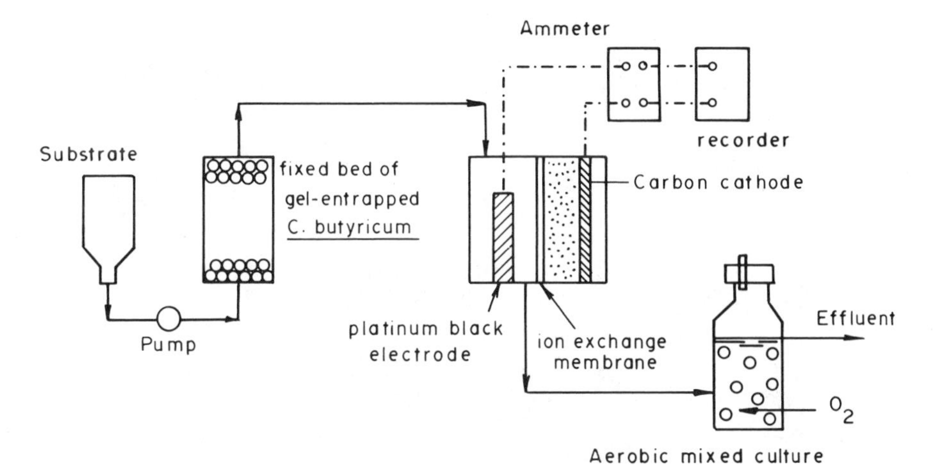

Figure 17.8 A fuel cell based on anaerobic production of H_2 by *C. butyricum* entrapped within polyacrylamide gel. H_2 produced from metabolism on an alcoholic wastewater as carbon source produced H_2, which was passed to the fuel cell, which generated a current of 8–9 mA for 20 days at an inlet BOD of 660 g m^{-3}. The anode chamber and cathode chamber of the fuel cell were separated by an anion exchange membrane. Effluent BOD (100 g m^{-3}) leaving the fuel cell passed directly to an aerobic biological reactor, employing a gel-immobilized aerobic mixed bacterial culture, to further reduce its organic concentration. (Reprinted with permission from Suzuki et al., 1978.)

and (2) the global reaction rate expression(s), i.e., the overall reaction rate (for reactant utilization or product formation) as a function of bulk reactor conditions, expressed on a reactor volume basis. The global reaction rate in turn consists of (1) the specific surface area reaction rate (i.e., the surface flux of reactant or product) evaluated at the biomass-bulk water interface, and (2) the reactor surface area to volume ratio. The exact form of the surface reaction rate is the intrinsic reaction rate expression as a function of local reaction conditions, integrated over the entire heterogeneous catalyst (or biofilm) volume. Both external and internal mass transfer resistance effects are incorporated within the resultant expression for the overall surface reaction rate (Figure 17.9). Consequently, once the intrinsic biological reaction rate is known as a function of local conditions, the design of any laboratory, pilot, or technical scale reactor is reduced to describing the mass transfer resistances and fluid contact patterns in the actual system. A summary of reaction rate design equations for the more pertinent types of reactor systems is presented in Table 17.4.

Problems in scaleup often result, not from changes in the intrinsic reaction rate, but rather from improper extrapolation of reactor mixing characteristics and mass transfer effects. Often times, this last procedural consideration is ignored in comparing two different reaction systems or in scaleup. For example, design criteria for fixed biofilm systems have been developed empirically without regard to the fundamental processes that contribute to their performance. For

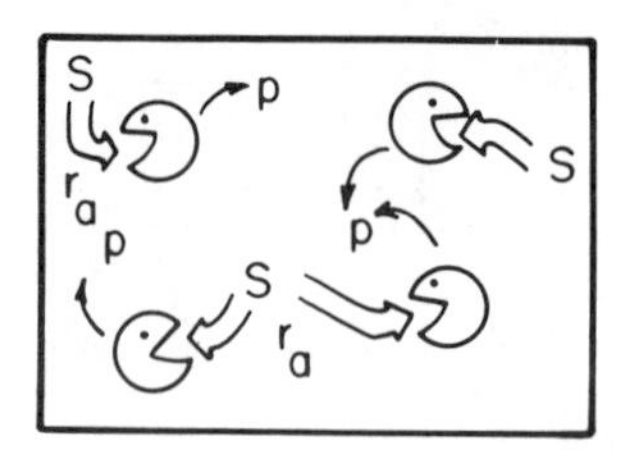

LOCAL BIOLOGICAL REACTION, r_a

$r_a(M_s/L^2 - t) = f$(S, temperature, pH, pE, cell concentration, . . .)

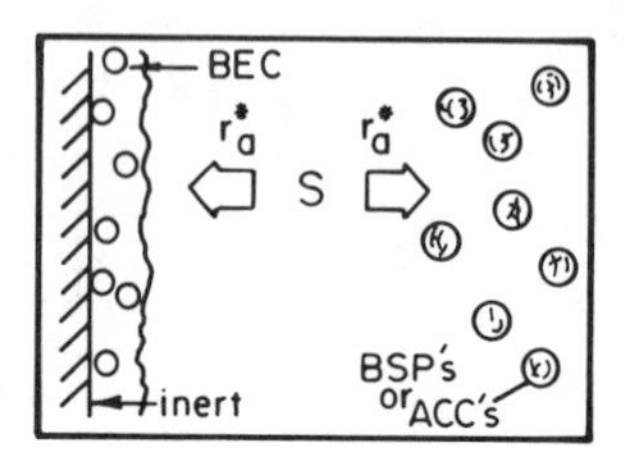

OVERALL SURFACE REACTION, r_a^*
(TOTAL SUBSTRATE FLUX TO SURFACE)

$$r_a^*(M_s/L^2 - t) = \eta\, r_a$$

η = correction for internal & external mass transfer effects.

= f(diffusivity, geometry, reaction rate, bulk fluid S concn.)

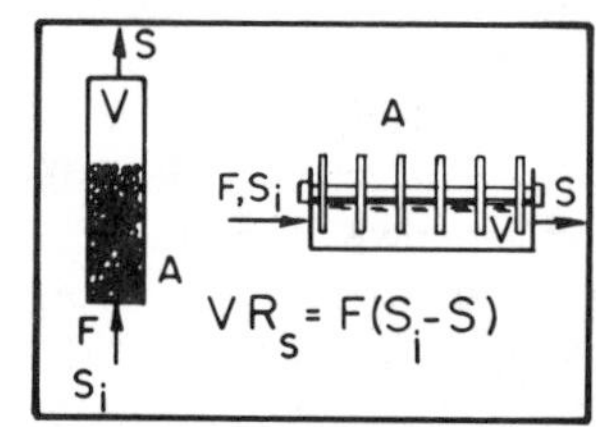

GLOBAL REACTION RATE, R_S

$$R_S(M_s/L^3 - t) = r_a^* A/V = \eta r_a A/V$$

MATERIAL BALANCE FOR S IN ANY TYPE OF REACTOR

$$V\begin{bmatrix}\text{RATE OF S}\\ \text{ACCUMULATION}\end{bmatrix} = \begin{bmatrix}\text{NET INFLOW}\\ \text{RATE OF S}\end{bmatrix} + \sum_{I=1}^{N}\left[v_{i,s} R_S V\right]$$

where $v_{i,s}$ = stoichiometric coefficient of S in the "i-th" reaction where the number of reactions involving S is "n".

Figure 17.9 Design equations or models for a reactor with heterogeneous reactions include two components: (1) a reactor design equation describing the flow patterns and residence time of the reactor, and (2) the global reaction rate expression(s), i.e., the overall reaction rate (for reactant utilization or product formation) as a function of bulk reactor conditions, expressed on a reactor volume basis. The global reaction rate in turn consists of (1) the specific surface area reaction rate (i.e., the surface flux of reactant or product) evaluated at the biomass-bulk water interface, and (2) the reactor surface area to volume ratio. The exact form of the surface reaction rate is the intrinsic reaction rate expression as a function of local reaction conditions, integrated over the entire heterogeneous catalyst (or biofilm) volume. Both external and internal mass transfer resistance effects are incorporated within the resultant expression for the overall surface reaction rate.

TABLE 17.4 A Summary of the More Common Types of Reactor Design Equations

Completely Mixed Reactors

(e.g., rotating biological contactors, agitated vessels with BSP, fluidized bed with recycle)

$$\Theta_{\text{CSTR}} = \frac{V}{Q} = \frac{S_i - S}{r_S}$$

where Θ = reactor mean residence time (t)
S_i = influent substrate concentration (ML^{-3})
S = effluent substrate concentration (ML^{-3})
r_S = overall substrate consumption rate ($\text{ML}^{-3}\text{t}^{-1}$)

Plug Flow Reactors

$$\Theta_{\text{PFR}} = \frac{V}{Q} = -\int_{S_i}^{S} \frac{1}{r_S}\, dS$$

Plug Flow Reactors with Recycle

$$\Theta_{\text{PFRR}} = -(\alpha + 1) \int_{(S_i+\Theta S)/(\alpha+1)}^{S} \frac{1}{r_S}\, dS$$

where α = recycle ratio, i.e., ratio of the volumetric flow rate recycled to reactor to the inlet volumetric flow rate

example, design criteria based on volumetric substrate loading, surface loading, or hydraulic loading predominate in the fixed film reactor literature, with no apparent consistency or rationale. Rittmann (1982) provides an outline for rigorously comparing the performance of various fixed biofilm systems.

Once the reactor volume has been determined to meet average and peak influent conditions, system-specific operational priorities must be addressed. For example, in either packed bed reactors (employing solid rock or plastic support media) or in expanded bed reactors (employing quartz or BSP), an operating protocol must include periodic backwashing of the bed to prevent biomass clogging. Also, in BSP or quartz systems where support material may deteriorate with time due to mechanical attrition or captured biomass mineralization, a periodic replacement scheme must be considered. Consequently, an economic analysis is required to optimize production time and replacement or cleaning time. Furthermore, incidental equipment for separating entrained biomass supports or BSP collection equipment must be included in the capital investment calculations.

Similar practical considerations must be made for artificially immobilized whole-cell systems. Growth of the captured cells is not encouraged, since desired conversions are carried out enzymatically. The reactivity of such a catalytic bed thus decreases with time. A practical operating protocol must consider a sequence of continuous operation interrupted periodically by either *in situ* cell regeneration or a catalyst replacement regime.

REFERENCES

Atkinson, B., "Consequences of aggregation," in K.C. Marshall (ed.), *Microbial Adhesion and Aggregation,* Springer-Verlag, Berlin, 1985.

Atkinson, B., G.M. Black, P.J.S. Lewis, and A. Pinches, *Biotechnol. Bioeng.,* **21**, 193 (1979).

Atkinson, B., G.M. Black, and A. Pinches, "Characteristics of solid supports and biomass support particles when used in fluidized beds," in P.F. Cooper, and B. Atkinson (eds.), *Biological Fluidized Bed Treatment of Water and Wastewater,* Ellis-Horwood Publishers, Chichester, 1981.

Atkinson, B., J.D. Cunningham, and A. Pinches, *Chem. Eng. Res. Des. Trans. Inst. Chem. Eng.,* **62**, 155 (1984).

Beeftink, H.H. and P. Staugaard, *Appl. Env. Microbiol.,* **52**, 1139–1146 (1986).

Beer, C.B., J.S. Jeris, and J.S. Mueller, "Biological denitrification using fluidized granular beds," Report to New York State Department of Environment & Conservation, Environmental Quality Research and Development Unit, June 1972.

Bousfield, D.W. and S.H. Hermanowicz, "Biomass distribution in a biological fluidized bed," in J.T. Bandy, Y.C. Wu, E.D. Smith, J.V. Basilico, and E.J. Opatken (eds.), *Proc. 2nd International Conference on Fixed Film Biological Processes,* Arlington, VA, 1984, pp. 206–226.

Cooper, P.F. and B. Atkinson (eds.), *Biological Fluidized Bed Treatment of Water and Wastewater,* Ellis-Horwood, Chichester, 1981.

Ganczarczyk, J.J. and A. Hutton, "Physical characteristics of PROPAK random media for biological filters," in J.T. Bandy, Y.C. Wu, E.D. Smith, J.V. Basilico, and E.J. Opatken (eds.), *Proc. 2nd International Conference on Fixed Film Biological Processes,* Arlington, VA, 1984, pp. 403–411.

Grady, C.P.L., "Modelling of biological fixed films: A state-of-the-art review," in Y.C. Wu, E.D. Smith, R.D. Miller, and E.J. Opatken (eds.), *Proc. 1st International Conference on Fixed Film Biological Processes,* Kings Island, OH, 1982, pp. 344–404.

Grady, C.P.L. and H.C. Lim, *Biological Wastewater Treatment,* Marcel Dekker, New York, 1980.

Heijnen, J.J., *Biological Industrial Wastewater Treatment Minimizing Biomass Production and Maximizing Biomass Concentration,* Delft University Press, 1984.

Hermanowicz, S.H. and J.J. Ganczarczyk, *Biotech. Bioeng.,* **25**, 1321–1330 (1983).

Hikuma, M., H. Suzuki, T. Yasuda, I. Karube, and S. Suzuki, *Europ. J. Appl. Microbiol. Biotechnol.,* **8**, 289 (1979).

Hikuma, M., T. Kabok, T. Yasuda, I. Karube, and S. Suzuki, *Anal. Chem.,* **52**, 1020 (1980).

Hogrefe, W., H. Grossenbacher, Y. Kido, A.M. Cook, and R. Hutter, "Biodegradation in a fluidized bed of real wastes containing *s*-triazines," in J.D. Bryers, G. Hamer, and M. Moo-Young (eds.), *Waste Treatment and Utilization—3, Conservation and Recycling,* **8**, 1985.

Hollo, J., J. Toth, R.P. Tengerdy, and J.E. Johnson, "Denitrification and removal of heavy metals from wastewater by immobilized microorganisms," in K. Venkatsubramanian, ed., *Immobilized Microbial Cells,* American Chemical Society Symposium Series, Vol. 106, Washington, 1979, Chapter 5.

Jewell, W.J. and M.S. Switzerbaum, *J. WPCF,* **53**, 482 (1981).

Karube, I., S. Suzuki, S. Kinoshita, and J. Mizuguchi, *J. Ind. Eng. Chem./Prod. Res. Develop.*, **10**, 160 (1971).

Karube, I., T. Matsunaga, and S. Suzuki, *J. Solid-Phase Biochem.* **2**(2), 97 (1977a).

Karube, I., T. Matsunaga, S. Mitsuda, and S. Suzuki, *Biotechnol. Bioeng.*, **19**, 1535 (1977b).

Karube, I., S. Kuriyama, T. Matsunaga, and S. Suzuki, *Biotechnol. Bioeng.*, **22**, 847 (1980).

Kennedy, J.F. and J.M.S. Cabral, "Immobilized living cells and their applications," in *Applied Biochemistry and Bioengineering,* Vol. 4, Academic Press, New York, 1983.

Kennedy, K.J. and L. van den Berg, "Effects of temperature and overloading on performance of anaerobic fixed film reactors," in *Proc. 36th Industrial Waste Conference,* Purdue University, Ann Arbor Science Publ., Ann Arbor, MI, 1981.

Klein, J., U. Hackel, and F. Wagner, "Phenol degradation by *Candida Tropicalis* whole cells entrapped in polymeric ionic networks," in K. Venkatsubramanian (ed.), *Immobilized Microbial Cells,* ACS Symposium Series 106, Washington, 1979, Chapter 7.

La Motta, E.J., R.F. Hickey, and J.F. Buydos, *J. Environ. Eng. Div. ASCE,* **108** (EE6), 1326 (1982).

Messing, R.A., *Biotechnol. Bioeng.*, **24**, 1115 (1982).

Mohan, R.R. and N.N. Li, *Biotechnol. Bioeng.*, **17**, 1137 (1982).

Nakajima, A., T. Horikoshi, and T. Sakaguchi, *Europ. J. Appl. Microbiol. Biotechnol.*, **16**, 88 (1982).

Nesaratnam, S.T. and F.H. Ghobrial, "Treatment of combined industrial/municipal wastewaters using lab scale CAPTOR[R] systems," in J.D. Bryers, G. Hamer, and M. Moo-Young (eds.), *Proc. Third International Conference on Waste Treatment and Utilization, Recycling and Conservation,* **8**, 1985.

Nickol, G.B., "Vinegar," in H.J. Peppler and D. Perlman (eds.), *Microbial Technology,* Academic Press, 1979, Chapter 6.

Nilsson, I. and S. Ohlson, *Europ. J. Appl. Microbiol. Biotechnol.*, **14**, 86 (1982).

Nilsson, I., S. Ohlson, L. Haggstrom, N. Molin, and K. Mosbach, *Europ. J. Appl. Microbiol. Biotechnol.*, **10**, 261 (1980).

Rittmann, B.E., *Biotechnol. Bioeng.*, **24**, 1341 (1982).

Rusten, B. and H. Odegaard, "Denitrification in a submerged biodisc system with raw sewage as carbon source," in Y.D. Wu, E.D. Smith, R.D. Miller, and E.J.D. Patken (eds.), *Proc. First International Conference on Fixed Film Biological Processes,* 20–23 April 1982.

Suzuki, S., I. Karube, and T. Matsunaga, *Biotechnol. Bioeng., Symp. No. 8*, 501 (1978).

Switzenbaum, M.S. and W.J. Jewell, *J. WPCF,* **52**, 1953 (1980).

Tait, S.J. and A.A. Friedman, *J. WPCF,* **52**(8), 2257, 1980.

Tanaka, H., S. Uzman, and I.J. Dunn, *Biotechnol. Bioeng.*, **23**, 1683 (1981).

van den Berg, L. and C.P. Lentz, "Performance and stability of anaerobic contact process as affected by waste composition, inoculation, and solids retention time," in *Proc. 35th Industrial Waste Conference, 13–15 May 1980,* Ann Arbor Science, Ann Arbor, MI, 1981, p. 496.

Walker, I. and E.P. Austin, "Use of plastic, porous supports in a pseudo-fluidized bed for effluent treatment," in P.F. Cooper and B. Atkinson (eds.), *Biological Fluidised Bed Treatment of Water and Wastewater,* Ellis-Horwood Publ., Chichester, 1981, Chapter 16.

Walker, I., P.F. Cooper, H.E. Crabtree, and R.P. Aldre, "Evaluation of the CAPTOR[R] process for uprating an overloaded sewage works," presented at Biological Systems Symposium for Institute of Chemical Engineers, Bradford, West Yorkshire, U.K., 1984.

Watanabe, Y., K. Nishidome, Ch. Thananteseth, and M. Ishiguro, "Kinetics and simulation of nitrification in a RBC," in Y.C. Wu, E.D. Smith, R.D. Miller, and E.J. Opatken (eds.), *Proc. 1st International Conference on Fixed Film Biological Processes,* Kings Island, OH, 1982, pp. 309–330.

Watanabe, Y., S. Masuda, K. Nishidome, and C. Wantawin, *Wat. Sci. Tech.,* **17**, 385 (1984).

Wetzel, E.D., A.T. Wallace, L.D. Benefield, and W.G. Characklis, "Inert media biomass support structures in aerated suspended growth systems: An innovative/alternative technology assessment," Final Report, Contract No. 68-03-1821, U.S. Environmental Protection Agency, September 1986.

18

BIOFILMS IN POROUS MEDIA

A. B. CUNNINGHAM

Montana State University, Bozeman, Montana

E. J. BOUWER

The Johns Hopkins University, Baltimore, Maryland

WILLIAM G. CHARACKLIS

Montana State University, Bozeman, Montana

SUMMARY

Biofilm accumulation in porous media is the net result of microbial cell adsorption, desorption, growth on surfaces, detachment and filtration. Accumulation of biofilm in porous media influences hydraulic conductivity and results in biotransformation of compounds, many of which are groundwater contaminants. The hydrodynamics of porous media indicates that pore size reduction as biofilm accumulates is likely to be the major cause of the decreased permeability under laminar flow that is characteristic of most groundwaters. This effect has been demonstrated in laboratory and field experiments. There is sufficient field and laboratory evidence that subsurface biofilms can bring about the biotransformation of many trace organic groundwater contaminants. Low biomass densities, improper environment, and poorly adapted microorganisms contribute to slow reaction rates. Competition among microbial species for available electron acceptors and redox conditions is an important factor influencing organic contaminant biotransformations. Stimulation of biotransformation, possibly by addition of primary substrate or inorganic nutrients, is attractive for the treatment of contaminated porous media, since contaminants may be destroyed rather than transferred to another environmental medium as encountered with physicochemical processes. Much more needs to be known about the rate-limiting processes for microbial cell accumulation and their biochemical capabilities under different environmental and physiological conditions characteristic of porous media in order to optimize the design and operation of subsurface contaminant treatment schemes.

18.1 INTRODUCTION

The dynamics of microbial populations in the subsurface are influenced by a complex and often heterogeneous set of environmental conditions. As a consequence, the biofilm processes occurring in flow in porous media are more difficult to monitor than in other common reactor types such as tanks, reservoirs, and pipelines. Subsurface biofilm accumulation, for example, is influenced by the nature of fluid and nutrient transport, which in a porous medium occurs along tortuous flow paths of varying dimension and geometry. Similarly, the wide distribution of pore velocities introduces considerable variation in the processes of cell adsorption, desorption, attachment, and detachment. Deposition of particulate material by filtration mechanisms must also be considered. An understanding of cause and effect relationships that influence these and other biofilm processes is essential for describing net subsurface biofilm accumulation.

Accumulation of biofilm in porous media is presently of interest from two distinct viewpoints: (1) analysis and mitigation of porous media biofouling, and (2) the *in situ* biodegradation of groundwater contaminants. This chapter presents basic concepts and research findings that address both topics, using the fundamentals of biofilm processes (Chapters 7 and 9) as a basis.

18.2 HYDRODYNAMICS OF POROUS MEDIA

The hydrodynamic principles presented here provide the necessary background for subsequent discussion of biofilm processes in porous media. For a more comprehensive treatment of flow through porous media, the reader should consult a standard groundwater text such as Bear (1979), H. Bouwer (1978), Freeze and Cherry (1979), or Todd (1980). These references also provide an overview of mass and momentum transport in porous media. The reader should also consult Chapter 9 for a discussion of the fundamental processes and equations that describe the transport of mass, momentum, and energy in biofilm systems. In porous media, mass transport determines the capacity of the flow regime to transport dissolved or suspended substances (i.e. tracers, chemicals, nutrients, and microbial cells), while momentum transport determines the resistance to flow through the porous matrix. Mass transport processes influence the net accumulation of biofilm and the rates of degradation of organic contaminants, and they control the migration of contaminated groundwater. Momentum transport processes are likewise of importance to the analysis of biofouling in packed bed filters and porous media.

18.2.1 Darcy's Law

Consider flow through a cylinder of cross section A filled with a granular porous medium (Figure 18.1). Analysis of this experimental system provides the basis for illustrating fundamental porous medium flow principles. If water is introduced into the cylinder until all the pores are filled, the inflow rate Q is equal to the outflow rate. If an arbitrary datum is chosen, the elevations of the manometer levels relative to the datum are h_1 and h_2. Then,

$$Q = \frac{-KA\,\Delta h}{\Delta L} \qquad [18.1]$$

where ΔL = distance between the manometer intakes (L)
$\Delta h = h_1 - h_2$ (L)
K = proportionality coefficient

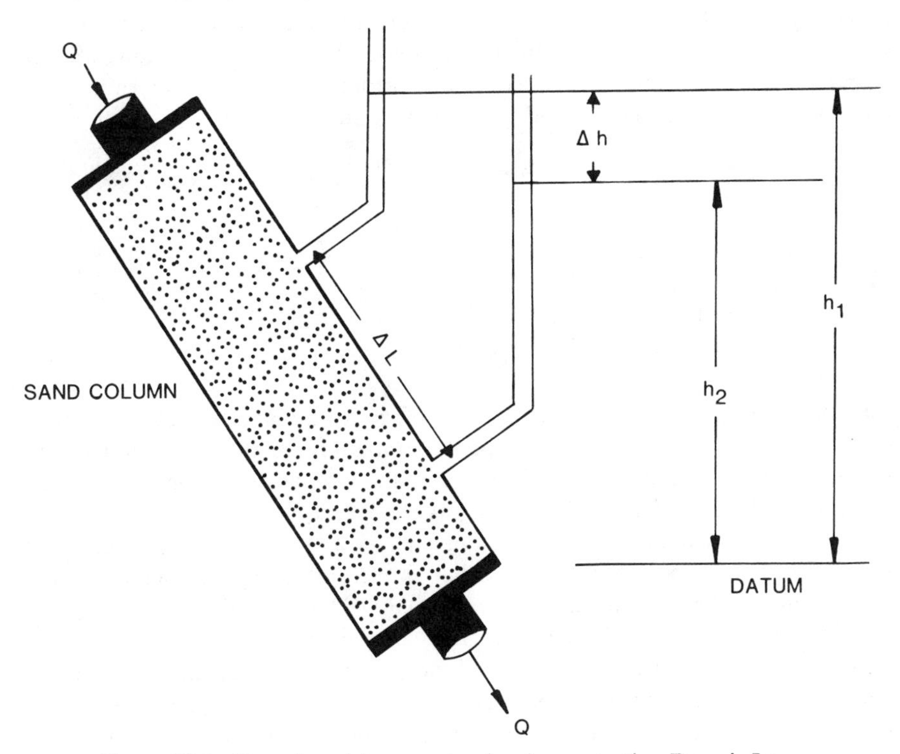

Figure 18.1 Experimental apparatus for demonstrating Darcy's Law.

Expressed in differential terms,

$$Q = -KA \frac{dh}{dL} \tag{18.2}$$

or

$$v = \frac{Q}{A} = -K \frac{dh}{dL} \tag{18.3}$$

where Q = volumetric flow rate (L^3t^{-1})
v = specific discharge (Lt^{-1})
K = hydraulic conductivity (Lt^{-1})
dh/dL = hydraulic gradient (—)

Preferred units for K and v are m day^{-1}, which result in values ranging from essentially zero for highly impermeable material (sandstone, clay, rock) to as much as 1000 m day^{-1} for very permeable media (gravel, fractured limestone). The negative sign indicates that the flow of water is in the direction of decreasing head. Equation 18.3 is known as Darcy's law and states that the specific discharge v equals the product of hydraulic conductivity and hydraulic gradient. Darcy's law is experimentally based, and its range of validity is determined by the magnitude of the porous medium Reynolds number

$$Re = \frac{vd\rho}{\eta} \tag{18.4}$$

where d = grain diameter (L)
η = fluid viscosity ($ML^{-1}t^{-1}$)
ρ = fluid mass density (ML^{-3})

d is the mean grain diameter, although alternative quantities such as d_{10} (the diameter such that 10% by weight of the grains are smaller) are used in the literature. According to Bear (1979), despite the various definitions for d, Darcy's law can be considered valid as long as Re does not exceed 1.

Darcy's law is of fundamental importance in the analysis of groundwater flow; it is also of importance in many other applications of porous medium flow. As indicated by Freeze and Cherry (1979), Darcy's law describes the flow of soil moisture and is used by soil physicists, agricultural engineers, and soil mechanics specialists. It describes the flow of oil and gas in deep geological formations and is used by petroleum reservoir analysts. It is relevant to the design of filters by chemical engineers. It has also been used by bioscientists to describe the flow of bodily fluids across porous membranes in the body.

18.2.2 Permeability

The hydraulic conductivity K is a function of porous medium properties and hydraulic properties; it provides a measure of the ability of a porous medium to conduct water. However, it is often useful to express the permeability of a porous medium as a property of the medium independent of the density and viscosity of the fluid. Thus, the intrinsic permeability is defined as

$$k = \frac{K\eta}{\rho g} \qquad [18.5]$$

where k = permeability (L^2)
g = acceleration due to gravity (Lt^{-2})

Expressing g in cm s^{-2}, ρ in g cm^{-3}, and K in cm s^{-1}, then k is expressed in cm^2. The customary unit of k is the darcy, which is 0.987×10^{-8} cm^2. Intrinsic permeability is mainly applied in the petroleum and natural gas industries, which deal with underground flow of fluids of varying viscosity and density.

Because water movement occurs through the pores and cracks of the aquifer material only, the actual pore velocity of the water is greater than the specific discharge (Figure 18.2). If the effects of tortuosity are neglected, the relation between the actual pore velocity v_p and the specific discharge v is

$$v_p = \frac{A}{A_{\text{eff}}} v \qquad [18.6]$$

where v_p = pore velocity (Lt^{-1})
A = total area normal to the flow direction (L^2)
A_{eff} = effective area available for flow (L^2)

or

$$v_p = \frac{v}{\alpha} \qquad [18.7]$$

where $\alpha = A_{\text{eff}}/A$ = porosity (—).

Equation 18.7 is theoretically correct only for the condition where the effective pore area is the same as the total pore area A_p. Also, if tortuosity is considered, the true microscopic velocities will be larger than v_p, since the fluid must travel along irregular flow paths that are longer than the linearly averaged path represented in v_p. However, according to H. Bouwer (1978), Eq. 18.7 reasonably estimates the actual velocity in most types of soil or aquifer materials.

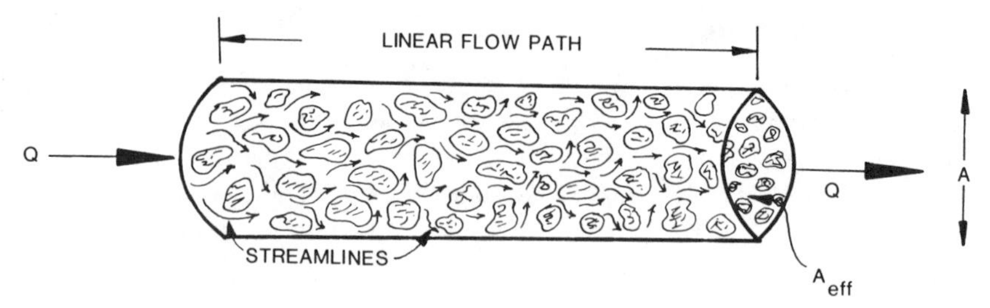

Figure 18.2 Streamline pattern representing tortuous flow through porous media.

18.2.3 Mass Transport

Consider the flux of a dissolved component into and out of a stationary porous medium control volume (Figure 18.2). The conservation of mass for this control volume can be written as

$$\begin{aligned} &\text{rate of accumulation of mass within control volume} \\ &\qquad = \text{net mass flux across all volume surfaces} \\ &\quad + \text{rate of change of mass within column volume due to reactions} \end{aligned} \qquad [18.8]$$

Two categories of components are of interest when considering biofilm processes in porous media: (1) nutrients, whose concentration will influence the cell growth rates along with the rates of contaminant degradation, and (2) suspended cells and solids, which contribute to biofilm accumulation and possible porous medium fouling.

Mass flux across the control surface occurs by advection and hydrodynamic dispersion, while chemical or biochemical transformations affect the mass concentration within the control volume. The combined effects of advection and dispersion, along with reaction of components with their surroundings, govern the variation in concentration with time and flow path position. Mass transport processes therefore have a fundamental influence on the net accumulation of subsurface biofilm.

18.2.3.1 Advection Advection is the component of mass transport resulting from the velocity of flow through porous media. Thus, the advective transport rate is typically taken as the average linear pore velocity (Eq. 18.7). In this context, advective transport is a macroscopic process that defines the linear rate of migration in the direction of flow for which v_p is measured. This representation of advection transport is considered adequate for describing the migration of tracers, nonreactive groundwater contaminants, and other components whose transport depends mainly on the bulk properties of the porous media. However, in the transport of nutrients, reactive chemicals, and microbial cells, it may be necessary to consider the true microscopic pore velocity distribution in order to adequately characterize the interaction of each component with microbial processes in porous media.

18.2.3.2 *Hydrodynamic Dispersion* Hydrodynamic dispersion, the other physical process influencing mass flux, results from mechanical mixing and molecular diffusion (as discussed in Chapter 9). Dispersion is the net result of three processes: (1) mixing due to the fluid velocity distribution within individual pore spaces, (2) mixing due to variation in true pore velocities among pore channels of different size and surface roughness, and (3) mixing due to convergence and divergence of individual pore channels. These three processes increase the spreading of a solute plume, in both longitudinal and transverse directions, from the path the plume would follow due solely to advection. Studies of contaminant plumes have shown that longitudinal dispersion greatly exceeds transverse dispersion (Freeze and Cherry 1979).

18.2.4 Momentum Transport

Conservation of fluid momentum through a control volume can be written as

$$\begin{aligned}&\text{rate of accumulation of momentum within control volume}\\&\quad = \text{net momentum flux across control surface}\\&\quad + \text{summation of forces acting on fluid within control volume}\end{aligned} \quad [18.9]$$

In steady state flow systems, momentum conservation relates the momentum transport rate to the individual forces acting on the fluid in motion. The forces of primary concern include (1) the force due to gravity, (2) viscous or "shear" forces due to fluid friction, and (3) forces due to pressure. In porous medium flow, the combined effects of these three forces equals the rate of momentum transfer from the fluid to the porous medium matrix. Frictional resistance to flow is measured by the permeability or the friction factor f (defined in the following section).

The discussion of frictional resistance in pipe flow systems presented in Section 9.5 (Chapter 9) can readily be extended to porous medium flow by reformulating Eq. 9.22 (Chapter 9) to allow for arbitrary conduit geometry. This is accomplished by defining the hydraulic radius R_H of the conduit cross section as

$$R_H = \frac{\text{cross-sectional area}}{\text{wetted perimeter}} \quad [18.10]$$

Since R_H for a circular conduit is equal to $d/4$, Eq. 9.22 can be written, after multiplying both sides by g, as

$$\frac{P_0 - P_L}{\rho g} = \frac{p_0 - p_L}{\rho g} + z_0 - z_L = \frac{f\,L\,V_m{}^2}{R_H\,2g} \qquad [18.11]$$

$(p_0 - p_L)/\rho g$ is the fluid energy per unit weight (head) that is dissipated along a path of length L of arbitrary geometry. The head loss h_L can be defined for porous media flow by assuming $v_m = v_p$ as

$$h_L = \frac{f_p L v_p^2}{R_H\,2g} \qquad [18.12]$$

where h_L = head loss (L)
f_p = porous medium friction factor (—)
R_H = hydraulic radius (L)

Defining an *average* hydraulic radius R_{Hm} for a unit volume of porous medium makes it more convenient to apply Eq. 18.12:

$$R_{Hm} = \frac{\text{volume of void region per unit volume}}{\text{total wetted surface per unit volume}} = \frac{\alpha}{a} \qquad [18.13]$$

According to Bird et. al. (1960),

$$a = a_v(1 - \alpha) \qquad [18.14]$$

where a_v = ratio of particle surface area to particle volume (L^{-1}).

For a sphere of diameter d,

$$a_v = \frac{6\pi d^2}{\pi d^3} = \frac{6}{d} \qquad [18.15]$$

Thus, the final expression for R_{Hm} becomes

$$R_{Hm} = \frac{\alpha d}{6(1 - \alpha)} \qquad [18.16]$$

If Eq. 18.7 and Eq. 18.16 are combined with Eq. 18.12, the resulting expression for the head loss along a porous medium flow path is

$$h_L = \frac{f_p L(1 - \alpha)v^2}{d\alpha^3 g} \qquad [18.17]$$

According to Tchobanoglous (1969), f_p is a dimensionless friction factor related to porosity and Reynolds number (Eq. 18.4) by

$$f_p = 150\left(\frac{1-\alpha}{Re}\right) + 1.75 \qquad [18.18]$$

A summary of alternative head loss formulas (after Tchobanoglous, 1969) is given in Table 18.1. All formulas in Table 18.1 express the head loss in terms of porous medium properties (d and α), flow path length (L), and specific discharge (v), as well as other parameters that take account of frictional resistance. The equations for friction factor (f_p) and drag coefficient (C_d) in Table 18.1 provide direct estimates of the frictional resistance in porous medium flow.

A more comprehensive treatment of frictional resistance (Fahien and Schriver, 1961) involves modification of Eq. 18.18 to allow for the probability that, for a given porous medium Reynolds number, flow could be turbulent through some pores while laminar in others. Accordingly dimensionless friction factor is defined by Fahien as

$$f_p^* = \frac{q f_{1L}}{Re' f_2} + \frac{(1-q)(f_{1T} + f_2 Re')}{Re' f_2} \qquad [18.19]$$

where $Re' = \dfrac{d\rho v_p}{(1-\alpha)\eta}$ (—)

$f_{1L} = \dfrac{136}{(1-\alpha)^{0.38}}$ (—)

$f_{1T} = \dfrac{29}{(1-\alpha)^{1.45}\,\alpha^2}$ (—)

$f_2 = \dfrac{1.87\,\alpha^{0.75}}{(1-\alpha)^{0.26}}$ (—)

q = laminar fraction of flow (—)

$1 - q$ = turbulent fraction of flow (—)

The laminar fraction of flow was determined experimentally as

$$q = \exp(-C_0 Re') \qquad [18.20]$$

where $C_0 = \dfrac{\alpha^2(1-\alpha)}{12.6}$ [18.21]

Fahien found that the friction factor f_p^* correlated with a modified porous medium Reynolds number Re^* given by

$$Re^* = \left(\frac{q}{(Re^*)_L} + \frac{(1-q)}{(Re^*)_T}\right)^{-1} \qquad [18.22]$$

TABLE 18.1 Formulas Governing the Flow of Clear Water through a Clean Granular Medium

Equations

CARMEN–KOZENY EQUATION

$$h_L = \frac{f_p (1 - \alpha) L v^2}{\phi \alpha^3 d g}$$

where

$$f_p = 150 \frac{1 - \alpha}{Re} + 1.75$$

$$Re = \frac{d v \rho}{\eta}$$

FAIR–HATCH EQUATION

$$h_L = k \nu \phi_1^2 \frac{(1 - \alpha)^2}{\alpha^3} \frac{L v}{d^2 g}$$

ROSE EQUATION

$$h_L = \frac{1.067}{\phi_2} C_d \frac{L v^2}{\alpha^4 d g}$$

where

$$C_d = \frac{24}{Re} + \frac{3}{Re} + 0.34$$

HAZEN EQUATION

$$h_L = \frac{1}{C} \left(\frac{60}{T + 10} \right) \frac{L}{d_{10}^2} v$$

Symbols

C = coefficient of compactness (600 to 1200)
C_d = coefficient of drag
d = grain diameter m
d_{10} = effective size gram diameter, mm
f_p = friction factor
g = gravity constant, 9.8 m s^{-2}
h_L = head loss, m
k = filtration coefficient: 5 based on sieve openings, 6 based on size of separation
L = depth, m
Re = Reynolds number
ϕ_1 = shape factor (varies between 60 and 77)
T = temperature, T
v = specific discharge, m s^{-1}
α = porosity
η = viscosity, N s m^{-2}
ν = kinematic visocity, m^2 s^{-1}
ρ = density, kg m^{-3}
ϕ_2 = shape factor (usually 1)

where

$$(Re^*)_L = \frac{Re'f_2}{f_{1L}}$$

and

$$(Re^*)_T = \frac{f_2Re'}{f_{1T}}$$

Experimental data validating the relationship between Eq. 18.19 and 18.22 are presented by Fahien (1983).

18.2.5 Effects of Biofilm Accumulation

Biofilm accumulation within porous media can substantially reduce the capacity for transport of mass and momentum. In flow systems with constant head gradient, excessive biofilm accumulation results in a decrease in specific discharge. In systems where the specific discharge is held constant, an increase in head gradient will result. This increased resistance to flow is due to reduction of the effective pore space caused by attached cells and their extracellular matrix (Figure 18.3). Reduced effective pore space results in decreased permeability and increased friction factor.

Laboratory experiments by Crawford (1987) demonstrate the relationship between biofilm accumulation and the hydraulic resistance of a porous medium as measured by both permeability and friction factor. The apparatus (Figure 18.4) used by Crawford measured the average thickness of a *P. aeruginosa* biofilm in a rectangular capillary reactor containing a single layer of 1 mm glass spheres. Periodic measurements of head gradient and specific discharge were used to determine the variation of permeability and friction factor with time. The increase in biofilm accumulation, reflected by the average biofilm thickness throughout the reactor, corresponds to a substantial decrease in permeability (Figure 18.5). The reduction in permeability continued after the average biofilm thickness had stabilized, indicating that the biofilm surface continued to develop in a progressively irregular manner (Figure 18.3).

Biofilm accumulation in Crawford's experiments also resulted in an increase in porous media friction factor. The development of a 75 μm biofilm on 1 mm glass spheres caused f_p to increase from an initial value of 8.1 up to 217 over a 10 day period (Figure 18.6). As with the permeability, f_p continued to increase after the biofilm thickness stabilized. Crawford's friction factor data are consistent with the general expression developed by Fahien using several types of porous media (Figure 18.7). This suggests that the presence of biofilm does not alter the observed relationship between friction factor and Reynolds number for flow through porous media.

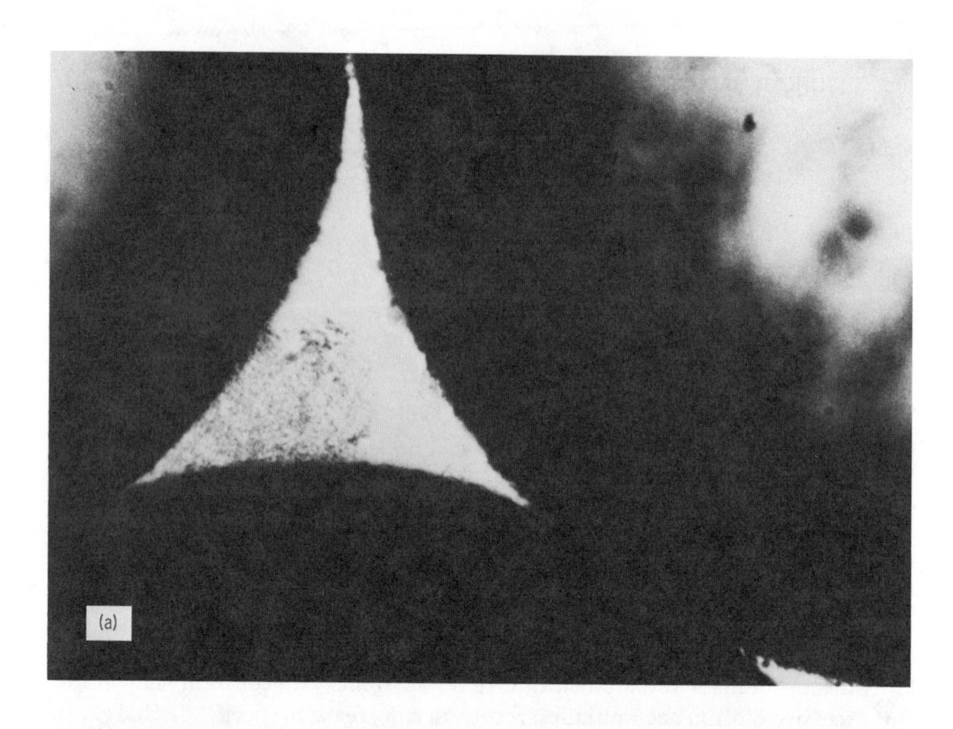

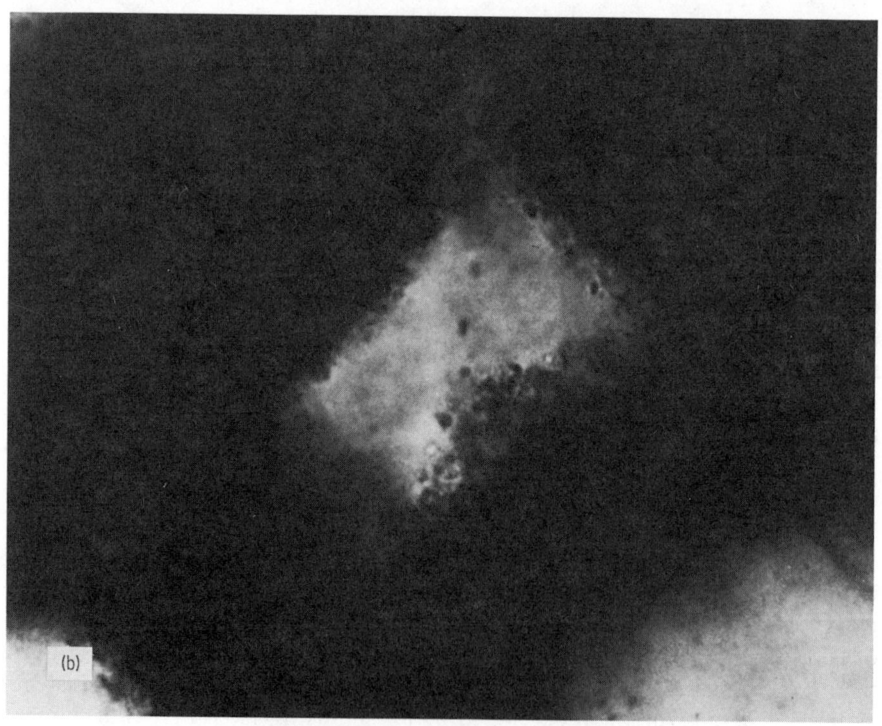

Figure 18.3 A *P. aeruginosa* biofilm accumulation on 1 mm ID glass spheres after 3 days (a) and 8 days (b) of continuous reactor operation. Substrate was glucose in a mineral salts medium. (Crawford, 1987.)

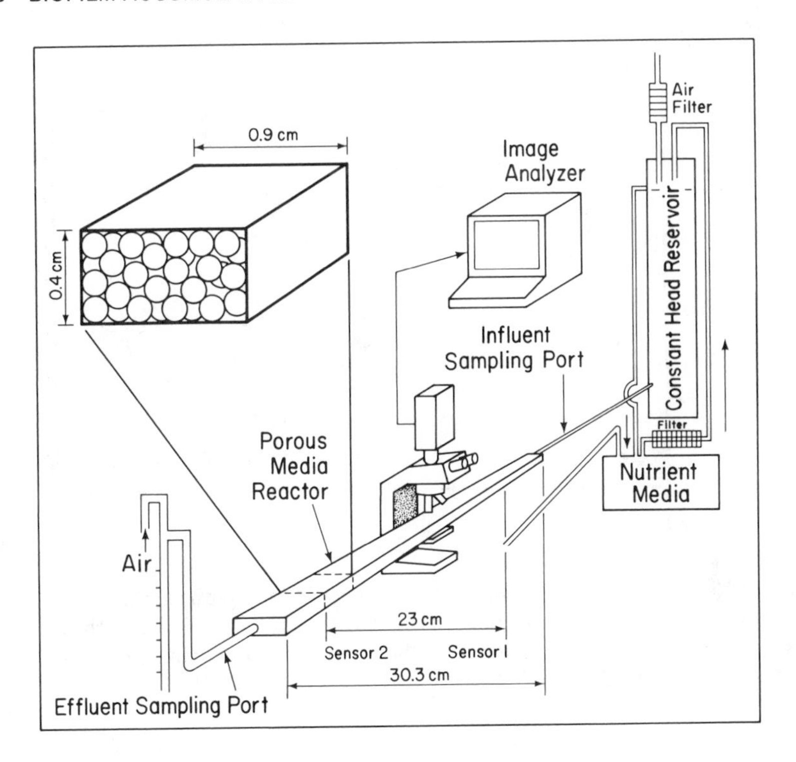

Figure 18.4 Image analysis system including the support equipment for providing a continuous flow of microorganisms to a rectangular capillary containing a single layer of a porous medium (see Chapter 3).

18.3 BIOFILM ACCUMULATION

Fundamental biofilm processes have been described previously (Chapter 7) in the context of their individual contributions to net accumulation of biofilm on a substratum. These processes, which include cell transport, adsorption, desorption, attachment, detachment, and cell growth, were defined for both laminar and turbulent flow conditions irrespective of flow path geometry. The presumption is that complex flow path geometry, although it may change the rate at which individual biofilm processes occur, does not alter their basic definition. However, in porous media the processes of attachment and detachment are complicated because of the potential for sloughing of biofilm fragments, which may be subsequently entrapped by downstream pore openings. Under these conditions, *filtration* must be considered as yet another biofilm process.

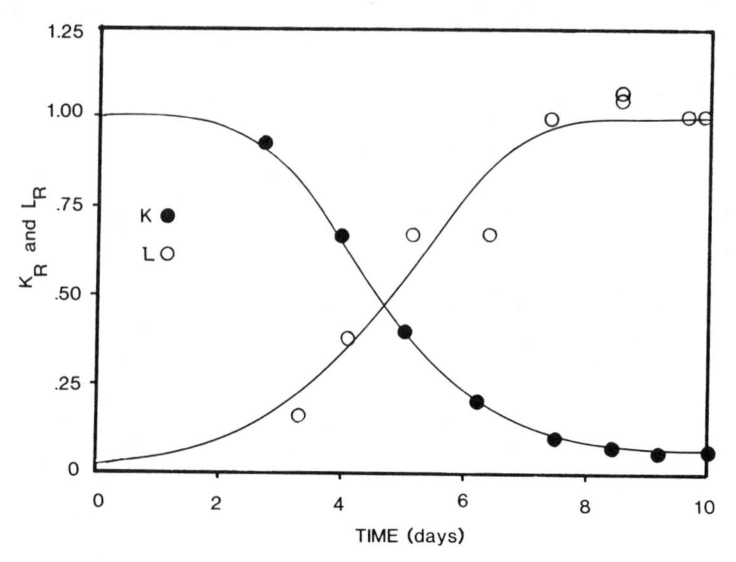

Figure 18.5 Relationship between biofilm thickness L_f and permeability K for *P. aeruginosa* biofilm accumulation on 1 mm glass spheres. L_R is the ratio of L_f to the maximum thickness, 75 μm. K_R is the ratio of K to the minimum permeability, 2.4×10^{-5} cm^2. (After Crawford, 1987.)

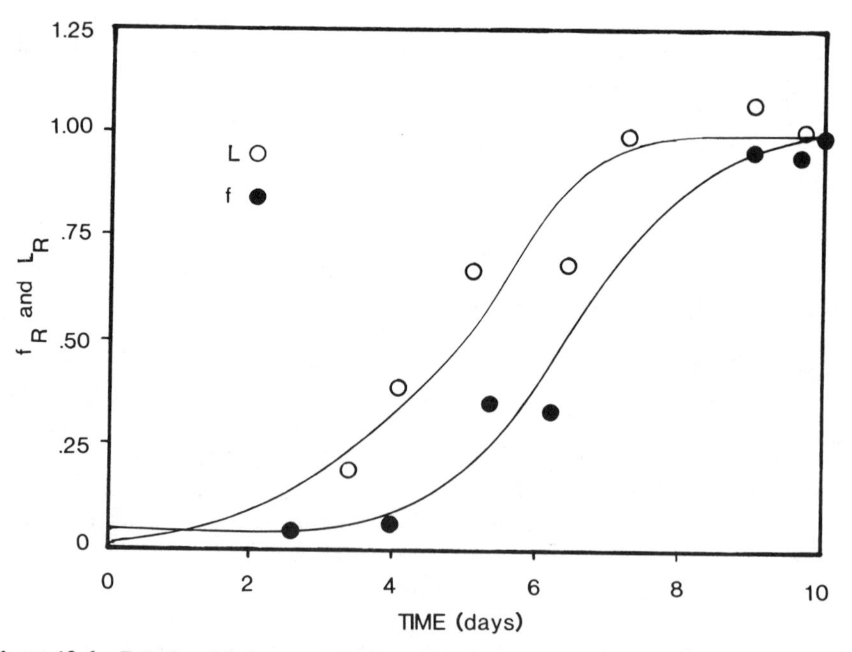

Figure 18.6 Relationship between biofilm thickness L_f and porous media friction factor f. f_R is the ratio of f to the maximum friction factor, 217. L_R is as defined in Figure 18.5.

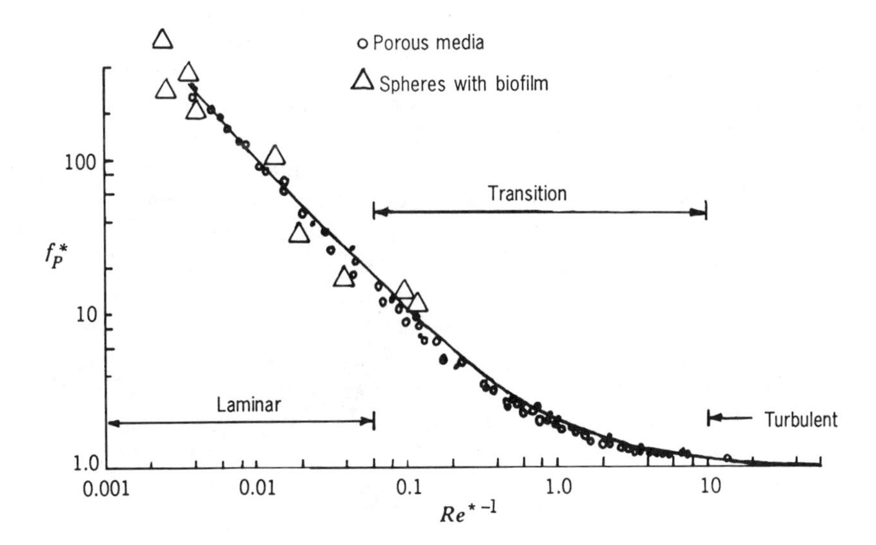

Figure 18.7 Porous medium friction factor f_p^* versus Reynolds number Re^* for a *P. aeruginosa* biofilm attached to 1 mm glass spheres. Data are compared with corresponding observations for clean porous media. (After Fahien, 1961.)

18.3.1 Transport

Filtration, which is defined as the removal of suspended material from flow passing through granular pore space, represents a variation of the attachment process unique to media in porous flow and, when active, can contribute substantially to net biofilm accumulation. The primary mechanism for biofilm filtration is mechanical straining, by which biofilm fragments sloughed at one location are entrapped by downstream pore spaces that are too small for them to pass through. Mathematical formulations to describe such particle deposition in porous media are discussed by Bouwer (1987).

Considering filtration as a separate process, the net rate of biofilm accumulation in porous medium flow can be written as

$$\text{accumulation} = \text{adsorption} + \text{attachment} + \text{filtration} + \text{growth} - \text{detachment} - \text{desorption} \quad [18.23]$$

The accumulation rate, as well as the rate of each contributing process, can be expressed in any consistent unit, such as biomass per unit time per unit volume of porous medium.

In porous media flow systems, biofilm accumulation will generally *decrease* with distance along the flow path especially if substrate concentration is depleted through the medium. In laboratory column experiments, such as reported by Raiders et al. (1986), maximum biofilm accumulation occurred immediately inside the column entrance and diminished rapidly within a few centimeters downstream. Similar observations have been reported from field investigations by Peterson and Oberdorfer (1985) with injection of treated wastewater, and van Beek (1984) with water well clogging. These laboratory and field observations illustrate the plug flow nature of biofilm accumulation in porous media (Figure 18.8). In a plug flow system, as discussed in Chapters 12 and 13, the concentration of substrate and oxygen (or other electron acceptors) will decrease in the downstream flow direction. Thus, cell growth will also diminish at some downstream point where substrate and/or oxygen become limiting to the growth process. The potential for attachment, along with sloughing and filtration, is likewise highest in upstream locations experiencing maximum cell growth rates.

Biofilm accumulation will likely be nonuniform or "patchy" along a porous medium flow path, particularly with low substrate concentration. This effect has been demonstrated by Cunningham et al. (1987), using a simulated porous media reactor (Figure 18.9). Experimental results (Figure 18.10) indicate the patchiness of a mixed culture biofilm. Here the major accumulations were ob-

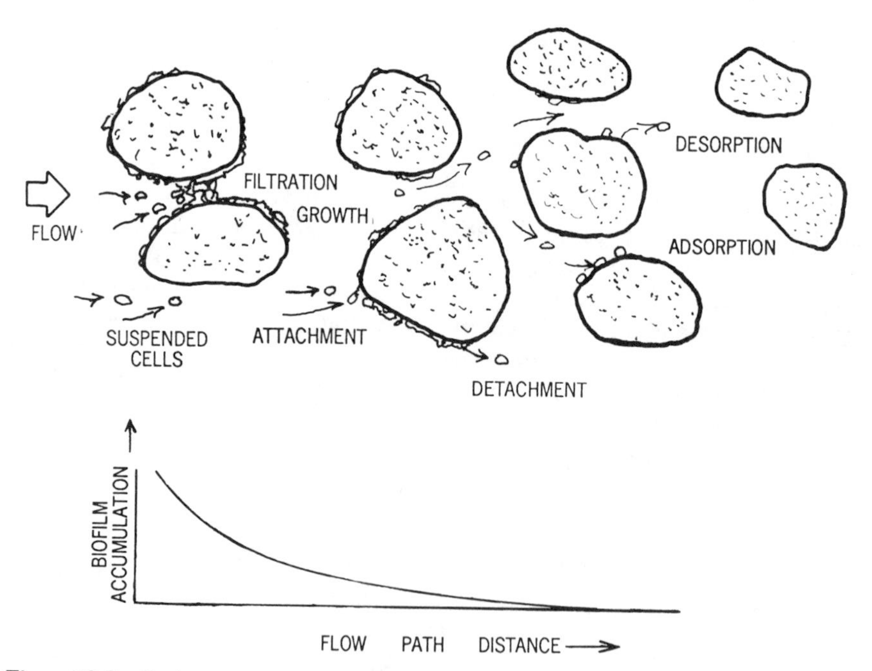

Figure 18.8 Various processes contribute to biofilm accumulation in saturated porous media, including adsorption, desorption, attachment, detachment, microbial growth, and filtration.

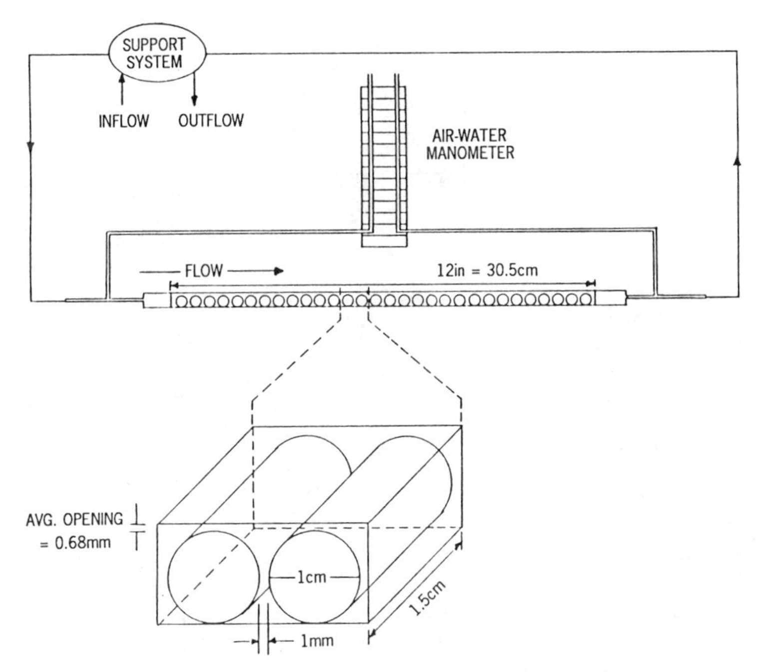

Figure 18.9 A schematic illustration of a continuous flow experimental system for simulating hydraulic resistance resulting from biofilm accumulation in porous media (Cunningham et al., 1987).

served on the upstream sides of the media particles. If accumulation increased indefinitely, the biofilm distribution would eventually form a continuous, albeit highly irregular, surface.

The independent processes that contribute to net accumulation of biofilm were investigated by Escher (1986), using continuous flow capillary tube reactors. Experiments were conducted with *Ps. aeruginosa* suspensions in various concentrations under laminar flow conditions. Image analysis permitted direct measurements of adsorption, desorption, and cell multiplication rates during early stages of colonization (Chapter 12). Typical observations (Figure 18.11a and b) illustrate the temporal behavior of each process.

Escher found that adsorption rates were a function of (1) suspended cell concentrations in the bulk fluid and (2) shear stress, and were independent of the biomass concentration at the substratum. Desorption rates were strongly dependent on adsorption rates and therefore were likewise governed by suspended cell concentration and shear stress. Growth-related processes, in contrast, depended only on microbial surface concentration. These experimental results provide considerable insight into the individual processes affecting net biofilm accumulation as well as the influence of environmental variables. However, additional experimentation involving tortuous flow paths is needed to better define the contributing biofilm processes in porous medium flow.

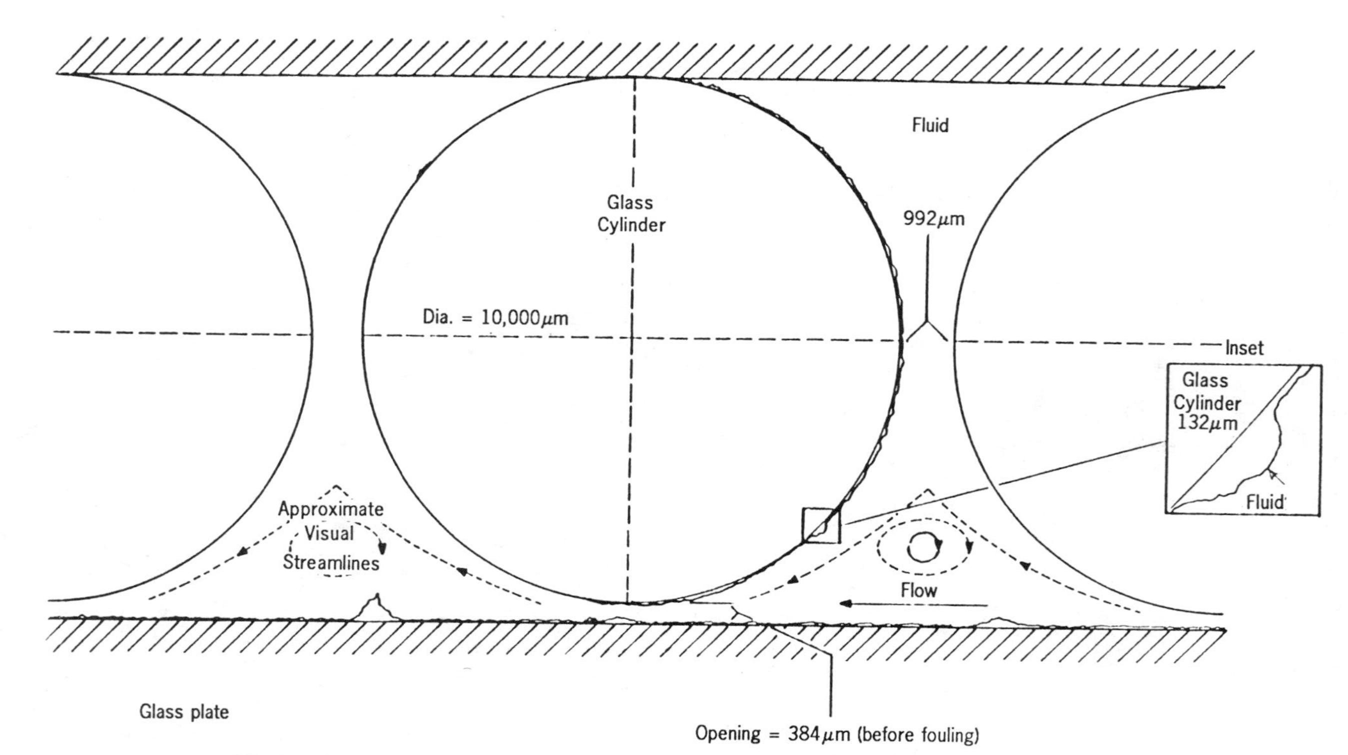

Figure 18.10 The spatial distribution of a mixed population biofilm accumulation in simulated porous media (Figure 18.9) is patchy (Cunningham et al., 1987).

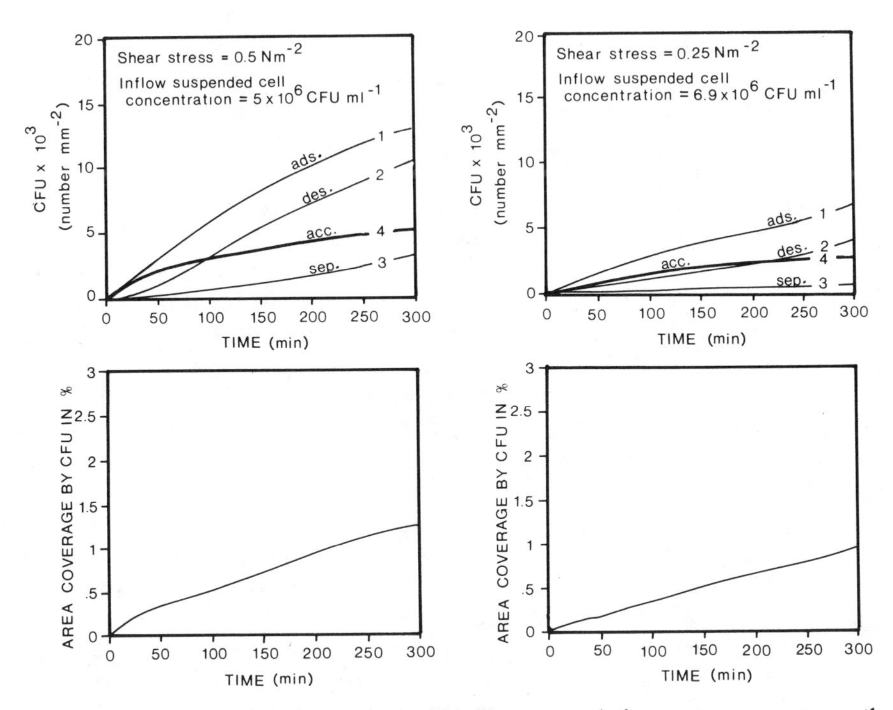

Figure 18.11 Cumulative observations of biofilm accumulation processes on a smooth substratum as a function of fluid shear stress and influent CFU (colony forming unit) concentration (Escher, 1986). The experimental system used (shown in Figure 18.4) permits the independent observation of CFU adsorption, desorption, and separation as well as the net accumulation of biofilm. 1 Adsorption, 2 desorption, 3 CFU separation, 4 net accumulation.

18.3.2 Biotransformation

In groundwater systems, many organic compounds are subject to transformation by microorganisms (Wilson et al., 1986a), and oxidation pathways for some chemical classes are fairly well known (Alexander, 1981). Biotransformation is one of the few processes that can bring about the destruction of organic contaminants in the subsurface. Light is not available for photolysis, water saturation reduces the opportunity for volatilization, and most abiotic transformations are slow. Using aseptic sampling techniques, diverse and active microbial populations have been measured in the subsurface, often at great depths (Wilson et al., 1983; Webster et al., 1985). Acridine orange direct counts of bacteria in aquifer material sampled at depths between 2 and 9 m ranged between 10^6 and 10^7 cells g^{-1} dry weight soil. ATP and tetrazolium dye reduction measurements indicated that between 1 and 10% of the total cell count was capable of respiration (Webster et al., 1985). The deep probe drilling project sponsored by the U.S. Department of Energy (1986) has identified considerable microbial activity at depths of

250 m. Highly transmissive strata tended to have greater active biomass due to greater influx of substrates and nutrients with the faster-moving groundwater. The diversity of biochemical reactions carried out by subsurface microorganisms appears similar to the capabilities of surface microorganisms.

Microorganisms growing as biofilms in the subsurface have an advantage over suspended species in that they can remain near the source of fresh substrate and nutrients contained in the groundwater that flows by them. The rate of biofilm growth, and thus the rate of biotransformation, is therefore strongly influenced by transport characteristics, including pore velocity distribution, dispersivity, molecular diffusivity, surface roughness, and other variables that affect the delivery rate of substrate and nutrients to the growing cells. Transformation rates are further affected if the biomass accumulation becomes so large as to *alter* the transport characteristics of the porous media. Thus, the processes of subsurface mass transport, biotransformation, and biofilm accumulation are highly interrelated.

Given the oligotrophic conditions in groundwater, subsurface microorganisms may need to utilize numerous different trace compounds to obtain sufficient energy to sustain themselves. The microorganisms are generally small (<1 μm) with high ratios of surface area to volume to aid in scavenging available substrates (Balkwill and Ghiorse, 1985). Growth rates can be extremely slow, with doubling times ranging from days to months. The oligotrophic conditions, combined with the high specific surface area of soil, favor microbial attachment to surfaces.

Modeling biomass accumulation for biofilms such as would be expected in an aquifer has been described in Chapters 11–13. In order for subsurface microorganisms to grow, sufficient substrate must be supplied to maintain a rate of growth that exceeds their rate of decay. Using Monod growth kinetics and first order decay, the net growth rate μ_{net} can be approximated by the following equation:

$$\mu_{net} = \underbrace{\frac{Yk_S}{K_S + S}}_{\text{growth}} - \underbrace{b}_{\text{decay}} \qquad [18.24]$$

where Y = yield ($M_X M_S^{-1}$)
k_S = maximum rate of substrate utilization ($M_S M_X^{-1} t^{-1}$)
S = substrate concentration (ML^{-3})
K_S = saturation coefficient (ML^{-3})
b = specific biomass decay coefficient (t^{-1})

In order to sustain a biofilm reaction in the long term, the substrate concentration must exceed a limiting value, S_{min}, where growth just balances decay:

$$S_{min} = \frac{bK_S}{Yk_S - b} \qquad [18.25]$$

If a single contaminant is present in an aquifer at a concentration below S_{min}, then there will be no net growth of microorganisms and the rate of contaminant biotransformation will not increase. Frequently, subsurface contaminants are present at concentrations that are in the ng/L and μg/L range, perhaps well below the S_{min}-values for microorganisms. Compounds present at trace concentrations may be consumed by microorganisms, however, if the population is supported with energy obtained from another substrate, i.e., the primary substrate, present in concentration above S_{min}. The trace organic contaminant is termed the secondary substrate, and the process is called secondary utilization.

18.3.2.1 Secondary Substrates The biofilm models described in Chapters 11–13 cannot be used for secondary substrates, since their utilization is inconsequential to biofilm growth. One approach to modeling trace contaminant utilization is to couple the biofilm mass, which is established by utilization of a primary substrate, with concentration and the intrinsic biodegradation kinetics of each secondary substrate. Laboratory studies with continuous flow columns containing glass beads and seeded with mixed bacterial cultures to simulate groundwater conditions have been conducted to evaluate the concept of secondary utilization and to determine the effect of electron acceptors on the biodegradation of a range of potentially hazardous organic chemicals commonly found in groundwater. The glass beads served as the substratum in aerobic studies; acetate at a concentration of 1 to 5 g m^{-3} was used as the primary substrate. The detention time in the column was about 0.33 h, and the column was seeded with settled domestic wastewater. Details of these aerobic column studies have been presented elsewhere (Bouwer and McCarty, 1982).

The time required for acclimation of the culture to the various trace organic substrates present and the fraction biologically transformed are presented in Table 18.2. With some, such as styrene and naphthalene, no acclimation period was required. With chlorinated benzenes, an acclimation time of a week or longer was required, depending upon the structure of the given compound. Carbon-14 tracer studies indicated these compounds were completely mineralized to carbon dioxide, i.e., no organic intermediates were found. However, the aerobic biofilm did not transform the halogenated aliphatic compounds after three years. The acetate primary substrate was sufficiently high in concentration to develop a deep biofilm, so that half-lives for the compounds were on the order of minutes.

Similar column systems were used for studies with anoxic biofilms under conditions of denitrification, sulfate respiration, and methanogenesis (Bouwer and Wright, 1988). The growth medium contained, as the primary substrate, 30 g m^{-3} acetate for denitrification and 250 g m^{-3} acetate for sulfate respiration and methanogenesis. The detention time in these anoxic columns was increased to 2.5 days. All of the halogenated aliphatic compounds studied were significantly transformed in an acetate-supported biofilm column under methanogenic conditions (Table 18.3). Many of these compounds were also transformed under sulfate-reducing conditions (Table 18.4). Less than two weeks of acclimation

TABLE 18.2 Trace Organics Removal in Aerobic Biofilm Column[a]

Compound	Acclimation Period (days)	Influent[b] (μg/L)	Removal[b] (%)
Chloroform	0	28 ± 4	0
Tetrachloroethylene	0	16 ± 3	0
1,1,1-Trichloroethane	0	10 ± 4	0
1,3-Dichlorobenzene	500	10 ± 2	71 ± 8
Chlorobenzene	20	10 ± 2	91 ± 3
1,2,4-Trichlorobenzene	40	9 ± 2	95 ± 3
1,2-Dichlorobenzene	20	10 ± 2	97 ± 1
1,4-Dichlorobenzene	10	10 ± 2	99 ± 1
Ethylbenzene	0	9 ± 2	>99
Naphthalene	0	14 ± 3	>99
Styrene	0	8 ± 1	>99

[a]20 min detention time.
[b]Mean ± 1 standard deviation.

TABLE 18.3 Removal of Trace Halogenated Compounds in Methanogenic Biofilm Column[a]

Compound	Acclimation Period (wk)	Influent[b] (μg/L)	Removal[b] (%)
Chlorinated benzenes	—	10 ± 2	0
Tetrachloroethylene	9–12	18 ± 3	86 ± 7
Chloroform		17 ± 4	95 ± 3
1,1,2,2-Tetrachloroethane		27 ± 1	97 ± 3
1,1,1-Trichloroethane		10 ± 2	>99
Carbon tetrachloride		7 ± 2	>99
Dibromochloropropane		17 ± 2	>99
Bromodichloromethane	2	13 ± 1	89 ± 8
Dibromochloromethane		34 ± 5	>99
Bromoform		18 ± 3	>99
Ethylene dibromide		20 ± 4	>99
Hexachloroethane		4 ± 1	>99

[a]2.5 day detention time.
[b]Mean ± 1 standard deviation.

time was required for onset of degradation for most of the trace organics. 1,1,1-Trichloroethane appeared more resistant and required ten months of acclimation time. A few of the halogenated aliphatic compounds (carbon tetrachloride, bromoform, bromodichloromethane, and hexachloroethane) were significantly transformed in the presence of an active denitrifying biofilm (Table 18.5). Sorption onto biomass may have been responsible for the minimal removals of 1,1,1-trichloroethane, ethylene dibromide, and dibromochloropropane.

TABLE 18.4 Removal of Trace Halogenated Compounds in Denitrifying Biofilm Column[a]

Compound	Acclimation Period (wk)	Influent[b] (μg/L)	Removal[b] (%)
Chloroform	—	28 ± 2	−13 ± 16
Chlorinated benzenes	—	10 ± 2	0
Tetrachloroethylene	—	7 ± 2	0 ± 40
Dibromochloropropane	—	37 ± 3	14 ± 11
Ethylene dibromide	—	30 ± 2	23 ± 7
1,1,1-Trichloroethane	—	27 ± 3	30 ± 7
Hexachloroethane	<2	4 ± 2	80 ± 17
Bromodichloromethane	<2	24 ± 2	90 ± 8
Bromoform	<2	36 ± 2	97 ± 3
Carbon tetrachloride	<2	17 ± 2	>99

[a]2.5 day detention time.
[b]Mean ± 1 standard deviation.

TABLE 18.5 Estimated Half-Lives for Trace Organic Substrates as a Function of Active Organism Concentration and Electron Acceptor

			Half-Life (days)		
Electron Acceptor	Trace Organics	Active biomass =	0.001	0.01	0.1 mg/L
O_2	Chlorobenzenes		280–60	28–6	3–0.6
NO_3^-	Haloaliphatics		350,000–1900	35,000–190	3500–19
SO_4^{-2}	Haloaliphatics		140,000–1000	14,000–100	1400–10
CO_2	Haloaliphatics		7000–290	700–29	70–3

Tetrachloroethylene appeared to persist under denitrification and sulfate respiration. Carbon tetrachloride transformation to chloroform, reported in batch studies (Bouwer and McCarty, 1983b), most likely explains the increase in chloroform concentration across the denitrifying and sulfate-reducing biofilm columns. The chlorinated benzenes that were degradable under aerobic conditions persisted under anoxic conditions.

Evidence was obtained for partial mineralization of chloroform, carbon tetrachloride, and ethylene dibromide to carbon dioxide using carbon-14 radiotracers (Bouwer and McCarty, 1983a). In contrast, transformation of tetrachloroethylene, 1,1,2,2-tetrachloroethane, bromoform, carbon tetrachloride, and ethylene dibromide to trichloroethylene, 1,1,2-trichloroethane, dibromomethane, chloroform, and ethylene, respectively, were evidence for reductive dehalogenation. The halogenated organic compounds served as electron acceptors, i.e., a halogen atom was removed and substituted with a hydrogen atom.

Reduction of halogenated aliphatic and aromatic contaminants is common in reducing subsurface environments. For example, tetrachloroethylene was converted to trichloroethylene, dichloroethylene, and vinyl chloride under methanogenic conditions (Vogel and McCarty, 1985). Reductive dechlorination of trichloroethylene occurred in anaerobic soil with production of 1,2-dichloroethylene (Kloepfer et al., 1985). Many chlorinated methane, ethane, and ethene compounds were reductively dehalogenated in microcosms containing mud from an aquifer recharge basin (Parsons et al., 1984). Microbial dehalogenation of haloaromatic substrates occurred in methanogenic environments (Suflita and Gibson, 1985). Complete dehalogenation of the aryl halides was required before substrates could be completely mineralized. 1,1,1-Trichloroethane can be converted chemically into 1,1-dichloroethylene and biologically into 1,1-dichloroethane and chloroethane (Vogel and McCarty, 1987).

The removals observed in the laboratory scale biofilm columns can provide valuable insight into relative rates of biotransformation in the subsurface environment (Bouwer and McCarty, 1985). At low substrate concentration, the rate of utilization approaches first order with respect to substrate concentration. For the case of low active organism concentration X and slow groundwater movement, an expression to estimate the half-life $t_{1/2}$ for an organic micropollutant is

$$t_{1/2} = \frac{\ln 2}{(k_S/K_S)X} \qquad [18.26]$$

The relationship between half-life and active biomass concentration in the presence of oxygen (aerobic respiration), nitrate (denitrification), sulfate (sulfate reduction), and carbon dioxide (methanogenesis) as electron acceptors is summarized in Table 18.5. The results indicate that biotransformation could be responsible for significant loss of organic micropollutants in subsurface environments, even with the very low indigenous microbial concentrations (0.001–0.01 g m^{-3}) that have been reported. Stimulation of growth to achieve an active biomass concentration of 0.1 g m^{-3}, perhaps by addition of suitable substrate and nutrients, may result in half-lives between days and months. The rate of degradation is markedly affected by the electron acceptor available. Other factors affecting the rate of biotransformation are pH, temperature, ionic strength, soil moisture content, presence of nutrients (such as nitrogen, phosphorus, calcium, and magnesium), presence of trace elements (such as iron, manganese, and cobalt), and presence of toxic components that may significantly inhibit microbial activity. Greater persistence for biodegradable contaminants may indicate that proper environmental conditions are not present or that the fraction of bacteria capable of a given transformation is not high.

18.3.2.2 Modeling Competitive and Sequential Reactions The laboratory data also indicate that the redox environment is an important factor affecting rates of biotransformation. The aromatic compounds degraded only when oxygen was available. The halogenated aliphatics were susceptible to transformation as conditions became more reducing. A variety of inorganic and organic compounds that could serve as energy substrates for microorganisms are generally present in the subsurface. Therefore, the spatial distribution of electron acceptors identifies regions favoring biotransformation of a particular organic micropollutant.

Microbial populations appear to preferentially utilize electron acceptors that provide the maximum free energy during respiration. Thus, the following four electron acceptors are used sequentially to oxidize an organic electron donor: oxygen, nitrate, sulfate, and carbon dioxide. Oxygen provides the most free energy to microorganisms during electron transfer, and carbon dioxide provides the least. Under aerobic conditions, there is competition between heterotrophs deriving energy from oxidation of reduced organic carbon and nitrifiers deriving energy from the oxidation of ammonia. As oxygen is depleted, anoxic conditions follow, with the sequential utilization of nitrate (denitrification), sulfate (sulfate respiration), and carbon dioxide (methanogenesis) as electron acceptors by heterotrophs.

The multiple electron acceptor biofilm model of Bouwer and Cobb (1987) can be used to simulate the microbial succession that would occur in groundwater with a mixture of primary substrates and electron acceptors. This model considers competitive and sequential microbiological processes of aerobic carbonaceous oxidation, nitrification, denitrification, sulfate respiration, and methanogenesis. By establishing the dominant electron acceptor condition in the direction of groundwater flow, the model simulates the secondary utilization of organic pollutants in redox regions favorable to their biotransformation.

A model simulation depicting movement of water containing organic carbon, ammonia, oxygen, nitrate, and sulfate through sandy soil appears in Figure 18.12. The various microbiological reactions can occur over a short distance in the soil. The capacity for aerobic oxidation of contaminants in groundwater is rather limited because reaeration is not possible. The possible generation of sulfides during sulfate reduction could be important in precipitating metal sulfides, thus immobilizing toxic metals, such as cadmium, chromium, and lead. In less polluted soil, the microbial reactions could extend over much greater distances, the dominant processes being aerobic carbonaceous oxidation, nitrification, and denitrification.

Model simulations of secondary utilization provided reasonable fits to measured data in a laboratory scale biofilm column reactor fed acetate as primary substrate, oxygen, nitrate, sulfate, and trace halogenated compounds. The influence of carbon tetrachloride on biomass accumulation, predicted from the steady state utilization of acetate under each electron acceptor condition, was coupled with rate parameters for carbon tetrachloride to obtain a model profile (Figure 18.13). A multiple solute biofilm model is necessary for assessing treatment strategies to clean and restore contaminated subsurface environments by biotransformation.

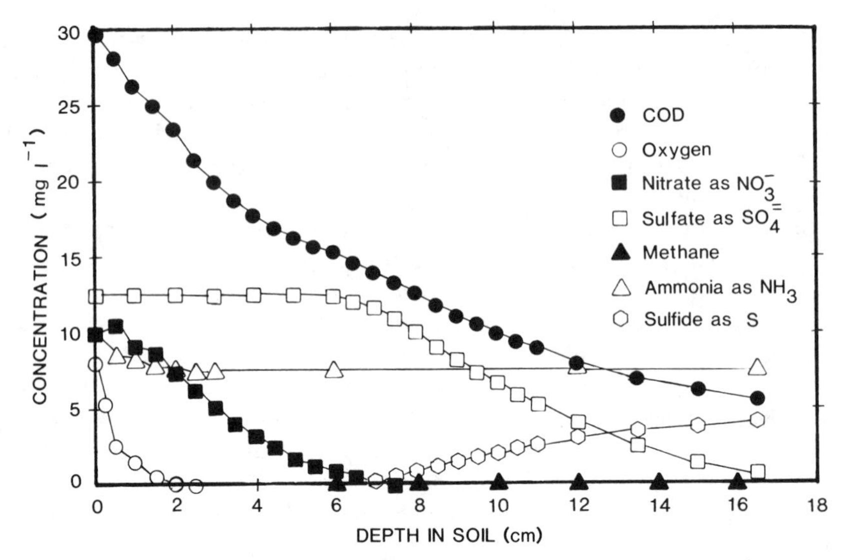

Figure 18.12 Model simulation of microbially mediated changes in chemical species with depth in soil (particle size 0.5 mm; Darcy velocity 25 cm day^{-1}).

18.4 FOULING OF POROUS MEDIA

Fouling is the accumulation of biological deposits or films on equipment surfaces that reduces performance and/or lifetime (Chapter 14). For flow in porous media, excessive accumulation of biofilm can reduce the effective pore area and, as a result, decrease the fluid transport capacity of the media. Mechanisms of porous medium fouling and various related measurement techniques will be presented in the context of laboratory and field investigations.

18.4.1 Fouling Mechanism

In flow systems experiencing both laminar and turbulent flow, fouling can occur due to (1) increased frictional resistance, (2) decreased flow path area, or both. If laminar flow prevails, frictional resistance becomes independent of surface roughness and fouling is due only to pore size reduction (clogging) resulting from biofilm accumulation. This effect has been demonstrated by Cunningham et al. (1987) in laboratory experiments using a simulated porous medium (Figure 18.9). The decrease in permeability was paralleled by a decrease in equivalent diameter of the pore openings (Figure 18.14). The permeability was a linear function of the equivalent diameter, so increased surface roughness due to biofilm accumulation was insignificant.

If the biofilm thickness remains small compared to the effective pore space, accumulation of biofilm will not affect the distribution of pore velocities within the porous medium. However, if biofilm occupies a significant fraction of the

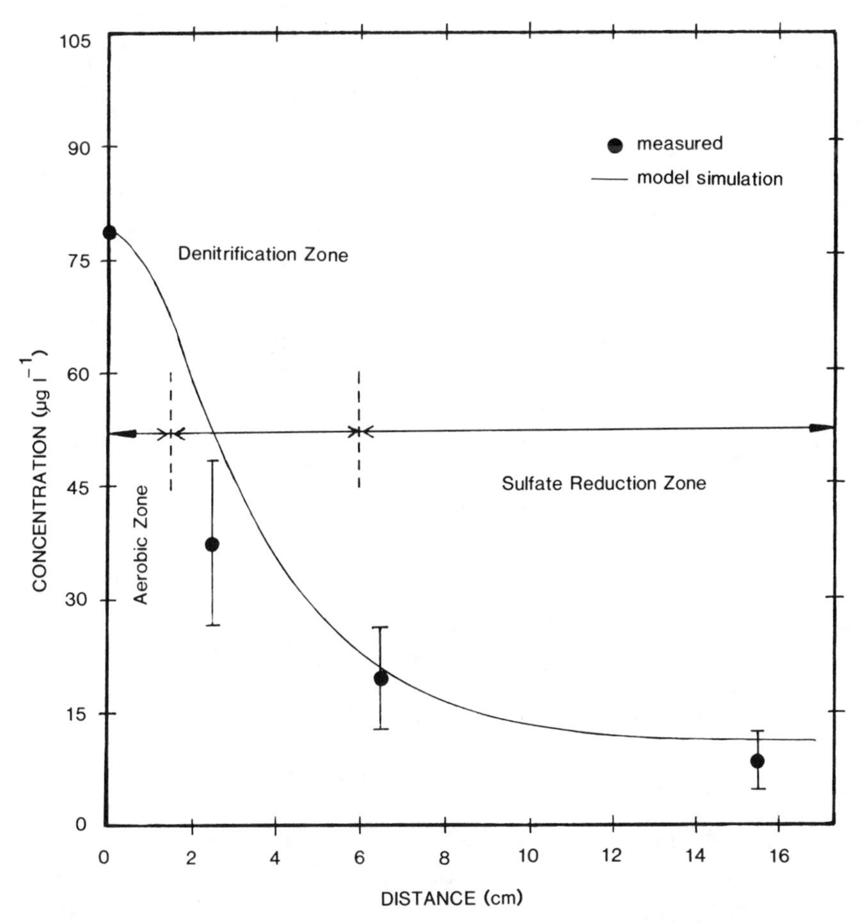

Figure 18.13 Model simulation of carbon tetrachloride transformation with depth in a laboratory scale biofilm column.

effective pore space, decreases in effective porosity and permeability of the formation will occur. Reduced pore velocities will affect biofilm accumulation and, depending on the magnitude of individual process rates, the net accumulation rate may subsequently increase or decrease. Thus, a *coupled relationship* exists between biofilm accumulation and porous media flow dynamics when a significant reduction in pore size occurs. An example of this phenomenon is reported by Torbati et al. (1986), based on a laboratory investigation of sandstone cores. *In situ* accumulation of microorganisms in Brea sandstone was observed to preferentially plug the large (59–69 μm) pore entrances (Figure 18.15). Because substrate flux was initially greater inside the large pore spaces, higher cell growth rates occurred, resulting in a significant reduction in pore space. Biofilm accumulation also resulted in a more homogeneous distribution of pore sizes (i.e., smaller peaks and valleys) than in the clean sandstone.

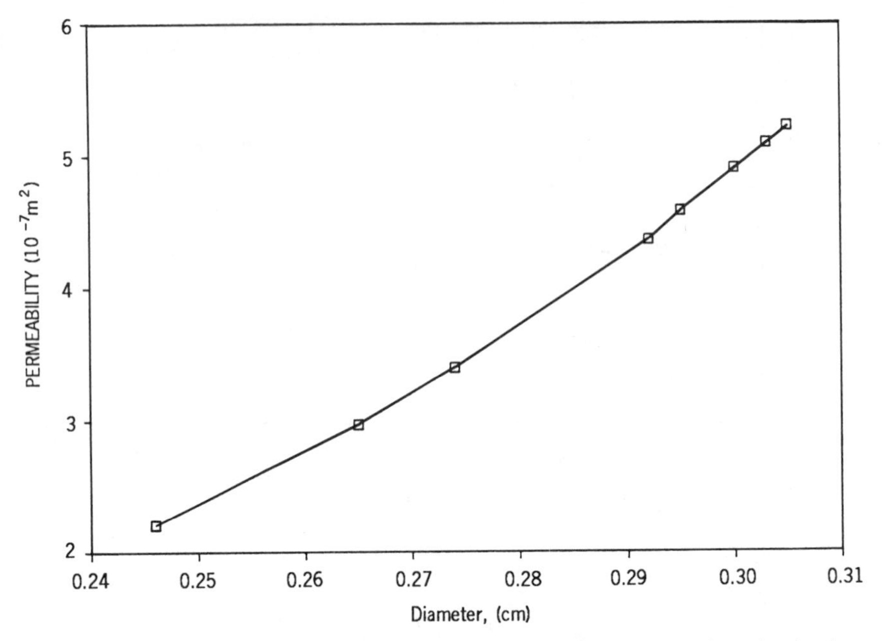

Figure 18.14 The relationship of permeability to equivalent diameter for mixed culture biofilm accumulation in a simulated porous medium reactor. The relationship indicates that reduced permeability is due to reduced pore space. (Crawford, 1987.)

18.4.2 Direct Measurement

Direct assessment of porous media fouling requires measurement of the reduction in effective pore space due to biofilm accumulation. However, methods for making *in situ* measurements of film thickness and density along tortuous flow paths are in the early stages of development. One such method has been reported by Bakke (1986), based on experiments with *Ps. aeruginosa* biofilms formed in horizontal capillary tubes under laminar flow conditions. Bakke measured the biofilm optical thickness and density *in situ* and nondestructively at various locations along the reactor. The optical thickness was subsequently translated into mechanical thickness using a geometric analysis of the microscope light path through the sample and the refractive indices of the sample and surrounding medium (Bakke and Ollson, 1986). The time progression of the biofilm thickness and spatial profiles within the reactor were also obtained by this method (Bakke, 1986). Similar optical methods for determining film thickness were used by Crawford (1987) to measure the biofilm thickness in a laboratory reactor (Figure 18.5).

18.4.3 Indirect Measurement

Although direct measurement of fouling deposits in porous media is difficult, more convenient methods can measure the extent of fouling. Consider the sys-

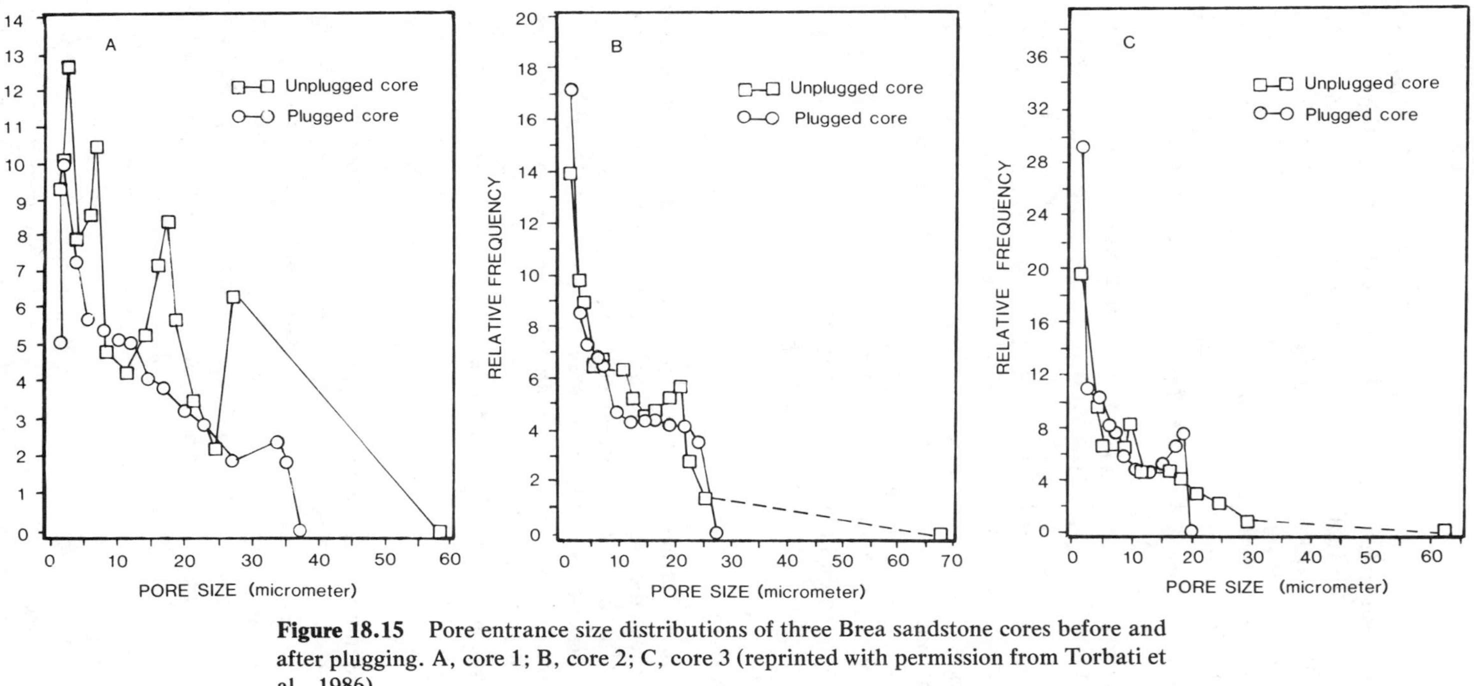

Figure 18.15 Pore entrance size distributions of three Brea sandstone cores before and after plugging. A, core 1; B, core 2; C, core 3 (reprinted with permission from Torbati et al., 1986).

tem described in Figure 18.1 (the basis for Eq. 18.3). If K is reduced due to fouling, a corresponding drop in v will result if dh/dL remains constant. Likewise, if v is held constant, a reduced K will result in an increase in dh/dL. In both laboratory and field investigations involving fouling of porous media, v and dh/dL are usually the variables that are measured directly. Thus, K can be computed from Eq. 18.3 and converted to k using Eq. 18.5. So the degree of fouling can be determined indirectly by measuring the specific discharge and pressure gradient and computing the corresponding reduction in permeability of the medium.

Investigations documenting permeability reduction resulting from biofilm accumulation have been reported by practitioners in petroleum recovery, disposal of treated wastewater, and recharge/recovery for water supply (Shaw et al., 1985; Raiders et al., 1986; Raleigh and Flock, 1965; Hart et al., 1960; Cerini et al., 1946; Jenneman et al., 1985; Okubo and Matsumoto, 1983). The reports cover a variety of porous media, including core samples from field sites as well as synthetic porous media (e.g., glass spheres). Pure cultures as well as mixed cultures were used as inoculum. Substantial reductions in permeability (65–99%) occurred in all cases. Raiders et al. (1986) report decreased permeability due to biofilm accumulation in effective pore space, in which extracellular polymeric substances play an important role in plugging.

18.4.4 Well Fouling

Fouling in porous formations, particularly in the vicinity of a well bore, can substantially reduce the performance of both discharge and injection wells. Injection wells are particularly sensitive to clogging because they can filter suspended solids from the water. Fouling due to accumulation of biofilm incrustation and corrosion products are common in both well types. The flow in the adjacent formation is *radial* for both injection and discharge wells. As a consequence, pore velocities, and hence nutrient supply rates, are greatest at the well casing and decrease with distance from the well. Thus, the greatest potential for fouling exists at the well screen and diminishes as the square of the distance away from the well.

If the groundwater contains Fe(II), the potential exists for fouling to occur inside the well screen as well as outside in the gravel pack and surrounding formation. The Fe(II) is readily oxidized to Fe(III) in the presence of oxygen as the groundwater is withdrawn. The conversion rate of Fe(II) to Fe(III) is accelerated by iron bacteria such as *Gallionella* and *Leptothrix*. As water is withdrawn from such a well, radial flow within the aquifer toward the well screen results in an increased nutrient supply to bacteria residing on or near the well screen. Under such conditions massive bacterial growth and production of insoluble Fe(III) complexes occurs, resulting in substantially decreased formation permeability. Figure 18.16a shows the biomass accumulation observable on a relief well screen in northern Mississippi. Wells in this region are extremely prone to iron bacterial fouling and are treated regularly with sodium hypochlorite and acid to main-

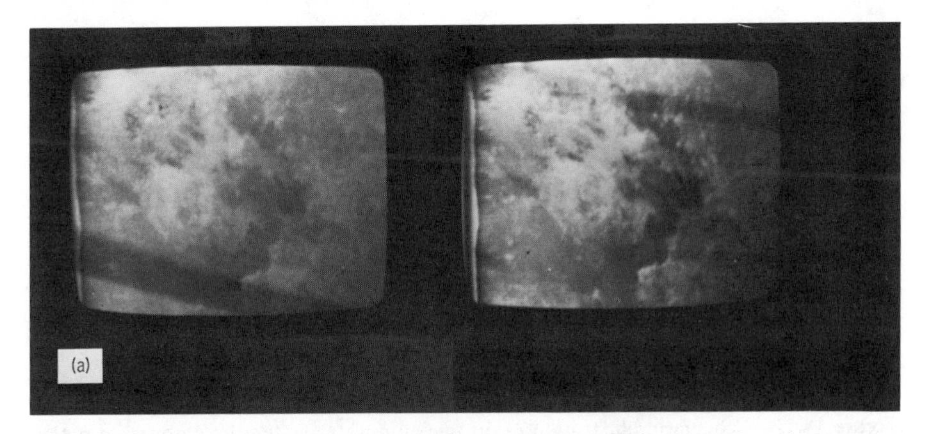

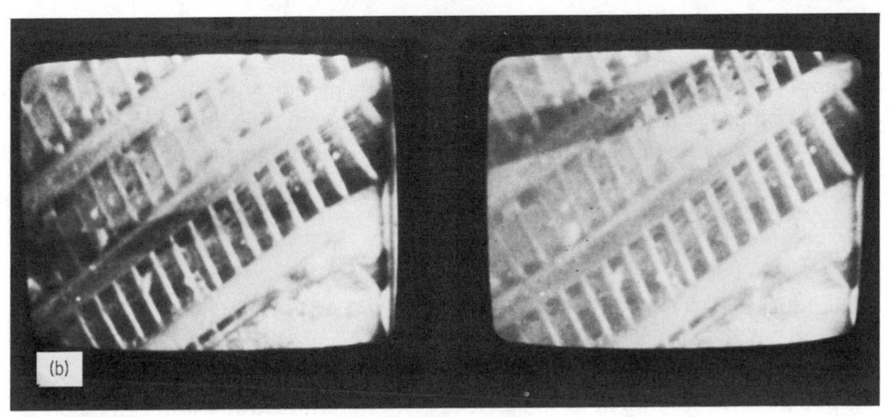

Figure 18.16 Photograph of bacterial fouling on a relief well screen before (a) and after (b) cleaning obtained with a down-hole video camera and monitor.

tain acceptable performance. Figure 18.16b illustrates the degree of biomass removal achieved by such treatment.

Biofilm accumulation in porous media is of great concern in petroleum recovery operations that involve injection of water into oil-bearing strata to flush out oil residing in the smaller pore spaces. During such operations, fouling due to filtration of suspended solids in the injection water may occur on the interior of the injection well screen. Fouling may be enhanced by additional microbial accumulation in the formation surrounding the well casing, thereby severely inhibiting the flow of injection fluids. However, if such fouling can be avoided, the controlled accumulation of biofilm can serve a useful purpose in the oil recovery process. The injected water migrates toward the recovery well(s) through the various strata constituting the oil bearing formation (Figure 18.17). Since fluid velocities are higher in the more permeable strata, the oil in these zones will be flushed first. Subsequently, deliberate plugging of the high permeability zone (to prevent injection water reaching the production well) is valuable and can be

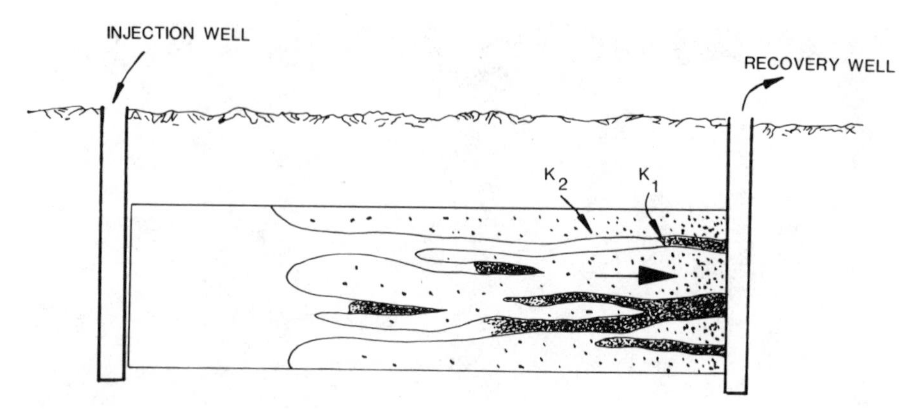

Figure 18.17 Schematic illustration of an injection water system for improving petroleum recovery.

accomplished by encouraging microbial growth in this zone by injecting cells and nutrients. If the plugging is successful, additional injection of fluid will facilitate removal of hydrocarbons from within surrounding zones of lower permeability. Laboratory results documenting preferential plugging of sandstone cores are reported by Torbati et al. (1986).

18.5 REMEDIATING SUBSURFACE CONTAMINATION

The injection of physiologically capable microorganisms for biotransformation has potential for treatment of subsurface contamination *in situ* and can result in contaminant removal, rather than transfer to another environmental medium as with land disposal, air stripping, or activated carbon treatment. Biorestoration schemes that have been proposed (see Figure 18.18) include: (1) stimulation of native bacteria by satisfying their needs for energy source, electron acceptor, and other growth factors to enhance transformation of contaminants, (2) addition of acclimated microorganisms obtained by enrichment culture or genetic manipulation, and (3) development of a localized zone of biological activity to intercept the groundwater flow and cleanse the water as it flows by the microorganisms (Wilson et al., 1986b). Substrate removal rates are important in determining whether or not a biodegradable secondary substrate can be removed biologically from groundwater. If the removal rates of a particular compound are too slow, it may pass through a biologically active zone with little or no removal. A great deal of development work on *in situ* biorestoration is needed before this can become a reliable mitigation process.

Evidence for subsurface biotransformation of trace halogenated organic compounds was obtained at the ground water recharge project in the Palo Alto, California baylands where reclaimed municipal wastewater was injected into a confined aquifer (Roberts et al., 1982). Within 50 days after stopping injection, the

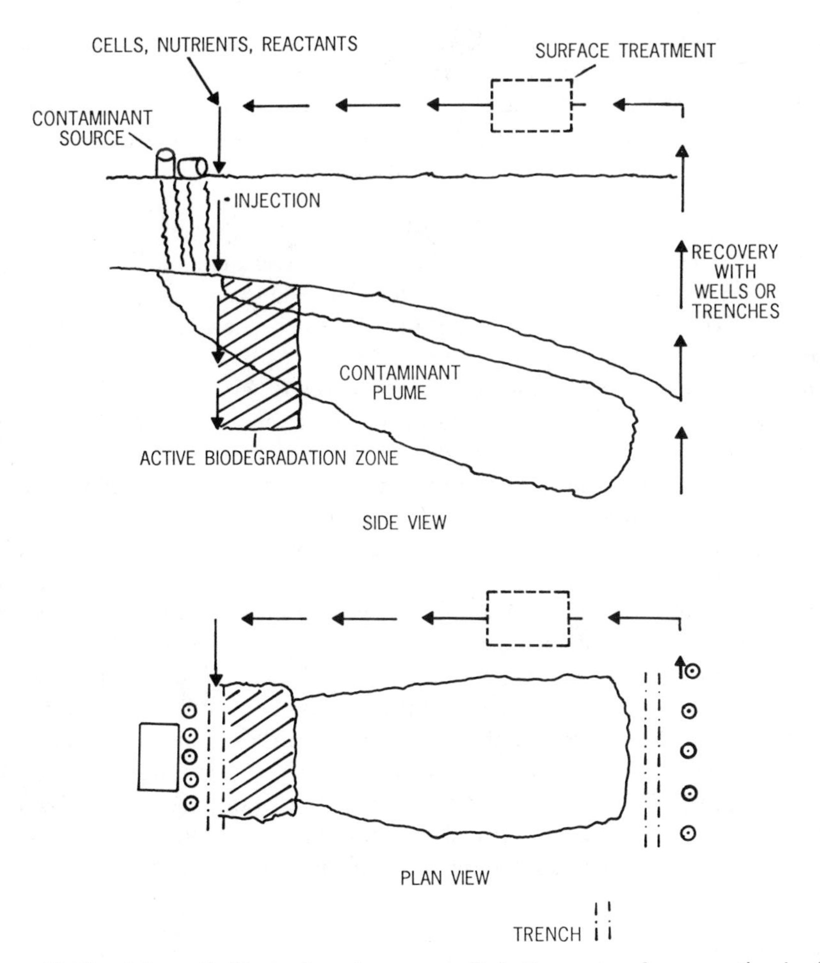

Figure 18.18 Schematic illustration of a recovery/injection system for promoting *in situ* biodegradation of groundwater contaminants.

concentrations of chloroform and other trihalomethanes at a nearby observation well were found to decrease significantly, yielding half-lives of 3–6 weeks. A much slower decline occurred in the concentrations of the chlorinated ethanes and ethenes, yielding half-lives of 5 to 9 months. Several halogenated aliphatic compounds were apparently degraded during soil percolation in the anoxic subsurface beneath wastewater infiltration basins in Phoenix, Arizona (Bouwer et al., 1984). Dune infiltration for more than 20 years in the Netherlands (Piet and Zoeteman, 1980) has resulted in no detectable haloforms after percolation through dune sand despite relatively high haloform concentrations (20–100 μg/L) in the influent water. The removal is attributed to biotransformation under anoxic conditions. Enhanced degradation of gasoline components occurred when oxygen, nitrogen, phosphorus, and other trace elements were added to aquifers (Raymond et al., 1975). Evidence for 1,4-dichlorobenzene degradation during infiltration from the River Glatt was found under aerobic conditions with an apparent half-life of 8 days (Schwarzenbach et al., 1983). Subsurface biological activity caused the aquifer to go anoxic in the summer months, and during this period 1,4-dichlorobenzene was observed to persist.

A complex set of transformations was observed for trace organics in a sanitary landfill leachate plume (Barker et al., 1986). 1,1,1-Trichloroethane and trichloroethylene were restricted to the immediate vicinity of the landfill, presumably by biotransformation. o-Xylene was rapidly lost from this system. In the strictly anaerobic segment of the plume, 1,2,4-trichlorobenzene and 1,4-dichlorobenzene were equally persistent, but downgradient in the less anaerobic region, the former appeared to be degraded more rapidly than the latter. More than 80% conversion of nine of ten halobenzoates, five of seven chlorophenols, and two phenoxyacetic acid herbicides was observed in methanogenic aquifer samples (Suflita and Gibson, 1985). These transformations were not observed in sulfate-reducing aquifers. Degradation of a number of alkylbenzenes in methanogenic aquifer material after long lag periods has been reported (Wilson et al., 1986b). Up to 40 weeks was needed for benzene, ethylbenzene, and o-xylene degradation, and 6 weeks for toluene degradation. Halogenated aliphatics, such as trichloroethylene, were also transformed with lag periods of a few weeks. Subsurface microorganisms isolated from aquifer material and groundwater collected at an abandoned creosoting site in Conroe, Texas could utilize anthracene, dibenzofuran, fluorene, and naphthalene as a sole carbon source (Borden et al., 1986). Hexachloroethane rapidly disappeared when it was introduced into an unconfined sand aquifer with a half-life of 40 days (Criddle et al., 1986). Laboratory scale studies confirmed biotransformation of hexachloroethane to tetrachloroethylene in microcosms containing unhomogenized aquifer material.

REFERENCES

Alexander, M., *Science*, **211**, 132–138 (1981).

Bakke, R., "Biofilm detachment," unpublished doctoral dissertation, Montana State University, Bozeman, Montana, 1986.

Bakke, R. and P.Q. Ollson, *J. Microbiol. Meth.*, **5**, 93–98 (1986).

Balkwill, D.L. and W.C. Ghiorse, *Appl. Environ. Microbiol.* **50**, 580–588 (1985).

Barker, J.F., J.S. Tessmann, P.E. Plotz, and M. Reinhard, *J. Contaminant Hydrology*, **1**, 171–189 (1986).

Bear, J., *Hydraulics of Groundwater*, McGraw-Hill, New York, 1979.

Bird, B.R., W.E. Stewart, and E.N. Lightfoot, *Transport Phenomena*, Wiley, New York, 1960.

Borden, R.C., P.B. Bedient, M.D. Lee, C.H. Ward, and J.T. Wilson, *Wat. Resources Res.*, **22**, 1983–1990 (1986).

Bouwer, E.J., *Wat. Res.*, **21**, 1489–1498 (1987).

Bouwer, E.J. and G.D. Cobb, *Wat. Sci. Tech.*, **19**, 769–779 (1987).

Bouwer, E.J. and P.L. McCarty, *Environ. Sci. Technol.*, **16**, 836–843 (1982).

Bouwer, E.J. and P.L. McCarty, *Appl. Environ. Microbiol.*, **45**, 1286–1294 (1983a).

Bouwer, E.J. and P.L. McCarty, *Appl. Environ. Microbiol.*, **45**, 1295–1299 (1983b).

Bouwer, E.J. and P.L. McCarty, *Biotech. Bioeng.*, **27**, 1564–1571 (1985).

Bouwer, E.J. and J.P. Wright, *J. Contaminant Hydrology*, **2**, 155–169 (1988).

Bouwer, E.J., P.L. McCarty, H. Bouwer, and R.C. Rice, *Wat. Res.*, **18**, 463–472 (1984).

Bouwer, H., *Groundwater Hydrology*, McGraw-Hill, New York, 1978.

Cerini, W.G., W.R. Battles, and P.H. Jones, *Petroleum Technology*, TIP 2028, (1946).

Crawford, D.J., "Hydraulic effects of biofilm accumulation in simulated porous media flow systems," unpublished master's thesis, Montana State University, Bozeman, MT, 1987.

Criddle, C.S., P.L. McCarty, M.C. Elliott, and J.F. Barker, *J. Contaminant Hydrology*, **1**, 133–142 (1986).

Cunningham, A.B., W.G. Characklis, and D. Crawford, "Modeling microbial fouling in porous media," in *Proc. National Water Well Conference*, Denver, CO, 1987.

Escher, A.R., "Bacterial colonization of a smooth surface: An analysis with image analyzer," unpublished doctoral dissertation, Montana State University, Bozeman, MT, 1986.

Fahien, R.W., *Fundamentals of Transport Phenomena*, McGraw-Hill, New York, 1983, pp. 459–465.

Fahien, R.W. and C.B. Schriver, "The effect of porosity and transition flow on pressure drop in packed beds," presented at AIChE Conference, Denver, CO, 1961.

Freeze, R.A. and J.A. Cherry, *Groundwater*, Prentice-Hall, Englewood Cliffs, NJ, 1979.

Hart, R.T., T. Fekete, and D.L. Flock, *Can. Mining Metallurgical Bull.*, **63**, 495–501 (1960).

Jenneman, G.E., M.J. McInerney, and R.M. Knapp, *Appl. Environ. Microbiol.*, **50**, 383–391 (1985).

Kloepfer, R.D., D.M. Easley, B.B. Haas, T.G. Delhi, D.E. Jackson, and C.J. Wuney, *Environ. Sci. Tech.*, **19**, 277–280 (1985).

Metcalf and Eddy, Inc. (G. Tchobanoglous), *Wastewater Engineering Treatment Disposal*, McGraw-Hill, New York, 1979, p. 242.

Okubo, T. and J. Matsumoto, *Wat. Res.*, **17**, 813–821 (1983).

Parsons, F., P.R. Wood, and J. DeMarco, *J. Am. Wat. Works Assoc.*, **76**, 56–59 (1984).

Peterson, F.L. and J.A. Oberdorfer, *Pacific Sci.*, **39**, 230–240 (1985).

Piet, G.J. and B.C.J. Zoeteman, *J. Am. Wat. Works Assoc.*, **72**, 400–404 (1980).

Raiders, R.A., M.J. McInerney, D.E. Revus, H.M. Torbati, R.M. Knapp, and G.E. Jenneman, *J. Ind. Microbiol.*, **1**, 195–203 (1986).

Raleigh, J.T. and D.L. Flock, *Petroleum Trans. Pet. and Nat. Gas Div.*, **68**, 201–206 (1965).

Raymond, R.L., J.O. Hudson, and V.W. Jamison, *Appl. Environ. Microbiol.*, **31**, 522–535 (1975).

Roberts, P.V., J. Schreiner, and G.D. Hopkins, *Water Res.*, **16**, 1025–1035 (1982).

Schwarzenbach, R.P., W. Giger, E. Hoehn, and J.K. Schneider, *Environ. Sci. Tech.*, **17**, 472-479 (1983).

Shaw, J.C., B. Bramhill, N.C. Wardlaw, and J.W. Costerton, *Appl. Env. Microbiol.*, **49**, 693-701 (1985).

Suflita, J.M. and S.A. Gibson, "Biodegradation of haloaromatic substrates in a shallow anoxic ground water aquifer," in N.N. Durham and A.E. Redelfs (eds.), *Proc. Second International Conference on Groundwater Quality Research*, 1985, pp. 30-32.

Tchobanoglous, G., "A study of the filtration of treated sewage effluent," unpublished doctoral dissertation, Stanford University, Palo Alto, CA, 1969.

Todd, D.K., *Groundwater Hydrology*, 2nd ed., Wiley, New York, 1980.

Torbati, H.M., R.A. Raiders, E.C. Donaldson, M.J. McInerney, G.E. Jenneman, and R.M. Knapp, *J. Ind. Microbiol.*, **1**, 227-234 (1986).

U.S. Dept. of Energy. "Microbiology of deep subsurface environments research program," Report No. DOE/ER-0280, 1986.

van Beek, C.G.E.M., *J. Am. Wat. Works Assoc.*, **76**, 66-72 (1984).

van Beek, C.G.E.M. and W.F. Kooper, *Groundwater*, **18**, 578-586 (1980).

Vogel, T.M. and P.L. McCarty, *Appl. Environ. Microbiol.*, **49**, 1080-1083 (1985).

Vogel, T.M. and P.L. McCarty, *J. Contaminant Hydrology*, **1**, 299-308 (1987).

Webster, J.J., G.J. Hampton, J.T. Wilson, W.C. Ghiorse, and F.R. Leach, *Ground Water*, **23**, 17-25 (1985).

Wilson, B.H., G.B. Smith, and J.F. Rees, *Environ. Sci. Technol.*, **20**, 997-1002 (1986a).

Wilson, J.T., J.F. McNabb, D.L. Balkwill, and W.C. Ghiorse, *Ground Water*, **21**, 134-142 (1983).

Wilson, J.T., L.E. Leach, M. Henson, and J.N. Jones, *Ground Water Monit. Rev.*, **6**, 56-64 (1986b).

19

BIOFILMS IN BIOTECHNOLOGY

JAMES D. BRYERS
Duke University, Durham, North Carolina

SUMMARY

Natural biofilm and support particle systems capitalize on the adsorption properties of the cells to stick to a substratum or to another cell, then grow and replicate, thus producing biologically active surfaces. Advantages of natural biofilm and biomass support particle systems include (1) the use of cheap, easily obtainable support material, (2) the ability to process large volumes of nutrient flow for long time periods, and (3) in microbial systems, the freedom to employ mixed cultures that can mediate multiple biological conversions. The major distinction between naturally and artificially immobilized cell systems is that the growth of cells is promoted in the former class, thus providing a constantly active system. Growth of artificially immobilized cells can cause support rupture and system contamination. Growth is only promoted periodically in some cases to regenerate decaying catalytic activity.

A major disadvantage of naturally formed biofilms is lack of control over the biofilm density and thickness, which can lead to excessive internal mass transfer limitations and potential biofilm sloughing. Biomass support systems promote control by defining the size and geometry of the biocatalyst. In both systems, slow-growing cells result in long reactor startup times.

Artificially immobilized whole cells indefinitely store a biological catalyst, with the convenience of short startup times at any location. This latter advantage has promoted the entire field of biosensor research and development. Despite the enormous volume of scientific literature on artificially immobilized whole cells for bioconversions, only three processes exist where such biocatalysts are employed at industrial scale: (1) L-aspartic acid production using nongrowing *E. coli*, (2) L-malic acid production using nongrowing *Brevibacterium ammoniagenes*, and (3) prednisolone production using a binary culture of *Curvularia lunata* and *Corynebacterium simplex*. This limited industrial use of artificially immobilized whole cells can be attributed to (1) carrier or immobilization reagent costs, (2) short active half-lives ($\approx$20–40 days), (3) toxic effects on cell viability, and (4) poor operational stability of the catalysts. The lower costs of alternative biofilm or suspended culture systems have, until now, economically outcompeted artificially immobilized systems. However, the growth of hybridomas or other mammalian cells, producing specialized products at exceptionally high per unit retail prices, will provide a more favorable incentive to use artificially immobilized cells. Consequently, one research area that should receive more attention will be the production of cheap, yet stable, immobilization materials.

19.1 INTRODUCTION

Biotechnology is the manipulation of cellular fragments (e.g., enzymes, organelles) or entire living cells to mediate desired biological reactions and the subsequent development of those reactions from laboratory scale procedure to commercial production. Progress in molecular biology, biochemistry, immunology, genetics, and microbiology in the past two decades now makes it possible to actually program living cells of virtually every type—microbial, plant, and mammalian—to generate a myriad of products ranging from simple molecules to complex proteins.

Biotechnology is the commercial exploitation of existing and novel biological techniques and processes.

Clearly, man has exploited biological processes for centuries in such activities as brewing, wine making, bread making, food preservation and modification (e.g., cheese, vinegar, soy sauce), and waste treatment. However until the 1970s, the performance, reactivity, and product specificity of a biological process was inherent to the specific biological catalyst (e.g., enzyme, whole cell) selected. The emergence of genetic manipulation, recombinant DNA technology, cell fusion, and advances in immunology now permit the orchestration of a desired metabolism, production of an exotic molecule, and improvement in system productivity/selectivity. Opportunities for commercialization of the "new" biology are both tantalizing and highly diverse (Table 19.1). The world market for biotechnologically derived products by the year 2000 is estimated at over $60 billion per annum (Table 19.2). In this chapter, I shall discuss the application of biofilms in the biotechnology sector, including microbial, plant, and mammalian cell biofilm systems.

TABLE 19.1 Prospects of Biotechnology

1. Human and Animal Health Care
   a. New diagnostic products based on enzymes, monoclonal antibodies, and other genetically engineered proteins
   b. Novel prophylactic products exemplified by new vaccines for the prevention of viral, bacterial, and protozoan diseases
   c. New therapeutic biologicals for treatment of cardiovascular and cerebrovascular disease, CNS diseases, and cancers
   d. Peptide hormonal substances that increase milk production in dairy cattle and stimulate growth
2. Human and Animal Nutrition
   a. Utilization of low-cost carbon sources for new microbial and enzymatic synthesis of amino acids, sugars, and edible fats and oils
   b. Use of large-scale fermentation of low-value feedstocks to produce nutritionally balanced protein
   c. Synthesis of vitamins and nutritional growth factors
3. Agricultural Chemicals and Related Products
   a. Biologically derived fungicides and herbicides
   b. Plant growth regulators
   c. Genetically manipulated plant cell clones to improve strain selection and fruit productivity
4. Environmental Improvement and Protection
   a. Microbial and enzymatic techniques for removing or eliminating toxic pollutants in industrial and municipal wastes
   b. Recovery of waste uranium
5. Natural Resource Recovery
   a. New biological processes for the recovery of metals and nonmetals from low-grade ores
   b. Increasing oil recovery by microbial enhanced methods
   c. Direct electricity generation by microbial processes

TABLE 19.2 Estimated World Markets for Biologically Derived Commodities by the Year 2000

Class	Sales (10^9 dollars/yr)
Chemicals	11
Energy	12
Food	12
Medicine	10
Miscellaneous (e.g., enhanced oil recovery, pollution monitoring and control, bioplastics, mineral leaching)	15
Total	60

19.2 MICROBIAL SYSTEMS

19.2.1 Natural Biofilms

Microbial biofilms accumulate as a consequence of the microbes' ability to adsorb, replicate, and metabolize on a substratum (Chapter 7). The adsorption of a cell to a substratum may be a complex process dependent upon physical and chemical characteristics of the cell, the substratum, and the environment. During nonspecific adsorption and subsequent cell growth, the cells remain viable and reactive. Consequently, biofilm-bound cells are generally in a growth mode, providing constant catalytic activity and the possibility of multiple biological reactions if mixed cultures are employed. However, the lack of control over the adsorption and growth processes is a disadvantage with natural biofilms. Slow biofilm accumulation rates can prolong startup of a reactor system, or excessive biofilm formation can create intolerable mass transfer limitations to system reactivity and, if ignored, catastrophic system calamities (e.g., excessive biofilm sloughing, clogging of packed beds).

19.2.1.1 Acetic Acid Production Vinegar (in French it literally means "sour wine"), by law, must be produced by the microbial oxidation of ethanol. One of the earliest applications of microbial biofilms in biotechnology was in the "quick vinegar" process invented in 1823 by Scheutzenbach. Wooden vats with perforated bottoms are packed with wood chips, and alcoholic solutions are trickled down through the packed bed, where an enriched biofilm of *Acetobacter* spp. develops. The biological conversion requires substantial amounts of oxygen and liberates a large quantity of heat. The process is depicted in Figure 19.1, where approximately 55 m^3 of beechwood chips (50 mm diameter) act as substratum for the bacterium.

19.2.1.2 Microbial Leaching Extraction of copper from ore deposits using acid solutions has been practiced for centuries, but the role of bacteria in metal dissolution was not verified until the late 1940s. Today, approximately 10–20% of the copper mined in the U.S. is extracted by microbial-assisted processing of low grade ores (Brierley, 1978 and 1982). There is also investment in extending microbial leaching to the recovery of other metals such as uranium, silver, cobalt, molybdenum, nickel, and gold (Lawrence and Bruynesteyn, 1983) as well as the microbial desulfurization of coal (Chandra et al., 1979; Kargi and Robinson, 1982; Gockay and Yurteri, 1983). Most microbial leaching depends upon microbial oxidization of metal sulfides. Aqueous environments in association with spent mineral produce very harsh conditions of low pH, high metal concentrations, and (often) elevated temperatures, which enrich for a microbial flora with very discriminating nutritional requirements (Table 19.3).

Dump leaching is the most common process used in the U.S. to extract copper microbially from spent ore ($\geq 0.4\%$ w/w copper). The process system consists of an aerobic, trickle packed bed system in which the spent ore serves as

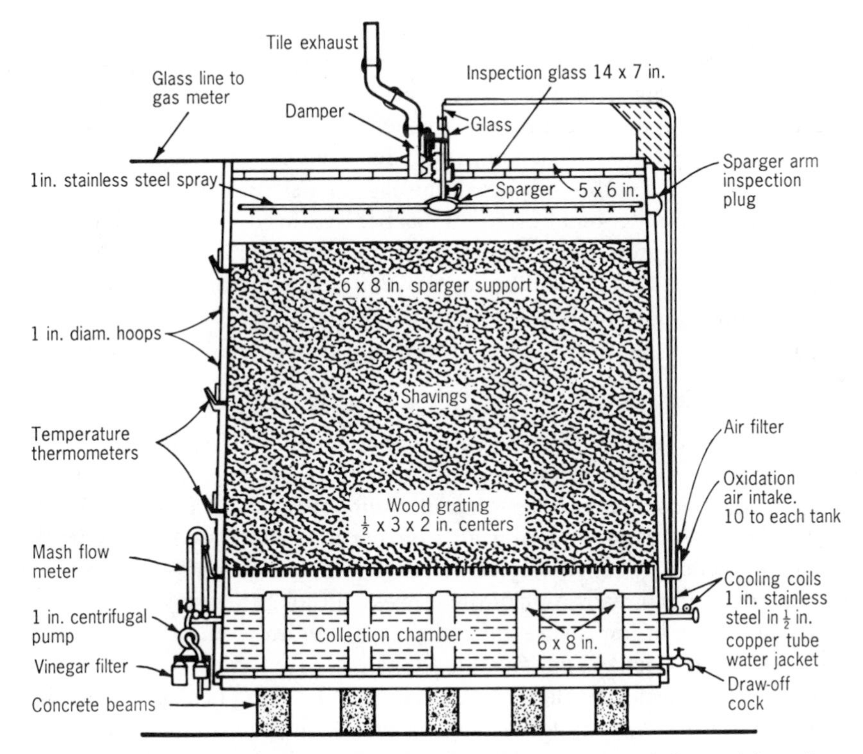

Figure 19.1 Vinegar production by alcohol oxidation employing biofilms of *Acetobacter* spp. (reprinted with permission from Bailey and Ollis, 1986).

TABLE 19.3 Bacteria Isolated from Metal Ore Environments

Organism	Growth Temperature	Energy Source	Carbon Source
Thiobacillus ferrooxidans	Mesophile	Fe^{+2}, V^{+4}, S^0	CO_2
Leptospirillum ferroxidans	Mesophile	Fe^{+2}	CO_2
Thiobacillus thiooxidans	Mesophile	S^0	CO_2
Sulfolobus spp.	Thermophile	S^0, organic C, Fe^{+2}	CO_2, organic C
Acidophilium cryptum	Mesophile	Organic C	Organic C

both substratum and substrate. To initiate the process, acidified water (pH = 1.5–3.0) is sprayed over the porous ore bed. Acidophilic bacteria, *Thiobacillus ferrooxidans,* actively oxidize the soluble ferrous iron and attack the sulfide minerals, releasing the soluble cupric ion. Elemental sulfur could accumulate on the ore, limiting the extent of leaching, if not for *Thiobacillus thiooxidans,* which oxidizes the reduced sulfur, thus maintaining the acidic environment.

Commercial production of uranium has been realized in a Canadian process where insoluble uranium ores are oxidized to soluble uranium species by the Fe^{+3} ion generated by *Thiobacillus ferrooxidans*. The microbe is also capable of directly oxidizing U(IV) to U(VI) for an energy source. At the Agnew Lake mine in Canada, approximately 7200 kg of uranium is generated per month in this way.

Desulfurization of coal has received considerable attention due to the obvious benefits of clean coal with respect to atmospheric pollution. Approximately 90% of the inorganic sulfur (sulfides) can be leached from various coal types using *Thiobacillus ferrooxidans* (Myerson and Kline, 1983). The accumulation of a *Thiobacillus ferrooxidans* biofilm on iron pyrite is illustrated in Figure 19.2 (Olsen and Kelley, 1986). Biological reaction and mass transfer rates presently limit the commercial realization of the process, although recent work has shown promising efforts in improving its efficiency (Chandra et al., 1979; Kargi and Robinson, 1982; Gockay and Yurteri, 1983).

Process engineering development of microbial leaching has been sorely lacking. Currently, most dump leach sites are piles of irregularly shaped ore rocks distributed over a large open pit. Pipes and spacers may be employed to facilitate aeration, but little other control can be exerted over the microbial process. Design and optimization of continuous efficient bioleaching systems appears limited by the lack of detailed kinetic and stoichiometric information. Although the cell–substrate interaction and the heterogeneous reaction system are complex, application of methodologies presented in Chapters 7–13 to microbial leaching would prove beneficial.

19.2.1.3 Polysaccharide Production Polysaccharides occur as microbial energy reserves and as structural material in cell walls and extracellular capsular layers. The production of extracellular polysaccharides is critical to the formation of the gel matrix surrounding cells in a biofilm (Chapter 4). A large number of extracellular polymers have gained commercial importance due to their ability to alter, at low concentrations, the rheological properties of aqueous solutions. Such polysaccharidic hydrocolloids are used extensively in the food, pharmaceutical, cosmetic, oil, paper, and textile industries as thickening agents and viscosity modifiers (Table 19.4).

Exploitation of microbial polymer production is limited in suspended cell culture by the high energy demands of mixing viscous reaction broths, by oxygen transfer limitations due to nonideal mixing in non-Newtonian solutions, and to separation of the polymer product from the microbial cells. The last is extremely important if the polysaccharide has food or pharmaceutical applications.

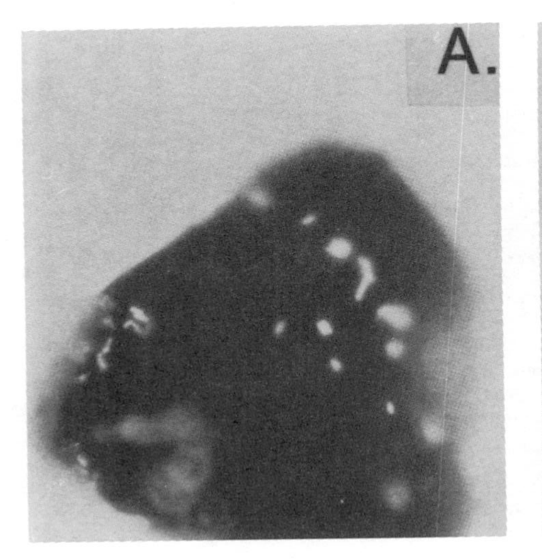

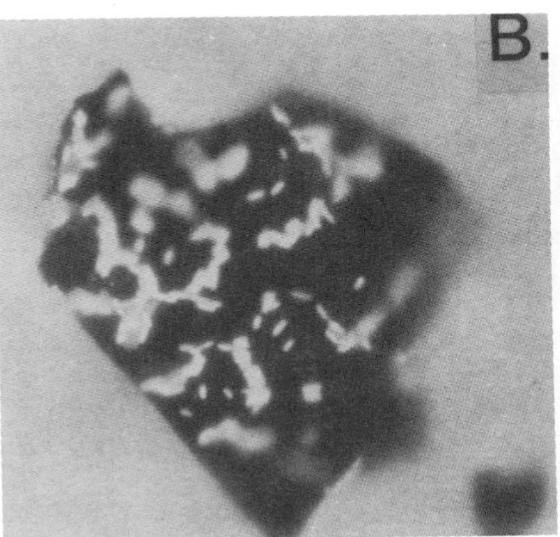

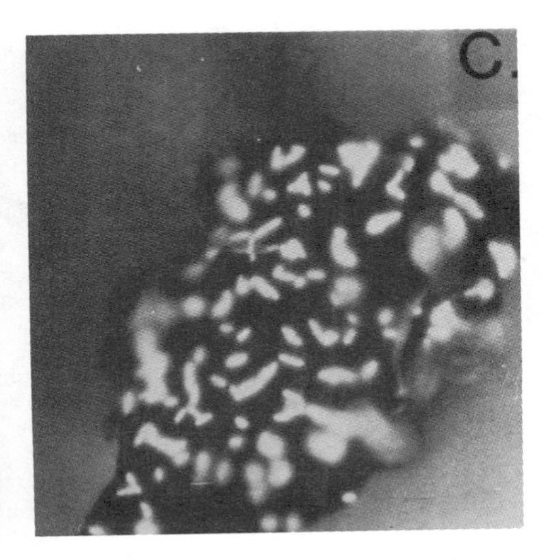

Figure 19.2 Biofilms of *Thiobacillus ferrooxidans* on iron pyrite result in accelerated mineral microbial bioleaching.

TABLE 19.4 Microbially Derived Commercial Extracellular Polysaccharides

Polysaccharide	Microorganism	Applications
Xanthan	*Xanthomonas campestris*	Viscosity modifier, gelling agent, enhanced oil recovery, food additive, cosmetic emulsifier
Dextran	*Aerobacter* spp. *Streptococcus bovis* *Streptococcus viridans* *Leuconostoc mesenteriodes*	Plasma substitutes (blood expanders), therapeutic burn treatment, polyelectrolytic, separation media
Alginates	*Pseudomonas aeruginosa* *Azotobacter vinelandii*	Thickening agent in dairy products, yogurt stabilizer
Gellan gum	*Pseudomonas elodea ATCC 31461*	Gel agent in microbiological media
Zanflo	*Erwinia tahitica*	Paint thickener
Polytran	*Sclerotium glucanicum*	Stabilizer for drilling muds; enhanced oil recovery, ceramic glazes
Alcaligenes spp.	*Alcaligenes* spp.	Suspending agent, emulsifier in oil industry, pesticide undefined polymers & fertilizer suspending agent (S130; S194; S198)
Curdlan	*Alcaligenes faecalis*	Gelling agent in cooked foods, support for enzymes
Pullulan	*Aureobasidium pullulans*	Packaging, flocculating agent
Unnamed heteroglycan	*Flavobacterium uliginosum* MP-55	Anticancer (tumor) therapeutic

One novel approach for exploiting the microorganism's tendency to overproduce extracellular polysaccharide during biofilm formation is offered by Robinson and Wang (1986). In this process, the mass transfer limitations inherent in biofilm systems are exploited, rather than avoided, in order to selectively retain the macromolecular product within the support. In typical *Xanthomonas campestris* fermentation, maximum product concentrations are apparently limited, not by feedback or other physiological inhibition, but by the corresponding decrease in mass transfer rates with increasing product concentration in the bulk phase. Robinson and Wang immobilized living *X. campestris* on celite, a highly porous, rigid support particle. Due to the low diffusivities of the macromolecule, xanthan was preferentially retained within the porous matrix; nutrients and other metabolites diffused rapidly to the cells, promoting further biofilm development. Retained polysaccharide concentrations were enhanced, so that the overall productivity was greater in the immobilized cell system than in a classical suspended culture (Figure 19.3).

19.2.2 Biomass Support Systems

Natural biofilm systems suffer from a lack of direct control over biofilm. Excessive biofilm formation can lead to such problems as: (1) unstable effluent suspended biomass concentrations, (2) excessive mass transfer limitations (Chapter 9), (3) packed bed hydraulic instabilities and potential bed plugging (Chapter 18), and (4) uncontrolled biofilm particle buoyancy in fluidized bed systems. To

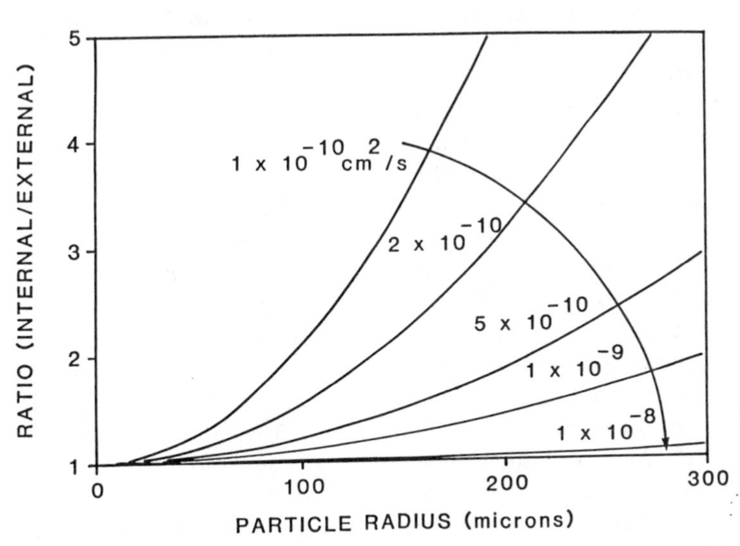

Figure 19.3 Predicted polysaccharide concentration ratios (internal/external) in the absence of nutrient transport limitations as a function of particle radius and effective xanthan gum diffusivity (Robinson and Wang, 1986).

alleviate the drawbacks, biomass support particles (BSP) were developed by Atkinson and coworkers in the late 1970s. BSP are three dimensional, randomly spun lattices or flexible porous sponges made in a variety of geometrical shapes, pore sizes, and overall dimensions. Details concerning BSP fabrication and physical properties are presented in Chapter 17 along with a detailed review of BSP wastewater treatment applications.

19.2.2.1 Ethanol Production in BSP Two reactor configurations, a fluidized bed and a gas-stirred circulating bed, were evaluated by Black et al. (1984) as alternative BSP systems. A single type of stainless steel BSP and three plastic foam BSP, of varying porosity (indicated as pores per linear inch—ppi) were employed to enhance overall ethanol productivity in the yeasts *Saccharomyces cerevisiae* and *Saccharomyces uvarum* (Table 19.5). Although no effort was made to optimize the process, the ethanol productivities were better in all cases than those for freely suspended cultures (Cysewski and Wilke, 1978). In terms of stable operation, there was no indication that either reactor geometry (fluidized bed or circulating bed) provided any distinct advantage. However, the stainless steel BSP exhibited lower biomass concentrations than any foam BSP, presumably due to the higher liquid recirculation rates required for fluidization of the particles.

19.2.2.2 Cellulase Production in a BSP Webb et al. (1986) report on the continuous production of the enzyme cellulase by the filamentous fungus *Trichoderma viride* QM 9123 captured in 6 mm diameter stainless steel BSP. The biocatalysts were fluidized in a 10 L spouted bed reactor that employs a liquid jet for fluidization and recirculation. Cellulase productivity for freely suspended cells (washout at $D = 0.012\ h^{-1}$) was much lower than for BSP-supported cells (Figure 19.4).

The high volumetric productivities for the BSP occur at lower specific activities than for suspended cells, which the authors attribute to diffusional limitations. Webb et al. (1986) suggest mass transfer limitations in the BSP lead to lower local growth rates, which, since cellulase is a secondary metabolite, induces its production. The lower growth rate also diverts substrate from cell mass production, and consequently a higher cellulase yield value results.

Atkinson (1985) indicates that mass transfer limitations can be advantageous in the case of substrate inhibition where the growth rate decreases with increasing substrate concentration. Then, mass transfer limitations result in an increase in overall reactivity.

Other factors distinguish immobilized cell activities from their dispersed counterparts. Many examples suggest that cells undergo a wide variety of modifications upon adsorption and growth at a surface; i.e., the optimum growth conditions (Williams and Munnecke, 1981), product yields (Navarro and Durrand, 1977), lag times (Hattori and Furusaka, 1959), and cell morphology (Jirku et al., 1980) can change. There have been claims that biofilm cells are more active, but such reports should be treated with caution, since oftentimes cell

TABLE 19.5 Performance Data for Ethanol Production Using Biomass Support Particles[a]

Reactor type: / Yeast strain:	Fluidized Bed / *S. uvarum* ATCC 26602	Circulating Bed / *S. cerevisiae* NCYC 1119
Reactor volume, m^{-3}	0.002	0.003
Dilution rate, h^{-1}	1.35	0.78
Influent glucose, kg m^{-3}	24.8	46.0
Oxygen conc., % sat at 30°C	20.0	>50.0
Number of BSP	8000[b]	4500[c]
Per particle biomass amount, mg	16	12.7 (30 ppi) 27.6 (45 ppi) 32.5 (60 ppi)
Biomass conc. in reactor, kg m^{-3}	63.6	39.1
Overall glucose uptake rate, kg m^{-3} h^{-1}	22.4	35.0
Overall specific glucose uptake rate, g g^{-1} h^{-1}	0.35	0.89
Overall ethanol production rate, kg m^{-3} h^{-1}	11.2	14.1
Ethanol yield, % of theoretical maximum	98	79

[a] Black et al. (1984).
[b] Stainless steel.
[c] 1500 each of the 30, 45, and 60 ppi BSP.

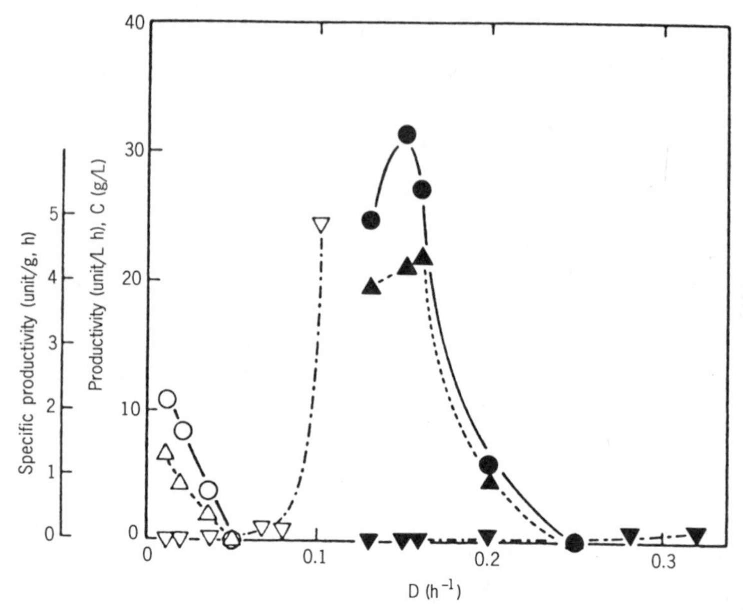

Figure 19.4 Effects of dilution rate on cellulase productivity for freely suspended and immobilized cell systems. (○), productivity for freely suspended cells; (△), specific productivity for freely suspended cells; (▽), glucose concentration for freely suspended cells; (●), reactor productivity for immobilized cells; (▲), specific productivity for immobilized cells; (▼), glucose concentration for immobilized cells.

numbers are not accurately monitored. Due to mass transfer limitations and the surface chemistry of any BSP, the microenvironment of the adsorbed cells is considerably different from the bulk water. However, the difference between biofilm cells and suspended cells is more than can be expected from differences in nutrient availability. Such deviations include (1) altered cell properties upon detachment and (2) different properties of daughter cells from mother cells although both are attached in the same manner (Hattori, 1972; Brodelius et al., 1979; Hattori and Furusaka, 1960; Larrson and Mosbach, 1979). Doran and Bailey (1986) observed changes in immobilized *Saccharomyces cerevisiae* cell composition (DNA, RNA, total protein, and reserve polysaccharides) and metabolism (increased specific ethanol production rate, increased specific glucose consumption rate, decreased specific growth rate), which they attributed to alteration of normal yeast cell wall synthesis and DNA replication that occurred upon attachment.

There are many mechanisms modifying the physiology of biofilm cells, including nutrient availability, membrane permeability changes, substratum electrochemical interactions with cell membrane transport systems, osmotic pressure, and shifts in metabolism due to extracellular polymer production.

19.2.2.3 *Protein Synthesis by a Recombinant Strain of* Escherichia coli *Using a Hollow Fiber Membrane Reactor* Inloes et al. (1983) propose a biomass support system with an alternative support geometry based upon hollow fiber (HF) membranes. Asymmetric HF membranes were originally developed for hemodiafiltration of uremic blood as an alternative to hemodialysis. Asymmetric cylindrical HF membrane reactors operate as a continuous plug flow system with cell and nutrient never directly contacting each other. This configuration theoretically eliminates the need for downstream cell separation. An asymmetric cylindrical HF membrane (Figure 19.5) consists of an ultramicroporous inner layer 0.1–0.5 μm thick surrounded by an open channel macroporous region. Pore diameters in the inner layer are on the order of nanometers, while outer surface pores are approximately 10 μm in diameter; the porosity varies between 60 and 90%.

Inloes et al. (1983) report on the growth of the recombinant bacterium *Escherichia coli* C600 (pBr 322), which had been genetically cloned to overproduce (about 50 times that of the wild type) the enzyme β-lactamase within a HF membrane tubule. A single HF membrane tubule was enclosed within a glass shell (Figure 19.6), and inoculum was introduced through the shell side of the fiber tube. Nutrients passed through the interior channel (lumen) and diffused radially outward; metabolic products diffused in the reverse direction and exited with the spent media. If required, gaseous nutrients could be either dissolved in the nutrient feed or supplied directly to the shell side. β-Lactamase activity for freely suspended cells and HF-supported cells indicate a marked improvement in volumetric enzyme productivity for the biofilm-bound cells (Table 19.6). Although preliminary helium gas leak testing indicated no breach between shell and lumen, contamination of the lumen inner surface did occur.

19.3 MAMMALIAN CELL SYSTEMS

Products derived from mammalian cells have proven to be extremely valuable in diagnostic and therapeutic medicine. Such products can be either the mammalian cell itself (e.g. β-islet cells, hepatocytes, bone marrow cells, lymphocytes) or cell-derived products, which include human growth factors, blood factors, interferons, proteases, hormones (e.g., insulin, parathyroid hormone), vaccines, and monoclonal antibodies.

Vaccines currently constitute the largest portion of the mammalian cell product array, with 1984 world sales over $95 million (Glacken et al., 1983). Animal vaccines are the greatest part of this market, with foot and mouth disease vaccine the major single product. Human vaccines produced from mammalian cells include polio, rubella, rabies, measles, and encephalitis. Unlike microbes, mammalian cells can replicate only a limited number of times. *Hybridoma* cells are composites, formed by the fusion of a particular mammalian cell and a cancer cell, thus allowing infinite replication. Monoclonal antibodies from hybridomas may prove to be the single most economically significant product of mam-

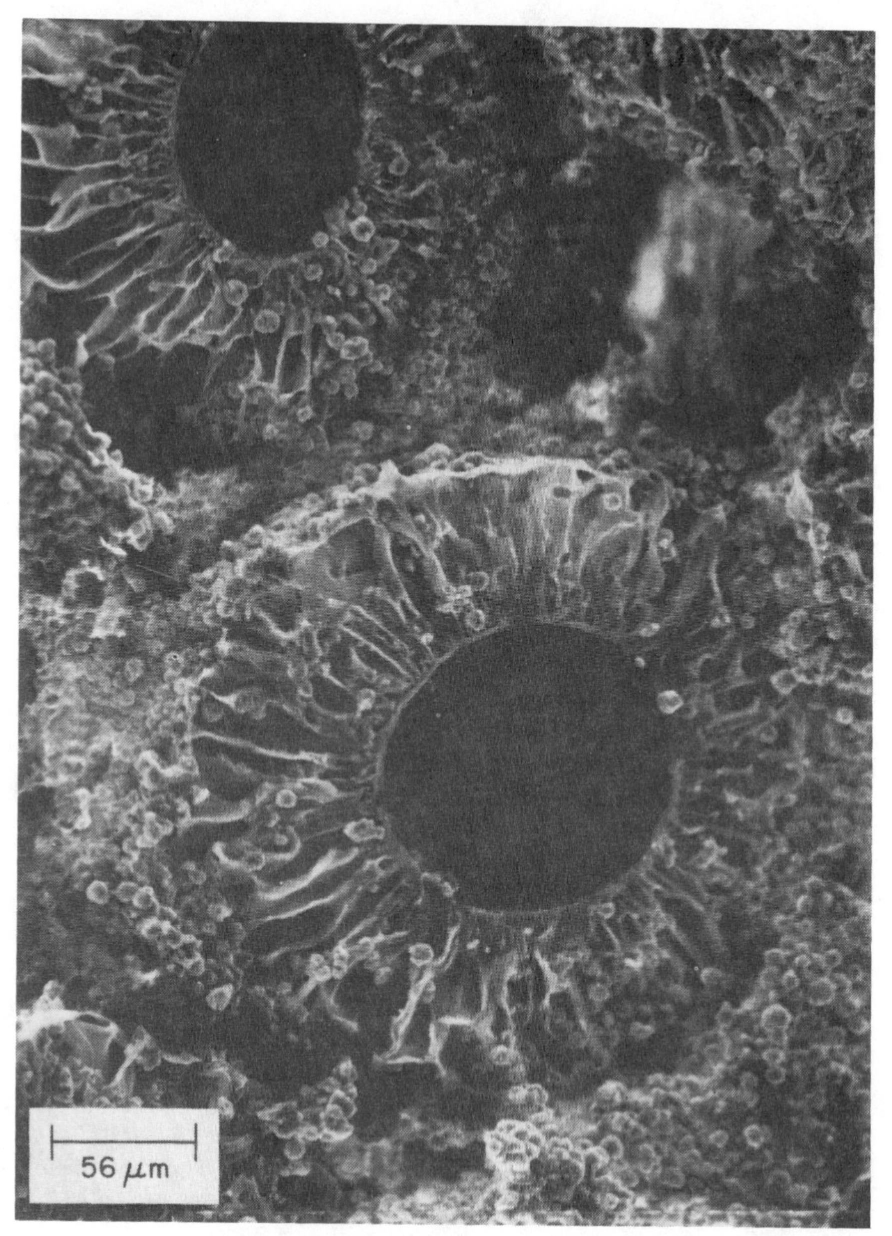

Figure 19.5 Scanning electron micrograph of an asymmetric hollow fiber.

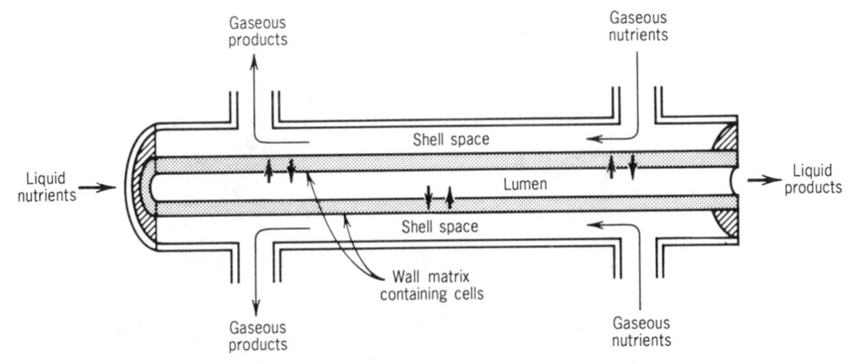

Figure 19.6 Asymmetric hollow fiber membrane reactor design (diffusion mode).

malian cells. Current applications of monoclonal antibodies include diagnostic medicine (detecting hormones and antigens associated with infectious diseases), tumor imaging, therapeutic medicine (antibody-targeted drugs in cancer treatment), and immunopurification.

Not every biological commodity can be produced microbially. Many proteins, such as interferon and insulin, have been successfully produced in bacteria via recombinant DNA technology. However, some proteins with complex tertiary or quaternary structure, or proteins that require posttranslational modification, may not be properly expressed in bacteria, i.e., mammalian cells are able to glycosylate properly, form correct disulfide linkages, and carry out other posttranslational folding that produces a functional protein molecule. Although some microbial physiologists and molecular geneticists argue that eventually microbes can be engineered to produce functional mammalian proteins, the investment costs may be prohibitive. Consequently, research and development in novel mammalian cell cultivation methods has increased dramatically over the past five years. Mass production of mammalian cells is not as simple as that of microorganisms, since mammalian cells are larger, more fragile, and nutritionally more discriminating. Mammalian cells are believed to be sensitive to damage from shear forces due to their lack of a rigid cell wall. However, recent studies indicate that anchorage-dependent mammalian cells are relatively insensitive to direct shear stress, up to 6 N m^{-2} (Stathopoulos and Hellums, 1985; Dewey et al., 1981) even though such cells are easily damaged by turbines or impellers. Such results imply that impaction, microturbulent eddies, and cavitation are more likely causes of mammalian cell damage than direct shear stress. Finally, the average doubling time of a mammalian cell is much greater than for microbes (20–50 h), making continuous long term aseptic operation expensive and difficult.

Cultivation requirements of mammalian cells are also related to the cells' specific anchorage dependency. Cell lines, such as lymphocytes, grow freely in suspension, while a large number of mammalian cells (e.g., fibroblasts, epithelial cells, and epidermal cells) must adsorb to a compatible substratum, followed by

TABLE 19.6 Comparison of β-Lactamase Reactor Productivities[a]

	Productivity[b]		
	Per Cell (u $cell^{-1}$ h^{-1})	Per Unit Total Volume (u ml^{-1} h^{-1})	Per Unit Void Volume (u ml^{-1} h^{-1})
Shake flask (1 day)	1×10^{-10}	0.084	—
HF membrane (<5 day)	2×10^{-11}	10	22
HF membrane (long term)	1×10^{-11}	8.0	23

[a] Inloes et al. (1983).
[b] u is specific activity of the enzyme.

tension and spreading, in order to grow. A third group of cells, such as HeLa cells, which are transformed anchorage-dependent cells, can grow either in suspension or anchored to a substratum. There is an obvious analogy between the heterogeneous reactions in a microbial biofilm and in anchorage-dependent mammalian cell development. However, development of mammalian cell biofilms does pose a series of unique problems.

Once a cell reaches a surface, it adsorbs, spreads out, becomes elongated, aligns with other cells in a pattern highly dependent upon prevailing fluid dynamics (Stathopoulos and Hellums, 1985; Dewey et al., 1981), and grows.

In contrast with microbial systems, no research has addressed the initial stages of mammalian cell transport and adsorption to an in vitro *substratum as a function of growth and fluid flow conditions.*

Cellular growth and accumulation at a substratum continue until cell to cell contact is established (termed "confluence"), forming a monolayer coverage of the substratum; cells will not overlap and will not continue to grow. At this stage, cells are removed from the substratum using a proteolytic agent (e.g., trypsin), and the cells are either harvested or exposed to a new, larger surface.

Mammalian cells capable of growth in free suspension can be cultivated in agitated vessels similar to microbial fermenters that are designed for low shear conditions (artificial immobilization of cells is an alternative approach to protect such cells from shear damage and is discussed in Section 19.5.2.3). The scaleup of suspension cell culture systems is similar to that of microbial suspension systems. However, the scaleup of anchorage-dependent mammalian cell has led to a variety of designs meant to increase the reactor surface area to volume ratio.

19.3.1 Anchorage-Dependent Mammalian Cell Systems

A variety of materials has been used as substratum for culturing anchorage dependent cells (Table 19.6; Reuveny, 1983). Glass has been recently replaced as the predominant substratum material by plastic with negatively charged surface functional groups that are produced by irradiation (McKeehan et al., 1981) or emulsion polymerization (Reuveny, 1983). Surface pretreatment agents have recently been commercialized that claim enhanced nonspecific cell attachment to any substratum. For example, CELL-TAK™, formulated from an adhesive protein secreted from the marine mussel *Mytilus edulis* (Waite and Tanzer, 1981), is reported to greatly enhance the adsorption of bacteria, yeast, and anchorage-dependent and -independent mammalian cells to a variety of substrata (Figure 19.7) without adverse effects on cell morphology, metabolism, or growth rate.

The most common devices for culturing anchorage-dependent cells are Petri dishes and roller bottles, neither of which provide the potential surface area nor the environmental control necessary for scaleup to commercial production. Consequently, a variety of heterogeneous reactor designs that increase the surface

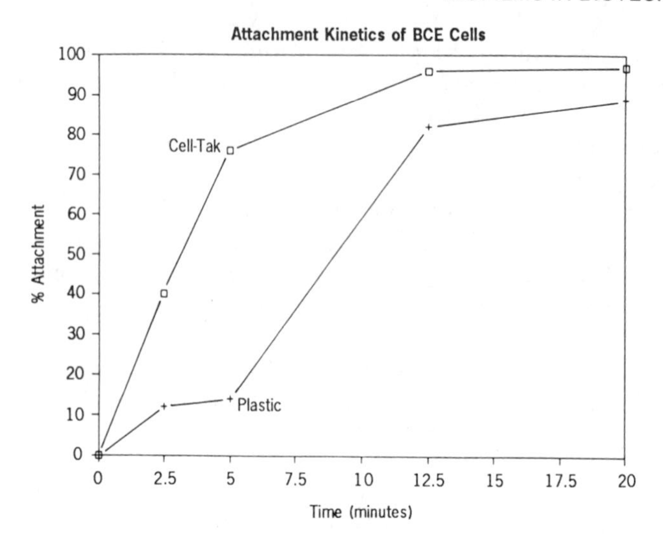

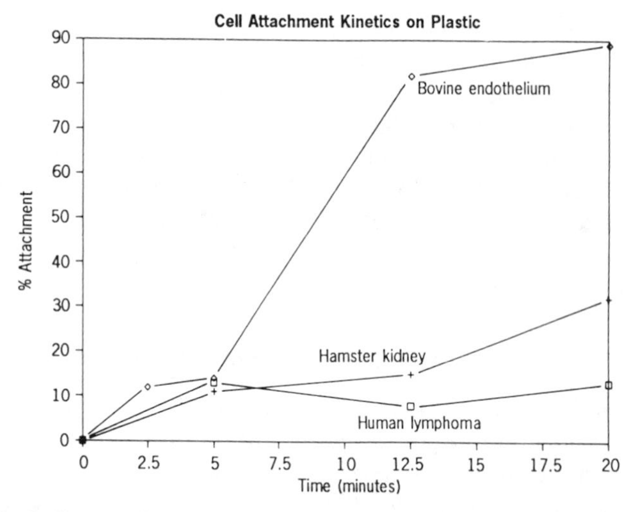

Figure 19.7 Influence of CELL-TAK on attachment of bovine corneal endothelial cells to various substrata.

area to volume ratio have been evaluated over the past twenty years, including: (1) the glass bead propagator (Wohler et al., 1972; Figure 19.8), (2) the multiple plate propagator (Weiss and Schleicher, 1968), (3) the spiral film reactor (House et al., 1972), (4) the Gyrogen^{TM} with rotating tubes (Girard et al., 1980), and (5) microcarrier beads (van Wezel, 1967; Clark and Hirtenstein, 1981; Figure 19.8).

The microcarrier developed by van Wezel (1967) was DEAE-Sephadex beads. Since this initial work, microcarriers with controlled lower surface charge densities have been developed using DEAE-Sephadex, gelatin-coated dextran beads,

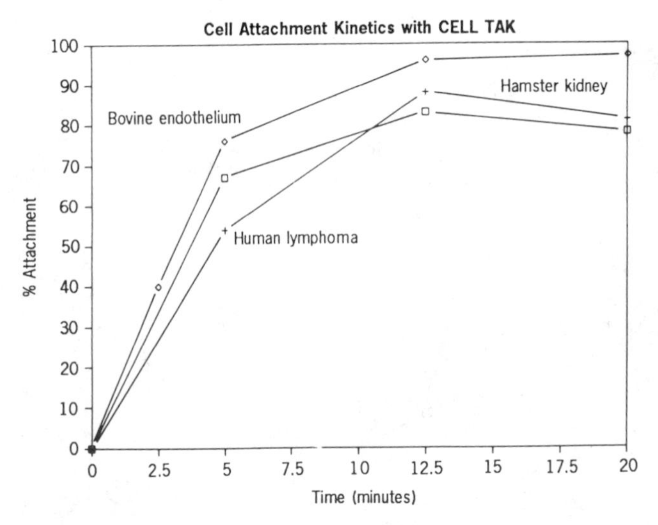

Figure 19.7 Continued

polyacrylamide, polystyrene, hollow glass, cellulose beads, and fluorocarbon droplets. Microcarriers (1) offer growth surface area to volume ratios over 10,000 m^{-1}, (2) permit simple cell separation from the medium, (3) provide a readily accessible and controlled homogeneous liquid phase environment, (4) support growth to over 10^{13} cells m^{-3}, (5) afford reduced medium requirements and increased yields, and (6) can be scaled to up to cubic meter proportions. Reviews by Reuveny (1983) and Pharmacia Fine Chemicals (1982) enumerate the variety of cell lines that can be propagated on microcarriers (Table 19.7). Microcarrier cell supports can be uniformly suspended in culture media, allowing the use of agitated vessels. The operation of such reactors depends on cell and product type. Where cell death follows product formation, batch operation is preferred. But where either the cell or a secreted protein is the desired product, fed batch or continuous operation is preferred. Cells are harvested by separating the microcarriers from the reactor contents followed by trypsin treatment to resuspend the cells. Use of cross-linked gelatin microcarriers provides an alternative method to harvest cells by enzymatic dissolution of the gelatin (Figure 19.9). Culture medium is delivered in a plug flow reactor mode; anchorage-dependent cells attach to glass surfaces and grow past confluence, where cells slough and become entrapped in the interstices of the bed (Figure 19.10). Recently, the effects of hydrodynamics (Croughan et al., 1987), inoculum requirements, (Hu et al., 1985), substratum surface properties (Varani et al., 1985), and reactor operation (Miller et al., 1986) on microcarrier performance have been addressed.

19.3.2 Mammalian Cell Support Systems

Despite obvious advantages, microcarriers do pose some problems. For example, the cell concentration within a reactor can be increased by increasing the

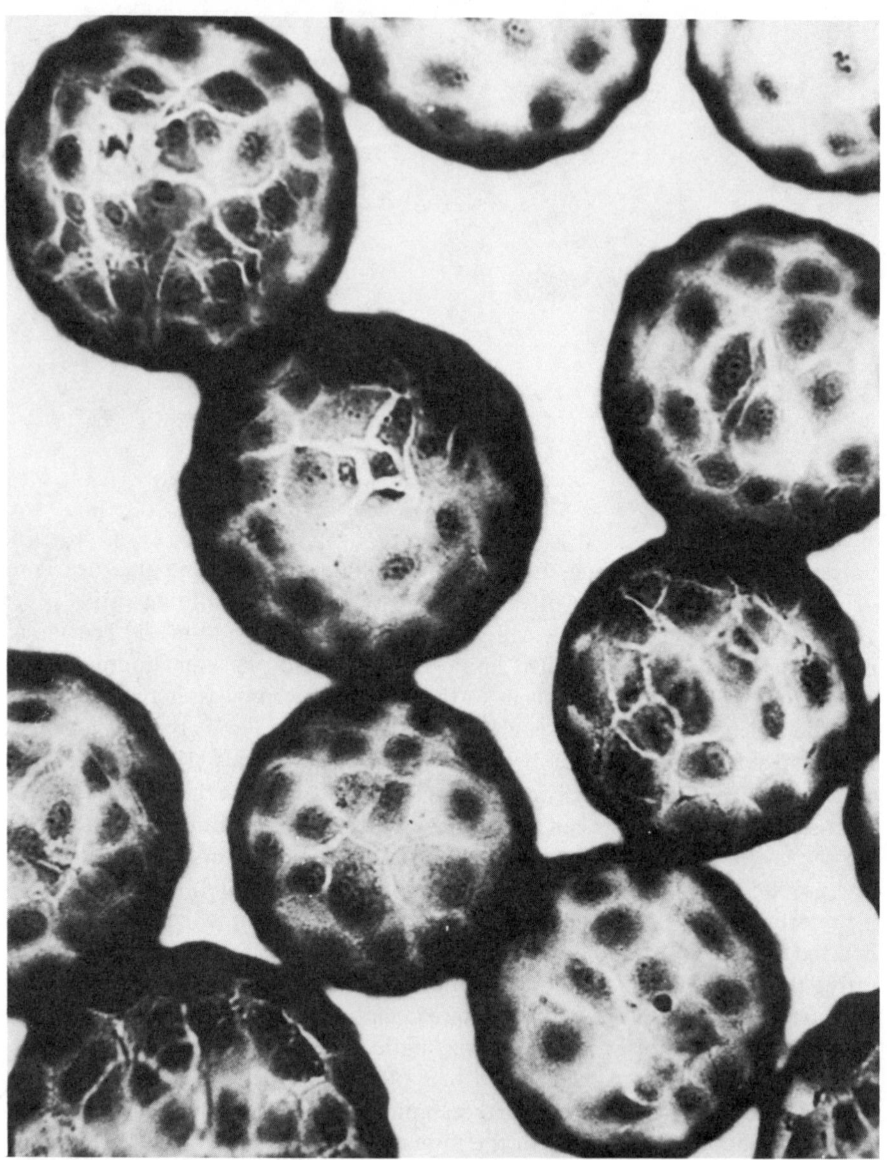

Figure 19.8 Human foreskin fibroblasts anchored to a glass bead. (Courtesy Biopolymers, Inc.)

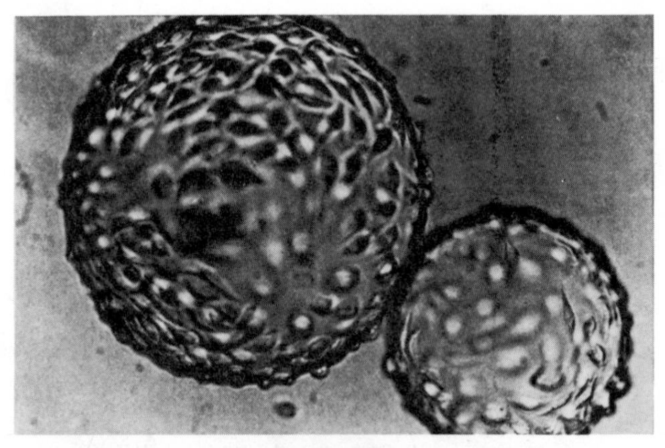

1. Start time = 0,0.01% crude collagenase.

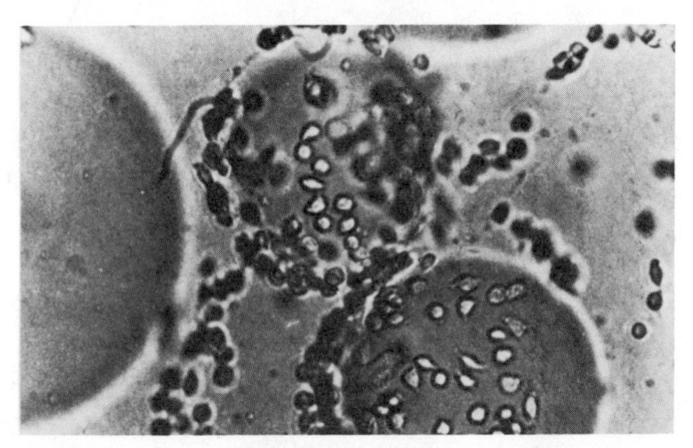

2. Time = 1.5 minutes. Cell release and partial microcarrier digestion achieved.

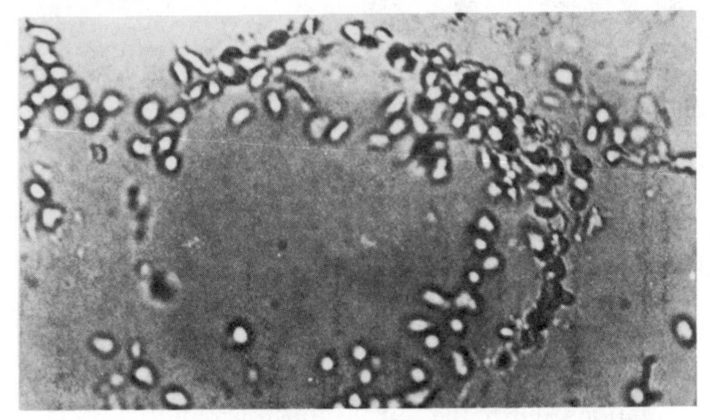

3. Time = 10 minutes. Complete microcarrier digestion. Cells ready for scale-up culture or alternative processing.

Figure 19.9 The harvesting progression for mureine L-cells from a Ventrex Ventregel gelatin microcarrier.

Figure 19.10 Photograph of a packed bed column reactor for continuous mammalian cell cultivation. Packing material is glass beads. (Courtesy of BIORESPONSE, Inc., Haywood, CA.)

microcarrier concentration. However, at some point, the carrier concentration is sufficiently high that shear forces and impaction can cause cell damage. To prevent such damage, agitation is decreased, which can reduce particle suspension. Increasing the surface area to volume ratio by decreasing the microcarrier size also has its limitations, due to the substratum area required per cell and physical requirements to keep the carriers in the reaction vessel. A second problem with microcarriers is that once cells reach confluence, additional growth causes cell sloughing into the medium, which is undesirable in long term production operation. To circumvent these problems, mammalian cell support systems (i.e., biomass support systems) of various geometries were developed to capture both anchorage-dependent and -independent cell lines. Such mammalian cell support systems include: (1) hollow fiber, (2) static maintenance, (3) ceramic matrix, and (4) sponge matrix particle reactors.

Hollow fiber (HF) reactors consist of a bundle of HF membrane tubules enclosed by an outer shell housing (Figure 19.11). Membrane construction has been discussed previously in Section 19.2.2.3. HF membrane reactors have been applied to both anchorage-dependent (Knazek et al., 1972) and anchorage-independent cell lines (Wiemann et al., 1983; Calabresi et al., 1981). Cells are inoculated into the shell space, where they become entrapped in the porous surface of the outer HF membrane. As in the case of microbial cultures, fresh medium is pumped from one end of the lumen to the other; nutrients diffuse through the membrane, supporting cell growth, while waste metabolites and desired products diffuse back into the medium, leaving the lumen. Judicious selection of membrane molecular weight exclusion properties permits selective concentration of desired products. Due to the eventual high cell concentrations in these units, gradients in nutrient and metabolite concentrations in the axial direction can occur in the lumen (Ku et al., 1981). Thus, scaleup of HF systems may require either arranging multiple units in parallel or some degree of medium recycling. Another drawback to the HF membrane reactor is the inability to either harvest or directly measure cell numbers (and their reactivity) in the sealed HF unit.

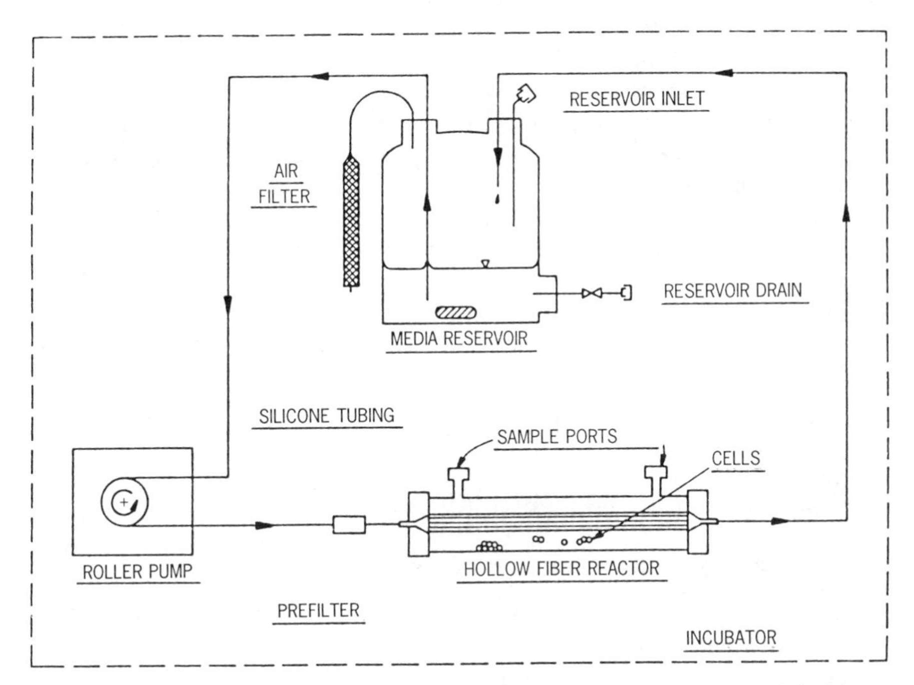

Figure 19.11 Flow diagram of the hollow fiber reactor system placed inside an incubator.

A *static maintenance* reactor does not attempt to grow cells at all, but operates on the basis that maximum product secretion occurs in some cell lines when the cells are not replicating or viable. In such a system, cells are actually cultivated elsewhere in suspension or on microcarriers, concentrated, and then mixed into a foam matrix material, which completely fills the interior of the reaction vessel. An array of porous tubes penetrate the foam material and deliver the needed nutrients and oxygen. Cells are maintained, by nutrient feed control, in an homeostatic environment to prevent cell proliferation.

Another mammalian cell support particle system is the *ceramic matrix support,* developed by Charles River Biotechnical Services. The OPTICORE™ matrix (Figure 19.12) can be fabricated either with an extremely smooth surface for anchorage-dependent cells or with a very rough surface for anchorage-independent cells. In contrast with the HF support systems, nutrient medium is delivered down each of the individual rectangular channels coming in direct contact with cells attached to the surfaces. Virtually no axial gradients have been reported (Lydersen et al., 1984). As with the HF membranes, direct measurement of cell numbers or cell reactivity within the ceramic matrix is currently not feasible.

Another approach to mammalian cell support particles is the *sponge matrix bead* (Runstadler and Cernek, 1987). The sponge matrix bead is fabricated from collagen, a natural biopolymer, cross-linked to render it insoluble at normal temperatures. At an elevated temperature, the liquid polymer is mixed with extremely fine metal particles (which provide a controlled buoyancy), and the polymer cast into shapes and dried at normal room temperature. The porosity of the beads (Figure 19.13) is produced by a proprietary process. Void fractions can range between 0.8 and 0.9. Design specifications for the porous collagen beads are: average diameter 500 μm, pore size 20–40 μm, and wet specific gravity 1.6–1.7. Inoculated beads (Figure 19.14), in a dense slurry, are fluidized in a column reactor by a recycled nutrient medium (Figure 19.15). Van Brunt (1986) reports that a 0.010 m^{-3} reaction vessel can be loaded with 6×10^7 beads, producing about 10^{14} cells m^{-3}. Currently, production figures are for a murine hybridoma producing a secreted monocolonal antibody and a recombinant CHO line producing tPA. Recovery of cells from the sponge beads would require denaturation of the polymer cross linking.

19.4 ARTIFICIALLY IMMOBILIZED CELLS

19.4.1 Immobilization Fundamentals

There are numerous methods available for artificially immobilizing whole cells, whether microbial, plant, or mammalian (Table 19.8). A synopsis of each technique will be given here; there are a number of excellent, detailed reviews on the topic (Kennedy and Cabral, 1983; Kennedy, 1979; Scott, 1987).

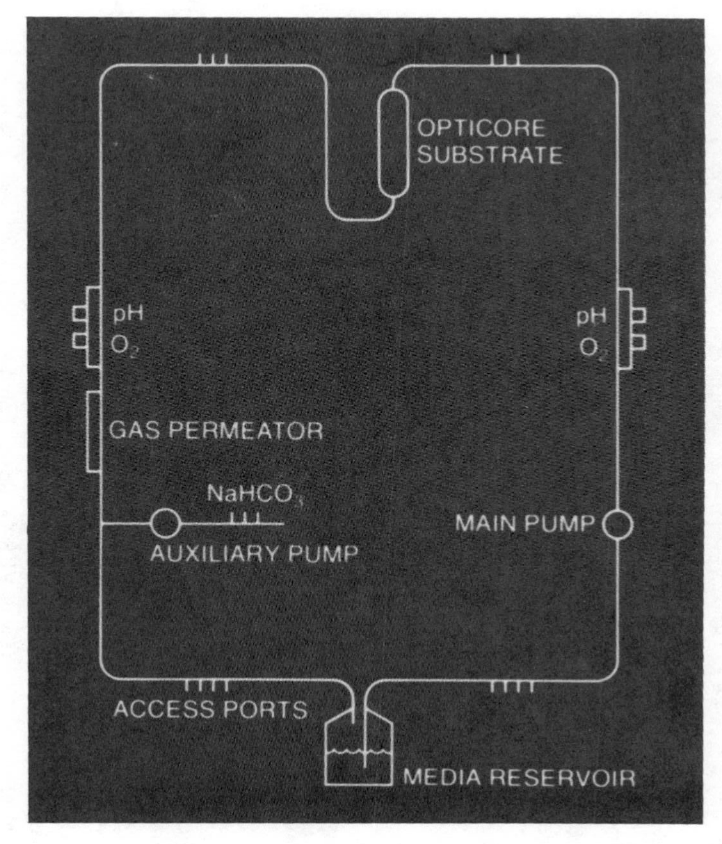

Figure 19.12 Diagram of the OPTICELL mammalian culture reactor system and the OPTI-CORE substratum cartridge (courtesy of Charles River Biotechnical Services, Inc., Wilmington, MA).

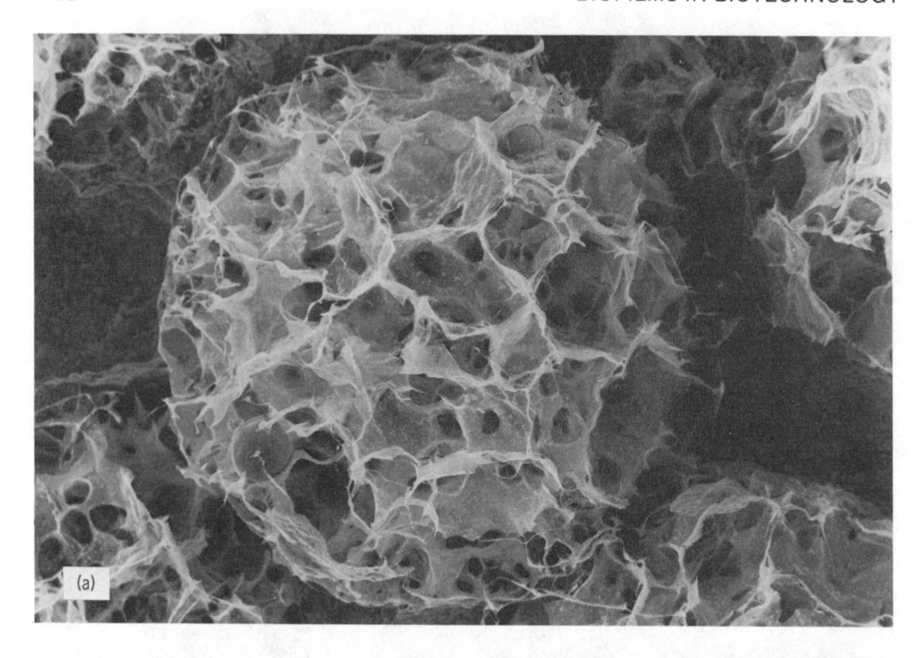

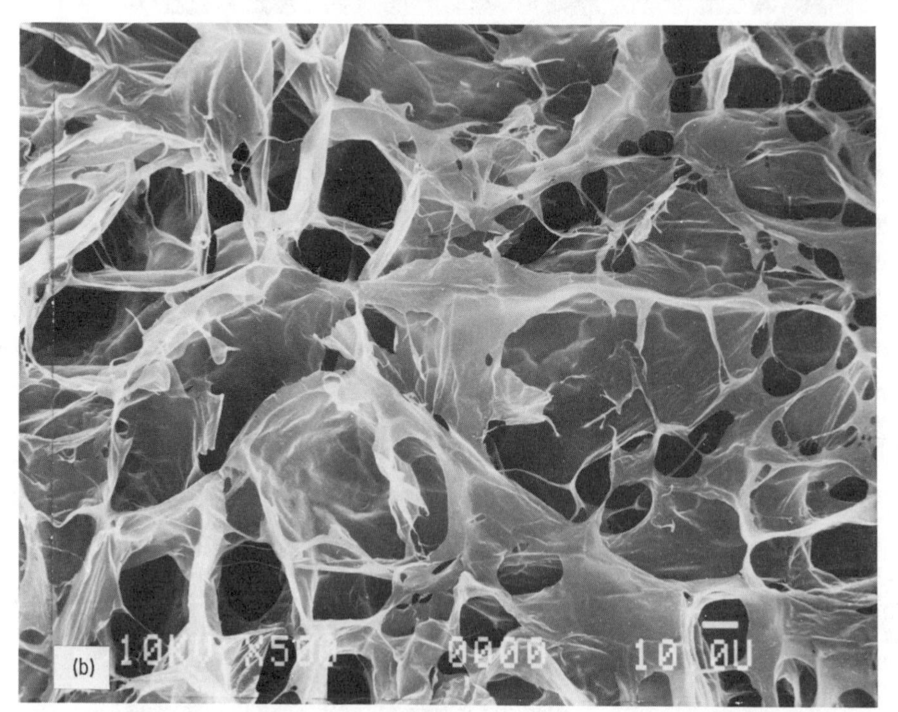

Figure 19.13 Scanning electron micrographs of a clean collagen porous sponge bead for mammalian cell cultivation: (a) $\approx 500\ \mu m$ diameter collagen microsphere; (b) leaflike collagen morphology. (Courtesy of VERAX Corp., Lebanon, NH).

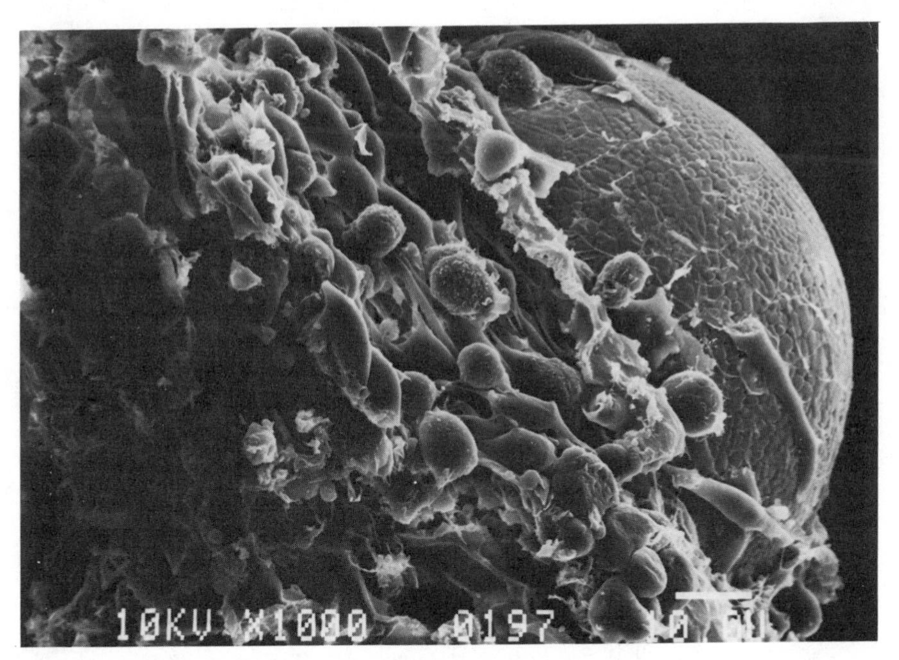

Figure 19.14 Scanning electron micrograph of anchorage-dependent cells populating collagen sponge beads (Courtesy of VERAX Corp., Lebanon, NH).

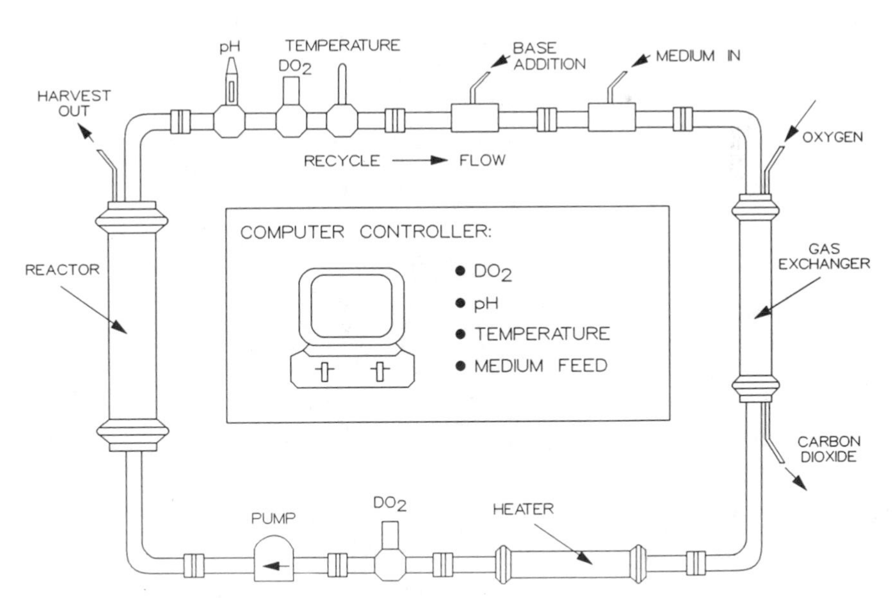

Figure 19.15 Schematic illustration of the continuous recycle reactor system using a fluidized bed of porous sponge beads for cultivation of hybridoma (Courtesy of VERAX Corp., Lebanon, NH).

TABLE 19.7 Commercially Available Microcarriers[a]

Type of MC	Registered Trade Name	Manufacturer	Matrix Composition	Charged Group	Capacity or Equivalent
Tertiary amino MCs	Biocarrier	Bio-Rad., USA	Polyacrylamide	Dimethylaminopropyl	1.4 meq g^{-1} dry materials
	Cytodex 1	Pharmacia, Sweden	Dextran	Diethylaminoethyl (DEAE)	1.5 meq g^{-1} dry materials
	Superbeads	Flow Labs., USA	Dextran	Diethylaminoethyl (DEAE)	2.0 meq g^{-1} dry dextran
Surface-charged MCs	Cytodex 2	Pharmacia, Sweden	Dextran	Trimethyl-2-hydroxy-amino propyl	0.6 meq g^{-1} dry materials
Collagen-coated MCs	Cytodex 3	Pharmacia, Sweden	Dextran	Collagen coated	60 μg collagen (cm^2 MC surface)$^{-1}$
Tissue-culture-treated polystyrene MCs	Biosilon	Nune, Denmark	Polystyrene	Negative charge (tissue culture treatment)	Surface charge of 2–20 × 10^{14} e cm^{-2}
	Cytospheres	Lux, USA	Polystyrene	Negative charge (tissue culture treatment)	Surface charge of 2–10 × 10^{14} e cm^{-2}

[a] Reuveni (1983).

19.4.1.1 Entrapment Techniques The most frequently employed entrapment technique is the capture of whole cells in a *polyacrylamide gel*. The procedure involves the polymerization of acrylamide monomers in an aqueous solution that contains suspended microorganisms. Acrylamide polymerization occurs by a free radical process in which linear chains of polymer are constructed. Inclusion of a bifunctional reagent with two unsaturated double bonds, which is susceptible to insertion in the polymer, yields cross-linking between linear polymer chains. Thus, the extent of cross-linking is a function of the relative amounts of monomer and cross-linking agent. The most common reaction system consists of (1) acrylamide (monomer), (2) N,N′-methylenebis(acrylamide) (cross-linking agent), and (3) persulfate ion plus β-dimethylaminopropionitrile (catalyst). Usually, the polymerization is carried out in an isotonic phosphate buffer solution. The gel containing cells can be easily granulated for use as column packing with the individual pellet porosity a function of the degree of cross-linking. The major disadvantage of this method is the toxicity of the acrylamide monomer, the cross-linking agent, and the catalyst.

Collagen gels have been used as enzyme, cell, and organelle immobilization material (Vieth and Venkatasubramanian, 1979). Pretreated cells (heated from 0 to 80°C) are added to a collagen dispersion, and the pH is then increased from slightly acidic (pH = 6.5) to very caustic (pH = 11.5). The resultant colloidal suspension is dried on a Mylar sheet, forming a flat membrane (2–109 mm thick), hardened in a tanning procedure using a 10% alkaline glutaraldehyde solution for 1–5 min and then washed thoroughly with water.

Investigations by Kennedy and coworkers (Kennedy et al., 1967; Kennedy, 1979, 1980) suggest a number of *hydrous metal oxides* can serve as immobilization supports. Ti^{+4} and Zr^{+4} chloride salts are added to an aqueous cell suspension, which results in pH-dependent formation of a gelatinous polymeric metal hydroxide precipitate where the metals are bridged by either hydroxyl or oxide groups. Using this technique, *Acetobacter* spp., *E. coli, Saccharomyces cerevisiae,* and *Serratia marcescens* have been immobilized without any disruption of their metabolism.

Entrapment in *agar gels* is a flexible process but little used, presumably due to the poor mechanical strength of the resultant gel. Toda and Shoda (1975) reported that whole cells (*S. cerevisiae*) can be captured in agar pellets by first suspending the desired microbe within a 2.5% agar solution and then injecting the hot solution (50°C), dropwise, into a cold solution of either toluene or tetrachloroethylene, forming nearly spherical pellets.

A promising alternative to agar gel is *K-carrageenan,* a seaweed-derived polysaccharide that can form a rigid gel at room temperature or upon contact with either metal ions, ammonium ions, or water-miscible organic solvents, such as methanol or acetone (Wada et al., 1979; Chibata, 1980). K-carrageenan has several advantages over agar gels, including a less toxic, milder treatment of the cells, superior structural properties, less leakage of cells during use, and higher effective nutrient and oxygen diffusion coefficients.

Another method of cell entrapment is the polymerization of polyelectrolytes by multivalent ions. *Ionic polymer network precipitation* proceeds by mixing either calcium alginate, carboxymethylcellulose, or styrene-maleic acid (all nontoxic) with the aqueous suspension of specific cells, then dispersing this mixture dropwise into a counterion solution. Drops precipitate to form uniform, spherical pellets with a high porosity. Calcium alginate entrapment was the first example of whole cell immobilization (Hackel et al., 1975).

One major disadvantage of calcium alginate pellets is that they are susceptible to dissolution due to extraction of the bound Ca^{+2} ion by certain chelating agents (e.g., phosphates) and other ions (e.g., K^{+2} and Mg^{+2}). Birnbaum et al. (1981) circumvented this problem by treating the alginate gels with polyamines prior to cross-linking. Lewandowski et al. (1987) immobilized both autotrophic nitrifiers and denitrifiers with calcium carbonate within a calcium alginate gel. Protons released due to microbial respiration reacted with the calcium carbonate, producing calcium ions, which internally stabilized the gel.

Mohan and Li (1974, 1975) demonstrated that intact whole cells (*Micrococcus denitrificans*) could be captured within a liquid surfactant membrane. This was achieved by adding a phosphate-buffered suspension of viable cells to a mixture of isoparaffin oil (85% v/v), surfactant (2% v/v), membrane strengthener (10% v/v), and anion transport facilitator, then stirring at ≈600 rpm at 18°C. Approximately 500–600 cells become entrapped on aqueous bubbles suspended within a larger oil droplet, itself colloidally stabilized by the surfactant, in a surrounding buffered medium.

19.4.1.2 Adsorption Techniques As an immobilization technique, adsorption of cells to a surface is a mild, simple, and sometimes nonspecific process that is not deleterious to the cell. Adsorption is critical in the formation of natural biofilms and has been discussed at length in Chapters 7 and 12. The reader should consult specific reviews for details (Kennedy and Cabral, 1983; Mosbach, 1981; Cheetam, 1979; Durand and Navarro, 1978).

Ion exchange resins have been employed to selectively separate and concentrate specific cells and to immobilize microorganisms for continuous by-product manufacture. From the extensive literature on bacteria and yeast adsorption to both cationic and anionic exchange materials (Rotman, 1960; Daniels, 1972), it is apparent that adsorption depends upon the chemical nature of the cell surface constituents that provide the local ionic sites for adsorption to a charged surface. The affinity of a specific species for a specific ion exchange material is not yet predictable. The main application of ion exchange cell adsorption is either chromatographic separation of microbial mixtures or the concentration and isolation of a given cell. Daniels (1972) and Kennedy and Cabral (1983) provide extensive reviews on the adsorption of microorganisms to ionic exchange resins.

19.4.1.3 Selective Binding by Bound Macromolecules Macromolecules immobilized on an inert surfaces can bind to certain sites on the surfaces of particular cells. This property has been used as a selective whole cell immobilization technique. The macromolecules used are naturally occurring proteins, usually lectins, obtained from either plants or fish. These proteins are capable of agglutinating certain cell lines. CELL-TAK (Section 19.3.1.1) is the first commercial adsorption promoter and is derived from the marine mussel, *Mytilus edulis*, for mammalian cell immobilization.

19.4.1.4 Covalent Binding Immobilization Although frequently employed to immobilize enzymes, covalent binding is less frequently employed to immobilize whole cells because of coupling agent toxicity (Chipley, 1974; Shimizu et al., 1975). Covalent binding directly links cells to an activated support. Linkages are by any reactive component of the cell surface, i.e., amino, carboxyl, thiol, hydroxl, imidazole, or phenol groups of proteins. The main advantages of covalent binding are that the resultant biocatalyst is free of the diffusion limitations inherent in entrapment techniques, and the immobilization process is stable, which minimizes cell leakage. Navarro and Durand (1977) successfully immobilized *Saccharomyces carlbergensis* on porous silica beads treated with s-aminopropyltriethoxysilane activated with glutaraldehyde. Messing et al. (1979) coupled cells of *Serratia marcescens, Saccharomyces cerevisiae,* and *Saccharomyces amurcae* by isocynate coupling to various silicate and ceramic carriers.

19.4.1.5 Comparison of Immobilization Techniques No one particular immobilization method clearly surpasses all others for whole cell catalyst preparation (Table 19.8). Each has advantages and disadvantages as an immobilization technique.

19.4.2 Applications of Artificially Immobilized Cells

Artificially immobilized whole cells are employed either in specific bioconversion reactions or for diagnostic analysis. Several reviews (Klein and Wagner, 1980; Durand and Navarro, 1978; Cheetam, 1979; Kennedy and Cabral, 1983) illustrate the many applications of artificially immobilized whole cells in both fields.

19.4.2.1 Microbial Systems Artificially immobilized cells permit the selection of type and concentration of biocatalyst to mediate a desired bioconversion. In general, artificial microbial biocatalysts have been applied to the following processes: the production of organic acids, amino acids, ethanol, wine, hydrogen, methane, antibiotics, enzymes, fine chemicals (e.g., NADP, enzyme CoA), and butanol; the degradation of phenolics, nitrate, urea, and malic acid; and the bioconversion of steroids, cellular nucleotides, and other specific commodities. In artificial immobilization, the ability to "locate" a cellular reaction has promoted the development of biosensors that have been used in a variety of analytical applications, as illustrated in Table 19.9.

A biosensor *consists of a biologically sensitive material in intimate contact with an appropriate transducing system that converts the biochemical reaction into an electrical signal, which can be processed.*

TABLE 19.8 Attributes of Different Artificial Immobilization Techniques[a]

Characteristic	Adsorption	Metal-Ion Precipitation	Covalent Binding	Entrapment
Preparation complexity	Simple	Simple	Difficult	Intermediate
Binding force	Weak	Intermediate	Strong	Intermediate
Activity retention	High	Intermediate	Low	Intermediate
Carrier regeneration	Possible	Possible	Rare	Impossible
Cost	Low	Low	High	Low
Stability	Low	Intermediate	High	High
General applicability	Yes	Yes	No	Yes
Protection from microbial attack	Yes	Yes	No	Yes
Maintains viability	Yes	Yes	No	Yes

[a] Kennedy and Cabral 1983.

TABLE 19.9 Examples of Microbial Biosensors[a]

Analysis (Immobilized Electrode)	Microbial Electrode + Other Electrode(s)	Immobilization Material
Antibiotics (Nystatin)	*Saccharomyces cerevisiae* + thermistor	Polyacrylamide gel
Cephalosporins	*Citrobacter freundii* + pH electrode	Collagen membrane
D-Glucose	*Pseudomonas fluorescens* + oxygen electrode	Collagen membrane
Sugar	*Brevibacterium lactogermentis* + oxygen electrode	Acetyl cellulose membrane
Ethanol	*Trichosporon brassilaw* + oxygen electrode	Agar gel
BOD	*Clostridium butylicum* + fuel cell *S. cerevisiae* + thermistor	Polyacrylamide gel
Nicotinic acid	*Lactobacillus arabinous* + pH electrode	Collagen membrane

[a] Kennedy and Cabral (1983).

The biological material, typically an enzyme, antibody, membrane component, organelle, or whole cell, is responsible for "recognition" of a desired component or class of components. Sensitivity and specificity of biosensors is dependent, to a large extent, on the biocatalyst–reactant pair. When the specific bioreaction occurs, a change in one or more physical–chemical parameters also occurs. For example, a change in proton concentration, the release or uptake of gases (O_2, CO_2, NH_3), a change in ions, the release or absorption of heat, a change in optical density, or electron transfer can be detected. For example, a microbial ammonia probe using nitrifying bacteria has been developed (Figure 17.6). These responses, in close proximity to a suitable transducer, result in a quantifiable electric signal. Thus, artificial immobilization is critical to biosensor fabrication. Different types of biosensors can also be classified according to the type of transducer system employed (Table 19.10).

19.4.2.2 Plant Cell Systems Heterogeneous plant cell systems have not been discussed in relation to either natural biofilm or biomass support systems, simply because of the lack of studies in these areas (Kargi and Rosenberg, 1987). Despite definite advantages of plant cell biofilms, all heterogeneous plant cell systems to date have been based on artificial immobilization, e.g., calcium alginate gel entrapment or covalent binding (Shuler et al., 1982; Brodelius and Mosbach, 1982; Brodelius, 1983). Plant cells are used for the production of a wide

TABLE 19.10 Biosensor Transducers and Applications

Transducer	Measurement Mode	Applications
Ion-selective electrode	Potentiometric	Ions in biological media, enzyme electrodes, enzyme immunosensors
Gas-sensing electrode	Potentiometric	Gases; enzyme, organelle, cell electrodes for substrates and inhibitors
Field-effect transistors	Potentiometric	Ions, gases, enzyme substrates
Optoelectronic, fiber-optic, and wavelength devices	Optical	pH, enzyme substrates, immunological compounds
Thermistors	Calorimetric	Enzyme, organelle, whole cell sensors for substrates, products, inhibitors, gases, pollutants, antibiotics, vitamins
Enzyme electrodes	Amperometric	Enzyme substrates
Conductimeter	Conductance	Enzyme substrates

variety of special commodities, including pharmaceuticals (codeine, scopolamine, vincristine, ajmalicine, and digoxin), biochemicals (pyrethrin, sallanin, rotenone, and other alleopathic compounds), flavors and fragrances (strawberry, vanilla, rose, and lemon), sweeteners (thaumatin and monellin), and food colorings (anthocyanin and saffron). A majority of these products are *secondary metabolites* formed by the plant cell when grown under suboptimal conditions (i.e., low growth rates, low nutrient concentrations). Plant cells can be grown in suspension culture but, like mammalian cells, have a propensity to associate with one another, forming large aggregates. Consequently, suspension culture systems cannot provide control over aggregate size without resorting to increased shear, which is detrimental to the cell itself. Other advantages of immobilized plant cells are (1) protection from shear stress damage and (2) better cell-cell contact, favoring the cell-cell differentiation essential for secondary metabolite production.

The reactor geometry most commonly used in plant cell cultivation consists of a column reactor packed with beads of gel-entrapped plant cells, receiving a high recycle flow of nutrient solution. Packed bed reactors are preferred, over air lift or fluidized bed reactors, since growth rates of plant cells are so slow it is unlikely that mass transfer of nutrients will be limiting unless cell concentrations are very high. Also, low internal nutrient concentrations also promote higher secondary metabolite formation.

Recent alternatives to artificial plant cell immobilization are hollow fiber reactors as used by Prenosil and Pedersen (1983) for polyphenol production using carrot cells, and flexible sponge biomass support particles as used by Park and Mavituna (1986) for the production of capsaicin (a flavor compound) using chili pepper plant cells, *Capsicum frutescens* (Figure 19.16).

19.4.2.3 Mammalian Cell Systems Artificially immobilized mammalian cells have bioconversion and diagnostic applications. Certain types of mammalian and hybridoma cells can be successfully cultivated in free suspension provided that the hydrodynamic conditions are sufficiently mild (Section 19.3). However, in conventionally agitated vessels, low shear conditions may lead to incomplete mixing and thus to external mass transfer limitations for nutrients and oxygen. One method to protect mammalian cells and hybridomas from shear damage is to create a rigid yet permeable artificial cell wall via *microencapsulation*. Microencapsulation has been used for drug delivery since 1980 (Lim and Sun). This method has been modified, using biocompatible materials and isotonic solutions, into a commercial method to mass produce monoclonal antibodies from entrapped hybridomas. In the ENCAPCEL™ process (Figure 19.17), hybridoma-producing monoclonal antibodies to B-cell lymphoma, are first suspended in a sodium alginate solution at 5×10^6 cells/ml. This solution is dripped into a Ca^{+2} ion solution, creating water-insoluble spheres. A semipermeable polymer membrane is layered onto the gel spheres, and the gel dissolved by a chelating agent. A microencapsulated solution containing viable hybridomas remains. Monoclonal antibodies secreted from the entrapped cells are re-

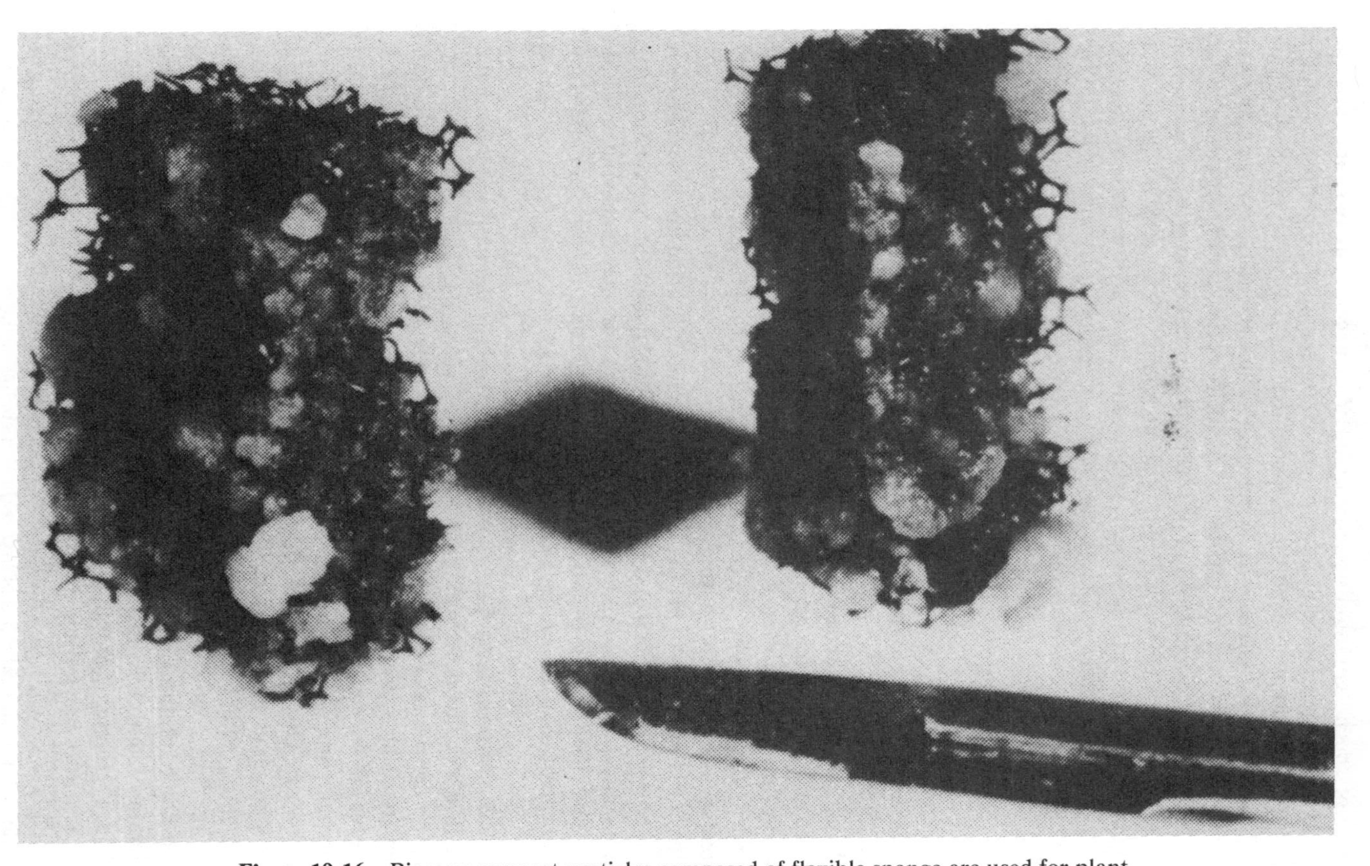

Figure 19.16 Biomass support particles composed of flexible sponge are used for plant cell cultivation (reprinted with permission from Park and Mavituna, 1986).

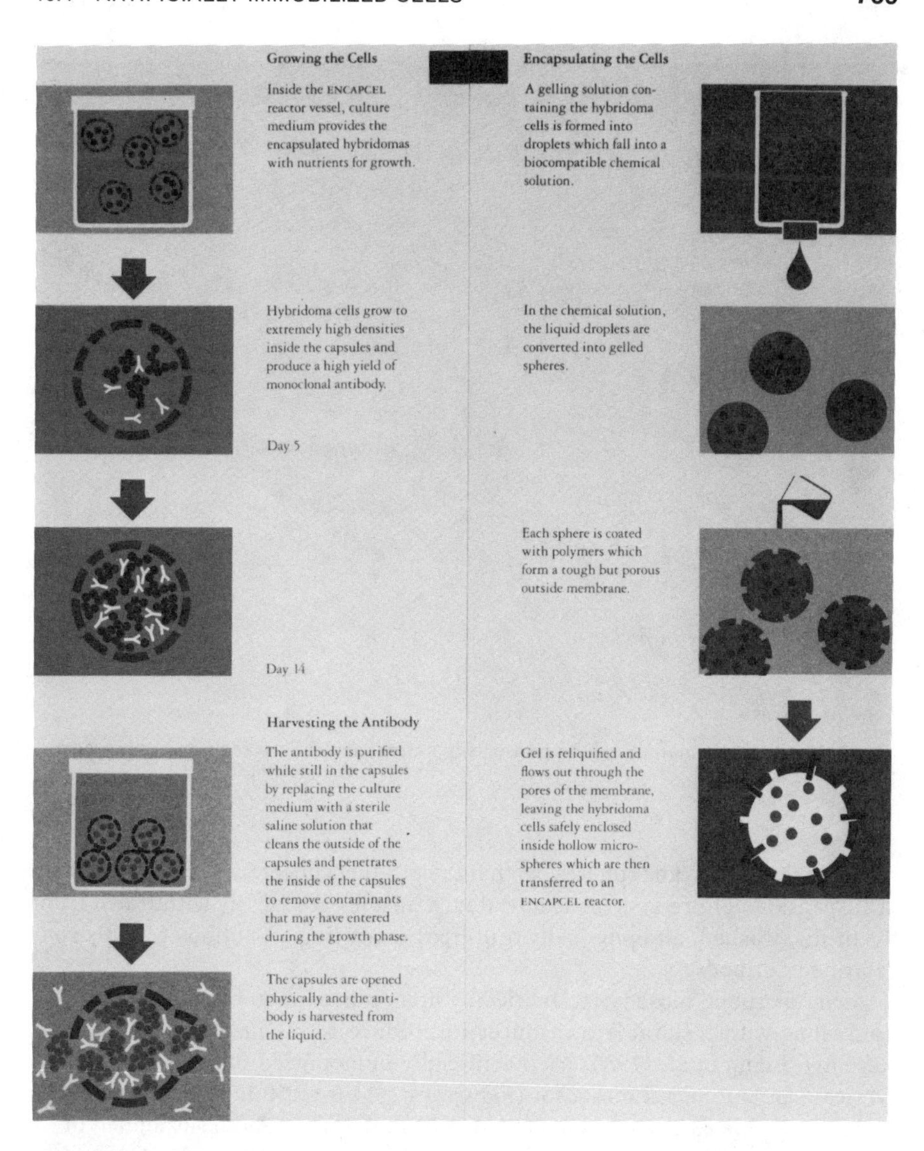

Figure 19.17 Schematic illustration of the ENCAPCEL process for microencapsulating and cultivating hybridomas (courtesy of DAMON Biotech, Inc., Needham Heights, MA).

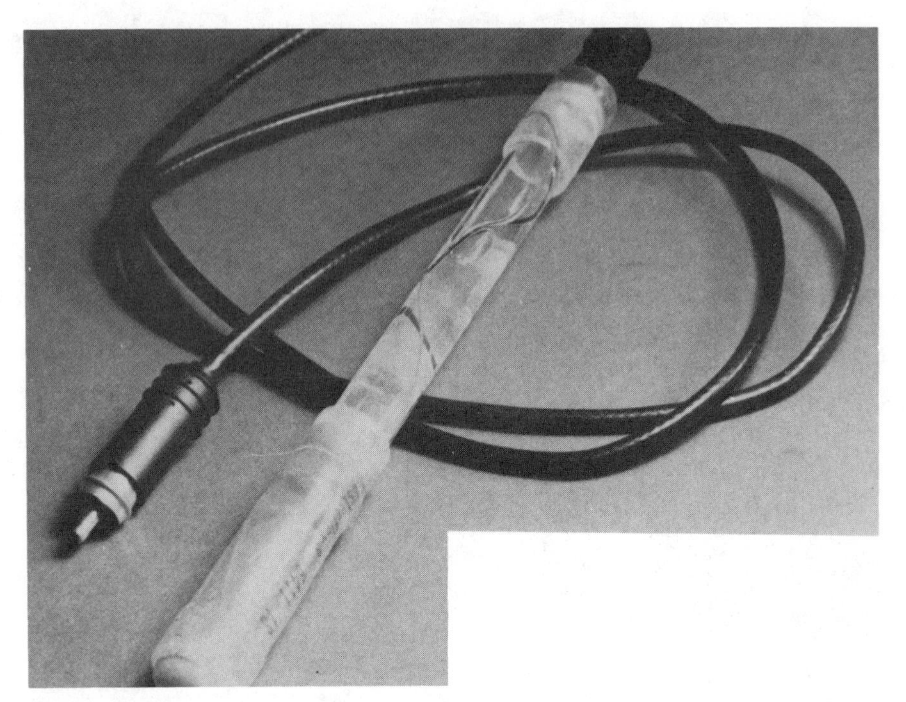

Figure 19.18 Mammalian cell biosensor for anticancer drug screening (Liang et al., 1986).

tained in the membrane spheres, so that the product concentration step required in suspension culture systems is obviated. Capsules are merely withdrawn from the media, washed, and physically ruptured, producing a solution of highly concentrated antibody.

As in microbial biosensors, artificially immobilized mammalian cells can be used along with a suitable transducer to produce a system-specific biosensor. Recently, Liang et al. (1986) used artificially immobilized mammalian cells to fabricate an automated biosensor (Figure 19.18) for antitumor drug screening. Mouse leukemia cells and human foreskin fibroblasts were both suspended in K-carrageenan solution and gelled onto the tip of an oxygen electrode. The antitumor drug mechlorethamine HCl was screened using the dynamic response of the oxygen uptake rate of the cells.

REFERENCES

Atkinson, B., in K.C. Marshall (ed.), *Microbial Adhesion and Aggregation*, Springer-Verlag, Berlin, 1985.

Atkinson, B., G.M. Black, P.J. Lewis, and A. Pinches, *Biotechnol. Bioeng.*, **21**, 193 (1979).

Bailey, J.E. and D.F. Ollis, *Biochemical Engineering,* 2nd ed., McGraw-Hill, 1986.

Birnbaum, S., D.O. Larsson, and K. Mosbach, in *Abstr. Commun. Eur. Congr. Biotechnol. 2nd,* 1981, p. 150.

Black, G.M., C. Webb, T.M. Matthews, and B. Atkinson, *Biotechnol. Bioeng.*, **26**, 134-141 (1984).

Brierley, C.L., *Crit. Rev. Microbiol.*, **6**, 207-262 (1978).

Brierley, C.L., *Sci. Am.*, **247**, 42-51 (1982).

Brodelius, P., *Ann. N.Y. Acad. Sci.*, 383-393 (1983).

Brodelius, P. and K. Mosbach, *J. Chem. Technol. Biotechnol.*, **32**, 330-337 (1982).

Brodelius, P.B., B. Deus, K. Mosbach, and M.H. Zenk, *FEBS Lett.*, **103**, 93 (1979).

Calabresi, P., L.L. McCarthy, D.L. Dexter, F.J. Cummings, and B. Rotman, *Proc. Amer. Assoc. Can. Res.*, **22**, 302 (1981).

Chandra, D., P. Roy, A.K. Mishra, J.N. Chakrabarti, and B. Sengupta, *Fuel,* **58**, 549 (1979).

Cheetham, P.S.J., *Enzyme Microb. Technol.*, **1**, 183 (1979).

Chibata, I., *Enzyme Eng.*, **5**, 393 (1980).

Chipley, J.R., *Microbios,* **10**, 115 (1974).

Clark, J.M. and M.D. Hirtenstein, *Ann. N.Y. Acad. Sci.*, **369**, 33-46 (1981).

Cooper, P.F. and B. Atkinson (eds.), *Biological Fluidized Bed Treatment of Water and Wastewater,* Ellis-Horwood, Chichester, U.K., 1981.

Croughan, M.S., J.-F. Hamel, and D.I.C. Wang, *Biotechnol. Bioeng.*, **29**, 130-141 (1987).

Cysewski, G.R. and C.R. Wilke, *Biotechnol. Bioeng.*, **20**, 1421 (1978).

Daniels, S.L., in E.D. Murray (eds.), *Dev. Ind. Microbiol.*, Pridemark Press, Baltimore, 1972, Chapter 19.

Dewey, C.F., S.R. Bussolari, M.A. Gimbrone, and P.F. Davies, *J. Biomech. Eng.*, **103**, 177-185 (1981).

Doran, P.M. and J.E. Bailey, *Biotechnol. Bioeng.*, **28**, 73-87 (1986).

Durand, G. and J.M. Navarro, *Process Biochem.*, **13**, 14 (1978).

Girard, H.C., M. Suteu, H. Erdom, and I. Gurham, *Biotechnol. Bioeng.*, **22**, 477 (1980).

Glacken, M.W., R.J. Fleischaker, and A.J. Sinskey, *Trends Biotech.*, **1**, 102-108 (1983).

Gockay, C.F. and R.N. Yurteri, *Fuel,* **62**, 1223 (1983).

Hackel, V., J. Klein, R. Megnet, and F. Wagner, *Eur. J. Appl. Microbiol.*, **1**, 291 (1975).

Hattori, R., *J. Gen. Appl. Microbiol.*, **18**, 319 (1972).

Hattori, T. and C. Furusaka, *Biochem. Biophys. Acta,* **31**, 581 (1959).

Hattori, T. and C. Furusaka, *J. Biochem.*, **48**, 831 (1960).

House, W., M. Shearer, and N.B. Maroudas, *Exp. Cell. Res.*, **71**, 293 (1972).

Hu, W.S., J. Meier, and D.I.C. Wang, *Biotechnol. Bioeng.*, **27**, 585-595 (1985).

Inloes, D.S., W.J. Smith, D.P. Taylor, S.N. Cohen, A.S. Michaels, and C.R. Robertson, *Biotechnol. Bioeng.*, **25**, 2653-2681 (1983).

Jirku, V., J. Turkova, and V. Krumphanzl, *Biotechnol. Lett.*, **2**, 509 (1980).

Kargi, F. and J.M. Robinson, *Biotechnol. Bioeng.*, **24**, 2115 (1982).

Kargi, F. and M.Z. Rosenberg, *Biotechnol. Prog.*, **3**(1), 1-8 (1987).

Kennedy, J.F., *ACS Symp. Ser.*, **106**, 119 (1979).

Kennedy, J.F., *Enzyme Microb. Technol.*, **2**, 164 (1980).

Kennedy, J.F. and J.M.S Cabral, *Appl. Biochem. Bioeng.*, **4**, 189 (1983).

Kennedy, J.F., S.A. Barker, and J.D. Humphreys, *J. Chem. Soc. Perkin Trans.*, **1**, 962 (1967).

Klein, J. and F. Wagner, *Enzyme Eng.*, **5**, 335 (1980).

Knazek, R.A., P.M. Gullino, P.O. Kohler, and R.L. Dedrick, *Science*, **178**, 65 (1972).

Ku, K., M.J. Kuo, J. Delente, B.S. Wildi, and J. Feder, *Biotechnol. Bioeng.*, **23**, 79 (1981).

Larrson, P.O. and K. Mosbach, *Biotechnol. Lett.*, **1**, 501 (1979).

Lawrence, R.W. and A. Bruynesteyn, *CIM Bull.*, **76**, 107 (1983).

Lewandowski, Z., R. Bakke, and W.G. Characklis, *Wat. Sci. Technol.*, **19**, 175-182 (1987).

Liang, B. S., X-M. Li, and H.Y. Wang, *Biotechnol. Progr.*, **2**(4), 187-191 (1986).

Lim, F. and A.M. Sun, *Science*, **210**, 908 (1980).

Lydersen, B.J., J. Putnam, E. Bognar, M. Patterson, G.G. Pugh, and L.A. Noll, in *Abst. Am. Chem. Soc. Ann. Meet.*, 1984.

McKeehan, W.L., K.A. McKeehan, and R.G. Ham, in *The Growth Requirements of Vertebrate Cells in vitro*, C. Waymouth, B.G. Ham, and D.G. Chapel, Cambridge University Press, Cambridge, UK., 1981, pp. 118-130.

Messing, R., R.A. Oppermann, and F. Kolot, *ACS Symp. Ser.*, **106**, 13, 1979.

Miller, S.J.O., M. Henrotte, and A.O.A. Miller, *Biotech. Bioeng.*, **28**, 1466-1473 (1986).

Mohan, R.R. and N.N. Li, *Biotechnol. Bioeng.*, **16**, 513 (1974).

Mohan, R.R. and N.N. Li, *Biotechnol. Bioeng.*, **17**, 1137 (1975).

Mosbach, K., in *Abstr. Commun. Eur. Congr. Biotechnol.*, 2nd, 1981, p. 48.

Myerson, A.S. and P.C. Kline, *Biotechnol. Bioeng.*, **25**, 1669 (1983).

Navarro, J.M. and G. Durand, *Eur. J. Appl. Microbiol.*, **4**, 243 (1977).

Olsen, G.J. and R.M. Kelley, *Biotechnol. Prog.*, **2**(1), 1-15 (1986).

Park, J.M. and F. Mavituna, in C. Webb, G.M. Black, and B. Atkinson (eds.), *Process Engineering Aspects of Immobilized Cell Systems, Inst. Chem. Eng.*, Rugby, Warwickshire, U.K., 1986, pp. 295-303.

Pharmacia Fine Chemicals, *Microcarrier Cell Culture*, Principles and Methods, Uppsala, Sweden, 1981.

Prenosil, J.G. and H. Pedersen, *Enzyme Microbiol. Technol.*, **5**, 323-331 (1983).

Reuveny, S., in *Adv. Biotechnol. Processes*, Vol. 2, A.R. Liss Publ., New York, 1983, pp. 1-32.

Robinson, D.K. and D.I.C. Wang, "A novel bioreactor system for polymer production," presented at Biochemical Engineering, V, July 1986.

Rotman, B., *Bact. Rev.*, **24**, 251 (1960).

Runstadler, P.W., and S.R. Cernek, VERAX Corporation Technical Memo TM200, Lebanon, NH (1988).

Scott, C.D., *Enzyme Microb. Technol.*, **9**, 66–73 (1987).

Shimizu, S., H. Morioka, Y. Tani, and K. Ogata, *J. Ferment. Technol.*, **53**, 77 (1975).

Shuler, M.L., O.P. Sahai, and G.A. Hallsbay, *Ann. N.Y. Acad. Sci.*, **413**, 373–382 (1982).

Stathopoulos, N.A. and J.D. Hellums, *Biotechnol. Bioeng.*, **27**, 1021–1026 (1985).

Toda, K., and H. Shoda, *Biotechnol. Bioeng.*, **17**, 481 (1975).

Van Brunt, J. *Bio/technology*, **4**, 505–510 (1986).

van Wezel, A.L., *Nature*, **216**, 64 (1967).

Varani, J., M. Dame, J. Rediske, T.F. Beals, and W. Hillegas, *J. Biolog. Standardiz.*, **13**, 67–76 (1985).

Vieth, W.R. and K. Venkatasubramanian, *ACS Symp. Ser.*, **106**, 1 (1979).

Wada, M., J. Kato, and I. Chibata, *Eur. J. Appl. Microbiol. Biotechnol.*, **8**, 241 (1979).

Waite, J.H. and M.L. Tanzer, *Science*, **212**, 1038–1040 (1981).

Webb, C., H. Fukunda, and B. Atkinson, *Biotechnol. Bioeng.*, **28**, 41–50 (1986).

Weiss, R.E. and J.B. Schleicher, *Biotechnol. Bioeng.*, **10**, 601 (1968).

Wiemann, M.C., E.D. Ball, M.W. Fanger, O.R. McIntyre, G. Berier, and P. Calabresi, *Clin. Res.*, **31**, S11A, 1983.

Williams, D. and D.M. Munnecke, *Biotechnol. Bioeng.*, **23**, 1813 (1981).

Wohler, W., H.W. Rudiger, and E. Passarge, *Exp. Cell Res.*, **74**, 571 (1972).

APPENDIX:

SYMBOLS AND NOTATION

Dimensions are defined according to the most fundamental measures: **A** for amperes, **L** for length, **M** for mass, **mol** for mole, **Q** for charge, **t** for time, **T** for temperature, and # for cell numbers. More specific definition for symbols and notation are formulated in individual chapters most often by incorporating superscripts and subscripts.

Symbol	Quantity name or names	Dimension
	Latin Characters	
A	Area	L^2
$\mathbf{a}$	Stoichiometric matrix	MM^{-1}
a_j	activity of component j	mol L^{-3}
B_j''	Mass (or cell) loading rate of j per unit area	$M_jL^{-2}t^{-1}$
B_j	Mass (or cell) loading rate of j per unit volume, or volumetric loading rate	$M_jL^{-3}t^{-1}$ (# $L^{-3}t^{-1}$)
b	Specific biomass loss rate	t^{-1} (from $M_XM_X^{-1}t^{-1}$)
C_j	Volumetric mass concentration of component j	M_jL^{-3}
C_{jk}	Volumetric mass concentration of j in phase (or compartment) k	$M_jL_k^{-3}$
C_{jk}''	Areal mass concentration of j in phase (or compartment) k	$M_jL_k^{-2}$
C_p	Heat capacity at constant pressure	$ML^2t^{-2}mol^{-1}T^{-1}$
D	Diffusion coefficient or diffusivity	L^2t^{-1}
D	Dilution rate	t^{-1} (from $L^3L^{-3}t^{-1}$)
d	Diameter	L
d_r	length of random run for motile organisms	L
E_a	Activation energy	$ML^2t^{-2}mol^{-1}$

E, E^0	Potential (standard potential)	$M L^2t^{-3} A^{-1}$ (volts)
e	Height of roughness elements	L
e_s	Equivalent sand roughness	L
F_A	Faraday constant = 9.652×10^4 coulombs g-equivalent^{-1}	charge $e^{-1}mol^{-1}$
F	Mass flow rate	Mt^{-1}
F_j	Force in the j-direction	MLt^{-2}
f	A ratio	—
f	Fanning friction factor	—
f_R	RotoTorque friction factor	—
G, G^0	Free energy (standard free energy)	ML^2t^{-2}
g	Gravitational acceleration	Lt^{-2}
g	Mean cell generation time	t
H	Height or depth	L
H	Enthalpy	ML^2t^{-2}
h	Heat transfer coefficient	$Mt^{-3}T^{-1}$
J_k	Mass flux of component k	$M_kL^{-2}t^{-1}$
k	Rate coefficient	variable
k_B	Boltzmann constant 1.38×10^{16} erg molecule^{-1} °K^{-1}	
k_M	Mass transfer coefficient	Lt^{-1}
k_T	Thermal conductivity	$MLt^{-3}T^{-1}$
K	Equilibrium constant or distribution coefficient	—
K_s	Half saturation coefficient	ML^{-3}
L	Length	L
L_f	Biofilm thickness	L
M	Molecular weight	$Mmol^{-1}$
N	Effectiveness factor (for example, correction of reaction rate for diffusion)	—
$\mathbf{n}$	inward normal unit vector on system surface	—
n	Reaction order	—
n	Number	#
n_e	Number of electrons	#
n_f	Cell number per unit biofilm volume	# L^{-3}
n_f''	Cell number per unit biofilm area	# L^{-2}
Nu	Nusselt number	—
P	Total pressure ($p + \rho gH$)	$ML^{-1}t^{-2}$
Pe	Peclet number	—
p	Pressure	$ML^{-1}t^{-2}$
Pr	Prandtl number	—
Q	Volumetric flow rate	L^3t^{-1}
Q_R	Volumetric recycle flow rate	L^3t^{-1}
Q_g	Gas volumetric flow rate	L^3t^{-1}
Q_H	Heat flow rate	ML^2t^{-3}
q_j	Reaction rate of j per unit biomass, i.e., specific reaction rate	$M_jM_X^{-1}t^{-1}$
$q_{j,max}$	Maximum reaction rate of j per unit biomass	$M_jM_X^{-1}t^{-1}$

Symbol	Definition	Units
q_{Hj}	Heat flux in the j-direction	Mt^{-3}
R	Conversion rate	Mt^{-1}
R	Ideal gas constant	$ML^2t^{-2}T^{-1}mol^{-1}$
R	Radius	L
$\mathbf{r}$	Reaction rate vector	(variable units)
r	Radial position	—
r''	Interfacial transfer or reaction rate	$ML^{-2}t^{-1}$
r	Reaction rate per unit volume	$ML^{-3}t^{-1}$
Re	Reynolds number	—
Re_r	Roughness Reynolds number	—
Re_R	Reynolds number for the RotoTorque	—
$\mathbf{S}$	Chemical state vector	ML^{-3}
S	Entropy	$ML^2t^{-2}T^{-1}$
S	Dissolved material concentration (typically, the substrate concentration)	ML^{-3}
S_{kj}	Dissolved material concentration of component k in phase (or compartment) j	$M_kL_j^{-3}$
S_i	Inlet dissolved material concentration	ML^{-3}
Sc	Schmidt number	—
St	Stanton number	—
Sh	Sherwood number	—
T	Temperature	T
t	Chronological or running time	t
Ta	Taylor number	—
T_q	Torque	ML^2t^{-2}
U	Overall heat transfer coefficient	$Mt^{-3}T^{-1}$
U	Mobility	—
U'	Apparent mobility	—
V	Volume	L^3
v	Velocity	Lt^{-1}
v	Superficial velocity	Lt^{-1} (from $L^3L^{-2}t^{-1}$)
W	Work	ML^2t^{-2}
X_i	Inlet particulate mass (or cell) volumetric concentration	ML^{-3}
X_j	Particulate mass (or cell) volumetric concentration of component j	M_jL^{-3} (# L^{-3})
X_j''	Particulate mass (or cell) areal concentration of component j	M_jL^{-2} (# L^{-2})
$Y_{j/k}$	Yield of component j from component k	$M_jM_k^{-1}$
Y_o	Observed or net biomass yield coefficient	M_xM^{-1}

Greek Characters

Symbol	Definition	Units
α	A ratio	—
	Specific lethality coefficient	variable
	angle of turn for motile organism	—
α_e	Sticking efficiency	—
β	A ratio	—

Γ	Gamma function	—
γ	activity coefficient	—
δ	Thickness	L
δ_v	Thickness of viscous sublayer	L
δ_t	Thickness of thermal sublayer	L
δ_m	Thickness of mass transfer sublayer	L
ϵ	Porosity	—
ϵ_i	Volume fraction of compartment *i*	—
ϵ_{sp}	Substratum-particle capture factor	—
Φ	Electric potential	$ML^2t^{-2}mol^{-1}$
ϕ_D	Volume fraction of gel	—
$\boldsymbol{\phi}$	Transport vector	Mt^{-1}
η	Dynamic viscosity	$ML^{-1}t^{-1}$
θ	Space time (volume of reaction phase divided by volumetric flow rate of phase entering the reactor)	t
θ	Mean hydraulic residence time	t
θ_X	Mean solids residence time	t
λ	Rate coefficient	$L^{-1}t^{-1}$
μ	Specific growth rate	t^{-1} (from $M_xM_x^{-1}t^{-1}$)
μ_{max}	Maximum specific growth rate	t^{-1}
μ_{obs}	Observed or net specific biomass growth rate	t^{-1}
$\bar{\mu}$	Electrochemical potential	$ML^2t^{-2}mol^{-1}$
μ^0	Standard electrochemical potential	$ML^2t^{-1}mol^{-1}$
ν	Kinematic viscosity	L^2t^{-1}
ν_{ij}	Stoichiometric coefficient	$M_iM_j^{-1}$
ρ	Density	ML^{-3}
σ_s	Velocity of sloughing	Lt^{-1}
σ_{ab}, σ_a, σ_b	Interfacial free energies	ML^2t^{-2}
τ	Momentum flux (hydrodynamic shear stress)	$ML^{-1}t^{-2}$
τ_w	Momentum flux at the wall	$ML^{-1}t^{-2}$
Ω	Rotational speed	t^{-1}
Ω_p	Penetration ratio	—

Subscripts

BIOFILM SYSTEM COMPARTMENTS

b	Bulk water compartment
f	Biofilm compartment
s	Solid phase of biofilm compartment
l	Liquid (water) phase of biofilm compartment
sb	Substratum compartment
g	Gas compartment

BIOFILM SYSTEM INTERFACES

0	Substratum—biofilm interface
1	Biofilm-bulk water interface
2	Air-bulk water interface

INDEX